Mathematical Statistics with Applications in R

Mathematical Statistics with Applications in R

Fourth Edition

Kandethody M. Ramachandran
Professor of Mathematics and Statistics
University of South Florida
Tampa, Florida

Chris P. Tsokos
Distinguished University Professor of Mathematics and Statistics
University of South Florida
Tampa, Florida

ELSEVIER

ACADEMIC PRESS
An imprint of Elsevier

Academic Press is an imprint of Elsevier
125 London Wall, London EC2Y 5AS, United Kingdom
50 Hampshire Street, 5th Floor, Cambridge, MA 02139, United States

ISBN: 978-0-443-27548-7

For Information on all Academic Press publications
visit our website at https://www.elsevier.com/books-and-journals

Publisher: *Linsley Peter*
Acquisitions Editor: Stephanie Xu
Editorial Project Manager: Prashant Mahajan
Publishing Services Manager: Deepthi Unni
Production Project Manager: Erragounta Saibabu Rao
Cover Designer: Raman Kumar

Typeset by MPS Limited, Chennai, India

Dedication

Dedicated to our families:

Usha, Vikas, Vishnu Vandhana, Vilas, Shrina, and Varsha Ramachandran
and
Debbie, Matthew, Jonathan, and Maria Tsokos

Contents

About the Authors

Dr. Kandethody M. Ramachandran is a Professor of Mathematics and Statistics at the University of South Florida (USF). He received his B.S. and M.S. degrees in Mathematics from Calicut University, India, and subsequently worked as a researcher at the Tata Institute of Fundamental Research (TIFR), Bangalore, in the Applied Mathematics Division. He earned his Ph.D. in Applied Mathematics from Brown University.

Dr. Ramachandran's research interests include applied probability, statistics, and data science. His scholarly publications span a variety of areas such as control of heavy-traffic queues, stochastic delay equations and control problems, stochastic differential games and their applications, reinforcement learning in game theory, software reliability modeling, applications of statistical methods to microarray data analysis, cybersecurity, streaming data analysis, and mathematical finance. He is the coauthor, with Chris Tsokos, of *Stochastic Differential Games: Theory and Applications* (Atlantis Press). He is also the inventor and coinvestigator on US patents related to his research interest.

Professor Ramachandran has been recognized for his excellence in teaching with the USF's Teaching Incentive Program (TIP) Award. He was an active member of the MEME Collaborative, an interdisciplinary partnership among mathematics, engineering, and mathematics education faculty that addresses challenges in mathematics education. He has also contributed to calculus reform efforts at the USF.

Dr. Ramachandran served as the Principal Investigator (PI) of a $2 million grant from the National Science Foundation (NSF) and as a Co-Principal Investigator (Co-PI) of a $1.4 million grant from the Howard Hughes Medical Institute (HHMI), both aimed at strengthening STEM education initiatives at the USF.

Dr. Chris P. Tsokos is a Distinguished University Professor Emeritus in the Department of Mathematics and Statistics at the USF.

Professor Tsokos received his B.S. in Engineering Science/Mathematics and his M.A. in Applied Mathematics from the University of Rhode Island and was recognized with the Distinguished Alumni Excellence Award in Science and Technology. He received his Ph.D. in Statistics and Probability from the University of Connecticut and was recognized with the Distinguished Alumni Excellence Award, along with the recognition of the State of Connecticut General Assembly Award for outstanding public philanthropic service to our country. Professor Tsokos has also served on the faculties at Virginia Polytechnic Institute and State University and the University of Rhode Island. Dr. Tsokos's research has extended into a variety of areas in the mathematical sciences, including stochastic systems, statistical models, reliability analysis, ecological systems, operations research, time series, Bayesian analysis, mathematical and statistical modeling of global warming, and both parametric and nonparametric survival analysis, cyber security, and financial systems, among others. He is the inventor/coinvestigator of more than nine US Patents in his research interests.

He is a recipient of many distinguished awards and honors, including Fellow of the American Statistical Association, the Royal Statistical Society, and the International Statistical Institute. Dr. Tsokos has served as an editor and reviewer for numerous international journals and has supervised more than 75 Ph.D. students who now hold influential positions in academia, industry, and government.

Throughout his distinguished career, Dr. Tsokos has been recognized for his excellence in teaching, research, and service to the profession. His dedication to integrating theory with practical application continues to inspire statisticians and data scientists worldwide.

Preface

Preface to Fourth Edition

In this fourth edition, we have preserved the core structure and strengths of the previous editions while incorporating substantial updates through new examples and exercises that reflect the rapid growth of statistics in emerging domains such as artificial intelligence (AI) and cryptocurrency analytics. Given the increasing prevalence of extreme observations in these fields, a new chapter on Extreme Value Theory (EVT) has been added to equip students and researchers with essential tools for modeling and inference in such contexts.

The content has been carefully revised and enhanced throughout, drawing on insightful feedback from students, instructors worldwide, and reviewers on platforms such as Amazon. These revisions improve the clarity, accuracy, and relevance of the matter. Chapter 15, on Statistical Modeling, now includes a range of real-world data applications, including analyses of global warming, Hurricane Katrina, national unemployment, brain cancer, rainfall variability, and prostate cancer.

Both the Instructor's Solution Manual and the Student's Solution Manual have been thoroughly updated to align with the new and revised content. The textbook datasets are available for download from the Elsevier website.

Preface to Third Edition

In the third edition, we kept much of the material the same as in the second edition. However, we have moved around a few materials to get a better flow of topics. We have added a few extra proofs such as the central limit theorem. Based on the comments from various instructors from around the globe, and from the reviews from the websites such as that of Amazon, we improved and corrected much of the material throughout the book. In addition, we have added a chapter on categorical data analysis and a section on empirical Bayes material in the Bayes chapter. In Chapter 14, we have included some real-world applications with actual data on global warming, Hurricane Katrina, national unemployment, brain cancer, rainfall data analysis, and prostate cancer.

We have updated both the Instructor's Solution Manual and the Student's Solution Manual. Textbook data will be available on the Elsevier website.

Preface to Second Edition

In the second edition, while keeping much of the material from the first edition, there are some significant changes and additions. Due to the popularity of R and its free availability, we have incorporated R-codes throughout the book. This will make it easier for students to do the data analysis. We have also added a chapter on goodness-of-fit tests and illustrated their applicability with several examples. In addition, we have introduced more probability distribution functions with real-world data-driven applications in global warming, brain and prostate cancer, national unemployment, and total rainfall. In this edition, we have shortened the point estimation chapter and merged it with interval estimation. In addition, many corrections and additions are made to reflect the continuous feedback we have obtained.

We have created a student companion website (http://booksite.elsevier.com/9780124171138) with solutions to selected problems and data on global warming, brain and prostate cancer, national unemployment, and total rainfall. We have also posted solutions to most of the problems on the instructor site (http://textbooks.elsevier.com/web/Manuals.aspx?isbn9780124171138).

Preface to First Edition

This textbook is of an interdisciplinary nature and is designed for a one- or two-semester course in probability and statistics, with basic calculus as a prerequisite. The book is primarily written to give a sound theoretical introduction to statistics while emphasizing applications. If teaching statistics is the main purpose of a two-semester course in probability and statistics, this textbook covers all the probability concepts necessary for the theoretical development of statistics in two chapters, and goes on to cover all major aspects of statistical theory in two semesters, instead of only a portion of statistical concepts. What is more, using the optional section on computer examples at the end of each chapter, the student can also simultaneously learn to utilize statistical software packages for data analysis. It is our aim, without sacrificing any rigor, to encourage students to apply the theoretical concepts they have learned. There are many examples and exercises concerning diverse application areas that will show the pertinence of statistical methodology to solving real-world problems. The examples with statistical software and projects at the end of the chapters will provide a good perspective on the usefulness of statistical methods. To introduce the students to modern and increasingly popular statistical methods, we have introduced separate chapters on Bayesian analysis and empirical methods.

One of the main aims of this book is to prepare advanced undergraduates and beginning graduate students in the theory of statistics with emphasis on interdisciplinary applications. The audience for this course is regular full-time students from mathematics, statistics, engineering, physical sciences, business, social sciences, materials science, and so forth. Also, this textbook is suitable for people who work in industry and in education as a reference on introductory statistics for a good theoretical foundation with clear indication of how to use statistical methods. Traditionally, one of the main prerequisites for this course is a semester of an introduction to probability theory. A working knowledge of elementary (descriptive) statistics is also a must. In schools where there is no statistics major, imposing such a background, in addition to calculus sequence, is very difficult. Most of the present books available on this subject contain full one-semester material for probability and then, based on those results, continue on to the topics in statistics. Also, some of these books include in their subject matter only the theory of statistics, whereas others take the cookbook approach of covering the mechanics. Thus, even with two full semesters of work, many basic and important concepts in statistics are never covered. This book has been written to remedy this problem. We fuse together both concepts in order for the student to gain knowledge of the theory and at the same time develop the expertise to use their knowledge in real-world situations.

Although statistics is a very applied subject, there is no denying that it is also a very abstract subject. The purpose of this book is to present the subject matter in such a way that anyone with exposure to basic calculus can study statistics without spending two semesters of background preparation. To prepare students, we present an optional review of the elementary (descriptive) statistics in Chapter 1. All the probability material required to learn statistics is covered in two chapters. Students with a probability background can either review or skip the first three chapters. It is also our belief that any statistics course is not complete without exposure to computational techniques. At the end of each chapter, we give some examples of how to use Minitab, SPSS, and SAS to statistically analyze data. Also, at the end of each chapter, there are projects that will enhance the knowledge and understanding of the materials covered in that chapter. In the chapter on the empirical methods, we present some of the modern computational and simulation techniques, such as bootstrap, jackknife, and Markov chain Monte Carlo methods. The last chapter summarizes some of the steps necessary to apply the material covered in the book to real-world problems. The first eight chapters have been class tested as a one-semester course for more than 3 years with five different professors teaching. The audience was junior- and senior-level undergraduate students from many disciplines who had had two semesters of calculus, most of them with no probability or statistics background. The feedback from the students and instructors was very positive. Recommendations from the instructors and students were very useful in improving the style and content of the book.

Aim and Objective of the Textbook

This textbook provides a calculus-based coverage of statistics and introduces students to methods of theoretical statistics and their applications. It assumes no prior knowledge of statistics or probability theory but does require calculus. Most books at this level are written with elaborate coverage of probability. This requires teaching one semester of probability and then continuing with one or two semesters of statistics. This creates a particular problem for nonstatistics majors from various disciplines who want to obtain a sound background in mathematical statistics and applications. It is our aim to introduce basic concepts of statistics with sound theoretical explanations. Because statistics is basically an interdisciplinary applied subject, we offer many applied examples and relevant exercises from different areas. Knowledge of using computers for data analysis is desirable. We present examples of solving statistical problems using Minitab, SPSS, and SAS.

Features

- During years of teaching, we observed that many students who do well in mathematics courses find it difficult to understand the concept of statistics. To remedy this, we present most of the material covered in the textbook with well-defined step-by-step procedures to solve real problems. This clearly helps the students to approach problem solving in statistics more logically.
- The usefulness of each statistical method introduced is illustrated by several relevant examples.
- At the end of each section, we provide ample exercises that are a good mix of theory and applications.
- In each chapter, we give various projects for students to work on. These projects are designed in such a way that students will start thinking about how to apply the results they learned in the chapter, as well as other issues they will need to know for practical situations.
- At the end of the chapters, we include an optional section on computer methods with Minitab, SPSS, and SAS examples with clear and simple commands that the student can use to analyze data. This will help the student to learn how to utilize the standard methods they have learned in the chapter to study real data.
- We introduce many of the modern statistical computational and simulation concepts, such as the jackknife and bootstrap methods, the EM algorithms, and the Markov chain Monte Carlo methods such as the Metropolis algorithm, the Metropolis–Hastings algorithm, and the Gibbs sampler. The Metropolis algorithm was mentioned in *Computing in Science and Engineering* as being among the top 10 algorithms having the "greatest influence on the development and practice of science and engineering in the 20th century."
- We have introduced the increasingly popular concept of Bayesian statistics and decision theory with applications.
- A separate chapter on the design of experiments, including a discussion on the Taguchi approach, is included.
- The coverage of the book spans most of the important concepts in statistics. Learning the material along with computational examples will prepare students to understand and utilize software procedures to perform statistical analysis.
- Every chapter contains discussion on how to apply the concepts and what are the issues related to applying the theory.
- A student's solution manual, instructor's manual, and data disk are provided.
- In the last chapter, we discuss some issues in applications to clearly demonstrate in a unified way how to check for many assumptions in data analysis and what steps one needs to follow to avoid possible pitfalls in applying the methods explained in the rest of this textbook.

Kandethody M. Ramachandran
Chris P. Tsokos

For additional information on topics covered in the book, visit the companion site:
https://www.elsevier.com/books-and-journals/book-companion/9780443275487

For qualified Instructors, additional information is available in the Instructor site:
https://educate.elsevier.com/9780443275487

Acknowledgments

We wish to express our heartfelt gratitude to our late colleague, coworker, and dear friend, Professor A.N.V. Rao, whose valuable ideas and encouragement greatly influenced the first edition of this book. We also thank Bong-jin Choi and Yong Xu for their assistance during the preparation of that edition.

In preparing the third edition, we were fortunate to receive generous help from A.K.M.R. Bashar, Jason Burgess, Doo Young Kim, Taysseer Sharaf, Bhikhari Tharu, Ram Kafle, Dr. Rebecca Wooten, and Dr. Olga Savchuk.

For the current edition, we are deeply indebted to many instructors around the world who offered insightful suggestions, and to numerous students whose comments helped make this book more engaging and student-friendly. We are especially thankful to Dr. Houston Sanders and Dr. Kim Jiwoong for identifying and helping to correct printing errors from the previous edition.

We also extend our appreciation to those who shared their thoughts through Amazon, Google Books, and other websites, especially Mansah Alkebu-lan for his encouraging words and Dr. Axel Boldt for his thoughtful and detailed review of the first edition. Our special thanks go to Dr. Lu Lu, Dr. Olga Savchuk, and many others who took the time to send comments and suggestions by email.

We are grateful to Vikas Ramachandran for preparing the Solution Manual for the fourth edition. Finally, our sincere thanks go to our students at the University of South Florida, whose curiosity, questions, and enthusiasm continue to inspire us.

Finally, we would like to express our sincere appreciation to the editorial team at Elsevier/Academic Press for their professional support and efficient handling of the publication process.

To all of them, we are truly thankful.

Kandethody M. Ramachandran
Chris P. Tsokos
Tampa, Florida

Flow Chart

In this flow chart, we suggest some options on how to use the book in a one-semester or two-semester course. For a two-semester course, we recommend coverage of the complete textbook. However, Chapters 1, 9, and 14 are optional for both one- and two-semester courses and can be given as reading exercises. For a one-semester course, we suggest the following options: A, B, C, D.

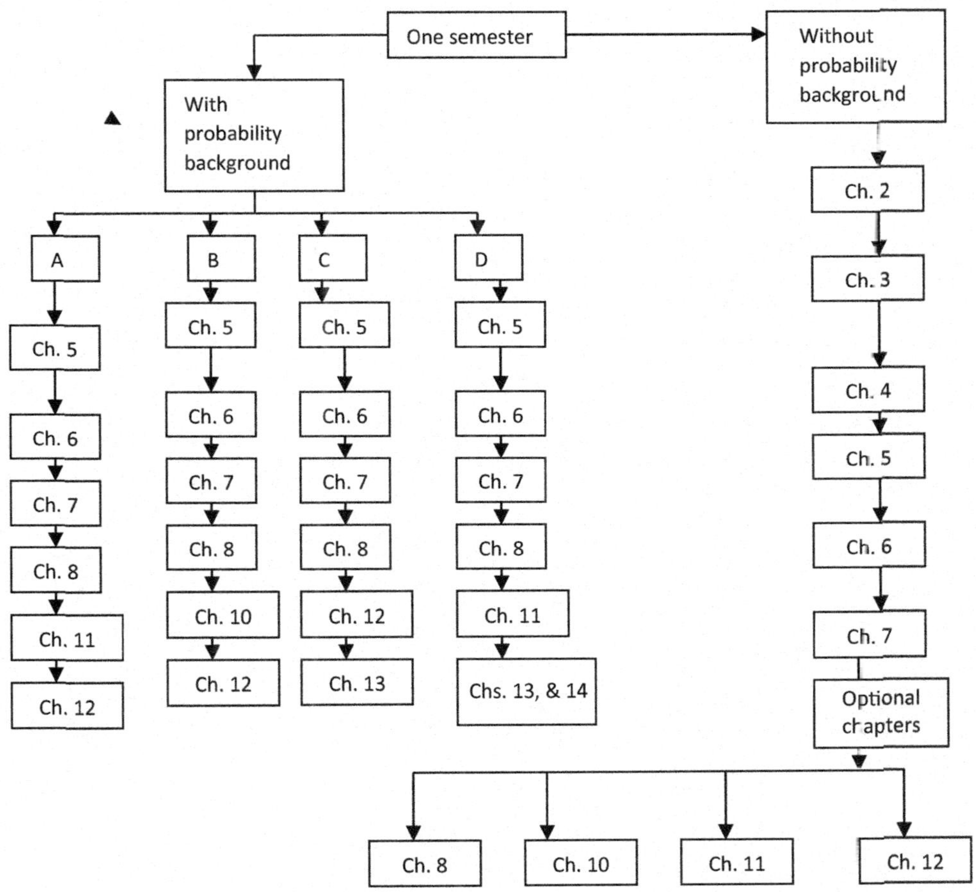

Chapter 1

Exploratory Data Analysis

Chapter outline

OBJECTIVE

Understand the data better and uncover hidden patterns by summarizing and visualizing the key characteristics of a data set.

Sir Ronald Aylmer Fisher
(*Source:* http://www.stetson.edu/~efriedma/periodictable/jpg/Fisher.jpg)

Mathematical Statistics with Applications in R. DOI: https://doi.org/10.1016/B978-0-443-27548-7.00001-3

Sir Ronald Fisher F.R.S. (1890–1962) was one of the leading scientists of the 20th century who laid the foundations for modern statistics. As a statistician working at the Rothamsted Agricultural Experiment Station, the oldest agricultural research institute in the United Kingdom, he also made major contributions to Evolutionary Biology and Genetics. The concept of randomization and the analysis of variance procedures that he introduced are now used throughout the world. In 1922 he gave a new definition of statistics. Fisher identified three fundamental problems in statistics: (1) specification of the type of population that the data came from, (2) estimation, and (3) distribution. His book *Statistical Methods for Research Workers* (1925) was used as a handbook for the methods for the design and analysis of experiments. Fisher also published the books titled *The Design of Experiments* (1935) and *Statistical* Tables (1947). While at the Agricultural Experiment Station, he conducted breeding experiments with mice, snails, and poultry, and the results he obtained led to theories about gene dominance and fitness that he published in *The Genetical Theory of Natural Selection* (1930).

1.1 INTRODUCTION

In today's society, decisions are made on the basis of data. Most scientific or industrial studies and experiments produce data, and the analysis of these data and drawing useful conclusions from them becomes one of the central issues. Statistics is an integral part of the quantitative approach to knowledge. The field of statistics is concerned with the scientific study of collecting, organizing, analyzing, and drawing conclusions from data. Statistics benefits all of us because of its ability to predict the future based on data we have previously gathered. Statistical methods help us transform data into information and knowledge. Statistical concepts enable us to solve problems in a diversity of contexts, add substance to decisions, and reduce guesswork. The discipline of statistics stemmed from the need to place knowledge management on a systematic evidence base. Earlier works on statistics dealt only with the collection, organization, and presentation of data in the form of tables and charts. In order to place statistical knowledge on a systematic evidence base, we require a study of the laws of probability. In mathematical statistics we create a probabilistic model and view the data as a set of random outcomes from that model. Advances in probability theory enable us to draw valid conclusions and to make reasonable decisions on the basis of data.

Statistical methods are used in almost every discipline, including agriculture, astronomy, biology, business, communications, economics, education, electronics, geology, health sciences, and many other fields of science and engineering, and can aid us in several ways. Modern applications of statistical techniques include statistical communication theory and signal processing, information theory, network security and denial of service problems, clinical trials, artificial and biological intelligence, quality control of manufactured items, software reliability, and survival analysis. The first of these is to assist us in designing experiments and surveys. We desire our experiment to yield adequate answers to the questions that prompted the experiment or survey. We would like the answers to be precise without involving a lot of expenditure. Statistically designed experiments facilitate development of robust products that are insensitive to changes in the environment and internal component variation. Another way that statistics assists us is in organizing, describing, summarizing, and displaying experimental data. This is termed *descriptive statistics*, or in general the area is known as Exploratory Data Analysis (EDA). A third use of statistics is in drawing inferences and making decisions based on data. For example, scientists may collect experimental data to prove or disprove an intuitive conjecture or hypothesis. Through the proper use of statistics, we can conclude whether the hypothesis is valid or not. In the process of solving a real-life problem using statistics, the following three basic steps may be identified. First, consistent with the objective of the problem, we identify the model using the appropriate statistical method. Then, we justify the applicability of the selected model to fulfill the aim of our problem. Last, we properly apply the related model to analyze the data and make the necessary decisions, which results in answering the question of our problem with minimum risk. Starting with Chapter 2, we will study the necessary background material to proceed with the development of statistical methods for solving real-world problems.

In the present chapter, we briefly review some of the basic concepts of EDA. A collection of numbers, or in the multivariate case, a rectangular array of numbers by itself, is not amenable for people to make inferences. EDA is a first and crucial step in the data analysis process. It involves investigating and summarizing the main characteristics of a data set, often using visual methods, before making any assumptions or building models. The goal of EDA is to understand the underlying structure of the data, identify patterns, spot anomalies and mistakes, and check for assumptions that will guide further analysis. We will mostly restrict EDA to univariate data. In multivariate cases, EDA will also involve determining relationships among explanatory variables and roughly assessing the direction and size of those relationships. We will present both graphical and nongraphical methods of EDA in this chapter. Graphical methods will involve summarizing data in diagrams or pictures, while nongraphical methods will involve the calculation of summary statistics. It should be noted that even in multivariate cases, it is better to perform univariate EDA on each of the component variables before performing multivariate EDA. Now, we proceed with some basic definitions and procedures.

1.1.1 Data Collection

One of the first problems that a statistician faces is obtaining the data and importing it into a suitable environment (such as R, SAS, Python). The inferences that we make depend critically on the data that we collect and analyze. Data collection involves the following important steps:

General procedure for data collection

1. Define the objectives of the problem and proceed to develop the experiment or survey.
2. Define the variables or parameters of interest.
3. Define the procedures of data collection and measuring techniques. This includes sampling procedures, sample size, and data-measuring devices (questionnaires, telephone interviews, etc.).

Example 1.1.1

We may be interested in estimating the average household income in a certain community. In this case, the parameter of interest is the average income of a typical household in the community. To acquire the data, we may send out a questionnaire or conduct a telephone interview. Once we have the data, we may first want to represent the data in graphical or tabular form to better understand its distributional behavior. Then we will use appropriate analytical techniques to estimate the parameter(s) of interest, in this case, the average household income.

Very often, a statistician is confined to data that have already been collected, possibly even collected for other purposes. This makes it very difficult to determine the quality of the data. Planned collection of data, using proper techniques, is much preferred.

1.1.2 Basic Concepts and Types of Data

Statistics is the science of data. This involves collecting, classifying, summarizing, organizing, analyzing, and interpreting data. It also involves model building. Suppose we wish to study household incomes in a certain neighborhood. We may decide to randomly select, say, 50 families and examine their household incomes. As another example, suppose we wish to determine the diameter of a rod, and we take 10 measurements of the diameter. When we consider these two examples, we note that in the first case, the population (the household incomes of all families in the neighborhood) really exists, whereas in the second, the population (set of all possible measurements of the diameter) is only conceptual. In either case, we can visualize the totality of the population values, of which our sample data are only a small part. Thus, we define a population to be the set of all measurements or objects that are of interest and a sample to be a subset of that population. The population acts as the sampling frame from which a sample is selected. Now we introduce some basic notions commonly used in statistics.

Definition 1.1.1. *A* **population** *is the collection or set of all objects or measurements that are of interest to the collector.*

Example 1.1.2

Suppose we wish to study the heights of all female students at a certain university. The population will be the set of the measured heights of all female students in the university. The population is not the set of all female students in the university.

In real-world problems it is usually not possible to obtain information on the entire population. The primary objective of statistics is to collect and study a subset of the population, called a sample, to acquire information on some specific characteristics of the population that are of interest.

Definition 1.1.2. *The* **sample** *is a subset of data selected from a population. The* **size** *of a sample is the number of elements in it.*

Example 1.1.3

We wish to estimate the percentage of defective parts produced in a factory during a given week (5 days) by examining 20 parts produced per day. The parts will be examined each day at randomly chosen times. In this case, "all parts produced during the week" is the population, and the (100) selected parts for 5 days constitute a sample.

Other common examples of sample and population are:

Political polls: The population will be all voters, whereas the sample will be the subset of voters we poll.
Laboratory experiment: The population will be all the data we could have collected if we were to repeat the experiment a large number of times (an infinite number of times) under the same conditions, whereas the sample will be the data actually collected by the one experiment.
Quality control: The population will be the entire batch of items produced, say, by a machine or by a plant, whereas the sample will be the subset of items we tested.
Clinical studies: The population will be all the patients with the same disease, whereas the sample will be the subset of patients used in the study.
Finance: All common stock listed in stock exchanges such as the New York Stock Exchange, the American Stock Exchanges, and over the counter is the population. A collection of 20 randomly picked individual stocks from these exchanges will be a sample.

The methods consisting mainly of organizing, summarizing, and presenting data in the form of tables, graphs, and charts are called *descriptive statistics*. The methods of drawing inferences and making decisions about the population using the sample are called *inferential statistics*. Inferential statistics use probability theory.

Definition 1.1.3. *A* **statistical inference** *is an estimate, a prediction, a decision, or a generalization about the population based on information contained in a sample.*

For example, we may be interested in the average indoor radiation level in homes built on reclaimed phosphate mine lands (many of the homes in west-central Florida are built on such lands). In this case, we can collect indoor radiation levels for a random sample of homes selected from this area and use the data to infer the average indoor radiation level for the entire region. In the Florida Keys, one of the concerns is that the coral reefs are declining because of the prevailing ecosystems. In order to test this, one can randomly select certain reef sites for study and, based on these data, infer whether there is a net increase or decrease in coral reefs in the region. Here, the inferential problem could be finding an estimate, such as in the radiation problem, or making a decision, such as in the coral reef problem. We will see many other examples as we progress through the book.

Types of Data

Data can be classified in several ways. We will give two different classifications, one based on whether the data are measured on a numerical scale or not and the other on whether the data are collected in the same time period or collected at different time periods.

Definition 1.1.4. Quantitative data *are observations measured on a numerical scale. Nonnumerical data that can only be classified into one of the groups of categories are said to be* **qualitative** *or* **categorical data**.

For the univariate quantitative data, EDA helps to make initial assessments about the population distribution using the observed sample values. EDA for categorical variables involves tabulation of the frequency of each category.

Example 1.1.4

Data on response to a particular therapy could be classified as no improvement, partial improvement, or complete improvement. These are qualitative data. The number of minority-owned businesses in Florida is quantitative data. The marital status of each person in a statistics class as married or not married is qualitative or categorical data. The number of car accidents in different US cities is quantitative data. The blood group of each person in a community as O, A, B, or AB is qualitative data.

Categorical data could be further classified as *nominal data* and *ordinal data.* Data characterized as nominals have data groups that do not have a specific order. An example of this could be state names, or names of the individuals, or courses by name. These do not need to be placed in any order. Data characterized as ordinal have groups that should be listed in a specific order. The order may be either increasing or decreasing. One example would be income levels. The data could have numeric values such as 1, 2, or 3 values such as high, medium, or low.

Definition 1.1.5. Cross-sectional data *is data collected on different elements or variables at the same point in time or for the same period of time.*

Example 1.1.5

The data in Table 1.1 below represent US federal support for the mathematical sciences in 1996, in millions of dollars (source: *AMS Notices*). This is an example of cross-sectional data, as the data are collected in one time period, namely, in 1996.

TABLE 1.1 Federal Support for the Mathematical Sciences, 1996.

Federal Agency	Amount
National Science Foundation	91.70
DMS	85.29
Other MPS	4.00
Department of Defense	77.30
AFOSR	16.70
ARO	15.00
DARPA	22.90
NSA	2.50
ONR	20.20
Department of Energy	16.00
University support	5.50
National laboratories	10.50
Total, all agencies	185.00

Definition 1.1.6. Time series data *are data collected on the same element or the same variable at different points in time or for different periods of time.*

Example 1.1.6

The data in Table 1.2 below represent US federal support for the mathematical sciences during the years 1995–1997, in millions of dollars (source: AMS Notices). This is an example of time series data, because they have been collected in different time periods, 1995 through 1997.

For an extensive collection of statistical terms and definitions, we can refer to many sources such as http://www.stats.gla.ac.uk/steps/glossary/index.html. We will give some other helpful Internet sources that may be useful for various aspects of statistics: http://www.amstat.org/ (American Statistical Association), http://www.stat.ufl.edu (University of Florida statistics department), http://www.statsoft.com/textbook/ (covers a wide range of topics, the emphasis is on techniques rather than concepts or mathematics), http://www.york.ac.uk/depts/maths/histstat/welcome.htm (some information about the history of statistics), http://www.isid.ac.in/ (Indian Statistical Institute), http://www.isi-web.org/30-statsoc/statsoc/282-nsslist (the International Statistical Institute), http://www.rss.org.uk/ (the Royal Statistical Society), and http://lib.stat.cmu.edu/ (an index of statistical software and routines). For energy-related statistics, refer to http://www.eia.doe.gov/. The Earth Observing System Data and Information System (https://earthdata.nasa.gov/about-eosdis) is one of the largest data sources for geological data. The Environmental Protection Agency (http://www.epa.gov/datafinder/) is another great source of data on environmental-related areas. If you want market data, YAHOO! Finance (http://finance.yahoo.com/) is a good source. There are various other useful sites that you could explore based on your particular need.

TABLE 1.2 US Federal Support for the Mathematical Sciences in Different Years.

Agency	1995	1996	1997
National Science Foundation	87.69	91.70	98.22
DMS	85.29	87.70	93.22
Other MPS	2.40	4.00	5.00
Department of Defense	77.40	77.30	67.80
AFOSR	17.40	16.70	17.10
ARO	15.00	15.00	13.00
DARPA	21.00	22.90	19.50
NSA	2.50	2.50	2.10
ONR	21.40	20.20	16.10
Department of Energy	15.70	16.00	16.00
University support	6.20	5.50	5.00
National laboratories	9.50	10.50	11.00
Total, all agencies	180.79	185.00	182.02

1.1.3 Sampling Schemes

In any statistical analysis, it is important that we clearly define the target population. The population should be defined in keeping with the objectives of the study. When the entire population is included in the study, it is called a *census* study, because data are gathered on every member of the population. In general, it is usually not possible to obtain information on the entire population because the population is too large to attempt a survey of all of its members, or it may not be cost-effective. A small but carefully chosen sample can be used to represent the population. A sample is obtained by collecting information from only some members of the population. A good sample must reflect all the characteristics (of importance) of the population. Samples can reflect the important characteristics of the populations from which they are drawn with differing degrees of precision. A sample that accurately reflects its population characteristics is called a *representative* sample. A sample that is not representative of the population characteristics is called a *biased* sample. The reliability or accuracy of conclusions drawn concerning a population depends on whether or not the sample is properly chosen so as to represent the population sufficiently well.

There are many sampling methods available. We mention a few commonly used simple sampling schemes. The choice between these sampling methods depends on (1) the nature of the problem or investigation, (2) the availability of good sampling frames (a list of all of the population members), (3) the budget or available financial resources, (4) the desired level of accuracy, and (5) the method by which data will be collected, such as questionnaires or interviews.

Definition 1.1.7. *A sample selected in such a way that every element of the population has an equal chance of being chosen is called a* **simple random sample**. *Equivalently each possible sample of size n has the same chance of being selected as any other subset of sample of size n.*

Example 1.1.7

For a state lottery, 52 identical ping-pong balls with a number from 1 to 52 painted on each ball are put in a clear plastic bin. A machine thoroughly mixes the balls, and then six are selected. The six numbers on the chosen balls are the six lottery numbers that have been selected by a simple random sampling procedure.

Some advantages of simple random sampling

1. Selection of sampling observations at random ensures against possible investigator biases.
2. Analytic computations are relatively simple, and probabilistic bounds on errors can be computed in many cases.
3. It is frequently possible to estimate the sample size for a prescribed error level when designing the sampling procedure.

Simple random sampling may not be effective in all situations. For example, in a US presidential election, it may be more appropriate to conduct sampling polls by state, rather than a nationwide random poll. It is quite possible for a candidate to get a majority of the popular vote nationwide and yet lose the election. We now describe a few other sampling methods that may be more appropriate in a given situation.

Definition 1.1.8. **A systematic sample** *is a sample in which every Kth element in the sampling frame is selected after a suitable random start for the first element. We list the population elements in some order (say alphabetical) and choose the desired sampling fraction.*

Steps for selecting a systematic sample

1. Number the elements of the population from 1 to N.
2. Decide on the sample size, say n, that we need.
3. Choose $K = N/n$.
4. Randomly select an integer between 1 and K.
5. Then take every Kth element.

Example 1.1.8

If the population has 1000 elements arranged in some order and we decide to sample 10% (i.e., $N = 1000$ and $n = 100$), then $K = 1000/100 = 10$. Pick a number at random between 1 and $K = 10$ inclusive, say 3. Then select elements numbered 3, 13, 23, ..., 993.

Systematic sampling is widely used because it is easy to implement. If the population elements are ordered, systematic sampling is a better sampling method. If the list of population elements is in random order to begin with, then the method is similar to simple random sampling. If, however, there is a correlation or association between successive elements, or if there is some periodic structure, then this sampling method may introduce biases. Systematic sampling is often used to select a specified number of records from a computer file.

Definition 1.1.9. *A sample obtained by stratifying (dividing into nonoverlapping groups) the sampling frame based on some factor or factors and then selecting some elements from each of the strata is called a* **stratified sample***. Here, a population with* N *elements is divided into* s *subpopulations. A sample is drawn from each subpopulation independently. The size of each subpopulation and sample sizes in each subpopulation may vary.*

A stratified sample is a modification of simple random sampling and systematic sampling and is designed to obtain a more representative sample but at the cost of a more complicated procedure. Compared to random sampling, stratified sampling reduces sampling error.

Steps for selecting a stratified sample

1. Decide on the relevant stratification factors (sex, age, race, income, etc.).
2. Divide the entire population into strata (subpopulations) based on stratification criteria. Sizes of strata may vary.
3. Select the requisite number of units using simple random sampling or systematic sampling from each subpopulation. The requisite number may depend on the subpopulation sizes.

Examples of strata might be males and females, undergraduate students and graduate students, managers and nonmanagers, or populations of clients in different racial groups such as African Americans, Asians, Whites, and Hispanics. Stratified sampling is often used when one or more of the strata in the population have a low incidence relative to the other strata. Through stratified random sampling, adequate representation of all subgroups can be ensured.

Example 1.1.9

In a population of 1000 children from an area school, there are 600 boys and 400 girls. We divide them into strata based on their parents' income, as shown in Table 1.3.

TABLE 1.3 Classification of School Children.

	Boys	Girls
Poor	120	240
Middle class	150	100
Rich	330	60

This is stratified data.

Example 1.1.10

Refer to Example 1.1.9. Suppose we decide to sample 100 children from the population of 1000 (i.e., 10% of the population). We also choose to sample 10% from each of the categories. For example, we would choose 12 (10% of 120) poor boys; 6 (10% of 60) rich girls, and so forth. This yields Table 1.4. This particular sampling method is called *proportional stratified sampling*.

TABLE 1.4 **Proportional Stratification of School Children.**

	Boys	Girls
Poor	12	24
Middle class	15	10
Rich	33	6

Some uses of stratified sampling

1. In addition to providing information about the whole population, this sampling scheme provides information about the subpopulations, the study of which may be of interest. For example, in a US presidential election, opinion polls by state may be more important in deciding on the Electoral College advantage than a national opinion poll.
2. Stratified sampling can be considerably more precise than a simple random sample, because the population is fairly homogeneous within each stratum, but there is a sizable variation between the strata.

Definition 1.1.10. *In* **cluster sampling**, *the sampling unit contains groups of elements called clusters instead of individual elements of the population. A cluster is an intact group naturally available in the field. Unlike the stratified sample, where the strata are created by the researcher based on stratification variables, the clusters naturally exist and are not formed by the researcher for data collection. Cluster sampling is also called* **area sampling**.

To obtain a cluster sample, first take a simple random sample of groups and then sample all elements within the selected clusters (groups). Cluster sampling is convenient to implement. When cost and time are important, cluster sampling may be used. However, because it is likely that units in a cluster will be relatively homogeneous, this method may be less precise than simple random sampling. The standard errors of estimates in cluster sampling are higher than in other sampling designs.

Example 1.1.11

Suppose we wish to select a sample of about 10% from all fifth-grade children of a county. We randomly select 10% of the elementary schools assumed to have approximately the same number of fifth-grade students and select all fifth-grade children from these schools. This is an example of cluster sampling, each cluster being an elementary school that was selected.

Definition 1.1.11. Multiphase sampling *involves the collection of some information from the whole sample and additional information either at the same time or later from subsamples of the whole sample. The multiphase or multistage sampling is basically a combination of the techniques presented earlier.*

Example 1.1.12

An investigator in a population census may ask basic questions such as sex, age, or marital status for the whole population, but only 10% of the population may be asked about their level of education or about how many years of mathematics and science education they had.

Errors in Sample Data

Irrespective of which sampling scheme is used, the sample observations are prone to various sources of error that may seriously affect the inferences about the population. Some sources of error can be controlled. However, others may be unavoidable because they are inherent in the nature of the sampling process. Consequently, it is necessary to understand the different types of errors for a proper interpretation and analysis of the sample data. The errors can be classified as *sampling errors* and *nonsampling errors*. Nonsampling errors occur in the collection, recording, and processing of sample data. For example, such errors could occur as a result of bias in the selection of elements of the sample, poorly designed survey questions, measurement and recording errors, incorrect responses, or no responses from individuals selected from the population. Sampling errors occur because the sample is not an exact representative of the population. Sampling error is due to the differences between the characteristics of the population and those of a sample from the population. For example, we are interested in the average test score in a large statistics class of size, say, 80. A sample of size 10 grades from this class resulted in an average test score of 75. If the average test for the entire 80 students (the population) is 72, then the sampling error is $75 - 72 = 3$.

Sample Size

In almost any sampling scheme designed by statisticians, one of the major issues is the determination of the sample size. In principle, this should depend on the variation in the population as well as on the population size and on the required reliability of the results, that is, the amount of error that can be tolerated. For example, if we are taking a sample of school children from a neighborhood with a relatively homogeneous income level to study the effect of parents' affluence on the academic performance of the children, it is not necessary to have a large sample size. However, if the income level varies a great deal in the feeding area of the school, then we will need a larger sample size to achieve the same level of reliability. In practice, another influencing factor is the available resources, such as money and time. In later chapters, we present some methods of determining sample size in statistical estimation problems.

The literature on sample survey methods is constantly changing with new insights that demand dramatic revisions in the conventional thinking. We know that representative sampling methods are essential to permit confident generalizations of results to populations. However, there are many practical issues that can arise in real-life sampling methods. For example, in sampling related to social issues, whatever the sampling method we employ, a high response rate must be obtained. It has been observed that most telephone surveys have difficulty in achieving response rates higher than 60%, and most face-to-face surveys have difficulty in achieving response rates higher than 70%. Even a well-designed survey may stop short of the goal of a perfect response rate. This might induce bias in the conclusions based on the sample we obtained. A low response rate can be devastating to the reliability of a study. We can obtain a series of publications on surveys, including guidelines on avoiding pitfalls, from the American Statistical Association (www.amstat.org). In this book, we deal mainly with samples obtained using simple random sampling.

EXERCISES 1.1

1.1.1 Give your own examples for qualitative and quantitative data. Also, give examples for cross-sectional and time series data.

1.1.2 Discuss how you will collect different types of data. What inferences do you want to derive from each of these types of data?

1.1.3 Refer to the data in Example 1.2.4. State a few questions that you can ask about the data. What inferences can you make by looking at these data?

1.1.4 Refer to the data in Example 1.2.5. Can you state a few questions that the data suggest? What inferences can you make by looking at these data?

1.1.5 Give your own examples for each of the sampling methods described in this section. Discuss the merits and limitations of each of these methods.

1.1.6 Using the information obtained from the publications of the American Statistical Association (www.amstat.org) or any other reference, write a short report on how to collect survey data and what the potential sources of error are.

1.2 GRAPHICAL REPRESENTATION OF DATA

The source of our statistical knowledge lies in the data. Once we obtain the sample data values, one way to become acquainted with them is through data visualization techniques such as displaying them in tables or graphically. Charts and graphs are very important tools in statistics because they communicate information visually and, in a way, they compress knowledge. Remember, our interest in the data lies with the story it tells. These visual displays may reveal the patterns of behavior of the variables being studied. In this chapter, we will consider one-variable data. The most common graphical displays are the *frequency table*, *pie chart*, *bar graph*, *Pareto chart*, and *histogram*. For example, in the business world, graphical representations of data are used as statistical tools for everyday process management and improvements by decision-makers (such as managers and frontline staff) to understand processes, problems, and solutions. The purpose of this section is to introduce several tabular and graphical procedures commonly used to summarize both qualitative and quantitative data. Tabular and graphical summaries of data can be found in reports, newspaper articles, websites, and research studies, among others.

Now we shall introduce some ways of graphically representing both qualitative and quantitative data. Bar graphs and Pareto charts are useful displays for qualitative data. With bar graphs, we can see how different things are distributed between separate categories. In practice, if there are too many categories, it may be helpful to compare only a limited number of categories or combine categories with very short bars into say, others, and draw the bar graphs.

Definition 1.2.1. *A graph of bars whose heights represent the frequencies (or relative frequencies) of respective categories is called a* **bar graph**.

Example 1.2.1

The data in Table 1.5 represent the percentages of price increases of some consumer goods and services for the period December 1990 to December 2000 in a certain city. Construct a bar chart for these data.

TABLE 1.5 Percentages of Price Increases of Some Consumer Goods and Services.

Medical care	83.3%
Electricity	22.1%
Residential rent	43.5%
Food	41.1%
Consumer price index	35.8%
Apparel and upkeep	21.2%

Solution

In the bar graph of Fig. 1.1, *we use MC for medical care El for electricity, RR for residential rent, Fd for food, CPI for consumer price index, and A & U for apparel and upkeep.*

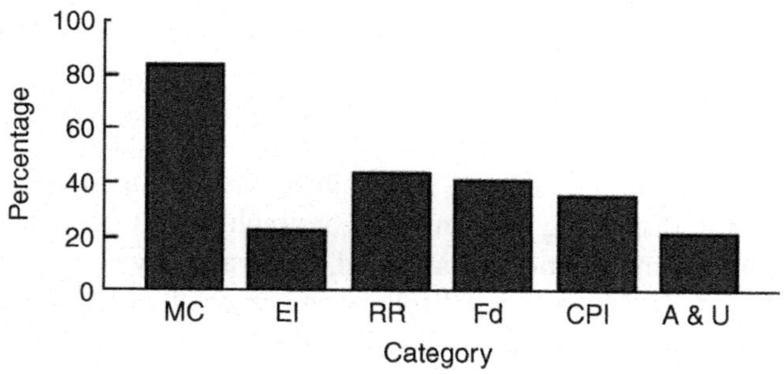

FIGURE 1.1 Percentage price increase of consumer goods.

Looking at Fig. 1.1, we can identify where the maximum and minimum responses are located, so that we can descriptively discuss the phenomenon whose behavior we want to understand.

For a graphical representation of the relative importance of different factors under study, one can use the *Pareto chart.* It is a bar graph with the height of the bars proportional to the contribution of each factor. The bars are displayed from the most numerous categories to the least numerous category, as illustrated by the following example. A Pareto chart helps in separating the few significant factors that have larger influence from the many trivial ones.

Example 1.2.2

For the data of Example 1.2.1, construct a Pareto chart.

Solution

First, rewrite the data in decreasing order. Then create a Pareto chart by displaying the bars from the most numerous category to the least numerous category.

Looking at Fig. 1.2, we can identify the relative importance of each category such as the maximum, the minimum, and the general behavior of the subject data.

Vilfredo Pareto (1848–1923), an Italian economist and sociologist, studied the distributions of wealth in different countries. He concluded that about 20% of people controlled about 80% of a society's wealth. This same distribution has been observed in other areas, such as quality improvement: 80% of problems usually stem from 20% of the causes. This phenomenon has been termed the Pareto effect or 80/20 rule. Pareto charts are used to display the Pareto principle, arranging data so that the few vital factors that are causing most of the problems reveal themselves. Focusing improvement efforts on these few causes will have a larger impact and be more cost-effective than undirected efforts.

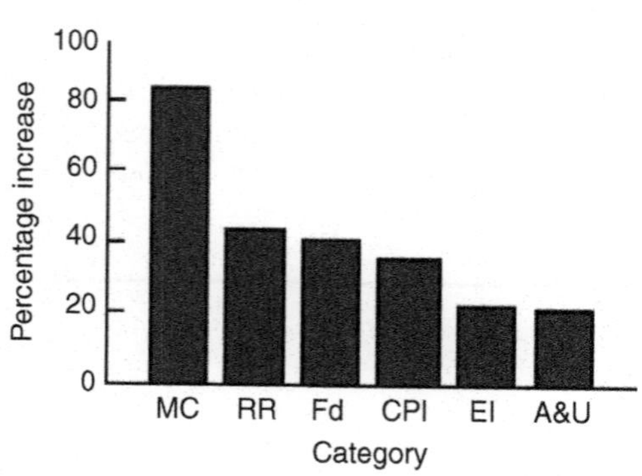

FIGURE 1.2 Pareto chart.

Pareto charts are used in business decision-making as a problem-solving and statistical tool that ranks problem areas, or sources of variation, according to their contribution to cost or to total variation.

Definition 1.2.2. *A circle divided into sectors that represent the percentages of a population or a sample that belongs to different categories is called a* **pie chart**.

Pie charts are especially useful for presenting categorical data. The pie "slices" are drawn such that they have an area proportional to the frequency. The entire pie represents all the data, whereas each slice represents a different class or group within the whole. Thus, we can look at a pie chart and identify the various percentages of interest and how they compare among themselves. Most statistical software can create 3D charts. Such charts are attractive; however, they can make pieces at the front look larger than they really are. In general, a two-dimensional view of the pie is preferable.

Example 1.2.3

The combined percentages of carbon monoxide (CO) and ozone (O_3) emissions from different sources are listed in Table 1.6. Construct a pie chart.

TABLE 1.6 Combined Percentages of CO and O_3 Emissions.

Transportation (T)	Industrial Process (I)	Fuel Combustion (F)	Solid Waste (S)	Miscellaneous (M)
63%	10%	14%	5%	8%

Solution

The pie chart is given in Fig. 1.3.

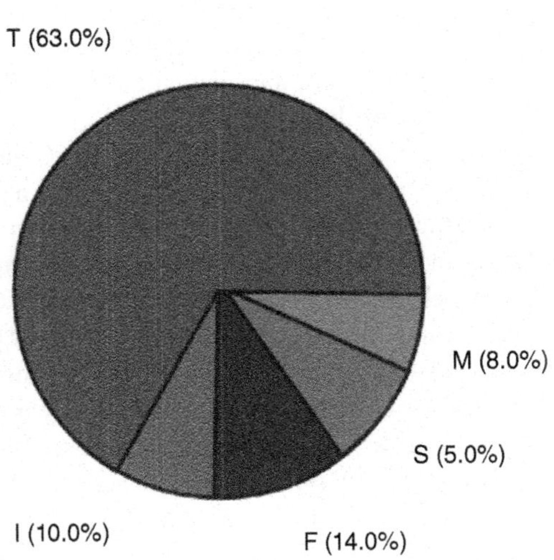

FIGURE 1.3 Pie chart for CO and O_3.

Definition 1.2.3. *A* **stem-and-leaf plot** *is a simple way of summarizing quantitative data and is well suited to computer applications. When data sets are relatively small, stem-and-leaf plots are particularly useful. In a stem-and-leaf plot, each data value is split into a "stem" and a "leaf." The "leaf" is usually the last digit of the number and the other digits to the left of the "leaf" form the "stem." Usually, there is no need to sort the leaves, although computer packages typically do. For more details, we refer the student to elementary statistics books. We illustrate this technique with an example.*

Example 1.2.4

Construct a stem-and-leaf plot for the 20 test scores given below.

78	74	82	66	94	71	64	88	55	80
91	74	82	75	96	78	84	79	71	83

Solution

At a glance, we see that the scores are distributed from the 50s through the 90s. We use the first digit of the score as the stem and the second digit as the leaf. The plot in Table 1.7 is constructed with stems in the vertical position.

TABLE 1.7 Stem-and-Leaf Display of 20 Exam Scores.

Stem	Leaves							
5	5							
6	6	4						
7	8	4	1	4	5	8	9	1
8	2	8	0	2	4	3		
9	4	1	6					

The stem-and-leaf plot condenses the data values into a useful display from which we can identify the shape and distribution of data, such as the symmetry, where the maximum and minimum are located with respect to the frequencies, and whether they are bell-shaped. This fact, that the frequencies are bell-shaped, will be of paramount importance as we proceed to study inferential statistics. Also, note that the stem-and-leaf plot retains the entire data set and can be used only with quantitative data. Examples 1.7.7 and 1.7.8 explain how to obtain a stem-and-leaf plot and histogram using SPSS. Refer to Section 1.7.3 for SAS commands to generate graphical representations of the data.

A *frequency table* is a table that divides a data set into a suitable number of categories (classes). Rather than retaining the entire set of data in a display, a frequency table essentially provides only a count of those observations that are associated with each class. Once the data are summarized in the form of a frequency table, a graphical representation can be given through bar graphs, pie charts, and histograms. Data presented in the form of a frequency table are called *grouped data*. A frequency table is created by choosing a specific number of classes in which the data will be placed. Generally, the classes will be intervals of equal length. The center of each class is called a *class mark*. The endpoints of each class interval are called class boundaries. Usually, there are two ways of choosing class boundaries. One way is to choose nonoverlapping class boundaries so that none of the data points will simultaneously fall in two classes. Another way is that for each class, except for the last, the upper boundary is equal to the lower boundary of the subsequent class. When forming a frequency table this way, one or more data values may fall on a class boundary. One way to handle such a problem is to arbitrarily assign it to one of the classes or to flip a coin to determine the class into which to place the observation at hand.

Definition 1.2.4. *Let f_i denote the frequency of the class i and let n be the sum of all frequencies. Then the* **relative frequency** *for the class i is defined as the ratio f_i/n. The* **cumulative relative frequency** *for the class i is defined by $\sum_{k=1}^{i} f_k/n$.*

The following example illustrates the foregoing discussion.

Example 1.2.5

The following data give the lifetime of 30 incandescent light bulbs (rounded to the nearest hour) of a particular type.

872	931	1146	1079	915	879	863	1112	979	1120
1150	987	958	1149	1057	1082	1053	1048	1118	1088
868	996	1102	1130	1002	990	1052	1116	1119	1028

Construct a frequency, relative frequency, and cumulative relative frequency table.

Solution

Note that there are n = 30 observations and that the largest observation is 1150 and the smallest one is 863, with a range of 287. We will choose six classes each with a length of 50.

Class	**Frequency f_i**	**Relative Frequency $\frac{f_i}{\Sigma f_i}$**	**Cumulative Relative Frequency $\Sigma_{k=1}^{i} \frac{f_k}{n}$**
50–900	4	4/30	4/30
900–950	2	2/30	6/30
950–1000	5	5/30	11/30
1000–1050	3	3/30	14/30
1050–1100	6	6/30	20/30
1100–1150	10	10/30	30/30

When data are quantitative in nature, and the number of observations is relatively large, and there are no natural separate categories or classes, we can use a histogram to simplify and organize the data. Since the classes are listed in order, histograms are great to identify the range and skew of the quantitative data.

Definition 1.2.5. *A* **histogram** *is a graph in which classes are marked on the horizontal axis, and either the frequencies, relative frequencies, or percentages are represented by the heights on the vertical axis. In a histogram, the bars are drawn adjacent to each other without any gaps.*

Histograms can be used only for quantitative data. In the univariate graphical EDA, histograms are used to get the sample's distributional characteristics. A histogram compresses a data set into a compact picture that shows the location of the mean and modes of the data and the variation in the data, especially the range. It identifies patterns in the data. This is a good aggregate graph of one variable. In order to obtain variability in the data, it is always good practice to start with a histogram of the data. The histogram is one of the simplest yet most informative tools in statistics. Looking at the histogram, we can obtain the shape of the distribution, such characteristics as symmetry, modality (unimodal, bimodal, or multimodal), and uniformity. Unlike qualitative data, quantitative data does not come with natural class separation points. For this, the following steps can be used as a general guideline to construct a frequency table and produce a histogram.

Guidelines FOR THE CONSTRUCTION OF A FREQUENCY TABLE AND HISTOGRAM

1. Determine the maximum and minimum values of the observations. The range, R = *maximum value – minimum value*.
2. Select from five to 20 classes that, in general, are non-overlapping intervals of equal length, so as to cover the entire range of data. The goal is to use enough classes to show the variation in the data, but not so many that there are only a few data points in many of the classes. The class width should be slightly larger than the ratio

$$\frac{\textit{Largest value Smallest value}}{\textit{Number of classes}}.$$

3. The first interval should begin a little below the minimum value, and the last interval should end a little above the maximum value. The intervals are called class intervals, and the boundaries are called class boundaries. The class limits are the smallest and the largest data values in the class. The class mark is the midpoint of a class.
4. None of the data values should fall on the boundaries of the classes.
5. Construct a table (frequency table) that lists the class intervals, a tabulation of the number of measurements in each class (tally), the frequency f_i of each class, and, if needed, a column with relative frequency, f_i/n, where n is the total number of observations.
6. Draw bars over each interval with heights being the frequencies (or relative frequencies).

Let us illustrate implementing these steps in the development of a histogram for the data given in the following example.

Example 1.2.6

The following data refer to a certain type of chemical impurity measured in parts per million in 25 drinking-water samples randomly collected from different areas of a county.

11	19	24	30	12	20	25	29	15	21
24	31	16	23	25	26	32	17	22	26
35	18	24	18	27					

(a) Make a frequency table displaying class intervals, frequencies, relative frequencies, and percentages.
(b) Construct a frequency histogram.

Solution

(a) *We will use five classes. The maximum and minimum values in the data set are 35 and 11. Hence the class width is* $(35-11)/5 = 4.8 \approx 5$. *Hence, we shall take the class width to be 5. The lower boundary of the first class interval will be chosen to be 10.5. With five classes, each of width 5, the upper boundary of the fifth class becomes 35.5. We can now construct a frequency table for the data.*

Class	**Class Interval**	f_i = Frequency	**Relative Frequency**	**Percentage**
1	10.5–15.5	3	3/25 = 0.12	12
2	15.5–20.5	6	6/25 = 0.24	24
3	20.5–25.5	8	8/25 = 0.32	32
4	25.5–30.5	5	5/25 = 0.20	20
5	30.5–35.5	3	3/25 = 0.12	12

(b) *We can generate a histogram as in* Fig. 1.4.

From the histogram, we should be able to identify the center (i.e., the location) of the data, spread of the data, skewness of the data, presence of outliers, presence of multiple modes in the data, and whether the data can be capped with a bell-shaped curve. These properties provide indications of the proper distributional model for the data.

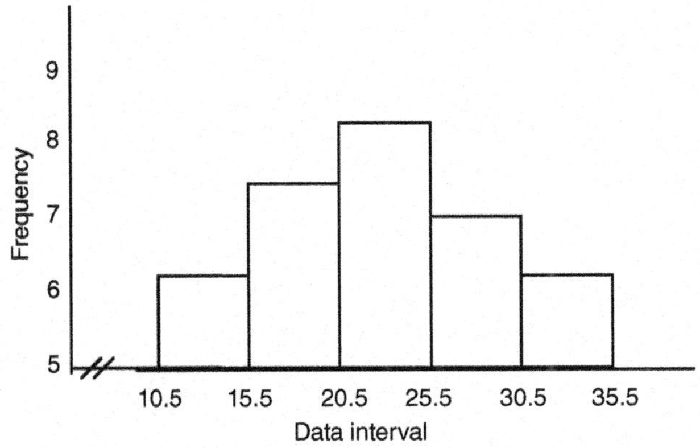

FIGURE 1.4 Frequency histogram of impurity data.

Another graphical tool used to compare the distribution of a data set to a theoretical distribution is **Quantile-Quantile plot (Q-Q plot)**. Q-Q plots are also used to compare two data sets. Since it is a bit more difficult, Q-Q plots will be discussed in Chapter 15.

EXERCISES 1.2

1.2.1. According to the recent US Federal Highway Administration *Highway Statistics*, the percentages of freeways and expressways in various road mileage-related highway pavement conditions are as follows:

Poor 10%, Mediocre 32%, Fair 22%, Good 21%, and Very good 15%.

(a) Construct a bar graph.

(b) Construct a pie chart.

1.2.2. More than 75% of all species that have been described by biologists are insects. Of the approximately 2 million known species, only about 30,000 are aquatic in any life stage. The data in Table 1.8 give the proportion of total species by insect order that can survive exposure to salt (source: http://entomology.unl.edu/).

TABLE 1.8 Percentage of Species by Insect.

Species	Percentage	Species	Percentage
Coleoptera	26%	Odonata	3%
Diptera	35%	Thysanoptera	3%
Hemiptera	15%	Lepidoptera	1%
Orthoptera	6%	Other	6%
Collembola	5%		

(a) Construct a bar graph.
(b) Construct a Pareto chart.
(c) Construct a pie chart.

1.2.3. The data in Table 1.9 are presented to illustrate the role of renewable energy consumption in the US energy supply in 2007 (source: http://www.eia.doe.gov/fuelrenewable.html). Renewable energy consists of biomass, geothermal energy, hydroelectric energy, solar energy, and wind energy.

TABLE 1.9 Renewable Energy Consumption.

Source	Percentage
Coal	22%
Natural gas	23%
Nuclear electric power	8%
Petroleum	40%
Renewable energy	7%

(a) Construct a bar graph.
(b) Construct a Pareto chart.
(c) Construct a pie chart.

1.2.4. A litter is a group of babies born from the same mother at the same time. Table 1.10 gives some examples of different mammals and their average litter size (source: http://www.saburchill.com/chapters/chap0032.html).

TABLE 1.10 Litter Size of Mammals.

Species	Litter Size
Bat	1
Dolphin	1
Chimpanzee	1
Lion	3
Hedgehog	5
Red fox	6
Rabbit	6
Black rat	11

(a) Construct a bar graph.
(b) Construct a Pareto chart.

1.2.5. The following data give the letter grades of 20 students enrolled in a statistics course.

A	B	F	A	C	C	D	A	B	F
C	D	B	A	B	A	F	B	C	A

(a) Construct a bar graph.
(b) Construct a pie chart.

1.2.6. According to the US Bureau of Labor Statistics (BLS), the median weekly earnings of full-time wage and salary workers by age for the third quarter of 1998 is given in Table 1.11.

TABLE 1.11 Weekly Wages and Salary Distribution by Age.

16–19 years	\$260
20–24 years	\$334
25–34 years	\$498
35–44 years	\$600
45–54 years	\$628
55–64 years	\$605
65 years and over	\$393

Construct a pie chart and bar graph for these data and interpret. Also, construct a Pareto chart.

1.2.7. The data in Table 1.12 is a breakdown of 18,930 workers in a town according to the type of work. Construct a pie chart and bar graph for these data and interpret.

1.2.8. The data in Table 1.13 represent the number (in millions) of adults and children living with HIV/AIDS by the end of 2000 according to the region of the world (source: http://w3.whosea.org/hivaids/factsheet.htm). Construct a bar graph for these data. Also, construct a Pareto chart and interpret.

1.2.9. The data in Table 1.14 give the life expectancy at birth, in years, from 1900 through 2000 (source: National Center for Health Statistics). Construct a bar graph for this data.

TABLE 1.12 Distribution of Workers by Type of Work.

Mining	58
Construction	1161
Manufacturing	2188
Transportation and public utilities	821
Wholesale trade	657
Retail trade	7377
Finance, insurance, and real estate	890
Services	5778
Total	18,930

TABLE 1.13 Number of People With HIV/AIDS.

Country	Adults and Children Living With HIV/AIDS (in Millions)
Sub-Saharan Africa	25.30
North Africa and Middle East	0.40
South and Southeast Asia	5.80
East Asia and Pacific	0.64
Latin America	1.40
Caribbean	0.39
Eastern Europe and Central Asia	0.70
Western Europe	0.54
North America	0.92
Australia and New Zealand	0.15

TABLE 1.14 Life Expectancy at Birth.

Year	Life Expectancy
1900	47.3
1960	69.7
1980	73.7
1990	75.4
2000	77.0

1.2.10. Dolphins are usually identified by the shape and pattern of notches and nicks on their dorsal fin. Individual dolphins are cataloged by classifying the fin based on the location of distinguishing marks. When a dolphin is sighted, its picture can then be compared to the catalog of dolphins in the area, and if a match is found, the dolphin can be recorded as resighted. These methods of mark-resight are for developing databases regarding the life history of individual dolphins. From these databases, we can calculate the levels of association between dolphins, population estimates, and general life history parameters such as birth and survival rates. The data in Table 1.15 represent frequently resighted individuals (as of January 2000) at a particular location (source: http://www.eckerd.edu/dolphinproject/biologypr.html).

TABLE 1.15 Number of Dolphin Resights by Type.

Hammer (adult female)	59
Mid Button Flag (adult female)	41
Luseal (adult female)	31
84 Lookalike (adult female)	20

Construct a bar graph for this data.

1.2.11. The data in Table 1.16 give death rates (per 100,000 population) for 10 leading causes in 1998 (source: National Center for Health Statistics, US Department of Health and Human Services).
(a) Construct a bar graph.
(b) Construct a Pareto chart.

1.2.12. In the fiscal year, a city collected $32.3 million in revenues. City spending for that year is expected to be nearly the same, with no tax increase projected.

Expenditure: Reserves 0.7%, capital outlay 29.7%, operating expenses 28.9%, debt service 3.2%, transfers 5.1%, personal services 32.4%.
Revenues: Property taxes 10.2%, utility and franchise taxes 11.3%, licenses and permits 1%, intergovernmental revenue 10.1%, charges for services 28.2%, fines and forfeits 0.5%, interest and miscellaneous 2.7%, transfers and cash carryovers 36%.
(a) Construct bar graphs for expenditure and revenues and interpret.
(b) Construct pie charts for expenditure and revenues and interpret.

TABLE 1.16 Death Rate by Cause.

Cause	Death Rate
Accidents and adverse effects	34.5
Chronic liver disease and cirrhosis	9.7
Chronic obstructive lung diseases and allied conditions	42.3
Cancer	199.4
Diabetes mellitus	23.9
Heart disease	268.0
Kidney disease	9.7
Pneumonia and influenza	35.1
Stroke	58.5
Suicide	10.8

1.2.13. Construct a histogram for the 24 examination scores given below:

78	74	82	66	94	71	64	88	55	80	73	86
91	74	82	75	96	78	84	79	71	83	78	79

1.2.14. The following table gives radon concentration in pCi/liter obtained from 40 houses in a certain area.

2.9	0.6	13.5	17.1	2.8	3.8	16.0	2.1	6.4	17.2
7.9	0.5	13.7	11.5	2.9	3.6	6.1	8.8	2.2	9.4
15.9	8.8	9.8	11.5	12.3	3.7	8.9	13.0	7.9	11.7
6.2	6.9	12.8	13.7	2.7	3.5	8.3	15.9	5.1	6.0

(a) Construct a stem-and-leaf display.
(b) Construct a frequency histogram and interpret.
(c) Construct a pie chart and interpret.

1.2.15. The following data give the mean of SAT Mathematics scores by state for 1999 for a randomly selected 20 states (source: *The World Almanac and Book of Facts 2000*).

558	503	565	572	546	517	542	605	493	499
568	553	510	525	595	502	526	475	506	568

(a) Construct a stem-and-leaf display and interpret.
(b) Construct a frequency histogram and interpret.
(c) Construct a pie chart and interpret.

1.2.16. A sample of 25 measurements is given here:

9	28	14	29	21	27	15	23	23	10
31	23	16	26	22	17	19	24	21	20
26	20	16	14	21					

(a) Make a frequency table displaying class intervals, frequencies, relative frequencies, and percentages.
(b) Construct a frequency histogram and interpret.

1.2.17. We may be interested in the changing demographics of the US population. The following table gives the demographics in 2010 (Overview of Race and Hispanic Origin: 2010, http://www.census.gov/prod/cen2010/briefs/c2010br-02.pdf). Table 1.17 gives a pretty good summary understanding.

TABLE 1.17 US Population Demographics.

Race/Ethnicity	Number	Percentage of US Population
White or European American	223,553,265	72.4%
Black or African American	38,929,319	12.6%
Asian American	14,674,252	4.8%
American Indian or Alaska Native	2,932,248	0.9%
Native Hawaiian or other Pacific Islander	540,013	0.2%
Some other race	19,107,368	6.2%
Two or more races	9,009,073	2.9%
Not Hispanic nor Latino	258,267,944	83.6%
Non-Hispanic White or European American	196,817,552	63.7%
Non-Hispanic Black or African American	37,685,848	12.2%
Non-Hispanic Asian	14,465,124	4.7%
Non-Hispanic American Indian or Alaska Native	2,247,098	0.7%
Non-Hispanic Native Hawaiian or other Pacific Islander	481,576	0.2%
Non-Hispanic some other race	604,265	0.2%
Non-Hispanic two or more races	5,966,481	1.9%
Hispanic or Latino	50,477,594	16.4%
White or European American Hispanic	26,735,713	8.7%

(Continued)

TABLE 1.17 (Continued)

Race/Ethnicity	Number	Percentage of US Population
Black or African American Hispanic	1,243,471	0.4%
American Indian or Alaska Native Hispanic	685,150	0.2%
Asian Hispanic	209,128	0.1%
Native Hawaiian or other Pacific Islander Hispanic	58,437	0.0%
Some other race Hispanic	18,503,103	6.0%
Two or more races Hispanic	3,042,592	1.0%
Total	308,745,538	100.0%

Draw a pie chart.

1.3 NUMERICAL ASSESSMENT OF UNIVARIATE DATA

In the previous section, we looked at some graphical and tabular techniques for describing a data set. We shall now consider some numerical characteristics of a set of measurements (quantitative data). Since the major aim of univariate EDA for quantitative data is to make rough assessments about population distribution using the sample data, the characteristics are its center, modality (how many peaks are in the pdf), spread, shape, and outliers. Suppose that we have a sample with values $x_1, x_2, \ldots, x_n$. There are many characteristics associated with this data set, for example, the central tendency and variability. A measure of the central tendency is given by the sample mean, median, or mode, and the measure of dispersion or variability is usually given by the sample variance or sample standard deviation or interquartile range.

Definition 1.3.1. *Let $x_1, x_2, \ldots, x_n$ be a set of sample values. Then the* **sample mean** *(or* **empirical mean***) $\overline{x}$ is defined by*

$$\overline{x} = \frac{1}{n} \sum_{i=1}^{n} x_i.$$

The **sample variance** *is defined by*

$$s^2 = \frac{1}{(n-1)} \sum_{i=1}^{n} (x_i - \overline{x})^2.$$

The **sample standard deviation** *is*

$$s = \sqrt{s^2}.$$

The sample variance s^2 and the sample standard deviation s are both measures of the variability or "scatteredness" of data values around the sample mean $\overline{x}$. The larger the variance, the more is the spread. We note that s^2 and s are both nonnegative. One question we may ask is "why not just take the sum of the differences $(x_i - \overline{x})$ as a measure of variation?" The answer lies in the following result which shows that if we add up all deviations about the sample mean, we always get a zero value.

Theorem 1.3.1. *For a given set of measurements $x_1, x_2, \ldots, x_n$, let $\overline{x}$ be the sample mean. Then*

$$\sum_{i=1}^{n} (x_i - \overline{x}) = 0.$$

Proof. Since $\overline{x} = (1/n) \sum_{i=1}^{n} x_i$, we have $\sum_{i=1}^{n} x_i = n\overline{x}$. Now,

$$\begin{aligned} \textstyle\sum_{i=1}^{n} (x_i - \overline{x}) &= \textstyle\sum_{i=1}^{n} x_i - \sum_{i=1}^{n} \overline{x} \\ &= n\overline{x} - n\overline{x} = 0. \end{aligned}$$

Thus, although there may be a large variation in the data values, $\sum_{i=1}^{n}(x_i - \overline{x})$ as a measure of spread would always be zero, implying no variability. So, it is not useful as a measure of variability.

Sometimes we can simplify the calculation of the sample variance s^2 by using the following computational formula:

$$s^2 = \frac{\left[\sum_{i=1}^{n} x_i^2 - \frac{1}{n}\left(\sum_{i=1}^{n} x_i\right)^2\right]}{(n-1)}.$$

If the data set has a large variation with some extreme values (called outliers), the mean may not be a very good measure of the center. For example, average salary may not be a good indicator of the financial well-being of the employees of a company if there is a huge difference in pay between support personnel and management personnel. In that case, one could use the median as a measure of the center, roughly 50% of data fall below and 50% above. The median is less sensitive to extreme data values.

Definition 1.3.2. *For a data set, the* **median** *is the middle number of the ordered data set. If the data set has an even number of elements, then the median is the average of the middle two numbers. The* **lower quartile** *is the middle number of the half of the data below the median, and the* **upper quartile** *is the middle number of the half of the data above the median. We will denote*

$$Q_1 = \text{lower quartile}$$
$$Q_2 = M = \text{middle quartile (median)}$$
$$Q_3 = \text{upper quartile.}$$

The difference between the quartiles is called the **interquartile range** *(IQR).*

$$IQR = Q_3 - Q_1.$$

A possible outlier (mild outlier) will be any data point that lies below

$$Q_1 - 1.5(IQR) \text{ or above } Q_3 + 1.5(IQR).$$

Thus, about 25% of the data lie below Q_1, and about 75% of the data lie below Q_3. Note that the IQR is unaffected by the positions of those observations in the smallest 25% or the largest 25% of the data.

Mode is another commonly used measure of central tendency. A mode indicates where the data tend to concentrate most.

Definition 1.3.3. Mode *is the most frequently occurring member of the data set. If all the data values are different, then by definition, the data set has no mode.*

Example 1.3.1

The following data give the time in months from hire to promotion to manager for a random sample of 25 software engineers from all software engineers employed by a large telecommunications firm.

5	7	229	453	12	14	18	14	14	483
22	21	25	23	24	34	37	34	49	64
47	67	69	192	125					

Calculate the mean, median, mode, variance, and standard deviation for this sample.

Solution

The sample mean is

$$\overline{x} = \frac{1}{n}\sum_{i=1}^{n} x_i = 83.28 \text{ months.}$$

To obtain the median, first arrange the data in ascending order:

5	7	12	14	14	14	18	21	22	23
24	25	34	34	37	47	49	64	67	69
125	192	229	453	483					

Now the median is the 13th number, which is 34 months.
Since 14 occurs most often (thrice), the mode is 14 months.
The sample variance is

$$s^2 = \frac{1}{n-1}\sum_{i=1}^{n}(x_i - \overline{x})^2$$
$$= \frac{1}{24}[(5 - 83.28)^2 + \cdots + (125 - 83.28)^2]$$
$$= 16,478.$$

and the sample standard deviation is, $s = \sqrt{s^2} = 128.36$ *months. Thus, we have a sample mean* $\overline{x} = 83.28$ *months, median = 34 months, and mode = 14 months. Note that the mean is very different from the other two measures of the center because of a few large data values. Also, the sample variance* $s^2 = 16{,}478$ *months, and the sample standard deviation* $s = 128.36$ *months.*

Example 1.3.2

For the data of Example 1.3.1, find lower and upper quartiles, median, and IQR. Check for any outliers.
Solution
Arrange the data in ascending order.

5	7	12	14	14	14	18	21	22	23
24	25	34	34	37	47	49	64	67	69
125	192	229	453	483					

Then the median M is the middle (13th) data value, $M = Q_2 = 34$. *The lower quartile is the middle number below the median,* $Q_1 = [(14\ 18)/2] = 16$. *The upper quartile,* $Q_3 = [(67\ 69)/2] = 68$.
The $IQR = Q_3 - Q_1 = 68 - 16 = 52$.
To test for outliers, compute

$$Q_1 - 1.5(IQR) = 16 - 1.5\,(52) = -\,62$$

and

$$Q_3 + 1.5(IQR) = 68 + 1.5(52) = 146.$$

Then all the data that falls above 146 are possible outliers. None is below −62. Therefore, the outliers are 192, 229, 453, and 483.

We have remarked earlier that the mean as a measure of central location is greatly affected by the extreme values or outliers. A robust measure of central location (a measure that is relatively unaffected by outliers) is the *trimmed mean*. For $0 \le \alpha \le 1$, a $100\alpha\%$ trimmed mean is found as follows: Order the data and then discard the lowest $100\alpha\%$ and the highest $100\alpha\%$ of the data values. Find the mean of the rest of the data values. We denote the $100\alpha\%$ trimmed mean by $\overline{x}_\alpha$. We illustrate the trimmed mean concept in the following example.

Example 1.3.3

For the data set representing the number of children in a random sample of 10 families in a neighborhood, find the 10% trimmed mean ($\alpha = 0.1$).

$$1 \quad 2 \quad 2 \quad 3 \quad 2 \quad 3 \quad 9 \quad 1 \quad 6 \quad 2.$$

Solution

Arrange the data in ascending order.

$$1 \quad 1 \quad 2 \quad 2 \quad 2 \quad 2 \quad 3 \quad 3 \quad 6 \quad 9.$$

The data set has 10 elements. Discarding the lowest 10% (10% of 10 is 1) and discarding the highest 10% of the data values, we obtain the trimmed data set as

$$1 \quad 2 \quad 2 \quad 2 \quad 2 \quad 3 \quad 3 \quad 6.$$

The 10% trimmed mean is

$$\bar{x}_{0.1} = \frac{1+2+2+2+2+3+3+6}{8} = 2.6.$$

Note that the mean for the data in the previous example, without removing any observations, is 3.1, which is different from the trimmed mean.

Although standard deviation is a more popular method, there are other measures of dispersion, such as average deviation or interquartile range. We have already seen the definition of interquartile range. The *average deviation* for a sample $x_1, \ldots, x_n$ is defined by

$$\text{Average deviation} = \frac{\sum_{i=1}^{n} |x_i - \bar{x}|}{n}.$$

Calculation of average deviation is simple and straightforward.

1.3.1 Numerical Measures for Grouped Data

When we encounter situations where the data are grouped in the form of a frequency table (see Section 1.2), we no longer have individual data values. Hence, we cannot use the formulas in Definition 1.3.1. The following formulas will give approximate values for $\bar{x}$ and s^2. Let the grouped data have l classes, with m_i being the midpoint and f_i being the frequency of class i, $i = 1, 2, \ldots, l$. Let $n = \sum_{i=1}^{l} f_i$.

Definition 1.3.4. *The* **mean** *for a sample of size n,*

$$\bar{x} = \frac{1}{n} \sum_{i=1}^{l} f_i m_i,$$

where m_i is the midpoint of the class i and f_i is the frequency of the class i.
Similarly, the **sample variance,**

$$s^2 = \frac{1}{n-1} \sum_{i=1}^{n} f_i (m_i - \bar{x})^2 = \frac{\sum m_i^2 f_i - \frac{\left(\sum_i f_i m_i\right)^2}{n}}{n-1}.$$

The following example illustrates how we calculate the sample mean for grouped data.

Example 1.3.4

The grouped data in Table 1.18 represents the number of children from birth through the end of the teenage years in a large apartment complex. Find the mean, variance, and standard deviation for these data:

TABLE 1.18 Number of Children and Their Age Group.

Class	0–3	4–7	8–11	12–15	16–19
Frequency	7	4	19	12	8

Here we use the usual convention that, until the child attains the next year, their age will be the previous year; for instance, until the child is 4 years old, we will say the child is 3 years old.

Solution

Note that even though the classes are given as disjoint, in actuality these are adjacent age intervals, like [0, 4), [4, 8), and so forth. When we take the class midpoint, we have to take this into account. For simplicity of calculation, we create Table 1.19.

TABLE 1.19 Summary Statistics for Number of Children.

Class	Interval	f_i	m_i	$m_i f_i$	$m_i^2 f_i$
0–3	[0, 4)	7	2	14	28
4–7	[4, 8)	4	6	24	144
8–11	[8, 12)	19	10	190	1900
12–15	[12, 16)	12	14	168	2352
16–19	[16, 20)	8	18	144	2592
		$n = 50$	$\sum m_i f_i = 540$	$\sum m_i^2 f_i = 7016$	

The sample mean is

$$\bar{x} = \frac{1}{n}\sum_i f_i m_i = \frac{540}{50} = 10.80.$$

The sample variance is

$$s^2 = \frac{\sum m_i^2 f_i - \frac{\left(\sum_i f_i m_i\right)^2}{n}}{n-1} = \frac{7016 - \frac{(540)^2}{50}}{49} = 24.1632650 \approx 24.16.$$

The sample standard deviation is $s = \sqrt{s^2} = 4.9156144 \approx 4.92.$

Using the following calculations, we can also find the *median for grouped data*. We only know that the median occurs in a particular class interval, but we do not know the exact location of the median. We will assume that the measures are spread evenly throughout this interval. Let

L = lower class limit of the interval that contains the median.
n = total frequency.
F_b = cumulative frequencies for all classes before the median class.

f_m = frequency of the class interval containing the median.
w = interval width of the interval that contains the median.

Then the median for the grouped data is given by

$$M = L + \frac{w}{f_m}(0.5n - F_b).$$

We proceed to illustrate with an example.

Example 1.3.5

For the data in Example 1.3.4, find the median.
Solution
First, we develop Table 1.20.

TABLE 1.20 Frequency Distribution for Number of Children.

Class	f_i	Cumulative f_i	*Cumulative f_i/n*
0–3	7	7	0.14
4–7	4	11	0.22
8–11	19	30	0.6
12–15	12	42	0.84
16–19	8	50	1.00

The first interval for which the cumulative relative frequency exceeds 0.5 is the interval that contains the median. Hence, the interval 8 to 11 contains the median. Therefore, $L = 8, f_m = 19, n = 50, w = 3$, and $F_b = 11$. Then, the median is

$$M = L + \frac{w}{f_m}(0.5n - F_b) = 8 + \frac{4}{19}((0.5)(50) - 11) = 10.947.$$

It is important to note that all the numerical measures we calculate for grouped data are only approximations to the actual values of the ungrouped data if they are available.

One of the uses of the sample standard deviation will be clear from the following result, which is based on the data following a bell-shaped curve. Such an indication can be obtained from the histogram or stem-and-leaf display.

Empirical rule

When the histogram of a data set is "bell-shaped" or "mound-shaped" and symmetric, the *empirical rule* states:

1. Approximately 68% of the data are in the interval $(\bar{x} - s, \bar{x} + s)$.
2. Approximately 95% of the data are in the interval $(\bar{x} - 2s, \bar{x} + 2s)$.
3. Approximately 99.7% of the data are in the interval $(\bar{x} - 3s, \bar{x} + 3s)$.

The bell-shaped curve is called a normal curve and is discussed later in Chapter 3. A typical symmetric bell-shaped curve is given by Fig. 1.5.

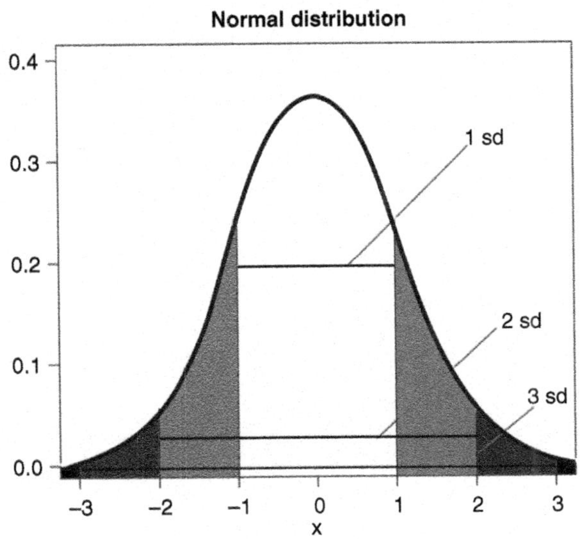

FIGURE 1.5 Bell-shaped curve.

1.3.2 Box Plots and Outlier Detection

The sample mean, or the sample standard deviation, focuses on a single aspect of the data set, whereas histograms and stem-and-leaf displays express rather general ideas about the data. A pictorial summary called a box plot (also called box-and-whisker plots) can be used to describe several prominent features of a data set, such as the center, the spread, the extent and nature of any departure from symmetry, and identification of outliers. Box plots are a simple diagrammatic representation of the five number summary: minimum, lower quartile, median, upper quartile, maximum.

Procedure to construct a box plot

1. Draw a vertical measurement axis and mark Q_1, Q_2 (median), and Q_3 on this axis as shown in Fig. 1.6, below. Let $IQR = Q_3 - Q_1$.
2. Construct a rectangular box whose bottom edge lies at the lower quartile, Q_1, and whose upper edge lies at the upper quartile, Q_3.
3. Draw a horizontal line segment inside the box through the median.
4. Extend the lines from each end of the box out to the farthest observation that is still within 1.5(IQR) of the corresponding edge. These lines are called *whiskers*.
5. Draw an open circle (or asterisks *) to identify each observation that falls between 1.5(IQR) and 3(IQR) from the edge to which it is closest; these are called *mild outliers*.
6. Draw a solid circle to identify each observation that falls more than 3(IQR) from the closest edge; these are called *extreme outliers*.

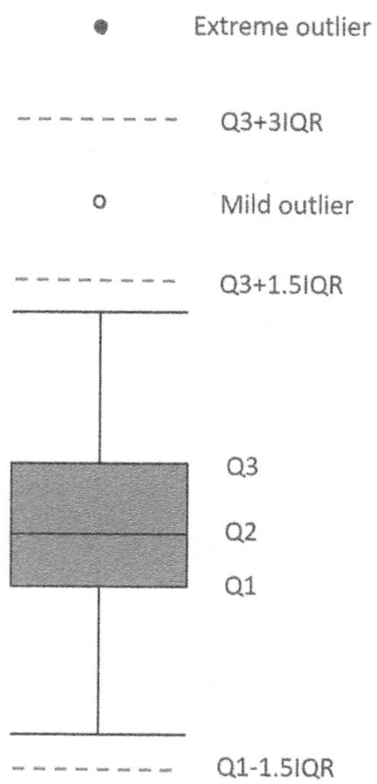

FIGURE 1.6 A typical box-and-whiskers plot.

We illustrate the procedure with the following example.

Example 1.3.6

The following data identify the time in months from hire to promotion to chief pharmacist for a random sample of 25 employees from a certain group of employees in a large corporation of drugstores.

5	7	229	453	12	14	18	14	14	483
22	21	25	23	24	34	37	34	49	64
47	67	69	192	125					

Construct a box plot. Does the data appear to be symmetrically distributed along the measurement axis?
Solution
Referring to Example 1.3.2, we find that the median, $Q_2 = 34$.

The lower quartile is $Q_1 = \frac{14+18}{2} = 16$.

The upper quartile is $Q_3 = \frac{67+69}{2} = 68$.

The interquartile range is $IQR = 68 - 16 = 52$.
To find the outliers, compute

$$Q_1 - 1.5(IQR) = 16 - 1.5(52) = -62$$

and

$$Q_3 + 1.5(IQR) = 68 + 1.5(52) = 146.$$

Using these numbers, we follow the procedure outlined earlier to construct the box plot shown by Fig. 1.7. *The * in the box plot represents an outlier. The first horizontal line is the first quartile, the second is the median, and the third is the third quartile.*

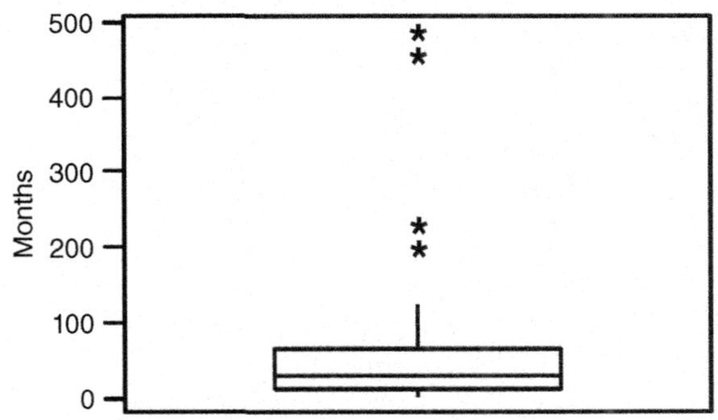

FIGURE 1.7 Box plot for months to promotion.

By examining the relative position of the median line (the middle line in Fig. 1.7), we can test the symmetry of the data. For example, in Fig. 1.7, the median line is closer to the lower quartile than the upper line, which suggests that the distribution is slightly nonsymmetric. Also, a look at this box plot alone shows the presence of two mild outliers and two extreme outliers. However, we have $Q_3 + 3(IQR) = 68 + 3(52) = 224$. Thus, in reality, there are three extreme outliers and one mild outlier.

A box plot is an effective tool to visualize an entire range of data. Box plots can tell us if the data is uniform or diverse and give us a broad overview of the data at hand that will help us ask more questions in a practical application as well as selection of analytical methods.

Skewness and kurtosis of a distribution are two additional univariate EDA descriptors. Asymmetry in distribution is measured by skewness. Kurtosis measures the peakedness of a distribution relative to normal distribution shape. Exact formulas to calculate these quantities and further discussions are given in Chapter 2.

EXERCISES 1.3

1.3.1. The prices of 12 randomly chosen homes in dollars (approximated to nearest thousand) in a growing region of Tampa in the summer of 2002 are given below.

176 105 133 140 305 215 207 210 173 150 78 96

Find the mean and standard deviation of the sampled home prices from this area.

1.3.2. The following is a sample of nine mortgage companies' interest rates for 30-year home mortgages, assuming 5% down.

7.625 7.500 6.625 7.625 6.625 6.875 7.375 5.375 7.500

(a) Find the mean and standard deviation and interpret.

(b) Find lower and upper quartiles, median, and interquartile range. Check for any outliers and interpret.

1.3.3. For four observations, it is given that the mean is 6, median is 4, and mode is 3. Find the standard deviation of this sample.

1.3.4. The data given below pertain to a random sample of disbursements of state highway funds (in millions of dollars), to different states.

1188	1050	2882	2802	780	1171	685
537	519	2523	316	1117	1578	261

(a) Find the mean, variance, and range for these data and interpret.
(b) Find lower and upper quartiles, median, and interquartile range. Check for any outliers and interpret.
(c) Construct a box plot and interpret.

1.3.5. Maximal static inspiratory pressure (PImax) is an index of respiratory muscle strength. The following data show the measure of PImax (cm H_2O) for 15 cystic fibrosis patients.

105	80	115	95	100	85	90	70
135	105	45	115	40	115	95	

(a) Find the lower and upper quartiles, median, and interquartile range. Check for any outliers and interpret.
(b) Construct a box plot and interpret.
(c) Are there any outliers?

1.3.6. Compute the mean, variance, and standard deviation for the data in Table 1.21 (assume that the data belongs to a sample).

TABLE 1.21 Class and Frequency.

Class	0–4	5–9	10–14	15–19	20–24
Frequency	5	14	15	10	6

1.3.7. **(a)** For any grouped data with l classes with group frequencies f_i, and class midpoints m_i, show that

$$\sum_{i=1}^{l} f_i\,(m_i - \bar{x}) = 0.$$

(b) Verify this result for the data given in Exercise 1.3.6.

1.3.8. **(a)** Given the sample values $x_1, x_2, \ldots, x_n$, show that

$$\sum_{i=1}^{n} (x_i - \bar{x})^2 = \sum_{i=1}^{n} x_i^2 - \frac{(\sum_{i=1}^{n} x_i)^2}{n}.$$

(b) Verify the result of part (a) for the data of Exercise 1.3.5.

1.3.9. The following are the closing prices of some securities that a mutual fund holds on a certain day:

10.25	5.31	11.25	13.13	18.00	32.56	37.06	39.00
43.25	45.00	40.06	28.56	22.75	51.50	47.00	53.50
32.00	25.44	22.50	30.00	24.75	53.37	51.38	26.00
53.50	29.87	32.00	28.87	42.19	37.50	30.44	41.37

(a) Find the mean, variance, and range for these data and interpret.
(b) Find lower and upper quartiles, median, and interquartile range. Check for any outliers.
(c) Construct a box plot and interpret.
(d) Construct a histogram.
(e) Locate on your histogram $\bar{x}$, $\bar{x} \pm s$, $\bar{x} \pm 2s$, and $\bar{x} \pm 3s$. Count the data points in each of the intervals $\bar{x} \pm s$, $\bar{x} \pm 2s$, and $\bar{x} \pm 3s$, and compare this with the empirical rule.

1.3.10. The radon concentration (in pCi/liter) data obtained from 40 houses in a certain area are given below.

2.9	0.6	13.5	17.1	2.8	3.8	16.0	2.1	6.4	17.2
7.9	0.5	13.7	11.5	2.9	3.6	6.1	8.8	2.2	9.4
15.9	8.8	9.8	11.5	12.3	3.7	8.9	13.0	7.9	11.7
6.2	6.9	12.8	13.7	2.7	3.5	8.3	15.9	5.1	6.0

(a) Find the mean, variance, and range for these data.
(b) Find lower and upper quartiles, median, and interquartile range. Check for any outliers.
(c) Construct a box plot.
(d) Construct a histogram and interpret.
(e) Locate on your histogram $\bar{x} \pm s$, $\bar{x} \pm 2s$, and $\bar{x} \pm 3s$. Count the data points in each of the intervals $\bar{x}$, $\bar{x} \pm s$, $\bar{x} \pm 2s$, and $\bar{x} \pm 3s$. How do these counts compare with the empirical rule?

1.3.11. A random sample of 100 households' weekly food expenditure represented by x from a particular city gave the following statistics:

$$\sum x_i = 11,000, \text{ and } \sum x_i^2 = 1,900,000.$$

(a) Find the mean and standard deviation for these data.
(b) Assuming that the food expenditure of the households of an entire city of 400,000 will have a bell-shaped distribution, how many households of this city would you expect to fall in each of the intervals, $\bar{x} \pm s$, $\bar{x} \pm 2s$, and $\bar{x} \pm 3s$?

1.3.12. The following numbers are the hours put in by 10 employees of a company in a randomly selected week:

40 46 40 54 18 45 34 60 39 42

(a) Calculate the values of the three quartiles and the interquartile range. Also, calculate the mean and standard deviation and interpret.
(b) Verify from this data set that $\sum_{i=1}^{10} (x_i - \bar{x}) = 0$.
(c) Construct a box plot.
(d) Does this data set contain any outliers?

1.3.13. For the following data:

6.3	2.9	4.5	1.1	1.8	4.0	1.2	3.1	2.0	4.0
7.0	2.8	4.3	5.3	2.9	8.3	4.4	2.8	3.1	5.6
4.5	4.5	5.7	0.5	6.2	3.7	0.9	2.4	3.0	3.5

(a) Find the mean, variance, and standard deviation.
(b) Construct a frequency table with five classes.
(c) Using the grouped data formula, find the mean, variance, and standard deviation for the frequency table constructed in part (b) and compare it to the results in part (a).

1.3.14. In order to assess the protective immunizing activity of various whooping cough vaccines, suppose that 30 batches of different vaccines are tested on groups of children. Suppose that the following data give immunity percentage in home exposure values (IPHE values).

85	51	41	90	91	40	39	69	45	47
42	12	70	38	97	34	94	77	88	91
79	90	43	40	89	85	71	30	25	21

(a) Find the mean, variance, and standard deviation and interpret.
(b) Construct a frequency table with five classes.
(c) Using the grouped data formula, find the mean, variance, and standard deviation for the table in part (b) and compare it to the results in part (a).

1.3.15. The grouped data in Table 1.22 give the number of births by age group of mothers between ages 10 and 39 in a certain state in 2000.

TABLE 1.22 Number of Births by Mother's Age Group.

Age of Mother	Number of Births
10–14	895
15–19	55,373
20–24	122,591
25–29	139,615
30–34	127,502
35–39	68,685

Find the median for this grouped data and interpret.

1.3.16. Table 1.23 gives the distribution of the masses (in grams) of 50 salmon from a single young cohort.

TABLE 1.23 Distribution of Salmon Mass.

Weight	155–164	165–174	175–184	185–194	195–204
Frequency	8	11	18	9	4

(a) Using the grouped data formula, find the mean, variance, and standard deviation.
(b) Find the median for this grouped data.

1.3.17. After a pollution accident, 180 dead fish were recovered from a stream. Table 1.24 gives their lengths measured to the nearest millimeter.

TABLE 1.24 Length of Dead Fish.

Length of fish (mm)	1–19	20–39	40–59	60–79	80–99
Frequency	38	31	59	45	7

(a) Using the grouped data formula, find the mean, variance, and standard deviation.
(b) Find the median for this grouped data and interpret.

1.4 DATA QUALITY EVALUATION

In any data-driven system—whether in artificial intelligence (AI), machine learning, or statistical modeling—the quality of data determines the reliability and interpretability of results.

Poor quality data can lead to biased models, misleading inferences, and incorrect decisions. Thus, Data Quality Evaluation is a foundational step in the data analysis pipeline, ensuring that the data used is accurate, consistent, complete, and relevant for the intended purpose. Data quality evaluation is the process of assessing the reliability of data by measuring its *accuracy*, *completeness*, *consistency*, *validity*, and *timeliness*.

Key Dimensions of Data Quality

Accuracy measures how closely the data values represent the true values. For example, if a sensor records temperature, accuracy indicates how close the recorded temperature is to the actual temperature. In order to assess accuracy, we can compare with ground truth or reference data set and calculate error measures such as Mean Absolute Error (MAE) or percentage deviation.

Completeness assesses whether all required data values are present. Missing or null values reduce the completeness of a data set. Completeness is assessed by

$$\text{Completeness} = 1 - \frac{\text{Number of missing values}}{\text{Total expected values}}.$$

Consistency checks whether data follow the same format, units, or definitions across sources or time. Inconsistent formats (e.g., "FL" vs. "Florida") cause integration and interpretation errors. Rule-based validation or schema comparison is used as an assessment method.

Validity ensures that data values fall within acceptable ranges or conform to domain rules.

For instance, a negative age or a date in the future is invalid. Range or constraint validation is the way to assess validation.

Timeliness measures whether the data are up-to-date. For example, outdated sales records may not reflect current trends. Timeliness can be measured using:

$$\text{Timeliness} = 1 - \frac{\text{Age of data}}{\text{Maximum allowable data age}}.$$

Uniqueness ensures that each record appears only once. Duplicate records can introduce bias. Uniqueness can be measured by

$$\text{Uniqueness} = 1 - \frac{\text{Duplicate records}}{\text{Total records}}.$$

Example 1.4.1

An AI model predicts student performance based on features such as attendance, homework score, and exam results, reproduced in Table 1.25.

TABLE 1.25 Student Class Record.

Student	Attendance (%)	Homework	Exam	Final Grade
1	95	90	88	89
2	85	88	NA	85
3	100	91	92	90
4	70	-10	80	78
5	105	85	86	84

Identify quality issues.

Solution

Looking at the data, we can identify the following quality issues. We have a missing value, Exam = NA for Student 2. This is a completeness issue. There is an invalid entry, Homework = − 10 for Student 4. Also, there is an out-of-range value, Attendance = 105 for Student 5. These are validity issues. These could have arisen from potential data entry errors.

We will not go into more details on data quality evaluation.

1.5 COMPUTERS AND STATISTICS

With present-day technology, we can automate most statistical calculations. For small sets of data, graphing calculators are sufficient for many basic calculations, such as finding means and standard deviations and creating simple charts. Students should learn how to perform statistical analysis using their handheld calculators. For deeper analysis and for large data sets, statistical software is necessary. Software also provides easier data entry and editing and much better graphics in comparison to calculators. There are many statistical packages available. Many such analyses can be performed with spreadsheet application programs such as Microsoft Excel, but a more thorough data analysis requires the use of more sophisticated software such as R, SAS, Python, and SPSS. For some students with programming abilities, packages such as MATLAB® may be more appealing. For very large data set and for complicated data analysis, one could use SAS. SAS is one of the most frequently used statistical packages. Many other statistical packages (such as Splus and StatXact) are available; the utilities and advantages of each are based on the specific application and personal taste. The software R is free software that is being increasingly used by statisticians and can be downloaded from http://www.r-project.org/, and many statistical tutorials for R are freely available on the Internet. For a good introduction to doing statistics with R, refer to the book by Peter Dalgaard, *Introductory Statistics with R (Second Edition)*, Springer, 2008, or its newer edition.

In this book, we will give some representative R, SPSS, and SAS commands at the end of each chapter just to get students started on the technology. These examples are by no means a tutorial for the respective software. For a more thorough understanding and use of technology, students should look at the users' manual that comes with the software or references given at the end of the book. The computer commands are designed to be illustrative, rather than completely efficient. In dealing with data analysis for real-world problems, we need to know which statistical procedure to use, how to prepare the data sets suitable for use in the particular statistical package, and finally how to interpret the results obtained. A good knowledge of theory supplemented with a good working knowledge of statistical software will enable students to perform sophisticated statistical analyses while understanding the underlying assumptions and the limitations of results obtained. This will prevent us from making misleading conclusions when using computer-generated statistical outputs.

1.6 CHAPTER SUMMARY

In this chapter, we dealt with some basic aspects of univariate EDA. First, we gave basic definitions of terms such as *population* and *sample*. Some sampling techniques were discussed. We learned about some graphical presentations in Section 1.4. In Section 1.5 we dealt with descriptive statistics, in which we learned how to find mean, median, and variance and how to identify outliers. A brief discussion of technology and statistics was given in Section 1.6. All the examples given in this chapter are for a univariate population, in which each measurement consists of a single value. Many populations are *multivariate*, where measurements consist of more than one value. For example, we may be interested in finding a relationship between blood sugar level and age or between body height and weight. These types of problems will be discussed in Chapter 8.

In practice, it is always better to run descriptive statistics as a check on one's data. The graphical and numerical descriptive measures can be used to verify that the measurements are sound and that there are no obvious errors due to collection or coding.

We now list some of the key definitions introduced in this chapter.

- Population
- Sample
- Statistical inference
- Quantitative data
- Qualitative or categorical data

- Cross-sectional data
- Time series data
- Simple random sample
- Systematic sample
- Stratified sample
- Proportional stratified sampling
- Cluster sampling
- Multiphase sampling
- Relative frequency
- Cumulative relative frequency
- Bar graph
- Pie chart
- Histogram
- Sample mean
- Sample variance
- Sample standard deviation
- Median
- Interquartile range
- Mode
- Mean
- Empirical rule
- Box plots

In this chapter, we have also introduced the following important concepts and procedures:

- General procedure for data collection
- Some advantages of simple random sampling
- Steps for selecting a stratified sample
- Procedures to construct frequency and relative frequency tables and graphical representations such as stem-and-leaf displays, bar graphs, pie charts, histograms, and box plots
- Procedures to calculate measures of central tendency, such as mean and median, as well as measures of dispersion, such as the variance and standard deviation for both ungrouped and grouped data
- Guidelines for the construction of frequency tables and histograms
- Procedures to construct a box plot

1.7 COMPUTER EXAMPLES

In this section, we give some examples of how to use SPSS and SAS for creating graphical representations of the data, as well as methods for the computation of basic statistics. Sometimes, the output obtained using a particular software package may not be exactly as explained in the book; they vary from one package to another and also depend on the particular software version. In fact, most of the outputs will not be shown in this book. It is important to obtain the explanation of output from the help menu of the particular software package for complete understanding. The "Computer Examples" sections of this book are not designed as manuals for the software, nor are they written in the most efficient way. The idea is only to introduce some basic procedures, so that the students can get started with applying the theoretical material they have seen in each of the chapters.

1.7.1 R Introduction and Examples

R is a free software for statistical computing and graphics that you can download from http://www.r-project.org/. Detailed help manuals are available from this site. In addition, you can get R help from numerous sources. One such book can be obtained at http://www.ecostat.unical.it/tarsitano/Didattica/LabStat2/Everitt.pdf. In this book, we are only introducing the reader to basic R-programming as a starting point. The R-commands are not optimal, nor are they comprehensive. If you don't have experience with R-program, we suggest that you start working with R-studio (https://www.rstudio.com/), which is much easier to work with its Windows interface.

How to input data?

Example 1.7.1 Using the following data:

66	74	79	80	69	77	78	65	79	81

We will make a single variable data set or vector named *x*. First manually and second using the *scan()* function for convenience.

R Code:

```
x=c(66,74,79,80,69,77,78,65,79,81);
OR
x=scan();
1: 66
2: 74
3: 79
4: 80
5: 69
6: 77
7: 78
8: 65
9: 79
10: 81
11:
```

Typing the comma can be time-consuming

This method allows you to type each number pressing enter between each entry designed with the number pad in mind. Notice the last entry is blank which ends the scan function.

Results: Both methods obtain the same output, which can be seen simply by typing *x* or *cat(x)* or *print(x)*. However, the scan method allows you to rapidly type your numbers into the variable using a numpad and enter key.

Importing a CSV file

It is common to import CSV (comma-separated value) files into R. This imports Example 7.7.1 data into variable *x*.

This example assumes your file is located on a D:\ drive. You may need to modify the path preceding the file name for the csv you wish to import.

R Code:

```
x=read.csv("D:/ch7_1.csv");
```

Results:

You should have obtained a variable containing the data from the CSV file; these files can be opened with Notepad to see their contents.

Exporting a CSV

It is common to export a CSV file of data you wish to save, back up (verb), or share.

Using R, we will export the following data:

```
Sample 1 (x): 1 2 3 4 5 6 7 8 9 10
```

This example is writing to the path C:\Users\Admin\Documents. Please modify the path to work on your computer.

R Code:

```
x=c(1:10);
write.csv(x,"C:\Users\Admin\Documents\myfile.csv");
```

Results: This should have created the specified file in the specified location. You can open this file with Notepad and should see the exported data.

Example 1.7.2 (Stem-and-leaf plot) **Using the following data, construct a stem-and-leaf plot.**

```
Sample X: 78 74 82 66 94 71 64 88 55 80 91 74 82 75 96 78 84 79 71 83
```

This assumes you've stored the data under variable x. Please modify your code appropriately.

R Code:

```
stem(x);
```

Output:

```
The decimal point is 1 digit(s) to the right of the |
5 | 5
6 | 46
7 | 11445889
8 | 022348
9 | 146
```

Example 1.7.3 (Histogram) **Using the following data, construct a histogram.**

```
Sample X: 25 37 20 31 31 21 12 25 36 27 38 16 40 32 33 24 39 26 27 19
```

This assumes you've stored the data into variable x. Please modify your code appropriately.

R Code:

```
hist(x);
```

Output:

Example 1.7.4 (Descriptive statistics) **Using the following data, generate descriptive statistics.**

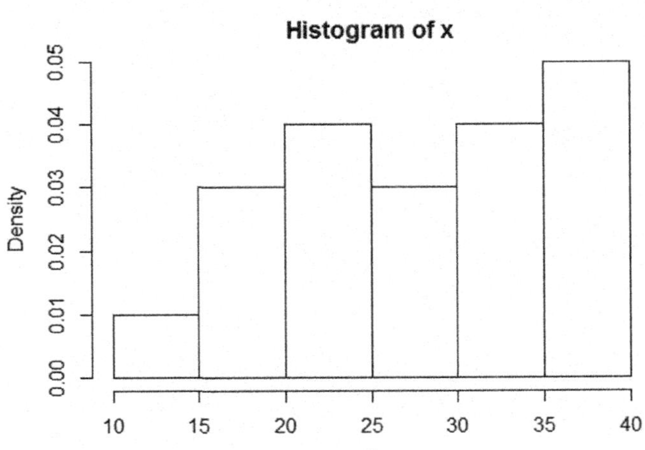

Sample X: 5 7 229 453 12 14 18 14 18 14 14 483 22 21 25 23 24 34 37 34 49 64 47 67 69 192 125

This assumes you've stored the data into variable *x*. Please modify your code appropriately.

R Code:

```
summary(x);
sd(x);       <- Standard Deviation
length(x);   <- Length of variable
```

Output:

```
Output:
Min. 1st Qu.  Median    Mean 3rd Qu.    Max.
5.00   18.00   34.00   83.28   67.00  483.00

128.3649   <- Standard Deviation

25   <- Length of variable
```

Example 1.7.5 (Box plot) Using the following data, create a box plot.

Sample X: 870 922 1146 1120 1079 905 888 865 1112 966 1150 977 958 1088 1139 1055 1082 1053 1048 1118 866 996 1102 1028 1130 1002 990 1052 1116 1109

This assumes you've stored the data into variable *x*. Please adjust your code appropriately.

R Code:

```
boxplot(x);
```

Output:

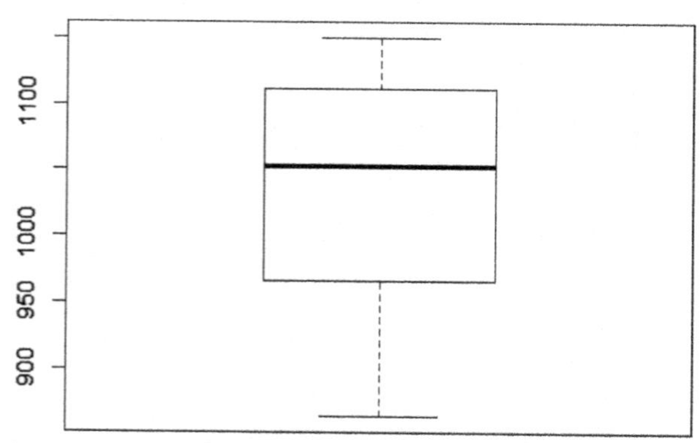

Example 1.7.6 (Test of Randomness) Using the following data, test whether or not the sample is random (details of this test are left undisclosed):

```
Sample X: 24 31 28 43 28 56 48 39 52 32 38 49 51 49 62 33 41 58 63 56
```

This test is known as "Runs test" and assumes you've stored the data into variable x. Please modify your code appropriately. Additionally, you will need to install the "lawstat" package to use this test.

R Code:

Installs and loads the required pacakge

```
install.packages('lawstat');

library('lawstat');

runs.test(x);
```

Output:

```
      Runs Test - Two sided
data: x
Standardized Runs Statistic = -1.3784, p-value = 0.1681
```

Using the methods of Chapter 6, we will see that since the p-value is not small, we cannot reject the hypothesis that the sample is random.

SPSS Examples

For SPSS, we will give only Windows commands. For all the pull-down menus, the sequence will be separated by the > symbol.

Example 1.7.7

Redo Example 1.7.1 with SPSS.

Solution

*After entering the data in **C1**,*

Analyze > Descriptive Statistics > Explore…>
At the ***Explore*** window select the variable and move to ***Dependent List***; then click ***Plots…***, select ***Stem-and-Leaf***, click ***Continue***, and click ***OK*** at the ***Explore*** Window

We will get the output with a few other things, including box plots along with the stem-and-leaf display, which we will not show here.

Example 1.7.8

Redo Example 1.7.2 with SPSS.

Solution

After entering the data:

Graphs > Histogram…>
At the ***Histogram*** window, select the variable and move to ***Variable***, and click ***OK***

We will get the histogram, which we will not display here.

Example 1.7.9

Redo Example 1.7.3 with SPSS.
Solution
Enter the data. Then:

Analyze > Descriptive Statistics > Frequencies…>
At the ***Frequencies*** window, select the variable(s); then open the ***Statistics*** window and check whichever boxes you desire under ***Percentile***, ***Dispersion***, ***Central Tendency***, and ***Distribution > continue > OK***

For example, if you select Mean, Median, Mode, Standard Deviation, and Variance, we will get the following output and more:

	Statistics	
	VAR00001	
N	*Valid*	*25*
	Missing	*0*
Mean		*83.2800*
Median		*34.0000*
Mode		*14.00*
Std. Deviation		*128.36488*
Variance		*16477.54333*

SAS Examples

We will now give some SAS procedures describing the numerical measures of a single variable. **PROC UNIVARIATE** will give mean, median, mode, standard deviation, skewness, kurtosis, and so forth. If we do not need median, mode, and so on, we could just as well use **PROC MEANS** in lieu of **PROC UNIVARIATE**. We can use the following general format in writing SAS programs with appropriate problem-specific modifications. There are many good online references as well as books available for SAS procedures. To get support on SAS, including many example codes, refer to the SAS support website: http://support.sas.com/. Another helpful site can be found at http://www.ats.ucla.edu/stat/sas/. There are many other sites that may suit your particular application.

General format of an SAS program

DATA give a name to the data set;
INPUT here we put variable names and column locations, if there are more than one variable;
CARDS; (also we can use DATALINES;)
Enter the data here;
TITLE 'here we include the title of our analysis';
PROC PRINT;
PROC name of procedure (such as PROC UNIVARIATE) goes here;
Options that we may want to include (such as the variables to be used) go here;
RUN;

After writing an SAS program, to execute it, we can go to the menu bar and select **run>submit**, or click the "**running man**" icon. On execution, SAS will output the results to the Output window. All the steps used to include time of execution, and any error messages will be given in the Log window.

In order to make the SAS output more manageable, we can use the following SAS command at the beginning of an SAS program:

```
options ls=80 ps=50;
```

ls stands for line size, and this sets each line to be 80 characters wide. ps stands for page size and allows 50 lines on each page. This reduces the number of unnecessary page breaks. In order to avoid date and number, we can use the option commands:

Options *nodate nonumber*;

Example 1.7.10

For the data of Example 1.7.3, use **PROC UNIVARIATE** to summarize the data.

Solution

*In the program editor window, type the following if you are entering the data directly. If you are using the data stored in a file, the comment line (with *) should be used instead of the input and data lines.*

```
Options nodate nonumber;
  DATA ex9;
  INPUT ex9 @@;
  DATALINES;
  5 7 229 453 12 14 18 14 14 483
  22 21 25 23 24 34 37 34 49 64
  47 67 69 192 125;
  PROC UNIVARIATE;
  TITLE;
  RUN;
```

In this case, we will get the following output:

```
The UNIVARIATE Procedure
Variable: ex9
Moments
N                         25                  Sum Weights          25
Mean                      83.28               Sum Observations     2082
Std Deviation             128.364884          Variance             16477.5433
Skewness                  2.45719194          Kurtosis             5.47138396
Uncorrected SS            568850              Corrected SS         395461.04
Coeff Variation           154.136508          Std Error Mean       25.6729767
Basic Statistical Measures
Location                              Variability
Mean              83.28000            Std Deviation                  128.36488
Median            34.00000            Variance                       16478
Mode              14.00000            Range                          478.00000
Interquartile Range                   49.00000
Tests for Location: Mu0=0
Test                      -Statistic-           -p            Value-
Student's t t             3.243878              Pr > |t|      0.0035
Sign            M         12.5                  Pr >= |M|     <.0001
Signed Rank S             162.5                 Pr >= |S|     <.0001
```

Quartiles (Definition 5)

Quartile	Estimate
100% Max	483
99%	483
95%	453
90%	229
75% Q3	67
50% Median	34
25% Q1	18
10%	12
5%	7
1%	5
0% Min	5

The UNIVARIATE Procedure
Variable: ex9
Extreme Observations

-Lowest-		-Highest-	
Value	Obs	Value	Obs
5	1	125	25
7	2	192	24
12	5	229	3
14	9	453	4
14	8	483	10

*We can observe from the previous output that **PROC UNIVARIATE** gives much information about the data, such as mean, standard deviation, and quartiles. If we do not want all these details, we could use the **PROC MEANS** command. In the previous code, if we replace **PROC UNIVARIATE** with the **PROC MEANS** statement, we will get the following:*

The MEANS Procedure
Analysis Variable: ex9

N	Mean	Std Dev	Minimum	Maximum
25 83.	2800000	128.3648836	5.0000000	483.0000000

The output is greatly simplified.

*If we use **PROC UNIVARIATE PLOT NORMAL**, this option will produce three plots: stem-and-leaf, box plot, and normal probability plot (this will be discussed later in the text). In order to obtain bar graphs at the midpoints of the class intervals, use the following commands:*

```
PROC CHART DATA=ex9;
  VBAR ex9;
```

If we want to create a frequency table, use the following:

```
PROC FREQ;
  table ex9;
  title 'Frequency tabulation';
```

Every PROC or procedure has its own name and options. We will use different PROCs as we need them. Always remember to enclose titles in single quotes. There are various other actions that we can perform for the data analysis using SAS. It is beyond the scope of this book to explain general and efficient SAS codes. For details, we refer to books dedicated to SAS, such as the book by Ronald P. Cody and Jeffrey K. Smith, *Applied Statistics and the SAS Programming Language*, 5th Edition, Prentice Hall, 2006. There are many websites that give SAS codes. One example with references for many aspects of SAS, including many codes, can be found at http://www.sas.com/service/library/onlinedoc/code.samples.html.

EXERCISES 1.7

1.7.1 The following data represent the lengths (to the nearest whole millimeter) of 80 shoots from seeds of a certain type planted at the same time.

75	72	76	76	72	74	71	75	77	72
74	71	76	76	76	72	71	73	73	71
72	72	75	70	74	74	78	74	76	79
75	76	73	73	71	72	79	74	77	72
76	70	72	75	78	72	69	75	72	71
77	79	76	73	75	73	72	75	74	78
73	77	73	77	70	74	66	74	73	77
75	79	75	70	72	73	80	73	78	75

Using one of the software packages (R, Minitab, SPSS, or SAS):

(a) Represent the data in a histogram.

(b) Find the summary statistics such as mean, median, variance, and standard deviation.

(c) Draw box plots and identify any outliers.

1.7.2 On a particular day, ask, "How many minutes did you exercise today?" The following were the responses of 30 randomly selected people:

15	30	25	10	30	15	10	45	20	22
18	0	45	12	15	10	17	30	30	15
10	30	20	8	18	30	27	33	15	0

Using one of the software packages (R, Minitab, SPSS, or SAS):

(a) Represent the data in a histogram.

(b) Find the summary statistics such as mean, median, variance, and standard deviation.

(c) Draw box plots and identify any outliers.

PROJECTS FOR CHAPTER 1

1A. World Wide Web and Data Collection

Statistical Abstracts of the United States is a rich source of statistical data (http://www.census.gov/prod/www/statistical-abstract-us.html). Pick any category of interest to you and obtain data (say, Income, Expenditures, and Wealth). Represent a section of the data graphically. Find mean, median, and standard deviation. Identify any outliers. There are many other sites, such as http://lib.stat.cmu.edu/datasets/ and http://it.stlawu.edu/~rlock/datasurf.html, that we can use for obtaining real data sets.

1B. Preparing a List of Useful Internet Sites

Prepare a list of Internet references for various aspects of statistical study.

1C. Dot Plots and Descriptive Statistics

From the local advertisements of apartments for rent, randomly pick 50 monthly rents for two-bedroom apartments. For this data, first draw a dot plot and then obtain descriptive statistics (use R, or any other statistical software).

1D. Importance of Statistics in Our Society

Write a short report on the importance of statistics in our modern society. Give different examples to illustrate your points. One of the interesting projects will be to study the role of Internet of Things (IoT)—a vast network of smart objects which work together in collecting and analyzing data and autonomously performing actions.

1E. Uses and Misuses of Statistics

"There are three types of lies—lies, damn lies, and statistics" Benjamin Disraeli

Write a short report on the uses and misuses of statistics.

1F. Download daily temperature, rainfall, or any other data from Kaggle (NOAA Global Historical Climatology Network Daily). Using R or other software

(a) Analyze the distribution of daily temperatures.
(b) Identify outliers (e.g., heatwaves).
(c) Examine central tendency and spread (mean, median, mode, variance).
(d) Plot histogram, box plot, and Q-Q plot and interpret.

Chapter 2

Probability Theory: Basic Concepts

Chapter outline

OBJECTIVE

In this chapter we will review some results from probability theory that are essential for the development of the statistical results of this book.

Andrei Nikolaevich Kolmogorov
(Source: *http://www.scholarpedia.org/article/Andrey_Nikolaevich_Kolmogorov*.)

Andrei Kolmogorov (1903–87) laid the mathematical foundations of probability theory and the theory of randomness. His monograph *Grundbegriffe der Wahrscheinlichkeitsrechnung*, published in 1933, introduced probability theory in a rigorous way from fundamental axioms. He later used probability theory to study the motion of the planets and the turbulent flow of air from a jet engine. He also made important contributions to stochastic processes, information theory, statistical mechanics, and nonlinear dynamics. Kolmogorov had numerous interests outside mathematics. In particular, he was interested in the form and structure of the poetry of the Russian author Pushkin.

2.1 INTRODUCTION

Probability theory provides a mathematical model for the study of randomness and uncertainty. The concept of probability occupies an important role in the decision-making process, whether the problem is one faced in business, in engineering, in government, in sciences, or just in one's own everyday life. Most decisions are made in the face of uncertainty. The mathematical models of probability theory enable us to make predictions about certain mass phenomena from the necessarily

Mathematical Statistics with Applications in R. DOI: https://doi.org/10.1016/B978-0-443-27548-7.00002-5

incomplete information derived from sampling techniques. It is the probability theory that enables one to proceed from descriptive statistics to inferential statistics. In fact, probability theory is the most important tool in statistical inference.

The origin of probability theory can be traced to the modeling of games of chance, such as dealing from a deck of cards or spinning a roulette wheel. The earliest results on probability arose from the collaboration of the eminent mathematicians Blaise Pascal and Pierre Fermant and a gambler, Chevalier de Méré. They were interested in what seemed to be contradictions between mathematical calculations and actual games of chance, such as throwing dice, tossing a coin, or spinning a roulette wheel. For example, in repeated throws of a die, it was observed that each number, 1–6, appeared with a frequency of approximately 1/6. However, if two dice are rolled, the sum of numbers showing on two dice, that is, 2–12, did not appear equally often. It was then recognized that, as the number of throws increased, the frequency of these possible results could be predicted by following some simple rules. Similar basic experiments were conducted using other games of chance, which resulted in the establishment of various basic rules of probability. Probability theory was developed solely to be applied to games of chance until the 18th century, when Pierre Laplace and Karl F. Gauss applied the basic probabilistic rules to other physical problems. Modern probability theory owes much to the 1933 publication *Foundations of Theory of Probability* by the Russian mathematician Andrei N. Kolmogorov. He developed probability theory from an axiomatic point of view. In the 21st century, probability is used in many real-life applications, such as to control the flow of traffic through a highway system or a computer network, to find the genetic makeup of individuals or populations, to model the spread of diseases, or the spread of information in a social network, and so forth. Governments routinely apply probabilistic methods in environmental regulations, and stock markets are perhaps the largest casinos in the world and cannot run without probability theory. Our objective in this chapter is to provide only a brief review of various definitions and facts from probability that are needed elsewhere in the text. Proofs are omitted in most cases. Many books are devoted solely to the study of probability theory, and we refer to them for further details and a deeper understanding.

2.2 RANDOM EVENTS AND PROBABILITY

Any process whose outcome is not known in advance but is random is termed an *experiment*. The term *experiment* is used here in a wider sense than the usual notion of a controlled laboratory testing situation. Thus, an experiment may include observing whether a fuse is defective or not, or the duration of time from start to end of rain in a particular place. Assume that the experiment can be repeated any number of times under identical conditions. Each repetition is called a *trial*. A (random) experiment satisfies the following three conditions: (1) the set of all possible outcomes is known in advance in each trial; (2) in any particular trial, it is not known which particular outcome will happen; and (3) the experiment can be repeated under identical conditions. We will now summarize some notations and concepts for our study of probability.

Basic definitions

1. The *sample space* associated with an experiment is the set consisting of all possible outcomes and is called the sure event in the experiment. A sample space is also referred to as a *probability space*. A sample space will be denoted by S.
2. An outcome in S is also called a *sample point*. An event A is a subset of outcomes in S, that is, $A \subset S$. We say that an event A occurs if the outcome of the experiment is in A.
3. The *null subset* ϕ of S is called an *impossible event*.
4. The event $A \cup B$ consists of all outcomes that are in A or in B, or in both.
5. The event $A \cap B$ consists of all outcomes that are both in A and B.
6. The event A^c (the complement of A in S) consists of all outcomes not in A, but in S.

Using these concepts, we can define the following. All events are considered to be subsets of S. For some more concepts from set theory, we refer to Appendix A1.

Definition 2.2.1. *Two events A and B are said to be* **mutually exclusive** *or* **disjoint** *if* $A \cap B = \phi$. *Mutually exclusive events cannot happen together.*

The mathematical definition of probability has changed from its earliest formulation as a measure of belief to the modern approach of defining through the axioms. We shall discuss four definitions of probability. We now give an informal definition of probability.

Informal definition of probability

Definition 2.2.2. *The* **probability** *of an event is a measure (number) of the chance with which we can expect the event to occur. We assign a number between 0 and 1 inclusive to the probability of an event. A probability of 1 means that we are 100% sure of the occurrence of an event, and a probability of 0 means that we are 100% sure of the nonoccurrence of the event. The probability of any event A in the sample space S is denoted by P(A).*

From this definition, we can see that $P(S) = 1$. The earliest approach to measuring uncertainty (in chance events) is the classical probability concept, which applies when all possible outcomes are equally likely or when the probabilities of outcomes are known.

Classical definition of probability

Definition 2.2.3. *If there are n equally likely possibilities, of which one must occur, and m of these are regarded as favorable to an event, or as "success," then the* **probability** *of the event or a "success" is given by m/n.*

Now we give steps that can be used to compute the probabilities of events using this classical approach.

Method of computing probability by the classical approach

A. *When all outcomes are equally likely*
   1. Count the number of outcomes in the sample space; say this is n.
   2. Count the number of outcomes in the event of interest, A, and say this is m.
   3. $P(A) = m/n$.

B. *When all outcomes are not equally likely*
   1. Let $\omega_1, \omega_2, \ldots, \omega_n$ be the outcomes of the sample space S. Let $P(\{\omega_i\}) = p_i$, $i = 1,2,\ldots,n$. In this case, the probability of each outcome, p_i, is assumed to be known.
   2. List all the outcomes in the event A, say, $\omega_i, \omega_j \ldots, \omega_m$.
   3. $P(A) = P(\{\omega_i\}) + P(\{\omega_j\}) + \ldots + P(\{\omega_m\}) = p_i + p_j + \ldots + p_m$, the sum of the probabilities of the outcomes in A.

EXAMPLE 2.2.1

A balanced die (with all outcomes equally likely) is rolled. Let A be the event that an even number occurs. Then, there are three favorable outcomes (2, 4, 6) in A, and the sample space has six elements, $\{1, 2, 3, 4, 5, 6\}$. Hence, $P(A) = 3/6 = 1/2$.

EXAMPLE 2.2.2

Suppose we toss two coins. Assume that all the outcomes are equally likely (fair coins).

(a) What is the sample space?
(b) Let A be the event that at least one of the coins shows up heads. Find $P(A)$.
(c) What would be the sample space if we knew that at least one of the coins showed up heads?

Solution

(a) *The sample space consists of four outcomes, namely S = {(H, H), (H, T), (T, H), (T, T)}.*

(b) *Event A has three outcomes: (H, H), (H, T), and (T, H). Therefore, P(A) = 3/4.*

(c) *Since we know that at least one of the coins showed up heads, the possible outcomes are (H, H), (H, T), and (T, H). The sample space now has only three outcomes {(H, H), (H, T), (T, H)}.*

The classical probability concept is not applicable in situations where the various possibilities cannot be regarded as equally likely. Suppose we are interested in whether or not it will rain on a given day with known meteorological conditions. Clearly, we cannot assume that the events of rain or no rain are equally likely. In such cases, one could use the so-called frequency interpretation of probability. The frequentist view is a natural extension of the classical view of probability. This definition was developed as the result of work by R. von Mises in 1936.

Frequency definition of probability

Definition 2.2.4. *The* **probability** *of an outcome (event) is the proportion of times the outcome (event) would occur in a long run of repeated experiments.*

For example, to find the probability of heads, H, using a biased coin, we would imagine the coin is repeatedly tossed. Let $n(H)$ be the number of times H appears in n trials. Then, the probability of heads is defined as $P(H) = \lim_{n\to\infty}(n(H)/n)$.

The frequency interpretation of probability is often useful. However, it is not complete. Because of the condition of repetition under identical circumstances, the frequency definition of probability is not applicable to every event. For a more complete picture, it makes sense to develop the probability theory through axioms. Now we will define probabilities axiomatically. This definition results from the 1933 studies of A. N. Kolmogorov.

Axiomatic definition of probability

Definition 2.2.5. *Let S be a sample space of an experiment. Probability P(.) is a real-valued function that assigns to each event A in the sample space S a number P(A), called the* **probability** *of A, with the following conditions satisfied:*

1. *It is nonnegative, $P(A) \geq 0$.*
2. *It is unity for a certain event. That is, $P(S) = 1$.*
3. *It is additive over the union of an infinite number of pairwise disjoint events, that is, if $A_1, A_2,...$form a sequence of pairwise mutually exclusive events (i.e., $A_i \cap A_j = \phi$, for $i \neq j$) in S, then $P\left(\cup_{i=1}^{\infty} A_i\right) = \sum_{i=1}^{\infty} P(A_i)$.*

From the previous three axioms, it can be shown that $P(\phi) = 0$, and if $A_1, A_2,...$ form a sequence of pairwise mutually exclusive events in S, then $P\left(\cup_{i=1}^{n} A_i\right) = \sum_{i=1}^{n} P(A_i)$ for a finite n. Also, we could verify that $0 \leq P(A) \leq 1$, for any event A. It is important to observe that the axioms do not tell us how to assign probabilities to events.

EXAMPLE 2.2.3

A die is loaded (not all outcomes are equally likely) such that the probability that the number i shows up is Ki, $i = 1,2,\ldots,6$, where K is a constant. Find

(a) the value of K.

(b) the probability that a number greater than 3 shows up.

Solution

(a) *Here the sample space S has six outcomes {1, 2...,6}. Hence, using axioms (2) and (3) we have,*

$$P(\{1\}) + P(\{2\}) + \ldots + P(\{6\}) = 1.$$

Since $P(i) = Ki$, we have

$$(K)(1) + (K)(2) + \ldots + (K)(6) = 1 \text{ or}$$
$$(K)(1 + 2 + \ldots + 6) = (K)(21) = 1.$$

Hence, $K = 1/21$.
The probability of, say, the number 5 showing up is 5/21.

(b) *Let A be the event that a number greater than 3 shows up. Then, the outcomes in A are {4, 5, 6} and they are mutually exclusive. Therefore,*

$$P(A) = P(\{4\}) + P(\{5\}) + P(\{6\})$$
$$= \frac{4}{21} + \frac{5}{21} + \frac{6}{21} = \frac{15}{21}.$$

The following properties help us in going beyond the axioms to actually compute various probabilities.

Some basic properties of probability

For two events A and B in S, we have the following:
1. $P(A^c) = 1 - P(A)$, where A^c is the complement of the set A in S.
2. If $A \subset B$, then $P(A) \leq P(B)$.
3. $P(A \cup B) = P(A) + P(B) - P(A \cap B)$.
In particular, if $A \cap B = \phi$, then $P(A \cup B) = P(A) + P(B)$.

EXAMPLE 2.2.4

In a large university, the freshman profile for one year's fall admission says that 40% of the students were in the top 10% of their high school class, and that 65% are White, of whom 25% were in the top 10% of their high school class. What is the probability that a freshman student selected randomly from this class was either in the top 10% of their high school class or is White?

Solution

Let E_1 be the event that a person chosen at random was in the top 10% of his or her high school class, and let E_2 be the event that the student is White. We are given $P(E_1) = 0.40$, $P(E_2) = 0.65$, and $P(E_1 \cap E_2) = 0.25$. Then, the event that the student chosen is White or was in the top 10% of his or her high school class is represented by $E_1 \cup E_2$. Thus,

$$P(E_1 \cup E_2) = P(E_1) + P(E_2) - P(E_1 \cap E_2)$$
$$= 0.40 + 0.65 - 0.25 = 0.80.$$

EXAMPLE 2.2.5

A subway station in a large city has 12 gates, six inbound (entering the subway station) and six outbound (exiting the subway station). The number of gates open in each direction is observed at a particular time of day. Assume that each outcome of the sample space is equally likely.

(a) Define a suitable sample space.
(b) What is the probability that at most one gate is open in each direction?
(c) What is the probability that at least one gate is open in each direction?
(d) What is the probability that the number of gates open is the same in both directions?
(e) What is the probability of the event that the total number of gates open is six?

Solution

(a) *We define the sample space to be the set of ordered pairs (x, y), where x is the number of inbound gates open, and y is the number of outbound gates open. For example, (4, 5) means four gates for inbound and five gates for outbound are open. (1, 0) means one gate is open in the inbound direction and no gate is open in the outbound direction.* Fig. 2.1 *represents the situation*

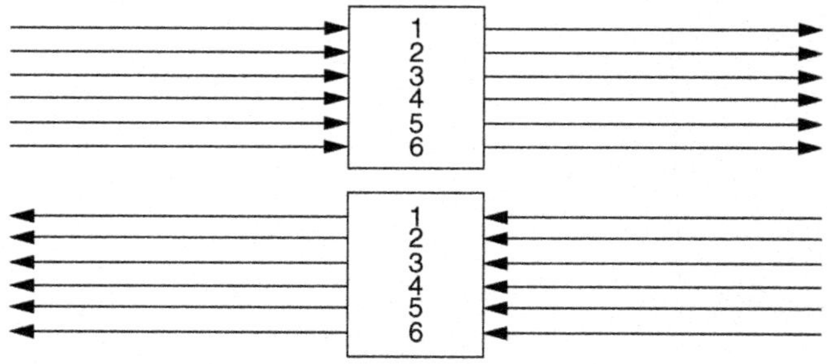

FIGURE 2.1 Inbound and outbound traffic.

$$s = \begin{Bmatrix} (0,0)(0,1)(0,2)(0,3)(0,4)(0,5)(0,6) \\ (1,0)(1,1)(1,2)(1,3)(1,4)(1,5)(1,6) \\ (2,0)(2,1)(2,2)(2,3)(2,4)(2,5)(2,6) \\ (3,0)(3,1)(3,2)(3,3)(3,4)(3,5)(3,6) \\ (4,0)(4,1)(4,2)(4,3)(4,4)(4,5)(4,6) \\ (5,0)(5,1)(5,2)(5,3)(5,4)(5,5)(5,6) \\ (6,0)(6,1)(6,2)(6,3)(6,4)(6,5)(6,6) \end{Bmatrix}.$$

We see that the sample space has 49 possible outcomes. We assume that these outcomes are equally likely.

(b) *Suppose that A is the event that at most one gate is open in each direction. Then,*

$$A = \{(0,0), (0,1), (1,0), (1,1)\}.$$

Hence,

$$P(A) = \frac{4}{49} = 0.082$$

(c) *Let B be the event that at least one gate is open in each direction. Then, B contains 36 elements. Hence,*

$$P(B) = \frac{36}{49} = 0.7347.$$

(d) *Let*

$$\begin{aligned} C &= \text{event that the number of open gates is the same both ways} \\ &= \{(0,0), (1,1), (2,2), (3,3), (4,4), (5,5), (6,6)\}. \end{aligned}$$

Then, $P(C) = \frac{7}{49} = 0.1428.$

(e) *Let*

$$\begin{aligned} D &= \text{the event that the total number of gates open is six} \\ &= \{(3,3), (2,4), (4,2), (5,1), (1,5), (6,0), (0,6)\}. \end{aligned}$$

Hence, $P(D) = 7/49$.

EXERCISES 2.2

2.2.1 Consider an experiment in which each of three cars exiting from a university main entrance turns right (R) or left (L). Assume that a car will turn right or left with equal probability of 1/2.

(a) What is the sample space S?

(b) What is the probability that at least one car will turn left?

(c) What is the probability that at most one car will turn left?

(d) What is the probability that exactly two cars will turn left?

(e) What is the probability that all three cars will turn in the same direction?

2.2.2 A coin is tossed three times. Define an appropriate sample space for the following cases:
(a) The outcome of each individual toss is of interest.
(b) Head appears for the first time.

2.2.3 A pair of six-sided balanced dice is rolled. What are the probabilities of getting the sum of the face values as follows?
(a) 8
(b) 6 or 9
(c) 3, 8, or 12
(d) Not an even number

2.2.4 An experiment has four possible outcomes: A, B, C, and D. Check whether the following assignments of probability are possible:
(a) $P(A) = 0.20$, $P(B) = 0.40$, $P(C) = 0.09$, $P(D) = 0.31$.
(b) $P(A) = 0.41$, $P(B) = 0.17$, $P(C) = 0.12$, $P(D) = 0.36$.
(c) $P(A) = 1/8$, $P(B) = 1/2$, $P(C) = 1/4$, $P(D) = 1/8$.

2.2.5 Suppose we toss two coins and suppose that each of the four points in the sample space $S = \{(H, H), (H, T), (T, H), (T, T)\}$ is equally likely. Let the events be $A = \{(H, H), (H, T)\}$ and $B = \{(H, H), (T, H)\}$. Find $P(A \cup B)$.

2.2.6 An urn contains 12 white, 5 yellow, and 13 black marbles. A marble is chosen at random from the urn, and it is noted that it is not one of the black marbles. What is the sample space in view of this knowledge? What is the probability that it is yellow?

2.2.7 Two fair dice are rolled, and face values are noted.
(a) What is the probability space?
(b) What is the probability that the sum of the numbers showing is 7?
(c) What is the probability that both dice show number 2?

2.2.8 In a city, 65% of people drink coffee, 50% drink tea, and 25% both. What is the probability that a person chosen at random will drink at least one coffee or tea? Will drink neither?

2.2.9 In a fruit basket, there are five mangos, of which two are spoiled. If we were to randomly pick two mangos:
(a) What would be our sample space?
(b) What is the probability that both mangos are good?
(c) What is the probability that no more than one mango is spoiled?

2.2.10 In a box, there are three slips of paper, with one of the letters A, C, T written on each slip. If the slips are drawn out of the box one at a time, what is the probability of obtaining the word *CAT?*

2.2.11 Suppose that the genetic makeup of the population of a city is as in Table 2.1.
An individual is considered to have the dominant characteristic if the person has the AA or Aa genetic trait. If we were to choose an individual from this city at random, what is the probability that this person has the dominant characteristic?

TABLE 2.1 Genetic Makeup of a Population.

Genetic makeup	AA	Aa	aa
Probability	p	$2q$	r

2.2.12 Using the axioms of probability, show that $P(\phi) = 0$, and if $A_1, \ldots, A_n$ are pairwise mutually exclusive, then $P(\cup_{i=1}^{n} A_i) = \sum_{i=1}^{n} P(A_i)$.

2.2.13 Using the axioms of probability, prove the following:
(a) If $A \subset B$, then $P(A) \le P(B)$.
(b) $P(A \cup B) = P(A) + P(B) - P(A \cap B)$. In particular, if $A \cap B = \phi$, then $P(A \cup B) = P(A) + P(B)$.

2.2.14 Using the axioms of probability, show that

$$P(A \cup B \cup C) = P(A) + P(B) + P(C) - P(A \cap B) - P(A \cap C) - P(B \cap C) + P(A \cap B \cap C).$$

2.2.15 Prove that

(a) $P(A \cap B) \geq P(A) + P(B) - 1$

(b) $P(\cup_{i=1}^{2} A_i) \leq \sum_{i=1}^{2} P(A_i)$

2.2.16 If A and B are mutually exclusive events, $P(A) = 0.17$ and $P(B) = 0.46$, find

(a) $P(A \cup B)$

(b) $P(A^c)$

(c) $P(A^c \cup B^c)$

(d) $P((A \cap B)^c)$

(e) $P(A^c \cap B^c)$

2.2.17 If $P(A) = 0.24$, $P(B) = 0.67$, and $P(A \cap B) = 0.09$, find

(a) $P(A \cup B)$

(b) $P((A \cup B)^c)$

(c) $P(A^c \cup B^c)$

(d) $P((A \cap B)^c)$

(e) $P(A^c \cap B^c)$

2.2.18 In a series of seven games, the first team to win four games wins the series. If the teams are evenly matched, what is the probability that the team that wins the first game will win the series?

2.2.19 In a survey, 1000 adults were asked if they would approve an increase in tax if the revenues went to build a football stadium. It was also noted whether the person lived in a city (C), suburb (S), or rural area (R) of the county. The results are summarized in Table 2.2.

TABLE 2.2 Survey Results for Opinion on a Tax Increase.

	Yes (For Tax Increase)	No (Against Tax Increase)
C	150	250
S	250	150
R	50	150

Define the following events:

A: person chosen is from the city

B: person disapproves tax increase

Find the following probabilities:

(i) $P(B)$, (ii) $P(A^c \cap B)$, and (iii) $P(A \cup B^c)$

2.2.20 A couple has two children. Suppose we know the elder child is a boy.

(a) Determine an appropriate sample space.

(b) Find the probability that both are boys.

2.2.21 A box contains three red and two blue flies. Two flies are removed with replacement. Let A be the event that both the flies are of the same color, and B be the event that at least one of the flies is red. Find (i) $P(A)$, (ii) $P(B)$, (iii) $P(A \cup B)$, and (iv) $P(A \cap B)$.

2.2.22 Prove that for any n,

$$P\left(\cup_{i=1}^{n} A_i\right) = \Sigma_{i=1}^{n} P(A_i) - \Sigma_{i_1<i_2} P(A_{i1} \cap A_{i2}) + \cdots$$
$$+(-1)^{m+1} \Sigma_{i_1<i_2<\ldots<i_m} P(A_{i1} \cap A_{i2} \cap \ldots \cap A_{im})$$
$$+\cdots + (-1)^{n+1} P(A_1 \cap A_2 \cap \ldots An).$$

The summation $\sum_{i_1<i_2<\ldots<i_m} P\left(A_{i_1} \cap A_{i_2} \cap \ldots \cap A_{i_m}\right)$ is taken over all of the $\binom{n}{m}$ subsets of size m from the set $\{1, 2,\ldots, n\}$.

2.2.23 A sequence of events $\{A_n, n \geq 1\}$ is said to be an increasing sequence if $A_1 \subset A_2 \subset \ldots \subset A_n \subset \ldots$, whereas it is said to be decreasing if $A_1 \supset A_2 \supset \ldots \supset A_n \supset \ldots$. If $\{A_n, n \geq 1\}$ is increasing sequence of events, then $\lim_{n\to\infty} A_n = \cup_{i=1}^{\infty} A_n$. Similarly, if $\{A_n, n \geq 1\}$ is decreasing sequence of events, then $\lim_{n\to\infty} A_n = \cap_{i=1}^{\infty} A_n$. Show that if $\{A_n, n \geq 1\}$ is either an increasing or a decreasing sequence of events, then $\lim_{n\to\infty} P(A_n) = P(\lim_{n\to\infty} A_n)$.

2.2.24 A group of n old people who needed a stick to support their walking attended a meeting about absent-mindedness. While entering the room, they kept their sticks in the entrance. After the meeting, none could recognize their own sticks, so they picked a stick at random. Also, with probability P and independently of others, they could break the stick on their way home. Find the probability that none had their own stick with them. Show that the probability is approximately $e^{-(1-p)}$.

2.3 COUNTING TECHNIQUES AND CALCULATION OF PROBABILITIES

In a sample space with a large number of outcomes, determining the number of outcomes associated with the events through direct enumeration could be tedious. In this section, we develop some counting techniques and use them in probability computations.

Multiplication principle

Theorem 2.3.1. *If the experiments $A_1, A_2, \ldots A_m$ contain, respectively, $n_1, n_2, \ldots n_m$ outcomes, such that for each possible outcomes of A_1 there are n_2 possible outcomes for A_2, and so on, then there are a total of $n_1, n_2, \ldots n_m$ possible outcomes for the composite experiment $A_1, A_2, \ldots A_m$.*

For $m = 2$ and $n_1 = 2$, $n_2 = 3$, the tree diagram in Fig. 2.2 illustrates the multiplication principle. If we count the total number of branches at the top of the tree, we get the total number of possible outcomes for the composite experiment. In Fig. 2.2, we can see that there are a total of six branches that represent all the possible outcomes of this experiment. The tree diagrams can be utilized for counting for any finite number of composite experiments.

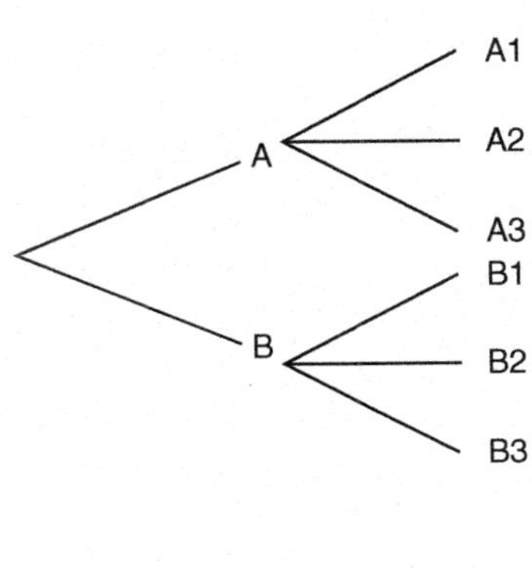

FIGURE 2.2 Tree diagram.

EXAMPLE 2.3.1

In how many different ways can a student club at a large university with 500 members choose its president and vice president?

Solution

The president can be chosen 500 ways, and the vice president can be chosen from the remaining 499 ways. Hence, by the multiplication principle, there are (500)(499) = 249,500 *ways in which the complete choice can be made.*

When a random sample of size k is taken with replacement from a total of n objects, the total number of ways in which the random sample of size k can be selected depends on the particular sampling method we employ. Here, we will consider four sampling methods: (i) sampling with replacement and the objects are ordered, (ii) sampling without replacement and the objects are ordered, (iii) sampling without replacement and the objects are not ordered, and (iv) sampling with replacement and the objects are not ordered.

(I) Sampling With Replacement and the Objects Are Ordered

When a random sample of size k is taken with replacement from a total of n objects and the objects being ordered, then there are nk possible ways of selecting k-tuples.

For example, (1) if a die is rolled four times, then the sample space will consist of 6^4 4-tuples. (2) If an urn contains nine balls numbered 1 to 9, and a random sample with replacement of size $k = 6$ is taken, then the sample space S will consist of 9^6 6-tuples.

(II) Sampling Without Replacement and the Objects Are Ordered

The symbol $n!$ (read n factorial) is defined as $n! = n(n - 1)...(2)(1)$. Clearly 1!=1. By definition, we take 0!= 1.

If r objects are chosen from a set of n distinct objects without replacement, any particular (ordered) arrangement of these objects is called a permutation. For example, *CDAB* is a permutation of the letters *ABCD*. The number of permutations of these four letters is 4! = 24, because the first position can be filled by any of the four letters, leaving only three possibilities for the second position, two for the third position, and only one for the fourth position, yielding the number of permutations to be 4.3.2.1 = 24.

Permutation of n objects taken m at a time

Theorem 2.3.2. *The number of permutations of m objects selected from a collection of n distinct objects is*

$$ {}_nP_m = \frac{n!}{(n-m)!} = n(n-1)(n-2)\ldots(n-m+1). $$

When a random sample of size k is taken without replacement from a total of n objects and the objects being ordered, we will apply the permutation formula.

Example 2.3.2

How many distinct three-digit numbers can be formed using the digits 2, 4, 6, and 8 if no digit can be repeated?

Solution

The number of distinct three-digit numbers will be the number of permutations of three numbers from the set of four numbers {2, 4, 6, 8}. *Hence, the number of distinct three-digit numbers will be* ${}_4P_3 = 4!/1! = 24$.

(III) Sampling Without Replacement and the Objects Are Not Ordered

Note that in a permutation, the order in which each object is selected becomes important. When the order of arrangement is not important—for example, if we do not distinguish between *AB* and *BA*—the arrangement is called a combination. We give the following result for the number of combinations.

Number of combinations of *n* objects taken *m* at a time

Theorem 2.3.3. *The number of ways in which m objects can be selected (without replacement) from a collection of n distinct objects is*

$$\binom{n}{m} = \frac{n!}{m!(n-m)!} = \frac{n(n-1)(n-2)\dots(n-m+1)}{m!},\ m = 0,1,2,\dots,n$$

The symbol $\binom{n}{m}$ is to be read as "n choose m." When a random sample of size k is taken without replacement from a total of n objects and the objects are not ordered, we will apply the combinations formula. An R command "*choose(n, m)*" (like choose(20, 10)) will calculate combinations.

EXAMPLE 2.3.3

How many different ways can the admissions committee of a statistics department choose four foreign graduate students from 20 foreign applicants and three US students from 10 US applicants?

Solution

The four foreign students can be chosen in $\binom{20}{4}$ *ways, and the three US students can be chosen in* $\binom{10}{3}$ *ways. Now, by the multiplication principle, the whole selection can be made in* $\binom{20}{4}\binom{10}{3} = 581{,}400$ *ways. "choose(20, 4)choose(10, 3)" will give the answer in R.*

(IV) Sampling With Replacement and the Objects Are Not Ordered

In obtaining an unordered sample of size k, with replacement, from a total of n objects, $(k - 1)$ replacements will be made before sampling ceases. Thus, n is increased by $(k - 1)$ so that sampling in this manner may be thought of as drawing an unordered sample of size k from a population of size $(n + k - 1)$. Hence, the number of possible samples can be obtained by using the formula

$$\binom{n+k-1}{k} = \frac{(n+k-1)!}{k!(n-1)!},\ k = 0,1,2,\ \dots.$$

EXAMPLE 2.3.4

An urn contains 15 balls numbered 1 to 15. If four balls are drawn at random, with replacement and without regard for order, how many samples are possible?

Solution

Using the previous formula, the number of possible samples is

$$\binom{15+4-1}{4} = \frac{18!}{4!14!} = 3060.$$

If we need to divide n objects into more than two groups, we can use the following result.

(cont'd)

Number of combinations of *n* objects into *m* classes

Theorem 2.3.4. *The number of ways that n objects can be grouped into m classes with n_i in the ith class, i = 1, 2,, m and $\sum_{i=1}^{m} n_i = n$ is given by*

$$\binom{n}{n_1 n_2 \ldots n_m} = \frac{n!}{n_1! n_2! \ldots n_m!}$$

In the foregoing theorem, the numbers $\binom{n}{n_1 n_2 \ldots n_m}$ are called *multinomial coefficients.*

We can use the previous computational technique to compute the probabilities of events of interest by using the frequency interpretation of probability. Suppose that there are a total of N possible outcomes for the experiment and let nA be the number of outcomes favoring an event A. Then, the probability of this event is $P(A) = n_A/N$. The following is a well-known problem that is called the birthday problem.

EXAMPLE 2.3.5

In a room there are *n* people. What is the probability that at least two of them have a common birthday?

Solution

Disregarding leap years, assume that every day of the year is equally likely to be a birthday. Let A be the event that there are at least two people with a common birthday. There are 365^n possible outcomes of which A^c can happen in 365 × 364 × (365 − n + 1) ways. Because event A can happen in many more ways, it is easier to calculate $P(A^c)$, that is, the probability that no two persons have the same birthday or equivalently, that they all have different birthdays. To count the number of n-tuples in A^c, because there are no common birthdays, we can use the method of choosing distinct objects without replacement for an ordered arrangement. Thus, there are 365 possibilities to choose the first person, 364 for the second person,...,(365 − (n − 1)) possibilities for the nth person. The product of these numbers gives the total number of elements in A^c. Thus,

$$P(A^c) = \frac{365 \times 364 \times \ldots \times (365 - n + 1)}{365^n}$$

and hence,

$$P(A) = 1 - \frac{365 \times 364 \times \ldots \times (365 - n + 1)}{365^n}.$$

For example, if n = 3, $P(A) = 1 - \frac{365 \times 364 \times 363}{365^3} = 0.0082$*, and if n = 40,*

$$P(A) = 1 - \frac{365 \times 364 \times \ldots \times (365 - 40 + 1)}{(365)^{40}} = 1 - 0.1087 = 0.89123.$$

That is, there is only a 0.82% chance of having a common birthday among three persons, whereas if n = 40, then P(A) = 0.89123—that is, the chance of having a common birthday among 40 persons increases to 89.1%. Thus, as the number of people increases, the chance of finding people with common birthdays also increases.

EXAMPLE 2.3.6

In a tank containing 10 fishes, there are three yellow and seven black fishes. We select three fishes at random.

(a) What is the probability that exactly one yellow fish gets selected?
(b) What is the probability that at most one yellow fish gets selected?
(c) What is the probability that at least one yellow fish gets selected?

Solution

Let A be the event that exactly one yellow fish gets selected, and B be the event that at most one yellow fish gets selected. There are $\binom{10}{3} = 120$ *ways to select three fish from 10.*

(a) *There are* $\binom{3}{1} = 3$ *ways to select a yellow fish and* $\binom{7}{2} = 21$ *ways to select two black fish. By the multiplication rule, the probability of selecting exactly one yellow fish is*

$$\frac{\binom{3}{1}\binom{7}{2}}{\binom{10}{3}} = \frac{3(21)}{120} = 0.525.$$

(b) *The probability that at most one yellow fish gets selected is the same as the probability of selecting none or one, which is*

$$\frac{\binom{3}{1}\binom{7}{2}}{\binom{10}{3}} + \frac{\binom{3}{0}\binom{7}{3}}{\binom{10}{3}} = 0.525 + 0.292 = 0.817.$$

(c) *The probability that at least one yellow fish gets selected is the same as 1 – P(none), which is* 1–0.292 = 0.708.

EXAMPLE 2.3.7

Refer to Example 2.3.3. Suppose that the admission committee decides to randomly choose seven graduate students from a pool of 30 applicants, of whom 20 are foreign and 10 are US applicants. What is the probability that a chosen seven will have four foreign students and three US students?

Solution

As in Example 2.3.3, the number of ways of selecting four foreign and three US students is

$$\binom{20}{4}\binom{10}{3} = 581{,}400.$$

The number of ways of selecting seven applicants out of 30 is

$$\binom{30}{7} = 2{,}035{,}800.$$

Hence, the probability that a randomly selected group of seven will consist of four foreign and three US students is

$$\frac{\binom{20}{4}\binom{10}{3}}{\binom{30}{7}} = \frac{581{,}400}{2{,}035{,}800} = 0.2856.$$

It should be noted that in the rapidly developing field of AI (artificial intelligence), search optimization, natural language processing, and machine learning, permutations and combinations provide a foundational toolset for tackling feature selection problems, enhancing model performance, probabilistic reasoning, and developing solutions that are both useful and efficient. The effectiveness of combinatorics lies in its ability to streamline the selection process, offering a rich collection of strategies for data scientists to navigate the vastness of possible feature sets and their arrangements.

EXERCISES 2.3

2.3.1 Determine the following:

$$\text{(i)} \binom{10}{2}, \text{(ii)} \binom{10}{0}, \text{(iii)} \binom{10}{9}, \text{(iv)} \binom{10}{2}\binom{10}{3}, \text{ and (v)} \binom{10}{2\ 3\ 5}.$$

2.3.2 A game in a state lottery selects four numbers from a set of numbers, $\{0,1,2,3,4,5,6,7,8,9\}$, with no number being repeated. How many possible groups of four numbers are possible?

2.3.3 A 10-bit binary word is a sequence of 10 digits, of which each may be either a 1 or a 0. How many 10-bit words are there?

2.3.4 Insulin, a peptide hormone built from 51 amino acid residues, is one of the smallest proteins known (note that proteins are made up of chains of amino acids) with a molecular weight of 5808 Da. Twenty amino acids are encoded by the standard genetic code; that is, proteins are built from a basic set of 20 amino acids. How many possible proteins of length 51 can be made with 20 amino acids for each position in the protein?

2.3.5 An examination is designed where the students are required to answer any 20 questions from a group of 25 questions. How many ways can a student choose the 20 questions?

2.3.6 How many different six-place license plates are possible if the first three places and the last place are to be occupied by letters and the fourth and fifth places are to be occupied by numbers?

2.3.7 In how many different ways can 15 tickets to a football game be distributed among a class of 30 students if each student gets at most one ticket?

2.3.8 How many different four-letter English words (with or without meaning) can be written using distinct letters from the alphabet?

2.3.9 DNA (deoxyribonucleic acid) is made from a sequence of four nucleotides (A, T, G, or C). Suppose a region of DNA is 40 nucleotides long. How many possible nucleotide sequences are there in this region of DNA?

2.3.10 Show that

(a) $\binom{n}{0} = \binom{n}{n} = 1.$

(b) $\binom{n}{m} = \binom{n-1}{m-1} + \binom{n-1}{m}, \quad 1 \le m \le n.$

(c) $\binom{n}{m} = \binom{n}{n-m}.$

2.3.11 A lot of 50 electrical components numbered 1–50 is drawn at random, one by one, and is divided among five customers.

(a) Suppose that it is known that components 3, 18, 12, 26, and 46 are defective. What is the probability that each customer will receive one defective component?

(b) What is the probability that one customer will have drawn five defective components?

(c) What is the probability that two customers will receive two defective components each, two none, and the other one?

2.3.12 A package of 15 apples contains two defective apples. Four apples are selected at random.

(a) Find the probability that none of the selected apples is defective.

(b) Find the probability that at least one of the selected apples is defective.

2.3.13 A homeowner wants to repaint her home and install new carpets (no store where she lives sells both paint and carpet). She plans to get the services from the stores where she buys the paint and carpet. Suppose there are 12 paint stores with painting services available and 15 carpet stores with installation services available in that city. In how many ways can she choose these two stores?

2.3.14 From an urn containing 15 white, 7 black, and 8 yellow balls a sample of 3 balls is drawn at random. Find the probability that

(a) All three balls are yellow.

(b) All three balls are of the same color.

(c) All three balls are of different colors.

2.3.15 Refer to Example 2.3.5. Compute (*A*) for (a) $n = 20$. (b) $n = 30$. Estimate n if you wish to have an approximately 50% chance of finding someone who shares your birthday.

2.3.16 A box of manufactured items contains 12 items, of which four are defective. If three items are drawn at random without replacement, what is the probability that
(a) The first one is defective, and the rest are good?
(b) Exactly one of the three is defective?

2.3.17 Five white and four black balls are arranged in a row. What is the probability that the end balls are of different colors?

2.3.18 Three numbers are chosen at random from the numbers $\{1, 2, ..., 9\}$. What is the probability that the middle number is 5?

2.3.19 In each of the following, find the number of elements in the resulting sample space.
(a) If a die is rolled five times, how many elements are there in the sample space?
(b) If 13 cards are selected from a deck of 52 playing cards without replacement, and the order in which the cards are drawn is important, how many elements are there in the sample space?
(c) Four players in a game of bridge are dealt 13 cards each from an ordinary deck of 52 cards. What is the total number of ways in which we can deal the 13 cards to the four players?
(d) If a football squad consists of 72 players, how many selections of 11-man teams are possible?

2.3.20 In the Florida Lotto, an urn contains balls numbered 1 to 53. From this urn, a machine chooses six balls at random and without replacement. The order in which the balls are selected does not matter. For a $1 bet, a player chooses six numbers. If all six numbers match with the six numbers chosen by the urn, the player wins the jackpot. What is the probability of winning the Florida Lotto jackpot?

2.3.21 The cells in our bodies receive half of their chromosomes from the father and the other half from the mother. So, for each pair of homologous chromosomes, one will be a paternal chromosome and one will be a maternal chromosome. We have 23 pairs of homologous chromosomes.
(a) How many possible combinations of paternal and maternal chromosomes are there?
(b) What is the probability of getting a gamete with nine paternal and 14 maternal chromosomes? Assume that any ordered combination is equally likely.

2.4 CONDITIONAL PROBABILITY, INDEPENDENCE, AND BAYES' RULE

If we know that an event has already occurred or we have some partial information about the event, then this knowledge may affect the probability of the event of interest. For example, if we were to guess on the probability of rain today, the answers will be different depending on whether we are sitting inside a windowless office or we are outside and can see the formation of heavy clouds. This leads to the idea of conditional probability.

Definition 2.4.1. *The* **conditional probability** *of an event A, given that an event B has occurred, denoted by P(A|B), is equal to*

$$P(A|B) = \frac{P(A \cap B)}{P(B)},$$

provided P(B) > 0.

It can be easily proved that conditional probability, $P(A|B)$ satisfies the axioms of probability, provided $P(B) > 0$, and hence a probability.

EXAMPLE 2.4.1

We toss two balanced dice and let *A* be the event that the sum of the face values of two dice is 8, and *B* be the event that the face value of the first one is 3. Calculate *P*(*A*|*B*).

Solution

The elements of events A and B are

$$A = \{(2,6), (6,2), (3,5), (5,3), (4,4)\}$$

and

$$B = \{(3,1), (3,2), (3,3), (3,4), (3,5), (3,6)\}.$$

Now, $A \cap B = \{(3, 5)\}$

$$P(A) = 5/36,\ P(B) = 6/36,\ \text{and}\ P(A \cap B) = 1/36.$$

Therefore,

$$P(A|B) = \frac{P(A \cap B)}{P(B)} = \frac{\frac{1}{36}}{\frac{6}{36}} = \frac{1}{6}.$$

It is important to note that the conditional probability $P(.|B)$, is a probability on B. It satisfies all the axioms of probability.

Some properties of conditional probability

1. If $E_2 \subset E_1$, then $P(E_2|A) \le P(E_1|A)$.
2. $P(E|A) = 1 - P(E^c|A)$.
3. $P(E_1 \cup E_2|A) = P(E_1|A) + P(E_2|A) - P(E_1 \cap E_2|A)$.
4. Multiplication law: $P(A \cap B) = P(B)P(A|B) = P(A)P(B|A)$.
   In general,

$$P(A_1 \cap A_2 \cap \ldots \cap A_n) = P(A_1)P\left(A_2|A_1\right)P\left(A_3|A_1 \cap A_2\right)\ldots P\left(A_n|A_1 \cap A_2 \cap \ldots \cap A_{n-1}\right).$$

EXAMPLE 2.4.2

A fruit basket contains 25 apples and oranges, of which 20 are apples. If two fruits are randomly picked in sequence, what is the probability that both the fruits are apples?

Solution

Let

$$A = \{\text{event that the first fruit is an apple}\}$$
$$B = \{\text{event that the second fruit is an apple}\}.$$

We need to find $P(A \cap B)$. *We have,*

$$P(A) = 20/25,\ P(B|A) = 19/24.$$

Now, using the multiplication principle for conditional probabilities,

$$P(A \cap B) = P(A)P(B|A) = \left(\frac{20}{25}\right)\left(\frac{19}{24}\right) = 0.633.$$

Hence, the probability that both the fruits are apples is 0.633.

Probability and statistics are proving to be very useful in the field of genetics. Genetics is the study of heredity—traits transmitted from parents to offspring. The starting point of the subject of genetics as presently known can be attributed to Gregor Mendel (1822–84), an Austrian monk. During the 1850s, Mendel was interested in plant breeding. He performed careful experiments with the garden peas, *Pisum sativum*, and uncovered the basic principles of genetic inheritance. Mendel discovered that traits are inherited in discrete units (known as genes). Mendel's law of independent segregation states that the parent transmits randomly one of its traits to the offspring. Geneticists use letters to represent alleles. An allele is an alternative form of a gene that is located at a specific position on a specific chromosome. Organisms have two alleles for each trait. A capital letter is used to represent a *dominant* trait, and a lowercase letter is used to represent a *recessive* trait. The combination pair of these traits that one inherits from parents is the genetic makeup. A dominant allele can be observed in the organism's appearance or physiology, whereas a recessive allele cannot be observed unless the individual has *two* copies of the recessive allele.

EXAMPLE 2.4.3

Suppose we are given a population with the following genetic distribution:

Alleles are randomly donated from parents to offspring. Assuming random mating, what is the probability that the mating is *Aa* × *Aa* and the offspring is *aa* (recessive trait)?

Genetic makeup	*AA*	*Aa*	*aa*
Probability	p	$2q$	r

Solution

Let B denote the event that the mating is Aa × Aa, and C denote the event that the offspring is aa. Then, we have $P(B) = 4q^2$. *Because the alleles are randomly donated from parents to offspring,* $P(C|B) = 1/4$. *Now, using the multiplication principle for conditional probabilities,*

$$P(B \cap C) = P(B)P(C|B) = (4q^2)\left(\frac{1}{4}\right) = q^2.$$

Hence, the probability that the mating is Aa × Aa and the offspring is of the recessive trait is q^2.

In order to compute probabilities similar to those in Example 2.4.3, we could use Table 2.3. The distributions of the progeny (zygotes) are the predicted values from Mendel's law.

TABLE 2.3 The Distribution of Zygotes.

Mating	Probability of Mating	Probability of Zygotes (Offspring)		
		AA	*Aa*	*aa*
AA × *AA*	p^2	1	0	0
AA × *Aa*	$2pq$	1/2	1/2	0
AA × *aa*	pr	0	1	0
Aa × *Aa*	$4q^2$	1/4	1/2	1/4
Aa × *aa*	$2qr$	0	1/2	1/2
aa × *aa*	r^2	0	0	1

If the occurrence of one event has no effect on the occurrence of another event, then those two events are said to be independent of each other. Thus, we have the following definition.

Definition 2.4.2. *Two events A and B with* $P(A) \neq 0$ *and* $P(B) \neq 0$ *are said to be* **independent** *if* $P(A|B) = P(A)$, *or* $P(B|A) = P(B)$. *Otherwise, A and B are* **dependent**.

As a consequence of the foregoing definition, two events A and B are independent if and only if $P(A \cap B) = P(A)P(B)$ and at least one of $P(A)$ or $P(B)$ is not zero. An alternative definition of the independence of two events A and B can be based on this equality. That is, two events A and B are said to be independent if

$$P(A \cap B) = P(A)P(B).$$

In this case it is not necessary to assume that at least one of $P(A)$ or $P(B)$ is not zero.

EXAMPLE 2.4.4

Suppose that we toss two fair dice. Let E_1 denote the event that the sum of the dice is 6 and E_2 denote the event that the first die equals 4. Then, $P(E_1 \cap E_2) = P\{(4, 2)\} = 1/36 \neq P(E_1)P(E_2) = 5/216$. Hence, E_1 and E_2 are dependent events.

Definition 2.4.3. *The k events $A_1, A_2, \ldots, A_k$ are* **mutually independent** *if for every $j = 2, 3, \ldots, k$ and every subset of distinct indices $i_1, i_2, \ldots, i_j$*

$$P(A_{i_1} \cap A_{i_2} \cap \ldots \cap A_{i_j}) = P(A_{i_1})P(A_{i_2})\ldots P(A_{i_j})$$

Mutually independent events will often be called independent. In particular, if $P\left(A_{i_j} \cap A_{i_m}\right) = P\left(A_{i_j}\right)P\left(A_{i_m}\right)$ for each $j \neq m$, then the events are called **pairwise independent**.

Now we will discuss computation of the probability $P(A_j|B)$ (called posterior probability) from the given prior probabilities $P(A_i)$ and conditional probabilities $P(B|A_i)$. First, we will state the total probability rule.

Law of total probability

Theorem 2.4.1. *Assume $S = A_1 \cup A_2 \cup \ldots \cup A_m$, where $P(A_i) > 0$, $i = 1, 2,, n$, and $A_i \cap A_j = \phi$ (null set) for $i \neq j$.* *Then, for any event B,*

$$P(B) = \sum_{i=1}^{n} P(A_i)P(B|A_i).$$

The set $A_1, A_2, \ldots, A_n$ given in Theorem 2.4.1 is called the partition of S.

EXAMPLE 2.4.5

Assume that a noisy channel independently transmits symbols, say 0's 60% of the time and 1's 40% of the time. At the receiver, there is a 1% chance of obtaining any particular symbol distorted. What is the probability of receiving a 1, irrespective of which symbol is transmitted?

Solution

Given

$$P(0) = P(\text{'0' is transmitted}) = 0.6$$

and

$$P(1) = P(\text{'1' is transmitted}) = 0.4.$$

Also, given that the probability that a particular symbol is distorted is 0.01; that is,

$$\begin{aligned} P(1|0) &= P(\text{1 is received}|\text{ 0 is transmiited}) \\ &= 0.01 = P(0|1) = P(\text{0 is recieved}|\text{ 1 is transmitted}). \end{aligned}$$

Hence, from the total probability rule, the probability of receiving a zero is

$$\begin{aligned} P(1) &= P(\text{received a 1}) = P(1|0)P(0) + P(1|1)P(1) \\ &= (0.01)(0.6) + (0.99)(0.4) = 0.402. \end{aligned}$$

Hence, irrespective of whether a 0 or 1 is transmitted, the probability of receiving a 1 is 0.402.

EXAMPLE 2.4.6

During an epidemic in a town, 40% of its inhabitants became sick. Of any 100 sick people, 10 will need to be admitted to an emergency ward. What is the probability that a randomly chosen person from this town will be admitted to an emergency ward?

Solution

Let

$$A = \{\text{the person is healthy}\}$$

and

$$B = \{\text{the person is admitted to an emergency ward}\}.$$

It is given

$$P(A^c) = 0.4.$$

Hence,

$$P(A) = 0.6.$$

We want to find P(B). Now, P(B|A) = 0, because a healthy person will not be admitted to an emergency ward. Also,

$$P(B|A^c) = \frac{10}{100} = 0.1.$$

Hence, by the total probability rule,

$$\begin{aligned} P(B) &= P(A)P(B|A) + P(A^c)P(B|A^c) \\ &= (0.6)(0) + (0.1)(0.4) = 0.04. \end{aligned}$$

Sometimes it is not possible to directly calculate the conditional probability that is needed, but other probabilities related to the probability in question are available. Bayes' rule shows how probabilities change in the light of information and how to calculate them. It is also an essential tool in the Bayesian inference. Bayes' theorem is named after an English clergyman, Reverend Thomas Bayes, who outlined the result in a paper published (posthumously) in 1763. This is one of those results that we can prove relatively easily. However, the implications of this result are profcund in statistics and many other applied fields; see Chapter 10.

Bayes' rule

Theorem 2.4.2. *Assume* $S = A_1 \cup A_2 \cup \ldots \cup A_n$, *where* $P(A_i) > 0$, $i = 1, 2, \ldots, n$ *and* $A_i \cap A_j = \phi$ *for* $i \neq j$. *Then, for any event B, with* $P(B) > 0$

$$P\left(A_j|B\right) = \frac{P\left(A_j\right)P\left(B|A_j\right)}{\sum_{i=1}^{n} P(A_i)P(B|A_i)}.$$

Proof. We have,

$$\begin{aligned} P(A_j|B) &= \frac{P(A_j \cap B)}{P(B)} \\ &= \frac{P(A_j \cap B)}{\sum_{i=1}^{n} P(A_i)P(B|A_j)}, \text{ by total probability rule for } P(B) \\ &= \frac{P(A_j)P(B|A_j)}{\sum_{i=1}^{n} P(A_i)P(B|A_i)}. \end{aligned}$$

In Bayes' theorem, the probabilities $P(A_i)$ are called the prior or a priori probabilities of the events A_i, and the conditional probability $P(A_j|B)$ is called the posterior probability of the event A_j. The events A_1 , A_n are sometimes called the states of nature.

EXAMPLE 2.4.7

Suppose a statistics class contains 70% male and 30% female students. It is known that in a test, 5% of males and 10% of females got an "A" grade. If one student from this class is randomly selected and observed to have an "A" grade, what is the probability that this is a male student?

Solution

Let A_1 denote that the selected student is a male, and A_2 denote that the selected student is a female. Here the sample space is $S = A_1 \cup A_2$. Let D denote that the selected student has an "A" grade. We are given $P(A_1) = 0.7$, $P(A_2) = 0.3$, $P(D|A_1) = 0.05$, and $P(D|A_2) = 0.10$. Then, by the total probability rule,

$$\begin{aligned} P(D) &= P(A_1)P(D|A_1) + P(A_2)P(D|A_2) \\ &= 0.035 + 0.030 = 0.065. \end{aligned}$$

Now by Bayes' rule,

$$\begin{aligned} P(A_1|D) &= \frac{P(A_1)P(D \mid A_1)}{P(A_1)P(D \mid A_1) + P(A_2)P(D \mid A_2)} \\ &= \frac{(0.7)(0.05)}{(0.065)} = \frac{7}{13} = 0.538. \end{aligned}$$

This shows that even though the probability of a male student getting an "A" grade is smaller than that for a female student, because of the larger number of male students in the class, a male student with an "A" grade has a greater probability of being selected than a female student with an "A" grade.

```
R-code: #Specify known probabilities
P_A1<-0.7
P_A2<-0.3
P_D_given_A1<-0.05
P_D_given_A2<-0.10
#Calculate P(D)
P_D<-P_A1*P_D_given_A1+P_A2*P_D_given_A2
P_D
#Calculate P(A1 given D)
P_A1_given__D<-P_A1*P_D_given_A1/(P_A1*P_D_given_A1+P_A2*P_D_given_A2)
P_A1_given__D
0.5384615
```

Steps to apply Bayes' rule

To find $P(A_1|D)$:

1. List all the probabilities, including conditional probabilities, given in the problem. That is, $P(A_1)$,..., $P(A_n)$ and $P(D|A_1)$,,$P(D|A_n)$.
2. Write the numerator as the product, $P(A_1)P(D|A_1)$.
3. Using the total probability rule, find the denominator probability by calculating $P(D) = \sum_{i=1}^{n} P(A_i)P\left(D|A_i\right)$, in the Bayes' rule.
4. The desired probability is $\frac{Numerator}{Denominator}$.

EXAMPLE 2.4.8

Suppose that three types of antimissile defense systems are being tested. From the design point of view, each of these systems has an equally likely chance of detecting and destroying an incoming missile within a range of 250 miles with a speed ranging up to nine times the speed of sound. However, in actual practice, it has been observed that the precision of these antimissile systems is not the same; that is, the first system will usually detect and destroy the target 10 of 12 times, the second will detect and destroy it 9 of 12 times, and the third will detect and destroy it 8 of 12 times. We have observed that a target has been detected and destroyed. What is the probability that the antimissile defense system was of the third type?

Solution

Let S_1, S_2, and S_3 be the events that the first, second, and third antimissile defense systems, respectively, are used. Also, let D be the event that the target has been detected and destroyed. We wish to find $P(S_3|D)$. Given that $P(S_1) = P(S_2) = P(S_3) = 1/3$, $P(D|S_1) = 10/12$, $P(D|S_2) = 9/12$, and $P(D|S_3) = 8/12$. By the total probability rule,

$$P(D) = P(S_1)P(D|S_1) + P(S_2)P(D|S_2) + P(S_3)P(D|S_3)$$
$$= \left(\frac{1}{3}\right)\left(\frac{10}{12}\right) + \left(\frac{1}{3}\right)\left(\frac{9}{12}\right) + \left(\frac{1}{3}\right)\left(\frac{8}{12}\right) = 0.75.$$

Now using the Bayes formula, we have

$$P(S_3|D) = \frac{P(S_3)P(D|S_3)}{P(D)} = \frac{(1/3)(8/12)}{0.75} = \frac{8}{27} = 0.2963.$$

If the target is destroyed, then the probability that the antimissile defense system was of the third type is 0.2963.

EXERCISES 2.4

2.4.1. Consider the portion of an electric circuit with three relays shown in Fig. 2.3. Current will flow from point a to point b if at least one of the relays closes properly when activated. The relays may malfunction and not close properly when activated. Suppose that the relays act independently of one another and close properly when activated with probability 0.9.

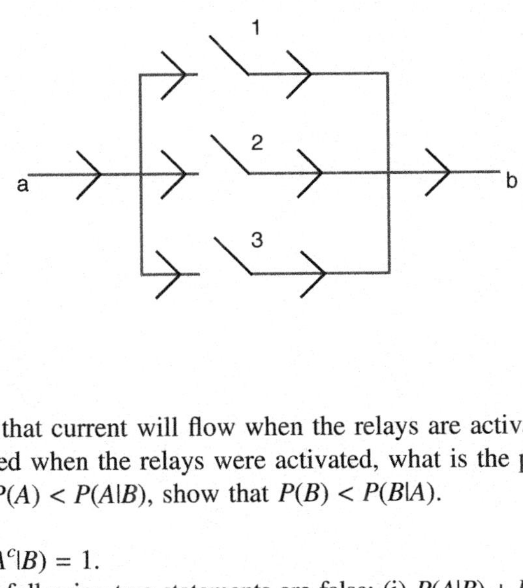

FIGURE 2.3 Electric circuit.

(a) What is the probability that current will flow when the relays are activated?
(b) Given that current flowed when the relays were activated, what is the probability that relay 1 functioned?

2.4.2. If $P(A) > 0$, $P(B) > 0$ and $P(A) < P(A|B)$, show that $P(B) < P(B|A)$.

2.4.3. If $P(B) > 0$,
(a) Show that $P(A|B) + P(A^c|B) = 1$.
(b) Show that in general the following two statements are false: (i) $P(A|B) + P(A|B^c) = 1$, (ii) $P(A|B) + P(A^c|B^c) = 1$.

2.4.4. If $P(B) = p$, $P(A^c|B) = q$, and $P(A^c \cap B^c) = r$, find (a) $P(A \cap B^c)$, (b) $P(A)$, and (c) $P(B|A)$.

2.4.5. If A and B are independent, show that so are (i) A^c and B, (ii) A and B^c, and (iii) A^c and B^c.

2.4.6. Show that two events A and B are independent if and only if $P(A \cap B) = P(A)P(B)$ when at least one of $P(A)$ or $P(B)$ is not zero.

2.4.7. A card is selected at random from an ordinary deck of 52 playing cards. If E is the event that the selected card is an ace and F is the event that it is a spade, show that E and F are independent events.

2.4.8. A fruit basket contains 30 apples, of which five are bad. If you pick two apples at random, what is the probability that both are good apples?

2.4.9. Two students are to be selected at random from a class with 10 girls and 12 boys. What is the probability that both will be girls?

2.4.10. Assume a population with the genetic distribution given in Example 2.4.3. Assume random mating. What is the probability that an offspring is *aa?*

2.4.11. One of the most common forms of colorblindness is a sex-linked hereditary condition caused by a defect on the X chromosome (one of the two chromosomes that determine gender). It is known that colorblindness is much more prevalent in males than in females. Suppose that 6% of males are colorblind, but only 0.75% of females are colorblind. In a certain population, 45% are male, and 55% are female. A person is randomly selected from this population.
(a) Find the probability that the person is colorblind.
(b) Find the probability that the person is colorblind given that the person is a male.

2.4.12. A survey asked a group of 400 people whether or not they were doing daily exercise. The responses by sex and physical activity are as in Table 2.4.

TABLE 2.4 Physical Activity Survey Results by Gender.

	Male	Female
Daily exercise	50	61
No daily exercise	177	112

A person is randomly selected.
(a) What is the probability that this person is doing daily exercise?
(b) What is the probability that this person is doing daily exercise if we know that this person is a male?

2.4.13. A laboratory blood test is 98% effective in detecting a certain disease if the person has the disease (sensitivity). However, the test also yields a "false positive" result for 0.5% of the healthy persons tested. (That is, if a healthy person is tested, then, with probability 0.005, the test result will show positive.) Assume that 2% of the population actually has this disease (prevalence). What is the probability that a person has the disease given that the test result is positive?

2.4.14. In order to evaluate the rate of error experienced in reading chest X-rays, the following experiment is done. Several people with known tuberculosis (TB) status (through other reliable tests) are subjected to chest X-rays. A technician who is unaware of this status reads the X-ray, and Table 2.5 gives the result. Here +X-ray means the technician concluded that the person has TB.

TABLE 2.5 Chest X-Ray Result for TB.

	Person without TB	Person with TB	Total
+X-ray	70	27	97
–X-ray	1883	20	1903
Total	1953	47	2000

Find (a) $P(TB| + X\text{-}ray)$, (b) $P(+X\text{-}ray|No\ TB)$, and (c) $P(No\ TB| - X\text{-}ray)$.

2.4.15. Each of the 12 ordered boxes contains 12 coins, consisting of pennies and dimes. The number of dimes in each box is equal to its order among the boxes; that is, box number 1 contains one dime and 11 pennies, box number 2 contains two dimes and 10 pennies, and so on. A pair of fair dice is tossed, and the total showing indicates which box is chosen to have a coin selected at random from it.

(a) Find the probability that a coin selected is a dime.
(b) It is observed that the selected coin is a penny. Find the probability that it came from box number 4.

2.4.16. Of 600 car parts produced, it is known that 350 are produced in one plant, 150 parts in a second plant, and 100 parts in a third plant. Also, it is known that the probabilities are 0.15, 0.2, and 0.25 that the parts will be defective if they are produced in the first, second, or third plants, respectively. What is the probability that a randomly picked part from this batch is not defective?

2.4.17. One class contains 5 girls and 10 boys, and the second class contains 13 boys and 12 girls. A student is randomly picked from the second class and transferred to the first one. After that, a student is randomly chosen from the first class. What is the probability that this student is a boy?

2.4.18. Consider that we have in an industrial complex two large boxes, each of which contains 30 electrical components. It is known that the first box contains 26 operable and 4 nonoperable components and that the second box contains 28 operable and 2 nonoperable components. Assume that the probability of making a selection from each of the boxes is the same.
(a) Find the probability that a component selected at random will be operable.
(b) Suppose the component chosen at random is operable. Find the probability that the component was chosen from box 1.

2.4.19. Urn 1 contains five white balls and three red balls. Urn 2 contains four white and six red balls. An urn is selected at random, and a ball is drawn at random from that urn. Find the probability that, if the ball selected is white, it came from urn 1.

2.4.20. An urn contains two white balls and two black balls. A number is randomly chosen from the set {1, 2, 3, 4}, and many balls are removed from the urn. Find the probability that the number i, $i = 1, 2, 3, 4$, was chosen if at least one white ball was removed from the urn.

2.4.21. A certain state groups its licensed drivers according to age into the following categories: (1) 16–25; (2) 26–45; (3) 46–65; and (4) over 65. Table 2.6 lists, for each group, the proportion of licensed drivers who belong to the group and the proportion of drivers in the group who had accidents.

TABLE 2.6 Accident Rate and Size by Age.

Group	Size	Accident Rate
1	0.250	0.086
2	0.257	0.044
3	0.347	0.056
4	0.146	0.098

(a) What proportion of licensed drivers had an accident?
(b) What proportion of those licensed drivers who had an accident were over 65?

2.4.22. It is known that a rare disease, K, is present only in 0.2% of the population. Performance of the test by a physician's diagnostic test for the presence or absence of the disease K is given in Table 2.7, where R^+ denotes the positive test result, and R^- denotes the negative result. Also, K^c denotes absence of the disease.

TABLE 2.7 Diagnostic Test Results for Disease K.

	R^+	R^-
K	0.98	0.02
Kc	0.01	0.99

(a) What is the probability that a patient has the disease, if the test result is positive?
(b) What is the probability that a patient has the disease, if the test result is negative?

2.4.23. A store has light bulbs from two suppliers, 1 and 2. The chance of supplier 1 delivering defective bulbs is 10%, whereas supplier 2 has a defective rate of 3%. Suppose 60% of the current supply of light bulbs came from supplier 1. If one of these bulbs is taken from the current supply and observed to be defective, find the probability that it came from supplier 2.

2.4.24. The quality control chart of a certain manufacturing company shows that 45% of the defective parts produced in the company are due to mechanical errors, and 55% were caused by human error. The defective parts caused by mechanical errors can be detected, with 95% accuracy rate, at an inspection station. The detection rate is only 80% if the defective parts are due to human error.
(a) Suppose a defective part was detected at the inspection station. What is the probability that this defective part is due to human error?
(b) Suppose that a customer returned a defective part that went undetected at the inspection station. What is the probability that the defective part is due to human error?

2.4.25. A circuit has three major components: A, B, and C. Component A operates independently of B and C. Components B and C are interdependent. It is known that component A works properly 85% of the time; component B, 90% of the time; and component C, 95% of the time. However, if component C fails, there is a 75% chance that B will also fail. Assume that at least two parts must operate for the circuit to function. What is the probability that the circuit will function properly?

2.4.26. Suppose that the data in Table 2.8 represent the approximate distribution of blood type frequency in percentage of the total population.

TABLE 2.8 Frequency of Blood Type.

Blood Type	O	A	B	AB
Frequency (%)	45	40	10	5

Assume that the blood types are distributed the same in both male and female populations. Also, assume that the blood types are independent of marriage.
(a) What is the probability that in a randomly chosen couple, the wife has type B blood and the husband has type O blood?
(b) It is known that a person with type B blood can safely receive transfusions only from persons with type B or type O blood. What is the probability a husband has type B or type O blood? If it is given that a woman has type B blood, what is the probability that her husband is an acceptable donor for her?

2.4.27. Suppose that there are 40 students in a statistics class and their blood type follows the percentage distribution given in Exercise 2.4.26.
(a) If we randomly select two students from this class, what is the probability that both will have the same blood type?
(b) If we randomly select two students from this class and it is observed that the first student's blood type is B, what is the probability that the second student's blood type is O?

2.4.28. A rare nonlethal disease (ND) that develops during adolescence is believed to be associated with a certain recessive genotype (aa) at a certain locus. It is known that in a population, 5% of adults have the disease. Suppose that among the adults with the disease ND, 85% have the aa genotype. Also, suppose that among the adults without the disease, 2% of them have the aa genotype. We have randomly selected an adult from this population.
(a) What is the probability that this person has the disease but not the aa genome type?
(b) What is the probability that this person has the aa genome type but not the disease ND?
(c) Given that this person has the aa genotype, what is the probability that this person has the disease ND?

2.4.29. (The gambler's ruin problem) Two gamblers, A and B, bet on the outcomes of successive flips of a coin. On each flip, if the coin comes up heads, A collects from B one unit, whereas if it comes up tails, A pays to B one unit. They continue to do this until one of them runs out of money. If it is assumed that the successive flips of the coin are independent and each flip results in a head with probability p, what is the probability that A winds up with all the money if A starts with i units and B starts with $N - i$ units?

2.4.30. From a life table the following probabilities are derived. Females can expect to live to an age of 60 years with a probability of 0.993. The probability drops to 0.956 for females with a life expectancy of 80 years. Given the woman is 60 years old, what is the probability that she lives to the age of 80 years?

2.5 RANDOM VARIABLES AND PROBABILITY DISTRIBUTIONS

An experiment may contain numerous characteristics that can be measured. However, in most cases, an experimenter will focus on some specific characteristics of the experiment. For example, a traffic engineer may focus on the number of vehicles traveling on a certain road or in a certain direction rather than the brand of vehicles or the number of passengers in each vehicle. In general, each outcome of an experiment can be associated with a number by specifying a rule of association. The concept of a random variable allows us to pass from the experimental outcomes to a numerical function of the outcomes, often simplifying the sample space (Fig. 2.4).

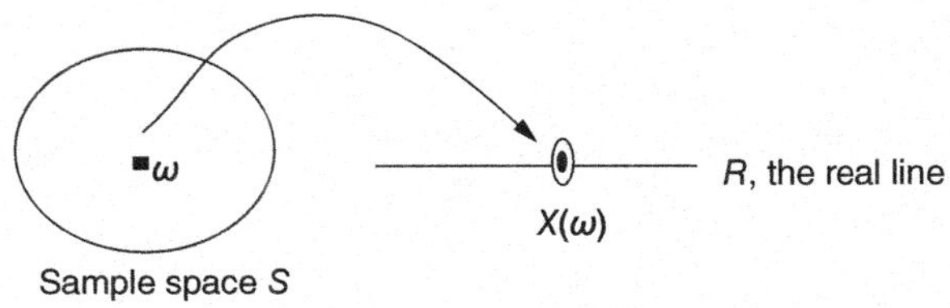

FIGURE 2.4 Random variable as a function.

Definition 2.5.1. *A* **random variable** (*r.v.*) *X is a function defined on a sample space, S, that associates a real number,* $X(\omega) = x$, *with each outcome* ω *in S.*

EXAMPLE 2.5.1

Two balanced coins are tossed and face values are noted. Then, the sample space $S = \{(H,H), (H,T), (T,H), (T,T)\}$. Define the random variable $X(\omega) = n$, where n is the number of heads and ω represents a simple event such as (H,H). Then,

$$X(\omega) = \begin{cases} 0, & \text{if}\omega = (T, T) \\ 1, & \text{if}\omega \in \{(H, T), (T, H)\} \\ 2, & \text{if}\omega = (H, H). \end{cases}$$

It can be noted that $X(\omega) = 0$ or 2 with probability 1/4 (w.p. 1/4) and $X(\omega) = 1$ (w.p. ½).

It is important to note that in the definition of a random variable, probability plays no role. However, as evidenced by the previous example, for each value or a set of values of the random variable, there are underlying collections of events, and through these events, one connects the values of random variables with probability measures.

The random variable is represented by a capital letter $(X, Y, Z,)$... and any particular real value of the random variable is denoted by the corresponding lowercase letter $(x, y, z,)$... We define two types of random variables, discrete and continuous. In this book, we will not deal with mixed random variables.

Definition 2.5.2. *A random variable X is said to be* **discrete** *if it can assume only a finite or countably infinite number of distinct values.*

Suppose an internet business firm had 1000 hits on a particular day. Let the random variable X be defined as the number of sales resulting on that day. Then, X can take values 0, 1,..., 1000. If we are to define a random variable as the number of telephone calls made from a large city on any given day, for all practical purposes, this can be assumed to take values 0, 1,... ∞.

EXAMPLE 2.5.2

In the tossing of three fair coins, let the random variable X be defined as X = number of tails. Then, X can assume values 0, 1, 2, and 3. We can associate these values with probabilities in the following way:

$$\begin{aligned} P(X=0) &= P(\{H,H,H\}) = 1/8 \\ P(X=1) &= P(\{H,H,T\} \cup \{H,T,H\} \cup \{T,H,H\}) = 3/8 \\ P(X=2) &= P(\{T,T,H\} \cup \{T,H,T\} \cup \{H,T,T\}) = 3/8 \\ P(X=3) &= P(\{T,T,T\}) = 1/8. \end{aligned}$$

We can write this in tabular form:

x	0	1	2	3
$p(x)$	1/8	3/8	3/8	1/8

Let X be a discrete random variable assuming values x_1, x_2, x_3, we have the following.

Definition 2.5.3. *The* **discrete probability mass function (pmf)** *of a discrete random variable X is the function*

$$p(x_i) = P(X = x_i),\ i = 1,2,3,....$$

A probability mass function (pmf) is more simply called a probability function (pf).
The **cumulative distribution function (cdf)** *F of the random variable X is defined by*

$$\begin{aligned} F(x) &= P(X \le x) \\ &= \Sigma_{\text{all} y \le x} p(y), \text{ for } -\infty < x < \infty. \end{aligned}$$

A cumulative distribution function is also called a **probability distribution function** *or simply the* **distribution function**.

The probability function $p(x)$ is nonnegative. In addition, because X must take on one of the values in $\{x_1, x_2, x_3 ...\}$, we have $\Sigma_{i=1}^{\infty} p(x_i) = 1$. Although the pmf $p(x)$ is defined only for a set of discrete values $x_1, x_2, x_3,...$, the cdf $F(x)$ is defined for all real numbers.

EXAMPLE 2.5.3

Suppose that a fair coin is tossed twice so that the sample space is $S = \{(H,H), (H,T), (T,H), (T,T)\}$. Let X be the number of heads.

(a) Find the probability function for X.
(b) Find the cumulative distribution function of X.

Solution

(a) *We have,*

$$1/4 = P(\{H,H\}) = P(\{H,T\}) = P(\{T,H\}) = P(\{T,T\}).$$

Hence, the pmf is given by

$$p(0) = P(X=0) = 1/4,\ p(1) = 1/2,\ p(2) = 1/4.$$

(b) *For example,*

$$\begin{aligned} F(1.5) &= P(X \le 1.5) = P(X = 0 \text{ or } 1) \\ &= P(X=0) + P(X=1) \\ &= \frac{1}{4} + \frac{1}{2} = \frac{3}{4}. \end{aligned}$$

Proceeding similarly, we obtain (as shown in Fig. 2.5*),*

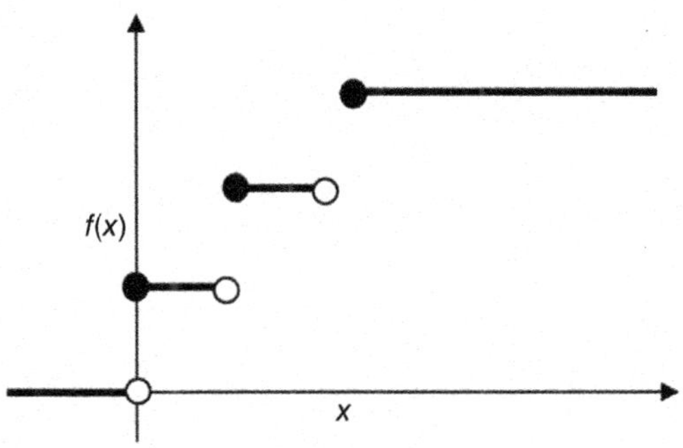

FIGURE 2.5 Graph of $F(x)$.

$$F(x) = \begin{cases} 0, & -\infty < x < 0 \\ 1/4, & 0 \le x < 1 \\ 3/4, & 1 \le x < 2 \\ 1, & 2 \le x < \infty. \end{cases}$$

We have seen that a discrete random variable assumes a finite or a countably infinite value. In contrast, we define a continuous random variable as one that assumes uncountably many values, such as the points on a real line. We now give the definition of a continuous random variable.

Definition 2.5.4. *Let X be a random variable. Suppose that there exists a nonnegative real-valued function:* $f: R \to [0, \infty)$ *such that for any interval* $[a, b]$,

$$P(X \in [a, b]) = \int_a^b f(t)dt.$$

Then, X is called a **continuous random variable**. *The function f is called the* **probability density function (pdf)** *of X. The* **cumulative distribution function (cdf)** *is given by*

$$F(x) = P(X \le x) = \int_{-\infty}^{x} f(t)dt.$$

For a given function f to be a pdf, it needs to satisfy the following two conditions: $f(x) \ge 0$ for all values of x, and $\int_{-\infty}^{\infty} f(x)dx = 1$.

Also, if f is continuous, then $\frac{dF(x)}{dx} = f(x)$, where $F(x)$ is the cdf. This follows from the fundamental theorem of calculus. If f is the pdf of a random variable X, then

$$P(a \le X \le b) = \int_a^b f(t)dt.$$

Fig. 2.6 represents $P(a \le X \le b)$.

As a result, for any real number a, $P(X = a) = 0$. Also,

$$P(a \le X \le b) = P(a < X \le b) = P(a \le X < b) = P(a < X < b).$$

If we have cdf $F(x)$, then we have,

$$P(a \le X \le b) = F(b) - F(a).$$

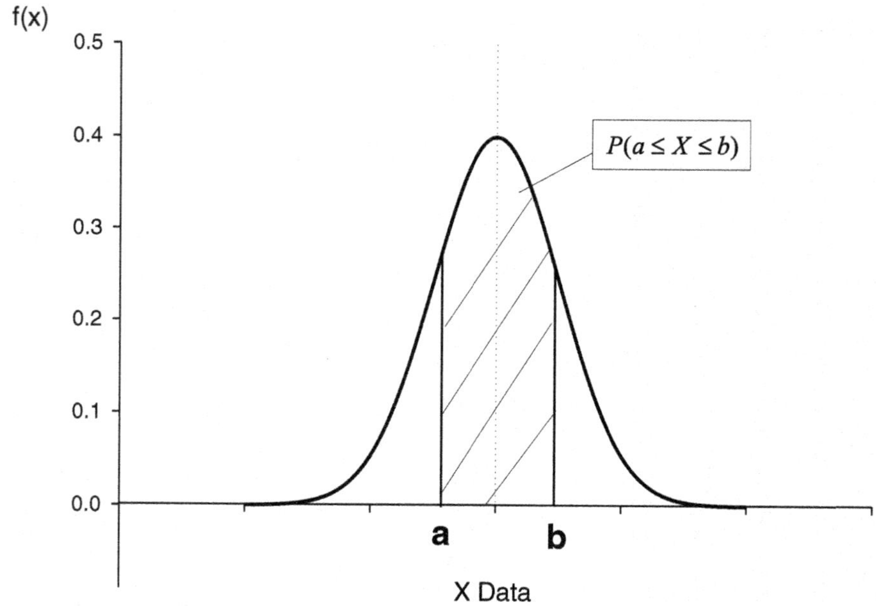

FIGURE 2.6 Probability as an area under a curve.

Some properties of distribution function

1. $0 \le F(x) \le 1$.
2. $\lim_{x \to -\infty} F(x) = 0$, and $\lim_{x \to \infty} F(x) = 1$.
3. F is a nondecreasing function, and right continuous.

EXAMPLE 2.5.4

Let the function

$$f(x) = \begin{cases} \lambda x e^{-x}, & x > 0 \\ 0, & \text{otherwise.} \end{cases}$$

(a) For what value of is f a pdf?
(b) Find $F(x)$.

Solution

(a) *First note that $f(x) \ge 0$. Now, for $f(x)$ to be a pdf, we need $\int_{-\infty}^{\infty} f(x)dx = 1$. Because $f(x) = 0$ for $x \le 0$, by integration by parts, we can get $\lambda = 1$ (see* Fig. 2.7*)*

$$1 = \int_{-\infty}^{\infty} f(x)dx = \int_{0}^{\infty} \lambda x e^{-x} dx$$

$$= \lambda \int_{-\infty}^{\infty} x e^{-x} dx = \lambda \left[-x e^{-x} \Big|_0^{\infty} + \int_0^{\infty} e^{-x} dx \right] (\text{usingintegrationbyparts})$$

$$= \lambda \left[0 - e^{-x} \Big|_0^{\infty} \right] = \lambda.$$

(b) *The cumulative distribution function is*

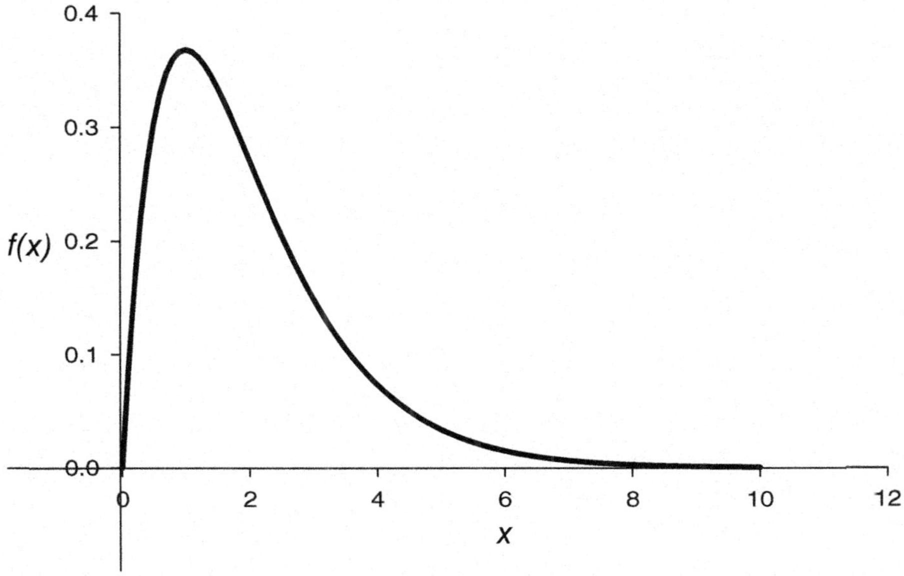

FIGURE 2.7 Graph of $f(x) = xe^{-x}$.

$$F(x) = \int_{-\infty}^{x} f(t)dt = \begin{cases} 0, & x < 0 \\ \int_0^x te^{-t}dt = 1 - (x+1)e^{-x}, & x \geq 0. \end{cases}$$

Fig. 2.8 *represents the cumulative distribution.*

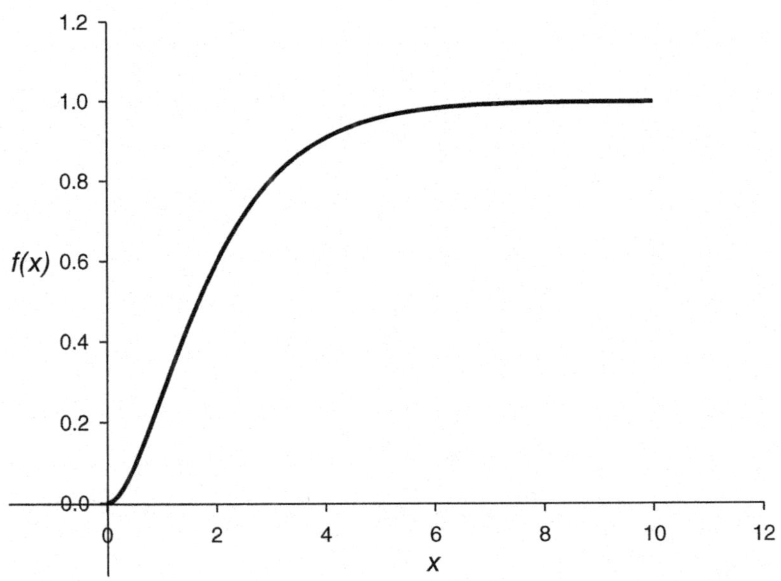

FIGURE 2.8 Graph of $F(x)$, $x \geq 0$.

EXAMPLE 2.5.5

Suppose that a large grocery store has shelf space for 150 cartons of fruit drink that are delivered on a particular day of each week. The weekly sales for a fruit drink show that the demand increases steadily up to 100 cartons and then levels off between 100 and 150 cartons. Let Y denote the weekly demand in hundreds of cartons. It is known that the pdf of Y can be approximated by

$$f(y) = \begin{cases} y, & 0 \le y \le 1 \\ 1, & 1 < y \le 1.5 \\ 0, & \text{elsewhere.} \end{cases}$$

(a) Find $F(y)$.
(b) Find $P(0 \le Y \le 0.5)$.
(c) Find $P(0.5 \le Y \le 1.2)$.

Solution

(a) *The graph of the density function f(y) is shown in* Fig. 2.9
*From the definition of cdf, we have (*Fig. 2.10*)*

$$F(y) = \int_{-\infty}^{y} f(t)dt = \begin{cases} 0, & y < 0 \\ \int_0^y t\,dt, & 0 \le y < 1 \\ \int_0^1 t\,dt + \int_1^y dt, & 1 \le y < 1.5 \\ \int_0^1 t\,dt + \int_1^{1.5} dt, & y \ge 1.5 \end{cases}$$

$$= \begin{cases} 0, & y < 0 \\ y^2/2, & 0 \le y < 1 \\ y - 1/2, & 1 \le y < 1.5 \\ 1 & y \ge 1.5. \end{cases}$$

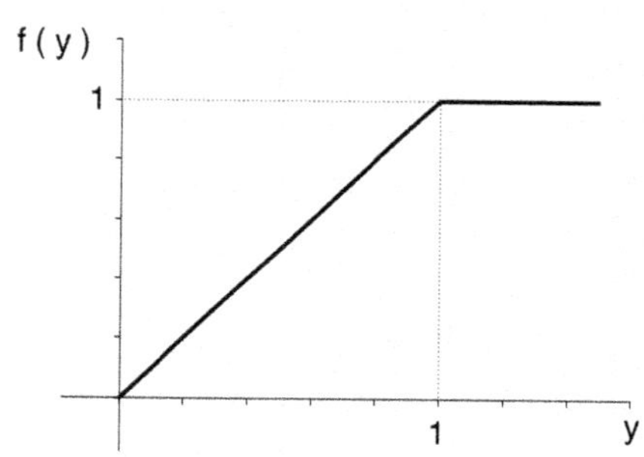

FIGURE 2.9 Graph of $f(y)$.

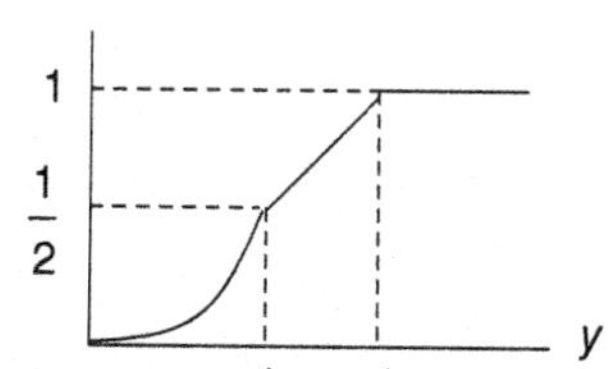

FIGURE 2.10 Graph of $F(y)$.

(b) *The probability*

$$P(0 \le Y \le 0.5) = F(0.5) - F(0)$$
$$= (0.5)^2/2 = 1/8 = 0.125.$$

(c)

$$P(0.5 \le Y \le 1.2) = F(1.2) - F(0.5)$$
$$= (1.2 - 1/2) - 0.125 = 0.575.$$

EXERCISES 2.5

2.5.1. The probability function of a random variable Y is given by $p(i) = \frac{c\lambda^i}{i!}$, $i = 0,1,2, ...,$ where is a known positive value and c is a constant.
(a) Find c.
(b) Find $P(Y = 0)$.
(c) Find $P(Y > 2)$.

2.5.2. Find k so that the function given by

$$p(x) = \frac{k}{x+1}, \ x = 1,2,3,4,$$

is a probability mass function. Graph the probability mass function and cumulative distribution function.

2.5.3. A random variable X has the following probability mass function:

x	−5	0	3	6
$p(x)$	0.2	0.1	0.4	0.3

Find the cumulative distribution function $F(x)$ and graph it.

2.5.4. The cumulative probability function of a discrete random variable X is given in the following table:

x	−1	0	2	5	6
$F(x)$	0.1	0.15	0.4	0.8	1

(a) Find $P(X = 2)$.
(b) Find $P(X > 0)$.

2.5.5. The cumulative distribution function $F(x)$ of a random variable X is given by

$$F(x) = \begin{cases} 0, & -\infty < x < -1 \\ 0.2, & -1 \le x < 3 \\ 0.8, & 3 \le x < 9 \\ 1, & x \ge 9. \end{cases}$$

Write down the values of the random variable X and the corresponding probabilities, $p(x)$.

2.5.6. The probability density function of a random variable X is given by

$$f(x) = \begin{cases} cx, & 0 < x < 4 \\ 0, & \text{otherwise}. \end{cases}$$

(a) Find c.
(b) Find the distribution function $F(x)$.
(c) Compute $P(1 < X < 3)$.

2.5.7. Let the function

$$f(x) = \begin{cases} cx^2, \ 0 < x < 3 \\ 0, \text{ otherwise.} \end{cases}$$

(a) Find the value of c so that $f(x)$ is a density function.
(b) Compute $P(2 < X < 3)$.
(c) Find the distribution function $F(x)$.

2.5.8. Suppose that Y is a continuous random variable whose pdf is given by

$$f(y) = \begin{cases} K(4y - 2y^2), & 0 < y < 2 \\ 0, & \text{elsewhere.} \end{cases}$$

(a) What is the value of K?
(b) Find $P(Y > 1)$.
(c) Find $F(y)$.

2.5.9. The random variable X has a cumulative distribution function

$$F(x) = \begin{cases} 0, & \text{for } x \le 0 \\ \frac{x^2}{1+x^2}, & \text{for } x > 0. \end{cases}$$

Find the probability density function of X.

2.5.10. A random variable X has a cumulative distribution function

$$F(x) = \begin{cases} 0, & \text{if } x \le 0 \\ ax + b, & \text{if } 0 \le x < 3 \\ 1, & \text{if } x \ge 3. \end{cases}$$

(a) Find the constants a and b.
(b) Find the pdf $f(x)$.
(c) Find $P(1 < X < 5)$.

2.5.11. The amount of time, in hours, that a machine functions before breakdown is a continuous random variable with pdf

$$f(t) = \begin{cases} \frac{1}{120} e^{-t/120}, & t \ge 0 \\ 0, & t < 0. \end{cases}$$

What is the probability that this machine will function between 98 and 145 hours before breaking down? What is the probability that it will function for fewer than 160 hours?

2.5.12. The length of time that an individual talks on a long-distance telephone call has been found to be of a random nature. Let X be the length of the talk; assume it to be a continuous random variable with a probability density function given by

$$f(x) = \begin{cases} \alpha e^{-(1/5)x}, \ x > 0 \\ 0, \text{ elsewhere.} \end{cases}$$

Find

(a) The value of α that makes $f(x)$ a probability density function.

(b) The probability that this individual will talk (i) between 8 and 12 minutes, (ii) less than 8 minutes, (iii) more than 12 minutes.

(c) Find the cumulative distribution function, $F(x)$.

2.5.13. Let T be the life length of a mechanical system. Suppose that the cumulative distribution of such a system is given by

$$F(t) = \begin{cases} 0, & t < 0 \\ 1 - \exp\left(-\frac{(t-y)^{\beta}}{\alpha}\right), & t \ge 0,\ \alpha > 0,\ \beta,\ y \ge 0. \end{cases}$$

Find the probability density function that describes the failure behavior of such a system.

2.6 MOMENTS AND MOMENT-GENERATING FUNCTIONS

One of the most useful concepts in probability theory is that of expectation of a random variable. The expected value may be viewed as the balance point of distribution of probability on the real line, or in common language, the average.

Definition 2.6.1. *Let X be a discrete random variable with pmf p(x). Then the* **expected value** *of X, denoted by E(X), is defined by*

$$\mu = E(X) = \sum_{\text{all}x} xp(x), \text{provided} \sum_{\text{all}x} |x|p(x) < \infty.$$

Now, we will define the expected value for a continuous random variable.

Definition 2.6.2. *The* **expected value** *of a continuous random variable X with pdf f(x) is defined by*

$$\mu = E(X) = \int_{-\infty}^{\infty} xf(x)dx, \text{ provided } \int_{-\infty}^{\infty} |x|f(x)dx < \infty.$$

The expected value of X is also called the *expectation* or mathematical expectation of X. We denote the expected value of X by μ.

EXAMPLE 2.6.1

Let

$$X = \begin{cases} 1, \text{ with a probability } 1/2 \\ 0, \text{ with a probability } 1/2. \end{cases}$$

Then, $E(X) = 1(1/2) + 0(1/2) = 1/2$.

EXAMPLE 2.6.2

Let X be a discrete random variable whose probability mass function is given in the following table:

x	−1	0	1	2	3	4	5
$p(x)$	$\frac{1}{7}$	$\frac{1}{7}$	$\frac{1}{14}$	$\frac{2}{7}$	$\frac{1}{14}$	$\frac{1}{7}$	$\frac{1}{7}$

Find $E(X)$.

Solution
By definition,

$$E(X) = \sum xp(x) = -1\left(\frac{1}{7}\right) + 0\left(\frac{1}{7}\right) + 1\left(\frac{1}{14}\right)$$
$$+2\left(\frac{2}{7}\right) + 3\left(\frac{1}{14}\right) + 4\left(\frac{1}{7}\right) + 5\left(\frac{1}{7}\right) = 2.$$

EXAMPLE 2.6.3

Let $X \geq 0$ be an integer-valued random variable such that $P(X = n) = p_n$. Show that $E(X) = \sum_{n=1}^{\infty} P(X \geq n)$.

Solution
Using the definition of expectation, and the fact that $(0)p_0 = 0$, *we have,*

$$\begin{aligned} E(X) &= \textstyle\sum_{n=1}^{\infty} np_n = 1p_1 + 2p_2 + 3p_3 + \cdots \\ &= p_1 + p_2 + p_3 + \cdots \\ &\quad + p_2 + p_3 + p_4 + \cdots \\ &\quad + p_3 + p_4 + \cdots \\ &= P(X \geq 1) + P(X \geq 2) + \cdots \\ &= \textstyle\sum_{n=1}^{\infty} P(X \geq n). \end{aligned}$$

Example 2.6.4

Suppose you are selling a car. Let $X_0, X_1, X_2, \ldots$ be the successive offers occurring at times 0, 1, 2,..., n, that you receive (assume that the offers are random, independent, and have the same distribution); see Fig. 2.11. Show that $E(N) = \infty$, where $N = \min\{n: X_n > X_0\}$, that is, the first time an offer exceeds the initial offer X_0 at time "0".

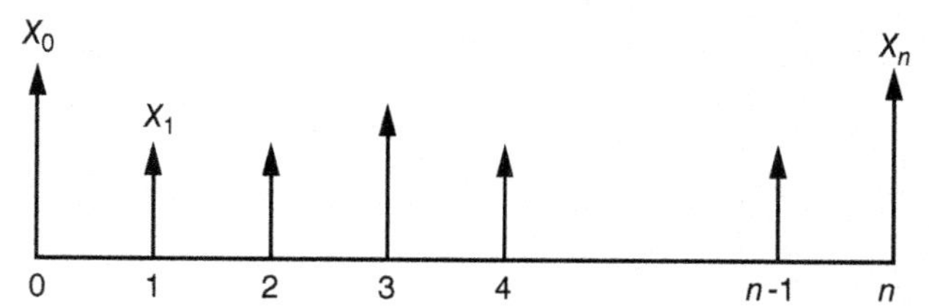

FIGURE 2.11 Size of successive offerings.

Solution
By definition,

$$\begin{aligned} P(N \geq n) &= P(X_0 \text{ is largest of } X_0, X_1, \ldots, X_{n-1}) \\ &= \frac{1}{n}, \text{ by symmetry,} \end{aligned}$$

as any of the $X'_i s$ *could be more than the rest. Hence, using Example 2.6.3,*

$$E(N) = \sum_{n=1}^{\infty} P(N \geq n) = \sum_{n=1}^{\infty} \frac{1}{n} = \infty.$$

You would expect to wait a long time to receive an offer better than the first one.

The variance is a measure of the spread or variability of values of a random variable around the mean.

Definition 2.6.3. *The* **variance** *of a random variable X is defined by*

$$\sigma^2 = Var(X) = E[(X - \mu)^2].$$

The square root of variance, denoted by σ, is called the **standard deviation.**

The next result shows how to obtain the expectation of a function of a random variable.

Expectation of function of a random variable

Theorem 2.6.1. *Let g(X) be a function of X, then the* **expected value** *of g(X) is*

$$E[g(X)] = \begin{cases} \sum_x g(x)p(x), \text{ if X is discrete} \\ \int_{-\infty}^{\infty} g(x)f(x)dx, \text{ if X is continuous,} \end{cases}$$

provided the sum or the integral exists.

We now give some properties of the expectation of a random variable.

Some properties of expected value and variance

Theorem 2.6.2. *Let c be a constant and let g(X), $g_1(X),\ldots, g_n(X)$ be functions of a random variable X such that E(g(X)) and $E(g_i(X))$ for i = 1, 2,, n exist. Then the following results hold:*

(a) $E(c) = c.$
(b) $E[cg(X)] = cE[g(X)].$
(c) $E[\sum_i g_i(X)] = \sum_i E[g_i(X)].$
(d) $Var(aX + b) = a^2 Var(X)$. In particular, $Var(aX) = a^2 Var(X)$.
(e) $Var(X) = E(X^2) - \mu^2.$

Proof. Proof of (a) through (d) will be given as an exercise. We will prove (e).

$$\begin{aligned} Var(X) &= E[(X - \mu)^2] \\ &= E(X^2 - 2X\mu + \mu^2) \\ &= E(X^2) - 2\mu E(X) + \mu^2 \\ &= E(X^2) - 2\mu^2 + \mu^2 \\ &= E(X^2) - \mu^2. \end{aligned}$$

EXAMPLE 2.6.5

A discrete random variable *X* is said to be uniformly distributed over the numbers 1, 2, 3,..., *n*, if

$$P(X = i) = \frac{1}{n}, i = 1,2, \ldots, n.$$

Find $E(X)$ and $Var(X)$.

Solution

By definition,

$$\begin{aligned} E(X) &= \Sigma_{i=n}^{n} x_i p_i \\ &= 1\left(\frac{1}{n}\right) + 2\left(\frac{1}{n}\right) + \cdots + n\left(\frac{1}{n}\right) \\ &= \frac{1}{n}\left[\frac{n(n+1)}{2}\right] = \frac{n+1}{2}. \end{aligned}$$

Similarly, using the summation formula $1^2 + 2^2 + \cdots + n^2 = \frac{n(n+1)(2n+1)}{6}$, *we get,*

$$\begin{aligned} E(X^2) &= 1^2\left(\frac{1}{n}\right) + 2^2\left(\frac{1}{n}\right) + \cdots + n^2\left(\frac{1}{n}\right) \\ &= \frac{1}{n}\left[\frac{n(n+1)(2n+1)}{6}\right] \\ &= \frac{(n+1)(2n+1)}{6}. \end{aligned}$$

Hence,

$$\begin{aligned} Var(X) &= E(X^2) - (EX)^2 \\ &= \frac{(n+1)(2n+1)}{6} - \left(\frac{n+1}{2}\right)^2 \\ &= \frac{n^2-1}{12}. \end{aligned}$$

EXAMPLE 2.6.6

To find out the prevalence of smallpox vaccine use, a researcher inquired into the number of times a randomly selected 200 people aged 16 and over in an African village had been vaccinated. He obtained the following figures: never, 17 people; once, 30; twice, 58; three times, 51; four times, 38; five times, 7. Assuming these proportions continue to hold exhaustively for the population of that village, what is the expected number of times those people in the village had been vaccinated, and what is the standard deviation?

Solution

Let X denote the random variable representing the number of times a person aged 16 or older in this village has been vaccinated. Then, we can obtain the following distribution:

x	0	1	2	3	4	5
$p(x)$	17/200	30/200	58/200	51/200	38/200	7/200

Then,

$$\begin{aligned} E(X) &= \Sigma\, xp(x) = \frac{1}{200}(0(17) + 1(30) + 2(58) + 3(51) + 4(38) + 5(7)) \\ &= 2.43. \end{aligned}$$

Also,

$$Var(X) = E(X^2) - (E(X))^2$$
$$= \sum x^2 p(x) - (2.43)^2 = 7.52 - (2.43)^2$$
$$= 1.6151.$$

Thus, the standard deviation is $\sqrt{1.6151} = 1.2709$.

EXAMPLE 2.6.7

Let Y be a random variable with pdf

$$f(y) = \begin{cases} \frac{3}{64} y^2 (4 - y), \ 0 \le y \le 4 \\ 0, \text{ elsewhere.} \end{cases}$$

(a) Find the expected value and variance of Y.
(b) Let $X = 300Y + 50$. Find $E(X)$ and $Var(X)$.
(c) Find $P(X > 750)$.

Solution

(a)

$$E(Y) = \int_{-\infty}^{\infty} y f(y) dy$$
$$= \frac{3}{64} \int_0^4 y y^2 (4 - y) dy$$
$$= 2.4,$$

and

$$Var(Y) = \int_0^4 (y - 2.4)^2 \frac{3}{64} y^2 (4 - y) dy$$
$$= 0.64.$$

(b) *Using the fact that* $Var(aY + b) = a^2 Var(Y)$, *we have*

$$Var(X) = (300)^2 Var(Y)$$
$$= 90{,}000(0.64) = 57{,}600.$$

(c)

$$P(X > 750) = P(300Y + 50 > 750)$$
$$= P\left(Y > \frac{7}{3}\right)$$
$$= \frac{3}{64} \int_{7/3}^{4} y^2 (4 - y) dy = 0.55339.$$

Example 2.6.8

Let $X_1, \ldots, X_n$ be independent, identically distributed (i.i.d) random variables with mean μ and variance σ^2. Let $\overline{X} = \frac{1}{n} \sum_{i=1}^{n} X_i$, find the mean of $S = \sum_{i=1}^{n} (X_i - \overline{X})^2$.

Solution

$$S = \Sigma_{i=1}^{n}(X_i - \overline{X})^2$$
$$= \Sigma_{i=1}^{n} X_i^2 - 2n\overline{X}^2 + n\overline{X}^2 = \Sigma_{i=1}^{n} X_i^2 - n\overline{X}^2.$$

Now, we know that
$E[X_i^2] = Var(X_i) + (E[X_i])^2 = \sigma^2 + \mu^2$, *and* $E[\overline{X}^2] = Var(\overline{X}) + (E[\overline{X}])^2 = \frac{\sigma^2}{n} + \mu^2$.
Thus,

$$E[S] = \Sigma_{i=1}^{n} E[X_i^2] - nE[\overline{X}^2]$$
$$= \Sigma_{i=1}^{n}(\sigma^2 + \mu^2) - n\left(\frac{\sigma^2}{n} + \mu^2\right)$$
$$= n\sigma^2 + n\mu^2 - \sigma^2 - n\mu^2 = (n-1)\sigma^2.$$

2.6.1 Skewness and Kurtosis

Even though the mean μ and the standard deviation σ are significant descriptive measures that locate the center and describe the spread or dispersion of the probability density function $f(x)$, they do not provide a unique characterization of the distribution. Two distributions may have the same mean and variance and yet could be very different, as in Fig. 2.12.

To better approximate the probability distribution of a random variable, we may need higher moments.

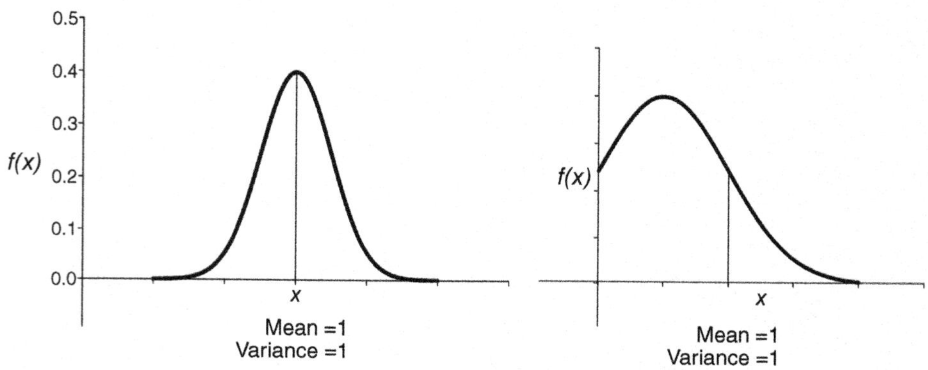

FIGURE 2.12 Same mean and variance.

Definition 2.6.4. *The* **kth moment about the origin** *of a random variable X is defined as* EX^k *and denoted by* μ'_k, *whenever it exists. The* **kth moment about its mean** *(also called the* **central kth moment***) of a random variable X is defined as* $E[(X-\mu)^k]$ *and denoted by* μ_k, $k = 2, 3, 4, \ldots$, *whenever it exists.*

In particular, we have $E(X) = \mu'_1 = \mu$, and $\sigma^2 = \mu_2$. We have seen earlier that the second moment about the mean (variance, σ^2) is used as a measure of dispersion about the mean.

Definition 2.6.5. *The standardized third moment about the mean,*

$$\alpha_3 = \frac{E(X-\mu)^3}{\sigma^3} = \frac{\mu_3}{\mu_2^{3/2}}$$

is called the **skewness** *of the distribution of X. The standardized fourth moment about the mean,*

$$\alpha_4 = \frac{E(X-\mu)^4}{\sigma^4}$$

is called the **kurtosis** *of the distribution.*

Skewness is used as a measure of the asymmetry (lack of symmetry) of a density function about its mean. Recall that a distribution, or data set, is symmetric if it looks the same to the left and right of the center point. Thus, for a symmetric distribution, $_3 = 0$. However, if $_3 = 0$, then we cannot say that the distribution is symmetric about the mean. For instance, if one tail is fat and the other tail is long, skewness does not obey such a simple rule. If $_3 > 0$, the distribution has a longer right tail, and if $_3 < 0$, the distribution has a longer left tail. Thus, the skewness of a normal distribution is zero. Kurtosis is a measure of whether the distribution is peaked or flat relative to a normal distribution. Kurtosis is based on the size of a distribution's tails. Positive kurtosis indicates too few observations in the tails, whereas negative kurtosis indicates too many observations in the tails of the distribution. Distributions with relatively large tails are called *leptokurtic*, and those with small tails are called *platokurtic*. A distribution that has the same kurtosis as a normal distribution is known as *mesokurtic*. It is known that the kurtosis for a standard normal distribution $_4 = 3$.

A sample of n values, $x_1, \ldots, x_n$ the skewness (g_1) and kurtosis (k_1) can be calculated using the following formulas.

$$g_1 = \frac{n}{(n-1)(n-2)} \sum_{i=1}^{n} \left(\frac{x_i - \overline{x}}{s}\right)^3$$

and

$$k_1 = \left[\frac{n(n+1)}{(n-1)(n-2)(n-3)} \sum_{i=1}^{n} \left(\frac{x_i - \overline{x}}{s}\right)^4\right] - \frac{3(n-1)^2}{(n-2)(n-3)}.$$

An important expectation is that the moment-generating function for a random variable, in a sense, contains all the moments for a random variable in one expression.

Definition 2.6.6. *For a random variable X, suppose that there is a positive number h such that for $-h < t < h$ the mathematical expectation $E(e^{tX})$ exists. The* **moment-generating function (mgf)** *of the random variable X is defined by*

$$M_X(t) = E(e^{tX}) = \begin{cases} \sum e^{tx} p(x), & \text{if discrete} \\ \int e^{tx} f(x)dx, & \text{if continuous.} \end{cases}$$

An advantage of the moment-generating function is its ability to give the moments. Recall that the Maclaurin series of the function *etx* is

$$e^{tx} = 1 + tx + \frac{(tx)^2}{2!} + \frac{(tx)^3}{3!} + \cdots + \frac{(tx)^n}{n!} + \cdots.$$

By using the fact that the expected value of the sum equals the sum of the expected values, the moment-generating function can be written as

$$M_X(t) = E[e^{tX}] = E\left[1 + tX + \frac{(tX)^2}{2!} + \frac{(tX)^3}{3!} + \cdots + \frac{(tX)^n}{n!} + \cdots\right]$$
$$1 + tE[X] + \frac{t^2}{2!}E[X^2] + \frac{t^3}{3!}E[X^3] + \cdots + \frac{t^n}{n!}E[X^n] + \cdots$$

Note that $M_X(0) = 1$ for all the distributions. Taking the derivative of $M_X(t)$ with respect to t, we obtain,

$$\frac{dM_X(t)}{dt} = M'_X(t) = E[X] + tE[X^2] + \frac{t^2}{2!}E[X^3]$$
$$+\frac{t^3}{3!}E[X^4] + \cdots + \frac{t^{(n-1)}}{(n-1)!}E[X^n] + \cdots$$

Evaluating this derivative at $t = 0$, all terms except $E[X]$ become zero. We have,

$$M'_X(0) = E[X].$$

Similarly, taking the second derivative of $M_X(t)$, we obtain,

$$M''_X(0) = E[X^2].$$

Continuing in this manner, from the nth derivative $M_X^{(n)}(t)$ with respect to t, we obtain all the moments to be

$$M_X^{(n)}(0) = E[X^n],\ n = 1,2,3,....$$

We summarize these calculations in the following theorem.

Theorem 2.6.3. *If $M_X(t)$ exists, then for any positive integer k,*

$$\frac{d^k M_X(t)}{dt^k}\Big|_{t=0} = M_X^{(k)}(0) = \mu'_k.$$

The usefulness of the foregoing theorem lies in the fact that, if the mgf can be found, the often difficult process of integration or summation involved in calculating different moments can be replaced by the much easier process of differentiation. The following examples illustrate this fact.

EXAMPLE 2.6.9

Let X be a random variable with pf

$$p(x) = \binom{n}{x} p^x (1-p)^{n-x},\ x = 0,1,2,\ ...,n.$$

(This random variable is called a binomial random variable, and the pmf is called a binomial distribution.) Show that $M_x(t) = [(1-p) + pe^t]^n$, for all real values of t. Also obtain the mean and variance of the random variable X.

Solution

The moment-generating function of X is

$$M_X(t) = E(e^{tX}) = \sum_{x=0}^{n} e^{tx}\binom{n}{x} p^x (1-p)^{n-x}$$
$$= \sum_{x=0}^{n}\binom{n}{x}(pe^t)^x (1-p)^{n-x}.$$

Using the binomial formula, we have,

$$M_X(t) = [pe^t + (1-p)]^n,\ -\infty < t < \infty.$$

The first two derivatives of $M_x(t)$ are

$$M'_X(t) = n[(1-p) + pe^t]^{(n-1)}(pe^t)$$

and

$$M''_X(t) = n(n-1)[(1-p) + pe^t]^{(n-2)}(pe^t)^2 + n[(1-p) + pe^t]^{(n-1)}(pe^t).$$

Thus,

$$\mu = E(X) = M'_X(0) = np$$

and

$$\sigma^2 = E(X^2) - \mu^2 = M''(0) - (np)^2$$
$$= n(n-1)p^2 + np - (np)^2 = np(1-p).$$

EXAMPLE 2.6.10

Let X be a random variable with pmf $f(x) = e^{-x}/(x!)$, $x = 0, 1, 2$, (Such a random variable is called a Poisson r.v. and the distribution is called a Poisson distribution with parameter). Find the mgf of X.

Solution

By definition,

$$M_X(t) = Ee^{tX} = \sum_{x=0}^{\infty} e^{tx} f(x)$$
$$= \sum_{x=0}^{\infty} e^{tx} \frac{e^{-\lambda}\lambda^x}{x!} = \sum_{x=0}^{\infty} e^{-\lambda} \frac{(e^t\lambda)^x}{x!}$$
$$= e^{-\lambda} \sum_{x=0}^{\infty} e^{\lambda e^t} \left[\frac{e^{-(\lambda e^t)}(\lambda e^t)^x}{x!}\right]$$
$$= e^{\lambda(e^t-1)} \sum_{x=0}^{\infty} \left[\frac{e^{-(\lambda e^t)}(\lambda e^t)^x}{x!}\right].$$

We observe that $e^{-(\lambda e^t)}(\lambda e^t)^x/x!$ *is a Poisson pf with parameter* e^t. *Hence,* $\sum_{x=0}^{\infty} \frac{e^{-(\lambda e^t)}(\lambda e^t)^x}{x!} = .1$ *Thus from (1),*

$$M_X(t) = e^{\lambda(e^t-1)}.$$

EXAMPLE 2.6.11

Let X be a random variable with pdf given by

$$f(x) = \begin{cases} \frac{1}{\beta} e^{-x/\beta}, & x > 0 \\ 0, & \text{otherwise.} \end{cases}$$

Find mgf $M_X(t)$.

Solution

By definition of mgf,

$$M_x(t) = \int_{\infty}^{-\infty} e^{tx} f(x)\,dx$$
$$= \int_{\infty}^{-\infty} e^{tx} \frac{1}{\beta} e^{-x/\beta} dx$$
$$= \frac{1}{\beta} \int_{\infty}^{-\infty} e^{-\left(\frac{1}{\beta}-t\right)x} dx, \left(t < \frac{1}{\beta}\right)$$
$$= \frac{1}{\beta}\left[-\frac{1}{((1/\beta)-t)} e^{\left(-\frac{1}{\beta}-t\right)} \Big|_{x=0}^{\infty}\right]$$
$$= \frac{1}{\beta}\frac{\beta}{1-\beta t} = \frac{1}{1-\beta t}, \; t < \frac{1}{\beta}.$$

EXAMPLE 2.6.12

Let X be a random variable with pdf $f(x) = (1/\sqrt{2\pi})e^{-x^2/2}$, $-\infty < x < \infty$. (We call such a random variable a standard normal random variable.) Find the mgf of X.

Solution

By the definition of mgf, we have

$$E(e^{tx}) = \frac{1}{\sqrt{2\pi}} \int_{-\infty}^{+\infty} e^{tx} e^{-x^2/2} dx$$
$$= \frac{1}{\sqrt{2\pi}} \int_{-\infty}^{+\infty} e^{-\frac{1}{2}(x^2-2tx)} dx$$
$$= \frac{1}{\sqrt{2\pi}} \int_{-\infty}^{+\infty} e^{-\frac{1}{2}(x^2-2tx+t^2)+\frac{t^2}{2}} dx$$
$$= \frac{1}{\sqrt{2\pi}} \int_{-\infty}^{+\infty} e^{-\frac{1}{2}(x-t)^2+\frac{t^2}{2}} dx$$
$$= e^{t^2/2} \frac{1}{\sqrt{2\pi}} \int_{-\infty}^{+\infty} e^{-\frac{1}{2}(x-t)^2} dx = e^{t^2/2}$$

as $1/\sqrt{2\pi} e^{-\frac{1}{2}(x-t)^2}$ *is a normal pdf with mean t and variance 1 and hence* $\frac{1}{\sqrt{2\pi}} \int_{-\infty}^{+\infty} e^{-\frac{1}{2}(x-t)^2} = 1$.
A random variable X with pdf

$$f(x) = (1/\sqrt{2\pi}) e^{-\frac{1}{2\sigma^2}(x-\mu)^2}, \quad -\infty < x < \infty$$

is called a normal random variable with mean μ *and variance* σ^2. *We will denote such random variables by* X: $N(\mu, \sigma^2)$.

Properties of the moment-generating function

1. The moment-generating function of X is unique in the sense that, if two random variables X and Y have the same mgf ($M_X(t) = M_Y(t)$, for t in an interval containing 0), then X and Y have the same distribution.
2. If X and Y are independent, then

$$M_{X+Y}(t) = M_X(t) M_Y(t).$$

That is, the mgf of the sum of two independent random variables is the product of the mgfs of the individual random variables. The result can be extended to "n" random variables.

3. Let $Y = aX + b$. Then,

$$M_Y(t) = e^{bt} M_X(at).$$

EXAMPLE 2.6.13

Find the mgf of $XN(\mu, \sigma^2)$.

Solution

Let Y: $N(0, 1)$ *and let* $X = \sigma Y + \mu$. *Then, by the foregoing property (3), and the Example 2.6.11, the mgf of X is*

$$M_X(t) = e^{\mu t} M_Y(\sigma t)$$
$$= e^{\mu t} e^{\frac{1}{2}\sigma^2 t^2} = e^{\mu t + \frac{1}{2}\sigma^2 t^2}.$$

EXAMPLE 2.6.14

Let X_1: $N\left(\mu_1, \sigma_1^2\right)$, X_2: $N\left(\mu_2, \sigma_2^2\right)$. Let X_1 and X_2 be independent. Find the mgf of $Y = X_1 + X_2$ and obtain the distribution of Y.

Solution

By property (2),

$$\begin{aligned} M_X(t) &= M_{X_1}(t) M_{X2}(t) \\ &= \left(e^{\mu_1 t + \frac{1}{2}\sigma_1^2 t^2}\right)\left(e^{\mu_2 t + \frac{1}{2}\sigma_2^2 t^2}\right) \\ & e^{(\mu_1+\mu_2)t + \frac{1}{2}\left(\sigma_1^2+\sigma_2^2\right)t^2}. \end{aligned}$$

This implies Y: $N\left(\mu_1 + \mu_2, \sigma_1^2 + \sigma_2^2\right)$.

This result can be generalized. If $X_1, \ldots, X_n$ are independent random variables such that X_i: $N\left(\mu_i, \sigma_1^2\right)$, $i = 1, \ldots, n$, then we can show that $\sum_{i=1}^n a_i X_i$: $N\left(\sum_{i=1}^n a_i \mu_i, \sum_{i=1}^n a_i^2 \sigma_i^2\right)$.

The moment-generating function often feels abstract, but it has real applications. The following example is an application to insurance risk modeling.

EXAMPLE 2.6.15

Insurance companies need to evaluate the probability that claims will exceed a certain threshold. Assume that the number of claims in a year, N, follows a Poisson distribution with parameter λ, and each claim size, X_i, $i = 1, \ldots, n$, is exponentially distributed with parameter θ. Also assume that X_i's are *i.i.d.*

(a) Find mean and variance of the total claim, $S = \sum_{i=1}^n X_i$.
(b) If the mean number of claims, $\lambda = 10$, and $\theta = 0.2$ (the mean single claim is $\frac{1}{\theta} = 5$), calculate the mean and variance of the total claim.

Solution

(a) *From Theorem 3.2.2, we have the moment-generating function of N is*

$$M_N(t) = \exp(\lambda(e^t - 1)).$$

Also, using Theorem 3.2.7 (with $\beta = \frac{1}{\theta}$*), we have the moment-generating function of X is*

$$M_X(t) = \frac{\theta}{\theta - t}, \ t < \theta.$$

First, we will show that $M_S(t) = M_N(\ln(M_X(t)))$.
Note that Nis a nonnegative integer-valued random variable. Since $\{X_i\}_{i=1}^{\infty}$ *are i.i.d. random variables with MGF* $M_X(t) = E\left[e^{tX}\right]$.
Now,

$$E\left[e^{tS} | N = n\right] = E\left[e^{t(X_1 + \ldots + X_n)}\right] = [M_X(t)]^n.$$

So,

$$M_S(t) = E\left[(M_X(t))^N\right].$$

Note that $(M_X(t))^N = e^{N \ln(M_X(t))}$. *Taking expectation,*

$$M_S(t) = E\left[e^{N \ln(M_X(t))}\right].$$

From this, we can calculate the moment-generating function of S as,

$$M_S(t) = M_N(\ln(M_X(t)))$$
$$= \exp(\lambda(M_X(t) - 1) = \exp(\lambda\left(\frac{\theta}{\theta - t} - 1\right)), \ t < \theta.$$

Now, differentiating $M_S(t)$,

$$M_S'(t) = M_S(t)\lambda M_X'(t).$$

Hence, at t=0,

$$E[S] = M_S'(0) = 1.\ \lambda M_X'(0) = \lambda E[X] = \lambda.\ \frac{1}{\theta}.$$

Similarly,

$$M_S''(0) = \lambda^2(E[X])^2 + \lambda E[X^2].$$

Thus,

$$Var(S) = M_S''(0) - (M_S'(0))^2 = (\lambda^2(E[X])^2 + \lambda E[X^2]) - \lambda^2(E[X])^2 = \lambda E[X^2]$$
$$= \lambda.\ \frac{2}{\theta^2} = \frac{2\lambda}{\theta^2}.$$

(b) *Using $\lambda = 10$, $\theta = 0.2$, we get*

$$E[S] = 10.\ \frac{1}{0.2} = 50, \text{ and } Var(S) = 500.\ Var(S) = 500.$$

Thus, the mean total claims is \$50, but the standard deviation is 22.36, which is rather large.

We will conclude this section by stating a result that will be useful in the proof of central limit theorem.

Theorem 2.6.4. *Let F_n be a sequence of cumulative distribution functions with the corresponding moment-generating functions M_n. Let F be a cdf with the mgf M. If $M_n(t) \to M(t)$ for all t in an open interval containing zero, then $F_n(x) \to F(x)$ for all x at which F is continuous.*

EXERCISES 2.6

2.6.1. Find $E(X)$ where X is the outcome when one rolls a six-sided balanced die. Find the mgf of X. Also, using the mgf of X, compute the variance of X.

2.6.2. The grades from a statistics class for the first test are given by

x_i	96	87	65	49	77	74	99	68	56	84
$p(x_i)$	3/15	2/15	1/15	1/15	2/15	1/15	1/15	1/15	1/15	2/15

(a) Find mean μ and variance σ^2.
(b) Find the mgf.

2.6.3. The cdf of a discrete random variable Y is given in the following table:

y	−1	0	2	5	6
$F(y)$	0.1	0.15	0.4	0.8	1

(a) Find $E(Y)$, $E(Y^2)$, $E(Y^3)$, and $Var(Y)$.
(b) Find the mgf of Y.

2.6.4. A discrete random variable X is such that

$$P(X = n) = \frac{2^{n-1}}{3^n},\ n = 1,2,\ \ldots,n,\ \ldots.$$

Show that $E(X) = 3$.

2.6.5. A discrete random variable X is such that

$$P(X = 2^n) = \frac{1}{2^n},\ n = 1,2,\ \ldots.$$

Show that $E(X) = \infty$. That is, X has no mathematical expectation.

2.6.6. Let X be a random variable with pdf $f(x) = kx^2$ where $0 \le x \le 1$.
(a) Find k.
(b) Find $E(X)$ and $Var(X)$.
(c) Find $M_X(t)$. Using the mgf, find $E(X)$.

2.6.7. Let X be a random variable with pdf $f(x) = ax^2 + b$, $0 \le x \le 1$. Find a and b such that $E(X) = 5/8$.

2.6.8. Given that X_1, X_2, X_3, and X_4 are independent random variables with mean 2, find $E(Y)$ and $E(Z)$ for

$$Y = 3X_4 - X_1 + \tfrac{1}{5}X_3$$
$$Z = X_2 + 7X_3 - 9X_1.$$

2.6.9. For a random variable X, prove (a)–(d) of Theorem 2.6.2.

2.6.10. Let ε (for "error") be a random variable with $E(\varepsilon) = 0$, and $Var(\varepsilon) = \sigma^2$. Define the random variable, $X = \mu + \varepsilon$, where μ is a constant. Find $E(X)$, $Var(X)$, and $E(\varepsilon^2)$.

2.6.11. A *degenerate random variable* is a random variable taking a constant value. Let $X = c$. Show that $E(X) = c$, and $Var(X) = 0$. Also, find the cumulative distribution function of the degenerate distribution of X.

2.6.12. Let $X \sim N(\mu, \sigma^2)$. Use the mgf to find $E(X^2)$ and $E(X^4)$.

2.6.13. Using Theorem 2.6.3, show that the mean and variance of the Poisson distribution, with parameter , is equal to .

2.6.14. Let X be a discrete random variable with a mass function

$$p(x) = \begin{cases} \frac{1}{x(x+1)}, & x = 1,2,\ldots, \\ 0, & \text{otherwise.} \end{cases}$$

Show that the moment-generating function does not exist for this random variable.

2.6.15. Let X be a random variable with a geometric pdf

$$f(x) = p(1-p)^{x-1},\ x = 1,2,3,\ldots.$$

(a) Find $E(X)$ and $Var(X)$.
(b) Show that $M_X(t) = \frac{pe^t}{1-(1-p)e^t}$, $t < -\ln(1-p)$.

2.6.16. Find $E(X)$ and $Var(X)$ for a random variable X with pdf $f(x) = \frac{1}{2}e^{-|x|}$, $-\infty < x < \infty$. Also, find the mgf of X.

2.6.17. The probability density function of the random variable X is given by

$$f(x) = \begin{cases} \frac{x^2}{2}, & 0 < x \le 1, \\ \frac{6x - 2x^2 - 3}{2}, & 1 < x \le 2, \\ \frac{(x-3)^2}{2}, & 2 < x \le 3, \\ 0, & \text{otherwise.} \end{cases}$$

Find the expected value of the random variable X.

2.6.18. Let the random variable X be normally distributed with mean 0 and variance σ^2. Show that $E(X^{(2k+1)}) = 0$, where $k = 0, 1, 2, .$

2.6.19. If the kth moment of a random variable exists, show that all moments of order less than k exist.

2.6.20. Suppose that the random variable X has an mgf

$$M_X(t) = \frac{\alpha}{\alpha - t}, \ t < \frac{1}{\alpha}.$$

Let the random variable Y have the following function for its probability density:

$$g(y) = \begin{cases} \alpha e^{-\alpha y}, & y > 0, \ \alpha > 0, \\ 0, & \text{otherwise.} \end{cases}$$

Can we obtain the probability density of the variable X with the foregoing information?

2.7 CHAPTER SUMMARY

In this chapter, we have introduced the concepts of random events and probability and how to compute the probabilities of events using counting techniques. We have studied the concepts of conditional probability, independence, and Bayes' rule. Random variables and distribution functions, moments, and moment-generating functions of random variables have also been introduced.

The following lists some of the key definitions introduced in this chapter.

- Sample space
- Mutually exclusive events
- Informal definition of probability
- Classical definition of probability
- Frequency interpretation of probability
- Axiomatic definition of probability
- Multinomial coefficients
- Conditional probability
- Mutually independent events
- Pairwise independent events
- Random variable (r.v.)
- Discrete random variable
- Discrete probability mass function
- Cumulative distribution function
- Continuous random variable
- Expected value
- kth moment about the origin
- kth moment about its mean
- Skewness and kurtosis
- Moment-generating function

The following important concepts and procedures have been discussed in this chapter:

- Method of computing probability by the classical approach
- Some basic properties of probability
- Computation of probability using counting techniques
- Four sampling methods:
  - Sampling with replacement and the objects are ordered
  - Sampling without replacement and the objects are ordered
  - Sampling without replacement and the objects are not ordered
  - Sampling with replacement and the objects are not ordered
- Permutation of n objects taken m at a time
- Combinations of n objects taken m at a time
- Number of combinations of n objects into m classes
- Some properties of conditional probability
- Law of total probability
- Steps to apply Bayes' rule
- Some properties of the distribution function
- Some properties of expected value
- Expectation of the function of a random variable
- Properties of moment-generating functions

2.8 COMPUTER EXAMPLES (OPTIONAL)

The three software packages, Software, SPSS, and SAS, that we are using in this book are not specifically designed for probability computations. However, the following examples are given to demonstrate that we will be able to use the software for some basic probability computations. We do not recommend using any of these three software packages for probability calculations; they are basically designed for statistical computations. There are many other software packages, such as Maple or MATLAB®, that can be used efficiently for probability computations.

2.8.1 Examples Using R

EXAMPLE 2.8.1 CALCULATING CUMULATIVE PROBABILITIES

Random variable X has the following distribution:

X	1	4	5	8	11
$p(x)$	0.2	0.2	0.1	0.15	0.35

Find $P(X \le 4)$. In this example we will use the which() statement to calculate the cumulative probability in R. However, there may be other methods available. Try using the which() statement by itself.

R Code:

```
x=c(1,4,5,8,11);
p=c(0.2,0.2,0.1,0.15,0.35);
sum(p[which(x<=4)]);
```

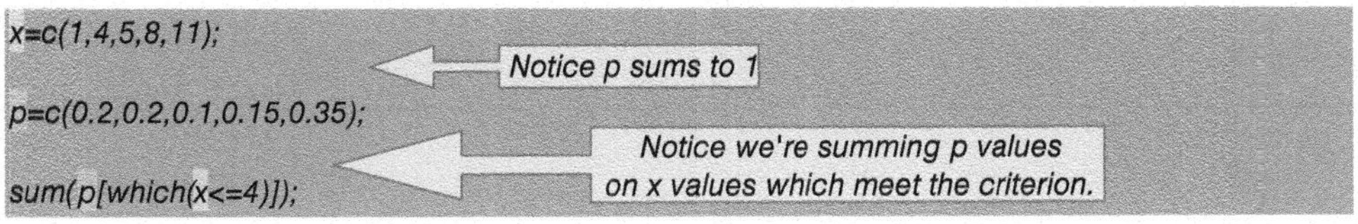

Output:
0.4

i.e, $P(X \leq 4) = 0.4$

EXAMPLE 2.8.2 EXPECTED VALUE

Using the data in the previous example (2.8.1) calculate $E(X)$ and $Var(X)$.

Since we're given the distribution, we can calculate it using the sum of the values multiplied by their probabilities.

R Code:

```
x=c(1,4,5,8,11);
p=c(0.2,0.2,0.1,0.15,0.35);
sum(xp);
sum(xxp)-sum(xp)^(2);
```

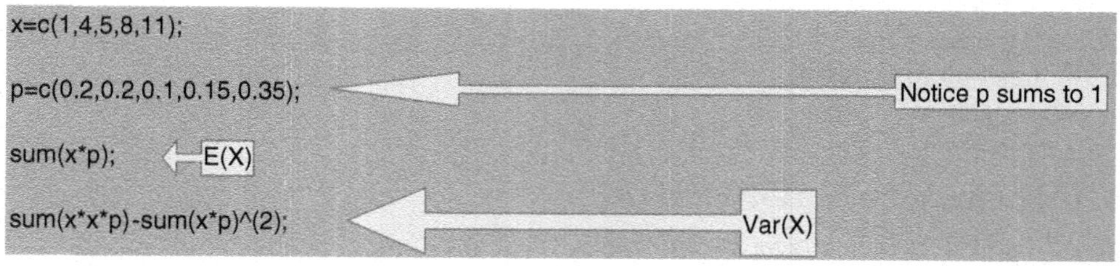

Output:

```
6.55
14.9475
```

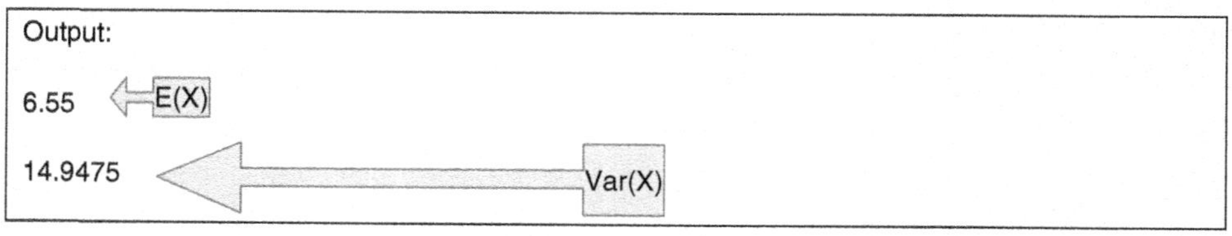

2.8.2 SPSS Examples

EXAMPLE 2.8.3

For the random variable X in Example 2.8.1, find $E(X)$.

Solution

In column 1, enter the x values and in column 2 enter the p(x) values. Then,

Transform > compute> in ***target variable:*** type a name, say, ***product***. Move ***var00001*** and ***var00002*** to ***Numeric Expression:*** field and put "" in between them as ***(var00001)(var00002)***. Then, use the ***SUM(., .)*** command to find the value of $E(X)$.

2.8.3 SAS Examples

EXAMPLE 2.8.4

A random variable X has the following distribution:

x	2	5	6	8	9
$p(x)$	0.1	0.2	0.3	0.1	0.3

Using SAS, find $E(X)$.

Solution

For discrete distributions where the random variable takes finite values, we can adapt the following procedure:

```
data evalue;
input x y n;
z=xyn;
cards;
2 .1 5
5 .2 5
6 .3 5
8 .1 5
9 .3 5
;
run;
proc means;
run;
```

We know that if proc means is used just for xy, that will give us $\frac{1}{n}\sum xp(x)$; hence, multiplying by n, the number of values X takes will give us $E(X) = \sum xp(x)$. We will get the following output:

The MEANS Procedure

Variable	N	Mean	Std Dev	Minimum	Maximum
X	5	6.0000000	2.7386128	2.0000000	9.0000000
Y	5	0.2000000	0.1000000	0.1000000	0.3000000
N	5	5.0000000	0	5.0000000	5.0000000
Z	5	6.5000000	4.8476799	1.0000000	13.5000000

From this, we can see that $E(X) = 6.5$. A direct way to find the expected value is by using "PROC IML."

```
options nodate nonumber;
/ Finding expected value of a random variable /
proc iml;
/ defining all the variables /
x={2 5 6 8 9}; / a row vector /
y={1 2 3 1 3}; / probabilities /
/ calculations /
z=xy';
/ print statements /
print "Display the vector x and probability y and the expected value";
print x y, z;
quit;
```

We will get the following output:

		X		
2	5	6	8	9
		Y		
0.1	0.2	0.3	0.1	0.3
		Z		
6.5				

PROJECTS FOR CHAPTER 2

2A. The Birthday Problem

The famous birthday problem is to find the smallest number of people one must ask to get an even chance that at least two people have the same birthday. To solve this, you can use the following steps.

Find the probability that in a group of k people, no two have the same probability. Let q be this probability. Then $p = 1 - q$ is the probability that at least two people have the same birthday. Ignoring leap years, take the sample space S as all sequences of length k with each element one of the 365 days in the year. Thus, there are 365^k elements in S.

(a) Find the total number of sequences with no common birthdays.

(b) Assuming that each sequence is equally likely, show that:

$$q = \frac{(365)(364)\ldots(365 - k + 1)}{365^k}.$$

(c) Write a computer program for calculating q for $k = 2$–50, and find the first k for which $p > 0.5$. This will give the least number of people we should ask to make it an even chance that at least two people will have the same birthday.

2B. The Hardy–Weinberg Law

Hereditary traits in offspring depend on a pair of genes, one each contributed by the father and the mother. A gene is either a dominant allele, denoted by A, or a recessive allele, denoted by a. If the genotype is AA, Aa, or aA, then the hereditary trait is A, and if the genotype is aa, then the hereditary trait is a. Suppose that the probabilities of the mother carrying the genotypes aa, aA (same as Aa), and AA are p, q, and r, respectively. Here $p + q + r = 1$. The same probabilities are true for the father.

(a) Assuming that the genetic contributions of the mother and father are independent and the matings are random, show that the respective probabilities for the first-generation offspring are

$$p_1 = (p + q/2)^2,\ q_1 = 2(r + q/2)(p + q/2),\ r_1 = (r + q/2)^2.$$

Also, find $P(\text{A})$ and $P(\text{a})$.

(b) The Englishman G. H. Hardy and the German W. Weinberg could show that the foregoing probabilities in a population stay constant for generations if certain conditions are fulfilled. This is known as the Hardy–Weinberg law. Under the conditions of part (a), using the induction argument, show that the Hardy–Weinberg law is satisfied, that is, $p_n = p_1$, $q_n = q_1$, and $r_n = r_1$ for all $n \geq 1$. The consequences of the Hardy–Weinberg law are that (i) no evolutionary change occurs through the process of sexual reproduction itself, and (ii) changes in allele and genotype frequencies can result only from additional forces on the gene pool of a species.

2C. Some Basic Probability Simulation

- Simulation imitates a real situation, and it models real situations by performing the experiment repeatedly. Repeated real experiments are time-consuming and expensive, and are difficult to calculate theoretically, whereas a computer simulation can take only seconds. For instance, think of a simple experiment of throwing a die 100 times and recording the up face. It will take a long time, whereas in R, the outcomes are obtained instantaneously. In simulations, often we have to make assumptions about situations being simulated, such as, there is an equal chance of producing a head or a tail. In this project, we will show a few simple examples and give a few more exercises. The idea of this project is to encourage students to explore probability simulation further.

We could simulate the tossing of a coin, say, 20 times by following commands.

```
n = 20

sample(c("Heads", "Tails"), n, rep = T)
```

Suppose we want to simulate tossing a die 100 times and observe up face each time. We could use the following R command.

```
Rolldie1 = function(n) sample(1:6, n, rep = T)

Rolldie1(100)
```

For another example, consider picking powerball numbers, where the powerball consists of choosing 5 numbers from 1:59 (without replacement) and 1 number (called the Powerball) chosen from 1:39. When people choose the numbers manually, people will usually choose 1:31 due to various birthdates. The following is a way of choosing a random combination, in which we give a higher probability for numbers 32:59. You can play around with this code to change various probabilities (code is based on the code in https://www.r-bloggers.com/picking-lotto-numbers/).

```
gen_lotto<-function(){
+       white<-seq(1:59)
+       red<-31:39
+       probs<-white
+       # Decrease probabilities for commonly chosen numbers
+       probs[probs<=31]<-1/(59)
+       probs[probs>=32]<-1/14
+       # We need 5 white
+       w<-sample(white,5,prob=probs)
+       # We need 1 PowerBall
+       r<-sample(red,1)
+       # Print results
+       cat(" White Balls:",w[order(w)],"\n","Powerball:",r)
+       # Make a good warning
+       cat("\n Remember, your odds of winning: \n","1 in 195,249,054")
+ }
> gen_lotto()
```

This will give you 5 numbers and a powerball! Good luck!

Exercises: Write and run R code for the following problems:

1. In 1000 coin tosses, what is the probability of having the same side come up 10 times in a row?
2. In ten coin tosses, what is the probability of having a different side come up with each throw, that is, that you never get two tails or two heads in a row?
3. Write code to generate numbers for the lottery you are interested in.

Think of a few other situations where simulation is appropriate.

Chapter 3

Probability Distributions

Chapter outline

OBJECTIVE

In this chapter, we will present some special distributions, joint distributions of several random variables, functions of random variables, and some important limit theorems.

Johann Carl Friedrich Gauss
(Source: *http://tobiasamuel.files.wordpress.com/2008/06/carl_friedrich_gauss.jpg*)

German mathematician and physicist Carl Friedrich Gauss (1777–1855) is sometimes called the "prince of mathematics." He was a child prodigy. At the age of 7, Gauss started elementary school, and his potential was noticed almost immediately. His teachers were amazed when Gauss summed the integers from 1 to 100 instantly. At the age of 24,

Mathematical Statistics with Applications in R. DOI: https://doi.org/10.1016/B978-0-443-27548-7.00003-7

Gauss published one of the most brilliant achievements in mathematics, *Disquisitiones Arithmeticae* (1801). In it, Gauss systematized the study of number theory. Gauss applied many of his mathematical insights in the field of astronomy, and by using the method of least squares, he successfully predicted the location of the asteroid Ceres in 1801. In 1820 Gauss made important inventions and discoveries in geodesy, the study of the shape and size of the earth. In statistics, he developed the idea of normal distribution. In the 1830s, he developed theories of non-Euclidean geometry and mathematical techniques for studying the physics of fluids. Although Gauss made many contributions to applied science, especially electricity and magnetism, pure mathematics was his first love. It was Gauss who first called mathematics "the queen of the sciences."

3.1 INTRODUCTION

In the previous chapter, we looked at the basic concepts of probability calculations, random variables, and their distributions. There are many special distributions that have useful applications in statistics. It is worth knowing the type of distribution that we can expect under different circumstances, because a better knowledge of the population will result in better inferential results. In the next section, we will discuss some of these distributions with some additional distributions presented in Appendix A3. We also briefly deal with joint distributions of random variables and functions of random variables. Limit theorems play an important role in statistics. We will present two limit theorems: the law of large numbers and the Central Limit Theorem.

3.2 SPECIAL DISTRIBUTION FUNCTIONS

Random variables are often classified according to their probability distribution functions. In any statistical analysis of quantitative data, it is important to know the form of the underlying probability distributions. There are certain basic probability distributions that are applicable in many diverse contexts and thus repeatedly arise in practice. A great variety of special distributions have been studied over the years. Also, new ones are frequently being added to the literature. It is impossible to give a comprehensive list of all probability distribution functions in this book. Many books and websites deal with the range of probability distribution functions. A good list of distributions can be obtained from http://www.causascientia.org/math_stat/Dists/Compendium.pdf. In this section, we will describe some of the commonly used probability distributions. In Appendix A3, we list some more distributions with their mean, variance, and moment-generating functions. First, we shall discuss some discrete probability distributions.

3.2.1 The Binomial Probability Distribution

The simplest distribution is the one with only two possible outcomes. For example, when a coin (not necessarily fair) is tossed, the outcomes are heads or tails, with each outcome occurring with some positive probability. These two possible outcomes may be referred to as "success" if heads occur and "failure" if tails occur. Assume that the probability of heads appearing in a single toss is p; then the probability of tails is $1 - p = q$. We define a random variable X associated with this experiment as taking value 1 with probability p if heads occur and value 0 if tails occur with probability q. Such a random variable X is said to have a *Bernoulli probability distribution*. That is, X is a Bernoulli random variable if for some p, $0 \le p \le 1$, the probability $P(X = 1) = p$ and $P(X = 0) = 1 - p$. The probability function of a Bernoulli random variable X can be expressed as

$$p(x) = P(X = x) = \begin{cases} p^x (1 - p)^{1-x}, & x = 0,\ 1 \\ 0, & otherwise. \end{cases}$$

Note that this distribution is characterized by single parameter p. It can be easily verified that the mean and variance of X are $E[X] = p$, $\mathrm{Var}(X) = pq$, respectively, and the moment-generating function is $MX(t) = pe^t + (1 - p)$.

Even when the experimental values are not dichotomous, reclassifying the variable as a Bernoulli variable can be helpful. For example, consider blood pressure measurements. Instead of representing the numerical values of blood pressure, if we reclassify the blood pressure as "high blood pressure" and "low blood pressure," we may be able to avoid dealing with a possible misclassification due to diurnal variation, stress, and so forth, and concentrate on the main issue, which would be: Is the average blood pressure is unusually high?

In a succession of Bernoulli trials, one is more interested in the total number of successes (whenever a 1 occurs in a Bernoulli trial, we term it a "success"). The probability of observing exactly k successes in n independent Bernoulli trials yields the binomial probability distribution. In practice, the binomial probability distribution is used when we are concerned with the occurrence of an event, not its magnitude. For example, in a clinical trial, we may be more interested in the number of survivors after treatment.

Definition 3.2.1. A **binomial experiment** *is one that has the following properties: (1) The experiment consists of n identical trials. (2) Each trial results in one of the two outcomes, called a success S and failure F. (3) The probability of success on a single trial is equal to p and remains the same from trial to trial. The probability of failure is 1 − p = q. (4) The outcomes of the trials are independent. (5) The random variable X is the number of successes in n trials.*
Earlier, we saw that the number of ways of obtaining x successes in n trials is given by

$$\binom{n}{x} = \frac{n!}{x!(n-x)!}.$$

Definition 3.2.2. *A random variable X is said to have a* **binomial probability distribution** *with parameters (n, p) if and only if*

$$P(X = x) = p(x) = \binom{n}{x} p^x q^{n-x}$$

$$= \begin{cases} \frac{n!}{x!(n-x)} & x = 0, 1, 2, \ldots, n, \ 0 \le p \le 1, \text{ and } q = 1 - p \\ 0 & \text{otherwise} \end{cases}$$

To show the dependence on n and p, we denote p(x) by b(x; n, p) and the cumulative probability distribution by

$$B(x, n, p) = \sum_{i=1}^{x} b(i; n, p).$$

The binomial probabilities have been tabulated and are given in the binomial table.
By the binomial theorem, we have

$$(p + q)^n = \sum_{x=0}^{n} \binom{n}{x} p^x q^{n-x}.$$

Because $(p + q) = 1$, we conclude that $\sum_{i=0}^{x} b(i; n, p) = \sum_{x=0}^{n} \binom{n}{x} p^x q^{n-x} = 1^n = 1$, for all $n \ge 1$ and $0 \le p \le 1$. Hence, $p(x)$ is indeed a probability mass function. The binomial probability distribution is characterized by two parameters: the number of independent trials n and the probability of success p. *Following* R *commands will help in binomial calculation. If we want the complete probability distribution, say for* $n = 10$, *and* $p = 0.2$, *use "dbinom(0:10, 10, 0.2)." Suppose we want* $P(X = 6)$, *use "dbinom(6, 10, 0.2)." If we want cumulative probability, say,* $P(X \le 3)$, *use "pbinom(3, 10, 0.2)."*

EXAMPLE 3.2.1

It is known that screws produced by a certain machine will be defective with probability 0.01, independently of each other. If we randomly pick 10 screws produced by this machine, what is the probability that at least two screws will be defective?

Solution

Let X be the number of defective screws out of 10. Then X can be considered as a binomial r.v. with parameters (10, 0.01). *Hence, using the binomial pmf* p(x), *given by definition 3.2.2, we define at least two screws will be defective as*

$$P(X \ge 2) = \sum_{x=2}^{10} \binom{10}{x} (0.01)^x (0.99)^{10-x}$$

$$= 1 - [P(X = 0) + P(X = 1)] = 0.004.$$

R-command: 1-pbinom(1,10,0.01).

In chapter 2, we introduced Mendel's law. In biology, the "gene frequencies and genotype ratios in a randomly breeding population remain constant from generation to generation" and is known as the *Hardy–Weinberg law*.

EXAMPLE 3.2.2

Suppose we know that the frequency of a dominant gene, A, in a population is equal to 0.2. If we randomly select eight members of this population, what is the probability that at least six of them will display the dominant phenotype? Assume that the population is sufficiently large that removing eight individuals will not affect the frequency and that the population is in Hardy–Weinberg equilibrium.

Solution

First, note that an individual can have the dominant gene, A, if the person has traits AA, aA, or Aa. Hence, if the gene frequency is 0.2, the probability that an individual is of genotype A is

$$P(A) = P(AA \cup Aa \cup aA) = P(AA) + 2P(Aa)$$
$$= (0.2)^2 + 2(0.2)(0.8) = 0.36.$$

Let X denote the number of individuals out of eight that display the dominant phenotype. Then X is binomial with $n = 8$, and $p = 0.36$. Thus the probability that at least six of them will display the dominant phenotype is

$$P(X \geq 6) = P(X = 6) + P(X = 7) + P(X = 8)$$
$$= \Sigma_{i=6}^{8} \binom{10}{i} (0.36)^i (0.64)^{10-i} = 0.029259.$$

R-command: *1-pbinom(5,8,0.36)*

For large n, calculations of the binomial probabilities are tedious. Many statistical software packages have binomial probability distribution commands. For the purpose of this book, we will use the binomial table that gives the cumulative probabilities $B(x, n, p)$ for $n = 2$ through $n = 20$ and $p = 0.05, 0.10, 0.15, \ldots, 0.90, 0.95$. If we need the probability of a single term, we can use the relation

$$P(X = x) = b(x;\ n,\ p) = B(x,\ n,\ p) - B(x - 1,\ n,\ p).$$

EXAMPLE 3.2.3

A manufacturer of inkjet printers claims that only 5% of their printers require repairs within the first year. If in a random sample of 18 of the printers, four required repairs within the first year, does this tend to refute or support the manufacturer's claim?

Solution

Let us assume that the manufacturer's claim is correct; that is, the probability that a printer will require repairs within the first year is 0.05. Suppose 18 printers are chosen at random. Let p be the probability that any one of the printers will require repairs within the first year. We now find the probability that at least four of these out of the 18 will require repairs during the first year. Let X represent the number of printers that require repair within the first year. Then X follows the binomial pmf with p = 0.05, n = 18. The probability that four or more of the 18 will require repair within the first year is given by

$$P(X \geq 4) = \sum_{x=4}^{18} \binom{18}{x} (0.05)^x (0.95)^{18-x}$$

or, using the binomial table,

$$\Sigma_{x=4}^{18}\ b(x;\ 18,\ 0.05) = 1 - B(3,\ 18,\ 0.05)$$
$$= 1 - 0.9891$$
$$= 0.0109.$$

This value (approximately 1.1%) is very small. We have shown that if the manufacturer's claim is correct, then the chances of observing four or more bad printers out of 18 are very small. But we did observe exactly four bad ones. Therefore we must conclude that the manufacturer's claim cannot be substantiated.

Mean, variance, and mgf of a binomial random variable

Theorem 3.2.1. *If X is a binomial random variable with parameters n and p, then*

$$E(X) = \mu = np$$

and

$$Var(X) = \sigma^2 = np(1 - p).$$

Also, the moment-generating function

$$M_X(t) = [pe^t + (1 - p)]^n.$$

Proof. We derive the mean and the variance. The derivation for mgf is given in Example 2.6.5. Using the binomial pmf, $p(x) = (n!/(x!(n-x)!))p^x q^{n-x}$, and the definition of expectation, we have

$$\begin{aligned}\mu = E(X) &= \sum_{x=0}^{n} xp(x) = \sum_{x=0}^{n} x\frac{n!}{x!(n-x)!}p^x(1-p)^{n-x}\\ &= \sum_{x=1}^{n} \frac{n!}{(x-1)!(n-x)!}p^x(1-p)^{n-x},\end{aligned}$$

because the first term in the sum is zero, as $x = 0$.

Let $i = x - 1$. When x varies from 1 through $n, i = (x - 1)$ varies from zero through $(n - 1)$. Hence,

$$\begin{aligned}\mu &= \sum_{i=0}^{n-1} \frac{n!}{i!(n-i-1)!}p^{i+1}(1-p)^{n-i-1}\\ &= np\sum_{i=0}^{n-1} \frac{(n-1)!}{i!(n-1-i)!}p^i(1-p)^{n-1-i}\\ &= np,\end{aligned}$$

because the last summand is that of a binomial pmf with parameter $(n - 1)$ and p, hence, equals 1.

To find the variance, we first calculate $E[X(X - 1)]$.

$$\begin{aligned}E[X(X-1)] &= \sum_{x=0}^{n} x(x-1)\frac{n!}{x!(n-x)!}p^x(1-p)^{n-x}\\ &= \sum_{x=2}^{n} \frac{n!}{(x-2)!(n-x)!}p^x(1-p)^{n-x},\end{aligned}$$

because the first two terms are zero. Let $i = x - 2$. Then,

$$\begin{aligned}E[X(X-1)] &= \sum_{i=0}^{n-2} \frac{n!}{i!(n-i-2)!}p^{i+2}(1-p)^{n-i-2}\\ &= n(n-1)p^2\sum_{i=0}^{n-2} \frac{(n-2)!}{i!(n-2-i)!}p^i(1-p)^n\\ &= n(n-1)p^2,\end{aligned}$$

because the last summand is that of a binomial pmf with parameter $(n - 2)$ and p thus equals 1.

Note that $E(X(X - 1)) = EX^2 - E(X)$, and so we obtain

$$\begin{aligned}\sigma^2 = Var(X) &= E(X^2) - [E(X)]^2\\ &= E[X(X-1)] + E(X) - [E(X)]^2\\ &= n(n-1)p^2 + np - (np)^2 = -np^2 + np\\ &= np(1-p).\end{aligned}$$

3.2.2 Poisson Probability Distribution

The Poisson probability distribution was introduced by the French mathematician Siméon-Denis Poisson in his book published in 1837, which was entitled *Recherches sur la probabilité des jugements en matières criminelles et matière civile* and dealt with the applications of probability theory to lawsuits, criminal trials, and the like. Consider a statistical experiment in which A is an event of interest. A random variable that counts the number of occurrences of A is called a *counting random variable*. The Poisson random variable is an example of a counting random variable. Here we assume that the numbers of occurrences in disjoint intervals is independent, and the mean of the number of occurrences is constant.

Definition 3.2.3. *A discrete random variable X is said to follow the* **Poisson probability distribution** *with parameter* $\lambda > 0$, *denoted by Poisson*(λ), *if*

$$P(X = x) = f(x;\ \lambda) = f(x) = \frac{e^{-\lambda}\lambda^x}{x!}, \quad x = 0,\ 1,\ 2,\ \ldots$$

The Poisson probability distribution is characterized by the single parameter λ, which represents the mean of a Poisson probability distribution. Thus, in order to specify the Poisson distribution, we only need to know the mean number of occurrences. This distribution is of fundamental theoretical and practical importance. Rare events are modeled by the Poisson distribution. For example, the Poisson probability distribution has been used in the study of telephone systems. The number of incoming calls into a telephone exchange during a unit time might be modeled by a Poisson variable, assuming that the exchange services a large number of customers who call more or less independently. Some other problems where Poisson representation can be used are the number of misprints in a book, radioactivity counts per unit time, the number of plankton (microscopic plant or animal organisms that float in bodies of water) per aliquot of seawater, or count of bacterial colonies per petri plate in a microbiological study. In stem cell research, the Poisson distribution is used to analyze the redundancy of clusters in the stem cell database. A Poisson probability distribution has a unique property that its mean equals its variance.

Mean, variance, and moment-generating function of a poisson random variable

Theorem 3.2.2. *If X is a Poisson random variable with parameter* λ, *then*

$$E(X) = \lambda$$

and

$$Var(X) = \lambda.$$

Also, the moment-generating function is

$$M_X(t) = e^{\lambda(e^t - 1)}.$$

The proof of this result is similar to that we used in Theorem 3.2.1 in this section. One needs to use the Maclaurin's expansion, $e^\lambda = \sum_{i=0}^{\infty} (\lambda^i/i!)$, in proving this result.

EXAMPLE 3.2.4

Let X be a Poisson random variable with $\lambda = 1/2$. Find

(a) $P(X = 0)$
(b) $P(X \geq 3)$

Solution

(a) *We have*

$$P(X = 0) = p(0) = \frac{e^{-1/2}(1/2)^0}{0!} = e^{-1/2} = 0.60653.$$

(b) *Here, we will use complementary event to compute the required probability. That is,*

$$\begin{aligned} P(X \geq 3) &= 1 - P(X \leq 2) = 1 - [p(0) + p(1) + p(2)] \\ &= 1 - \left[e^{-1/2} + \frac{e^{-1/2}(1/2)}{1!} + \frac{e^{-1/2}(1/2)^2}{2!}\right] \\ &= 1 - 0.98561 = 0.01439. \end{aligned}$$

R-command *"ppois(2, lambda=1/2, lower=FALSE)."*

When n is large and p small, the binomial probabilities are often approximated by Poisson probabilities. In these situations, where performing the factorial and exponential operations required for direct calculation of binomial probabilities is a lengthy and tedious process, and tables are not available, Poisson approximation is more feasible. The following theorem states this result.

Poisson approximation to the binomial probability distribution

Theorem 3.2.3. *If X is a binomial r.v. with parameters n and p, then for each value* $x = 0, 1, 2, \ldots$ and as $p \to 0$, $n \to \infty$ *with* $np = \lambda$ *constant,*

$$\lim_{n\to\infty}\binom{n}{x}p^x(1-p)^{n-x} = \frac{e^{-\lambda}\lambda^x}{x!}.$$

The proof of this result is similar to that we used in Theorem 3.2.1. In the present context, the Poisson probability distribution is sometimes referred to as "the distribution of rare events" because of the fact that p is quite small when n is large. Usually, if $p \le 0.1$ and $n \ge 40$, we could use the Poisson approximation in practice. In general, another rule of thumb is to use the Poisson approximation for binomial problems when $n > 50$, and $np < 5$.

EXAMPLE 3.2.5

If the probability that an individual suffers an adverse reaction from a particular drug is known to be 0.001, determine the probability that out of 2000 individuals, (a) exactly three and (b) more than two individuals will suffer an adverse reaction.

Solution

Let Y be the number of individuals who suffer an adverse reaction. Then Y is binomial with $n = 2000$ *and* $p = 0.001$. *Because n is large and p is small, we can use the Poisson approximation with* $\lambda = np = 2$.

(a) *The probability that exactly three individuals will suffer an adverse reaction is*

$$P(Y=3) = \frac{2^3 e^{-2}}{3!} = 0.18.$$

That is, there is approximately an 18% chance that exactly three individuals out of 2000 will suffer an adverse reaction.

(b) *The probability that more than two individuals will suffer an adverse reaction is*

$$P(Y>2) = 1 - P(Y=0) - P(Y=1) - P(Y=2)$$
$$= 1 - 5e^{-2} = 0.323.$$

Similarly, there is approximately a 32.3% chance that more than two individuals will have an adverse reaction.
R-command: *"ppois(2, lambda = 2, lower = FALSE)"*

Now, we will discuss some continuous distributions. As we mentioned earlier, if X is a continuous random variable with pdf $f(x)$, then

$$P(a \le X \le b) = \int_a^b f(x)dx.$$

3.2.3 Uniform Probability Distribution

The uniform probability distribution is used to generate random numbers from other distributions, and also is useful as a "first guess" if no other information about a random variable X is known, other than that it is between a and b. Also, in real-world problems that have uniform behavior in a given interval, we can characterize the probabilistic behavior of such a phenomenon by the uniform distribution. See Fig. 3.1, below,

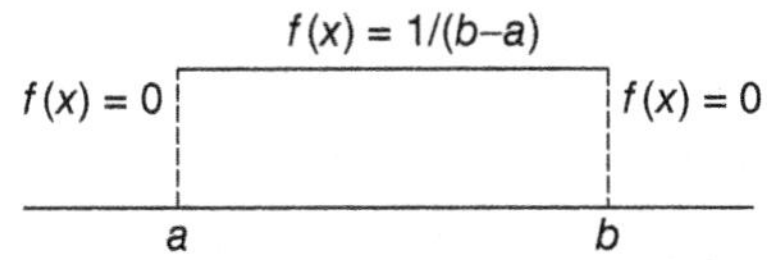

FIGURE 3.1 Uniform probability density.

Definition 3.2.4. *A random variable X is said to have a* **uniform probability distribution** *on (a, b), denoted by U(a, b), if the density function of X is given by*

$$f(x) = \begin{cases} \frac{1}{b-a}, & a \le x \le b, \\ 0, & \text{otherwise.} \end{cases}$$

The cumulative distribution function is given by

$$F(x) = \int_{-\infty}^{x} \frac{1}{b-a} dx = \begin{cases} 0, & x < a \\ \frac{x-a}{b-a}, & a \le x < b \\ 1, & x \ge b. \end{cases}$$

EXAMPLE 3.2.6

If X is a uniformly distributed random variable over (0, 10), calculate the probability that (a) $X < 3$, (b) $X > 6$, and (c) $3 < X < 8$.

Solution

(a)

$$P(X < 3) = \int_{0}^{3} \frac{1}{10} dx = \frac{3}{10}.$$

(b)

$$P(X > 6) = \int_{6}^{10} \frac{1}{10} dx = \frac{4}{10}.$$

(c)

$$P(3 < X < 8) = \int_{3}^{8} \frac{1}{10} dx = \frac{1}{2}.$$

Mean, variance, and moment-generating function of a uniform random variable

Theorem 3.2.4. *If X is a uniformly distributed random variable on (a, b), then*

$$E(X) = \frac{a+b}{2},$$

and

$$Var(X) = \frac{(b-a)^2}{12}.$$

Also, the moment-generating function is

$$M_X(t) = \begin{cases} \frac{e^{tb} - e^{ta}}{t(b-a)}, & t \ne 0 \\ 1, & t = 0. \end{cases}$$

Proof. We will obtain the mean and the variance and leave the derivation of the moment-generating function as an exercise. By definition, we have

$$E(X) = \int_{-\infty}^{\infty} x \frac{1}{b-a} dx$$
$$= \int_a^b x \frac{1}{b-a} dx = \frac{1}{b-a}\left(\frac{x^2}{2}\Big|_a^b\right)$$
$$= \frac{a+b}{2}.$$

Also,

$$E(X^2) = \int_a^b x^2 \frac{1}{b-a} dx = \frac{1}{b-a}\left(\frac{x^3}{3}\Big|_a^b\right)$$
$$= \frac{1}{3}\frac{b^3-a^3}{b-a}$$
$$= \frac{1}{3}(b^2 + ab + a^2) \text{ as } b^3 - a^3 = (b-a)(b^2 + ab + a^2).$$

Thus,

$$Var(X) = E(X^2) - (E(X))^2$$
$$= \frac{1}{3}(b^2 + ab + a^2) - \frac{(a+b)^2}{4}$$
$$= \frac{1}{12}(b-a)^2.$$

EXAMPLE 3.2.7

The melting point, X, of a certain solid may be assumed to be a continuous random variable that is uniformly distributed between the temperatures 100°C and 120°C. Find the probability that such a solid will melt between 112 °C and 115 °C.

Solution

The probability density function is given by

$$f(x) = \begin{cases} \frac{1}{20}, & 100 \le x \le 120 \\ 0, & \text{otherwise.} \end{cases}$$

Hence,

$$P(112 \le X \le 115) = \int_{112}^{115} \frac{1}{20} dx = \frac{3}{20} = 0.15.$$

Thus, there is a 15% chance of this solid melting between 112°C and 115°C.

3.2.4 Normal Probability Distribution

The single most important distribution in probability and statistics is the normal probability distribution. The density function of a normal probability distribution is bell-shaped and symmetric about the mean. The normal probability distribution was introduced by the French mathematician **Abraham de Moivre** in 1733. He used it to approximate probabilities associated with binomial random variables when n is large. This was later extended by **Laplace** to the so-called **Central Limit Theorem**, which is one of the most important results in probability. **Carl Friedrich Gauss**, in 1809, used the normal distribution to solve the important statistical problem of combining observations. Because Gauss played such a prominent role in determining the usefulness of the normal probability distribution, the normal probability distribution is often called the *Gaussian distribution*. Gauss and Laplace noticed that measurement errors tend to follow a bell-shaped curve, a normal probability distribution. Today, the normal probability distribution arises repeatedly in diverse areas of applications. For example, in biology, it has been observed that the normal probability distribution fits data on the heights and weights of human and animal populations, among others.

We should also mention here that almost all basic statistical inference is based on the normal probability distribution. The question that often arises is, when do we know that our data follows the normal distribution? To answer this question, we have specific statistical procedures that we study in later chapters, but at this point, we can obtain some constructive indications of whether the data follows the normal distribution by using descriptive statistics. That is, if the histogram of our data can be capped with a bell-shaped curve (Fig. 3.2), if the stem-and-leaf diagram is fairly symmetrical with respect to its center, and/or by invoking the empirical rule "backwards," we can obtain a good indication whether our data follow the normal probability distribution.

Definition 3.2.5. *A random variable X is said to have a* **normal probability distribution** *with parameters* μ and σ^2, *if it has a probability density function given by*

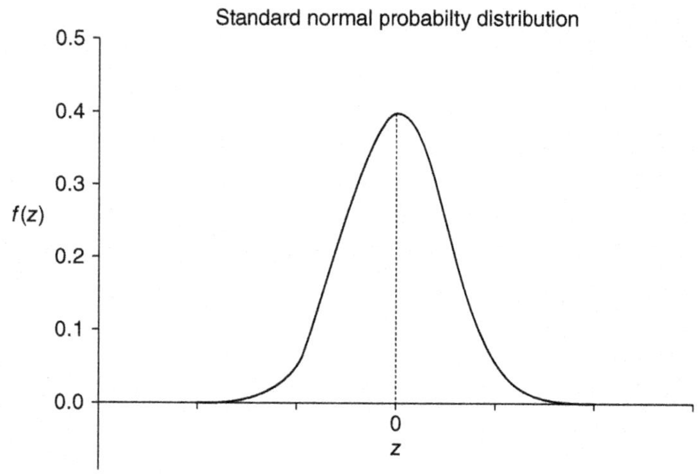

FIGURE 3.2 Standard normal density function.

$$f(X) = \frac{1}{\sqrt{2\pi}\sigma}e^{-(x-\mu)^2/2\sigma^2}, \ -\infty < x < \infty, \ -\infty < \mu < \infty, \ \sigma > 0.$$

If $\mu = 0$, and $\sigma = 1$, we call it a standard normal random variable.

For any normal random variable with mean μ and variance σ^2, we use the notation $X \sim N(\mu, \sigma 2)$. When a random variable X has a standard normal probability distribution, we will write $X \sim N(0, 1)$ (X is a normal with mean 0 and variance 1). The probabilities for a standard normal probability distribution are given in the normal table.

Mean, variance, and mgf of a normal random variable

Theorem 3.2.5. *If $X \sim N(\mu, \sigma^2)$, then $E(X) = \mu$ and Var $(X) = \sigma^2$. Also, the moment-generating function is*

$$M_X(t) = e^{t\mu + \frac{1}{2}t^2\sigma^2}.$$

If $X \sim N(\mu, \sigma^2)$, then the *z-transform* (or *z*-score) of X, $Z = \frac{X-\mu}{\sigma} \sim N(0, 1)$. This fact will be used in calculating probabilities for normal random variables. The normal table given in appendix is based on the standard normal distribution. Note that for the continuous random variables, the probabilities of strict and nonstrict inequalities are the same, that is, $P(X > a) = P(X \geq a)$.

EXAMPLE 3.2.8

(a) For $Z \sim N(0, 1)$, calculate $P(Z \geq 1.13)$.
(b) For $X \sim N(5, 4)$, calculate $P(-2.5 < X < 10)$.

Solution

(a) Using the normal table,

$$P(Z \geq 1.13) = 0.5 - 0.3708 = 0.1292.$$

The shaded part in the graph represents the $P(Z \geq 1.13)$.

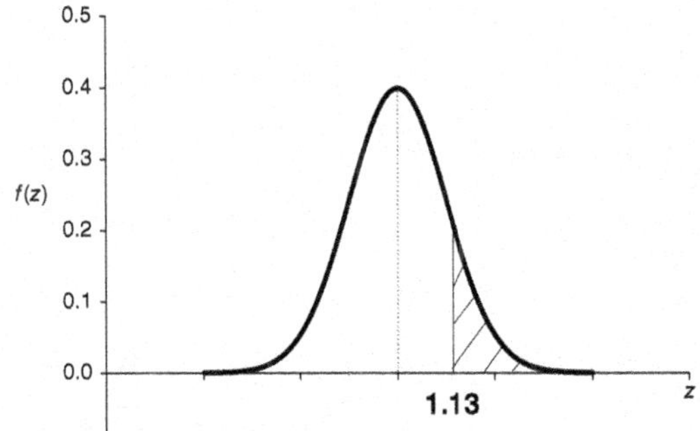

R-code: pnorm(1.13, mean=0, sd=1, lower.tail=FALSE)

(b) Using the z-transform, we have

$$
\begin{aligned}
P(-2.5 < X < 10) &= P\left(\frac{-2.5-5}{2} < Z < \frac{10-5}{2}\right) \\
&= P(-3.75 < Z < 2.5) \\
&= P(-3.75 < Z < 0) + P(0 < Z < 2.5) \\
&= 0.9938.
\end{aligned}
$$

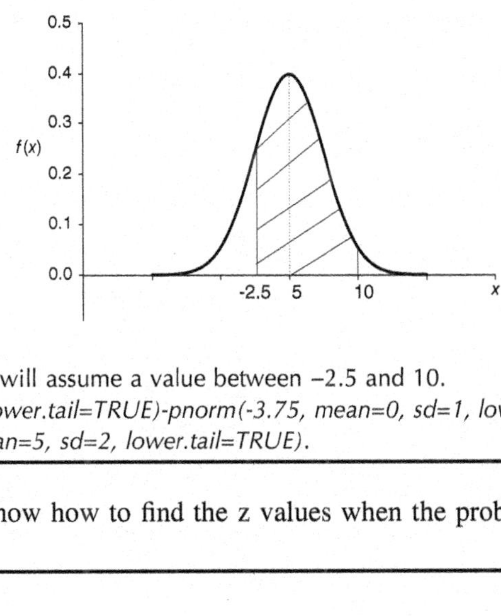

That is, we are 99.38% certain the Z will assume a value between -2.5 and 10.

R-code: *pnorm(2.5, mean=0, sd=1, lower.tail=TRUE)-pnorm(-3.75, mean=0, sd=1, lower.tail=TRUE)* or *pnorm(10, mean=5, sd=2, lower.tail=TRUE)-pnorm(-2.5, mean=5, sd=2, lower.tail=TRUE).*

In the following example, we will show how to find the z values when the probabilities are given.

EXAMPLE 3.2.9

For a standard normal random variable Z, find the value of z_0 such that

(a) $P(Z > z_0) = 0.25$.
(b) $P(Z < z_0) = 0.95$.
(c) $P(Z < z_0) = 0.12$.
(d) $P(Z > z_0) = 0.68$.

Solution

(a) *From the normal table, and using the fact that the shaded area in the Figure is 0.25, we have* $P(Z > z_0) = 0.5 - P(0 \leq Z \leq z_0) = 0.25$. *Thus,* $P(0 \leq Z \leq z_0) = 0.25$ *and hence, we obtain* $z_0 \approx 0.675$.

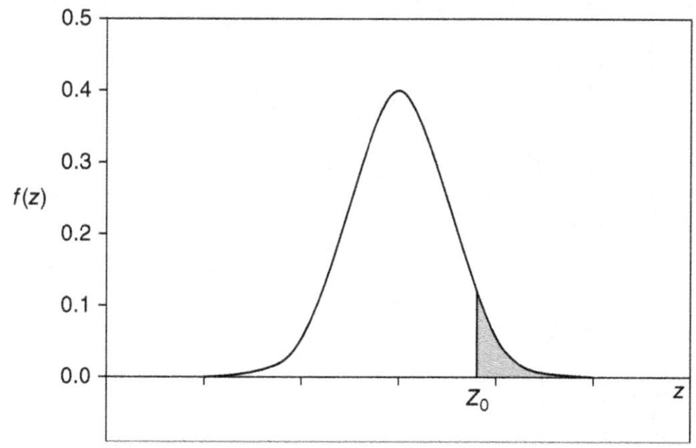

(b) *Because the* $P(Z < z_0) = 1 - P(Z \geq z_0) = 0.95 = 0.5 + 0.45$. *From the normal table,* $z_0 = 1.645$.
(c) *From the normal table,* $z_0 = -1.175$.
(d) *Using the normal table, we have* $P(Z > z_0) = 0.5 + P(0 < Z < z_0) = 0.68$.

This implies, $P(z_0 < Z < 0) = 0.18$. *From the normal table,* $z_0 = -0.465$.

EXAMPLE 3.2.10

The scores of an examination are assumed to be normally distributed with $\mu = 75$ and $\sigma^2 = 64$. What is the probability that a student score chosen at random will be greater than 85?

Solution

Let X be a randomly chosen score from the exam scores. Then, $X \sim N(75, 64)$.

$$P(X > 85) = P\left(\frac{X-75}{8} > \frac{85-75}{8} = 1.25\right)$$
$$= P(Z > 1.25) = 0.1056.$$

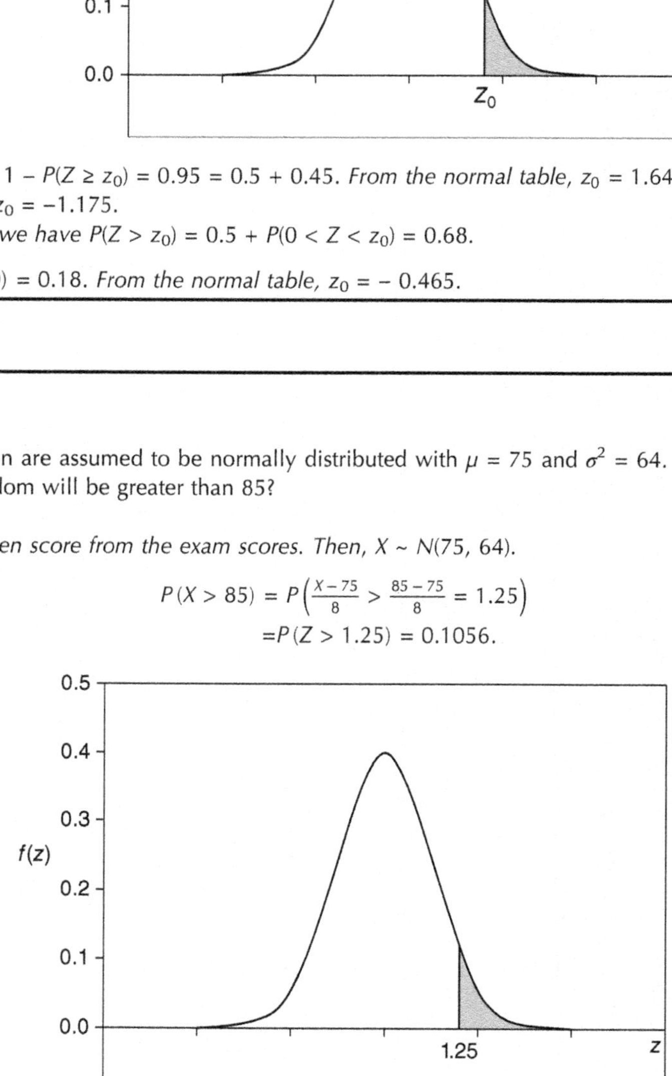

Thus, there is about a 10.56% chance that the score will be greater than 85.

In practice, whenever a large number of small effects are present and *acting additively*, it is reasonable to assume that observations will be normal. When the number of data is small, it is risky to assume a normal distribution without proper testing. Apart from histogram, box-plot, and stem-and-leaf-displays, one of the most useful tools for assessing normality is a quantile-quantile or QQ plot. This is a scatterplot with the quantiles of the scores on the horizontal axis and the expected normal scores on the vertical axis. The expected normal scores are calculated by taking the z-scores of $(r_i - 0.5)/n$, where r_i is rank of the ith observation in increasing order. The steps in constructing a QQ plot are as follows: First, we sort the data in ascending order. If the plot of these scores against the expected normal scores is a straight line, then the data can be considered normal. Any curvature of the points indicates departures from normality. This procedure for obtaining a normal plot (QQ plot is similar to normal plot for a normal distribution) is described in Project 4C. Fig. 3.3 shows a normal probability plot.

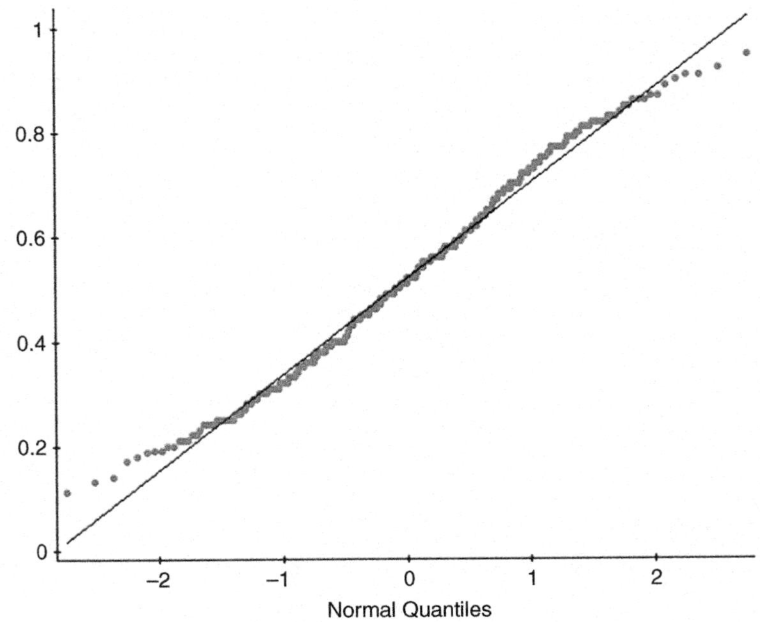

FIGURE 3.3 Normal probability plot.

If plotted points do not fit the line well, but bend away from it in places, the distribution may be nonnormal. The shapes in Fig. 3.4 will give some indication of the distribution of the data.

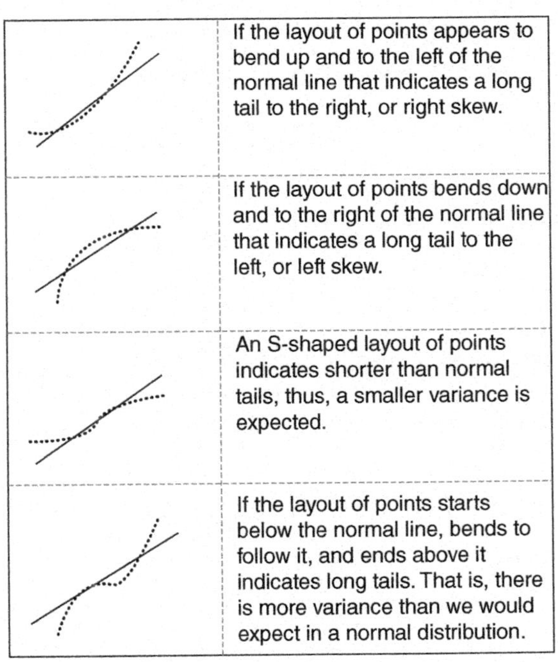

FIGURE 3.4 Shapes indicating distribution behavior of the data.

Almost all of the statistical software packages include a procedure for obtaining the graph of a normal probability plot that can be used to test the normality of data. Errors in the measurements can also act in a *multiplicative* (rather than additive) manner. In that case, the assumption of normality is not justified.

A closely related distribution to the normal distribution is the log-normal distribution. A variable might be modeled as log-normal if it can be thought of as the multiplicative effect of many small independent factors. This distribution arises in physical problems when the domain of the variate, X, is greater than zero and its histogram is markedly skewed. If a random variable Y is normally distributed, then exp(Y) has a *log-normal distribution*. Thus, the natural logarithm of a log-normally distributed variable is normally distributed. That is, if X is a random variable with a log-normal distribution, then ln(X) is normally distributed. Most biological evidence suggests that the growth processes of living tissue proceed by multiplicative, not additive, increments. Thus, the measures of body size should at most follow a log-normal rather than a normal distribution. Also, the sizes of plants and animals are approximately log-normal. The log-normal distribution is also useful in modeling claim sizes in the insurance industry.

The probability density function of a log-normal random variable, X, is given as

$$f(x) = \begin{cases} \frac{1}{x\sigma_y\sqrt{2\pi}} e^{-(\ln x-\mu_y)^2/2\sigma_y^2}, & x > 0,\ \sigma_y > 0,\ -\infty < \mu_y < \infty \\ 0, & \text{otherwise.} \end{cases}$$

where μy and σy are the mean and standard deviation of $Y = \ln(X)$. These parameters are related to the parameters of the random variable X as follows:

$$\mu_y = \ln\left(\sqrt{\frac{\mu_x^4}{\mu_x^2 + \sigma_x^2}}\right), \quad \sigma_y = \ln\left(\sqrt{\frac{\mu_x^2 + \sigma_x^2}{\mu_x^2}}\right).$$

We can verify that the expected value X is

$$E(X) = e^{\mu_y + (\sigma_y^2/2)}$$

and the variance is

$$\text{Var}(X) = (e^{\sigma_y^2} - 1)e^{2\mu_y + \sigma_y^2}.$$

The question of when the log-normal distribution is applicable in a given physical problem after a certain amount of data has been obtained can be answered by creating a normal probability plot of ln(X) and testing for normality. Thus, if the natural logarithms of the data show normality, a log-normal distribution may be more appropriate.

If X is log-normally distributed with parameters μ_y and σ_y, and $0 < a < b$, then with $Y = \ln(X)$,

$$\begin{aligned} P(a \le X \le b) &= P(\ln a \le Y \le \ln b) \\ &= P\left(\frac{\ln a - \mu_y}{\sigma_y} \le \frac{Y - \mu_y}{\sigma_y} \le \frac{\ln b - \mu_y}{\sigma_y}\right) \\ &= P(a' \le Z \le b'), \end{aligned}$$

where $Z \sim N(0, 1)$. This probability can be obtained from the standard normal table.

Example 3.2.11

In an effort to establish a suitable height for the controls of a moving vehicle, information was gathered about X, the amounts by which the heights of the operators vary from 60 inches, which is the minimum height. It was verified that the data that was collected followed the log-normal distribution by normal probability plot of $Y = \ln X$. Assume that $\mu_x = 6$ *in.* and $\sigma_x = 2$ *in.*

(a) What percentage of the operators would have a height less than 65.5 in.?

(b) If an operator is chosen at random, what is the probability that his or her height will be between 64 and 66 in.?

Solution

(a) Here, X = 65.5–60 = 5.5. Also,

$$\mu_y = \ln\left(\sqrt{\frac{\mu_x^4}{\mu_x^2 + \sigma_x^2}}\right) = \ln\sqrt{\frac{6^4}{6^2 + 2^2}} = 1.74,$$

and

$$\sigma_y = \ln\left(\sqrt{\frac{\mu_x^2 + \sigma_x^2}{\mu_x^2}}\right) = \ln\sqrt{\frac{6^2 + 2^2}{6^2}} = 0.053.$$

Thus,

$$P(X \le 5.5) = P(Y \le \ln 5.5) = P\left(Z \le \frac{(\ln 5.5) - 1.74}{0.053}\right)$$
$$= P(Z \le -0.67) = 0.2514.$$

Hence, about 25.14% of the heights of the operators vary from 60 inches.

(b) *Similar to part (a), we get*

$$P(4 \le X \le 6) = P(\ln 4 \le Y \le \ln 6)$$
$$= P\left(\frac{(\ln 4) - 1.74}{0.053} \le Z \le \frac{(\ln 6) - 1.74}{0.053}\right)$$
$$= P(-6.67 \le Z \le 0.98) = 0.8365.$$

Thus 83.65% of the heights of the operators will be between 64 and 66 inches.

3.2.5 Gamma Probability Distribution

The gamma probability distribution has been applied in various fields. For example, in engineering, gamma probability distribution has been employed in the study of system reliability. We describe the gamma function before we introduce the gamma probability distribution. The *gamma function*, denoted by $\Gamma(a)$, is defined as

$$\Gamma(a) = \int_0^\infty e^{-x} x^{a-1} dx, \quad a > 0.$$

It can be shown using the integration by parts that for $a > 1$, $\Gamma(a) = (a-1)\Gamma(a-1)$. In particular, if n is a positive integer, $\Gamma(n) = (n-1)!$.

Definition 3.2.6. *A random variable X is said to possess a* **gamma probability distribution** *with parameters* $\alpha > 0$ and $\beta > 0$ *if it has the pdf given by*

$$f(x) = \begin{cases} \frac{1}{\beta^\alpha \Gamma(\alpha)} x^{\alpha-1} e^{-x/\beta}, & \text{if } x > 0 \\ 0, & \text{otherwise.} \end{cases}$$

The gamma density has two parameters, α and β. We denote this by *Gamma*(α, β). The parameter α is called a *shape parameter*, and β is called a *scale parameter*. Changing α changes the shape of the density, whereas varying β corresponds to changing the units of measurement (such as changing from seconds to minutes). Varying these two parameters will generate different members of the gamma family. If we take α to be a positive integer, we get a special case of the gamma probability distribution, known as the *Erlang distribution.* This is used extensively in queuing theory to model waiting times. Fig. 3.5 gives an indication of how α and β influence the shape and scale of $f(x)$.

Mean, variance, and mgf of a gamma random variable

Theorem 3.2.6. *If X is a gamma random variable with parameters* $\alpha > 0$ *and* $\beta > 0$, *then*

$$E(X) = \alpha\beta \quad \text{and} \quad Var(X) = \alpha\beta^2.$$

Also the moment-generating function is

$$M_X(t) = \frac{1}{(1 - \beta t)^\alpha}, \quad t < \frac{1}{\beta}.$$

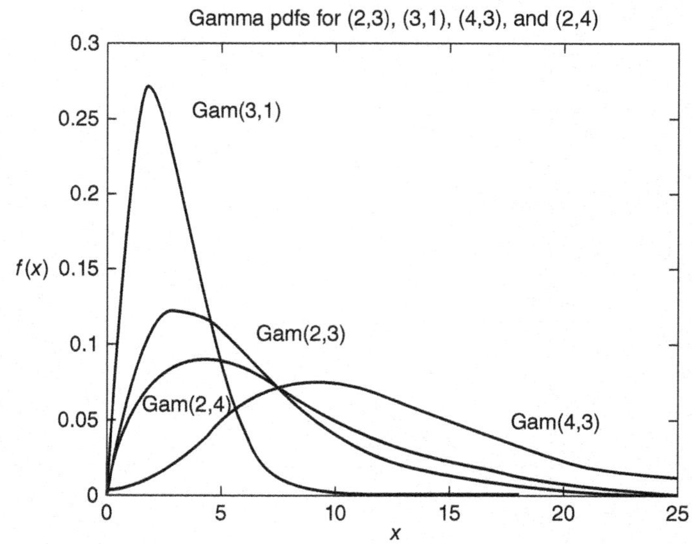

FIGURE 3.5 Gamma pdfs for different parametes.

EXAMPLE 3.2.12

The daily consumption of aviation fuel in millions of gallons at a certain airport can be treated as a gamma random variable with $\alpha = 3, \beta = 1$.

(a) What is the probability that on a given day the fuel consumption will be less than 1 million gallons?
(b) Suppose the airport can store only 2 million gallons of fuel. What is the probability that the fuel supply will be inadequate for a given day?

Solution

(a) *Let X be the fuel consumption in millions of gallons on a given day at a certain airport. Then,* $X \sim \Gamma(\alpha = 3, \beta = 1)$ *and*

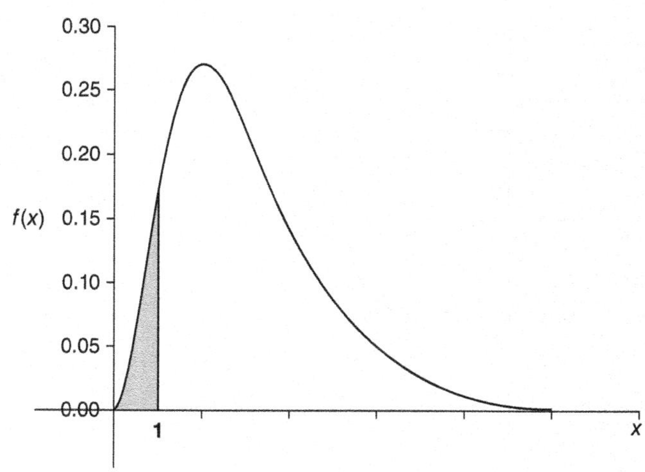

$$f(x) = \frac{1}{\Gamma(3)(1^3)} x^{3-1} e^{-x} = \frac{1}{2} x^2 e^{-x}, \quad x > 0.$$

Hence, using integration by parts, we obtain

$$P(X < 1) = \frac{1}{2}\int_0^1 x^2 e^{-x} dx = 1 - \frac{5}{2e} = 0.08025.$$

Thus there is about an 8% chance that on a given day, the fuel consumption will be less than 1 million gallons.

(b) *Because the airport can store only 2 million gallons, the fuel supply will be inadequate if the fuel consumption X is greater than 2. Thus,*

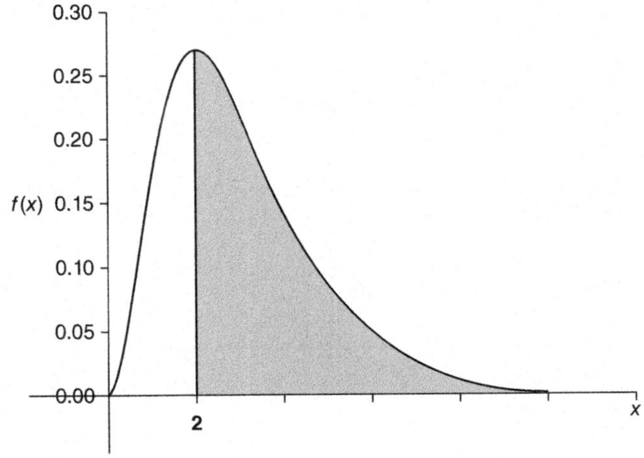

$$P(X > 2) = \frac{1}{2}\int_2^\infty x^2 e^{-x} dx = 0.677.$$

We can conclude that there is about 67.7% chance that the fuel supply of 2 million gallons will be inadequate on a given day. So, if the model is right, the airport needs to store more than 2 million gallons of fuel.

We now describe two special cases of gamma probability distribution. In the pdf of the gamma, we let $\alpha = 1$, we get the pdf of an **exponential random variable**.

Definition 3.2.7. *A random variable X is said to have an* **exponential probability distribution** *with parameter β if the pdf of X is given by*

$$f(x) = \begin{cases} \frac{1}{\beta} e^{-x/\beta}, & \beta > 0;\ 0 \le x < \infty \\ 0, & \text{otherwise.} \end{cases}$$

Exponential random variables are often used to model the lifetimes of electronic components such as fuses, for reliability and survival analysis, among others. The exponential distribution (Fig. 3.6) is also used in developing models of insurance risks. The exponential distribution is related to the Poisson distribution. When the events can occur more than once within a given unit of time and the time elapsed between two consecutive occurrences is exponentially distributed and independent of previous occurrences of the events, then the random variable defined by the number of occurrences has a Poisson distribution. A graph of the exponential pdf with $\beta = 3$ is given below in Fig. 3.6.

Mean, variance, and mgf of an exponential random variable

Theorem 3.2.7. *If X is an exponential random variable with parameters $\beta > 0$, then*

$$E(X) = \beta \quad and \quad \mathrm{Var}(X) = \beta^2.$$

Also the moment-generating function is

$$M_X(t) = \frac{1}{(1 - \beta t)}, \quad t < \frac{1}{\beta}.$$

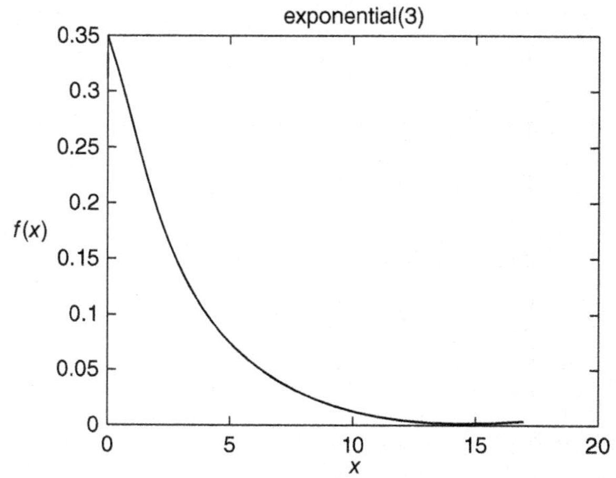

FIGURE 3.6 Probability density function for exponential r.v.

EXAMPLE 3.2.13

The time, in hours, during which an electrical generator is operational is a random variable that follows the exponential distribution with $\beta = 160$. What is the probability that a generator of this type will be operational for

(a) Less than 40 hours?
(b) Between 60 and 160 hours?
(c) More than 200 hours?

Solution

Let X denote the random variable corresponding to time (in hours) during which the generator is operational. Then the probability density function of X is given by

$$f(x) = \begin{cases} \frac{1}{160} e^{-\left(\frac{x}{160}\right)}, & x \geq 0 \\ 0, & \text{otherwise.} \end{cases}$$

Thus we have the following:

(a) $P(X \leq 40) = \int_0^{40} \frac{1}{160} e^{-(x/160)} dx = 0.22119$. *There is about a 22.1% chance that a generator of this type will be operational for less than 40 hours.*
(b) $P(60 \leq X \leq 160) = \int_{60}^{160} \frac{1}{160} e^{-(x/160)} dx = 0.3194$. *Hence, there is about a 31.94% chance that a generator of this type will be operational between 60 and 160 hours.*
(c) $P(X > 200) = \int_{200}^{\infty} \frac{1}{160} e^{-(x/160)} dx = 0.2865$. *The chance that the generator will last more than 200 hours is about 28.65%.*

Another special case of the gamma probability distribution that is useful in statistical inference problems is the chi-square distribution.

Definition 3.2.8. *Let n be a positive integer. A random variable, X, is said to have a* **chi-square** (χ^2) **distribution** *with n degrees of freedom if and only if X is a gamma random variable with parameters* $\alpha = n/2$ and $\beta = 2$. *We denote this by* $X \sim \chi^2(n)$.

Hence, the probability density function of a chi-square distribution with n degrees of freedom is given by

$$f(x) = \begin{cases} \frac{1}{\Gamma\left(\frac{n}{2}\right) 2^{n/2}} x^{(n/2)-1} e^{-x/2}, & 0 \leq x < \infty \\ 0, & \text{otherwise.} \end{cases}$$

Fig. 3.7 illustrates the dependence of the chi-square distribution on n.
The mean and variance of a chi-square random variable follow directly from Theorem 3.2.6.

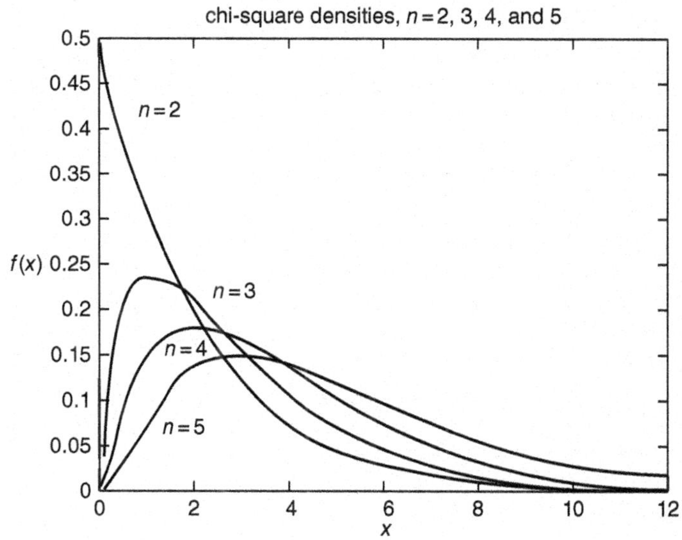

FIGURE 3.7 Chi-square pdfs for different degrees of freedom.

Mean, variance, and mgf of a chi-square random variable

Theorem 3.2.8. *If X is a chi-square random variable with n degrees of freedom, then E(X) = n and Var(X) = 2n. Also, the moment-generating function is given by*

$$M_X(t) = \frac{1}{(1-2t)^{n/2}}, \quad t < \frac{1}{2}.$$

Another class of distributions that plays a crucial role in Bayesian statistics is the beta distribution. The beta distribution is used as a prior distribution for binomial or geometric proportions. Also, for a random phenomenon that occurs between 0 and 1, a random variable X is said to have a *beta distribution* with parameters α and β if and only if the density function of X is given by

$$f(x) = \begin{cases} \frac{x^{\alpha-1}(1-x)^{\beta-1}}{B(\alpha,\beta)}, & \alpha, \beta > 0;\ 0 \le x \le 1 \\ 0, & \text{otherwise,} \end{cases}$$

where $B(\alpha, \beta) = \int_0^1 x^{\alpha-1}(1-x)^{\beta-1}dx$. It can be shown that $B(\alpha, \beta) = \frac{\Gamma(\alpha)\Gamma(\beta)}{\Gamma(\alpha+\beta)}$, and that $E(X) = \frac{\alpha}{\alpha+\beta}$ and $Var(X) = \frac{\alpha\beta}{(\alpha+\beta)^2(\alpha+\beta+1)}$.

Given a set of data, one of the questions we may have is: "How do we know which distribution to use in a given physical problem?" There is no simple and direct answer to this question. One intuitive way is to construct a histogram from the information at hand; from the shape of this histogram, we obtain a visual view of whether the random variable follows a particular distribution, such as gamma distribution. In Chapter 11, we discuss statistical tests called goodness-of-fit tests that will identify the pdf of a given data with a high degree of accuracy. Once we decide that it follows a particular distribution, then the parameters of this distribution, such as α and β, must be statistically estimated. In Chapter 5, we discuss how to estimate these parameters.

EXERCISES 3.2

3.2.1 A fair coin is tossed 10 times. Let X denote the number of heads obtained. Find the following probabilities:

(a) $P(X = 7)$
(b) $P(X \le 7)$
(c) $P(X > 0)$
(d) $E(X)$ and $Var(X)$

3.2.2 Let X be a Poisson random variable with $\lambda = 1/3$. Find

(a) $P(X = 0)$
(b) $P(X \geq 4)$.

3.2.3 For a standard normal random variable Z, find the value of z_0 such that
(a) $P(Z > z_0) = 0.05$
(b) $P(Z < z_0) = 0.88$
(c) $P(Z < z_0) = 0.10$
(d) $P(Z > z_0) = 0.95$.

3.2.4 Let $X \sim N(12, 5)$. Find the value of x_0 such that
(a) $P(X > x_0) = 0.05$
(b) $P(X < x_0) = 0.98$
(c) $P(X < x_0) = 0.20$
(d) $P(X > x_0) = 0.90$.

3.2.5 Let $X \sim N(10, 25)$. Compute the following probabilities:
(a) $P(X \leq 20)$
(b) $P(X > 5)$
(c) $P(12 \leq X \leq 15)$
(d) $P(|X - 12| \leq 15)$.

3.2.6 A quarterback on a football team has a pass completion rate of 0.62. If, in a given game, he attempts 16 passes, what is the probability that he will complete
(a) 12 passes?
(b) More than half of his passes?
(c) Interpret your result.
(d) Out of the 16 passes, what is the expected number of completions?

3.2.7 A consulting group believes that 70% of the people in a certain county are satisfied with their health coverage. Assuming that this is true, find the probability that in a random sample of 15 people from the county:
(a) Exactly 10 are satisfied with their health coverage and interpret.
(b) Not more than 10 are satisfied with their health coverage and interpret.
(c) What is the expected number of people out of 15 who are satisfied with their health coverage?

3.2.8 A man fires at a target. The probability of him hitting it each time is 0.40 and is independent of other tries.
(a) What is the probability that the man will hit the target at least once if he fires six times?
(b) How many times must he fire at the target so that the probability of hitting it at least once is greater than 0.77?
(c) Interpret your findings.

3.2.9 A certain electronics company produces a particular type of vacuum tube. It has been observed that, on average, three tubes of 100 are defective. The company packs the tubes in boxes of 400. What is the probability that a certain box of 400 tubes will contain
(a) r defective tubes?
(b) At least k defective tubes?
(c) At most one defective tube?
(d) Interpret your answers to (a), (b), and (c).

3.2.10 Suppose that, on average, in every two pages of a book there is one typographical error, and that the number of typographical errors on a single page of the book is a Poisson r.v. with $\lambda = 1/2$. What is the probability of at least one error on a certain page of the book? Interpret your result.

3.2.11 Show that the probabilities assigned by the Poisson probability distribution satisfy the requirements that $0 \leq p(x) \leq 1$, for all x and $\sum_{all\ x} p(x) = 1$.

3.2.12 In determining the range of an acoustic source using the triangulation method, the time at which the spherical wave front arrives at a receiving sensor must be measured accurately. Measurement errors in these times can be modeled as possessing uniform probability distribution from −0.05 to 0.05 microseconds. What is the probability that a particular arrival time measurement will be in error by less than 0.01 microsecond? What does your answer mean?

3.2.13 The hardness of a piece of ceramic is proportional to the firing time. Assume that a rating system has been devised to rate the hardness of a ceramic piece and that this measure of hardness is a random variable and is distributed uniformly between 0 and 10. If the hardness in the interval [5, 9] is desirable for kitchenware, what is the probability that a piece chosen at random will be suitable for kitchen use?

3.2.14 A receiver receives a string of 0's and 1's transmitted from a certain source. The receiver used a majority rule. That is, if the receiver acquires five symbols, xxxxx, x is 0 or 1, of which three or more are 1s, it decides that a 1 was transmitted. The receiver is correct only 85% of the time. What is $P(W)$, the probability of a wrong decision if the probabilities of receiving 0's and 1's are equally likely? What can you conclude from your result?

3.2.15 The efficiency X of a certain electrical component may be assumed to be a random variable that is distributed uniformly between 0 and 100 units. What is the probability that X is

(a) Between 60 and 80 units?

(b) Greater than 90 units?

(c) Interpret (a) and (b).

3.2.16 The *reliability function* of a system or a piece of equipment at time t is defined by

$$R(t) = P(T \geq t) = 1 - F(t),$$

where T, the failure time, is a random variable with a known distribution. A certain vacuum tube has been observed to fail uniformly over the interval $[t_1, t_2]$.

(a) Determine the reliability of such a tube at time t, $t_1 \leq t \leq t_2$.

(b) If $180 \leq t \leq 220$, what is the reliability of such a tube at 200 hours?

(c) The *failure or hazard rate function* $\rho(t)$ is defined by

$$\rho(t) = \frac{f(t)}{1 - F(t)} = \frac{f(t)}{R(t)} = \frac{\frac{-dR(t)}{dt}}{R(t)}.$$

Calculate the failure rate of this vacuum tube. Interpret your result.

3.2.17 An electrical component was studied in the laboratory, and it was determined that its failure rate was approximately equal to $\frac{1}{\beta} = 0.05$. What is the reliability of such a component at 10 hours?

3.2.18 Suppose that the life length of a mechanical component is normally distributed.

(a) If $\sigma = 3$ and $\mu = 100$, find the reliability of such a system at 105 hours.

(b) What should be the expected life of the component if it has reliability of 0.90 for 120 hours?

3.2.19 A geologist defines granite as a rock containing quartz, feldspar, and small amounts of other minerals, provided that it contains not more than 75% quartz. If all the percentages are equally likely, what proportion of granite samples that the geologist collects during his lifetime will contain from 50% to 65% quartz?

3.2.20 For a normal random variable with pdf,

$$f(x) = \frac{1}{\sqrt{2\pi}\,\sigma} e^{-(x-\mu)^2/2\sigma^2}, \qquad \infty < x < \infty, \; -\infty < \mu < \infty, \; \sigma^2 > 0,$$

show that $\int_{-\infty}^{\infty} f(x)dx = 1$. [Hint: use polar coordinates.]

3.2.21 A professor in a large statistics class has a grading policy such that only the 15% of the students with the highest scores will receive grade A. The mean score for this class is 72 with a standard deviation of 6. Assuming that all the grades for this class follow a normal probability distribution, what is the minimum score that a student in this class has to get to receive an A grade?

3.2.22 The scores, X, of an examination may be assumed to be normally distributed with $\mu = 70$ and $\sigma^2 = 49$. What is the probability that

(a) A score chosen at random will be between 80 and 85?

(b) A score will be greater than 75?

(c) A score will be less than 90?

(d) Interpret the meaning of (a), (b), and (c).

3.2.23 Suppose that the diameters of golf balls manufactured by a certain company are normally distributed with $\mu = 1.96$ in. and $\sigma = 0.04$ in. A golf ball will be considered defective if its diameter is less than 1.90 in. or greater than 2.02 in. What is the percentage of defective balls manufactured by the company? What did the answer indicate?

3.2.24 Suppose that the arterial diastolic blood pressure readings in a population follow a normal probability distribution with mean 80 mm Hg and standard deviation 6.2 mm Hg. Suppose it is recommended that a physician be consulted if an individual has an arterial diastolic blood pressure reading of 90 mm Hg or more. If an individual is randomly picked from this population, what is the probability that this individual needs to consult a physician? Discuss the meaning of your result.

3.2.25 In a certain pediatric population, systolic blood pressure is normally distributed with mean 115 mm Hg and standard deviation of 10 mm Hg. Find the probability that a randomly selected child from this population will have

(a) A systolic pressure greater than 125 mm Hg.

(b) A systolic pressure less than 95 mm Hg.
(c) A systolic pressure below which 95% of this population lies.
(d) Interpret (a), (b), and (c).

3.2.26 A physical fitness test was given to a large number of college freshmen. In part of the test, each student was asked to run as far as they could in 10 minutes. The distance each student ran in miles was recorded and can be considered to be a random variable, say, X. The data showed that the random variable X followed the log-normal distribution with $\mu_y = 0.35$ and $_{\sigma}y = 0.5$, where $Y = \ln X$. A student is considered physically fit if they are able to run 1.5 miles in the time allowed. What percentage of college freshmen would be considered physically fit if we consider only this part of the test?

3.2.27 An experimenter is designing an experiment to test tetanus toxoid in guinea pigs. The survival of the animal following the dose of the toxoid is a random phenomenon. Past experience has shown that the random variable that describes such a situation follows the log-normal distribution with $\mu_y = 0$ and $\sigma_y = 0.65$. As a requirement of good design, the experimenter must choose doses at which the probability of surviving is 0.20, 0.50, and 0.80. What three doses should the experimenter choose?

3.2.28 Show that $\Gamma(1) = 1$ and for $a > 1$, prove that $\Gamma(a) = (a - 1)\Gamma(a - 1)$.

3.2.29 **(a)** Find the moment-generating function for a gamma probability distribution with parameter $\alpha > 0$ and $\beta > 0$. [Hint: In the integral representation of $E(e^{tX})$, change the variable t to $u = (1 - \beta t)x/\beta$, with $(1 - \beta t) > 0$.]
(b) Using the mgf of a gamma probability distribution, find $E(X)$ and $Var(X)$.

3.2.30 Let X be an exponential random variable. Show that, for numbers $a > 0$ and $b > 0$,

$$P(X > a + b|X > a) = P(X > b).$$

This property of the exponential distribution is called the memoryless property of the distribution.

3.2.31 A random variable X is said to have a *beta distribution* with parameters α and β if and only if the density function of X is

$$f(x) = \begin{cases} \frac{x^{\alpha-1}(1-x)^{\beta-1}}{B(\alpha,\beta)}, & \alpha, \beta > 0;\ 0 \le x \le 1 \\ 0, & \text{otherwise} \end{cases}$$

where

$$B(\alpha, \beta) = \int_0^1 x^{\alpha-1}(1 - x)^{\beta-1}dx.$$

(a) Show that $B(\alpha, \beta) = \frac{\Gamma(\alpha)\Gamma(\beta)}{\Gamma(\alpha+\beta)}$.

(b) Show that $E(X) = \frac{\alpha}{\alpha+\beta}$ and $Var(X) = \frac{\alpha\beta}{(\alpha+\beta)^2(\alpha+\beta+1)}$.

3.2.32 The daily proportion of major automobile accidents across the United States can be treated as a random variable having a beta distribution with $\alpha = 6$ and $\beta = 4$. Find the probability that, on a certain day, the percentage of major accidents is less than 80% but greater than 60%. Interpret your answer.

3.2.33 Suppose that network breakdowns occur randomly and independently of each other at an average rate of three per month.
(a) What is the probability that there will be just one network breakdown during December? Interpret.
(b) What is the probability that there will be at least four network breakdowns during December? Interpret.
(c) What is the probability that there will be at most seven network breakdowns during December? Interpret.

3.2.34 Let X be a random variable denoting the number of events occurring in the time interval $(0, t)$. Show that X has a gamma probability distribution with parameters n and λ.

3.2.35 In order to etch an aluminum tray successfully, the pH of the acid solution used must be between 1 and 4. This acid solution is made by mixing a fixed quantity of etching compound in powder form with a given volume of water. The actual pH of the solution obtained by this method is affected by the potency of the etching compound, by slight variations in the volume of water used, and perhaps by the pH of the water. Thus the pH of the solution varies. Assume that the random variable that describes the random phenomenon is gamma distributed with $\alpha = 2$ and $\beta = 1$.
(a) What is the probability that an acid solution made by the foregoing procedure will satisfactorily etch a tray?

(b) What would the answer to part (a) be if $\alpha = 1$ and $\beta = 2$?

3.2.36 If $X_i \sim Pois(\lambda_i)$, $i = 1, 2, \ldots, k$, are independent, and $\lambda = \sum_{i=1}^{k} \lambda_i$, then,

$$Y = \sum_{i=1}^{n} X_i \sim Pois(\lambda).$$

3.2.37 If $X_i \sim Exp(\beta)$, $i = 1, 2, \ldots, k$ are independent, then show that

$$Y = \sum_{i=1}^{k} X_i \sim Gamma(k, \beta).$$

3.2.38 In order to evaluate a new release of a database management system, a database administrator runs a benchmark program several times and measures the time to completion in seconds. Assuming that the distribution of times is normal with a mean of 95 seconds and a standard deviation of 10 seconds, what proportion of measurement times will fall below 85 seconds?

3.2.39 For COVID-19 vaccine recipients, suppose it is known that 90% of people have at least one side effect at the first or second dose, find the probability that more than 60 of the patients in a random sample of 100 will experience at least one side effect.

3.2.40 In order to have an idea of how much bandwidth to provide to different websites, website hosting companies use Poisson distribution to model the number of expected visitors per hour that websites will receive. Suppose a website receives an average of 30 visitors per hour, find the probability that the website receives more than 50 visitors in a given hour.

3.3 JOINT PROBABILITY DISTRIBUTIONS

We have thus far confined ourselves to studying one-dimensional or univariate random variables and their properties. In many practical situations, we are required to deal with several, not necessarily independent random variables. For example, we might be interested in a study involving the weights and heights (W, H) of a certain group of people. In this situation, we need the two random variables (W, H), and it is likely that these two are related. Then it becomes important to study the joint effect of these random variables, which will lead to finding the joint probability distribution. In this section, we will confine our studies to two random variables and their joint probability distribution, which is called a *bivariate probability distribution*. We consider the random variables to be either both discrete or both continuous. We now define joint distribution of two random variables.

Definition 3.3.1.

(a) *Let X and Y be random variables. If both X and Y are discrete, then*

$$p(x, y) = P(X = x, Y = y),$$

is called the **joint probability function** *(joint pmf) of X and Y.*

(b) *If both X and Y are continuous, then f(x, y) is called the* **joint probability density function** *(joint pdf) of X and Y if and only if*

$$P(a \le X \le b, c \le Y \le d) = \int_a^b \int_c^d f(x, y)dydx.$$

In order to reduce repetitions, most of the time, we may use $f(.)$ in place of $p(.)$ for discrete and continuous cases. The reader can use $p(.)$ in the case of discrete distributions.

Example 3.3.1

A probability class contains 10 African American, 8 Hispanic American, and 15 White students. If 12 students are randomly selected from this class, and if X = number of Black students, and Y = number of White students, find the joint probability function of the bivariate random variable (X, Y).

Solution

There are total of 33 students. The number of ways in which x African American, and y White students can be picked (which means, the remaining $12 - (x + y)$ students are Hispanic American) can be obtained using the multiplication principle, that is,

$$\binom{10}{x}\binom{15}{x}\binom{8}{12-x-y}.$$

The number of ways to pick 12 students from 33 students is $\binom{33}{12}$. *Hence, the joint probability mass function is*

$$p(x, y) = P(X = x, Y = y) = \frac{\binom{10}{x}\binom{15}{y}\binom{8}{12-x-y}}{\binom{33}{12}},$$

where $0 \le x \le 10$, $0 \le y \le 12$, and $4 \le x + y \le 12$. *The last constraint is needed because there are only eight Hispanic Americans, so the combined minimum number of whites and African Americans should be at least 4.*

We follow the notation: $\Sigma_{x,y}$ to denote $\Sigma_x\Sigma_y$. The joint distribution of two random variables has to satisfy the following conditions:

Properties of Joint probability distribution

Theorem 3.3.1. *If X and Y are two random variables with joint probability function* $f(x, y)$, *then*

(i) *If X and Y are discrete, then* $p(x, y) \ge 0$, *for all x and y, and*
$\Sigma_{x,y}\, p(x, y) = 1$,
where the sum is over all values (x, y) that are assigned nonzero probabilities.

(ii) *If X and Y are continuous, then* $f(x, y) \ge 0$ *for all x and y, and*

$$\iint f(x, y)dxdy = 1.$$

Given the joint probability distribution (pdf or pmf), the probability distribution function of a component random variable can be obtained through the marginals.

Definition 3.3.2. *The* **marginal pdf (pmf)** *of X denoted by* $f_X(x)(p_X(x)$ *in case of discrete) (or f(x), when there is no confusion) is defined by*

$$f_X(x) = \begin{cases} \int_{-\infty}^{\infty} f(x, y)dy, & \text{if } X \text{ and } Y \text{ are continuous,} \\ \Sigma_{\text{all } y}\, p(x, y), & \text{if } X \text{ and } Y \text{ are discrete.} \end{cases}$$

Similarly, the **marginal pdf** *of Y denoted by* $f_Y(y)$ *is defined by*

$$f_Y(y) = \begin{cases} \int_{-\infty}^{\infty} f(x, y)dx, & \text{if } X \text{ and } Y \text{ are continuous,} \\ \Sigma_{\text{all } x}\, p(x, y), & \text{if } X \text{ and } Y \text{ are discrete.} \end{cases}$$

Note that we can obtain marginal probabilities of X, that is,

$$P(a \le X \le b) = \begin{cases} \int_a^b f_X(x)dx, & \text{if } X \text{ and } Y \text{ are continuous,} \\ \Sigma_{\text{all } y} f_X(x), & \text{if } X \text{ and } Y \text{ are discrete,} \end{cases}$$

where the summation is over all values of X from a to b.

Example 3.3.2

Find the marginal probability density function of the random variables X and Y, if their joint probability function is given by Table 3.1.

TABLE 3.1 Joint pmf of X and Y.

	Y				
X	−2	0	1	4	Sum
−1	0.2	0.1	0.0	0.2	0.5
3	0.1	0.2	0.1	0.0	0.4
5	0.1	0.0	0.0	0.0	0.1
Sum	0.4	0.3	0.1	0.2	1.0

Find the marginal densities of X and Y.

Solution

By definition, the marginal pmf of X is given by the column sums (summands over y for fixed x), and the marginal pmf of Y are obtained by the row sums. Hence,

x_i	−1	3	5	otherwise	y_i	−2	0	1	4	otherwise
$f_X(x_i)$	0.5	0.4	0.1	0	$f_y(y_i)$	0.4	0.3	0.1	0.2	0

Using the joint probability distribution and the marginals, we can now introduce the conditional probability distribution function.

Definition 3.3.3. ***The*** **conditional probability distribution** *of the random variable X given Y is given by*

$$f(x|y) = f(x|Y = y)$$
$$= \begin{cases} \dfrac{f(x, y)}{f_Y(y)}, & \text{if } X \text{ and } Y \text{ are continuous, } f_Y(y) \neq 0, \\ \dfrac{P(X = x,\ Y = y)}{f_Y(y)}, & \text{if } X \text{ and } Y \text{ are discrete.} \end{cases}$$

We note that both the marginal probability densities of X and Y, as well as the conditional pdf, must satisfy the two important conditions of a pdf.

We know that two events A and B are independent if $P(A \cap B) = P(A)P(B)$. It is usually more convenient to establish independence through the probability functions. Hence, we define independence for a bivariate probability distribution as follows.

Definition 3.3.4. *Let X and Y have a joint pmf or pdf, f (x, y). Then X and Y are* **independent** *if and only if*

$$f(x, y) = f_X(x) f_Y(y), \quad \text{for all } x \text{ and } y.$$

That is, for independent random variables, the joint pdf is the product of the marginals.

EXAMPLE 3.3.3

Let

$$f(x, y) = \begin{cases} 3x, & 0 \le y \le x \le 1, \\ 0, & \text{otherwise.} \end{cases}$$

(a) Find $P\left(X \le \frac{1}{2}, \frac{1}{4} < Y < \frac{3}{4}\right)$.
(b) Find the marginals $f_X(x)$ and $f_Y(y)$.
(c) Find the conditional $f(x|y)$ $(0 < y < 1)$. Also compute $f\left(x|Y = \frac{1}{2}\right)$.
(d) Are X and Y independent?

Solution

(a) *The domain of the function* $f(x, y)$ *is given in* Fig. 3.8. *The required probability* $P\left(X \le \frac{1}{2}\frac{1}{4} < Y < \frac{3}{4}\right)$ *is the volume over the area of the shaded region as shown by* Fig. 3.9. *That is,*

$$P\left(X \le \frac{1}{2}, \frac{1}{4} < Y < \frac{3}{4}\right) = \int_{1/4}^{1/2}\int_{1/4}^{x} 3x\,dy\,dx$$
$$= \int_{1/4}^{1/2} 3x\left(x - \frac{1}{4}\right)dx$$
$$= \left(\frac{3x^3}{3} - \frac{3x^2}{8}\right)\Big|_{1/4}^{1/2}$$
$$= \frac{5}{128}.$$

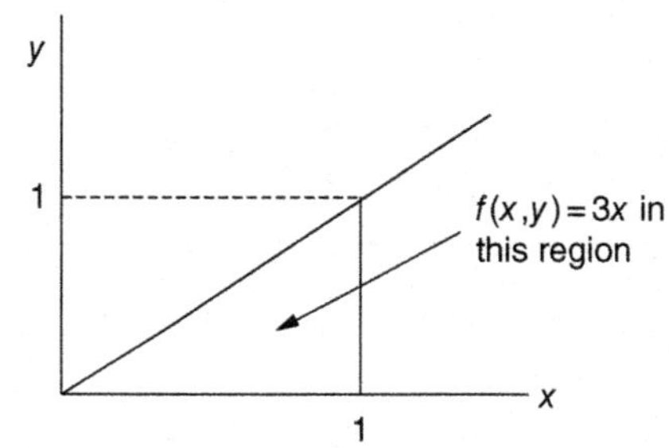

FIGURE 3.8 **Domain of** $f(x,y)$.

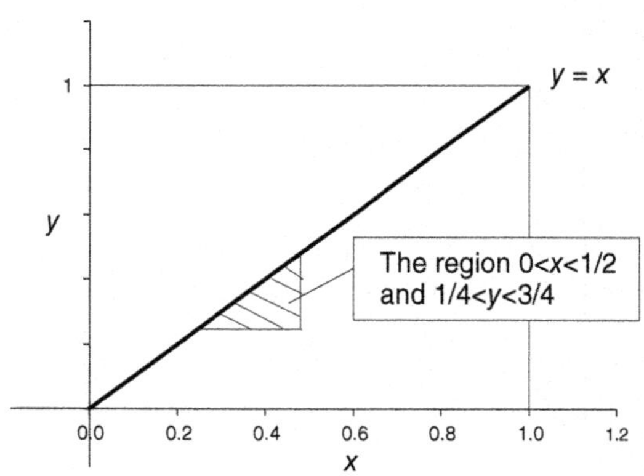

FIGURE 3.9 Region of integration.

(b) *To find the marginals, we note that for each* x, y *varies from* 0 *to* x $(0 < y < x)$. *Therefore*

$$f_X(x) = \int_0^x 3x\,dy = 3x\left(y|_0^x\right) = 3x^2, \ 0 < x < 1.$$

Similarly, for each y, x *varies from* y *to* 1.

$$f_Y(y) = \int_y^1 3x dx = \frac{3x^2}{2}\Big|_y^1 = \frac{3}{2} - \frac{3y^2}{2}$$
$$= \frac{3}{2}\left(1 - y^2\right), \; 0 < y < 1.$$

(c) *Using the definition of conditional density*

$$f\left(x|y\right) = \frac{f(x, y)}{f_Y(y)} = \frac{3x}{\frac{3}{2}\left(1 - y^2\right)} = \frac{2x}{1 - y^2}, \quad y \le x \le 1.$$

Thus, from this, we have

$$f\left(x|y = \frac{1}{2}\right) = \frac{2x}{\left(1 - \left(\frac{1}{2}\right)^2\right)} = \frac{8}{3}x, \quad \frac{1}{2} \le x \le 1.$$

(d) *To check for independence of X and Y,*

$$f_X(1) f_Y\left(\frac{1}{2}\right) = (3)\left(\frac{9}{8}\right) = \frac{27}{8} \ne 3 = f\left(1, \frac{1}{2}\right).$$

Hence, X and Y are not independent.

Recall that in the case of a univariate random variable X, with probability function $f(x)$, the expected value of X is

$$E(X) = \begin{cases} \sum_x xf(x), & \text{if } \sum_x |x| f(x) < \infty, \text{ for discrete r.v.} \\ \int xf(x)dx, & \text{if } \int |x| f(x)dx < \infty, \text{ for continuous r.v.} \end{cases}$$

Now we proceed to define similar concepts for a bivariate probability distribution.

Definition 3.3.5. *Let $f(x, y)$ be the joint probability function, and let $g(x, y)$ be such that $\sum_{x,y} |g(x, y)| f(x, y) < \infty$ in the discrete case, or $\int_{-\infty}^{\infty} \int_{-\infty}^{\infty} |g(x, y)| f(x, y) dxdy < \infty$, in the continuous case. Then the* **expected value** *of $g(X, Y)$ is given by*

$$Eg(X, Y) = \begin{cases} \sum_{x,y} g(x, y) f(x, y), & \text{if } X, Y \text{ are discrete,} \\ \int_{-\infty}^{\infty} \int_{-\infty}^{\infty} g(x, y) f(x, y) dxdy, & \text{if } X, Y \text{ are continuous.} \end{cases}$$

In particular,

$$E(XY) = \begin{cases} \sum_{x,y} xyf(x, y), & \text{if } X, Y \text{ are discrete,} \\ \int_{-\infty}^{\infty} \int_{-\infty}^{\infty} xyf(x, y)dxdy, & \text{if } X, Y \text{ are continuous.} \end{cases}$$

The following properties of mathematical expectation are easy to verify:

Properties of expected value

1. $E(aX + bY) = aE(X) + bE(Y)$.
2. If X and Y are independent, then $E(XY) = E(X)E(Y)$. However, the converse is not necessarily true.

EXAMPLE 3.3.4

Let $f(x, y) = 3x$, $0 \le y \le x \le 1$.

(a) Find $E(4X - 3Y)$,
(b) Find $E(XY)$.

Solution

(a) $E(X) = \int x f_X(x)dx$ *and* $E(Y) = \int y f_Y(y)dy$.
Recall that earlier (Example 3.3.3) we have computed $f_X(x) = 3x^2$, $(0 < x < 1)$ *and* $f_Y(y) = 2(1 - y^2)$, $0 \le y \le 1$. *Using these results, we have*

$$E(X) = \int_0^1 x3x^2dx = \frac{3}{4},$$

and

$$E(Y) = \int_0^1 y\frac{3}{2}(1 - y^2)dy = \frac{3}{8}.$$

Hence,

$$E(4X - 3Y) = 3 - \frac{9}{8} = \frac{15}{8}.$$

(b) $E(XY) = \int_0^1 \int_0^x xy(3x)dydx = \frac{3}{10}$.

Conditional expectations are defined in the same manner as univariate expectations, except that the conditional density is utilized in place of the unconditional density function.

Definition 3.3.6. *Let X and Y be jointly distributed with pmf or pdf f (x, y). Let g be a function of x. Then the* **conditional expectation** *of g(x) given Y = y is*

$$E(g(X)|y) = E(g(X)|Y = y)$$
$$= \begin{cases} \Sigma_{\text{all } x}\, g(x)f(x|y), & \text{if } X, Y \text{ are discrete}, \\ \int g(x)f(x|y)dx, & \text{if } X, Y \text{ are continuous}, \end{cases}$$

and

$$Var(X|y) = E[(X - E(X|y))^2|y]$$
$$= E[X^2|y] - [E(X|y)]^2.$$

Note that $E(g(X)|y)$ is a function of y. If we let Y range over all of its possible values, the conditional expectation $E(g(X)|Y)$ can be thought of as a function of the random variable Y. We will then be able to find the mean and variance of $E(g(X)|Y)$, as given in the following theorem, the proof of which is left as an exercise:

Theorem 3.3.2. *Let X and Y be two random variables. Then*

(a) $E(X) = E[E(X|Y)]$.
(b) $Var(X) = E[Var(X|Y)] + Var[E(X|Y)]$.

We can define the conditional variance, $Var(Y|X) = E([Y - E(Y|X)]^2|X)$.

EXAMPLE 3.3.5

Let X and Y be two random variables with joint density function given by

$$f(x, y) = \begin{cases} x^2 + \frac{xy}{3}, & 0 \le x \le 1 \text{ and } 0 \le y \le 2 \\ 0, & \text{otherwise.} \end{cases}$$

Find the conditional expectation, $E\left(X|Y = \frac{1}{2}\right)$.

Solution

First, we will find the conditional density, $f(x|y)$. *The marginal pdf is given by*

$$f_Y(y) = \int_0^1 \left(x^2 + \frac{xy}{3}\right)dx = \frac{1}{3} + \frac{1}{6}y, \quad 0 < y < 2.$$

Therefore

$$f(x|y) = \frac{f(x, y)}{f_Y(y)} = \frac{x^2 + \frac{xy}{3}}{\frac{1}{6}y + \frac{1}{3}}, \quad 0 \le x \le 1.$$

Hence,

$$f\left(x|Y = \frac{1}{2}\right) = \frac{x^2 + \frac{x}{6}}{\frac{1}{12} + \frac{1}{3}} = \frac{12}{5}\left(x^2 + \frac{x}{6}\right).$$

Thus

$$E\left(X|Y = \frac{1}{2}\right) = \int_0^1 xf\left(x|Y = \frac{1}{2}\right)dx$$
$$= \int_0^1 x\frac{12}{5}\left(x^2 + \frac{x}{6}\right)dx = \frac{11}{15} = 0.733.$$

EXAMPLE 3.3.6

Let the joint density of two random variables X and Y is given by

$$f(x, y) = \begin{cases} \frac{1}{4}(2x + y), & 0 \le x \le 1,\ 0 \le y \le 2 \\ 0, & \text{otherwise.} \end{cases}$$

(a) Find $f_X(x)$ and $f_Y(y)$.
(b) Find $Var(X)$
(c) Find $E(X|Y)$, and $Var(X|Y)$.

Solution

(a) We have

$$f_X(x) = \int_{-\infty}^{\infty} f(x, y)dy = \int_0^2 \frac{1}{4}(2x + y)dy$$
$$= \frac{1}{4}(4x + 2), \quad 0 \le x \le 1.$$

Similarly, the marginal pdf of X, $f_Y(y) = \int_0^1 \frac{1}{4}(2x + y)dx = \frac{1}{4}(1 + y), \quad 0 \le y \le 2.$

(b) To find the variance, we have

$$E(X) = \int_0^1 \frac{1}{4}x(4x + 2)dx = \frac{7}{12},$$

and

$$E(X^2) = \int_0^1 \frac{1}{4}x^2(4x + 2)dx = \frac{5}{12}.$$

Thus the variance of X is

$$\text{var}(X) = E(X^2) - [E(X)]^2$$
$$= \frac{5}{12} - \left(\frac{7}{12}\right)^2 = \frac{11}{144}.$$

(c) First, we will find the conditional density of X given that $Y = y$, that is,

$$f_{X|Y}(x|y) = \frac{f(x, y)}{f_Y(y)} = \frac{\frac{1}{4}(2x + y)}{\frac{1}{4}(1 + y)}$$
$$= \frac{(2x + y)}{(1 + y)}, \quad 0 \le x \le 1,\ 0 \le y \le 2.$$

Then the conditional expectation is given by

$$E[X|Y] = \int_0^1 x\frac{(2x+y)}{(1+y)}dx = \frac{1}{1+y}\int_0^1 (2x^2 + xy)dx$$
$$= \frac{1}{1+y}\left(\frac{2}{3} + \frac{1}{2}y\right) = \left(\frac{1}{6}\right)\frac{(4+3y)}{(1+y)}.$$

For the conditional variance, we also need to find

$$E[X^2|Y] = \int_0^1 x^2\frac{(2x+y)}{(1+y)}dx = \frac{1}{1+y}\int_0^1 (2x^3 + x^2y)dx$$
$$= \left(\frac{1}{6}\right)\left(\frac{3+2y}{1+y}\right).$$

Thus

$$Var[X|Y] = E(X^2|Y) - \left[E(X|Y)\right]^2$$
$$= \left(\frac{1}{6}\right)\left(\frac{3+2y}{1+y}\right) - \left(\frac{1}{36}\right)\frac{(4+3y)^2}{(1+y)^2}$$
$$= \frac{3y^2+6y+2}{36(1+y)^2}.$$

3.3.1 Covariance and Correlation

We will now define the covariance and correlation coefficient of two random variables.

Definition 3.3.7.

(i) *The* **covariance** *between two random variables X and Y is defined by*

$$\sigma_{XY} = Cov(X,\ Y) = E(X - \mu_X)(Y - \mu_Y) = E(XY) - \mu_X\mu_Y,$$

where $\mu_X = E(X)$ and $\mu_Y = E(Y)$.

(ii) *The* **correlation coefficient,** $\rho_{X,Y}$ *is defined by*

$$\rho_{X,Y} = \frac{Cov(X,\ Y)}{\sqrt{Var(X)\,Var(Y)}}.$$

Correlation is the measure of the ***linear relationship*** *between the random variables X and Y. If $Y = aX+b$ ($a \neq 0$), then $|\rho_{X,Y}| = 1$. If the dependence of X and Y needs to be specified, we will use the notation,* $\rho_{X,Y}$ or $\rho(X,\ Y)$.

From the definition of the covariance of X and Y, we note that if small values of X, for which $(X - \mu X) < 0$, tend to be associated with small values of Y, for which $(Y - \mu Y) < 0$, and similarly large values of X with large values of Y, then Cov $(X, Y) \equiv E[(X - \mu X)(Y - \mu Y)]$ can be expected to be positive. On the other hand, if small values of X tend to be associated with large values of Y and vice versa so that $(X - \mu X)$ and $(Y - \mu Y)$ are of opposite signs, then $Cov(X, Y) < 0$. Thus covariance can be thought of as a signed measure of the variation of Y relative to X. If X and Y are independent, then it follows from the definition of covariance that $Cov(X, Y) = 0$. The correlation coefficient of X and Y is a dimensionless quantity that measures the linear relationship between the random variables X and Y.

Properties of covariance and the correlation coefficient

(a) $-1 \leq \rho_{X,Y} \leq 1$.

(b) If X and Y are independent, then $\rho = 0$. The converse is not true.

(c) If $Y = aX + b$, then

$$\rho_{X,Y} = \begin{cases} 1, & \text{if } a > 0, \\ -1, & \text{if } a < 0. \end{cases}$$

Note that $Cov(X, X) = Var(X)$.

(d) If $U = a_1X + b_1$ and $V = a_2Y + b_2$, then

(i) $Cov(U, V) = a_1a_2Cov(X, Y)$, and

(ii) $\rho_{U,V} = \begin{cases} \rho_{X,Y}, & \text{if } a_1a_2 > 0 \\ -\rho_{X,Y}, & \text{otherwise.} \end{cases}$

(e) $Var(aX + bY) = a^2Var(X) + b^2Var(Y) + 2abCov(X, Y)$. In particular, if X and Y are independent, then $Var(aX + bY) = a^2Var(X) + b^2Var(Y)$.

(f) If $X_1, \ldots, X_n$ are independent, then $Var(\sum_{i=1}^{n} X_i) = \sum_{i=1}^{n} Var(X_i)$.

EXAMPLE 3.3.7

The joint probability density function of the random variables X and Y is given by

$$f(x, y) = \begin{cases} \frac{1}{64}e^{-y/8}, & 0 \le x \le y \le \infty \\ 0, & \text{otherwise.} \end{cases}$$

Find the covariance of X and Y.

Solution

We can use the formula, $\text{Cov}(X, Y) = E(XY) - E(X)E(Y)$. *Now, using integration by parts (three times), we will obtain*

$$E(XY) = \int_0^\infty \int_0^y (xy)\frac{1}{64}e^{-y/8}dxdy$$
$$= \frac{1}{64}\int_0^\infty ye^{-y/8}\left(\int_0^y xdx\right)dy$$
$$= \frac{1}{128}\int_0^\infty y^3 e^{-y/8}dy = 192.$$

We can also obtain

$$E(X) = \int_0^\infty \int_0^y x\frac{1}{64}e^{-y/8}dxdy = 8,$$

and

$$E(Y) = \int_0^\infty \int_0^y y\frac{1}{64}e^{-y/8}dxdy = 16.$$

Thus $\text{Cov}(X, Y) = 192 - (8)(16) = 64$.

Next, we will define the moment-generating function for the bivariate probability distributions.

Definition 3.3.8. *Let X and Y be jointly distributed. Then the joint* **moment-generating function** *is defined by*

$$M_{(X,Y)}(t_1, t_2) = E\left(e^{t_1 X + t_2 Y}\right)$$
$$= \begin{cases} \sum_y \sum_x e^{t_1 x + t_2 y} f(x, y), & \text{if } X \text{ and } Y \text{ are discrete} \\ \int_{-\infty}^{\infty}\int_{-\infty}^{\infty} e^{t_1 x + t_2 y} f(x, y)dxdy, & \text{if } X \text{ and } Y \text{ are continuous.} \end{cases}$$

EXERCISES 3.3

3.3.1 An experiment consists of drawing four objects from a container without replacement, which holds eight operable, six defective, and 10 semi-operable objects. Let X be the number of operable objects drawn and Y the number of defective objects drawn.

(a) Find the joint probability function of the bivariate random variable (X, Y).

(b) Find $P(X = 3, Y = 0)$.

(c) Find $P(X < 3, Y = 1)$.

(d) Give a graphical presentation of (a), (b), and (c).

3.3.2 Let

$$f(x, y) = \begin{cases} \frac{1}{50}(x^2 + 2y), & x = 0, 1, 2, 3, \text{ and } y = x + 3, \\ 0, & \text{otherwise.} \end{cases}$$

Show that $f(x, y)$ satisfies the conditions of a probability mass function.

3.3.3 **Let**

$$f(x, y) = c(1 - x)(1 - y), \quad -1 \le x \le 1, \quad -1 \le y \le 1.$$

Find the value of c that makes $f(x, y)$ the joint probability density function of the random variable (X,Y).

3.3.4 **Let**

$$f(x, y) = xe^{-xy}, \quad x \ge 0, \quad y \ge 1.$$

Is $f(x, y)$ a joint probability density function? If not, find the proper constant to multiply with $f(x, y)$ so that it will be a probability density.

3.3.5 Find the marginal probability mass functions of the random variables X and Y, if their joint probability mass function is given in Table 3.2.

TABLE 3.2 Joint pmf of X and Y.

		Y		
X	–2	0	1	4
–1	0.3	0.1	0.0	0.2
3	0.0	0.2	0.1	0.0
5	0.1	0.0	0.0	0.0

3.3.6 Find the marginal density functions of the random variables X and Y if their joint probability density function is given by

$$f(x, y) = \begin{cases} \frac{1}{5}(3x - y), & 1 \le x \le 2, \ 1 \le y \le 3, \\ 0, & \text{otherwise.} \end{cases}$$

3.3.7 Determine the conditional probability $P(X = -1|Y = 0)$ for the random variables defined in problem 3.3.5.

3.3.8 Find k so that $f(x, y) = kxy$, $1 \le x \le y \le 2$ will be a probability density function. Also find (i) $P\left(X \le \frac{3}{2}, Y \le \frac{3}{2}\right)$, and (ii) $P\left(X + Y \le \frac{3}{2}\right)$.

3.3.9 The random variables X and Y have a joint probability density function

$$f(x, y) = \begin{cases} \frac{8}{9}xy, & 1 \le x \le y \le 2, \\ 0, & \text{elsewhere.} \end{cases}$$

Find:

(a) The marginal of **X**.

(b) $P(1.5 < X < 1.75, Y > 1)$.

3.3.10 The joint pdf of X and Y is

$$f(x, y) = \begin{cases} \frac{1}{28}(4x + 2y + 1), & 0 \le x \le 2, \ 0 \le y \le 2 \\ 0, & \text{elsewhere.} \end{cases}$$

Find (a) $f_X(x)$ and $f_Y(y)$, and (b) $f(y|x)$.

3.3.11 Find the joint mgf of the random variables (X, Y) defined in problem 3.3.9.

3.3.12 The joint density of a random variable (X, Y) is given by

$$f(x, y) = \begin{cases} \frac{x^3y^3}{16}, & 0 \le x \le 2, \ 0 \le y \le 2 \\ 0, & \text{elsewhere.} \end{cases}$$

(a) Find marginals of X and Y, and (b) find f(y|x).

3.3.13 The joint probability mass function of a discrete random variable (X, Y) is given by

$$f(x, y) = \begin{cases} \left[\frac{6xy}{n(n+1)(2n+1)}\right]^2, & x, y = 1, 2, \ldots, n, \\ 0, & \text{otherwise.} \end{cases}$$

Find (a) f(x|y), and (b) f(y|x).[Hint:$\sum_{i=1}^{n} i^2 = (n(n+1)(2n+1))/6$].

3.3.14 Consider bivariate random variables with the pmf

$f(x, y) = \binom{n}{x} y^{x+\alpha-1}(1-y)^{n-x+\beta-1}$, for $x = 0, 1, \ldots, n$, and $0 < y \le 1$. Verify that

$$f(x|y) \propto \binom{n}{x} y^x (1-y)^{n-x},$$

and

$$f(y|x) \propto y^{x+\alpha-1}(1-y)^{n-x+\beta-1}.$$

3.3.15 The joint mass function of the discrete random variable (X,Y) is given by Table 3.3.

TABLE 3.3 Joint density of X and Y.

	Y		
X	1	2	3
1	$\frac{1}{6}$	$\frac{1}{6}$	$\frac{1}{6}$
2	$\frac{1}{6}$	$\frac{1}{12}$	$\frac{1}{12}$
3	$\frac{1}{12}$	$\frac{1}{12}$	0

(a) Find $E(XY)$.
(b) Find $Cov(X, Y)$.
(c) Find the correlation coefficient ρ_{XY}.

3.3.16 The joint probability function of the continuous random variable (X, Y) is given by

$$f(x, y) = \begin{cases} \frac{1}{28}(4x + 2y + 1), & 0 \le x < 2, \quad 0 \le y < 2, \\ 0, & \text{otherwise.} \end{cases}$$

(a) Find $E(XY)$.
(b) Find $Cov(X, Y)$.
(c) Find the correlation coefficient ρ_{XY}.

3.3.17 Let X and Y be random variables and $U = aX + b$, $V = cY + d$, where a, b, c, d are constants. Show that

$$\rho_{U,V} = \begin{cases} \rho_{X,Y}, & \text{if } ac > 0 \\ -\rho_{X,Y}, & \text{otherwise.} \end{cases}$$

3.3.18 Let X and Y be random variables, and let $Y = aX + b$, where a and b are constants. Show that (a) $\rho_{XY} = 1$, if a > 0, and (b) $\rho_{XY} = -1$, if a < 0.

3.3.19 If $\left|\rho_{XY}\right| = 1$, then prove that $P(Y = aX + b) = 1$.

3.3.20 Let X and Y be two random variables with a joint probability density function

$$f(x, y) = \begin{cases} 8xy, & 0 \le x \le y \le 1 \\ 0, & \text{otherwise.} \end{cases}$$

(a) Find the conditional expectation, $E(X|Y = \frac{3}{4})$.
(b) Find Cov(X, Y).

3.3.21 Let X and Y be two random variables with a joint density function

$$f(x, y) = \begin{cases} e^{-y}, & 0 \le x \le y \\ 0, & \text{otherwise.} \end{cases}$$

(a) Find the conditional expectation, $E(X|Y = y)$.
(b) Find $Cov(X, Y)$.
(c) Are X and Y independent? Why?

3.3.22 Let

$$f(x, y) = \frac{c}{(1 + x^2)\sqrt{1 - y^2}}, \qquad -\infty < x < \infty, \qquad -1 < y < 1.$$

Find the c that makes $f(x, y)$ the probability density function of the random variable (X, Y). Determine whether X and Y are independent.

3.3.23 If the random variables X and Y are independent and have equal variances, what is the coefficient of correlation between the random variables X and $aX + Y$, where a is constant?

3.3.24 Suppose you recorded, for four sample plots, the number of mango trees planted (X) and weekly harvest (pounds) from each plot (Y). The data is given as

X	3	5	7	9
Y	5	10	12	24

(a) Calculate (i) sample mean, sample variance, and sample covariance.
(b) Calculate sample correlation.
(c) Let B denote the event each mango is ripe on a given day. Assume that $B \sim$ *Bernoulli* (p) and that each mango ripens independently of the others with the same probability. Find the MGF of B. Also, using MGF with $p =$ 0.4, calculate the mean and variance.
(d) For this data, write and run R-code to calculate mean, variance, covariance, correlation, and MGF.

3.4 FUNCTIONS OF RANDOM VARIABLES

In this section, we will discuss the methods of finding the probability distribution of a function of a random variable X. We are given the distribution of X, and we are required to find the distribution of $g(X)$. There are many physical problems that call for the derivation of the distribution of a function of a random variable. The following is one of the classical examples. The velocity V of a gas molecule (Maxwell–Boltzmann law) behaves as a gamma-distributed random variable. We would like to derive the distribution of $E = mV^2$, the kinetic energy of the gas molecule. Because the value of the velocity is the outcome of a random experiment, so is the value of E. This is a problem of finding the distribution of a function of a random variable $E = g(V)$. We now illustrate various methods for finding the distribution of $g(X)$ by means of examples.

3.4.1 Method of Distribution Functions

Basically, the *method of distribution functions* is as follows. If X is a random variable with pdf $f_X(x)$ and if Y is some function of X, then we can find the cdf $F_Y(y) = P(Y \le y)$ directly by integrating $f_X(x)$ over the region for which $\{Y \le y\}$. Now, by differentiating $F_Y(y)$, we get the probability density function $f_Y(y)$ of Y. In general, if Y is a function of random variables $X_1, \ldots, X_n$, say $g(X_1, \ldots, X_n)$, then we can summarize the method of distribution function as follows:

Procedure to find cdf of a function of r.v. using the method of distribution functions

1. Find the region $\{Y \le y\}$ in the $(x_1, \ldots, x_n)$ space, that is find the set of $(x_1, \ldots, x_n)$ for which $g(x_1, \ldots, x_n) \le y$.
2. Find $F_Y(y) = P(Y \le y)$ by integrating $f(x_1, x_2, \ldots, x_n)$ over the region $\{Y \le y\}$.
3. Find the density function $f_Y(y)$ by differentiating $F_Y(y)$.

EXAMPLE 3.4.1

Let $X \sim N(0,1)$. Using the cdf of X, find the pdf of X^2.

Solution

Let $Y = X_2$. Note that the marginal pdf of X is

$$f_X(x) = \frac{1}{\sqrt{2\pi}} e^{-x^2/2}, \quad -\infty < x < \infty.$$

Then the cumulative distribution function of Y for a given $y \ge 0$ is

$$\begin{aligned} F(y) &= P(Y \le y) = P(X^2 \le y) \\ &= P(-\sqrt{y} \le X \le \sqrt{y}) \\ &= \int_{-\sqrt{y}}^{\sqrt{y}} \frac{1}{\sqrt{2\pi}} e^{-x^2/2} dx \\ &= 2\int_0^{\sqrt{y}} \frac{1}{\sqrt{2\pi}} e^{-x^2/2} dx, \quad \text{(by the symmetry of } e^{-x^2/2}). \end{aligned}$$

Hence, by differentiating $F(y)$ with respect to y, we obtain the marginal probability density function of Y, that is,

$$\begin{aligned} f_Y(y) &= \frac{2}{\sqrt{2\pi}} e^{-y/2} \frac{1}{2\sqrt{y}} \\ &= \begin{cases} \frac{1}{\sqrt{2\pi}} y^{-1/2} e^{-y/2}, & 0 < y < \infty \\ 0, & \text{otherwise.} \end{cases} \end{aligned}$$

This is a χ^2-distribution with 1 degree of freedom.

The same method can be used for discrete cases.

EXAMPLE 3.4.2

Suppose that the random variable X has a Poisson probability distribution, that is,

$$f(x) = \begin{cases} \frac{e^{-\lambda} \lambda^x}{x!}, & x = 0, 1, 2, \ldots \\ 0, & \text{otherwise.} \end{cases}$$

Find the cumulative distribution function of $Y = aX + b$.

Solution

The cdf of Y is given by

$$\begin{aligned} F(y) &= P(Y \le y) = P(aX + b \le y) \\ &= P\left(X \le \frac{y-b}{a}\right) = \sum_{x=0}^{\left[\frac{y-b}{a}\right]} \frac{e^{-\lambda}\lambda^x}{x!}, \end{aligned}$$

where $[x]$ is the largest integer less than or equal to x. Therefore

$$F(y) = \begin{cases} 0, & y < b \\ \sum_{x=0}^{\left[\frac{y-b}{a}\right]} \frac{e^{-\lambda}\lambda^x}{x!}, & y \geq b. \end{cases}$$

It should be noted here that the pmf, $f_Y(y)$ of Y, can be found from the equation

$$f_Y(y) = F_Y(y) - F_Y(y-1), \quad \text{for } y = an + b, \quad n = 0, 1, 2, \ldots.$$

The multivariate case (in particular, the bivariate case), though it is more difficult, can be handled similarly.

3.4.2 The pdf of $Y = g(X)$, where g is Differentiable and Monotone Increasing or Decreasing

We now consider the distribution of a random variable $Y = g(X)$, where X is a continuous random variable with pdf $f_X(x)$. Assume that g is differentiable and the inverse function g^{-1} of g exists. Let $X = g^{-1}(Y)$. Let $f_X(x)$ be the probability density function of X. Then the density function of Y can be obtained using the method we have discussed. Thus

$f_Y(y) = f_X(g^{-1}(y)) \cdot \frac{d}{dy} g^{-1}(y)$.

This is a special case of the transformation method, which is explained later in Subsection 3.4.4.

Example 3.4.3

Let $X \sim N(0, 1)$. Find the pdf of $Y = e^X$.

Solution

Here, $g(x) = e^x$, and hence, $g^{-1}(y) = \ln(y)$. Thus $\frac{d}{dy} g^{-1}(y) = \frac{1}{y}$.
Also,

$$f_X(x) = \frac{1}{\sqrt{2\pi}} e^{-x^2/2}, \quad -\infty < x < \infty.$$

Therefore the pdf of Y is

$$f_Y(y) = \begin{cases} \frac{1}{y\sqrt{2\pi}} e^{-[\ln(y)]^2/2}, & y > 0, \\ 0, & \text{otherwise.} \end{cases}$$

3.4.3 Probability Integral Transformation

Let X be a continuous random variable, with pdf $f_X(x)$ and cdf $F(x)$. Let $Y=F(X)$. Then,

$$P(Y \leq y) = P(F(X) \leq y) = P(X \leq F^{-1}(y))$$
$$= \int_{-\infty}^{F-1(y)} f_x(x)dx = F_x(x)|_{-\infty}^{F^{-1}(y)} = y.$$

Hence,

$$f(y) = \begin{cases} 1, & 0 < y < 1 \\ 0, & \text{otherwise.} \end{cases}$$

Thus Y has a $U(0,1)$ distribution. The transformation $Y=F(X)$ is called a *probability integral transformation*. It is interesting to note that irrespective of the pdf of X, Y is always uniform in (0, 1).

EXAMPLE 3.4.4

Let X be a normal random variable with mean μ and variance σ^2. Thus

$$f(x) = \frac{1}{\sqrt{2\pi}\sigma} e^{-(x-\mu)/2\sigma^2}, \quad -\infty < x < \infty, \quad -\infty < \mu < \infty, \quad \text{and } \sigma^2 > 0.$$

Let $Y = \int_0^X \frac{1}{\sqrt{2\pi}\sigma} e^{-(u-\mu)/2\sigma^2} du$. Then $Y = F(X)$, where F is the cdf of a standard normal random variable. Therefore Y is uniform on (0, 1). That is,

$$f(y) = \begin{cases} 1, & \text{if } 0 < y < 1 \\ 0, & \text{otherwise.} \end{cases}$$

3.4.4 Functions of Several Random Variables: Method of Distribution Functions

We now discuss the distribution of Y, when Y is a function of several random variables, $Y = g(X_1,\ldots,Xn)$.

EXAMPLE 3.4.5

Let $X_1, \ldots, X_n$ be continuous iid random variables with pdf $f(x)$ (cdf $F(x)$). Find the pdfs of

$$Y_1 = \min(X_1, \ldots, X_n) \quad \text{and} \quad Y_n = \max(X_1, \ldots, X_n).$$

Solution

For the random variable Y_1, we have

$$\begin{aligned} 1 - F_{Y_1}(y) &= P(Y_1 > y) \\ &= P(X_1 > y, X_2 > y, \ldots, X_n > y) \\ &= P(X_1 > y)P(X_2 > y) \ldots P(X_n > y) \text{ (because of independence)} \\ &= (1 - F(y))^n. \end{aligned}$$

This implies that

$$F_{Y_1}(y) = 1 - (1 - F(y))^n$$

and

$$f_{Y_1}(y) = n(1 - F(y))^{n-1} f(y).$$

Consider Y_n. Its cdf is given by

$$F_{Y_1}(y) = P(Y_n \le y) = (F(y))^n.$$

This implies that

$$f_{Y_n}(y) = n(F(y))^{n-1} f(y).$$

3.4.5 Transformation Method

A simple generalization of the method of distribution functions to functions of more than one variable is the *transformation method*. We illustrate the method for bivariate distributions. The method is similar for the multivariate case. Let the joint pdf of (X,Y) be $f(x,y)$. Let $U = g_1(X, Y)$; $V = g_2(X, Y)$. The mapping from (X, Y) to (U, V) is assumed to be one-to-one and onto. Hence, there are functions h_1 and h_2 such that

$$x = h_1^{-1}(u, v),$$

and

$$y = h_2^{-1}(u, v).$$

Define the Jacobian of the transformation J by

$$J = \begin{vmatrix} \frac{\partial x}{\partial u} & \frac{\partial x}{\partial v} \\ \frac{\partial y}{\partial u} & \frac{\partial y}{\partial v} \end{vmatrix}.$$

Then the joint pdf of U and V is given by

$$f_{(U,V)}(u, v) = f(h_1^{-1}(u, v), h_2^{-1}(u, v))|J|.$$

EXAMPLE 3.4.6

Let X and Y be independent random variables with common pdf $f(x) = e^{-x}$, $(x > 0)$. Find the joint pdf of $U = X/(X + Y)$, $V = X + Y$.

Solution

We have $U = X/(X + Y) = X/V$. Hence, $X = UV$ and $Y = V - X = V - UV = V(1 - U)$. *Thus the Jacobian is given by*

$$J = \begin{vmatrix} v & u \\ -v & 1-u \end{vmatrix}.$$

Then $|J| = v(1 - u) + uv = v(>0)$. *Note that* $0 \le u \le 1$, $0 < v < \infty$, *and*

$$\begin{aligned} f(u, v) &= f(h_1^{-1}(u, v), h_2^{-1}(u, v))|J| \\ &= e^{-uv}e^{-v(1-u)}v \\ &= ve^{-v}, \quad 0 \le u \le 1, \ 0 < v < \infty, \end{aligned}$$

The joint pdf of UV.

Suppose we want the marginal pdfs $f_V(v)$ and $f_U(v)$, that is,

$$F_V(v) = \int_0^1 ve^{-v}du = ve^{-v}, \quad 0 < v < \infty$$

and

$$f_U(u) = \int_0^\infty ve^{-v}dv = 1, \quad 0 \le u \le 1.$$

Sometimes the expressions for two variables, U and V, may not be given. Only one expression is available. In that case, call the given expression of X and Y as U, and define $V = Y$. Then, we can use the previous method to first find the joint density and proceed to find the marginal to obtain the pdf of U. The following example demonstrates this method:

EXAMPLE 3.4.7

Let X and Y be independent random variables uniformly distributed on [0, 1], Find the pdf of $X + Y$.

Solution

Let

$$\begin{aligned} U &= X + Y, \\ V &= Y, \\ f(x, y) = 1, &\quad 0 \le x \le 1, \ 0 \le y \le 1, \\ X &= U - V, \\ Y &= V, \end{aligned}$$

and

$$J = \begin{vmatrix} 1 & -1 \\ 0 & 1 \end{vmatrix} = 1.$$

Thus we have the joint pdf (U, V),

$$f(u, v) = \begin{cases} 1, & 0 \le u - v \le 1, \\ 0, & \text{otherwise.} \end{cases} \quad 0 \le v \le 1,$$

Note that because V *is the variable we introduced; to get the pdf of* U*, we just need to find the marginal pdf from the joint pdf.* *From* Fig. 3.10, *the regions of integration are* $0 \le u \le 1$, and $0 \le u \le 2$. *That is,* Fig. 3.11

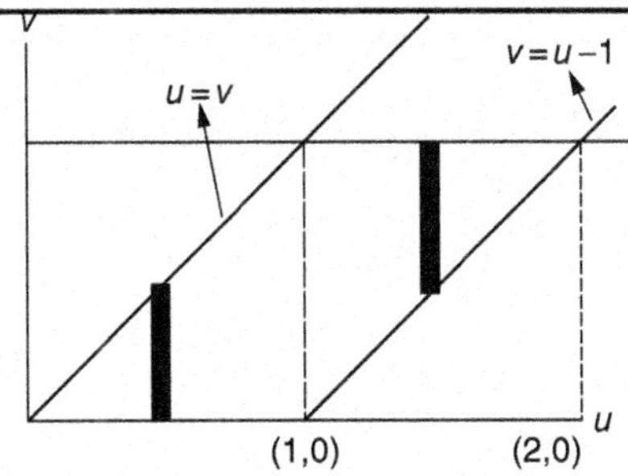

FIGURE 3.10 The regions of integration.

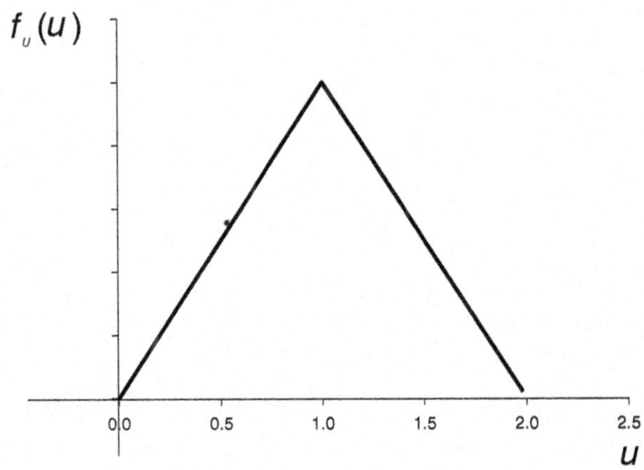

FIGURE 3.11 Graph of $f_U(u)$.

$$f_U(u) = \int f(u, v)dv = \int 1dv$$
$$= \begin{cases} \int_0^u dv = u, & 0 \le u \le 1 \\ \int_{u-1}^1 dv = 2 - u, & 0 \le u \le 2 \end{cases}$$

EXERCISES 3.4

3.4.1. Let X be a uniformly distributed random variable over $(0, a)$, $a > 0$. Find the pdf of $Y = cX + d$ *for a constant* $c > 0$.

3.4.2. The joint pdf of (X, Y) is

$$f(x, y) = \frac{1}{\theta^2}e^{-\frac{x+y}{\theta}}, \quad x, y > 0, \ \theta > 0.$$

Find the pdf of $U = X - Y$.

3.4.3. Let $f(x,y)$ be the probability density function of the continuous random variable (X, Y). If $U = XY$, show that the probability density function of U is given by

$$f_U(u) = \int_{-\infty}^{\infty} f\left(\frac{u}{v}, v\right)\left|\frac{1}{v}\right|dv.$$

3.4.4. The joint pdf of X and Y is

$$f(x, y) = \theta e^{-(x+\theta y)}, \quad \theta > 0, \ x > 0.$$

Find the pdf of XY.

3.4.5. If the joint pdf of (X, Y) is

$$f(x, y) = \frac{1}{2\pi\sigma_1\sigma_2} e^{-\frac{1}{4\sigma_1^2\sigma_2^2}(x^2+y^2)}, \quad -\infty < x < \infty, -\infty < y < \infty; \sigma_1, \ \sigma_2 > 0 \text{ find the pdf of } X^2 + Y^2.$$

3.4.6. Let $X_1, \ldots, X_n$ be independent and identically distributed random variables with pdf $f(x) = (1/\theta)e^{-x/\theta}$, $x > 0, \theta > 0$. Find the pdf of $\Sigma_{i=1}^{n} X_i$.

3.4.7. Let $f(x,y)$ be the pdf of the continuous random variable (X, Y). If $U = X + Y$, then show that the probability density function of U is given by

$$f_U(u) = \int_{-\infty}^{\infty} f(u - v, v)dv.$$

3.4.8. Let X be uniformly distributed over $(-2, 2)$ and $Y = X^2$. Find the $\text{Cov}(X, Y)$. Are X and Y independent?

3.4.9. Let $X \sim N(\mu,\sigma^2)$. Show that

(a) $Z = \frac{(X-\mu)}{\,}$ is $N(0, 1)$.
(b) $U = \frac{(X \not\subset \mu)^2}{\sigma^2}$ is $\chi^2(1)$.

3.4.10. Let $X \sim N(\mu,\sigma^2)$. Find the pdf of $Y = e^X$.

3.4.11. The probability density of the velocity, V, of a gas molecule, according to the Maxwell–Boltzmann law, is given by

$$f(v, \beta) = \begin{cases} cv^2 e^{-\beta v^2}, & v > 0, \\ 0, & \text{elsewhere}, \end{cases}$$

where c is an appropriate constant and β depends on the mass of the molecules and the absolute temperature. Find the density function of the kinetic energy E, which is given by $E = g(V) = \frac{1}{2}mV^2$.

3.4.12. Let X and Y be two independent random variables, each normally distributed, with parameters (μ_1, σ_1^2), and (μ_2, σ_2^2), respectively. Show that the probability density function of $U = X/Y$ is given by

$$f_U(u) = \frac{\sigma_1\sigma_2}{\pi(\sigma_1^2 + \sigma_2^2 u^2)}, \quad -\infty < u < \infty.$$

3.4.13. Let

$$f(x, y) = \frac{1}{2\pi\sigma^2} e^{-(1/2\sigma^2)(x^2+y^2)}, \quad -\infty < x, y < \infty$$

be the joint pdf of (X, Y). Let

$$U = \sqrt{X^2 + Y^2} \quad \text{and} \quad V = \tan^{-1}\left(\frac{Y}{X}\right), \quad 0 \le V \le 2\pi.$$

Find the joint pdf of (U, V).

3.4.14. Let the joint pdf of (X, Y) be given by

$$f(x, y) = \begin{cases} \beta^{-2} e^{-\{(x+y)/\beta\}}, & x, y > 0, \ \beta > 0, \\ 0, & \text{elsewhere}. \end{cases}$$

Let $U = \frac{X-Y}{2}$ and $V = Y$. Find the joint pdf of (U, V).

3.4.15. Let X and Y be independent and identically distributed random variables with pdf

$$f(x) = \begin{cases} \frac{1}{2} e^{-x/2}, & x \ge 0, \\ 0, & \text{otherwise}. \end{cases}$$

Find the distribution of $(X - Y)/2$.

3.4.16. If X and Y are independent and chi-square distributed random variables with n_1 and n_2 degrees of freedom, respectively, obtain the joint distribution of (U, V), where $U = X + Y$ and $V = X/Y$.

3.4.17. Suppose the proportion of ripe mangos on a tree at inspection, U, is uniform on [0, 1]. A quality-control metric converts proportion to a waiting time until first defect by a transformation $T = -\ln(U)$. Find the distribution of T and compute $P(T > 2)$ and $E[T]$.

3.5 LIMIT THEOREMS

Limit theorems play a very important role in the study of probability theory and in its applications. In Chapter 2, we discussed that the frequency interpretation of probability depends on the long-run proportion of times the outcome (event) would occur in repeated experiments. Also, in Section 3.2, we learned that some binomial probabilities can be computed using either the Poisson probability distribution or the normal probability distribution using the limiting arguments. Many random variables that we encounter in nature have distributions close to the normal probability distribution. These modeling simplifications are possible because of various limit theorems. In this section, we discuss the law of large numbers and the Central Limit Theorem.

First, we introduce Chebyshev's theorem, which is a useful result for proving the limit theorems. It gives a lower bound for the area under a curve between two points that are on opposite sides of the mean and are equidistant from the mean. The strength of this result lies in the fact that we need not know the probability distribution of the underlying population, other than its mean and variance. This result was developed by the Russian mathematician Pafnuty Chebyshev (1821–1894).

Chebyshev's theorem

Theorem 3.5.1. *Let the random variable X have a mean μ and standard deviation σ. Then, for $K > 0$, a constant,*

$$P(|X - \mu| < K\sigma) \geq 1 - \frac{1}{K^2}.$$

Proof. We will work with the continuous case. By definition of the variance of X, we have

$$\begin{aligned}\sigma^2 &= E(X-\mu)^2 = \int_{-\infty}^{\infty}(x-\mu)^2 f(x)dx \\ &= \int_{-\infty}^{\mu-K\sigma}(x-\mu)^2 f(x)dx + \int_{\mu-K\sigma}^{\mu+K\sigma}(x-\mu)^2 f(x)dx + \int_{\mu+K\sigma}^{\infty}(x-\mu)^2 f(x)dx \\ &\geq \int_{-\infty}^{\mu-K\sigma}(x-\mu)^2 f(x)dx + \int_{\mu+K\sigma}^{\infty}(x-\mu)^2 f(x)dx.\end{aligned}$$

Note that $(x-\mu)^2 \geq K^2\sigma^2$ for $x \leq \mu - K\sigma$ or $x \geq \mu + K\sigma$. The equation above can be rewritten as

$$\begin{aligned}\sigma^2 &\geq K^2\sigma^2\left[\int_{-\infty}^{\mu-K\sigma} f(x)dx + \int_{\mu+K\sigma}^{\infty} f(x)dx\right] \\ &= K^2\sigma^2\left[P\{X \leq \mu - K\sigma\} + P\{X \geq \mu + K\sigma\}\right] \\ &= K^2\sigma^2 P\{|X-\mu| \geq K\sigma\}.\end{aligned}$$

This implies that

$$P\{|X-\mu| \geq K\sigma\} \leq \frac{1}{K^2}$$

or

$$P(|X-\mu| < K\sigma) \geq 1 - \frac{1}{K^2}.$$

We can also write Chebyshev's theorem as

$$P\{|X-\mu| \geq \varepsilon\} \leq \frac{E[(X-\mu)^2]}{\varepsilon^2} = \frac{Var(X)}{\varepsilon^2}, \quad \text{for some } \varepsilon > 0.$$

Equivalently,

$$P\{|X - \mu| \geq K\sigma\} \leq \frac{1}{K^2}.$$

In other words, Chebyshev's inequality states that the probability that a random variable X differs from its mean by at least K standard deviations is less than or equal to $1/K^2$, ($K \geq 2$, for $K = 1$, the result is obvious.)

In statistics, if we do not have any idea of the population probability distribution, Chebyshev's theorem is used in the following manner. For any data set (regardless of the shape of the distribution), at least $(1 - (1/k^2))100\%$ of observations will lie within $k(\geq 1)$ standard deviations of the mean. For example, at least $(1 - (1/2^2))$ $100\% = 75\%$ of the data will fall in the interval $(\bar{x} - 2s,\ \bar{x} + 2s)$ and at least 88.9% of the observations will lie within three standard deviations of the mean. If the population distribution is bell-shaped, we have a better result than Chebyshev's theorem, namely, the empirical rule that states the following: (i) approximately 68% of the observations lie within one standard deviation of the mean, (ii) approximately 95% of the observations lie within two standard deviations of the mean, and (iii) approximately 99.7% of the observations lie within three standard deviations of the mean.

EXAMPLE 3.5.1

A random variable X has a mean of 24 and a variance of 9. Obtain a bound on the probability that the random variable X assumes values between 16.5 to 31.5.

Solution

From Chebyshev's theorem,

$$P\{\mu - K\sigma < X < \mu + K\sigma\} \geq 1 - \frac{1}{K^2}.$$

Equating $\mu + K\sigma$ to 31.5 and $\mu - K\sigma$ to 16.5 with $\mu = 24$ and $\sigma = \sqrt{9} = 3$, we obtain $K = 2.5$. Hence,

$$P\{16.5 < X < 31.5\} \geq 1 - \frac{1}{(2.5)^2} = 0.84.$$

EXAMPLE 3.5.2

Let X be a random variable that represents the systolic blood pressure of the population of 18- to 74-year-old men in the United States. Suppose that X has mean 129 mm Hg and standard deviation of 19.8 mm Hg.

(a) Obtain a bound on the probability that the systolic blood pressure of this population will assume values between 89.4 and 168.6 mm Hg.

(b) In addition, assume that the distribution of X is approximately normal. Using the normal table, find $P(89.4 \leq X \leq 168.6)$. *Compare this with the empirical rule.*

Solution

(a) *Because we are given only the mean and standard deviation, and no probability distribution is specified, we can use Chebyshev's theorem. We have*

$$P\{\mu - K\sigma < X < \mu + K\sigma\} \geq 1 - \frac{1}{K^2}.$$

Equating $\mu + K\sigma$ to 168.6 and $\mu - K\sigma$ to 89.4 with $\mu = 129$ and $\sigma = 19.8$, we obtain $K = 2$. Hence,

$$P\{89.4 \leq X \leq 168.6\} \geq 1 - \frac{1}{(2)^2} = 0.75.$$

(b) *Because X is normally distributed with a mean of 129 and a standard deviation of 19.8, using the z-score, we get*

$$P(89.4 \leq X \leq 168.6) = P\left(\frac{89.4 - 129}{19.8} \leq Z \leq \frac{168.6 - 129}{19.8}\right)$$
$$= P(-2 \leq Z \leq 2) = 0.9544.$$

Hence, approximately 95.44% of this population will have systolic blood pressure values between 89.4 and 168.6 mm Hg. This compares well with the 95% value from the empirical rule.

We could use Chebyshev's inequality to prove the following result, which is called the weak law of large numbers, in which the convergence is in probability. A stronger version of this result is called the strong law of large numbers, in which convergence is with probability one. Because in this book, we do not introduce various modes of convergence, we will not discuss the strong law of large numbers. The law of large numbers states that if the sample size n is large, the sample mean rarely deviates from the mean of the distribution of X, which in statistics is called the population mean.

Law of large numbers

Theorem 3.5.2. *Let $X_1,\ldots,X_n$ be a set of pairwise independent random variables with $E(X_i) = \mu$, and Var $(X_i) = \sigma^2$. Then for any $c > 0$,*

$$P\{\mu - c \le \overline{X} \le \mu + c\} \ge 1 - \frac{\sigma^2}{nc^2}$$

and as $n \to \infty$, the probability approaches 1. Equivalently,

$$P\left(\left|\frac{S_n}{n} - \mu\right| < \varepsilon\right) \to 1$$

as $n \to \infty$. Here, $\overline{X} = \frac{1}{n}\sum_{i=1}^{n} X_i$ and $S_n = \sum_{i=1}^{n} X_i$.

Proof. Because $X_1, \ldots, X_n$ are independent and identically distributed (iid) random variables (random sample), we know that $Var(Sn) = n\sigma^2$, and $Var(S_n/n) = \sigma^2/n$. *Also,* $E(Sn/n) = \mu$. By Chebyshev's theorem, for any $\varepsilon > 0$,

$$P\left(\left|\frac{Sn}{n} - \mu\right| \ge \varepsilon\right) \le \frac{\sigma^2}{n\varepsilon^2}.$$

Thus, for any fixed ε,

$$P\left(\left|\frac{Sn}{n} - \mu\right| \ge \varepsilon\right) \to 0$$

as $n \to \infty$. Equivalently,

$$P\left(\left|\frac{Sn}{n} - \mu\right| < \varepsilon\right) \to 1$$

as $n \to \infty$.

Thus, without any knowledge of the probability distribution function of S_n, the (weak) law of large numbers states that the sample mean, $\overline{X} = S_n/n$, will differ from the population mean by less than an arbitrary constant, $\varepsilon > 0$, with probability that tends to 1 as n tends to ∞. Because of this, the law of large numbers is also called the "law of averages." This result basically states that we can start with a random experiment whose outcome cannot be predicted with certainty, and by taking averages, we can obtain an experiment in which the outcome can be predicted with a high degree of accuracy. The law of large numbers in its simplest form for the Bernoulli random variables was introduced by Jacob Bernoulli toward the end of the 16th century. This result in generality was first proved by the Russian mathematician A. Khintchine in 1929. The result is widely used in its applications to insurance, statistics, and the study of heredity.

EXAMPLE 3.5.3

Let $X_1, \ldots, X_n$ be iid Bernoulli random variables with parameter p. Verify the weak law of large numbers.

Solution

For Bernoulli random variables, we know that $E(Xi) = p$, and $Var(Xi) = p(1 - p)$. Thus, by Chebyshev's theorem,

$$P\{p - c \le \overline{X} \le p + c\} = P\left\{\left|\frac{Sn}{n} - p\right| \le c\right\} \ge 1 - \frac{\sigma^2}{nc^2}$$

$$= 1 - \frac{p(1-p)}{nc^2} \to 1, \quad \text{as } n \to \infty.$$

This verifies the weak law of large numbers.

Example 3.5.4

Consider n rolls of a balanced die. Let X_i be the outcome of the ith roll, and let $S_n = \sum_{i=1}^{n} X_i$. Show that, for any $\varepsilon > 0$,

$$P\left(\left|\frac{Sn}{n} - \frac{7}{2}\right| \geq \varepsilon\right) \to 0$$

as $n \to \infty$.

Solution

Because the die is balanced, $E(Xi) = 7/2$. *By the law of large numbers, for any* $\varepsilon > 0$,

$$P\left(\left|\frac{Sn}{n} - \frac{7}{2}\right| \geq \varepsilon\right) \to 0$$

as $n \to \infty$, *or equivalently,*

$$P\left(\left|\frac{Sn}{n} - \frac{7}{2}\right| < \varepsilon\right) \to 1$$

as $n \to \infty$.

One of the most important results in probability theory is the **Central Limit Theorem.** This theorem basically states that the z-transform of the sample mean is asymptotically standard normal. The amazing thing about the Central Limit Theorem is that no matter what the shape of the original distribution is, the (sampling) distribution of the mean approaches a normal probability distribution. We state one version of the Central Limit Theorem. In a restricted case, the proof uses the idea that the moment-generating functions of Z_n converge to the moment-generating function of the standard normal random variable. The general proof is a little bit more involved. Because proof of the Central Limit Theorem is available in most probability books, we will not give the proof here.

Central limit theorem (clt)

Theorem 3.5.3. *If* $X_1,..., X_n$ *is a random sample from an infinite population with mean* $\mu < \infty$ *and variance* $\sigma^2 < \infty$, *then the limiting distribution of* $Z_n = (\overline{X} - \mu)/(\sigma/\sqrt{n})$ *as* $n \to \infty$ *is the standard normal probability distribution. That is,*

$$\lim_{n\to\infty} P(Z_n \leq z) = \frac{1}{\sqrt{2\pi}} \int_{-\infty}^{z} e^{-t^2/2}dt.$$

If $S_n = \sum_{i=1}^{n} X_i$, then we can rewrite Z_n as

$$Z_n = \frac{\overline{X} - \mu}{\sigma/\sqrt{n}} = \frac{n(\overline{X} - \mu)}{n\sigma/\sqrt{n}},$$

$$= \frac{S_n - n\mu}{\sigma\sqrt{n}}, \quad \text{since } n\overline{X} = \textstyle\sum_{i=1}^{n} X_i.$$

Then, the CLT states that $Z_n = (S_n - n\mu)/\sigma\sqrt{n}$ is approximately $N(0, 1)$ for large n.

The CLT basically says that when we repeat an experiment a large number of times, the average (almost always) follows a Gaussian distribution.

Proof: We will prove the result with $\mu = 0$, thus $Z_n = S_n/\sigma\sqrt{n}$. For $\mu \neq 0$, take $U_n = X_n - \mu$ for each n. Let $Y_n = \sum_{i=1}^{n} U_i$. Then we have

$$\lim_n P\left(\frac{S_n - n\mu}{\sigma\sqrt{n}} \leq x\right) = \lim_n P\left(\frac{Y_n}{\sigma\sqrt{n}} \leq x\right).$$

Hence, it suffices for the result in the case $\mu = 0$.

Because S_n is a sum of independent random variables $M_{S_n}(t) = [M(t)]^n$, and hence,

$$M_{Z_n}(t) = \left[M\left(\frac{t}{\sigma\sqrt{n}}\right)\right]^n.$$

Using Example 2.6.11, it is enough to show that $\lim_{n\to\infty} \ln M(t/\sigma\sqrt{n}) = t^2/2$. Let $x = 1/\sqrt{n}$ in this limit; thus we need to find

$$\lim_{x\to 0} \frac{\ln M(tx/\sigma)}{x^2} = \frac{t^2}{2} \text{ (using L'Hopitals'rule a couple of times).}$$

Thus the proof.

EXAMPLE 3.5.5

Let $X_1, X_2, \ldots$ are iid random variables such that

$$X_i = \begin{cases} 1, & \text{with probability} p, \\ 0, & \text{with probability } 1-p. \end{cases}$$

Show that $Z_n = (S_n - np)/\sqrt{npq}$ is approximately normal for large n, where $S_n = \Sigma_{i=1}^{n} X_i$, and $q = 1 - p$.

Solution

We know that

$$E(X) = p,\ E(X^2) = p,\ Var(X) = p - p^2 = pq.$$

Hence, by the CLT, the limiting distribution of $Z_n = (S_n - np)/\sqrt{npq}$ *as* $n \to \infty$ *is the standard normal probability distribution.*

EXAMPLE 3.5.6

A soft-drink vending machine is set so that the amount of drink dispensed is a random variable with a mean of 8 ounces and a standard deviation of 0.4 ounces. What is the approximate probability that the average of 36 randomly chosen fills exceeds 8.1 ounces?

Solution

From the CLT, $((\overline{X} - 8)/(0.4/\sqrt{36})) \sim N(0, 1)$. *Hence, from the normal table,*

$$P\{\overline{X} > 8.1\} = P\left\{Z > \frac{8.1 - 8.0}{\frac{0.4}{\sqrt{36}}}\right\}$$
$$= p\{Z > 1.5\} = 0.0668.$$

Thus there is approximately 6.68% chance that the chosen fills exceed 8.1 ounces.

EXAMPLE 3.5.7

Numbers in decimal form are often approximated by the closest integers. Suppose n numbers $X_1, \ldots, X_n$ are approximated by their closest integers $J_1, J_2, \ldots, J_n$. Let $U_i = X_i - J_i$. Assume that U_i are uniform on (–0.5, 0.5) and that $U'_i s$ are independent.

(a) Show that $\frac{\Sigma_{i=1}^{n} U_i}{\sqrt{n/12}} \sim N(0, 1)$ as $n \to \infty$.

(b) For $n = 300$, find $P\left\{\frac{-5}{\sqrt{300/12}} \le \frac{\Sigma_{i=1}^{n} U_i}{\sqrt{300/12}} \le \frac{5}{\sqrt{300/12}}\right\}$.

(c) For $n = 300$, find the value of a such that $P\{-a \le \Sigma U_i \le a\} = 0.95$

(d) For $n = 10^6$, find *a such that* $P\left\{-a \le \Sigma_{i=1}^{10^6} U_i \le a\right\} = 0.99$.

Solution

(a) *Because* $U'_i s$ *are uniform in (–0.5, 0.5),* $\Sigma Ui = 0$, $Var(Ui) = 1/12$. *Let,* $S_n = \Sigma_{i=1}^{n} X_i$, *and* $K_n = \Sigma_{i=1}^{n} J_i$. *Then,*

$$P\{|S_n - K_n| \le a\} = P\{-a \le \Sigma(X_i - J_i) \le a\}$$
$$= P\{-a \le \Sigma U_i \le a\}.$$

By the CLT, $\frac{\Sigma_{i=1}^{n} U_i - 0}{\sqrt{n/12}} \sim N(0, 1)$ as $n \to \infty$.

(b) *For n = 300; a = 5. Using the normal table,*

$$P\left\{\frac{-5}{\sqrt{300/12}} \le \frac{\sum_{i=1}^{n} U_i}{\sqrt{300/12}} \le \frac{5}{\sqrt{300/12}}\right\} = 0.68.$$

(c) *Now,*

$$0.95 = P\{-a \le \sum U_i \le a\}$$
$$= P\left\{\frac{-a}{\sqrt{300/12}} \le Z \le \frac{a}{\sqrt{300/12}}\right\}.$$

From the normal table, we get $\frac{a}{\sqrt{300/12}} = 1.96$. *This implies a = 9.8.*

(d) *We have*

$$0.99 = P\left\{-a \le \sum_{i=1}^{10^6} U_i \le a\right\}$$
$$= P\left\{\frac{-a}{\sqrt{10^6/12}} \le Z \le \frac{a}{\sqrt{10^6/12}}\right\}.$$

Now, using the normal table, we have $a/\sqrt{10^6/12} = 2.58$. *Hence, a = 745.*

EXAMPLE 3.5.8

A casino has a coin, suspected to be biased. Estimate p (probability of heads) such that they can be 99% confident that their estimate (say, $\hat{p}$) is within 0.01 of p (unknown). What is the minimum number of times we need to toss this coin?

Solution

Set

$$X_j = \begin{cases} 1, & \text{if } H \text{ appear in } j\text{'th toss} \\ 0, & \text{if } T \text{ appear in } j\text{'th toss} \end{cases}.$$

Suppose we decided to use $\hat{p} = \frac{\sum X_i}{n}$, *that is,* $\left(\frac{\#\,Heads}{n}\right)$, as an estimate of p.

We want $P\left\{\left|\bar{X} - p\right| < 0.01\right\} = 0.99.$

Because $Y = \sum_{i=1}^{n} X_i \sim Bin(n, p)$, *we have E(Y)=np, Var(Y)= npq. By the CLT,* $(\bar{X} - p)/\sqrt{pq/n} \sim N(0, 1)$. *Now,*

$$0.99 = P\left\{\frac{-0.01}{\sqrt{pq/n}} < \frac{\bar{X} - p}{\sqrt{pq/n}} < \frac{0.01}{\sqrt{pq/n}}\right\}$$
$$= P\left\{\frac{-0.01}{\sqrt{pq/n}} < Z < \frac{0.01}{\sqrt{pq/n}}\right\}.$$

Using the normal table, $(0.01/\sqrt{pq/n}) = 2.58$, *this implies that* $\sqrt{n} \ge (2.58\sqrt{pq}/0.01)$.
Because the maximum of $pq = 1/4$, *it is sufficient that,*

$$\sqrt{n} = \frac{(2.58)(\sqrt{(1/4)})}{0.01} = 129.$$

Hence, $n = (129)^2 = 16{,}641$, *and we should choose the sample size* $n \ge 16{,}641$. Thus the coin must be tossed at least 16,641 times to determine if the coin is not fair.

Note that the method used in Example 3.5.8 can be used to estimate any unknown probability, not just an unfair coin. Also, the fact that $pq = 1/4$ is maximum can be shown by calculus: let $f(p) = pq = p(1 - p) = p - p^2$. $0 = f'(p) = 1 - 2p$ implies $p = 1/2$, thus $q = 1/2$.

The CLT is extremely important in statistics because it says that we can approximate the distribution of certain statistics without much of the knowledge about the underlying probability distribution of those statistics for a relatively "large" sample size. How large the n should be for this normal approximation to work depends on the distribution of the original distribution. A rule of thumb is that the sample size n must be at least 30. We deal with these issues in Chapter 4.

EXERCISES 3.5

3.5.1 Let X be a random variable with probability density function

$$f(x) = \begin{cases} 630x^4(1-x)^4, & 0 < x < 1, \\ 0, & \text{otherwise.} \end{cases}$$

(a) Obtain the lower bound given by Chebyshev's inequality for $P\{0.2 < X < 0.8\}$.
(b) Compute the exact probability, $P\{0.2 < X < 0.8\}$.

3.5.2 Suppose that the number of cars arriving in 1 hour at a busy intersection is a Poisson probability distribution with $\lambda = 100$. Find, using Chebyshev's inequality, a lower bound for the probability that the number of cars arriving at the intersection in 1 hour is between 70 and 130.

3.5.3 Prove Chebyshev's inequality for the discrete case.

3.5.4 Suppose that the number of cars arriving at a busy intersection in a given 20-minute interval in a large city has a Poisson distribution with mean 120. Determine a lower bound for the probability that the number of cars arriving in a given 20-minute period will be between 100 and 140 using Chebyshev's inequality.

3.5.5 Find the smallest value of n in a binomial distribution for which we can assert that

$$P\left(\left|\frac{X_n}{n} - p\right| < 0.1\right) \geq 0.90.$$

3.5.6 How large should the size of a random sample be so that we can be 90% certain that the sample mean $\overline{X}$ will not deviate from the true mean by more than $\sigma/2$?

3.5.7 Let a fair coin be tossed n times and let S_n be the number of heads that turn up. Show that the fraction of heads, S_n/n, will be near to 1/2 for large n. What can we conclude if the coin is not fair?

3.5.8 Suppose that a failure of certain component follows the distribution $f(x) = p^x(1-p)^x$ for $x = 0$, 1, and zero, elsewhere. How many components must one test in order that the sample mean $\overline{X}$ will lie within 0.4 of the true state of nature with probability at least as great as 0.95?

3.5.9 Let $X_1, \ldots, X_n$ be a sequence of mutually independent random variables, with probability distribution

$$P(X_i = \sqrt{i}) = \frac{1}{2} \quad \text{and} \quad P(X_i = -\sqrt{i}) = \frac{1}{2}.$$

Show that this sequence of random variables does not satisfy the conditions of the law of large numbers.

3.5.10 Give proof of the CLT.

3.5.11 Let $X_1, \ldots, X_n$ be a sequence of independent Poisson-distributed random variables, with parameter λ. Let $S_n = \Sigma_{i=1}^n X_i$. Show that

$$Z_n = ((S_n - n\lambda)/\sqrt{n\lambda}) \sim N(0, 1).$$

3.5.12 Let $X_1, \ldots, X_n$ be a sequence of independent uniformly distributed over [0,1) random variables. Let $S_n = \Sigma_{i=1}^n X_i$. Show that $Z_n = ((S_n - n\lambda)/\sqrt{n\lambda}) \sim N(0, 1)$.

3.5.13 Suppose that 2500 customers subscribe to a telephone exchange. There are 80 trunk lines available. Any one customer has the probability of 0.03 of needing a trunk line on a given call. Consider the situation as 2500 trials with probability of "success" $p = 0.03$. What is the approximate probability that the 2500 customers will "tie up" the 80 trunk lines at any given time?

3.5.14 Suppose a group of people have an average IQ of 122 with a standard deviation of 2. Obtain a bound on the probability that IQ values of this group will be between 104 and 140.

3.5.15 Let X be a random variable that represents the diastolic blood pressure (DBP) of the population of 18- to 74-year-old men in the United States who are not taking any corrective medication. Suppose that X has mean 80.7 mm Hg and standard deviation 9.2.

(a) Obtain a bound on the probability that the DBP of this population will assume values between 53.1 and 108.3 mm Hg.

(b) In addition, assume that the distribution of X is approximately normal. Using the normal table, find $P(53.1 \le X \le 108.3)$. Compare the exact value of the probability with the lower bound obtained in part (a).

3.5.16 Color blindness appears in 2% of the people in a certain population. How large must a random sample be in order to be 99% certain that a color-blind person is included in the sample?

3.5.17 A shirt manufacturer knows that, on average, 2% of his product will not meet quality specifications. Find the greatest number of shirts constituting a lot that will have, with probability 0.95, fewer than five defectives.

3.5.18 A drug manufacturer receives a shipment of 10,000 calibrated "eyedroppers" for administering the Sabin poliovirus vaccine. If the calibration mark is missing on 500 droppers, which are scattered randomly throughout the shipment, what is the probability that, at most, two defective droppers will be detected in a random sample of 125?

3.5.19 Let $X_1, \ldots, X_n$ be a random sample, each with mean μ_1 and standard deviation σ_1. Also, let $Y_1, \ldots, Y_m$ be a random sample, each with mean μ_2 and a standard deviation σ_2. Assume that both samples are from the normal populations. Verify that $(\overline{X} - \overline{Y}) \sim N\left(\mu_1 - \mu_2, \frac{1}{n}\sigma_1^2 + \frac{1}{m}\sigma_2^2\right)$.

3.5.20 Let $X1 \ldots X_n$ be a random sample, each with mean μ_1 and standard deviation σ_1. Also, let $Y_1, Y_2, \ldots, Y_n$ be a random sample independent of $X_1 \ldots X_n$, each with mean μ_2 and a standard deviation σ_2. Prove that the random variable

$$V_n = \frac{(\overline{X} - \overline{Y}) - (\mu_1 - \mu_2)}{\sqrt{\frac{\sigma_1^2 + \sigma_2^2}{n}}}$$

satisfies the conditions of Theorem 3.5.3 and hence V_n is asymptotically normal.

3.5.21 A random sample size of 150 is taken from an infinite population having mean $\mu = 8$ and variance $\sigma^2 = 4$. What is the probability that $\overline{X}$ will be between 7.5 and 10?

3.5.22 A machine that is used to fill bottles with soda has been observed to have a true standard deviation in the amounts of fill of approximately $\sigma = 1.25$ ounces. However, the mean ounces of fill μ may change from day to day, because of a change of operator or adjustments in the machine. If $n = 55$ observations on ounces of fill are taken on a given day, find the probability that the sample mean will be within 0.5 ounce of the true population mean. State any assumptions.

3.5.23 The times spent by customers coming to a certain gas station to fill up can be viewed as independent random variables with a mean of 3 minutes and a variance of 1.5 minutes. Approximate the probability that a random sample of 75 customers in this gas station will spend a total time less than 3 hours. Interpret your results and state any assumptions.

3.5.24 Refer to Exercise 4.4.3. Find the number of customers, m, such that the probability that all the m customers can fill up in less than 3 hours is approximately 0.2.

3.5.25 In the mathematics department of a certain university, in a particular semester, 1250 students took the elementary algebra final examination. The mean was 69% with a standard deviation of 5.4%. If a random sample of 60 students is selected from this population, what is the probability that the average score of this sample will be at most 75.08? Interpret your results and state any assumptions.

3.5.26 For a newborn full-term infant, the weight appropriate for gestational age is assumed to be normally distributed with $\mu = 3025$ grams and $\sigma = 165$ grams. Compute the probability that a random sample of 50 infants born at full-term results in a sample mean of less than 3500 grams.

3.5.27 A small persimmon grower in North Florida measures individual fruit weights. Individual fruit weights are skewed. Suppose they are well-modeled by an exponential distribution with mean $\mu = 220$ grams (so the standard deviation $\sigma = 220$ grams as well). For a random sample of $n = 50$ fruits, find $P(\overline{X} > 240 \text{ g})$. Use the CLT.

3.6 CHAPTER SUMMARY

In this chapter, we studied some special probability distribution functions that we encounter in practice. It should be noted that we discussed only a few of the important probability distributions. There are many other discrete and continuous distributions that will be useful and appropriate in particular applications. Some of them are given in Appendix A3. A larger list of probability distributions can be found at http://www.causascientia.org/math_stat/Dists/Compendium.pdf, among many other places. For more than one random variable, we learned the behavior of joint probability distributions. We also learned how to find the probability (mass) density function and cumulative distribution for the functions of a random variable. Limit theorems are a crucial part of probability theory. We have introduced Chebyshev's inequality, the law of large numbers, and the CLT for random variables.

We now list some of the key definitions introduced in this chapter:

- Bernoulli probability distribution
- Binomial experiment
- Poisson probability distribution
- Probability distribution
- Normal (or Gaussian) probability distribution
- Standard normal random variable
- Gamma probability distribution
- Exponential probability distribution
- Chi-square (χ^2) distribution
- Joint probability density function
- Bivariate probability distributions
- Marginal pdf
- Conditional probability distribution
- Independence of two r.v.s
- Expected value of a function of bivariate r.v.s
- Conditional expectation
- Covariance
- Correlation coefficient

In this chapter, we have also learned the following important concepts and procedures:

- Mean, variance, and moment-generating function (mgf) of a binomial random variable
- Mean, variance, and mgf of a Poisson random variable
- Poisson approximation to the binomial probability distribution
- Mean, variance, and mgf of a uniform random variable
- Mean, variance, and mgf of a normal random variable
- Mean, variance, and mgf of a gamma random variable
- Mean, variance, and mgf of an exponential random variable
- Mean, variance, and mgf of a chi-square random variable
- Properties of expected value
- Properties of the covariance and correlation coefficient
- Procedure to find the cdf of a function of r.v. using the method of distribution functions
- The pdf of $Y = g(X)$, where g is differentiable and monotone increasing or decreasing
- The pdf of $Y = g(X)$, using the probability integral transformation
- The transformation method to find the pdf of $Y = g(X_1, \ldots, X_n)$
- Chebyshev's theorem
- Law of large numbers
- Central Limit Theorem (CLT)

3.7 COMPUTER EXAMPLES (OPTIONAL)

3.7.1 The R-Examples

EXAMPLE 3.7.1 PDFS AND CDFS IN R

R contains functions for many distribution functions with a logical format to access each. This example will translate to other distributions such as Poisson and normal; however, the examples will be with the binomial. Specifically in R there are four command prefixes (***p,q,r,d***); ***p*** will return the probabilities (cumulative) while ***q*** returns values (quantile, the inverse c. d.f.) ***r*** generates random values from the distribution, and ***d*** returns the density. In the case of R and these functions, everything is cumulative, and you will need to adjust for this when seeking non-cumulative probabilities. Using the *help()* function is recommended because each distribution takes different arguments. "for example, *help(pbinom)*"

R-code:

```
pbinom(c(0:5),5,0.4);
pbinom(3,5,0.4)-pbinom(2,5,0.4);
qbinom(0.5,5,0.4);
```

```
pnorm(4.2,4,2);
qnorm(0.5,4,2);
```

Output:

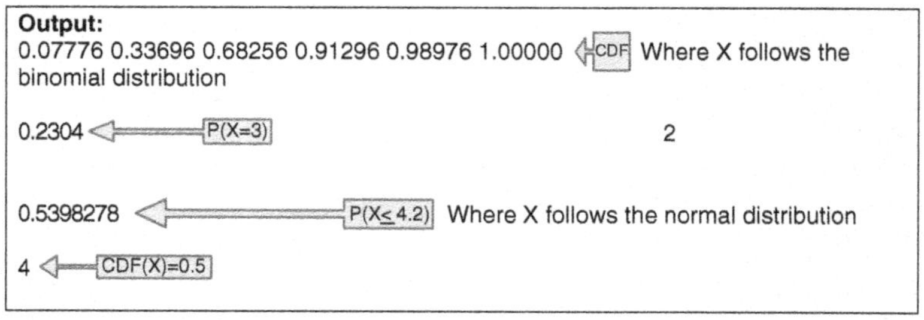

EXAMPLE 3.7.2 Binomial Experiment

A manufacturer of a color printer claims that only 5% of their printers require repairs within the first year. If out of a random sample of 18 of their printers, four required repairs within the first year, does this tend to refute or support the manufacturer's claim?

R Code:

```
1-pbinom(3,18,0.05);
```

Output:

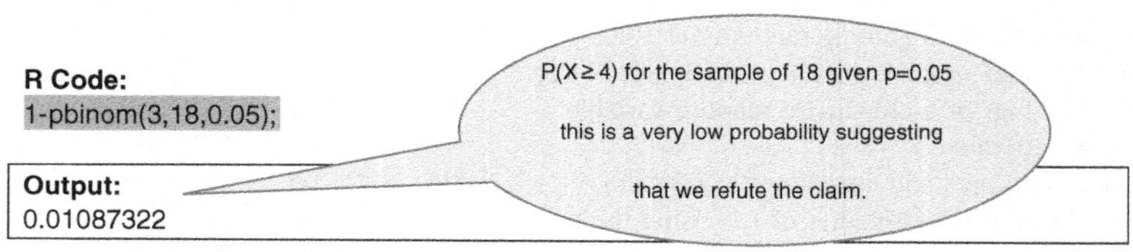

EXAMPLE 3.7.3 Binomial Experiment

Suppose that a certain drug to treat a disease has a success rate of $p = 0.65$. This drug is given to $n = 500$ patients with the disease.

(a) What is the probability that 335 or fewer show improvement?
(b) What is the probability that more than 320 show improvement?
(c) What is the probability that exactly 300 show improvement?
(d) What is the probability that the number of improvements lies in the interval (300,350)?

R Code:

```
pbinom(335,500,0.65);
1-pbinom(320,500,0.65);
pbinom(300,500,0.65)-pbinom(299,500,0.65);
pbinom(349,500,0.65)-pbinom(300,500,0.65);
```

Output:

```
0.8375342
0.6648447
```

```
0.002462253
0.9784924
```

Output:
0.8375342 ← P(X< 335)
0.6648447 ← P(X>320)
0.002462253 ← P(X=300)
0.9784924 ← P(300 < X < 350)

3.7.2 SPSS Examples

EXAMPLE 3.7.4

For the data of Example 3.7.1, using SPSS, find $P(X \leq 3)$.

Solution

*Enter numbers 1 through 18 in **C1**. Then use the following.*

Transform>Compute> type in the ***Target Variable: y >*** Use the scroll bar beside the Functions box to find ***CDF.BINOM(q,n, p)>*** Highlight it and use the up button to load it into the ***Numeric Expression:*** box. Set ***q*** to ***3*** (success, the x-value), ***n*** to ***18*** (total trials) and ***p*** to ***0.05*** (probability of success)> ***OK***

In the second column, we will get the y-values as 0.99. Hence, $P(X \leq 3) = 0.99$.

We can use this procedure for many other distributions.

3.7.3 SAS Examples

Sometimes, we can use computer calculations to find out the exact probability of a certain event in lieu of approximations. For example, when n is large in a binomial experiment, we can use the normal approximation to calculate the probabilities. The following example shows how to calculate binomial probabilities using SAS codes.

EXAMPLE 3.7.5

Suppose that a certain drug to treat a disease has a success rate of $p = 0.65$. This drug is given to $n = 500$ patients with the disease.

(a) What is the probability that 335 or fewer show improvement?
(b) What is the probability that more than 320 show improvement?
(c) What is the probability that exactly 300 show improvement?
(d) What is the probability that the number of improvements lies in the interval (300,350)?

Solution

Let X = number of patients showing improvement. Then X is a binomial random variable with parameters $n = 500$ and $p = 0.65$.

(a) *The first three lines in the following code are comment lines. In general, it is always helpful to include the comment lines to explain about the program.*

```
/*This program can be used to compute probability*/
/* that a Binomial variable with parameters p*/
/*and n is less than or equal to x*/
data binomial;
        p = 0.65;
        n = 500;
        x = 335;
        y = probbnml(p,n,x);
cards;
```

```
proc print;
run;
```

(b) *To calculate P(X > 320), we can use the following.*

```
data binomial;
        p = 0.65;
        n = 500;
        x = 320;
        y = probbnml(p,n,x);
        z = 1 - y;
cards;
proc print;
run;
```

(c) *To find P(X = 300), we can use the following:*

```
data binomial;
        p = 0.65;
        n = 500;
        x1 = 300;
        y1 = probbnml(p,n,x1);
        x2 = 299;
        y2 = probbnml(p,n,x2);
        z = y1 - y2;
cards;
proc print;
run;
```

(d) *To find P(300 <X< 350), use the following:*

```
data binomial;
        p = 0.65;
        n = 500;
        x1 = 300;
        y1 = probbnml(p,n,x1);
        x2 = 349;
        y2 = probbnml(p,n,x2);
        z = y2 - y1;
cards;
proc print;
run;
```

Similar procedures could be used to calculate probabilities for other distributions.

In order to *test for normality* of a given data set using a normal probability plot, we can use PROC UNIVARIATE (see Chapter 1 for explanation) in the following manner. A normal plot is called qqplot in SAS.

```
proc univariate data=K noprint; /*Specify the name of data set as K*/ qqplot standard;
run;
quit;
```

Note that this avoids printing all the standard outputs due to the univariate command, and we get only the QQ plot. If we need a straight line in the plot, we can modify the commands as follows:

```
proc univariate data=K noprint; /*Specify the name of data set as B*/ qqplot standard/ normal (mu=m,
sigma=s);
run;
quit;
```

PROJECTS FOR CHAPTER 3

3A. Mixture Distribution

In statistical modeling, if the data are contaminated by outliers or if the samples are drawn from a population formed by a mixture of two populations, one could use mixture distributions. Mixture distributions are used frequently in medical applications, such as microarray analyses. Suppose a random variable X has pdf $f_1(x)$ with probability p_1 and pdf $f_2(x)$ with probability p_2, where $p_1 + p_2 = 1$. Then we say that the r.v. X has a *mixture distribution*. This can be thought of as observing a Bernoulli random variable Z that is equal to 1 with probability p_1 and 2 with probability p_2. Thus

$$X = \begin{cases} X_1 \sim f_1(x), & \text{if } Y = 1, \\ X_2 \sim f_2(x), & \text{if } Y = 2. \end{cases}$$

(a) Show that the pdf of X is given by $f(x) = p_1f_1(x) + p_2f_2(x)$.
(b) If (μ_1, σ_1^2) and (μ_2, σ_2^2) are means and variances of $f_1(x)$ and $f_2(x)$, respectively, show that

$$\mu = E(X) = p_1\mu_1 + p_2\mu_2,$$

and

$$\sigma^2 = Var(X) = p_1\sigma_1^2 + p_2\sigma_1^2 + p_1\mu_1^2 + p_2\mu_2^2 - (p_1\mu_1 + p_2\mu_2)^2.$$

3B. Generating Samples from Exponential and Poisson Probability Distribution

(a) Generate a sample from $\frac{1}{\theta}e^{-x/\theta}$ (θ is chosen). Let $Y_1, Y_2,\ldots,Y_n$ be a sample from a $U(0, 1)$ distribution. Let $F(x) = 1-e^{-x/\theta}$ (cdf of exponential). Then $Y = F(x)$ is uniform. $y_j = 1-e^{-x/\theta}$ implies $x_j = -\theta \ln(1 - y_i) = -\theta \ln u_i$, where $u_1,u_2, \ldots, u_n$ is a sample from $U(0, 1)$. Then $X_1,\ldots,X_n$ is a sample from an exponential distribution with parameter θ.
(b) Suppose we want to generate a sample from a Poisson probability distribution with parameter λ. $X_1,\ldots,X_n$ is a sample from an exponential distribution with parameter $1/\lambda$ till $\sum_{i=1}^{n} X_i$ just exceeds 1. Then $y_n(n - 1)$ is a sample values form a Poisson probability distribution with parameter λ.

EXERCISE 3B

Let $u_1, u_2,\ldots,u_n$ be a sample from $U(0, 1)$. Show that

(i) $X = -2\sum_{i=1}^{n} \ln(u_i) \sim \chi^2_{2n}$,
(ii) $X = -\beta\sum_{i=1}^{\alpha} \ln(u_i) \sim gamma(\alpha, \beta)$, and
(iii) $X = \frac{\sum_{i=1}^{\alpha} \ln(u_i)}{\sum_{i=1}^{\alpha+\beta} \ln(u_i)} \sim beta(\alpha, \beta)$.
(iv) Search the Internet and create a list of transformations that uses a uniform random variable to generate random variables from other distributions. Discuss the computational efficiency of such methods.

3C. Coupon Collector's Problem

Suppose there are n distinct colors of coupons. Each color of the coupon is equally likely to occur. When a complete set of coupons with each color represented is assembled, you win a prize. Let X = # coupons for a complete set. Find (a) Distribution of X, (b) $E(X)$, and (c) $Var(X)$.

3D. Recursive Calculation of Binomial and Poisson Probabilities

A simple way to calculate binomial probabilities is as follows: For a given n and p, evaluate $b(0,n,p)$ and then apply the recursive relationship

$$b(x + 1, n, p) = b(x, n, p)\frac{p(n - x)}{(1 - p)(x + 1)},$$

to obtain other binomial probabilities.

(a) Derive this recursion formula.
(b) For $n = 15$, $p = 0.4$, using the recursive formula, compute all other probabilities starting from x=0.

The following recursive formulas are very useful in calculating successive Poisson probabilities:

$$f(x-1,\ \lambda) = f(x,\lambda)\tfrac{x}{\lambda},$$

and

$$f(x+1,\ \lambda) = \frac{e^{-\lambda}\lambda^{x+1}}{(x+1!)} = f(x,\lambda)\frac{\lambda}{x+1}.$$

For example, if $\lambda = 2.5$, we know that $f(0,2.5) = e^{-2.5} = 0.08208$. Using this, calculate (c) $f(1, 2.5)$ and $f(2, 2.5)$.

3E. Simulation of Poisson Approximation of Binomial

Write and run R-code with various n and p to see how the errors compare as n increases and p decreases, by calculating actual binomial probabilities as well as Poisson probabilities with $\lambda = np$.

3F. Generating a Large Amount of Random Data Using R

Although Projects 3B and 3E can be used for random sample generation, it is tedious for a large amount of data generation. Now we will give some R commands for generating n random samples from few common distributions.

runif(n, min = a, max = b) will generate n random values from $U(a, b)$. For instance, to generate 100 random samples from uniform (0,2), use the command *runif(100, 0, 2)*.

Similarly, *rnorm (n, mean = a, sd = b)* will generate n random values from normal distribution with mean a and standard deviation b. Thus *rnorm (100, mean = 4, sd = 2)* will generate 100 random values from a normal distribution with mean 4 and standard deviation of 2. Similarly, *rbinom(n, m,p)* will generate n random values from a binomial distribution with parameters m and p. For instance, to generate 100 sample values from binomial with $m = 10$, and $p = 0.5$, use *rbinom (100, 10, 0.5)*. Similarly, *rexp(n, rate)* will generate n random samples from exponential distribution with specified rate. Thus *rexp(100, 1/1000)* will generate 100 exponential values with rate $\lambda = 1/1000$. Similarly, we can generate random samples from any distribution.

EXERCISE 3F

For each of the generated random samples, write an R-code to create a histogram overlapped by the corresponding density function.

3G. Demonstrate CLT Empirically by Simulating Sum/Averages from Skewed Distributions

In Project 3F, you have learned to generate samples from various distribution. In this project, we will use those samples to demonstrate the Central Limit Theorem. Consider an exponential distribution with parameter $\lambda = 1$. Clearly this is highly skewed distribution. Here true mean is $\frac{1}{\lambda} = 1$, and true variance is $\frac{1}{\lambda^2} = 1$. Follow the procedure:

1. Draw many samples from Exponential(1).
2. Compute sample means for different sample sizes n.
3. Standardize them:

$$Z = \frac{\overline{X}_n - \mu}{\sigma/\sqrt{n}}.$$

Then, the probability distribution should approximate $N(0, 1)$.

Write a run an R-code to demonstrate CLT in this case with different n, such as $n = 5,\ 30,\ 100,$ and 1000.

Chapter 4

Sampling Probability Distributions

Chapter outline

OBJECTIVE

In this chapter, we will study the probability distributions of various sample statistics, such as the sample mean and the sample variance, and illustrate their usefulness.

Abraham de moivre
(*source*: *http://en.wikipedia.org/wiki/file:abraham_de_moivre.jpg*)

Abraham de Moivre (1667–1754) was a French mathematician known for his work on the normal distribution and probability theory. He is famous for de Moivre's formula, which links complex numbers and trigonometry. He fled France and went to England to escape the persecution of Protestants. In England, he wrote a book on probability theory, titled The Doctrine of Chances. This book was very popular among gamblers. The normal distribution was first introduced by de Moivre in an article in 1733 in the context of approximating certain binomial distributions for large n, and this approximation result is now called the Theorem of de Moivre–Laplace.

Mathematical Statistics with Applications in R. DOI: https://doi.org/10.1016/B978-0-443-27548-7.00004-9

4.1 INTRODUCTION

Sampling probability distributions play a very important role in statistical analysis and decision-making. We begin by studying the distribution of a statistic computed from a random sample. Based on the probabilistic foundation of Chapters 2 and 3, the present study marks the beginning of our learning of statistics beyond the descriptive phase. Because a sample is a set of random variables $X_1, \ldots, X_n$, it follows that a sample statistic that is a function of the sample is also random. We call the probability distribution of a sample statistic its sampling distribution. Sampling distributions provide the link between probability theory and statistical inference. The ability to determine the distribution of a statistic is a critical part in the construction and evaluation of statistical procedures. It is important to observe that there is a difference between the distribution of population from which the sample was taken and the distribution of the sample statistic. In general, a population has a distribution called a population distribution, which is usually unknown, whereas a statistic has a sampling distribution, which is usually different from the population probability distribution. The sampling distribution of a statistic provides a theoretical model of the relative frequency histogram for the likely values of the statistic that one would observe through repeated sampling. Even though some of the terms in this section have already been defined in Chapter 1, we now present these definitions in terms of random variables. These abstractions are introduced to develop scientifically based methods of analyzing the data, and one should always keep in mind the underlying population.

Definition 4.1.1. *A* **sample** *is a set of observable random variables $X_1, \ldots, X_n$. The number n is called the* **sample size**.

In most of the inferential procedures that we study in this book, we are dealing with random samples. We call the random variables $X_1, \ldots, X_n$ identically distributed if every X_i has the same probability distribution.

Definition 4.1.2. *A* **random sample** *of size n from a population is a set of n independent and identically distributed (iid) observable random variables $X_1, \ldots, X_n$.*

Note that in a sample (not a random sample), X_is need not be independent or identically distributed. For the results of this book to be applicable, it is important to ensure that the selection of a sample is at least approximately random. The significance of random sampling is that the probability distribution of a statistic can be easily derived. Random sampling helps us to control systematic biases. For a finite population, one can serially number the elements of the population and then select a random sample with the help of a table of random digits. One of the simplest ways to select a random sample of finite size is to use a table of random numbers. When the population size is very large, such a method can become very taxing and sometimes practically impossible. However, there are excellent computer programs for generating random samples from large populations, and these programs can be used. Now we define a statistic.

Definition 4.1.3. *A function T of observable random variables $X_1, \ldots, X_n$ that does not depend on any unknown parameters is called a* **statistic**.

The sample mean $\overline{X} = (1/n)\sum_{i=1}^{n} X_i$ is a function of $X_1, \ldots, X_n$. The sample median and sample variance S^2 are also examples of statistics. It is important to observe that even with random sampling, there is sampling variability or error. That is, if we select different samples from the same population, a statistic will take different values in different samples. Thus, a sample statistic is a random variable, and hence it has a probability distribution. In order for us to study the behavior of the phenomenon a sample statistic represents, we must identify its probability distribution.

Definition 4.1.4. *The probability distribution of a sample statistic is called the* **sampling distribution**.

We can illustrate these definitions with the following example with a finite population and a finite sample size. In this case, we take all possible samples of size n from a population of size N.

EXAMPLE 4.1.1

Let the population consist of the numbers {1, 2, 3, 4, 5}. Consider all possible samples consisting of three numbers randomly chosen without replacement from this population. Obtain the distribution of the sample mean.

Solution

Disregarding the order, it is clear that there are $\binom{5}{3} = 10$ *equally likely possible samples of size 3. They are (1,2,3), (1,2,4), (1,2,5), (1,3,4), (1,3,5), (1,4,5), (2,3,4), (2,3,5), (2,4,5), and (3,4,5). Calculating the sample mean,* $\overline{X}$*, for each of the samples, we will get the sampling distribution of* $\overline{X}$ *as*

$\overline{x}$	$\frac{2}{1}$	$\frac{7}{3}$	$\frac{8}{3}$	$\frac{3}{1}$	$\frac{10}{3}$	$\frac{11}{3}$	$\frac{4}{1}$
$p(\overline{x})$	$\frac{1}{10}$	$\frac{1}{10}$	$\frac{2}{10}$	$\frac{2}{10}$	$\frac{2}{10}$	$\frac{1}{10}$	$\frac{1}{10}$

For example, in the table, $P(\overline{X} = 8/3) = 2/10$ *because the two samples (1,2,5) and (1,3,4) both give an* $\overline{x} = 8/3$*, which is an estimate of the population mean,* μ*.*

In general, sampling distributions are theoretical distributions that consist of possibly an infinite number of sample statistics taken from an infinite number of randomly selected samples of a fixed sample size. For example, if a sample of size $n = 30$ were taken from a large population an infinite number of times, the combined means taken from all the samples would make up the sampling distribution of the mean. Every sample statistic has a sampling distribution. The next result states that if one selects a random sample from a population with mean μ and variance σ^2, then regardless of the form of the population distribution, one can obtain the mean and standard deviation of the statistic $\overline{X}$ in terms of the mean and standard deviation of the population. This is explained in the following result.

Theorem 4.1.1. Let $X_1, \ldots, X_n$ *be a random sample of size n from a population with mean* μ *and variance* σ^2*. Then* $E(\overline{X}) = \mu$ and $Var(\overline{X}) = \sigma^2/n$.

Proof. The mean and variance of $\overline{X}$ is given by,

$$E(\overline{X}) = E\left(\frac{1}{n}\sum_{i=1}^{n} X_i\right) = \frac{1}{n}\sum_{i=1}^{n} E(X_i)$$

$$= \frac{1}{n}\sum_{i=1}^{n}\mu = \frac{1}{n}n\mu = \mu,$$

and

$$Var(\overline{X}) = Var\left(\frac{1}{n}\sum_{i=1}^{n} X_i\right)$$

$$= \frac{1}{n^2}\sum_{i=1}^{n} Var(X_i) \text{ (because } X'_i\text{ s are independent and}$$

$$Var(aX_i) = a^2 Var(X_i))$$

$$= \frac{1}{n^2}n\sigma^2 = \frac{\sigma^2}{n}.$$

We denote $E(\overline{X}) = \mu_{\overline{X}}$ and $Var(\overline{X}) = \sigma_{\overline{X}}^2$. Note that from the previous theorem, $\mu_{\overline{X}} = \mu$ and $\sigma_{\overline{X}} = \sigma/\sqrt{n}$. Here, $\sigma_{\overline{X}}$ is called the standard error of the mean. It is important to notice that the variance of each of the random variables $X_1, X_2, \ldots, Xn$ is σ^2, whereas the variance of the sample mean $\overline{X}$ is σ^2/n, which is smaller than the population variance σ^2 for $n \geq 2$.

The implication of Theorem 4.1.1 is that the sample means become more and more reliable as an estimate of μ as the sample size is increased, as we would expect. From Chebyshev's inequality,

$P(|\overline{X} - \mu_{\overline{X}}| < k\sigma_{\overline{X}}) \geq 1 - \frac{1}{k^2}$. Let $\varepsilon = (k\sigma/\sqrt{n})$. Then $k = (\varepsilon\sqrt{n})/\sigma$. Since $\mu_{\overline{X}} = \mu$, the above inequality can be written as

$$P(|\overline{X} - \mu| < \varepsilon) \geq 1 - \frac{\sigma^2}{n\varepsilon^2}.$$

Thus, for any $\varepsilon > 0$, the probability that the difference between $\overline{X}$ and μ less than ε can be made arbitrarily close to 1 by choosing the sample size n is sufficiently large. We illustrate this result in the following example.

EXAMPLE 4.1.2

A particular brand of drink has an average of 12 ounces per can. As a result of randomness, there will be small variations in how much liquid each bottle really contains. It has been observed that the amount of liquid in these bottles is normally distributed with $\sigma = 0.8$ ounce. A sample of 10 bottles of this brand of soda is randomly selected from a large lot of bottles, and the amount of liquid, in ounces, is measured in each. Find the probability that the sample mean will be within 0.5 ounce of 12 ounces.

Solution

Let $X_1, X_2, \ldots, X_{10}$ denote the ounces of liquid measured for each of the bottles. We know that X_{is} are normally distributed with mean $\mu = 12$ and variance $\sigma^2 = 0.64$. From Theorem 4.1.1, $\overline{X}$ possesses a normal distribution (actually, for the normality part, we use Corollary 4.2.2) with a mean 12 and variance $\sigma 2/n = 0.64/10 = 0.064$. We find that

$$
\begin{aligned}
P(|\overline{X} - 12| \leq 0.5) &= P(-0.5 \leq (\overline{X} - 12) \leq 0.5) \\
&= P\left(-\frac{0.5}{\sigma/\sqrt{n}} \leq \frac{\overline{X} - 12}{\sigma/\sqrt{n}} \leq \frac{0.5}{\sigma/\sqrt{n}}\right) \\
&= P\left(-\frac{0.5}{0.253} \leq Z \leq \frac{0.5}{0.253}\right) \\
&= P(-1.97 \leq Z \leq 1.97) \\
&= 0.9512. \text{ (using standard normal table).}
\end{aligned}
$$

Hence, the chance is about 95% that the mean amount of drink in any 10 bottles randomly chosen will be between 11.5 and 12.5 ounces.

4.1.1 Finite Population

Let $\{c_1, c_2, \ldots, c_N\}$ be a finite population. Then, the population mean $\mu = (1/N)\sum_{i=1}^{N} c_i$ and the population variance $\sigma^2 = (1/N)\sum_{i=1}^{N}(c_i - \mu)^2$. The following theorem for the sample mean and variance is stated without proof.

Theorem 4.1.2. *If $X_1, \ldots, X_n$ is a sample of size n (chosen without replacement) from a population $\{c_1, c_2, \ldots, c_N\}$, then*

$$
E(\overline{X}) = \mu
$$

$$
Var(\overline{X}) = \frac{\sigma^2}{n}\left(\frac{N-n}{N-1}\right).
$$

We remark here that the sample in the theorem is not a random sample and X_is are not iid random variables. The factor $(N - n)/(N - 1)$ in the foregoing theorem is often called the ***finite population correction factor***. It is close to 1 unless the sample amounts to a significant portion of the population. Note that sampling without replacement causes dependence among the X_is. However, if the sample size n is small relative to the population size N, the population correction factor is approximately 1. Hence, we will not use the finite population correction factor in the derivation of a sampling distribution unless it is absolutely necessary.

EXAMPLE 4.1.3

Obtain the mean and variance of $\overline{X}$ in Example 4.1.1.

Solution

First note that for the population in Example 4.1.1, the population mean is $\mu = (1/N)\sum_{i=1}^{N} c_i = 3$ and the population variance is $\sigma^2 = (1/N)\sum_{i=1}^{N}(c_i - \mu)^2 = 2$. Applying the probability distribution of $\overline{X}$ given in Example 4.3.1, we obtain

$$
\begin{aligned}
E(\overline{X}) &= 2\left(\frac{1}{10}\right) + \frac{7}{3}\left(\frac{1}{10}\right) + \frac{8}{3}\left(\frac{2}{10}\right) + 3\left(\frac{2}{10}\right) + \frac{10}{3}\left(\frac{2}{10}\right) \\
&+ \frac{11}{3}\left(\frac{1}{10}\right) + 4\left(\frac{1}{10}\right) \\
&= 3,
\end{aligned}
$$

and

$$
\begin{aligned}
Var(\overline{X}) = E(\overline{X}^2) - E(\overline{X}^2) &= 2^2\left(\frac{1}{10}\right) + \left(\frac{7}{3}\right)^2\left(\frac{1}{10}\right) + \left(\frac{8}{3}\right)^2\left(\frac{2}{10}\right) \\
&+ 3^2\left(\frac{2}{10}\right) + \left(\frac{10}{3}\right)^2\left(\frac{2}{10}\right) + \left(\frac{11}{3}\right)^2\left(\frac{1}{10}\right) + 4^2\left(\frac{1}{10}\right) - 3^2 \\
&= \frac{2}{3} \times \frac{1}{2} = 0.3333.
\end{aligned}
$$

This is the same as $\frac{\sigma^2}{n}\left[\frac{(N-n)}{(N-1)}\right]$. In this case, we observe that the variance of $\overline{X}$ is precisely one-sixth of the original variance.

EXAMPLE 4.1.4

Let $X_1,\ldots,X_n$ be a random sample from a population with mean μ and variance σ^2. Consider the sample variance

$$S^2 = \frac{1}{n-1}\sum_{i=1}^{n}(X_i - \overline{X})^2.$$

Show that $E(S^2) = \sigma^2$.

Solution

It can be shown that (see Exercise 1.5.8)

$$\frac{1}{n-1}\sum_{i=1}^{n}(X_i - \overline{X})^2 = \frac{\sum_{i=1}^{n}X_i^2 - n\overline{X}^2}{n-1}.$$

Hence,

$$E(S^2) = E\left(\frac{\sum_{i=1}^{n}X_i^2 - n\overline{X}^2}{n-1}\right) = \frac{1}{n-1}\sum_{i=1}^{n}E(X_i^2) - \frac{n}{n-1}E(\overline{X}^2).$$

Using the fact that $E(X^2) = Var(X) + \mu^2$ and Theorem 4.1.1, we have

$$\begin{aligned} E(S^2) &= \frac{1}{n-1}n\left(\sigma^2 + \mu^2\right) - \frac{n}{n-1}\left(\frac{\sigma^2}{n} + \mu^2\right) \\ &= \left(\frac{n}{n-1} - \frac{1}{n-1}\right)\sigma^2 + \left(\frac{n}{n-1} - \frac{n}{n-1}\right)\mu^2 \\ &= \sigma^2. \end{aligned}$$

This shows that the expected value of the sample variance is the same as the variance of the population under consideration.

EXERCISES 4.1

4.1.1. Let the population be given by the numbers {–2, –1, 0, 1, 2}. Take all random samples of size 3.
- **(a)** Without replacement, obtain the following in each case.
  - **(i)** The sampling distribution of the sample mean.
  - **(ii)** The sampling distribution of the sample median.
  - **(iii)** The sampling distribution of the sample standard deviation.
  - **(iv)** The mean and variance of the sample mean.
- **(b)** How many samples of size 3 can we get if we sample with replacement?

4.1.2.
- **(a)** How many different samples of size $n = 2$ can be chosen from a finite population of size 12 if the sampling is without replacement?
- **(b)** *What* is the probability of each sample in part (a), if each sample of size 2 is equally likely?
- **(c)** Find the value of the finite population correction factor.

4.1.3. Let the population be given by {1, 2, 3}. Let $p(x) = 1/3$ for $x = 1, 2, 3$. Take samples of size 3 with replacement.
- **(a)** Calculate μ and σ^2.
- **(b)** *Obtain* the sampling distribution of the sample mean.
- **(c)** Obtain the mean and variance of the sample mean.

4.1.4. Find the value of the finite population correction factor for
- **(a)** $n = 8$ and $N = 60$.
- **(b)** $n = 8$ and $N = 1000$.
- **(c)** $n = 15$ and $N = 60$.

4.1.5. For a random sample $X_1, \ldots, X_n$, let $(S')^2 = (1/n)\sum_{i=1}^{n}(X_i - \overline{X})^2$. Find $E[(S')^2]$. Compare this with $E(S^2)$.

4.1.6. For a random sample $X_1, \ldots, X_n$ with mean μ and variance σ^2, let $T_n = \sum_{i=1}^{n} X_i$, the sample total. Show that $E(Tn) = n\mu$ and $Var(T_n) = n\sigma^2$.

4.1.7. A particular brand of sugar is sold in 5-lb packages. The weight of sugar in these packages can be assumed to be normally distributed with mean $\mu = 5$ lb and standard deviation $\sigma = 0.2$ lb. What is the probability that the mean weight of sugar in 15 randomly selected packages will be within 0.2 lb of 5 lb?

4.1.8. A random sample of size 50 is taken from an infinite population having the mean $\mu = 15$ and standard deviation $\sigma = 4$. What is the probability that $\overline{X}$ will be between 13.5 and 16.6?

4.1.9. The distribution of heights of all students in a large university has a normal distribution with a mean of 66 inches and a standard deviation of 2 inches. What is the probability that the mean height of 26 randomly selected students from this university will be more than 67 inches?

4.1.10. An image-encoding algorithm, when used to encode images of a certain size, uses a mean of 110 milliseconds with a standard deviation of 15 milliseconds. What is the probability that the mean time (in milliseconds) for encoding 50 randomly selected images of this size will be between 104 milliseconds and 115 milliseconds?

4.1.11. (3.5.11.) Let $X_1, \ldots, X_n$ be independent discrete random variables identically distributed as

$$f(x_i) = \begin{cases} 0.2, & x_i = 0,\ 1,\ 2,\ 3,\ 4, \\ 0, & \text{otherwise.} \end{cases}$$

Using CTL, find the approximate value of $P(\overline{X}_{100} > 2)$, where $\overline{X}_{100} = (1/100)\Sigma_{i=1}^{100} X_i$.

4.1.12. A population of disk drives manufactured by a certain company runs with mean seek time of 10 milliseconds with standard deviation of 1 millisecond. What proportion of the samples of size 250 would you expect to result in a mean less than 9.9 milliseconds?

4.1.13. Suppose that the national norm of a science test for 12th graders in a particular year has a mean of 215 and a standard deviation of 35.

(a) A random sample of 55 12th graders is selected. What is the probability that this group will average more than 225?

(b) A random sample of 200 12th graders is selected. What is the probability that this group will average over 225?

(c) A random sample of 35 12th graders is selected. What is the probability that this group will average over 225?

(d) How does the sample size influence the probability?

4.1.14. Scores on the Wechsler Adult Intelligence Scale for the 20 to 34 age group are approximately normally distributed with mean equal to 110 and standard deviation equal to 25. If we select 100 people at random, what is the probability that this group will have an average score of 116 or above?

4.1.15. It is known that a healthy human body has an average temperature of 98.6°F, with a standard deviation of 0.95°F. Sixty healthy humans are selected at random. What is the probability that their average temperature is at least 98.8°F?

4.1.16. (3.5.19).A random sample of size 100 is taken from a population with mean 1 and variance 0.04. Find the probability that the sample mean is between 0.99 and 1.

4.1.17. (3.5.20). The lifetime X (in hours) of a certain electrical component has the pdf $f(x) = (1/3)e–(1/3)x$, $x > 0$. If a random sample of 36 is taken from these components, find $P(\overline{X} < 2)$.

4.2 SAMPLING DISTRIBUTIONS ASSOCIATED WITH NORMAL POPULATIONS

The sampling distribution of a statistic will depend upon the population distribution from which the samples are taken. In this section, we discuss the sampling distributions of some statistics that are based on a random sample drawn from a normal distribution. These statistics are used in many statistical procedures that are very important in solving real-world problems. The following result establishes the distribution of a linear combination of independent normal random variables.

Theorem 4.2.1. *Let $X_1, \ldots, X_n$ be independent random variables with the distribution of X_i being normal with mean μ_i and variance σ_i^2. Let $a_1, a_2, \ldots, a_n$ be real constants. Then the distribution of $Y = \Sigma_{i=1}^n a_i X_i$ is normal with mean $\mu_Y = \Sigma_{i=1}^n a_i \mu_i$ and variance $\sigma_Y^2 = \Sigma_{i=1}^n a_i^2 \sigma_i^2$.*

Proof. The moment-generating function of Y is given by

$$\begin{aligned} M_Y(t) &= E e^{(\Sigma_{i=1}^n a_i X_i)t} \\ &= \Pi_i E e^{(a_i X_i)t} \quad \text{[by independence of } X'_i\, s] \\ &= \Pi_i E e^{(a_i t) X_i} \\ &= \Pi_i M_{X_i}(a_i t) \quad \text{[using the definition of mgf)} \\ &= \Pi_i\, e^{(a_i \mu_i t + (1/2) a_i^2 \sigma_i^2 t^2)} \quad \text{[using mgf of a normal]} \\ &= e^{[(\Sigma_i a_i \mu_i) t + (1/2)(\Sigma_i a_i^2 \sigma_i^2) t^2]}, \end{aligned}$$

which is the mgf of a normal random variable with mean $\Sigma_i a_i\mu_i$ and variance $\Sigma_i a_i^2\sigma_i^2$.

In Theorem 4.2.1 let $a_i = 1/n$, $\mu_i = \mu$, and $\sigma_1^2 = \sigma^2$, we obtain the following result, which provides the distribution of the sample mean.

Corollary 4.2.2. *Let $X_1, \ldots, X_n$ be a random sample of size n from a normal population with mean μ and variance σ^2. Then,*

$$\overline{X} = (1/n)\sum_{i=1}^{n} X_i$$

is normally distributed with mean $\mu_{\overline{X}} = \mu$ and variance $\sigma_{\overline{X}}^2 = \sigma^2/n$.

Recall that we have used the notation $X \sim N(\mu, \sigma^2)$ to mean that the random variable X is normally distributed with mean μ and variance σ^2. From Corollary 4.2.2, $\overline{X}\sim N(\mu, \sigma^2/n)$ and hence by the z-transformation we obtain the standard normal random variable, $Z = (\overline{X} - \mu)/(\sigma/\sqrt{n})\sim N(0, 1)$.

EXAMPLE 4.2.1

A company that manufactures cars claims that the gas mileage for its new line of hybrid cars, on average, is 60 miles per gallon with a standard deviation of 4 miles per gallon. A random sample of 16 cars yielded a mean of 57 miles per gallon. If the company's claim is correct, what is the probability that the sample mean is less than or equal to 57 miles per gallon? Comment on the company's claim about the mean gas mileage per gallon of its cars. What assumptions did you make?

Solution

Let X represent the gas mileage for the new car (in miles per gallon). If the company's claim is true, then from Corollary 4.2.2, $\overline{X}$ is normally distributed with mean $\mu = 60$ and variance $\sigma^2/n = 16/16 = 1$. Hence,

$$\begin{aligned} P(\overline{X} \le 57) &= P\left(\frac{\overline{X} - 60}{1} \le \frac{57 - 60}{1}\right) \\ &= P(Z \le -3) \approx 1 - 0.999 \\ &= 0.001. \end{aligned}$$

Therefore, if the company's claim is correct, it is very unlikely that the mean value of the random sample of 16 cars will be 57 miles per gallon. Because the mean is indeed 57 miles per gallon, we conclude that the company's claim is very likely not true. Here we have assumed that the sample of 16 measurements comes from a normal population, so that we could apply the results of Corollary 4.2.2.

Now we introduce some distributions that can be derived from a normal distribution. These distributions play a very important role in inferential statistics.

4.2.1 Chi-Square Distribution

A chi-square distribution is used in many inferential problems, for example, in inferential problems dealing with the variance. Recall that the chi-square distribution is a special case of a gamma distribution with $\alpha = n/2$ and $\beta = 2$. If n is a positive integer, then the parameter n is called the ***degrees of freedom***. However, if n is not an integer, but $\beta = 2$, we still refer to this distribution as a chi-square. The mgf of a χ^2– random variable is $M(t) = (1–2t)–n/2$. The mean and variance of a chi-square distribution are $\mu = n$ and $\sigma^2 = 2n$, respectively. That is, the mean of a $\chi^2(n)$ random variable is equal to its degree of freedom, and the variance is twice the degree of freedom. We now give some useful results for χ^2 – random variables.

Theorem 4.2.3. *Let $X_1, \ldots, X_k$ be independent χ^2– random variables with $n_1, \ldots, n_k$ degrees of freedom, respectively. Then the sum $V = \Sigma_{i=1}^{k} X_i$ is chi-square distributed with $n_1 + n_2 + \ldots + n_k$ degrees of freedom.*

Proof. The mgf of V is

$$M_V(t) = \prod_{i=1}^{k} (1 - 2t)^{-n_i/2} = (1 - 2t)^{-\left(\sum_{i=1}^{k} n_i\right)/2}.$$

This implies that $V\sim\chi^2(\Sigma_{i=1}^{k} n_i)$.

Our next result states that the difference of two chi-square random variables is a chi-square random variable, given by the following theorem. The proof is left as an exercise.

Theorem 4.2.4. *Let X_1 and X_2 be independent random variables. Suppose that X_1 is χ^2 with n_1 degrees of freedom, whereas $Y = X_1 + X_2$ is chi-square with n degrees of freedom, where $n > n_1$. Then $X_2 = Y - X_1$ is a chi-square random variable with $n - n_1$ degrees of freedom.*

The following result shows that we can generate a chi-square random variable from a gamma random variable.

Theorem 4.2.5. *If a random variable X has a gamma probability distribution with parameters α and β, then*

$$Y = \frac{2X}{\beta} \sim \chi^2(2\alpha).$$

Proof. Recall that the mgf of the gamma random variable X is $(1 - \beta t)^{-\alpha}$. Thus

$$\begin{aligned} M_Y(t) &= M_{\frac{2X}{\beta}}(t) = E\left(e^{\frac{2X}{\beta}t}\right) \\ &= E\left(e^{X\left(\frac{2}{\beta}t\right)}\right) = M_X\left(\frac{2}{\beta}t\right) \\ &= (1 - 2t)^{-\alpha} = (1 - 2t)^{-\frac{2\alpha}{2}}. \end{aligned}$$

Hence, $Y \sim \chi^2(2\alpha)$.

The following result states that by squaring a standard normal random variable, we can generate a chi-square random variable, with one degree of freedom.

Theorem 4.2.6. *If X is a standard normal random variable, then X^2 is chi-square random variable with 1 d.f.*

Proof. Because $X \sim N(0, 1)$, the moment-generating function of X^2 is

$$M_{X^2}(t) = \int_{-\infty}^{\infty} e^{tx^2} \frac{1}{\sqrt{2\pi}} e^{-x^2/2} dx = (1 - 2t)^{-1/2}.$$

This implies that $X^2 \sim \chi^2(1)$. Fig. 4.1 gives the probability densities of the random variables X and X^2.

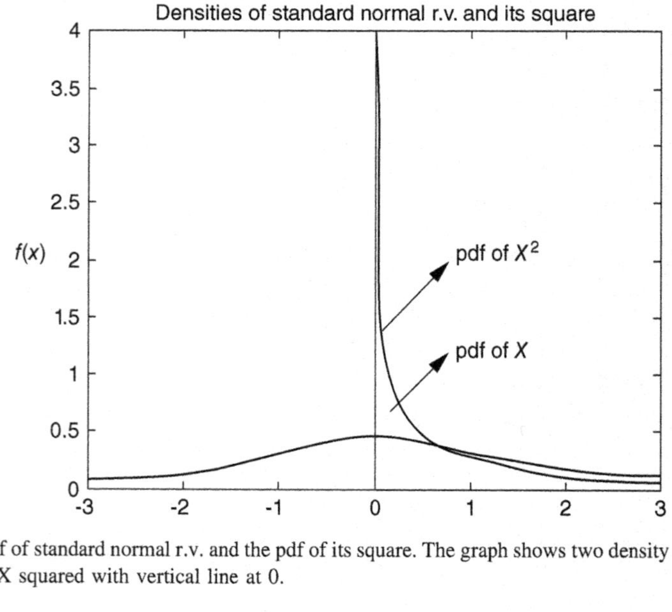

FIGURE 4.1 pdf of standard normal r.v. and the pdf of its square. The graph shows two density curves labeled pdf of X and pdf of X squared with vertical line at 0.

The following result is a direct consequence of Theorems 4.2.3 and 4.2.6. This result illustrates how to obtain a random sample from chi-square distribution if we have a random sample of n measurements from a normal population.

Theorem 4.2.7. *Let the random sample $X_1, \ldots, Xn$ be from a $N(\mu,\sigma^2)$ distribution. Then $Z_i = (X_i - \mu)/\sigma$, $i = 1, \ldots, n$ are independent standard normal random variables and*

$$\sum_{i=1}^{n} Z_i^2 = \sum_{i=1}^{n} \left(\frac{X_i - \mu}{\sigma}\right)^2$$

has a χ^2-distribution with n degrees of freedom. In particular, if $X_1, \ldots, Xn$ are independent standard normal random variables, then $Y^2 = \sum_{i=1}^{n} X_i^2$ is chi-square distributed with n degrees of freedom.

If $X \sim \chi^2(n)$, then from the chi-square table, we can compute the values of $\chi^2_\alpha(n)$ such that

$$P\left(X > \chi^2_\alpha(n)\right) = \alpha,$$

as shown by Fig. 4.2.

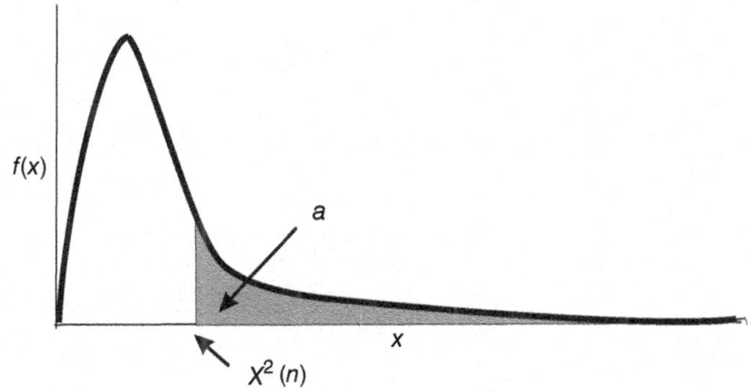

FIGURE 4.2 Chi-square probability density.

For example, if $X \sim \chi^2(15)$, to find $\chi^2_{0.95}(15)$ look in the chi-square table with the row labeled 15 d.f. and the column headed $\chi^2_{0.950}$ and obtain the value as 7.26094. Thus, with 15 degrees of freedom, $P(X > 7.261) = 0.95$. Also, if X is a chi-square random variable with 11 degrees of freedom, from the chi-square table we have $\chi^2_{0.05}(11) = 19.675$. Therefore, $P(X > 19.675) = 0.05$.

EXAMPLE 4.2.2

Let the random variables $X_1, X_2, \ldots, X_5$ be from an $N(5,1)$ distribution. Find a number a such that

$$P\left(\sum_{i=1}^{5} (X_i - 5)^2 \le a\right) = 0.90.$$

Solution

By Theorem 4.2.7, $\sum_{i=1}^{5} Z_i^2 = \sum_{i=1}^{5} \left(\frac{X_i - 5}{1}\right)^2 = \sum_{i=1}^{5} (X_i - 5)^2$ has a chi-square distribution with 5 degrees of freedom. Because the upper tail area is 0.10, looking at the chi-square table with 5 d.f. and the column corresponding to $\chi^2_{0.10}$, we obtain a = 9.23635. Thus,

$$P\left(\sum_{i=1}^{5} (X_i - 5)^2 \le 9.23635\right) = 0.90.$$

EXAMPLE 4.2.3

Suppose that X is a χ^2 – random variable with 20 degrees of freedom. Use the chi-square table to obtain the following:

(a) Find x_0 such that $P(X > x_0) = 0.95$.
(b) Find $P(X \le 12.443)$.

Solution

(a) *For 20 degrees of freedom, using the chi-square table, we have*

$$P(X > 10.851) = 0.95.$$

Hence, x0 = 10.851.

(b) *From the chi-square table,*

$$P(X \le 12.443) = 0.10.$$

The following result gives the probability distribution for a function of the sample variance S^2.

Theorem 4.2.8. *If $X_1, \ldots, Xn$ is a random sample from a normal population with the mean μ and variance σ^2, then*

(a) *the random variable*

(b) $\frac{\sum_{i=1}^{n}(X_i - \overline{X})^2}{\sigma^2} = \frac{(n-1)S^2}{\sigma^2}$,
has a chi-square distribution with $(n-1)$ degrees of freedom.

(c) $\overline{X}$ and S^2 are independent.

Proof. We will only prove part (a). For part (b), we will give some comments on the proof.

(a) We know from Theorem 4.2.7 that $(1/\sigma^2)\sum_{i=1}^{n}(X_i - \mu)^2$ has a chi-square distribution with n degrees of freedom. Thus

$$\frac{1}{\sigma^2}\sum_{i=1}^{n}(X_i - \mu)^2 = \frac{1}{\sigma^2}\sum_{i=1}^{n}(X_i - \overline{X} + \overline{X} - \mu)^2$$

$$= \frac{1}{\sigma^2}\left[\sum_{i=1}^{n}(X_i - \overline{X})^2 + \sum_{i=1}^{n}(\overline{X} - \mu)^2\right]$$

$$\left(\text{Since } 2\sum_{i=1}^{n}(X_i - \overline{X})(\overline{X} - \mu) = 0\right)$$

$$= \frac{(n-1)S^2}{\sigma^2} + \left(\frac{\overline{X} - \mu}{\sigma/\sqrt{n}}\right)^2.$$

The left-hand side of this equation has a chi-square distribution with n degrees of freedom. Also, since $(\overline{X} - \mu)|(\sigma|\sqrt{n}) \sim N(0, 1)$ by Theorem 4.2.6, we have $[(\overline{X} - \mu)|(\sigma|\sqrt{n})]^2 \sim \chi^2(1)$. Now from Theorem 4.2.4, $(n-1)\, S^2/\sigma^2 \sim \chi 2\, (n-1)$.

(b) We will accept the result of part (b) without proof here. A rigorous proof depends on the geometric properties of a multivariate normal distribution, which is beyond the scope of this book. A proof based on moment-generating functions is relatively straightforward where, essentially, we can first show that the random variable $\overline{X}$ and the vector of random variables $(X_1 - \overline{X}, \ldots, X_n - \overline{X})$ are independent. Because S^2 is a function of the vector$(X_1 - \overline{X}, \ldots, X_n - \overline{X})$, it is then independent of $\overline{X}$.

EXAMPLE 4.2.4

Let $X_1, X_2, \ldots, X_{10}$ be a random sample from a normal distribution with $\sigma^2 = 0.8$. Find two positive numbers a and b such that the sample variance S^2 satisfies

$$P(a \le S^2 \le b) = 0.90.$$

Solution

Because $\frac{(n-1)S^2}{\sigma^2} \sim \chi^2(n-1)$, *we have*

$$P(a \le S^2 \le b) = P\left(\frac{(n-1)a}{\sigma^2} \le \frac{(n-1)S^2}{\sigma^2} \le \frac{(n-1)b}{\sigma^2}\right).$$

The desired values can be found by setting the upper tail area and lower tail area, each equal to 0.05. Using the chi-square table with n − 1 = 9 degrees of freedom, we have

$$\frac{(n-1)b}{\sigma^2} = \frac{9b}{0.8} = 16.919 = \chi^2_{0.05,9},$$

which implies $b = (16.919) \times (0.8)/9) = 1.50$. *Similarly,*

$$\frac{(n-1)a}{\sigma^2} = \frac{9a}{0.8} = 3.325 = \chi^2_{0.95,9}$$

So we have $a = ((3.325) \times (0.8)/9) = 0.295$.

Hence,

$$P(0.295 \le S^2 \le 1.50) = 0.90.$$

It is important to note that this is not the only interval that would satisfy

$$P(a \le S^2 \le b) = 0.90$$

but it is a convenient one.

EXAMPLE 4.2.5

A fruit-drink company wants to know the variation, as measured by the standard deviation, of the amount of juice in 16-ounce cans. From past experience, it is known that $\sigma^2=2$. The company statistician decides to take a sample of 25 cans from the production line and compute the sample variance. Assuming that the sample values may be viewed as a random sample from a normal population, find a value of b such that $P(S^2 > b) = 0.05$.

Solution

To find the necessary probability, use the fact that $(n-1)S^2/\sigma^2 \sim \chi^2(n-1)$, *with* $n = 25$,

$$0.05 = P(S^2 > b) = P\left(\frac{24S^2}{2} > \frac{24b}{2}\right)$$
$$= P(\chi^2 > c).$$

From the chi-square table, we obtain $c = 36.4151$. *Hence,* $b = \frac{2}{24}c = \frac{2}{24}(36.4151) = 3.03$ *and*

$$P(S^2 > 3.03) = 0.05.$$

SUMMARY OF CHI-SQUARE DISTRIBUTION

Let $X_1, \ldots, X_n$ be iid $N(\mu, \sigma^2)$ random variables. Then

1. $\overline{X}$ has $N(\mu, \sigma^2/n)$ distribution,
2. $(n-1)S^2/\sigma^2$ has a chi-square distribution with $(n-1)$ degrees of freedom, and
3. $\overline{X}$ and S^2 are independent.
4. A χ^2– random variable has a mean equal to its degrees of freedom and a variance equal to twice its degrees of freedom.

Student t-Distribution

Let the random variables $X_1, \ldots, X_n$ follow a normal distribution with mean μ and variance σ^2. If σ is known, then we know that $\sqrt{n}((\overline{X} - \mu)/\sigma)$ is $N(0, 1)$. However, if σ is not known (as is usually the case), then it is routinely replaced by the sample standard deviation s. If the sample size is large, one could suppose that $s \approx \sigma$ and apply the Central Limit Theorem and obtain that $\sqrt{n}((\overline{X} - \mu)/S)$ is approximately an $N(0, 1)$. However, if the random sample is small, then the distribution of $\sqrt{n}((\overline{X} - \mu)/S)$ is given by the so-called *Student t-distribution* (or simply *t*-distribution). This was originally developed by W. S. Gosset in 1908. Because his employers, the Guinness brewery, would not permit him to publish this important work in his own name, he used the pseudonym "Student." Thus, the distribution is known as the Student *t*-distribution.

Definition 4.2.2. *If Y and Z are independent random variables, Y has a chi-square distribution with n degrees of freedom, and* $Z \sim N(0, 1)$, *then*

$$T = \frac{Z}{\sqrt{Y/n}}$$

is said to have a (Student) ***t-distribution*** *with n degrees of freedom. We denote this by* $T \sim T_n$.

The probability density of the random variable T with n degrees of freedom is given by

$$f(t) = \frac{\Gamma\left(\frac{n+1}{2}\right)}{\sqrt{\pi n}\,\Gamma\left(\frac{n}{2}\right)}\left(1 + \frac{t^2}{n}\right)^{-\frac{n+1}{2}}, \quad -\infty < t < \infty.$$

Fig. 4.3 illustrates the behavior of the t-distributions for $n = 2, 10, 20$, and 30. It is clear from Fig. 4.3 that as n becomes larger and larger, it is almost impossible to distinguish the graphs. It can be shown that the t-distribution tends to a standard normal distribution as the degrees of freedom (equivalently, the sample size n) tend to infinity. In fact, the standard normal distribution provides a good approximation to the t-distribution for sample sizes of 30 or more. We will use this approximation in the statistical inference problems for $n \geq 30$.

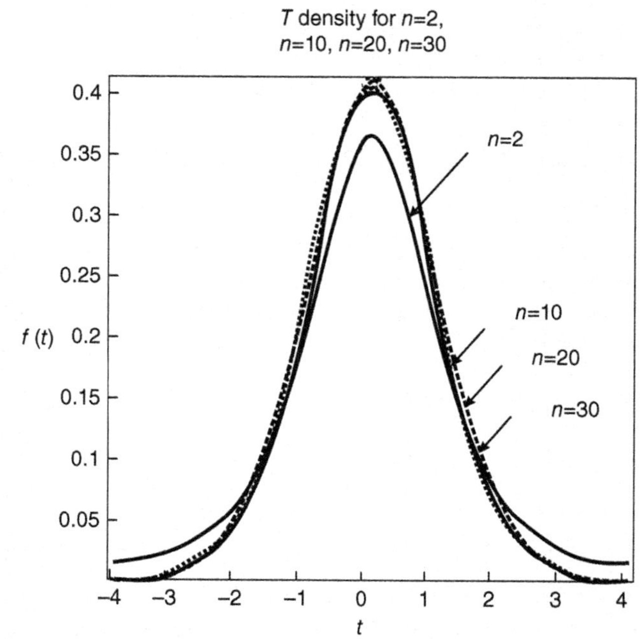

FIGURE 4.3 The Student t-distribution. The graph depicts multiple density curves labeled n equals 2, 10, 20, 30 with peaks near 0 and varying spreads.

The t-density is symmetric about zero, and then we have $E(T) = 0$. If $n > 2$, it can be shown that $Var(T) = n/(n-2)$. The value of $t_{\alpha,n}$ is such that $P(t > t_{\alpha,n}) = \alpha$ (the shaded area in Fig. 4.4) is obtained from the t-table. For example, if a random variable X has a t-distribution with 9 degrees of freedom and $\alpha = 0.01$, then $t_{0.01,9} = 2.821$.

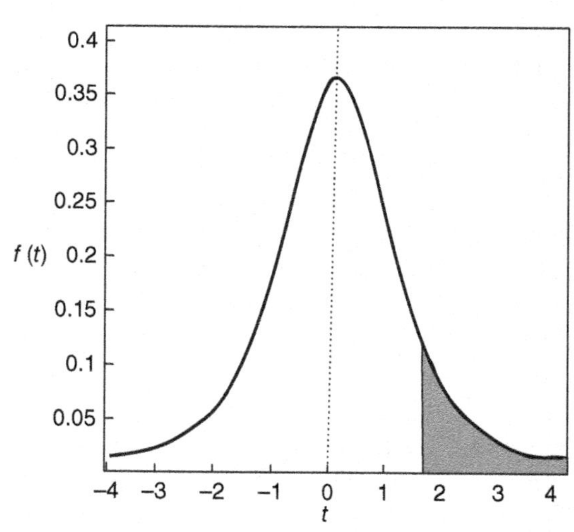

FIGURE 4.4 Probability of t-distribution.

If we have a random sample from a normal population, the following result involving a t-distribution is useful in applications.

Theorem 4.2.9. *If $\overline{X}$ and S^2 are the mean and the variance of a random sample of size n from a normal population with the mean μ and variance σ^2, then*

$$T = \frac{\overline{X} - \mu}{S/\sqrt{n}},$$

has a t-distribution with (n – 1) degrees of freedom.

Proof. By Corollary 4.2.2,

$$Z = \frac{\overline{X} - \mu}{\sigma/\sqrt{n}} \sim N(0, 1).$$

By Theorem 4.2.8, we have

$$Y = \frac{(n-1)S^2}{\sigma^2} = \frac{1}{\sigma^2}\sum_{i=1}^{n}(X_i - \overline{X})^2 \sim \chi^2(n-1).$$

Hence,

$$T = \frac{\frac{\overline{X} - \mu}{(\sigma/\sqrt{n})}}{\sqrt{\frac{(n-1)S^2}{\sigma^2(n-1)}}} \sim \frac{Z}{\sqrt{\frac{\chi^2(n-1)}{n-1}}}.$$

Also, $\overline{X}$ and S^2 are independent. Thus, Y and Z are independent, and by definition 4.2.2, T follows a *t*-distribution with $(n-1)$ degrees of freedom.

How can we distinguish between given degrees of freedom and the degrees of freedom from a sample? For the *t*-distribution, if n is given as the degrees of freedom, we will just use *n*. However, if a random sample of size *n* is given, then the corresponding degrees of freedom will be $(n-1)$, as given in Theorem 4.2.9.

The assumption that the sample comes from a normal population is not that onerous. In practice, it is necessary to check that the sampled population is approximately bell-shaped and not too much skewed. Construction of the normal-scores plot or histogram is a way to check for approximate normality. See Project 4C.

EXAMPLE 4.2.6

A manufacturer of fuses claims that with 20% overload, the fuses will blow in less than 10 minutes on average. To test this claim, a random sample of 20 of these fuses was subjected to a 20% overload, and the times it took them to blow had a mean of 10.4 minutes and a sample standard deviation of 1.6 minutes. It can be assumed that the data constitute a random sample from a normal population. Do they tend to support or refute the manufacturer's claim?

Solution

Given $\bar{y} = 10.4$, $s = 1.6$, $n = 20$, *and* $\mu = 10$. *Hence,*

$$t = \frac{\bar{y} - \mu}{s/\sqrt{n}} = \frac{10.4 - 10}{1.6/\sqrt{20}} = 1.118.$$

The degree of freedom is n – 1 = 19. From the t-table, the probability that t exceeds 1.328 is 0.10, and because the observed value of t = 1.118 is less than $t_{0.10,19} = 1.328$ *and 0.10 is a pretty large probability, we conclude that the data tend to agree with the manufacturer's claim.*

We will study the problems of the foregoing type in Chapter 6, where we will be learning about hypothesis testing. Prior to Gosset's work on the *t*-distribution, a very large number of observations were necessary for the design and analysis of experiments. Today, the use of the *t*-distribution often makes it possible to draw reliable conclusions from samples as small as 15 to 30 experimental units, provided that the samples are representative of their populations and that normality could reasonably be assumed or justified for the population. Example 4.2.7 suggests that we need to be careful about the use of *t*-distribution. It not only depends on sample size, but also the knowledge deviation.

EXAMPLE 4.2.7

The human gestation period—the period of time between conception and labor—is approximately 40 weeks (280 days), measured from the first day of the mother's last menstrual period. For a newborn full-term infant, the appropriate length for gestational age is assumed to be normally distributed with $\mu = 50$ centimeters and $\sigma = 1.25$ centimeters. Compute the probability that a random sample of 20 infants born at full term results in a sample mean greater than 52.5 centimeters.

Solution

Let X be length (measured in centimeters) of a newborn full-term infant. Then $\overline{X} \sim N(50, 1.56/20)$. Note that even though the sample size is small, since σ is known, we do not use the t-distribution and instead use a normal distribution. Hence,

$$P(\overline{X} > 52.5) = P\left(z > \frac{52.5 - 50}{1.56/\sqrt{20}} = 7.17\right) \approx 0.$$

Thus, the probability of such an occurrence is negligible.

In the previous example, it should be noted that $P(\overline{X} > 52.5) \approx 0$ does not imply that the probability of observing a newborn full-term infant with a length greater than 52.5 centimeters is zero. In fact, with 19 degrees of freedom, $P(X > 52.5) = P(Z > 2) is \approx 0.0228$.

4.2.2 *F*-Distribution

The F-distribution was developed by Fisher to study the behavior of two variances from random samples taken from two independent normal populations. In applied problems, we may be interested in knowing whether the population variances are equal or not, based on the response of the random samples. Knowing the answer to such a question is also important in selecting the appropriate statistical methods to study their true means.

Definition 4.2.3. *Let U and V be chi-square random variables with n1 and n2 degrees of freedom, respectively. Then, if U and V are independent,*

$$F = \frac{U/n_1}{V/n_2}$$

is said to have an **F-distribution** *with n_1 numerator degrees of freedom and n_2 denominator degrees of freedom. We denote this by $F \sim F(n_1, n_2)$.*

The pdf for a random variable $X \sim F(n_1, n_2)$ is given by

$$f(x) = \begin{cases} \frac{\Gamma((n_1+n_2)/2)}{\Gamma(n_1/2)\Gamma(n_2/2)} \left(\frac{n_1}{n_2}\right)^{n_1/2} x^{\frac{n_1}{2}-1} \left(1 + \frac{n_1}{n_2}x\right)^{-(n_1+n_2)/2}, & x > 0 \\ 0, & \text{elsewhere.} \end{cases}$$

A graph of $f(x)$ for various values of n is given in Fig. 4.5.

To find $F_\alpha(n_1, n_2)$ such that $P(F > F_\alpha(n_1, n_2)) = \alpha$ (shaded area in Fig. 4.6), we use the F-table. For example, if F has 3 numerator and 6 denominator degrees of freedom, then $F_{0.01}(3, 6) = 9.78$.

If we know $F_\alpha(n_1, n_2)$, it is possible to find $F_{1-\alpha}(n_2, n_1)$ by using the identity

$$F_{1-\alpha}(n_2, n_1) = 1/F_\alpha(n_1, n_2).$$

Using this identity, we can obtain $F_{0.99}(6, 3) = 1/F_{0.01}(3, 6) = 1/9.78 = 0.10225$.

When we need to compare the variances of two normal populations, we will use the following result.

Theorem 4.2.10. *Let two independent random samples of size n_1 and n_2 be drawn from two normal populations with variances σ_1^2, σ_2^2, respectively. If the variances of the random samples are given by S_1^2, S_2^2, respectively, then the statistic*

$$F = \frac{S_1^2/\sigma_1^2}{S_2^2/\sigma_2^2} = \frac{\sigma_2^2 S_1^2}{\sigma_1^2 S_2^2}$$

has the F-distribution with $(n_1 - 1)$ numerator and $(n_2 - 1)$ denominator degrees of freedom.

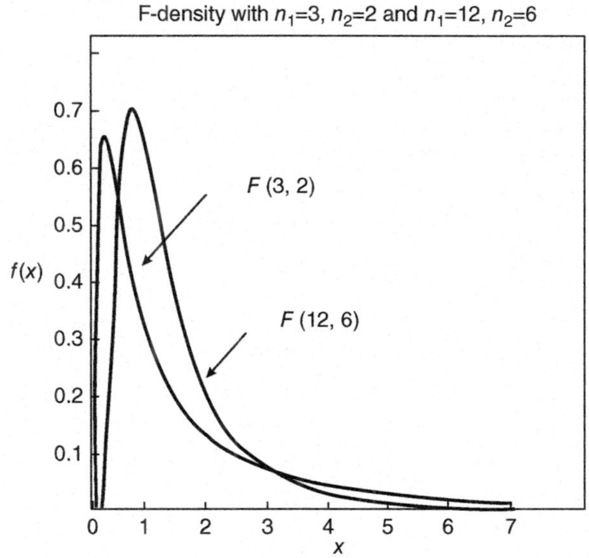

FIGURE 4.5 pdfs of the F-distribution.

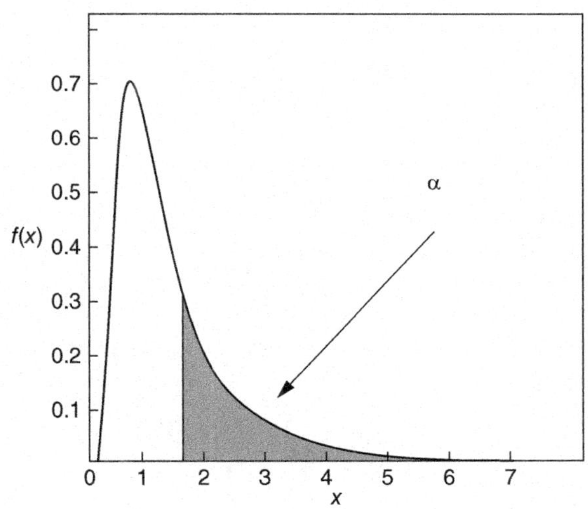

FIGURE 4.6 pdf of F-distribution.

Proof. From Theorem 4.2.9, we know that

$$U = \frac{(n_1 - 1)S_1^2}{\sigma_1^2} \sim \chi^2(n_1 - 1)$$

and

$$V = \frac{(n_2 - 1)S_2^2}{\sigma_2^2} \sim \chi^2(n_2 - 1).$$

Also, U and V are independent. From Definition 4.2.3, $F \sim F(n_1 - 1, n_2 - 1)$.

Corollary 4.2.11. *If* $\sigma_1^2 = \sigma_2^2$, *then*

$$F = \frac{S_1^2}{S_2^2} \sim F(n_1 - 1, n_2 - 1).$$

When $\sigma_1^2 = \sigma_2^2$, we refer to them as two populations that are homogeneous with respect to their variances.

EXAMPLE 4.2.8

Let S_1^2 denote the sample variance for a random sample of size 10 from Population I and let S_2^2 denote the sample variance for a random sample of size 8 from Population II. The variance of Population I is assumed to be three times the variance of Population II. Find two numbers a and b such that $P(a \le S_1^2/S_2^2 \le b) = 0.90$, assuming S_1^2 to be independent of S_2^2.

Solution

From the problem, we can assume that $\sigma_1^2 = 3\sigma_2^2$ *with* $n_1 = 10$ *and* $n_2 = 8$*. Thus, we can write*

$$\frac{S_1^2/\sigma_1^2}{S_2^2/\sigma_2^2} = \frac{S_1^2/3\sigma_2^2}{S_2^2/\sigma_2^2} = \frac{S_1^2}{3S_2^2},$$

this has F-distribution with $n_1 - 1 = 9$ *numerator and* $n_2 - 1 = 7$ *denominator degrees of freedom. Using the F-table,* $F_{0.05}(9, 7) = 3.68$*. Now to find* $F_{0.95}$ *such that*

$$P\left(\frac{S_1^2}{3S_2^2} < F_{0.95}\right) = 0.05.$$

We proceed as follows:

$$P\left(\frac{S_1^2}{3S_2^2} < F_{0.95}\right) = P\left(\frac{3S_2^2}{S_1^2} > \frac{1}{F_{0.95}}\right) = 0.05.$$

Indexing $v_1 = 7$ *and* $v_2 = 9$ *in the F-table, we have* $1/F_{0.95}(7, 9) = F_{0.05}(9, 7) = 3.68$ or $F_{0.95} = 1/3.68 = 0.2717$*. Hence, the entire probability statement is given by*

$$P\left(0.2717 \le \frac{S_1^2}{3S_2^2} \le 3.68\right) = P\left(0.815 \le \frac{S_1^2}{S_2^2} \le 11.04\right) = 0.90.$$

Thus $a = 0.815$ *and* $b = 11.04$.

When we need to find values not given in t-table (or chi-square, F-tables), and those values are not available in the corresponding table, what can we do? Most of the time, it is easier to use statistical programs to obtain so-called p-values (introduced in Chapter 6). In similar situations, most of the values given in the solution manual are obtained in this way. If the software is not available and we need to find these values from one of these (t-, chi-square, or F-) tables, then, either using linear interpolation or transformations, we can obtain approximate values. We will illustrate this only for linear interpolation. Given two points (x_1, y_1) and (x_2, y_2) on a line, any other point (x, y) on the line satisfies the following equation

$$y = y_1 + \frac{y_2 - y_1}{x_2 - x_1}(x - x_1).$$

Using this relationship, let us say we want to find the t-value for $\alpha = 0.15$ with 6 degrees of freedom. In our table, we have t-values for $\alpha = 0.1$ *and* $\alpha = 0.25$. Then

$$\begin{aligned} t_{0.15,6} &\approx 1.439756 + \frac{(0.717558 - 1.439756)}{(0.25 - 0.1)}(0.15 - 0.1) \\ &= 1.199023. \end{aligned}$$

We will use this method for finding critical values of t-, chi-square, and F-values that are not available in the table.

EXERCISES 4.2

4.2.1. Let Y have a chi-square distribution with 15 degrees of freedom.

(a) Find y_0 if $P(Y \le y_0) = 0.025$.

(b) Find a and b if $P(a < Y < b) = 0.95$.

(c) Find the probability, $P(Y \ge 22.307)$.

4.2.2. Let Y have a chi-square distribution with 7 degrees of freedom.

(a) Find y_0 if $P(Y > y_0) = 0.025$

(b) Find a and b if $P(a < Y < b) = 0.90$

(c) Find the probability, $P(Y > 1.239)$.

4.2.3. The time to failure T of a microwave oven has an exponential distribution with pdf

$$f(t) = \frac{1}{2}e^{-t/2}, \quad t > 0.$$

If three such microwave ovens are chosen and $\overline{T}$ is the mean of their failure times, find the following:

(a) Distribution of $\overline{T}$.
(b) $P(\overline{T} > 2)$.

4.2.4. Let $X_1, X_2, \ldots, X_{10}$ be a random sample from a standard normal distribution. Find the numbers a and b such that

$$P\left(a \le \sum_{i=1}^{10} X_i^2 \le b\right) = 0.95.$$

4.2.5. Let $X_1, X_2, \ldots, X_5$ be a random sample from the normal distribution with mean 55 and variance 223. Let

$$Y = \sum_{i=1}^{5} (X_i - 55)^2/223$$

and

$$U = \sum_{i=1}^{5} (X_i - \overline{X})^2/223.$$

(a) Find the distribution of the random variables Y and Z.
(b) Are Y and Z independent?
(c) Find **(i)** $P(0.554 \le Y \le 0.831)$, and **(ii)** $P(0.297 \le U \le 0.484)$.

4.2.6. Let X and Y be independent chi-square random variables with 14 and 5 degrees of freedom, respectively. Find

(a) $P(|X - Y| \le 11.15)$,
(b) $P(|X - Y| \ge 3.8)$.

4.2.7. A particular type of vacuum-packed coffee packet contains an average of 16 ounces. It has been observed that the number of ounces of coffee in these packets is normally distributed with $\sigma = 1.41$ ounces. A random sample of 15 of these coffee packets is selected, and the observations are used to calculate s. Find the numbers a and b such that $P(a \le S^2 \le b) = 0.90$.

4.2.8. An optical firm buys glass slabs to be ground into lenses, and it is known that the variance of the refractive index of the glass slabs is to be no more than 1.04×10^{-3}. The firm rejects a shipment of glass slabs if the sample variance of 16 pieces selected at random exceeds 1.15×10^{-3}. Assuming that the sample values may be looked on as a random sample from a normal population, what is the probability that a shipment will be rejected even though $\sigma^2 = 1.04 \times 10^{-3}$?

4.2.9. Assume that T has a t-distribution with 8 degrees of freedom. Find the following probabilities.

(a) $P(T \le 2.896)$.
(b) $P(T \le -1.860)$.
(c) The value of a such that $P(-a < T < a) = 0.99$.

4.2.10. Assume that T has a t-distribution with 15 degrees of freedom. Find the following probabilities.

(a) $P(T \le 1.341)$.
(b) $P(T \ge -2.131)$.
(c) The value of a such that $P(-a < T < a) = 0.95$.

4.2.11. A psychologist claims that the mean age at which female children start walking is 11.4 months. If 20 randomly selected female children are found to have started walking at a mean age of 12 months with a standard deviation of 2 months, would you agree with the psychologist's claim? Assume that the sample came from a normal population.

4.2.12. Let U_1 and U_2 be independent random variables. Suppose that U_1 is χ^2 with v_1 degrees of freedom, while $U = U_1 + U_2$ is chi-square with v degrees of freedom, where $v > v_1$. Then prove that U_2 is a chi-square random variable with $v - v_1$ degrees of freedom.

4.2.13. Show that if $X \sim \chi^2\,(v)$, then $EX = v$ and $Var\,(X) = 2v$.

4.2.14. Let $X_1, \ldots, X_n$ be a random sample with $X_i \sim \chi^2\,(1)$, for $i = 1, \ldots, n$. Show that the distribution of

$$Z = \frac{\overline{X} - 1}{\sqrt{2/n}}$$

as $n \to \infty$ is standard normal.

4.2.15. Find the variance of S^2, assuming the sample $X_1, X_2, \ldots, X_n$ is from $N(\mu, \sigma^2)$.

4.2.16. Let X_1, $X2$, ...,Xn be a random sample from an exponential distribution with parameter θ. Show that the random variable $2\theta^{-1}(\sum_{i=1}^{n} X_i) \sim \chi^2(2n)$.

4.2.17. Let X and Y be independent random variables from an exponential distribution with common parameter $\theta = 1$. Show that X/Y has an F-distribution. What is the number for degrees of freedom?

4.2.18. Prove that if X has a t-distribution with n degrees of freedom, then $X^2 \sim F\,(1, n)$.

4.2.19. Let X be F distributed with 9 numerator and 12 denominator degrees of freedom. Find

(a) $P(X \le 3.87)$.

(b) $P(X \le 0.196)$.

(c) The value of a and b such that $P\,(a < Y < b) = 0.95$.

4.2.20. Prove that if $X \sim F(n_1,n_2)$, then $1/X \sim F(n_2,n_1)$.

4.2.21. Find the mean and variance of $F\,(n_1, n_2)$ random variable.

4.2.22. Let $X_{11}, X_{12}, \ldots, X_{1n_1}$ be a random sample with sample mean $\overline{X}_1$ from a normal population with mean μ_1 and variance σ_1^2, and let $X_{21}, X_{22}, \ldots, X_{2n_2}$ be a random sample with sample mean $\overline{X}_2$ from a normal population with mean μ_2 and variance σ_2^2. Assume the two samples are independent. Show that the sampling distribution of $(\overline{X}_1 - \overline{X}_2)$ is normal with mean $\mu_1 - \mu_2$ and variance $\sigma_1^2/n_1 + \sigma_2^2/n_2$.

4.2.23. Let $X_1, X_2, \ldots, X_{n1}$ be a random sample from a normal population with mean μ_1 and variance σ^2, and $Y_1, Y_2, \ldots, Y_{n2}$ be a random sample from an independent normal population with mean μ_2 and variance σ^2. Show that

$$T = \frac{(\overline{X} - \overline{Y}) - (\mu_1 - \mu_2)}{\sqrt{\frac{(n_1 - 1)S_1^2 + (n_2 - 1)S_2^2}{n_1 + n_2 - 2}\left(\frac{1}{n_1} + \frac{1}{n_2}\right)}} \sim T_{(n_1+n_2-2)}.$$

4.2.24. Show that a t-distribution tends to a standard normal distribution as the degrees of freedom tend to infinity.

4.2.25. Show that the mgf of a χ^2 random variable with n degrees of freedom is $M(t)=(1-2t)^{-n/2}$. Using the mgf, show that the mean and variance of a chi-square distribution are n and $2n$, respectively.

4.2.26. Let the random variables $X_1, X_2, \ldots, X_{10}$ be normally distributed with mean 8 and variance 4. Find a number a such that

$$P\left(\sum_{i=1}^{10}\left(\frac{X_i - 8}{2}\right)^2 \le a\right) = 0.95.$$

4.2.27. Let $X^2 \sim F(1,n)$. Show that $X \sim t(n)$.

4.3 ORDER STATISTICS

In practice, the random variables of interest may depend on the relative magnitudes of the observed variable. For example, we may be interested in the maximum mileage per gallon of a particular class of cars. In this section, we study the behavior of ordering a random sample from a continuous distribution.

Definition 4.3.1. *Let $X_1, \ldots, X_n$ be a random sample from a continuous distribution with pdf $f(x)$. Let $Y_1, \ldots, Y_n$ be a permutation of $X_1, \ldots, X_n$ such that*

$$Y_1 \le Y_2 \le \cdots \le Y_n.$$

Then the ordered random variables $Y_1, \ldots, Y_n$ are called the **order statistics** *of the random sample $X_1, \ldots, X_n$. Here Y_k is called the* **kth order statistic**. *Because of continuity, the equality sign could be ignored.*

Remark. Although X_i's are iid random variables, the random variables Y_i's are neither independent nor identically distributed.

Thus, the minimum of X_i's is

$$Y_1 = min\ (X_1, \ \ldots, X_n)$$

and the maximum is

$$Y_n = max\ (X_1, \ \ldots, X_n).$$

The order statistics of the sample $X_1, X_2, \ldots, X_n$ can also be denoted by $X_{(1)}, X_{(2)}, \ldots, X_{(n)}$ where

$$X_{(1)} < X_{(2)} < \ldots < X_{(n)}.$$

Here, $X_{(k)}$ is the kth order statistic and is equal to Y_k in Definition 4.3.1. One of the most commonly used order statistics is the median, the value in the middle position in the sorted order of the values.

EXAMPLE 4.3.1

(i) The range $R = Y_n - Y_1$ is a function of order statistics.
(ii) The sample median M equals Y_{m+1} if $n = 2m + 1$.

Hence, the sample median M is an order statistic, when n is odd. If n is even then the sample median can be obtained using the order statistic, $M = (1/2)\ [Y_{n/2} + Y_{(n/2)+1}]$.

The following result is useful in determining the distribution of functions of more than one order statistics.

Theorem 4.3.1. *Let $X_1, \ldots, X_n$ be a random sample from a population with pdf $f(x)$. Then the joint pdf of order statistics $Y_1, \ldots, Y_n$ is*

$$f\,(y_1, \ldots, y_n) = \begin{cases} n!f\,(y_1)f\,(y_2)\ldots f\,(y_n), & \text{for } y_1 < \cdots < y_n \\ 0, & \text{otherwise.} \end{cases}$$

The pdf of the kth order statistic is given by the following theorem.

Theorem 4.3.2. *The pdf of Y_k is*

$$f_k\,(y) = f_{Y_k}\,(y) = \frac{n!}{(k-1)!(n-k)!} f\,(y)\,(F\,(y))^{k-1}(1 - F\,(y))^{n-k},$$

for $-\infty < y < \infty$, where $F(y)=P(X_i \le y)$ is the cdf of X_i.

In particular, the pdf of Y_1 is $f_1\,(y) = nf\,(y)\,[1 - F(y)]^{n-1}$ and the pdf of Y_n is $f_n(y) = nf(y)[F(y)]^{n-1}$. In the following example, we will derive pdf for Y_n.

EXAMPLE 4.3.2

Let $X_1, \ldots, X_n$ be a random sample from $U[0,1]$. Find the pdf of the kth order statistic Y_k.

Solution

Since the pdf of X_i is $f(x)=1, 0 \le x \le 1$, the cdf is $F(x)=x,\ 0 \le x \le 1$. Using Theorem 4.3.2, the pdf of the kth order statistic Y_k reduces to

$$f_k(y) = \frac{n!}{(k-1)!(n-k)!} y^{k-1}(1-y)^{n-k}, \quad 0 \le y \le 1$$

which is a beta distribution with $\alpha = k$ and $\beta = n - k + 1$.

The next example gives the so-called extreme (i.e., largest) value distribution, which is the distribution of the order statistic Y_n.

EXAMPLE 4.3.3

Find the distribution of the nth order statistic Y_n of the sample $X_1, \ldots, X_n$ from a population with pdf $f(x)$.

Solution

Let the cdf of Y_n be denoted by $F_n(y)$. Then

$$F_n(y) = P(Y_n \le y) = P\left(\max_{1\le i\le n} X_i \le y\right)$$
$$= P(X_1 \le y, \ldots, X_n \le y) = [F(y)]^n \text{ (by independence)}.$$

Hence, the pdf $f_n(y)$ of Y_n is

$$f_n(y) = \frac{d}{dy}[F(y)]^n = n[F(y)]^{n-1}\frac{d}{dy}F(y)$$
$$= n[F(y)]^{n-1}f(y).$$

In particular, if $X_1, \ldots, X_n$ is a random sample from $U[0, 1]$, then the cumulative extreme value distribution is given by

$$F_n(y) = \begin{cases} 0, & y < 0 \\ y^n, & 0 \le y \le 1 \\ 1, & y > 1. \end{cases}$$

EXAMPLE 4.3.4

A string of 10 light bulbs is connected in series, which means that the entire string will not light up if any one of the light bulbs fails. Assume that the lifetimes of the bulbs, $\tau_1, \ldots, \tau_{10}$, are independent random variables that are exponentially distributed with mean 2. Find the distribution of the life length of this string of light bulbs.

Solution

Note that the pdf of τ_i is $f(t)=2e^{-2t}$, $0 < t < \infty$, and the cumulative distribution of τ_i is $F_{\tau i}(t) = 1 - e^{-2t}$. Let T represent the lifetime of this string of light bulbs. Then,

$$T = \min(\tau_1, \ldots, \tau_{10}).$$

Thus,

$$F_T(t) = 1 - \left[1 - F_{\tau_i}(t)\right]^{10}.$$

Hence, the density of T is obtained by differentiating $F_T(t)$ with respect to t, that is,

$$f_T(t) = 10 f_{\tau i}(t)[1 - F_{\tau i}(t)]^9$$
$$= \begin{cases} 2(10)e^{-2t}(e^{-2t})^9 = 20e^{-20t}, & 0 < t < \infty \\ 0, & \text{otherwise.} \end{cases}$$

The joint pdf of the order statistics is given by the following result.

Theorem 4.3.3. *Let $X_1, \ldots, X_n$ be a random sample with continuous probability density function $f(x)$ and a distribution function $F(x)$. Let $Y_1, \ldots, Y_n$ be the order statistics. Then for any $1 \le i < k \le n$ and $-\infty < x \le y < \infty$, the joint pdf of Y_i and Y_k is given by*

$$f_{Y_i,Y_k}(x, y) = \frac{n!}{(i-1)!(k-i-1)!(n-k)!}[F(x)]^{i-1}$$
$$\times [F(y) - F(x)]^{k-i-1}[1 - F(y)]^{n-k} f(x) f(y)$$

EXAMPLE 4.3.5

Let $X_1, \ldots, X_n$ be a random sample from $U[0,1]$. Find the joint pdf of Y_2 and Y_5.

Solution

Taking $i = 2$ and $k = 5$ in Theorem 4.3.3, we get the joint pdf of Y_2 and Y_5 as

$$f_{Y_2,Y_5}(x, y) = \frac{n!}{(2-1)!(5-2-1)!(n-5)!}[F(x)]^{2-1}$$
$$[F(y) - F(x)]^{5-2-1} \times [1 - F(y)]^{n-5} f(x) f(y)$$
$$= \begin{cases} \frac{n!}{2(n-5)!} x(y-x)^2(1-y)^{n-5}, & 0 < x \le y < 1 \\ 0, & \text{otherwise.} \end{cases}$$

EXERCISES 4.3

4.3.1 The lifetime X of a certain electrical fuse has the following probability density function

$$f(x) = \begin{cases} \frac{1}{10} e^{-x/10}, & x > 0 \\ 0, & \text{otherwise.} \end{cases}$$

Suppose two such fuses are in series and operate independently in a system. Find the pdf of the lifetime Y of the system. (The system will work only if both of the fuses operate.)

4.3.2 Suppose that the time between two telephone calls at an office, in minutes, is uniformly distributed on the interval [0, 20]. If there were 15 calls, **(i)** what is the probability that the longest time interval between calls is less than 15 minutes? **(ii)** What is the probability that the shortest time interval between calls is greater than 5 minutes?

4.3.3 Let X_1, X_2, X_3 be three random variables of discrete type. Let X_1, X_2 take values 0, 1, and X_3 take values 1, 2, 3. What are the values of the order statistics Y_1, Y_2, Y_3?

4.3.4 Let $X_1, \ldots, X_{10}$ be a random sample from $U[0, 1]$. Find the joint density of Y_2 and Y_7, where Y_i, $i = 1, 2, \ldots, 10$ are order statistics of $X_1, \ldots, X_{10}$.

4.3.5 Let $X_1, \ldots, X_n$ be a random sample from exponential distribution with a mean of θ. Show that $Y_1 = \min(X_1, X_2, \ldots, X_n)$ has an exponential distribution with mean θ/n. Also, find the pdf of $Y_n = \max(X_1, X_2, \ldots, X_n)$.

4.3.6 A string of 10 light bulbs is connected in parallel, which means that the entire string will fail to light up only if all 10 of the light bulbs fail. Assume that the lifetimes of the bulbs, $\tau_1, \ldots, \tau_{10}$, are independent random variables that are exponentially distributed with mean θ. Find the distribution of the lifetimes of this string of light bulbs.

4.3.7 Let $X_1, \ldots, X_n$ be a random sample from the uniform distribution $f(x) = 1/2$, $0 \le x \le 2$. Find the probability density function for the range $R = (X_{(n)} - X_{(1)})$.

4.3.8 Given a sample of 25 observations from a distribution with pdf

$$f(x) = \begin{cases} e^{-x}, & x > 0 \\ 0, & \text{otherwise} \end{cases}$$

let M be the sample median. Compute $P(M \ge b)$.[Hint: Note that M is the 13th order statistic.]

4.3.9 Let $X_1, \ldots, X_n$ be a random sample from a normal population with mean 10 and variance 4. What is the probability that the largest observation is greater than 10?

4.3.10 Let $X_1, \ldots, X_n$ be a random sample from an exponential population with parameter θ. Let $Y_1, \ldots, Y_n$ be the ordered random variables.

(a) Show that the sampling distributions of Y_1 and Y_n are given by

$$f_1(y_1) = \begin{cases} \frac{n}{\theta} e^{-ny_1/\theta}, & \text{if } y_1 > 0 \\ 0, & \text{otherwise,} \end{cases}$$

and

$$f_n(y_n) = \begin{cases} \frac{n}{\theta} e^{-y_n/\theta} [1 - e^{-y_n/\theta}]^{n-1}, & \text{if } y_n > 0 \\ 0, & \text{otherwise.} \end{cases}$$

(b) Let $n = 2l + 1$. Show that the sampling distribution of the median, M, is given by

$$f(m) = \begin{cases} \frac{n!}{(l!)^2\theta} e^{-m(l+1)/\theta}[1 - e^{-m/\theta}]^l, & \text{for } m > 0 \\ 0, & \text{otherwise.} \end{cases}$$

4.3.11 Let $X_1, \ldots, X_n$ be a random sample from a beta distribution with $\alpha = 2$ and $\beta = 3$. Find the joint pdf of Y_1 and Y_n.
4.3.12 Let $X_1, \ldots, X_n$ be a random sample from a geometric distribution with pmf

$$p_i = P(X = i) = pq^{i-1}, \ i = 1, 2, \ldots, 0 < p < 1, \ q = 1 - p.$$

Show that

$$P(Y_k = y) = \Sigma_{i=k}^{n} \binom{n}{i} q^{(y-1)(n-i)} \{q^{n-i}[1 - q^y]^i - [1 - q^{y-1}]^i\},$$

$$y = 1, 2, \ldots .$$

4.4 THE NORMAL APPROXIMATION TO THE BINOMIAL DISTRIBUTION

We know that a binomial random variable Y, with parameters n and $p = P$(success), can be viewed as the number of successes in n trials and can be written as

$$Y = \sum_{i=1}^{n} X_i$$

where,

$$X_i = \begin{cases} 1 & \text{with probability } p \\ 0 & \text{with probability } (1 - p). \end{cases}$$

The fraction of successes in n trials is

$$\frac{Y}{n} = \frac{1}{n} \sum_{i=1}^{n} X_i = \overline{X}.$$

Hence, Y/n is a sample mean. Since $E(X_i) = p$ and $Var(X_i) = p(1 - p)$, we have

$$E\left(\frac{Y}{n}\right) = E\left(\frac{1}{n} \sum_{i=1}^{n} X_i\right) = \frac{1}{n} np = p$$

and

$$Var\left(\frac{Y}{n}\right) = \frac{1}{n^2} \sum_{i=1}^{n} Var(X_i) = \frac{p(1-p)}{n}.$$

Because $Y = n\overline{X}$, by the Central Limit Theorem, Y has an approximate normal distribution with mean $\mu = np$ and variance $\sigma^2 = np(1 - p)$. Because the calculation of the binomial probabilities is cumbersome for large sample sizes n, the normal approximation to the binomial distribution is widely used. A useful rule of thumb for use of the normal approximation to the binomial distribution is to make sure n is large enough if $np \geq 5$ and $n(1 - p) \geq 5$. Otherwise, the binomial distribution may be so asymmetric that the normal distribution may not provide a good approximation. Other rules, such as $np \geq 10$ and $n(1 - p) \geq 10$, or $np(1 - p) \geq 10$, are also used in the literature. Because all of these rules are only approximations, for consistency's sake we will use $np \geq 5$ and $n(1 - p) \geq 5$ to test for largeness of sample size in the normal approximation to the binomial distribution. If need arises, we could use the more stringent condition $np(1 - p) \geq 10$.

Recall that discrete random variables take no value between integers, and their probabilities are concentrated at the integers as shown in Fig. 4.7. However, the normal random variables have zero probability at these integers; they have nonzero probability only over intervals. Because we are approximating a discrete distribution with a continuous distribution, we need to introduce a correction factor for continuity which is explained below.

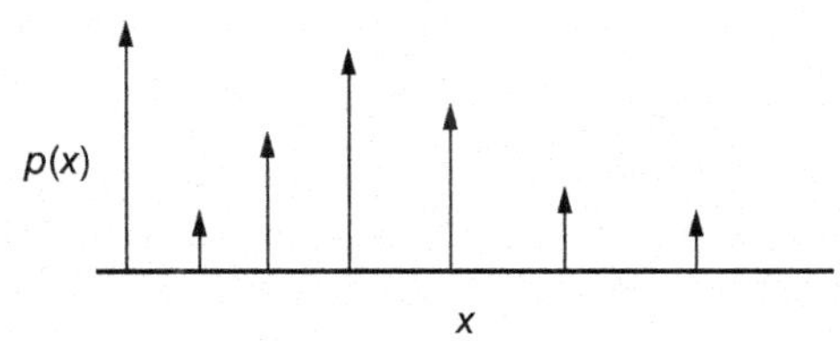

FIGURE 4.7 Probability function of discrete r.v. X. The diagram depicts discrete probability p of x with vertical spikes of varying heights across x representing the corresponding probabilities.

CORRECTION FOR CONTINUITY FOR THE NORMAL APPROXIMATION TO THE BINOMIAL DISTRIBUTION

(a) To approximate $P(X \le a)$ or $P(X > a)$, the correction for continuity is $(a + 0.5)$, that is,

$$P(X \le a) = P\left(Z < \frac{(a + 0.5) - np}{\sqrt{np(1 - p)}}\right)$$

and

$$P(X > a) = P\left(Z > \frac{(a + 0.5) - np}{\sqrt{np(1 - p)}}\right).$$

(b) To approximate $P(X \ge a)$ or $P(X < a)$, the correction for continuity is $(a - 0.5)$, that is,

$$P(X \ge a) = P\left(Z > \frac{(a - 0.5) - np}{\sqrt{np(1 - p)}}\right)$$

and

$$P(X < a) = P\left(Z < \frac{(a - 0.5) - np}{\sqrt{np(1 - p)}}\right).$$

(c) To approximate $P(a \le X \le b)$, treat ends of the intervals separately, calculating two distinct z-values according to steps (a) and (b), that is,

$$P(a \le X \le b) = P\left(\frac{(a - 0.5) - np}{\sqrt{np(1 - p)}} < Z < \frac{(b + 0.5) - np}{\sqrt{np(1 - p)}}\right).$$

(d) Use the normal table to obtain the approximate probability of the binomial event.

The shaded area in Fig. 4.8 represents the continuity correction for $P(X = i)$.

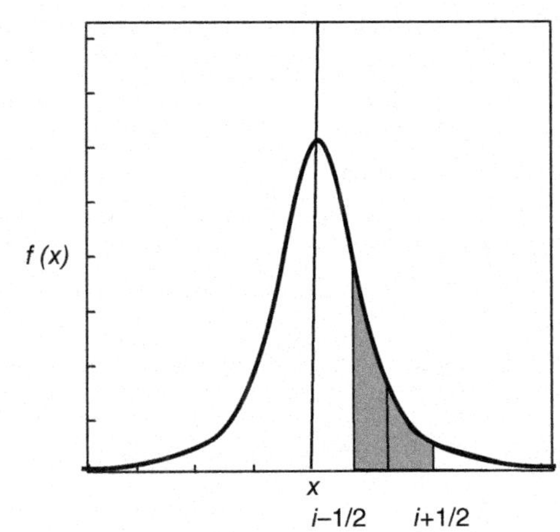

FIGURE 4.8 Continuity correction for $P(X = i)$.

EXAMPLE 4.4.2

A study of parallel interchange ramps revealed that many drivers do not use the entire length of parallel lanes for acceleration, but seek, as soon as possible, a gap in the major stream of traffic to merge. At one site on Interstate Highway 75, 46% of drivers used less than one-third of the lane length available before merging. Suppose we monitor the merging pattern of a random sample of 250 drivers at this site.

(a) What is the probability that fewer than 120 of the drivers will use less than one-third of the acceleration lane length before merging?
(b) What is the probability that more than 225 of the drivers will use less than one-third of the acceleration lane length before merging?

Solution

First we check for adequacy of the sample size:

$$np = (250)(0.46) = 115 \text{ and } n(1-p) = (250)(1-0.46) = 135.$$

Both are greater than 5. Hence, we can use the normal approximation. Let X be the number of drivers using less than one-third of the lane length available before merging. Then X can be considered to be a binomial random variable. Also,

$$\mu = np = (250)(0.46) = 115.0$$

and

$$\sigma = \sqrt{np(1-p)} = \sqrt{250(0.46)(0.54)} = 7.8804.$$

Thus,

(a) $P(X < 120) = P\left(Z < \frac{119.5 - 115}{7.8804} = 0.57103\right) = 0.7157$, *that is, we are approximately 71.57% certain that fewer than 120 drivers will use less than one-third of the acceleration length before merging.*

(b) $P(X > 225) = P\left(Z > \frac{225.5 - 115}{7.8804} = 14.02213\right) \approx 0$, *that is, there is almost no chance that more than 225 drivers will use less than one-third of the acceleration lane length before merging.*

EXERCISES 4.4

4.4.1 Suppose X is a binomial random variable with $n = 20$ and $p = 0.2$. Find the probability that $X \le 10$ using binomial tables and compare this to the corresponding value found from normal approximation.

4.4.2 Using normal approximation, find the probability of obtaining at least 90 heads in 150 tosses of a fair coin. Is the normal approximation valid? Why?

4.4.3 A car rental company finds that each day, 6% of the persons making reservations will not show up. If the rental company reserves for 215 persons with only 200 automobiles, what is the probability that an automobile will be available for every person who shows up holding a reservation? (Use the normal approximation.)

4.4.4 The president of the United States is thought to have a positive approval rating of 58% of the people at a certain time. In a random sample of 1200 people, what is the approximate probability that the number of positive approvals will be at least 750? Interpret your results and state any assumptions.

4.4.5 In the United States, sudden infant death syndrome (SIDS) is one of the leading causes of post neonatal deaths (those occurring between the ages of 28 days and 1 year). Thus far, the most significant risk factor discovered for SIDS is placing babies to sleep in a prone position (on their stomachs). Suppose the rate of death due to SIDS is 0.00103 per year. In a random sample of 5000 infants between the ages of 28 days and 1 year, what is the approximate probability that the number of SIDS-related deaths will be at least 10? Interpret your results and state any assumptions.

4.4.6 Let X and Y be independent binomial random variables with parameters (n, p_1) and (m, p_2), respectively.

(a) Find $E\left(\frac{X}{n} - \frac{Y}{m}\right)$.

(b) Find $Var\left(\frac{X}{n} - \frac{Y}{m}\right)$.

(c) Show that $\left(\frac{X}{n} - \frac{Y}{m}\right) \sim N\left(E\left(\frac{X}{n} - \frac{Y}{m}\right), Var\left(\frac{X}{n} - \frac{Y}{m}\right)\right)$, for large m and n.

4.5 CHAPTER SUMMARY

In this chapter, we learned about sampling distributions. In sampling distributions associated with normal populations, we have seen that we can generate chi-square, t-, and F-distributions. In Section 4.3, we dealt with order statistics. Then, in Section 4.4, we looked at large sample approximations such as the normal approximation to the binomial distribution. In the following section, we will give Software examples to show how the idea of sampling distribution can be explored using statistical software.

We will now list some of the key definitions introduced in this chapter.

- Sampling distribution
- Sample and sample size
- Random sample
- Statistic
- Standard error
- Finite population correction factor
- Degrees of freedom
- t-distribution
- F-distribution
- Order statistics

In this chapter, we have also presented the following important concepts and procedures:

- Sampling distribution associated with normal distribution
- Results on chi-square distribution
- Results on Student t-distribution
- Results on F-distribution
- Derivation of probability density functions for order statistics
- Large sample approximations
- Normal approximation to the binomial
- Correction for continuity for the normal approximation to the binomial distribution

4.6 COMPUTER EXAMPLES

4.6.1 Examples Using R

Note: For the following problems you're generating random samples, and your answers will vary!

Example 4.6.1 Generating Normal Random Samples

Create three samples of size 30 from a standard normal distribution and draw histograms for each sample.

Notice the last two arguments are the mean and standard deviation of the distribution 0, 1. Additionally, plot a density curve over the histogram. Only one output is shown for this example.

R Code:

```
sample1=rnorm(30,0,1);
sample2=rnorm(30,0,1);
sample3=rnorm(30,0,1);
hist(sample1,prob=T);
lines(density(sample1),col="red");
hist(sample2,prob=T);
lines(density(sample2),col="red");
hist(sample3,prob=T);
lines(density(sample3),col="red");
```

Output:

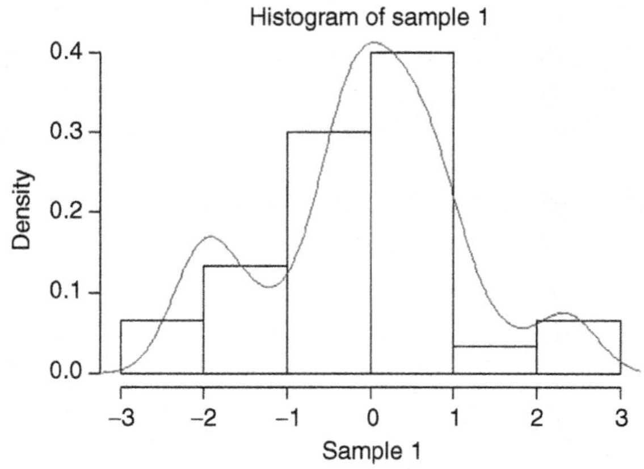

Example 4.6.2 Generating a normal random sample

Generate 50,000 observations from a normal distribution with mean 30 and standard deviation 8. Obtain summary statistics for this data and draw a graph.

R Code:

```
sample=rnorm(50000,30,8);
summary(sample);
sd(sample);
hist(sample,prob=T);
lines(density(sample),col="red");
```

Output:

```
   Min. 1st Qu  Median  Mean 3rd Qu  Max.
-0.08056 24.62000 30.01000 30.03000 35.42000 60.82000
7.981699
```

Standard deviation

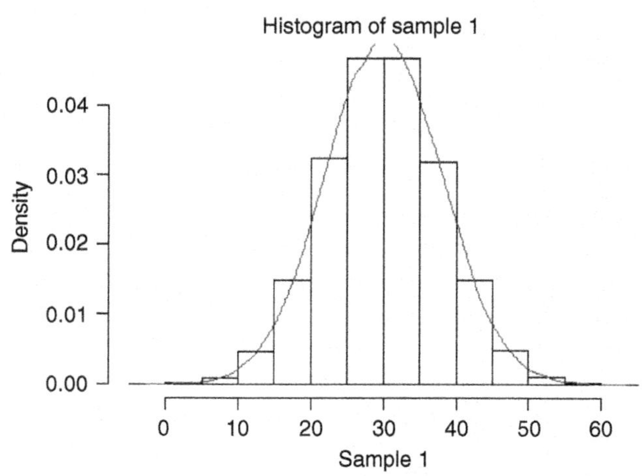

Example 4.6.3 Generating a random Exponential sample

From an exponential distribution, draw 10,000 samples, each sample of size 15. Compute the mean of each sample and draw a chart for the means. This will be an approximate sampling distribution of $\overline{x}$ for a fixed sample of size 15.

R Code:

```
samples_means=c();     ##Creates an empty array for us to store the means in.
for(i in 1:10000) {    ## This for loop repeats the code inside it change variable i over the range
```

```
  sample=rexp(15,3);     ##Generates a random sample of 15 from an exponential.
  mean=mean(sample); ## calculates the mean of that sample.
  samples_means=c(sample_mean, mean);## store the mean inside our array for later use.
}
hist(samples_means,prob=T);##Use previous methods to check the distribution of the means.
lines(density(samples_means),col="red");
summary(samples_means);
sd(samples_means);
```

Output:

No output given for this particular problem, please see the graph generated by R.
You have stored the samples_means in this variable. Use previous analysis methods on this variable.

4.6.2 SPSS Examples

If we have the full version of SPSS, we can write code that can be used to simulate a sampling distribution with different values of p. However, with the student version, it is not easy to simulate. Therefore, we will not give SPSS examples in this chapter.

4.6.3 SAS Examples

EXAMPLE 4.6.4

Generate 50,000 observations from a normal distribution with mean 30 and standard deviation 8. Obtain summary statistics for these data and draw a graph.

Solution

We could use the following program.

```
title '50000 Obs Sample from a Normal Distribution';
title2 'with Mean=30 and Standard Deviation=8';
data normaldat;
  do n=1 to 50000;
      X=8*rannor(55)+30;
      output;
  end;
run;
proc univariate data=normaldat;
  var x;
run;
proc chart;
    vbar x / midpoints=6 to 54 by 2;
    format x msd.;
run;
```

*In the foregoing program, rannor(**55**), the number 55 is just a seed number to obtain the same series of random numbers each time we run the program. If we use "0", each time we run the program we will get a different set of random numbers. We will not give the output.*

EXAMPLE 4.6.5

From an exponential distribution, draw 10,000 samples, each sample of size 15. Compute the mean of each sample and draw a chart for the means. This will be an approximate sampling distribution of $\overline{X}$ for a fixed sample of size 15.

Solution

Use the following program.

```
title '10000 Sample Means with 15 Obs per Sample';
title2 'Drawn from an Exponential Distribution';
```

```
data sample15;
  do Sample=1 to 10000;
      do n=1 to 15;
        X=ranexp(3);
          output;
      end;
    end;
  proc means data=sample 15 noprint;
  output out=mean 15 mean=Mean;
  var x;
  by sample;
run;
proc chart data=mean 15;
    vbar mean/axis=1800
        midpoints=0.10 to 2.05 by.1;
run;
proc univariate data=mean4 noextrobs=0 normal
        mu0=1;
  var mean;
run;
```

This will produce an approximate sampling distribution of $\overline{X}$. We will not give the output.

4.7 PROJECTS FOR CHAPTER 4

4.7.1 A Method to Obtain Random Samples From Different Distributions

Most of the statistical software packages contain a random number generator that produces approximations to random numbers from the uniform distribution U [0, 1]. To simulate the observation of any other continuous random variables, we can start with uniform random numbers and associate these to the distribution we want to simulate. For example, suppose we wish to simulate an observation from the exponential distribution

$$F(x) = 1 - e^{-0.5x}, \quad 0 < x < \infty.$$

First, produce the value of y from the uniform distribution. Then solve for x from the equation

$$y = F(x) = 1 - e^{-0.5x}.$$

So $x = [-\ln(1 - y)]/0.5$ is the corresponding value of the exponential random variable. For instance, if $y = 0.67$, then $x = [-\ln(1 - y)]/0.5 = 2.2173$. If we wish to simulate a sample from the distribution F from the different values of y obtained from the uniform distribution, the procedure is repeated for each new observation x.

(a) Simulate 10 observations of a random variable having exponential distribution with mean and standard deviation both equal to 2.

(b) Select 1500 random samples of size $n = 10$ measurements from a population with an exponential distribution with mean and standard deviation both equal to 2. Calculate the sample mean for each of these 1500 samples and draw a relative frequency histogram. Based on Theorems 4.1.1 and 4.4.1, what can you conclude?

It should be noted that in general, if $Y \sim U(0, 1)$ random variable, then we can show that $X = -\frac{\ln Y}{\lambda}$ will give an exponential random variable with parameter λ. Uniform random variables could also be used to generate random variables from other distributions. For example, let U_i's be iid U[0, 1] random variables. Then,

$$X = -2\sum_{i=1}^{v} \ln(U_i) \sim \chi^2_{2v},$$

and

$$Y = -\beta\sum_{i=1}^{\alpha} \ln(U_i) \sim Gamma\ (\alpha, \beta).$$

Of course, these transformations are useful only when ν and α are integers. More efficient methods based on Monte Carlo simulations, such as MCMC methods, are discussed in Chapter 13.

4.7.2 Simulation Experiments

When the derivation via probability rules is too difficult or complicated to be carried out, one can use simulation experiments to obtain information about a statistic's sampling distribution. The following characteristics of the experiment must be specified:

(i) The population distribution (normal with $\mu = 10$ and $\sigma = 2$, exponential with $\lambda = 5$, etc.)
(ii) The sample size n and the statistic of interest ($\overline{X}$, S, etc.)
(iii) The number of replications k (such as $k = 300$)

Then, using a computer program, obtain k different random samples, each of size n, from the designated population distribution. Calculate the value of the statistic for each of the k replications. Construct a histogram for this k statistic. This histogram gives the approximate sampling distribution of the statistic. The larger the value of k, the better will be the approximation.

(a) For your simulation study, use the population distribution as normal with $\mu = 3.4$ and $\sigma = 1.2$.
(b) For $n = 8$, perform $k = 500$ replications and draw a histogram for values of the sample means. Repeat the experiment with $n = 15$, $n = 25$, and $n = 35$ and draw the histograms. Based on this exercise, you will be able to intuitively verify the result that $\overline{X}$ based on a large n tends to be closer to μ than does $\overline{X}$ based on a small n.
(c) Repeat the experiment of part (a) with different values of k, such as $k = 200$, $k = 750$, and $k = 1000$.
(d) Repeat the simulation study with different distributions, such as exponential distribution.

4.7.3 A Test for Normality

Many statistical procedures require that the population be at least approximately normal. Therefore, a procedure is needed for checking that the sampled data could have come from a normal distribution. There are many procedures, such as the normal-score plot, or Lilliefors test for normality, available in statistics for this purpose. We will describe the *normal-score plot*, which is an effective way to detect deviations from normality. The *normal scores* consist of values of z that divide the axes into equal probability intervals. For a sample of size 4, the normal scores are $-z_{0.20} = -0.84$, $-z_{0.40} = -0.25$, $z_{0.40} = -0.25$, and $z_{0.20} = 0.84$.

Steps to construct a normal plot

1. Rearrange the n data points in ascending order.
2. Obtain the n normal scores.
3. Plot the kth largest observation, versus the kth normal score, for all k.
4. If the data were from a standard normal distribution, the plot would resemble a 45 degree line through the origin.
5. If the observations were from normal (but not from standard normal), the pattern should still be a straight line. However, the line need not pass through the origin or have a slope 1.

In applications, a minimum of 15 to 20 observations is needed to reach a more accurate conclusion.

4.8 EXERCISES

1. For different observations, construct normal plots and check for normality of the corresponding populations.
2. Using software (such as Software), generate 15 observations each from the following distributions: **(a)** Normal (2, 4), **(b)** Uniform (0, 1), **(c)** Gamma (2, 4), and **(d)** Exponential (2).

For each of these data sets, draw a probability plot and note the geometry of the plots.

4.8.1 Simulation for Comparing Estimator Performance

Compare sampling distributions of different estimators, namely, mean, median, and trimmed mean. Do the following steps.

1. Simulate a skewed population, such as an exponential distribution.

2. Calculate each of the estimators for the simulated sample.
3. For each estimator, repeat sampling 10,000 times.
4. Plot sampling distributions side-by-side.
5. Compare bias and variance.

What insight can you derive from this project?

Chapter 5

Statistical Estimation

Chapter outline

OBJECTIVE

In this chapter, we will study some statistical methods to find estimators of population parameters and study their properties. This will include methods of finding point estimation as well as interval estimation of the unknown population parameters.

Mathematical Statistics with Applications in R. DOI: https://doi.org/10.1016/B978-0-443-27548-7.00005-0

C.R. Rao
(Source: http:http://www.science.psu.edu/alert/Rao6–2007.htm)

Calyampudi Radhakrishna (C.R.) Rao (1920–2023) is a statistician whose work has influenced not just statistics but such diverse fields as anthropology, biometry, demography, economics, genetics, geology, and medicine. Several statistical terms and equations are named after Rao. He has worked with many other famous statisticians such as Blackwell, Fisher, and Neyman and has had dozens of theorems named after him. Rao earned an M.A. in mathematics and another M.A. in statistics, both in India, and earned his Ph.D. and Sc.D. at Cambridge University. The following was stated in the preface in the 1991 special issue of the *Journal of Quantitative Economics* in Rao's honor: "Dr. Rao is a very distinguished scientist and a highly eminent statistician of our time. His contributions to statistical theory and applications are well known, and many of his results, which bear his name, are included in the curriculum of courses in statistics at bachelor's and master's level all over the world. He was an inspiring teacher and guided the research work of numerous students in all areas of statistics. His early work had greatly influenced the direction of statistical research during the last four decades. One of the purposes of this special issue is to recognize Dr. Rao's own contributions to econometrics and we acknowledge his major role in the development of econometric research in India. The importance of statistics can be summarized in Rao's own words: 'If there is a problem to be solved, seek statistical advice instead of appointing a committee of experts. Statistics can throw more light than the collective wisdom of the articulate few'" (http://www.finse.uio.no/events/international-workshops/introduction-to-estimation/)

5.1 INTRODUCTION

In statistical analysis, estimation of population parameters plays a very significant role. In most applied problems, a certain numerical characteristic of the physical phenomenon may be of interest; however, its value may not be observable directly. Instead, suppose it is possible to observe one or more random variables, the distribution of which depends on the characteristic of interest. Our objective will be to develop methods that use the observed values of random variables (sample data) in order to gain information about the unknown and unobservable characteristics of the population.

In studying a real-world phenomenon, we begin with a random sample of size n taken from the totality of a population. In estimation theory, it is assumed the observations are random with a probability distribution that depends on some parameters of interest. The initial step in statistically analyzing the data is to be able to identify the probability distribution that characterizes this information. Since the parameters of a distribution are its defining characteristics, it becomes necessary to know the parameters. In the present chapter, we shall assume that the form of the population distribution is known (such as a binomial, normal, etc.), but the parameters of the distribution (p for a binomial; μ and σ^2 for a normal, etc.) are unknown. We shall estimate these parameters using the data from our random sample. It is extremely important to have the best possible estimate of the population parameter(s). Having such estimates will lead to a better and more accurate statistical analysis.

For example, due to phosphate mining in Florida, we may be interested in estimating the average radioactivity from both uranium and radium in a clay settling area of a mining site. Suppose that a random sample of ten such sites resulted in a sample average of 40 pCi/g (picocuries/gram) of radioactivity. We may use this value as an estimate of the average radioactivity for all of the settling areas of mining sites in Florida. We may also want to know a range of values of radioactivity with a certain degree of confidence. Since many Florida crops are grown on clay settling areas, these types of estimates are important for accessing the risks associated with radioactivity due to eating food from the crops grown on these clay settling areas.

There are two types of estimators, namely, point estimators and interval estimators. First, we will introduce statistical point estimation methods, discuss their properties, and illustrate their usefulness with a number of applications. Point estimation gives a single "best guess" for the parameter(s) of interest. The importance of point estimates lies in the fact that many statistical formulas are based on them. For example, point estimates of the population mean and standard deviation are needed in the calculation of confidence intervals and in many formulas for hypothesis testing. These topics will be covered subsequently. In general, the point estimates will differ from the true parameter values by varying amounts depending on the sample values obtained. In addition, the point estimates do not convey any measure of reliability. To deal with these issues, we will also introduce so-called interval estimation, or confidence intervals.

5.2 THE METHODS OF FINDING POINT ESTIMATORS

Let $X_1 \ldots X_n$ be independent and identically distributed (iid) random variables (in statistical language, a random sample) with a pdf or pmf $f(x, \theta_1,\ldots, \theta_l)$, where $\theta_1, \ldots, \theta_l$ are the unknown population parameters (characteristics of interest). For example, a normal pdf has parameters μ (the mean) and σ^2 (the variance). The actual values of these parameters are not known. The problem in point estimation is to determine statistics $g_i(X_1,\ldots, X_n)$, $i = 1,\ldots, l$, which can be used to estimate the value of each of the parameters, that is, to assign an appropriate value for the parameters $\theta = (\theta_1,\ldots, \theta_l)$ based on observed sample data from the population. These statistics are called *estimators* for the parameters, and the values calculated from these statistics using particular sample data values are called *estimates* of the parameters. Estimators of θ_i are denoted by $\widehat{\theta_i}$, where $\widehat{\theta_i} = g_i(X_1, \ldots,X_n)$, $i = 1, \ldots l$. Observe that the estimators are random variables. As a result, an estimator has a distribution (which we called the sampling distribution in Chapter 4). When we actually run the experiment and observe the data, let the observed values of the random variables be $X_1,\ldots, X_n$ be $x_1, \ldots, x_n$; then, $\widehat{\theta}(X_1, \ldots,X_n)$ is an estimator, and its value $\widehat{\theta}(x_1, \ldots,x_n)$ is an estimate. For example, in case of the normal distribution, the parameters of interest are $\theta_1 = \mu$, and $\theta_2 = \sigma^2$, that is, $\boldsymbol{\theta} = (\mu, \sigma^2)$. If the estimators of μ and σ^2 are $\overline{X} = (1/n)\sum_{i=1}^{n} X_i$ and $S^2 = (1/(n-1))\sum_{i=1}^{n}(X_i - \overline{X})^2$, respectively, then, the corresponding estimates are $\overline{x} = (1/n)\sum_{i=1}^{n} x_i$ and $s^2 = (1/n-1)\sum_{i=1}^{n}(x_i - \overline{x})^2$, the mean and variance corresponding to the particular observed sample values. In this book, we use capital letters such as $\overline{X}$ and S^2 to represent the estimators and lowercase letters such as $\overline{x}$ and s^2 to represent the estimates.

There are many methods available for estimating the true value(s) of the parameter(s) of interest. Three of the more popular methods of estimation are the *method of moments*, the *method of maximum likelihood*, and *Bayes' method.* A very popular procedure among econometricians to find a point estimator is the generalized method of moments. In this chapter, we study only the method of moments and the method of maximum likelihood for obtaining point estimators and some of their desirable properties. In Chapter 10, we shall discuss Bayes' method of estimation.

There are many criteria for choosing a desired point estimator. Heuristically, some of them can be explained as follows. An estimator, $\hat{\theta}$, is unbiased if the mean of its sampling distribution is the parameter θ. The bias of $\hat{\theta}$ is given by $B = E(\hat{\theta}) - \theta$. The estimator has sufficiency property if it fully uses all the sample information. Minimal sufficient statistics are those that are sufficient for the parameter and are functions of every other set of sufficient statistics for those same parameters. A method due to Lehmann and Scheffé can be used to find a minimal sufficient statistic. In addition, the estimator is said to satisfy the consistency property if the sample estimator has a high probability of being close to the population value θ for a large sample size. The concept of efficiency is based on comparing variances of the different unbiased estimators. If there are two unbiased estimators, it is desirable to have the one with a smaller variance. However, some of these properties will not be discussed in this book.

How do we find a good point estimator with desirable properties? To answer this question, we will study two methods of finding point estimators, namely, the *method of moments* and the *method of maximum likelihood.*

5.2.1 The Method of Moments

One of the oldest methods for finding point estimators is the method of moments. This is a very simple procedure for finding an estimator for one or more population parameters. Let $\mu'_k = E[X^k]$ be the kth moment about the origin of a random variable X, whenever it exists. Let $m'_k = (1/n)\sum_{i=1}^{n} X_i^k$ be the corresponding kth sample moment. Then, the estimator of μ'_k by the method of moments is m'_k. The method of moments is based on matching the sample moments with the corresponding population (distribution) moments and is founded on the assumption that sample moments should provide good estimates of the corresponding population moments. Because the population moments

$\mu'_k = h_k(\theta_1, \theta_2, \dots, \theta_l)$ are often functions of the population parameters, we can equate corresponding population and sample moments and solve for these parameters in terms of the moments.

Method of moments

Choose as estimates those values of the population parameters that are solutions of the equations $\mu'_k = m'_k$, $k = 1,2, \dots, l$. Here μ'_k is a function of the population parameters.

For example, the first population moment is $\mu'_1 = E(X)$, and the first sample moment is $\overline{X} = \sum_{i=1}^{n} X_i/n$. Hence, the moment estimator of μ'_1 is $\overline{X}$. If $k = 2$, then the second population and sample moments are $\mu'_2 = E(X^2)$ and $m'_2 = (1/n)\sum_{i=1}^{n} X_i^2$, respectively. Basically, we can use the following procedure in finding point estimators of the population parameters using the method of moments.

The method of moments procedure

Suppose there are l parameters to be estimated, say $\theta = (\theta_1, \dots, \theta_l)$.

1. Find l population moments, μ'_k, $k = 1,2, \dots, l$. μ'_k will contain one or more parameters $\theta_1, \dots, \theta_l$.
2. Find the corresponding l sample moments, m'_k, $k = 1,2, \dots, l$. The number of sample moments should equal the number of parameters to be estimated.
3. From the system of equations, $\mu'_k = m'_k$, $k = 1,2, \dots, l$, solve for the parameter $\theta = (\theta_1, \dots, \theta_l)$; this will be a moment estimator of $\hat{\theta}$.

The following examples illustrate the method of moments for population parameter estimation.

Example 5.2.1

Let $X_1, \dots, X_n$ be a random sample from a Bernoulli population with parameter p.

(a) Find the moment estimator for p.

(b) Tossing a coin 10 times and equating heads to value 1 and tails to value 0, we obtained the following values:

0 1 1 0 1 0 1 1 1 0

Obtain a moment estimate for p, the probability of success (head).

Solution

(a) *For the Bernoulli random variable,* $\mu'_k = E[X] = p$, *so we can use* m'_1 *to estimate* p. *Thus,*

$$m'_1 = \hat{p} = \frac{1}{n}\sum_{i=1}^{n} X_i.$$

Let

$$Y = \sum_{i=1}^{n} X_i.$$

Then, the method of moments estimator for p is $\hat{p} = Y/n$. That is, the ratio of the total number of heads to the total number of tosses will be an estimate of the probability of success.

(b) *Note that this experiment results in Bernoulli random variables. Thus, using part (a) with Y = 6, we get the moment estimate of p is $\hat{p} = \frac{6}{10} = 0.6$.*

We would use this value $\hat{p} = 0.6$, to answer any probabilistic questions for the given problem. For example, what is the probability of exactly obtaining 8 heads out of 10 tosses of this coin? This can be obtained by using the binomial formula (or R-command: pbinom(8,10, 0.6)-pbinom(7,10, 0.6)), with $\hat{p} = 0.6$, that is,

$$P(X = 8) = \binom{10}{8}(0.6)^8(0.4)^{10-8} = 0.1209324.$$

In Example 5.2.1, we used the method of moments to find a single parameter. We demonstrate in Example 5.2.2 how this method is used for estimating more than one parameter.

Example 5.2.2

Let $X_1,\ldots,X_n$ be a random sample from a gamma probability distribution with parameters α and β. Find moment estimators for the unknown parameters α and β.

Solution

For the gamma distribution (see Section 3.2.5),

$$E[X] = \alpha\beta \quad \text{and} \quad E[X^2] = \alpha\beta^2 + \alpha^2\beta^2.$$

Because there are two parameters, we need to find the first two moment estimators. Equating sample moments to the distribution (theoretical) moments, we have

$$\frac{1}{n}\sum_{i=1}^{n} X_i = \overline{X} = \alpha\beta, \text{ and } \quad \frac{1}{n}\sum_{i=1}^{n} X_i^2 = \alpha\beta^2 + \alpha^2\beta^2.$$

Solving for α and β, we obtain the estimates as $\alpha = (\overline{x}/\beta)$ and $\beta = [\{(1/n)\Sigma_{i=1}^{n} x_i^2 - \overline{x}^2\}/\overline{x}]$. Therefore, the method of moments estimators for α and β are

$$\hat{\alpha} = \frac{\overline{X}}{\hat{\beta}}$$

and

$$\hat{\beta} = \frac{\frac{1}{n}\Sigma_{i=1}^{n} X_i^2 - \overline{X}^2}{\overline{X}} = \frac{\Sigma_{i=1}^{n}(X_i - \overline{X})^2}{n\overline{X}},$$

which implies that

$$\hat{\alpha} = \frac{\overline{X}}{\hat{\beta}} = \frac{\overline{X}^2}{\frac{1}{n}\Sigma_{i=1}^{n} X_i^2 - \overline{X}^2} = \frac{\overline{X}^2}{\Sigma_{i=1}^{n}(X_i - \overline{X})^2}.$$

Thus, we can use these values in the gamma pdf to answer questions concerning the probabilistic behavior of the r.v. X.

The following example shows that once we find the moments estimator theoretically, the estimate can be obtained by simply substituting sample statistics into the formulas.

Example 5.2.3

Let the distribution of X be $N(\mu, \sigma^2)$.

(a) For a given sample of size n, use the method of moments to estimate μ and σ^2.
(b) The following data (rounded to the third decimal digit) were generated using Software from a normal distribution with mean 2 and a standard deviation of 1.5.

3.163	1.883	3.252	3.716	−0.049	−0.653	0.057	2.987
4.098	1.670	1.396	2.332	1.838	3.024	2.706	0.231
3.830	3.349	−0.230	1.496				

Obtain the method of moments estimates of the true mean and the true variance.

Solution

(a) *For the normal distribution,* $E(X) = \mu$, and because $Var(X) = EX^2 - \mu^2$, we have the second moment as $E(X^2) = \sigma^2 + \mu^2$. *Equating sample moments to distribution moments, we have*

$$\frac{1}{n}\sum_{i=1}^{n} X_i = \mu'_1 = \mu$$

and

$$\mu'_2 = \frac{1}{n}\sum_{i=1}^{n} X_i^2 = \sigma^2 + \mu^2.$$

Solving for μ *and* σ^2, *we obtain the moment estimators as*

$$\hat{\mu} = \overline{X}$$

and

$$\hat{\sigma}^2 = \frac{1}{n}\sum_{i=1}^{n} X_i^2 - \overline{X}^2 = \frac{1}{n}\sum_{i=1}^{n} (X_i - \overline{X})^2.$$

(b) *Because we know that the estimator of the mean is* $\hat{\mu} = \overline{X}$ *and the estimator of the variance is* $\hat{\sigma}^2 = (1/n)\Sigma_{i=1}^{n} X_i^2 - \overline{X}^2$, *using the data, the estimates are* $\hat{\mu} = 2.005$, and $\hat{\sigma}^2 = 6.12 - (2.005)^2 = 2.1$. *Notice that the true mean is 2 and the true variance is 2.25, which we used to simulate the data.*

In general, using the population pdf, we evaluate the lower-order moments, finding expressions for the moments in terms of the corresponding parameters. Once we have the population (theoretical) moments, we equate them to the corresponding sample moments to obtain the moment estimators.

Example 5.2.4

Let $X_1,\ldots,X_n$ be a random sample from a uniform distribution on the interval $[a, b]$. Obtain method of moments estimators for a and b.

Solution

Here a and b are treated as parameters. That is, we only know that the sample comes from a uniform distribution on some interval, but we do not know from which interval. Our interest is to estimate this interval. The pdf of a uniform distribution is

$$f(x) = \begin{cases} \frac{1}{b-a}, & a \le x \le b \\ 0, & \text{otherwise.} \end{cases}$$

Hence, the first two population moments are

$$\mu_1 = E(X) = \int_a^b \frac{x}{b-a}dx = \frac{a+b}{2} \quad \text{and} \quad \mu_2 = E(X^2) = \int_a^b \frac{x^2}{b-a}dx = \frac{a^2+ab+b^2}{3}.$$

The corresponding sample moments are

$$\hat{\mu}_1 = \overline{X} \text{ and } \hat{\mu}_2 = \frac{1}{n}\sum_{i=1}^{n} X_i^2.$$

Equating the first two-sample moments to the corresponding population moments, we have

$$\hat{\mu}_1 = \frac{a+b}{2} \text{ and } \hat{\mu}_2 = \frac{a^2+ab+b^2}{3}$$

and solving for a and b results in the moment estimators of a and b,

$$\hat{a} = \hat{\mu}_1 - \sqrt{3\left(\hat{\mu}_2 - \hat{\mu}_1^2\right)} \text{ and } \hat{b} = \hat{\mu}_1 + \sqrt{3\left(\hat{\mu}_2 - \hat{\mu}_1^2\right)}.$$

In Example 5.2.4, if $a = -b$, that is, $X_1,\ldots,X_n$ is a random sample from a uniform distribution on the interval $(-b, b)$, the problem reduces to a one-parameter estimation problem. However, in this case $E(X_i) = 0$, so the first moment cannot be used to estimate b. It becomes necessary to use the second moment. For the derivation, see Exercise 5.2.3.

It is important to observe that the method of moments estimators need not be unique. The following is an example of the nonuniqueness of moment estimators.

Example 5.2.5

Let $X_1,\ldots, X_n$ be a random sample from a Poisson distribution with parameter $\lambda > 0$. Show that both $(1/n)\sum_{i=1}^{n} X_i$ and $(1/n)\sum_{i=1}^{n} X_i^2 - ((1/n)\sum_{i=1}^{n} X_i)^2$ are moment estimators of λ.

Solution

We know that $E(X) = \lambda$, *from which we have a moment estimator of* λ as $(1/n)\sum_{i=1}^{n} X_i$. *Also, because we have* $Var(X) = \lambda$, *equating the second moment, we can see that*

$$\lambda = E(X^2) - (EX)^2,$$

so that

$$\hat{\lambda} = \frac{1}{n}\sum_{i=1}^{n} X_i^2 - \left(\frac{1}{n}\sum_{i=1}^{n} X_i\right)^2.$$

Thus,

$$\hat{\lambda} = \frac{1}{n}\sum_{i=1}^{n} X_i$$

and

$$\hat{\lambda} = \frac{1}{n}\sum_{i=1}^{n} X_i^2 - \left(\frac{1}{n}\sum_{i=1}^{n} X_i\right)^2.$$

Both are moment estimators of λ. Thus, the moment estimators may not be unique. We generally choose $\overline{X}$ as an estimator of λ, for its simplicity.

It is important to note that, in general, we have as many moment conditions as the parameters. In Example 5.2.5, we have more moment conditions than parameters, because both the mean and variance of the Poisson random variables are the same. Given a sample, this results in two different estimates of a single parameter. One of the questions could be, can these two estimators be combined in some optimal way? This is done by the so-called generalized method of moments. We will not deal with this topic. The method of moments often provides estimators when other methods fail to do so or when estimators are harder to obtain, as in the case of the gamma distribution. Compared to other methods, method of moments estimators are easier to compute and have some desirable properties that we will discuss in ensuing section.

5.2.2 The Method of Maximum Likelihood

Now we will present an important method for finding estimators of parameters proposed by geneticist/statistician Sir Ronald A. Fisher around 1922 called the method of maximum likelihood. Even though the method of moments is intuitive and easy to apply, it usually does not yield "good" estimators. The method of maximum likelihood is intuitively appealing, because we attempt to find the values of the true parameters that would most accurately represent the behavior of the data that we have observed. For most cases of practical interest, the performance of the maximum likelihood estimators are optimal for large enough data. This is one of the most versatile methods for fitting parametric statistical models to the data. First, we define the concept of a likelihood function.

Definition 5.2.1 *Let $f(x_1,...,x_n;\theta)$, $\theta \in \Theta \subseteq \mathbb{R}^k$, be the joint probability (or density) function of n random variables $X_1,..., X_n$* with sample values $x_1,...,x_n$. *The* **likelihood function** *of the sample is given by*

$$L(\theta; x_1, \ldots, x_n) = f(x_1, \ldots, x_n; \theta), \ [\ = L(\theta), \textit{for a briefer notation}].$$

We emphasize that L is a function of θ for a fixed sample values.

The likelihood of a set of parameter values θ, given $x_1, \ldots, x_n$, is equal to the probability of those observed outcomes for the given parameter values. If $X_1,...,X_n$ are discrete iid random variables with probability function $p(x, \theta)$, then, the likelihood function is given by

$$\begin{aligned} L(\theta) &= P(X_1 = x_1, \ldots, X_n = x_n) \\ &= \prod_{i=1}^{n} P(X_i = x_i), \text{ (by multiplication rule for independent random variables)} \\ &= \prod_{i=1}^{n} p(x_i, \theta) \end{aligned}$$

and in the continuous case, if the density is $f(x, \theta)$, then the likelihood function is

$$L(\theta) = \prod_{i=1}^{n} f(x_i, \theta).$$

It is important to note that the likelihood function, although it depends on the observed sample values $x = (x_1,...,x_n)$, is to be regarded as a function of the parameter θ. In the discrete case, $L(\theta; x_1,..., x_n)$ gives the probability of observing $x = (x_1,...,x_n)$, for a given θ. Thus, the likelihood function is a statistic, depending on the observed sample $x = (x_1, \ldots, x_n)$.

Example 5.2.6

Let $X_1,\ldots,X_n$ be iid $N(\mu, \sigma^2)$ random variables. Let $x_1,\ldots,x_n$ be the sample values. Find the likelihood function.

Solution

The density function for the normal variable is given by $f(x) = \frac{1}{\sigma\sqrt{2\pi}} \exp\left(-\frac{(x-\mu)^2}{2\sigma^2}\right)$ *Hence, the likelihood*

$$L(\mu, \sigma^2) = \prod_{i=1}^{n} \frac{1}{\sigma\sqrt{2\pi}} \exp\left(-\frac{(x_i-\mu)^2}{2\sigma^2}\right) = \frac{1}{(2\pi)^{n/2}\sigma^n} \exp\left(-\frac{\sum_{i=1}^{n}(x_i-\mu)^2}{2\sigma^2}\right).$$

A statistical procedure should be consistent with the assumption that the best explanation of a set of data is provided by an estimator $\hat{\theta}$, which will be the value of the true parameter θ that maximizes the likelihood function. This value of θ will be called the maximum likelihood estimator and will be denoted by $\hat{\theta}$. The goal of maximum likelihood estimation is to find the parameter value(s) that makes the observed data most likely.

Definition 5.2.2 *The* **maximum likelihood estimators (MLEs)** *are those values of the parameters that maximize the likelihood function with respect to the parameter θ. That is,*

$$L(\hat{\theta}; x_1 \ldots,x_n) = \max_{\theta\in\Theta} L(\theta; x_1, \ldots, x_n),$$

where Θ is the set of possible values of the parameter θ.

The method of maximum likelihood extends to the case of several parameters. Let $X_1,\ldots, X_n$ be a random sample with joint pmf (if discrete) or pdf (if continuous)

$$L(\theta_1,\ldots,\theta_m; x_1 \ldots,x_n) = f(x_1, x_2, \ldots,x_n; \theta_1, \theta_2, \ldots,\theta_m)$$

where the values of the parameters $\theta_1, \ldots, \theta_m$ are unknown and $x_1, \ldots, x_n$ are the observed sample values. Then, the maximum likelihood estimates $\hat{\theta}_1, \ldots ,\hat{\theta}_m$ are those values of the $\theta'_i s$ that maximize the likelihood function, so that

$$f(x_1,\ldots,x_n; \hat{\theta}_1,\ldots,\hat{\theta}_m) \geq f(x_1,\ldots,x_n; \theta_1,\ldots,\theta_m)$$
for all allowable $\theta_1,\ldots,\theta_m$.

Note that the likelihood function conveys to us how feasible the observed sample is as a function of the possible parameter values. Maximum likelihood estimates give the parameter values for which the observed sample is most likely to have been generated from. In general, the maximum likelihood method results in the problem of maximizing a function of single or several variables. Hence, in most situations, the methods of calculus can be used. In deriving the MLEs, however, there are situations where the techniques developed are more problem specific. Sometimes we need to use numerical methods, such as Newton's method to obtain approximate estimates of the true parameters.

In order to find an MLE, we need only to compute the likelihood function and then maximize that function with respect to the parameter of interest. In many cases, it is easier to work with the natural logarithm (ln) of the likelihood function, called the *log-likelihood function*. Because the natural logarithm function is increasing, the maximum value of the likelihood function, if it exists, will occur at the same point as the maximum value of the log-likelihood function. We now summarize the calculus-based procedure to find MLEs.

Procedure to find an MLE

1. Define the likelihood function, $L(\theta)$.
2. Often it is easier to take the natural logarithm (ln) of $L(\theta)$.
3. When applicable, differentiate ln $L(\theta)$ with respect to θ, and then equate the derivative to zero.
4. Solve for the parameter θ, and we will obtain$\hat{\theta}$.
5. Check whether it is a maximizer or global maximizer.

Example 5.2.7

Suppose $X_1, \ldots, X_n$ are a random sample from a geometric distribution with parameter p, $0 \le p \le 1$. Find MLE of p.

Solution

For the geometric distribution, the pmf is given by

$$f(x, p) = p(1-p)^{x-1}, \; 0 \le p \le 1, \; x = 1,2,3, \ldots.$$

Hence, the likelihood function is

$$L(p) = \prod_{i=1}^{n}\left[p(1-p)^{x-1}\right] = p^n(1-p)^{-n+\sum_{i=1}^{n} x_i}.$$

Taking the natural logarithm of L(p), we get

$$\ln L = n \ln p + \left(-n + \sum_{i=1}^{n} x_i\right)\ln(1-p).$$

Taking the derivative with respect to p, gives us

$$\frac{d \ln L}{dp} = \frac{n}{p} - \frac{(-n + \sum_{i=1}^{n} x_i)}{(1-p)}.$$

Equating $\frac{d \ln L(p)}{dp}$ *to zero results in*

$$\frac{n}{p} - \frac{(-n + \sum_{i=1}^{n} x_i)}{(1-p)} = 0.$$

Solving for p, we have

$$p = \frac{n}{\sum_{i=1}^{n} x_i} = \frac{1}{\overline{x}}.$$

Thus, we obtain a maximum likelihood estimator of p as

$$\hat{p} = \frac{n}{\sum_{i=1}^{n} X_i} = \frac{1}{\overline{X}}.$$

We remark that $(1/\overline{X})$ *is the maximum likelihood estimator of p. It can be shown that* $\hat{p}$ *is a global maximum.*

Example 5.2.8

(a) Suppose $X_1, \ldots, X_n$ are random samples from a Poisson distribution with parameter λ. Find MLE of λ.

(b) Traffic engineers use the Poisson distribution to model light traffic. This is based on the rationale that when the rate is approximately constant in light traffic, the distribution of counts of cars in a given time interval is considered to follow the Poisson pmf. The following data shows the number of vehicles that turn left in 15 randomly chosen 5-minute intervals at a specific intersection. Calculate the maximum likelihood estimate.

10	17	12	6	12	11	9	6
10	8	8	16	7	10	6	

Solution (a)

We have the probability mass function

$$p(x) = \frac{\lambda^x e^{-\lambda}}{x!}, \ x = 0,1,2, \ \ldots, \lambda > 0.$$

Hence, the likelihood function is

$$L(\lambda) = \prod_{i=1}^{n} \frac{\lambda^{x_i} e^{-\lambda}}{x_i!} = \frac{\lambda^{\sum_{i=1}^{n} x_i} e^{-n\lambda}}{\prod_{i=1}^{n} x_i!}.$$

Then, taking the natural logarithm, we have

$$\ln L(\lambda) = \sum_{i=1}^{n} x_i \ln \lambda - n\lambda - \sum_{i=1}^{n} \ln(x_i!)$$

and differentiating with respect to λ results in

$$\frac{d \ln L(\lambda)}{d\lambda} = \frac{\sum_{i=1}^{n} x_i}{\lambda} - n$$

and

$$\frac{d \ln L(\lambda)}{d\lambda} = 0, \text{ implies } \frac{\sum_{i=1}^{n} x_i}{\lambda} - n = 0.$$

That is,

$$\lambda = \frac{\sum_{i=1}^{n} x_i}{n} = \bar{x}.$$

Hence, the MLE of λ is

$$\hat{\lambda} = \bar{X}.$$

(b) *From part (a), we have the estimate as*

$$\hat{\lambda} = \bar{x} = 9.8,$$

That is, approximately 10 vehicles per 5 minutes turn left at this intersection.

It can be verified that the second derivative is negative and, hence, we really have a maximum.

Sometimes the derivatives method cannot be used for finding the MLEs. For example, the likelihood is not differentiable for a given range space. In this case, we need to make use of the special structures available in a specific situation to solve the problem. The following example illustrates such an approach.

Example 5.2.9

Let $X_1,\ldots, X_n$ be a random sample from $U(0, \theta)$, $\theta > 0$. Find the MLE of θ.

Solution

Note that the pdf of the uniform distribution is

$$f(x) = \begin{cases} \frac{1}{\theta}, & 0 \le x \le \theta \\ 0, & \text{otherwise.} \end{cases}$$

$L(\theta)$

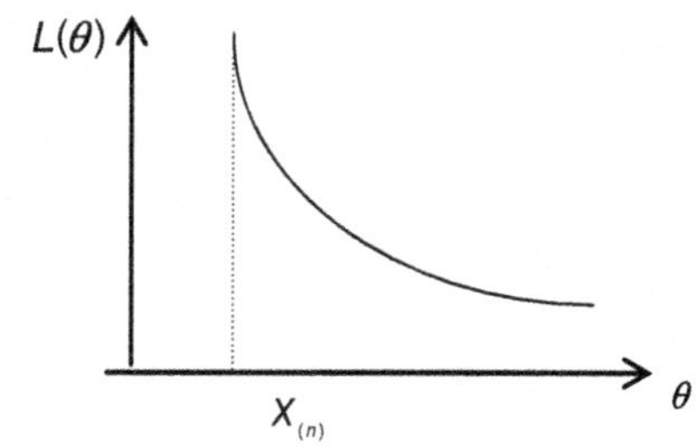

FIGURE 5.1 Likelihood function for uniform probability distribution.

Hence, the likelihood function is given by

$$L(\theta, x_1, x_2, \ldots, x_n) = \begin{cases} \frac{1}{\theta^n}, & 0 \le x_1, x_2, \ldots, x_n \le \theta \\ 0, & \text{otherwise.} \end{cases}$$

When $\theta \ge \max(x_i)$, the likelihood is $(1/\theta^n)$, which is positive and decreasing as a function of θ (for fixed n). However, for $\theta < \max(x_i)$, the likelihood drops to 0, creating a discontinuity at the point $\max(x_i)$ (this is the minimum value of θ that can be chosen, which still satisfies the condition $0 \le x_i \le \theta$), and Fig. 5.1 shows that the maximum occurs at this point. Hence, we will not be able to find the derivative. Thus, the MLE is the largest order statistic,

$$\hat{\theta} = \max(X_i) = X_{(n)}.$$

In the previous example, because $E(X) = (\theta/2)$, we can see that $\theta = 2E(X)$. Hence, the method of moments estimator for θ is $\hat{\theta} = 2\overline{X}$. Sometimes the method of moments estimator can give meaningless results. To see this, suppose we observe values 3, 5, 6, and 18 from a $U(0, \theta)$ distribution. Clearly, the maximum likelihood estimate of θ is 18, whereas the method of moments estimate is 16, which is not quite acceptable, because we have already observed a value of 18.

As mentioned earlier, if the unknown parameter θ represents a vector of parameters, say $\theta = (\theta_1,\ldots, \theta_l)$, then the MLEs can be obtained from the solutions of the system of equations

$$\frac{\partial}{\partial \theta} \ln L(\theta_1, \ldots, \theta_n) = 0, \text{ for } i = 1, \ldots, l.$$

These are called the *maximum likelihood equations*, and the solutions are denoted by $(\hat{\theta}_1, \ldots, \hat{\theta}_l)$.

Example 5.2.10

Let $X_1,\ldots,X_n$ be $N(\mu, \sigma^2)$.

(a) If μ is unknown and $\sigma^2 = \sigma_0^2$ is known, find the MLE for μ.
(b) If $\mu = \mu_0$ is known and σ^2 is unknown, find the MLE for σ^2.
(c) If μ and σ^2 are both unknown, find the MLE for $\theta = (\mu, \sigma^2)$.

Solution

In order to avoid notational confusion when taking the derivative, let $\theta = \sigma^2$. Then, the likelihood function is

$$L(\mu, \theta) = (2\pi\theta)^{-n/2} \exp\left(-\frac{\sum_{i=1}^n (x_i - \mu)^2}{2\theta}\right)$$

or

$$\ln L(\mu, \theta) = -\frac{n}{2}\ln(2\pi) - \frac{n}{2}\ln\theta - \frac{\sum_{i=1}^n (x_i - \mu)^2}{2\theta}.$$

(a) *When $\theta = \theta_0 = \sigma_0^2$ is known, the problem reduces to estimating the only one parameter, μ. Differentiating the log-likelihood function with respect to μ, we have*

$$\frac{\partial}{\partial\mu}(\ln L(\mu, \theta_0)) = \frac{2\sum_{i=1}^n (x_i - \mu)}{2\theta_0}.$$

Setting the derivative equal to zero and solving for μ results in

$$\sum_{i=1}^n (x_i - \mu) = 0.$$

Thus

$$\sum_{i=1}^n x_i = n\mu \text{ or } \mu = \bar{x}.$$

Thus the MLE of μ *is* $\hat{\mu} = \bar{X}$.

(b) *When $\mu = \mu_0$ is known, the problem reduces to estimating the only parameter $\sigma^2 = \theta$. Differentiating the log-likelihood function with respect to θ,*

$$\frac{\partial \ln L(\mu, \theta)}{\partial\theta} = \frac{-n}{2\theta} + \frac{\sum_{i=1}^n (x_i - \mu)^2}{2\theta^2}.$$

Setting the derivative equal to zero and solving for θ, we get

$$\hat{\theta} = \hat{\sigma}^2 = \frac{\sum_{i=1}^n (X_i - \mu_0)^2}{n}.$$

(c) *When both μ and θ are unknown, we need to differentiate with respect to both μ and θ individually; that is,*

$$\frac{\partial \ln L(\mu, \theta)}{\partial\mu} = \frac{2\sum_{i=1}^n (x_i - \mu)}{2\theta}$$

and

$$\frac{\partial \ln L(\mu, \theta)}{\partial \theta} = \frac{-n}{2\theta} + \frac{\sum_{i=1}^{n}(x_i - \mu)^2}{2\theta^2}.$$

Setting the derivatives equal to zero and solving simultaneously, we obtain

$$\hat{\mu} = \overline{X},$$

and

$$\hat{\sigma}^2 = \hat{\theta} = \frac{\sum_{i=1}^{n}(X_i - \overline{X})^2}{n} = S'^2.$$

Note that in (a) and (c), the estimates for μ are the same; however, in (b) and (c), the estimates for σ^2 are different.

If we want to simulate some data from the normal distribution with mean 6 and standard deviation 3 and use it to obtain MLE of the population mean and variance using 1000 samples, we can use the following R-code:

```
# Simulate some data
set.seed(123)
x <- rnorm(1000, mean = 6, sd = 3) # true mean = 6, sd = 3
# MLE of mean: sample mean
mle_mu <- mean(x)
# MLE of variance: use divisor n (not n-1)
mle_sigma2 <- sum((x - mle_mu)^2) / length(x)
# Print results
cat("MLE of mean (mu):", mle_mu, "\n")
cat("MLE of variance (sigma^2):", mle_sigma2, "\n")
# MLE of mean: sample mean
mle_mu <- mean(x)
MLE of mean (mu): 6.048384
MLE of variance (sigma^2): 8.842279
```

Note that we can easily modify the code by inputting real data.

At times, the MLEs may be hard to calculate. It may be necessary to use numerical methods to obtain approximate values of the estimate. The following example gives one such case.

Example 5.2.11

Let $X_1,\ldots,X_n$ be a random sample from a population with gamma distribution and parameters α and β. Find MLEs for the unknown parameters α and β.

Solution

The pdf for the gamma distribution is given by

$$f(x) = \begin{cases} \frac{x^{\alpha-1}e^{-x/\beta}}{\Gamma(\alpha)\beta^{\alpha}}, & x > 0,\ \alpha > 0,\ \beta > 0 \\ 0, & \text{otherwise.} \end{cases}$$

The likelihood function is given by

$$L = L(\alpha, \beta) = \frac{1}{(\Gamma(\alpha)\beta^{\alpha})^n} \prod_{i=1}^{n} x_i^{\alpha-1} e^{-\sum_{i=1}^{n} x_i/\beta}.$$

Taking the logarithms gives

$$\ln L = -n \ln \Gamma(\alpha) - n\alpha \ln \beta + (\alpha - 1) \sum_{i=1}^{n} \ln x_i - \sum_{i=1}^{n} \frac{x}{\beta}.$$

Now taking the partial derivatives with respect to α and β and setting both equal to zero, we have

$$\frac{\partial}{\partial \alpha} \ln L = -n\frac{\Gamma'(\alpha)}{\Gamma(\alpha)} - n \ln \beta + \sum_{i=1}^{n} \ln x_i = 0$$

$$\frac{\partial}{\partial \beta} \ln L = -n\frac{\alpha}{\beta} + \sum_{i=1}^{n} \frac{x_i}{\beta^2} = 0.$$

Solving the second one to get β in terms of α, we have

$$\beta = \frac{\overline{x}}{\alpha}.$$

Substituting this β in the first equation, we have to solve

$$-n\frac{\Gamma'(\alpha)}{\Gamma(\alpha)} - n \ln \frac{\overline{x}}{\alpha} + \sum_{i=1}^{n} \ln x_i = 0$$

for $\alpha > 0$. There is no closed-form solution for α and β. In this case, one can use numerical methods such as the Newton–Raphson method to solve for α and then use this value to find β.

There are many references available on the Web (such as http://www.mn.uio.no/math/tjenester/kunnskap/kompendier/num_opti_likelihoods.pdf) explaining the Newton–Raphson method for the gamma distribution.

In only a few cases are we able to obtain a simple form for the maximum likelihood equation that can be solved by setting the first derivative to zero. Often, we cannot write an equation that can be differentiated to find the MLE parameter estimates. This is especially true in the situation where the model is complex and involves many parameters. Evaluating the likelihood exhaustively for all values of the parameters becomes almost impossible, even with modern computers. This is why so-called *optimization* algorithms have become indispensable to statisticians. The purpose of an optimization algorithm is to find, as fast as possible, the set of parameter values that make the observed data most likely. There are many such algorithms available. We describe the Newton–Raphson method in Project 5F, and another powerful algorithm, known as the EM algorithm, is given in Section 13.4.

We have been introduced to several classical discrete and continuous *pdf*, such as the Binomial, Poisson, Gaussian (normal), Gamma, and exponential pdf, among others. Note that when we use one of these *pdf* to study a given set of data, we refer to it as parametric analysis, because each of the classical *pdf* contains at least one parameter that plays a major role in the shape of the probability distribution that characterizes the behavior of the phenomenon of interest.

5.2.3 Some Additional Probability Distributions

Now, we will introduce some additional probability distributions that play a major role in analyzing data, or information, in health science, environmental science, engineering, business, and economics, among many other important areas in our society. We shall study **the three parameter gamma *pdf*** and the **Weibull *pdf***. The **Rayleigh *pdf*** and the **power exponential *pdf*** are other examples that will be in Chapter 14. Each of these *pdf*s will be applied to real data: cancer data, hurricane data, global warming data, and environmental (rainfall) data in Chapter 15.

In Example 5.2.11, we have studied the 2-parameter gamma probability distribution (*pdf*); here we shall introduce the 3-parameter version that is useful when we analyze data that exhibits positive skewness. The ***3-parameter gamma*** *pdf* is given by

$$f(x) = \frac{1}{\beta^{\alpha}\Gamma_{(\alpha)}}(x - \gamma)^{\alpha-1} \ \exp - \frac{(x - \gamma)}{\beta},$$

where $x > \gamma$, $\beta > 0$, and $\Gamma_{(a)} = \int_0^\infty x^{\alpha-1}e^{-x}dx$.
The corresponding cumulative distribution function (cdf) is given by

$$\begin{aligned} F(x) &= P(X \le x) \\ &= \int_\gamma^x \frac{1}{\beta^\alpha \Gamma_{(a)}}(y-\gamma)^{\alpha-1}\exp-\frac{(y-\gamma)}{\beta}dy \\ &= \Gamma_{\frac{x-\gamma}{\beta}}(\alpha).\ \frac{1}{\Gamma(\alpha)}. \end{aligned}$$

The expected value is given by

$$E(X) = \int_0^\infty xf(x)dx = \gamma + \alpha\beta.$$

Note when the location parameter $\gamma = 0$, we obtain the 2-parameter gamma (pdf).

Example 5.2.12

Given a random sample $X_1, \ldots, X_n$ from a 3-parameter gamma distribution, obtain the MLEs of the parameters.

Solution

The likelihood function is given by

$$L(\alpha, \beta, \gamma) = \pi_{i=1}^n f(x)$$

$$= \left(\frac{1}{\beta^\alpha \Gamma_{(\alpha)}}\right)^n \sum_{i=1}^n (x_i - \gamma)^{\alpha-1} \pi_{i=1}^n \exp - \left(\frac{x_i - \gamma}{\beta}\right),$$

where $\alpha > 0$, $\beta > 0$, and $\gamma \in \mathbb{R}$. Then, the log-likelihood function $\ell(\alpha, \beta, \gamma)$ of $L(\alpha, \beta, \gamma)$ is given by

$$\ell(\alpha, \beta, \gamma) = -n\,\alpha \ln \beta - n \ln \Gamma_{(\alpha)} + (\alpha - 1)\sum_{i=1}^n \ln(x_i - \gamma) - \sum_{i=1}^n \frac{x_i - \gamma}{\beta}.$$

The maximum likelihood estimate (MLE) can be obtained by setting $\frac{\partial \ell}{\partial \alpha} = 0$, $\frac{\partial \ell}{\partial \beta} = 0$ and $\frac{\partial \ell}{\partial \gamma} = 0$.

That is,

$$\frac{\partial \ell}{\partial \beta} = -\frac{n\alpha}{\beta} + \frac{\sum_{i=1}^n (x_i - \gamma)}{\beta^2} = 0,$$

which results in the MLE of β to be

$$\widehat{\beta} = \frac{\sum_{i=1}^n (x_i - \widehat{\gamma})}{n\widehat{\alpha}}, \tag{1}$$

$$\frac{\partial \ell}{\partial \alpha} = -n \ln \beta - n\frac{\Gamma'_{(a)}}{\Gamma_{(a)}} + \sum_{i=1}^n \ln(x_i - \gamma) = 0.$$

Substituting $\widehat{\beta}$ in the above expression, we have

$$\ln \widehat{\alpha} - \frac{\Gamma'_{(\widehat{\alpha})}}{\Gamma_{(\widehat{\alpha})}} = \ln\left[\frac{1}{n}\sum_{i=1}^n (x_1 - \widehat{\gamma})\right] - \frac{1}{n}\sum_{i=1}^n \ln(x_i - \widehat{\gamma}), \tag{2}$$

where $\frac{\Gamma'(a)}{\Gamma(a)}$ is called the ***digamma function****, which is defined as the logarithmic derivative of the gamma function. Now*

$$\frac{\partial \ell}{\partial \gamma} = -(\alpha - 1)\sum_{i=1}^{n}\frac{1}{(x_i-\gamma)} + \sum_{i=1}^{n}\frac{1}{\beta} = 0$$

which reduces to

$$\sum_{i=1}^{n}\frac{1}{x_i-\hat{\gamma}} = \frac{n}{\hat{\beta}(\hat{\alpha}-1)}. \tag{3}$$

Thus, we can proceed to numerically solve (1), (2), and (3) to obtain (numerically) an approximate MLE $\hat{\alpha}$, $\hat{\beta}$, and $\hat{\gamma}$ so that we can apply the subject pdf to real data.

We can also use the cumulative probability distribution of the 3-parameter gamma *pdf* to obtain the quantile, x_p, for which $F(x_p) = 1 - p$, that is,

$$F(x_p) = \frac{\Gamma_{x_p} - \gamma^{(\alpha)}}{\beta}\cdot\frac{1}{\Gamma_{(\alpha)}} = 1 - p.$$

Substitute the *MLE* for α, β, and γ, that is, $\hat{\alpha}$, $\hat{\beta}$, and $\hat{\gamma}$ and, proceed to obtain approximate estimates of x_p.

The Weibull probability distribution is very important in characterizing the behavior of health, engineering, and environmental data, among others. The ***Weibull*** *pdf* is given by

$$f(x) = \frac{\alpha}{\beta}\left(\frac{x-\gamma}{\beta}\right)^{\alpha-1}\exp\left[-\left(\frac{x-\gamma}{\beta}\right)^{\alpha}\right],$$

where $x > 0$, the shape parameter α is greater than zero, the scale parameter β is $\beta > 0$, and the location parameter γ is $x > \gamma$. The cumulative probability distribution of the Weibull pdf is given by

$$F(x) = P(X \le x) = \int_{\gamma}^{x}\frac{\alpha}{\beta}\left(\frac{t-\gamma}{\beta}\right)^{\alpha-1}\exp\left[-\left(\frac{t-\gamma}{\beta}\right)^{\alpha}\right]dt$$
$$= 1 - \exp\left[-\left(\frac{x-\gamma}{\beta}\right)^{\alpha}\right].$$

When $\gamma = 0$, the subject pdf is reduced to a 2-parameter Weibull, and it is commonly used because of the difficulty in estimating the 3-parameter pdf.

Example 5.2.13

For a random sample $X_1, \ldots, X_n$ drawn from the 3-parameterWeibull *pdf*, obtain MLEs for the parameters.

Solution

The likelihood function, $L(\alpha, \beta, \gamma)$ is given by

$$L(\alpha,\beta,\gamma) = \alpha^n\beta^{-n\alpha}\left\{\prod_{i=1}^{n}(x_i-\gamma)\right\}^{\alpha-1}\exp\left\{-\beta^{-\alpha}\sum_{i=1}^{n}(x_i-\gamma)^{\alpha}\right\}$$

and the log-likelihood function $\ell(\alpha, \beta, \gamma)$ *of* $L(\alpha, \beta, \gamma)$ *is given by*

$$\ell(\alpha, \beta, \gamma) = n \ln \alpha - n\alpha \ln \beta + (\alpha - 1)$$

$$\cdot \sum_{i=1}^{n} \ln(x_i - \gamma) - \beta^{-\alpha} \sum_{i=1}^{n} (x_i - \gamma)^{\alpha}$$

Setting $\frac{\partial \ell}{\partial \alpha} = 0$, $\frac{\partial \ell}{\partial \beta} = 0$ and $\frac{\partial \ell}{\partial \gamma} = 0$ *and taking the partial derivatives and substituting* $\alpha = \hat{\alpha}$, $\beta = \hat{\beta}$, and $\gamma = \hat{\gamma}$ *and simplifying the resulting expression, we have*

$$\hat{\alpha} + \sum_{i=1}^{n} \ln(x_i - \hat{\gamma}) = \frac{n \sum_{i=1}^{n} (x_i - \hat{\gamma})^{\hat{\alpha}} \ln(x_i - \hat{\gamma})}{\sum_{i=1}^{n} (x_i - \hat{\gamma})^{\hat{\alpha}}},$$

$$\frac{n\hat{\alpha} \sum_{i=1}^{n} (x_i - \hat{\gamma})^{\hat{\alpha}-1}}{\sum_{i=1}^{n} (x_i - \hat{\gamma})^{\hat{\alpha}}} = (\hat{\alpha} - 1) \sum_{i=1}^{n} \frac{1}{x_i - \hat{\gamma}}$$

and

$$\hat{\beta} = \left\{ \frac{1}{n} \sum_{i=1}^{n} (x_i - \hat{\gamma})^{\hat{\alpha}} \right\}^{\frac{1}{\hat{\alpha}}}$$

The above equation cannot be analytically solved without further restrictions, so that we cannot obtain exact values for $\hat{\alpha}$, $\hat{\beta}$, and $\hat{\gamma}$. *However, there are software packages that we can use to obtain approximate estimates of the subject parameters.*

One of the solutions for Example 5.2.13 is given in http://math.ut.ee/acta/12/Bartkute-Sakalauskas.pdf. Thus we can see from the previous examples that even though MLEs are elegant estimators, sometimes it is not easy or possible to obtain explicit forms. For these estimates to perform parametric analysis on a given set of data that represents a real-world phenomenon of interest, we will need numerical approximations.

We can use the cumulative probability distribution function $F(x)$ to the **quantile** x_p for which $F(x_p) = 1 - p$, which reduces to

$$x_p = \gamma + \beta(-\ln p)^{\frac{1}{\alpha}}.$$

Thus, using the MLE of the parameters, we have

$$\hat{x}_p = \hat{\gamma} + \hat{\beta}(-\ln p)^{\frac{1}{\hat{\alpha}}}.$$

The following graphs illustrate how Weibull pdf varies with the shape parameter α (Fig. 5.2A) and with scale parameter β (Fig. 5.2B).

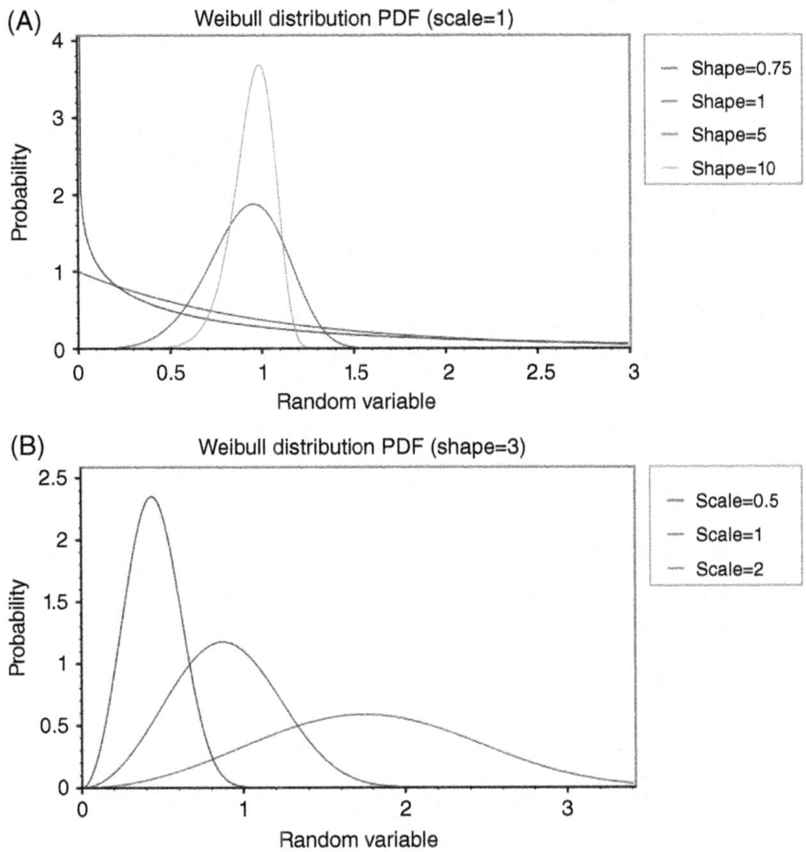

FIGURE 5.2 (A) Weibull distribution with different shape parameters. (B) Weibull distribution with different scale parameters. Two line graphs showing Weibull distribution PDFs with varying shape and scale parameters.

The ***exponential power or error probability distribution*** is usually applicable in characterizing continuous data that is very non-symmetric with respect to its mean. It has been shown to be useful in analyzing environmental, engineering, and health data, among others. It is characterized by three parameters that offer the flexibility of addressing different skewness behaviors. Let X be a continuous random variable that characterizes the behavior of a certain problem of interest; the power exponential or error pdf is given by

$$f(x) = \lambda \left(e^{1-e^{\lambda x^k}}\right) e^{\lambda x^k} x^{k-1}, \; x > 0. \; \lambda > 0, \; k > 0,$$

where λ and k are location and shape parameters, respectively.

The cumulative probability distribution function of the random variable X that follows the exponential power pdf is given by

$$F(x) = 1 - e^{1-e^{\lambda x^k}}, \; x > 0, \; \lambda > 0, \; k > 0.$$

The population mean and variance of X are mathematically intractable. Obtaining MLE analytically is difficult.

The ***Rayleigh distribution*** characterizes the behavior of a continuous random variable that represents many real-world problems. This pdf arises when, in a two-dimensional vector (e.g., wind velocity data as measured by an anemometer and wind range that consists of speed value and direction), each of the components is normally distributed, are not correlated, and have equal variance. Let X be a continous random variable that assumes such data. The Rayleigh pdf of the random variable X is given by

$$f(x; \sigma) = \frac{x}{\sigma^2} e^{\left(\frac{-x^2}{2\sigma^2}\right)}, \; x > 0,$$

where the scale parameter $\sigma > 0$. The pdf of various values of parameters is given in Fig. 5.3.

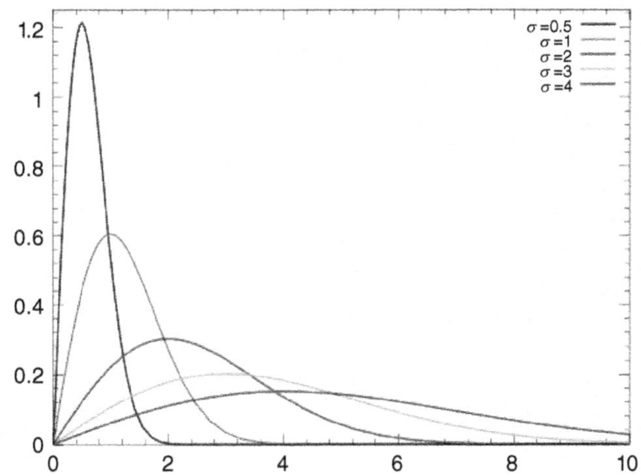

FIGURE 5.3 Rayleigh pdf for various values of σ. Line graph of Rayleigh probability density for sigma values 0.5, 1, 2, 3 and 4, showing peaks and decay trends.

The cdf is given by

$$P(X \le x) = \frac{1}{\sigma^2}\int_0^x \frac{t}{\sigma^2} e^{\left(\frac{-t^2}{2\sigma^2}\right)} dt = 1 - e^{\frac{-x^2}{2\sigma^2}},\ x > 0,\ \sigma > 0.$$

The expected value and the variance are given by

$$E(X) = \sigma\sqrt{\frac{\pi}{2}} = 1.25\sigma$$

and

$$Var(X) = \frac{4-\pi}{2}\sigma^2 = 0.429\sigma^2.$$

For a random sample $X_1, \ldots, X_n$ from Rayleigh pdf, we can verify that the MLE of σ is given by

$$\hat{\sigma} = \left[\frac{1}{2n}\sum_{i=1}^{n} X_i^2\right].$$

Sometimes, it may be necessary to estimate the function of a parameter. The following invariance property of MLEs is very useful in those cases.

Theorem 5.2.1 *Let $h(\theta)$ be a one-to-one function of θ. If $\hat{\theta} = (\hat{\theta}_1, \ldots, \hat{\theta}_l)$ is the MLE of $\theta = (\theta_1, \ldots, \theta_l)$, then the MLE of a function $h(\theta) = (h_1(\theta), \ldots, h_k(\theta))$ of these parameters is $h(\hat{\theta}) = (h_1(\hat{\theta}), \ldots, h_k(\hat{\theta}))$ for $1 \le k \le l$.*

As a consequence of the invariance property, in Example 5.2.10, we can obtain the estimator of the true standard deviation as $\hat{\sigma} = \sqrt{\hat{\sigma}^2} = \sqrt{(1/n)\Sigma_{i=1}^{n}(Xi - \overline{X})^2}$.

It is also known that, under very general conditions on the joint distribution of the sample and for a large sample size n, the MLE $\hat{\theta}$ is approximately the minimum variance unbiased estimator (this concept is introduced in the next section) of θ.

Both methods of moments and maximum likelihood methods are powerful techniques to find a point estimator. MLEs can be shown to be asymptotically optimal in the sense of smallest standard errors. However, MLEs require more assumptions. This means the better method may depend on the particular data setting.

EXERCISES 5.2

5.2.1 Let $X_1,\ldots,X_n$ be a random sample of size n from the geometric distribution for which p is the probability of success.
(a) Use the method of moments to find a point estimator for p.
(b) Use the following data (simulated from a geometric distribution) to find the moment estimate for p:

2	5	7	43	18	19	16	11	22
4	34	19	21	23	6	21	7	12

How will you use this information? [The pmf of a geometric distribution is $f(x) = p(1-p)^{x-1}$, for $x = 1, 2$, Also, $\mu = 1/p$.]

5.2.2 Let $X_1, \ldots, X_n$ be a random sample of size n from the exponential distribution whose pdf (by taking $\theta = 1/\beta$ in Definition 2.3.7) is

$$f(x, \theta) = \begin{cases} \theta e^{-\theta x}, & x \geq 0 \\ 0, & x < 0. \end{cases}$$

(a) Use the method of moments to find a point estimator for θ.
(b) Find MLE of θ.
(c) Using the invariance property, obtain an MLE of the variance.
(d) The following data represent the time intervals between the emissions of beta particles.

0.9	0.1	0.1	0.8	0.9	0.1	0.1	0.7	1.0	0.2
0.1	0.1	0.1	2.3	0.8	0.3	0.2	0.1	1.0	0.9
0.1	0.5	0.4	0.6	0.2	0.4	0.2	0.1	0.8	0.2
0.5	3.0	1.0	0.5	0.2	2.0	1.7	0.1	0.3	0.1
0.4	0.5	0.8	0.1	0.1	1.7	0.1	0.2	0.3	0.1

Assuming the data follow an exponential distribution, obtain a moment estimate for the parameter θ. Interpret.

5.2.3 Let $X_1,\ldots,X_n$ be a random sample from a uniform distribution on the interval $(\theta-1, \theta+1)$.
(a) Find a moment estimator for θ.
(b) Use the following data to obtain a moment estimate for θ:

11.72	12.81	12.09	13.47	12.37

5.2.4 The probability density of a one-parameter Weibull distribution is given by

$$f(x) = \begin{cases} 2\alpha x e^{-\alpha x^2}, & x > 0, \alpha > 0 \\ 0, & \text{otherwise.} \end{cases}$$

(a) Using a random sample of size n, obtain a moment estimator for α.
(b) Assuming that the following data are from a one-parameter Weibull population,

1.87	1.60	2.36	1.12	0.15
1.83	0.64	1.53	0.73	2.26

obtain a moment estimate of α.

5.2.5 Let $X_1,\ldots, X_n$ be a random sample from the truncated exponential distribution with pdf

$$f(x) = \begin{cases} e^{-(x-\theta)}, & x \geq \theta \\ 0, & \text{otherwise.} \end{cases}$$

(a) Find the method of moments estimate of θ.
(b) Show that the MLE of θ is $\min(X_{i;})$.

5.2.6 Let $X_1,\ldots, X_n$ be a random sample from a distribution with pdf

$$f(x) = \begin{cases} \frac{1+\alpha x}{2}, & -1 \leq x \leq 1, \text{ and } -1 \leq \alpha \leq 1 \\ 0, & \text{otherwise.} \end{cases}$$

Find the moment estimators for α.

5.2.7 Let $X_1,\ldots, X_n$ be a random sample from a population with pdf

$$f(x) = \begin{cases} \frac{2\alpha^2}{x^3}, & x \geq \alpha \\ 0, & \text{otherwise.} \end{cases}$$

Find a method of moments estimator for α.

5.2.8 Let $X_1,\ldots, X_n$ be a random sample from a negative binomial distribution with pmf

$$p(x, r, p) = \binom{x+r-1}{r-1} p^x (1-p)^x,\ 0 \leq p \leq 1,\ x = 0,1,2, \ldots.$$

Find method of moments estimators for r and p. [Here $E[X] = r(1-p)/p$ and $E[X^2] = r(1-p)(r - rp + 1)/p^2$.]

5.2.9 Let $X_1,\ldots, X_n$ be a random sample from a distribution with pdf

$$f(x) = \begin{cases} (\theta+1)x^\theta,\ 0 \leq x \leq 1;\ \theta > -1 \\ 0, \quad \text{otherwise.} \end{cases}$$

Use the method of moments to obtain an estimator of θ.

5.2.10 Let $X_1,\ldots, X_n$ be a random sample from a distribution with pdf

$$f(x) = \begin{cases} \frac{2\beta - 2x}{\beta^2}, & 0 < x < \beta \\ 0, & \text{otherwise.} \end{cases}$$

Use the method of moments to obtain an estimator of β.

5.2.11 Let $X_1,\ldots, X_n$ be a random sample with common mean μ and variance σ^2. Obtain a method of moments estimator for σ.

5.2.12 Let $X_1,\ldots, X_n$ be a random sample from the beta distribution with parameters α and β. Find the method of moments estimator for α and β.

5.2.13 Let $X_1, X_2,\ldots, X_n$ be a random sample from a distribution with unknown mean μ and variance σ^2. Show that the method of moments estimators for μ and σ^2 are, respectively, the sample mean $\overline{X}$ and$S'^2 = (1/n)\sum_{i=1}^{n}(X - \overline{X})^2$. Note that $S'^2 = [(n-1)/n]S^2$ where S^2 is the sample variance.

5.2.14 Let $X_1, \ldots, X_n$ be a random sample recorded as heads or tails resulting from tossing a coin n times with unknown probability p of heads. Find the MLE $\hat{p}$ of p. Also, using the invariance property, obtain an MLE for $q = 1 - p$. How would you use the results you have obtained?

5.2.15 Let X be a random variable representing the time between successive arrivals at a checkout counter in a supermarket. The values of X in minutes (rounded to the nearest minute) are

1	2	3	7	11	4	13
12	7	3	2	11	7	2

Assume that the pdf of X is $f(x) = (1/\theta)e^{-(x/\theta)}$. Use this data to find the MLE of θ. How can you use this estimate you have just derived? Also, find the method of moment estimate.

5.2.16 The pdf of a random variable X is given by

$$f(x) = \begin{cases} \frac{2x}{\alpha^2} e^{-x^2/\alpha^2}, & x > 0,\ \alpha > 0 \\ 0, & \text{otherwise.} \end{cases}$$

Using a random sample of size n, obtain the MLE $\hat{\alpha}$ for α.

5.2.17 The pdf of a random variable X is given by

$$P(X = n) = \frac{1}{n!} \exp(\alpha n - e^{\alpha}),\ n = 0,1,2, \ldots.$$

Using a random sample of size n, obtain the MLE $\hat{\alpha}$ for α.

5.2.18 Let $X_1, \ldots, X_n$ be a random sample from a two-parameter Weibull distribution with pdf

$$f(x) = \begin{cases} \frac{\alpha}{\beta^{\alpha}} x^{\alpha-1} e^{-(x/\beta)^{\alpha}}, & x \geq 0 \\ 0, & \text{otherwise.} \end{cases}$$

Obtain maximum likelihood equations, and indicate how to obtain the MLEs of α and β.

5.2.19 Let $X_1, \ldots, X_n$ be a random sample from a Rayleigh distribution with pdf

$$f(x) = \begin{cases} \frac{x}{\alpha} e^{-x^2/2\alpha}, & x > 0 \\ 0, & \text{otherwise.} \end{cases}$$

Find the MLEs of α.

5.2.20 Let $X_1, \ldots, X_n$ be a random sample from a two-parameter exponential population with density

$$f(x, \theta, \upsilon) = \frac{1}{\theta} e^{-\frac{(x-\upsilon)}{\theta}}, \text{ for } x \geq \upsilon,\ \theta > 0.$$

Find MLEs for θ and υ when both are unknown.

5.2.21 Let $X_1, \ldots, X_n$ be a random sample from the shifted exponential distribution with

$$f(x) = \begin{cases} \lambda e^{-\lambda(x-\theta)}, & x \geq \theta \\ 0, & \text{otherwise.} \end{cases}$$

Obtain the MLEs of θ and λ.

5.2.22 Let $X_1,\ldots, X_n$ be a random sample on [0, 1] with pdf

$$f(x) = \frac{\Gamma(2\theta)}{\Gamma(\theta)^2}[x(1-x)]^{\theta-1},\ \theta > 0.$$

What equation does the maximum likelihood estimate of θ satisfy?

5.2.23 Let $X_1,\ldots, X_n$ be a random sample with pdf

$$f(x) = \begin{cases} (\alpha+1)x^{\alpha}, & 0 \le x \le 1 \\ 0, & \text{otherwise.} \end{cases}$$

Find the MLE of α.

5.2.24 Let $X_1,\ldots, X_n$ be a random sample from a uniform distribution with pdf

$$f(x) = \begin{cases} \frac{1}{3\theta+2}, & 0 \le x \le 3\theta+2 \\ 0, & \text{otherwise.} \end{cases}$$

Obtain the MLE of θ.

5.2.25 Let $X_1,\ldots, X_n$ be a random sample from a Cauchy distribution with pdf

$$f(x) = \frac{1}{\pi[1+(x-\beta)^2]},\quad -\infty < x < \infty.$$

Obtain maximum likelihood equations, and indicate how to obtain the MLE for β.

5.2.26 The following data represents the amount of leakage of a fluorescent dye from the bloodstream into the eye in patients with abnormal retinas:

1.6	1.4	1.2	2.2	1.8	1.7
1.8	6.3	2.4	2.3	18.9	22.8

Assuming that these data come from a normal distribution, find the maximum likelihood estimate of (μ, σ).

5.2.27 Let $X_1,\ldots, X_n$ be a random sample from a population with gamma distribution and parameters α and β. Show that the MLE of $\mu = \alpha\beta$ is the sample mean $\hat{\mu} = \overline{X}$.

5.2.28 The lifetimes X of a certain brand of component used in a machine can be modeled as a random variable with pdf $f(x) = (1/\theta)\, e^{-(x/\theta)}$. The reliability $R(x)$ of the component is defined as $R(x) = 1 - F(x)$. Suppose $X_1, X_2,\ldots,X_n$ are the lifetimes of n components randomly selected and tested. Find the MLE of $R(x)$.

5.2.29 Using the method explained in Project 4A, generate 20 observations of a random variable having an exponential distribution with mean and standard deviation both equal to 2. What is the maximum likelihood estimate of the population mean? How much is the observed error?

5.2.30 Let $X_1,\ldots,X_n$ be a random sample from a Pareto distribution (named after the economist Vilfredo Pareto) with shape parameter a. The density function is given by

$$f(x) = \begin{cases} \frac{a}{x^{a+1}}, & x \ge 1 \\ 0, & \text{otherwise.} \end{cases}$$

(The Pareto distribution is a skewed, heavy-tailed distribution. Sometimes it is used to model the distribution of incomes.) Show that the maximum likelihood estimator of a is

$$\hat{a} = \frac{n}{\sum_{i=1}^{n} \ln(X_i)}.$$

5.2.31 Let $X_1,\ldots,X_n$ be a random sample from $N(\theta, \theta)$, $0 < \theta < \infty$. Find the maximum likelihood estimate of θ.

5.3 SOME DESIRABLE PROPERTIES OF POINT ESTIMATORS

Two different methods of finding estimators for population parameters have been introduced in the preceding section. We have seen that it is possible to have several estimators for the same parameter. For a practitioner of statistics, an important question is going to be which of many available sample statistics, such as mean, median, smallest observation, or largest observation, should be chosen to represent all of the samples? Should we use the method of moments estimator, the maximum likelihood estimator, or an estimator obtained through some other method such as the least squares (we will see this method in Chapter 7)? Now we introduce some common ways to distinguish between them by looking at some desirable properties of these estimators.

5.3.1 Unbiased Estimators

It is desirable to have the property that the expected value of an estimator of a parameter is equal to the true value of the parameter. Such estimators are called unbiased estimators.

Definition 5.3.1. *A point estimator* $\hat{\theta}$ *is called an* **unbiased estimator** *of the parameter* θ *if* $E(\hat{\theta}) = \theta$ *for all possible values of* θ. *Otherwise* $\hat{\theta}$ *is said to be* **biased**. *Furthermore, the* **bias** *of* $\hat{\theta}$ *is given by*

$$B = E(\hat{\theta}) - \theta.$$

Note that the bias is nothing but the expected value of the (random) error, $E(\hat{\theta} - \theta)$. Thus, the estimator is unbiased if the bias is 0 for all values of θ. The bias occurs when a sample does not accurately represent the population from which the sample is taken. It is important to observe that in order to check whether $\hat{\theta}$ is unbiased, it is not necessary to know the value of the true parameter. Instead, one can use the sampling distribution of $\hat{\theta}$. We demonstrate the basic procedure through the following example.

Example 5.3.1

Let $X_1, \ldots, X_n$ be a random sample from a Bernoulli population with parameter p. Show that the method of moments estimator is also an unbiased estimator.

Solution

We can verify that the moment estimator of p is

$$\hat{p} = \frac{\sum_{i=1}^{n} X_i}{n} = \frac{Y}{n}.$$

Because for binomial random variables, E (Y) = np, it follows that

$$E(\hat{p}) = E\left(\frac{Y}{n}\right) = \frac{1}{n}E(Y) = \frac{1}{n}\cdot np = p.$$

Hence, $\hat{p} = Y/n$ *is an unbiased estimator for p.*

In fact, we have the following result, which states that the sample mean is always an unbiased estimator of the population mean.

Theorem 5.3.1 *The mean of a random sample $\overline{X}$ is an unbiased estimator of the population mean μ.*

Proof. Let $X_1,\ldots,X_n$ be random variables with mean μ. Then, the sample mean is $\overline{X} = (1/n)\Sigma_{i=1}^{n} X_i$.

$$E\overline{X} = \frac{1}{n}\sum_{i=1}^{n} EX_i = \frac{1}{n}\cdot n\mu = \mu.$$

Hence, $\overline{X}$ is an unbiased estimator of μ.

How is this interpreted in practice? Suppose that a data set is collected with n numerical observations $x_1,\ldots,x_n$. The resulting sample mean may be either less than or greater than the true population mean, μ (remember, we do not know this value). If the sampling experiment was repeated many times, then the average of the estimates calculated over these repetitions of the sampling experiment will equal the true population mean.

If we have to choose among several different estimators of a parameter θ, it is desirable to select one that is unbiased. The following result states that the sample variance $S^2 = (1/n - 1)\Sigma_{i=1}^{n}(X_i - \overline{X})^2$ is an unbiased estimator of the population variance σ^2. This is one of the reasons why in the definition of the sample variance, instead of dividing by n, we divide by $(n - 1)$.

Theorem 5.3.2 *If S^2 is the variance of a random sample from an infinite population with finite variance σ^2, then S^2 is an unbiased estimator for σ^2.*

Proof. Let $X_1,\ldots,X_n$ be random sample with variance $\sigma^2 < \infty$. We have

$$E(S^2) = \frac{1}{n-1} E\sum_{i=1}^{n}(X_i - \overline{X})^2 = \frac{1}{n-1} E\left[\sum_{i=1}^{n}\{(X_i - \mu) - (\overline{X} - \mu)\}^2\right]$$

$$= \frac{1}{n-1}\left[\sum_{i=1}^{n} E\{X_i - \mu\}^2 - nE\{\overline{X} - \mu\}^2\right].$$

Because $E\{(X_i - \mu)^2\} = \sigma^2$ and $E\{(\overline{X} - \mu)^2\} = \sigma^2/n$, it follows that

$$E(S^2) = \frac{1}{n-1}\left[\sum_{i=1}^{n}\sigma^2 - n\frac{\sigma^2}{n}\right] = \sigma^2.$$

Hence, S^2 is an unbiased estimator of σ^2.

It is important to observe the following:

1. S^2 is not an unbiased estimator of the variance of a finite population.
2. Unbiasedness may not be retained under functional transformations; that is, if $\hat{\theta}$ is an unbiased estimator of θ, it does not follow that $f(\hat{\theta})$ is an unbiased estimator of $f(\theta)$.
3. MLEs or moment estimators are not, in general, unbiased.
4. In many cases, it is possible to alter a biased estimator by multiplying by an appropriate constant to obtain an unbiased estimator.

The following example will show that unbiased estimators need not be unique.

Example 5.3.2

Let $X_1,\ldots,X_n$ be a random sample from a population with finite mean μ. Show that the sample mean $\overline{X}$ and $\frac{1}{3}\overline{X} + \frac{2}{3}X_1$ are both unbiased estimators of μ.

Solution

By Theorem 1, $\overline{X}$ is unbiased. Now

$$E\left[\frac{1}{3}\overline{X} + \frac{2}{3}X_1\right] = \frac{1}{3}\mu + \frac{2}{3}\mu = \mu.$$

Hence $\frac{1}{3}\overline{X} + \frac{2}{3}X_1$ is also an unbiased estimator of μ.

How many unbiased estimators can we find? In fact, the following example shows that if we have two unbiased estimators, there are infinitely many unbiased estimators.

Example 5.3.3

Let $\hat{\theta}_1$ and $\hat{\theta}_2$ be two unbiased estimators of θ. Show that

$$\hat{\theta}_3 = a\hat{\theta}_1 + (1 - a)\hat{\theta}_2, \ 0 \le a \le 1$$

is an unbiased estimator of θ. Note that $\hat{\theta}_3$ is a convex combination of $\hat{\theta}_1$ and $\hat{\theta}_2$. In addition, assume that $\hat{\theta}_1$ and $\hat{\theta}_2$ are independent, $Var(\hat{\theta}_1) = \sigma_1^2$ and $Var(\hat{\theta}_2) = \sigma_2^2$. How should the constant a be chosen in order to minimize the variance of $\hat{\theta}_3$?

Solution

We are given that $E(\hat{\theta}_1) = \theta$ and $E(\hat{\theta}_2) = \theta$. Therefore,

$$\begin{aligned} E(\hat{\theta}_3) &= E[a\hat{\theta}_1 + (1 - a)\hat{\theta}_2] = aE\hat{\theta}_1 + (1 - a)E\hat{\theta}_2 \\ &= a\theta + (1 - a)\theta = \theta. \end{aligned}$$

Hence $\hat{\theta}_3$ is unbiased. By independence,

$$\begin{aligned} Var(\hat{\theta}_3) &= Var[a\hat{\theta}_1 + (1 - a)\hat{\theta}_2] \\ &= a^2 Var(\hat{\theta}_1) + (1 - a)^2 Var(\hat{\theta}_2) \\ &= a^2\sigma_1^2 + (1 - a)^2\sigma_2^2. \end{aligned}$$

To find the minimum,

$$\frac{d}{da}Var(\hat{\theta}_3) = 2a\sigma_1^2 - 2(1 - a)\sigma_2^2 = 0$$

gives us

$$a = \frac{\sigma_2^2}{\sigma_1^2 + \sigma_2^2}.$$

Because $\frac{d^2}{da^2}V(\hat{\theta}_3) = 2\sigma_1^2 + 2\sigma_2^2 > 0$, $V(\hat{\theta}_3)$ has a minimum at this value of 'a'. Thus if $\sigma_1^2 = \sigma_2^2$, then $a = 1/2$.

Example 5.3.4

Let $X_1, \ldots, X_n$ be a random sample from a population with pdf

$$f(x) = \begin{cases} \frac{1}{\beta}e^{-x/\beta}, & x > 0 \\ 0, & \text{otherwise.} \end{cases}$$

Show that the method of moments estimator for the population parameter β is unbiased.

Solution

From Section 5.2, we have seen that the method of moments estimator for β is the sample mean $\overline{X}$, and the population mean is β. Because $E(\overline{X}) = \mu = \beta$, the method of moments estimator for the population parameter β is unbiased.

As we have seen, there can be many unbiased estimators of a parameter θ. Which one of these estimators can we choose? If we have to choose an unbiased estimator, it will be desirable to choose the one with the least variance. If an estimator is biased, then we should prefer the one with low bias as well as low variance. Generally, it is better to have an estimator that has low bias as well as low variance. This leads us to the following definition.

Definition 5.3.2. *The* **mean square error** *of the estimator* $\hat{\theta}$, *denoted by* $MSE(\hat{\theta})$, *is defined as*

$$MSE(\hat{\theta}) = E(\hat{\theta} - \theta)^2.$$

Through the following calculations, we will now show that the MSE is a measure that combines both bias and variance.

$$\begin{aligned} MSE(\hat{\theta}) &= E(\hat{\theta} - \theta)^2 = E[(\hat{\theta} - E(\hat{\theta})) + (E(\hat{\theta}) - \theta)]^2 \\ &= E\left[(\hat{\theta} - E(\hat{\theta}))^2 + (E(\hat{\theta}) - \theta)^2 + 2(\hat{\theta} - E(\hat{\theta}))(E(\hat{\theta}) - \theta)\right] \\ &= E(\hat{\theta} - E(\hat{\theta}))^2 + E(E(\hat{\theta}) - \theta)^2 + 2E(\hat{\theta} - E(\hat{\theta}))(E(\hat{\theta}) - \theta) \\ &= Var(\hat{\theta}) + [E(\hat{\theta}) - \theta]^2. \end{aligned}$$

Letting $B = E(\hat{\theta}) - \theta$, we get

$$MSE(\hat{\theta}) = Var(\hat{\theta}) + B^2.$$

B is called the ***bias*** of the estimator. Also, $E(\hat{\theta} - E(\hat{\theta}))(E(\hat{\theta}) - \theta) = 0$.

Because the bias is zero for unbiased estimators, it is clear that $MSE(\hat{\theta}) = Var(\hat{\theta})$. Mean square error measures, on average, how close an estimator comes to the true value of the parameter. Hence, this could be used as a criterion for determining when one estimator is "better" than another. However, in general, it is difficult to find $\hat{\theta}$ to minimize $MSE(\hat{\theta})$. For this reason, most of the time, we look only at unbiased estimators in order to minimize $Var(\hat{\theta})$. This leads to the following definition.

Definition 5.3.3. *The unbiased estimator* $\hat{\theta}$ *that minimizes the mean square error is called the* **minimum variance unbiased estimator (MVUE)** of θ.

Example 5.3.5

Let X_1, X_2, X_3 be a sample of size $n = 3$ from a distribution with unknown mean μ, $-\infty < \mu < \infty$, where the variance σ^2 is a known positive number. Show that both $\hat{\theta}_1 = \overline{X}$ and $\hat{\theta}_2 = [(2X_1 + X_2 + 5X_3)/8]$ are unbiased estimators for μ. Compare the variances of $\hat{\theta}_1$ and $\hat{\theta}_2$.

Solution

We have

$$E(\hat{\theta}_1) = E(\overline{X}) = \frac{1}{3} \cdot 3\mu = \mu,$$

and

$$\begin{aligned} E(\hat{\theta}_2) &= \frac{1}{8}[2EX_1 + EX_2 + 5EX_3] \\ &= \frac{1}{8}[2\mu + \mu + 5\mu] = \mu. \end{aligned}$$

Hence both $\hat{\theta}_1$ and $\hat{\theta}_2$ are unbiased estimators.
However,

$$Var(\hat{\theta}_1) = \frac{\sigma^2}{3},$$

whereas

$$Var(\hat{\theta}_2) = Var\left(\frac{2X_1 + X_2 + 5X_3}{8}\right)$$
$$= \frac{4}{64}\sigma^2 + \frac{1}{64}\sigma^2 + \frac{25}{64}\sigma^2 = \frac{30}{64}\sigma^2.$$

Because $Var(\hat{\theta}_1) < Var(\hat{\theta}_2)$, we see that $\overline{X}$ is a better unbiased estimator in the sense that the variance of $\overline{X}$ is smaller.

It is important to observe that the MLEs are not always unbiased, but it can be shown that for such estimators the bias goes to zero as the sample size increases.

5.3.2 Sufficiency

In the statistical inference problems on a parameter, one of the major questions is: Can a specific statistic replace the entire data without losing pertinent information? Suppose $X_1, \ldots, X_n$ is a random sample from a probability distribution with unknown parameter θ. In general, statisticians look for ways of reducing a set of data so that these data can be more easily understood without losing the meaning associated with the entire collection of observations. Intuitively, a statistic U is a sufficient statistic for a parameter θ if U contains all the information available in the data about the value of θ. For example, the sample mean may contain all the relevant information about the parameter μ, and in that case $U = \overline{X}$ is called a sufficient statistic for μ. An estimator that is a function of a sufficient statistic can be deemed a "good" estimator, because it depends on fewer data values. When we have a sufficient statistic U for θ, we need to concentrate only on U because it exhausts all the information that the sample has about θ. That is, knowledge of the actual n observations does not contribute anything more to the inference about θ.

Definition 5.3.4. *Let $X_1, \ldots, X_n$ be a random sample from a probability distribution with unknown parameter θ. The statistic $U = g(X_1, \ldots, X_n)$ is then said to be* **sufficient** *for θ if the conditional pdf or pf of $X_1, \ldots, X_n$ given $U = u$ does not depend on θ for any value of u. An estimator of θ that is a function of a sufficient statistic for θ is said to be a* **sufficient estimator** *of θ.*

Example 5.3.6

Let $X_1, \ldots, X_n$ be iid Bernoulli random variables with parameter θ. Show that $U = \Sigma_{i=1}^{n} X_i$ is sufficient for θ.

Solution

The joint probability mass function of $X_1, \ldots, X_n$ is

$$f(X_1, \ldots, X_n; \theta) = \theta^{\sum_{i=1}^{n} X_i}(1-\theta)^{n-\sum_{i=1}^{n} X_i}, \; 0 \leq \theta \leq 1.$$

Because $U = \Sigma_{i=1}^{n} X_i$ we have

$$f(X_1, \ldots, X_n; \theta) = \theta^U (1-\theta)^{n-U}, \; 0 \leq U \leq n.$$

Also, because $U \sim B(n, \theta)$, we have

$$f(u; \theta) = \binom{n}{u}\theta^U (1-\theta)^{n-U}.$$

Also

$$f(x_1, \ldots, x_n | U = u) = \frac{f(x_1, \ldots, x_n, u)}{f_U(u)} = \begin{cases} \frac{f(x_1, \ldots, x_n)}{f_U(u)}, & u = \sum x_i \\ 0, \text{ otherwise.} \end{cases}$$

Therefore

$$f(x_1, \ldots, x_n | U = u) = \begin{cases} \frac{\theta^u (1-\theta)^{n-u}}{\binom{n}{u} \theta^u (1-\theta)^{n-u}} = \frac{1}{\binom{n}{u}} & \text{if } u = \sum x_i \\ 0, \text{ otherwise.} \end{cases}$$

which is independent of θ. Therefore, U is sufficient for θ.

Example 5.3.7

Let $X_1, \ldots, X_n$ be a random sample from $U(0, \theta)$. That is,

$$f(x) = \begin{cases} \frac{1}{\theta}, & \text{if } 0 < x < \theta \\ 0, \text{ otherwise.} \end{cases}$$

Show that $U = \max\limits_{1 \le i \le n} X_i$ is sufficient for θ.

Solution

The joint density or the likelihood function is given by

$$f(x_1, \ldots, x_n; \theta) = \begin{cases} \frac{1}{\theta^n}, & \text{if } 0 < x_1, \ldots, x_n < \theta \\ 0, \text{ otherwise.} \end{cases}$$

The joint pdf $f(x_1, \ldots, x_n; \theta)$ can be equivalently written as

$$f(x_1, \ldots, x_n; \theta) = \begin{cases} \frac{1}{\theta^n}, & \text{if } x_{min} > 0,\ x_{max} < \theta \\ 0, \text{ otherwise.} \end{cases}$$

Now, we can compute the pdf of U.

$$\begin{aligned} F(u) &= P(U \le u) = P(X_1, \ldots, X_n \le u) \\ &= \prod_{i=1}^{n} P(X_i \le u) \text{(because of independence)} \\ &= \prod_{i=1}^{n} \left(\int_0^u \frac{1}{\theta} dx \right) = \frac{u^n}{\theta^n}, \ 0 < u < \theta. \end{aligned}$$

The pdf of U may now be obtained as

$$f(u) = \frac{d}{du} F(u) = \frac{n u^{n-1}}{\theta^n}, \ 0 < u < \theta$$

Moreover,

$$f(x_1, \ldots, x_n | u) = \begin{cases} \frac{f(x_1, \ldots, x_n, u)}{f_U(u)} = \frac{f(x_1, \ldots x_n)}{f_U(u)}, & \text{if } u = x_{max} \text{ and} x_{min} > 0 \\ 0, \text{ otherwise.} \end{cases}$$

Using the expressions for $f(x_1,\ldots,x_n)$ and $f_U(u)$, we obtain

$$f(x_1, \ldots, x_n | u = u) = \begin{cases} \frac{1/\theta^n}{nu^{n-1}/\theta^n} = \frac{1}{nu^{n-1}}, & \text{if } u = x_{max} \text{ and } x_{min} > 0 \\ 0, & \text{otherwise,} \end{cases}$$

$f(X_1,\ldots,X_n|U)$ is a function of u and x_{min}, which is independent of θ. Hence $U = \max_{1\le i\le n} X_i$ is sufficient for θ.

The outcome $X_1,\ldots,X_n$ is always sufficient, but we will exclude this trivial statistic from consideration. In the previous two examples, we were given a statistic and asked to check whether it was sufficient. It can often be tedious to check whether a statistic is sufficient for a given parameter based directly on the foregoing definition. If the form of the statistic is not given, how do we guess what the sufficient statistic is? Now think of working out the conditional probability by hand for each of our guesses! In general, this will be a tedious way to go about finding sufficient statistics. Fortunately, the Neyman–Fisher factorization theorem makes it easier to spot a sufficient statistic. The following result will give us a convenient way of verifying sufficiency of a statistic through the likelihood function.

Neyman–Fisher Factorization Criteria

Theorem 5.3.3. *Let U be a statistic based on the random sample $X_1,\ldots,X_n$. Then, U is a sufficient statistic for θ if and only if the joint pdf (or pf) $f(x_1,\ldots,x_n;\theta)$ (which depends on the parameter θ) can be factored into two non-negative functions.*

$$f(x_1, \ldots, x_n; \theta) = g(u, \theta)h(x_1, \ldots, x_n), \text{ for all } x_1, \ldots, x_n,$$

where $g(u, \theta)$ is a function only of u and θ and $h(x_1, \ldots, x_n)$ is a function of only $x_1, \ldots, x_n$ and not of θ.

Proof. (Discrete case.) We will only give the proof in the discrete case, even though the result is also true for the continuous case. First suppose that $U(X_1,\ldots,X_n)$ is sufficient for θ. Then, $X_1 = x_1, X_2 = x_2,\ldots, X_n = x_n$ if and only if $X_1 = x_1$, $X_2 = x_2,\ldots, X_n = x_n$ and $U(X_1,\ldots,X_n) = U(x_1,\ldots,x_n) = u(say)$. Therefore

$$\begin{aligned} f(x_1,\ldots,x_n;\theta) &= P_\theta(X_1 = x_1, X_2 = x_2,\ldots,X_n = x_n \text{ and } U = u) \\ &= P_\theta(X_1 = x_1, X_2 = x_2,\ldots,X_n = x_n|U = u)P_\theta(U = u). \end{aligned}$$

Because U is assumed to be sufficient for θ, the conditional probability $P_\theta(X_1 = x_1, \ldots, X_n = x_n|U = u)$ does not depend on θ. Let us denote this conditional probability by $h(x_1,\ldots,x_n)$. Clearly $P_\theta(U = u)$ is a function of u and θ. Let us denote this by $g(u, \theta)$.

It now follows from the equation above that

$$f(x_1, \ldots, x_n; \theta) = g(u, \theta)h(x_1, \ldots, x_n)$$

as was to be shown.

To prove the converse, assume that

$$f(x_1, \ldots, x_n; \theta) = g(u, \theta)h(x_1, \ldots, x_n).$$

Define the set A_u by

$$A_u = \{(x_1,\ldots,x_n): U(x_1,\ldots,x_n) = u\}.$$

That is, A_u is the set of all $(x_1,\ldots,x_n)$ such that U maps it into u. We note that A_u does not depend on θ. Now

$$
\begin{aligned}
&P_\theta(X_1 = x_1, X_2 = x_2, \ldots, X_n = x_n | U = u)\\
&= \frac{P_\theta(X_1 = x_1, X_2 = x_2, \ldots X_n = x_n \text{ and } U = u)}{P_\theta(U = u)}\\
&= \begin{cases} \dfrac{P_\theta(X_1 = x_1, X_2 = x_2, \ldots, X_n = x_n \text{ and } U = u)}{P_\theta(U = u)}, & \text{if}(x_1, \ldots, x_n) \in A_u \\ 0, & \text{if}(x_1, \ldots, x_n) \notin A_u. \end{cases}
\end{aligned}
$$

If $(x_1,\ldots, x_n) \notin A_u$, then, clearly,

$$f(x_1, \ldots, x_n; \theta) = P_\theta(X_1 = x_1, X_2 = x_2, \ldots, X_n = x_n | U = u)$$

which is independent of θ.
If $(x_1,\ldots,xn) \in A_u$, then, using the factorization criterion, we obtain

$$
\begin{aligned}
&P_\theta(X_1 = x_1, X_2 = x_2, \ldots, X_n = x_n | U = u)\\
&= \frac{P_\theta(X_1 = x_1, X_2 = x_2, \ldots X_n = x_n)}{P_\theta(U = u)}\\
&= \frac{f(x_1, \ldots, x_n; \theta)}{P_\theta(U = u)} = \frac{g(u, \theta)h(x_1, \ldots, x_n)}{\sum_{(x_1, \ldots, x_n) \in A_u} g(u, \theta)h(x_1, \ldots, x_n)}\\
&= \frac{g(u, \theta)h(x_1, \ldots, x_n)}{g(u, \theta)\sum_{(x_1, \ldots, x_n) \in A_u} h(x_1, \ldots, x_n)} = \frac{h(x_1, \ldots, x_n)}{\sum_{(x_1, \ldots, x_n) \in A_u} h(x_1, \ldots, x_n)}.
\end{aligned}
$$

Therefore, the conditional distribution of $X_1,\ldots, X_n$ given U does not depend on θ, proving that U is sufficient.
One can use the following procedure to verify that a given statistic is sufficient. This procedure is based on factorization criteria rather than using the definition of sufficiency directly.

Procedure to Verify Sufficiency

1. Obtain the joint pdf or pf $f_\theta(x_1, \ldots, x_n)$.
2. If necessary, rewrite the joint pdf or pf in terms of the given statistic and parameter so that one can use the factorization theorem.
3. Define the functions g and h, in such a way that g is a function of the statistic and parameter only and h is a function of the observations only.
4. If step 3 is possible, then the statistic is sufficient. Otherwise, it is not sufficient.

In general, it is not easy to use the factorization criterion to show that a statistic U is *not* sufficient. We now give some examples using the factorization theorem.

Example 5.3.8

Let $X_1,\ldots,X_n$ denote a random sample from a geometric population with parameter p. Show that $\overline{X}$ is sufficient for p.

Solution

For the geometric distribution, the pf is given by

$$f(x, p) = \begin{cases} p(1 - p)^{x-1}, & x \geq 1 \\ 0, & \text{otherwise.} \end{cases}$$

Hence the joint pf is

$$f(x_1,\ldots,x_n;\, p) = p^n(1-p)^{-n+\sum_{i=1}^{n} x_i} = \begin{cases} p^n(1-p)^{n\bar{x}-n}, & \text{if } x_1,\ldots,x_n \geq 1 \\ 0, & \text{otherwise.} \end{cases}$$

Take

$$g(\bar{x}, p) = p^n(1-p)^{n\bar{x}-n} \text{ and } h(x_1, \ldots, x_n) = \begin{cases} 1, & \text{if } x_i \geq 1 \\ 0, & \text{otherwise.} \end{cases}$$

Thus $\bar{X}$ *is sufficient for* p.

Example 5.3.9

Let $X_1,\ldots,X_n$ denote a random sample from a $U(0, \theta)$ with pdf

$$f_\theta(x) = \begin{cases} \frac{1}{\theta}, & 0 < x < \theta, \theta > 0 \\ 0, & \text{otherwise.} \end{cases}$$

Show that $X_{(n)} = \max_{1 \leq i \leq n} X_i$ is sufficient for θ, using the factorization theorem.

Solution

The likelihood function of the sample is

$$f_\theta(x_1, \ldots, x_n) = \begin{cases} \frac{1}{\theta^n}, & \text{if } 0 < x_1, \ldots, x_n < \theta, \\ 0, & \text{otherwise.} \end{cases}$$

We can now write $f_\theta(x_1,\ldots,x_n)$ as

$$f_\theta(x_1, \ldots, x_n) = h(x_1, \ldots, x_n) g\left(\theta, x_{(n)}\right), \text{ for all } x_1, \ldots, x_n$$

where

$$h(x_1, \ldots, x_n) = \begin{cases} 1, & \text{if } x_1, \ldots, x_n > 0 \\ 0, & \text{otherwise} \end{cases}$$

and

$$g\left(\theta;\, x_{(n)}\right) = \begin{cases} \frac{1}{\theta^n}, & \text{if } 0 < x_{(n)} < \theta, \\ 0, & \text{otherwise.} \end{cases}$$

From the factorization theorem, we now conclude that $X_{(n)}$ *is sufficient for* θ*. In the next definition, we introduce the concept of joint sufficiency.*

To show that a statistic $U = U(X)$ is not sufficient, either we can show that the factorization does not hold, or the conditional distribution of the data given the statistic depends on the parameter θ. For instance, let X_1, X_2 be random variables from a uniform distribution, $U(0, \theta)$. *Take the statistic as* $U(X) = X_1$. *It can be shown that the conditional distribution of* X_2 *given* $X_1 = x_1$ *depends on* θ *even after observing* X_1. *Thus,* $U = U(X)$ *is not sufficient for* θ. *This can be shown in the following way:*

The joint distribution is

$$f_\theta(x_1, x_2) = \begin{cases} \frac{1}{\theta^2}, \text{ if } 0 < x_1, x_2 < \theta, \\ 0, \text{ otherwise.} \end{cases}$$

Since $X_1 \sim U(0, \theta)$, the marginal pdf is

$$f_{X_1}(x_1) = \begin{cases} \frac{1}{\theta}, & 0 < x_1 < \theta \\ 0, & \textit{otherwise} \end{cases}$$

The conditional pdf of X_2 given $X_1 = x_1$ is

$$f(x_2|x_1) = \frac{f(x_1, x_2)}{f_{X_1}(x_1)} = \frac{\frac{1}{\theta^2}}{\frac{1}{\theta}} = \frac{1}{\theta}, \textit{ for } 0 < x_2 < \theta,$$

which depends on the parameter θ.

Definition 5.3.5. *Two statistics U_1 and U_2 are said to be* **jointly sufficient** *for the parameters θ_1* and *θ_2 if the conditional distribution of $X_1, \ldots, X_n$ given U_1* and *U_2 does not depend on θ_1* or *θ_2. In general, the statistic $\mathbf{U} = (U_1, \ldots, U_n)$ is jointly sufficient for $\boldsymbol{\theta} = (\theta_1, \ldots, \theta_n)$ if the conditional distribution of $X_1, \ldots, X_n$ given $\mathbf{U}$ is free of $\boldsymbol{\theta}$.*

Now we state the factorization criteria for joint sufficiency analogous to the single population parameter case.

The Factorization Criteria for Joint Sufficiency

Theorem 5.3.4. *The two statistics U_1 and U_2 are jointly sufficient for θ_1 and θ_2 if and only if the likelihood function can be factored into two non-negative functions,*

$$f(x_1, \ldots, x_n; \theta_1, \theta_2) = g(u_1, u_2; \theta_1, \theta_2) h(x_1, \ldots, x_n)$$

where $g(u_1, u_2; \theta_1, \theta_2)$ is only a function of u_1, u_2; θ_1 and θ_2, and $h(x_1, \ldots, x_n)$ is free of θ_1 or θ_2.

Example 5.3.10

Let $X_1, \ldots, X_n$ be a random sample from $N(\mu, \sigma^2)$.

(a) If μ is unknown and $\sigma^2 = \sigma_0^2$ is known, show that $\overline{X}$ is a sufficient statistic for μ.
(b) If $\mu = \mu_0$ is known and σ^2 is unknown, show that $\Sigma_{i=1}^n (X_i - \mu_0)^2$ is sufficient for σ^2.
(c) If μ and σ^2 are both unknown, show that $\Sigma_{i=1}^n X_i$ and $\Sigma_{i=1}^n X_i^2$ are jointly sufficient for μ and σ^2.

Solution

The likelihood function of the sample is

$$L = \frac{1}{(2\pi)^{n/2}\sigma^n} \exp\left[-\frac{\Sigma_{i=1}^n (X_i - \mu)^2}{2\sigma^2}\right]$$

$$= \frac{1}{(2\pi)^{n/2}\sigma^n} \exp\left[\frac{1}{2\sigma^2}\left(\Sigma_{i=1}^n x_i^2 - 2\mu \Sigma_{i=1}^n x_i + n\mu^2\right)\right]$$

$$= (2\pi)^{-n/2}\sigma^{-n} \exp\left(-\frac{\Sigma_{i=1}^n x_i^2}{2\sigma^2}\right) \exp\left(\frac{2\mu n\overline{x}}{2\sigma^2}\right) \exp\left(-\frac{n\mu^2}{2\sigma^2}\right).$$

(a) *When $\sigma^2 = \sigma_0^2$ is known, use the factorization criteria, with*

$$g(\bar{x}, \mu) = \exp\left(\frac{2n\mu\bar{x} - n\mu^2}{2\sigma_0^2}\right)$$

and

$$h(x_1, \ldots, x_n) = (2\pi)^{-n/2}\sigma^{-n}\exp\left(-\frac{\sum_{i=1}^n x_i^2}{2\sigma^2}\right).$$

Therefore, $\bar{X}$ is sufficient for μ.

(b) *When $\mu = \mu_0$ is known, let*

$$g\left(\sum_{i=1}^n (X_i - \mu)^2, \sigma^2\right) = \sigma^{-n}\exp\left|-\frac{\sum_{i=1}^n (x_i - \mu)^2}{2\sigma^2}\right|$$

and

$$h(x_1, \ldots, x_n) = \frac{1}{(2\pi)^{n/2}}.$$

Thus, $\sum_{i=1}^n (X_i - \mu)^2$ is sufficient for σ^2.

(c) *When both μ and σ^2 are unknown, use*

$$g\left(\sum_{i=1}^n x_i, \sum_{i=1}^n x_i^2, \mu, \sigma^2\right) = \sigma^{-n}\exp\left|-\frac{\sum_{i=1}^n x_i^2 - 2\mu\sum_{i=1}^n x_i + n\mu^2}{2\sigma^2}\right|$$

and

$$h(x_1, \ldots, x_n) = \frac{1}{(2\pi)^{n/2}}.$$

Hence $\sum_{i=1}^n X_i$ and $\sum_{i=1}^n X_i^2$ are jointly sufficient for μ and σ^2.

Example 5.3.11

Suppose that we have a random sample $X_1, \ldots, X_n$ from a discrete distribution given by

$$f_\theta(x) = C(\theta)2^{-x/\theta},\ x = \theta, \theta + 1, \theta + 2, \ldots; \theta > 0$$

where $C(\theta) > 0$ is a normalizing constant. Using the factorization theorem, find a sufficient statistic for θ.

Solution

The joint density function $f(x_1, \ldots, x_n; \theta)$ of the sample $X_1, \ldots, X_n$ is

$$f(x_1, \ldots, x_n; \theta) = \begin{cases} C(\theta)2^{-\sum_{i=1}^n (x_i/\theta)}, & x_1, x_2, \ldots, x_n \text{ are integers} \geq \theta \\ 0, \text{ otherwise.} \end{cases}$$

The function $f(x_1,\ldots,x_n;\theta)$ can be written as

$$f(x_1,\ldots,x_n;\theta) = h(x_1,\ldots,x_n)C(\theta)2^{-\sum_{i=1}^{n}(x_i/\theta)}g_1(\theta, x_{(1)})$$

where $x(1) = \min_i(x_1,\ldots,x_n)$, and

$$h(x_1, x_2, \ldots, x_n) = \begin{cases} 1, \text{ if } x_j - x_{(1)} \geq 0 \text{ is an integer for } j = 1,2,\ldots,n \\ 0, \text{ otherwise} \end{cases}$$

and

$$g_1(\theta, x_{(1)}) = \begin{cases} 1, \text{ if } x_{(1)} \geq \theta \\ 0, \text{ otherwise.} \end{cases}$$

Thus

$$f(x_1,\ldots,x_n;\theta) = h(x_1,\ldots,x_n)g\left(\theta, \sum x_i, x_{(1)}\right)$$

where $g(\theta, \Sigma x_i, x_1) = C(\theta)2^{-\Sigma_{i=1}^{n}(x_i/\theta)}g_1(\theta, x_{(1)})$. Using the factorization theorem, we conclude that $(\Sigma x_i, x_{(1)})$ is jointly sufficient for θ. This result shows that even for a single parameter, we may need more than one statistic for sufficiency.

When using the factorization criterion, one has to be careful in cases where the range space depends on the parameter.

Using the factorization criterion, we can prove the following result, which says that if we have a unique maximum likelihood estimator, then that estimator will be a function of the sufficient statistic.

Theorem 5.3.5. *If* U *is a sufficient statistic for* θ, *the maximum likelihood estimator of* θ, *if unique, is a function of* U.

Proof. Because U is sufficient, by Theorem 5.4.1, the joint pdf can be factored as

$$f(x_1,\ldots,x_n;\theta) = g(u,\theta)h(x_1,\ldots,x_n).$$

This depends on θ only through the statistic U. To maximize L we need to maximize $g(U,\theta)$.

Many common distributions, such as Poisson, Normal, gamma, and Bernoulli, are members of the *exponential family* of probability distributions. The exponential family of distributions has density functions of the form

$$f(x;\theta) = \begin{cases} \exp[k(x)c(\theta) + S(x) + d(\theta)], \text{ if } x \in B \\ 0, \ x \notin B \end{cases}$$

where B does not depend on the parameter θ.

Example 5.3.12

Write the following in exponential form:

(a) $\dfrac{e^{-\lambda}\lambda^x}{x!}$

(b) $p^x(1-p)^{1-x}$

(c) $\dfrac{1}{\sqrt{2\pi}}e^{-(x-\mu)^2/2}$

Solution

(a) We have

$$\frac{e^{-\lambda}\lambda^x}{x!} = \exp[x \ln \lambda - \ln x! - \lambda].$$

Here, $k(x) = x$, $c(\lambda) = \ln \lambda$, $S(x) = -\ln(x!)$, and $d(\lambda) = -\lambda$.

(b) *Similarly,*

$$p^x(1-p)^{1-x} = \exp\left[x \ln\left(\frac{p}{1-p}\right) + \ln(1-p)\right], \; x = 0 \text{ or } 1.$$

(c) *This is the standard normal density.*

$$\frac{1}{\sqrt{2\pi}}e^{-(x-\mu)^2/2} = \exp\left[x\mu - \frac{x^2}{2} - \frac{\mu^2}{2} - \frac{1}{2}\ln(2\pi)\right], \; -\infty < x < \infty.$$

Note that in the previous example, for each of the cases, $\Sigma_{i=1}^{n} X_i$ is a sufficient statistic for the parameter. In the next result, we give a generalization of this fact.

Theorem 5.3.6. *Let $X_1,\ldots,X_n$ be a random sample from a population with pdf or pmf of the exponential form*

$$f(x;\theta) = \begin{cases} \exp[k(x)c(\theta) + S(x) + d(\theta)], \text{ if } x \in B \\ 0, \; x \notin B \end{cases}$$

where B *does not depend on the parameter θ. The statistic $\Sigma_{i=1}^{n} k(X_i)$ is sufficient for θ.*

Proof. The joint density

$$\begin{aligned} f(x_1,\ldots,x_n;\theta) &= \exp\left[c(\theta)\sum_{i=1}^{n} k(x_i) + \sum_{i=1}^{n} S(x_i) + nd(\theta)\right] \\ &= \left\{\exp\left[c(\theta)\sum_{i=1}^{n} k(x_i) + nd(\theta)\right]\right\}\left\{\exp\left[\sum_{i=1}^{n} S(x_i)\right]\right\}. \end{aligned}$$

Using the factorization theorem, the statistic $\Sigma_{i=1}^{n} k(X_i)$ is sufficient.

It does not follow that every function of a sufficient statistic is sufficient. However, any one-to-one function of a sufficient statistic is also sufficient. Every statistic need not be sufficient. When they do exist, sufficient estimators are very important, because if one can find a sufficient estimator, it is ordinarily possible to find an unbiased estimator based on the sufficient statistic. Actually, the following theorem shows that if one is searching for an unbiased estimator with minimal variance, it has to be restricted to functions of a sufficient statistic.

Rao–Blackwell Theorem

Theorem 5.3.7. *Let $X_1, \ldots, X_n$ be a random sample with joint pf or pdf $f(x_1, \ldots, x_n; \theta)$, and let $U = (U_1, \ldots, U_n)$ be jointly sufficient for $\theta = (\theta_1, \ldots, \theta_n)$. If T is any unbiased estimator of $k(\theta)$, and if $T^* = E(T \mid U)$, then:*

(a) *T^* is an unbiased estimator of $k(\theta)$*

(b) *T^* is a function of U, and does not depend on θ*

(c) *$Var(T^*) \le Var(T)$ for every θ, and $Var(T^*) < Var(T)$ for some θ unless $T^* = T$ with probability 1*

Proof.

(a) By the property of conditional expectation and by the fact that T is an unbiased estimator of $k(\theta)$,

$$E(T^*) = E(E(T|U)) = E(T) = k(\theta).$$

Hence T^* is an unbiased estimator of $k(\theta)$.

(b) Because U is sufficient for θ, the conditional distribution of any statistic (hence for T), given U, does not depend on θ. Thus, $T^* = E(T|U)$ is a function of U.

(c) From the property of conditional probability, we have the following:

$$\begin{aligned} Var(T) &= E(Var(T|U)) + Var(E(T|U)) \\ &= E(Var(T|U)) + Var(T^*). \end{aligned}$$

Because $Var(T|U) \geq 0$ for all u, it follows that $E(Var(T|U)) \geq 0$. Hence $Var(T^*) \leq Var(T)$. We note that $Var(T^*) = Var(T)$ if and only if $Var(T|U) = 0$ or T is a function of U, in which case $T^* = T$ (from the definition of $T^* = E(T|U) = T$).

In particular, if $k(\theta) = \theta$, and T is an unbiased estimator of θ, then $T^* = E(T|U)$ will typically give the MVUE of θ. If T is the sufficient statistic that best summarizes the data from a given distribution with parameter θ, and we can find some function g of T such that $E(g(T)) = \theta$, it follows from the Rao–Blackwell theorem that $g(T)$ is the UMVUE for θ.

EXERCISES 5.3

5.3.1 Let $X_1,\ldots, X_n$ be a random sample from a population with density

$$f(x) = \begin{cases} e^{-(x-\theta)}, & \text{for } x > \theta \\ 0, & \text{otherwise.} \end{cases}$$

(a) Show that $\overline{X}$ is a biased estimator of θ.

(b) Show that $\overline{X}$ is an unbiased estimator of $\mu = 1 + \theta$.

5.3.2 The mean and variance of a finite population $\{a_1,\ldots,a_N\}$ are defined by

$$\mu = \frac{1}{N}\sum_{i=1}^{N} a_i \text{ and } \sigma^2 = \frac{1}{N}\sum_{i=1}^{N}(a_i - \mu)^2.$$

For a finite population, show that the sample variance S^2 is a biased estimator of σ^2.

5.3.3 For an infinite population with finite variance σ^2, show that the sample standard deviation S is a biased estimator for σ. Find an unbiased estimator of σ. [We have seen that S^2 is an unbiased estimator of σ^2. From this exercise, we see that the function of an unbiased estimator need not be an unbiased estimator.]

5.3.4 Let $X_1,\ldots,X_n$ be a random sample from an infinite population with finite variance σ^2. Define

$$S'^2 = \frac{1}{n}\sum_{i=1}^{n}(X_i - \overline{X})^2.$$

Show that S'^2 is a biased estimator for σ^2 and that the bias of S'^2 is $-\frac{\sigma^2}{n}$. Thus S'^2 is negatively biased and so on average underestimates the variance. Note that S'^2 is the MLE of σ^2.

5.3.5 Let $X_1,\ldots,X_n$ be a random sample from a population with the mean μ. What condition must be imposed on the constants $c_1, c_2,\ldots,c_n$ so that

$$c_1X_1 + c_2X_2 + \cdots + c_nX_n$$

is an unbiased estimator of μ?

5.3.6 Let $X_1,\ldots,X_n$ be a random sample from a geometric distribution with parameter θ. Find an unbiased estimate of θ.

5.3.7 Let $X_1,\ldots,X_n$ be a random sample from $U\ (0,\ \theta)$ distribution. Let $Y_n = \max\{X_1,\ldots,X_n\}$. We know (from Example 5.3.4) that $\widehat{\theta}_1 = Y_n$ is a maximum likelihood estimator of θ.

(a) Show that $\widehat{\theta}_2 = 2\overline{X}$ is a method of moments estimator.

(b) Show that $\widehat{\theta}_1$ is a biased estimator, and $\widehat{\theta}_2$ is an unbiased estimator of θ.

(c) Show that $\widehat{\theta}_3 = \frac{n+1}{n}\widehat{\theta}_1$ is an unbiased estimator of θ.

5.3.8 Let $X_1,\ldots,X_n$ be a random sample from a population with mean μ and variance 1. Show that $\hat{\mu}^2 = \overline{X}^2$ is a biased estimator of μ^2, and compute the bias.

5.3.9 Let $X_1,\ldots,X_n$ be a random sample from an $N\ (\mu,\ \sigma^2)$ distribution. Show that the estimator $\hat{\mu} = \overline{X}$ is the MVUE for μ.

5.3.10 Let $X_1,\ldots,X_{n_1}$ be a random sample from an $N\ (\mu_1,\ \sigma^2)$ distribution, and let $Y_1,\ldots,\ Y_{n2}$ be a random sample from a $N\ (\mu_2,\sigma^2)$ distribution. Show that the pooled estimator

$$\hat{\sigma}^2 = \frac{(n_1 - 1)S_1^2 + (n_2 - 1)S_2^2}{n_1 + n_2 - 2}$$

is unbiased for σ^2, where S_1^2 and S_2^2 are the respective sample variances.

5.3.11 Let $X_1,\ldots,X_n$ be a random sample from an $N\ (\mu,\ \sigma^2)$ distribution. Show that the sample median, M, is an unbiased estimator of the population mean μ. Compare the variances of $\overline{X}$ and M. [Note: For the normal distribution, the mean, median, and mode all occur at the same location. Even though both $\overline{X}$ and M are unbiased, the reason we usually use the mean instead of the median as the estimator of μ is that $\overline{X}$ has a smaller variance than M.]

5.3.12 Let $X_1,\ldots,X_n$ be a random sample from a Poisson distribution with parameter λ. Show that the sample mean $\overline{X}$ is sufficient for λ.

5.3.13 Let $X_1,\ldots,X_n$ be a random sample from a population with density function

$$f_\sigma(x) = \frac{1}{2\sigma}\exp\left(-\frac{|x|}{\sigma}\right), \quad -\infty < X < \infty,\ \sigma > 0.$$

Find a sufficient statistic for the parameter σ.

5.3.14 Show that if $\widehat{\theta}$ is a sufficient statistic for the parameter θ and if the maximum likelihood estimator of θ is unique, then the maximum likelihood estimator is a function of this sufficient statistic $\widehat{\theta}$.

5.3.15 Let $X_1,\ldots,X_n$ be a random sample from an exponential population with parameter θ. Show that $\Sigma_{i=1}^n X_i$ is sufficient for θ. Also show that $\overline{X}$ is sufficient for θ.

5.3.16 The following is a random sample from an exponential distribution.

1.5	3.0	2.6	6.8	0.7	2.2	1.3	1.6	1.1	6.5
0.3	2.0	1.8	1.0	0.7	0.7	1.6	3.0	2.0	2.5
5.7	0.1	0.2	0.5	0.4					

(a) What is an unbiased estimate of the mean?

(b) Using part (a) and these data, find two sufficient statistics for the parameter θ.

5.3.17 Let $X_1,\ldots,X_n$ be a random sample from a one-parameter Weibull distribution with pdf

$$f(x) = \begin{cases} 2\alpha x e^{-\alpha x^2}, & x > 0 \\ 0, & \text{otherwise.} \end{cases}$$

(a) Find a sufficient statistic for α.
(b) Using part (a), find an UMVUE for α.

5.3.18 Let $X_1,\ldots,X_n$ be a random sample from a population with density function

$$f(x) = \begin{cases} \frac{1}{\theta}, & -\frac{\theta}{2} \le x \le \frac{\theta}{2}, \ \theta > 0 \\ 0, & \text{otherwise.} \end{cases}$$

Show that $\left(\min_{1\le i\le n} X_i, \max_{1\le i\le n} X_i\right)$ is sufficient for θ.

5.3.19 Let $X_1,\ldots,X_n$ be a random sample from a $G(1, \beta)$ distribution.
(a) Show that $U = \sum_{i=1}^{n} X_i$ is a sufficient statistic for β.
(b) The following is a random sample from a $G(1,\beta)$ distribution.

0.3	3.4	0.4	1.8	0.7	1.0	0.1	2.3	3.7	2.0
0.3	3.7	0.1	1.3	1.2	3.3	0.2	1.3	0.6	0.4

Find a sufficient statistic for β.

5.3.20 Show that X_1 is not sufficient for μ, if $X_1,\ldots,X_n$ is a sample from $N(\mu, 1)$.

5.3.21 Let $X_1,\ldots,X_n$ be a random sample from the truncated exponential distribution with pdf

$$f(x) = \begin{cases} e^{\theta - x}, & x > 0 \\ 0, & \text{otherwise.} \end{cases}$$

Show that $X_{(1)} = \min(X_i)$ is sufficient for θ.

5.3.22 Let $X_1,\ldots,X_n$ be a random sample from a distribution with pdf

$$f(x) = \begin{cases} \theta x^{\theta-1}, & 0 < x < 1, \ \theta > 0 \\ 0, & \text{otherwise.} \end{cases}$$

Show that $U = X_1,\ldots,X_n$ is a sufficient statistic for θ.

5.3.23 Let $X_1,\ldots,X_n$ be a random sample from a Rayleigh distribution with pdf

$$f(x) = \begin{cases} \frac{2x}{\alpha} e^{-x^2/\alpha}, & x > 0 \\ 0, & \text{otherwise.} \end{cases}$$

Show that $\sum_{i=1}^{n} X_i^2$ is sufficient for the parameter α.

5.4 A METHOD OF FINDING THE CONFIDENCE INTERVAL: PIVOTAL METHOD

In the previous sections, we studied methods for finding point estimators for the population parameters. In general, the estimates will differ from the true parameter values by varying amounts depending on the sample values obtained. In addition, the point estimates do not convey any measure of reliability.

Now, we discuss another type of estimation, called an *interval estimation*. Although point estimators are useful, interval estimators convey more information about the data that are used to obtain the point estimate. The purpose of using an interval estimator is to have some degree of confidence of securing the true parameter. For an interval estimator of a single parameter θ, we will use the random sample to find two quantities L and U such that $L < \theta < U$ with some probability. Because L and U depend on the sample values, they will be random. This interval (L, U) should have two properties: (1) $P(L < \theta < U)$ is high, that is, the true parameter θ is in (L, U) with high probability, and (2) the length of the interval (L, U) should be relatively narrow on the average.

In summary, interval estimation goes a step beyond point estimation by providing, in addition to the estimating interval (L, U), a measure of one's confidence in the accuracy of the estimate. Interval estimators are called *confidence intervals*, and the limits are called U and L, the *upper* and *lower confidence limits*, respectively. The associated levels of confidence are determined by specified probabilities. The width of the confidence interval reflects the amount of variability inherent in the point estimate. Thus, our objective is to find a narrow interval with high probability of enclosing the true parameter, θ. We will restrict our attention to single parameter estimation.

The probability that a confidence interval will contain the true parameter θ is called the *confidence coefficient*. The confidence coefficient gives the fraction of the time that the constructed interval will contain the true parameter, under repeated sampling.

Let L and U be the lower and upper confidence limits for a parameter θ based on a random sample $X_1,\ldots, X_n$. Both L and U are functions of the sample. We can write the interval estimate of θ as

$$P(L \leq \theta \leq U) = 1 - \alpha,$$

and we read it as we are $(1 - \alpha)$ 100% confident that the true parameter θ is, located in the interval (L, U). The number $1 - \alpha$ is the confidence coefficient, and the interval (L, U) is referred to as a $(1 - \alpha)$ 100% *confidence interval* ($(1 - \alpha)$ 100% ***CI***) for θ. Thus, if we want a 95% confidence interval for, say, population mean μ, then $\alpha = 0.05$. Note that for the discrete random variables, we may not be able to find a lower bound L and an upper bound U such that the probability, $P(L \leq \theta \leq U)$, is exactly $(1 - \alpha)$. In such a case, we can choose L and U such that $P(L \leq \theta \leq U) \geq 1 - \alpha$.

How do we find the confidence interval? For this, we use the error structure of the point estimator to obtain this interval. For instance, we know that the sample mean, $\overline{X}$, is a point estimate (MLE or unbiased estimator) of the population mean μ. In this case, we also know that the standard error of $\overline{X}$ is $\sigma/\sqrt{n}$. If the sample came from a normal population, then for a 95% confidence interval for the mean, multiply the standard error by 1.96, and then add and subtract this product from the sample mean. From this we can also observe that, if everything else remains the same, the size of the confidence interval reduces as the sample size increases.

Example 5.4.1

As part of a promotion, the management of a large health club wants to estimate average weight loss for its members within the first 3 months after joining the club. They took a random sample of 45 members of this health club and found that they lost an average of 13.8 pounds within the first 3 months of membership with a sample standard deviation of 4.2 pounds. Find a 95% confidence interval for the true mean. What if a random sample of 200 members of this health club also resulted in the same sample mean and sample standard deviation?

Solution

Here a point estimate of the true mean μ *is the sample mean* $\bar{x} = 13.8$ *pounds. Because* $n = 45$ *is large enough, we can use the Central Limit Theorem and use approximate normality for the distribution of* $\overline{X}$ *with mean* μ *and the approximate standard error* $(4.2/\sqrt{45}) = 0.626$. *Thus a 95% confidence interval is* $13.8 \pm (1.96)(0.626)$, *resulting in the interval* $(12.57, 15.03)$. *Thus, on average, with 95% confidence, one can expect the true mean to lie in this interval.*

For $n = 200$, *the standard error is* $(4.2/\sqrt{200}) \approx 0.297$. *Thus a 95% confidence interval is* $13.8 \pm (1.96)(0.297)$ *resulting in the interval* $(13.22, 14.38)$. *Thus the more sample values (i.e., the more information) we have, the tighter (smaller width) the interval.*

The previous example was built on our knowledge of the sampling distribution of the sample mean. What if the sampling distribution of the statistic we are interested in is not readily available? More generally, our success in building confidence intervals for an estimate of a parameter depends on identifying a quantity known as the pivot. We now describe this method.

The *pivotal method* is a general method of constructing a confidence interval using a pivotal quantity. This relies on our knowledge of sampling distributions. Here we have to find a pivotal quantity with the following two characteristics:

(i) It is a function of the random sample (a statistic or an estimator $\hat{\theta}$) and the unknown parameter θ, where θ is the only unknown quantity.
(ii) It has a probability distribution that does not depend on the parameter θ.

Suppose that $\hat{\theta} = \hat{\theta}(X)$ is a point estimate of θ, and let $p(\hat{\theta}, \theta)$ be the pivotal quantity. Now, for a given value of α, $(0 < \alpha < 1)$, and constants a and b, with $(a < b)$, let

$$P(a \le p(\hat{\theta}, \theta) \le b) = 1 - \alpha.$$

Hence, given $\hat{\theta}$, the inequality is solved for θ to obtain a region of θ values, usually an interval corresponding to the observed $\hat{\theta}$-value. This will be a desired confidence interval.

From (i) and (ii), it is important to note that the pivotal quantity depends on the parameter, but its distribution is independent of the parameter. Let $X_1, \ldots, X_n$ be a random sample, and let $\hat{\theta}$ be a reasonable point estimate of θ. For instance, $\hat{\theta}$ could be the maximum likelihood (or some other) estimator of θ. In general, finding a pivotal quantity may not be easy. However, if $\hat{\theta}$ is the sample mean $\overline{X}$ or sample variance S^2, we could find a pivotal quantity with known sampling distributions. Suppose $p(\hat{\theta}, \theta)$ is a pivotal quantity with known probability distribution that is independent of θ. (Usually, the probability distribution of the pivotal quantity will be standard normal, t, χ^2, or F-distribution.) The following are some of the standard pivotal quantities: If the sample $X_1, \ldots, X_n$ is from $N(\mu, \sigma^2)$, with μ unknown and σ known, let $\overline{X}$ be the sample mean. Then, the pivot is $(\overline{X} - \mu)/(\sigma/\sqrt{n})$, which has an $N(0, 1)$ distribution (see comments after Corollary 4.2.2).

With μ unknown and σ unknown, then the pivot is $(\overline{X} - \mu)/(S/\sqrt{n})$, which has a t-distribution with $(n - 1)$ degrees of freedom (see Theorem 4.2.9). If n is large, using CLT, the distribution of the pivot is approximately $N(0, 1)$.

If σ^2 is unknown, then the pivot is $(n - 1)S^2/\sigma^2$, which has a χ^2-distribution with $(n - 1)$ degrees of freedom (see Theorem 4.2.8).

The following examples illustrate the pivotal method.

Example 5.4.2

Suppose we have a random sample $X_1, \ldots, X_n$ from $N(\mu, 1)$. Construct a 95% confidence interval for μ.

Solution

Here the confidence coefficient is 0.95. We know that the maximum likelihood estimator of μ is $\overline{X}$, which has an $N(\mu, 1/n)$ distribution. Note that this distribution depends on the unknown value of μ and hence $\overline{X}$ cannot be a pivot. However, taking the z-transform of $\overline{X}$, we obtain the pivotal quantity as

$$Z = \frac{\overline{X} - \mu}{\sigma/\sqrt{n}} = \frac{\overline{X} - \mu}{1/\sqrt{n}},$$

which has an $N(0, 1)$ distribution that is a function of the sample measurements and does not depend on μ. Hence this Z can be taken as a pivot $p(\hat{\theta}, \theta)$. Now, to find a and b such that $P(a \le Z) = p(\hat{\theta}, \theta) \le b) = 0.95$. One such choice is to find the value of a such that $P(-a \le Z \le a) = 0.95$. From the normal table,

$$P(-z_{\alpha/2} \le Z \le z_{\alpha/2}) = 0.95,$$

where $z_{\alpha/2}$ represents the value of z with tail area $\alpha/2$. This implies $a = z_{\alpha/2} = 1.96$. Hence

$$P(-1.96 \le Z \le 1.96) = 0.95,$$

or, using the definition of Z and solving for μ, we obtain

$$P\left(\overline{X} - \frac{1.96}{\sqrt{n}} \le \mu \le \overline{X} + \frac{1.96}{\sqrt{n}}\right) = 0.95.$$

Hence, a 95% confidence interval for μ is $(\overline{X} - (1.96/\sqrt{n}), \overline{X} + (1.96/\sqrt{n}))$. Thus the lower confidence limit L is $\overline{X} - (1.96/\sqrt{n})$, and the upper confidence limit U is $\overline{X} + (1.96/\sqrt{n})$.

From the derivation of Example 5.4.1, it follows that

$$P\left(|\overline{X} - \mu| < z_{\alpha/2}\frac{\sigma}{\sqrt{n}}\right) = 1 - \alpha.$$

Thus for a normal population with known variance σ^2, if $\overline{X}$ is used as an estimator of the true mean μ, the probability that the error will be less than $z_{\alpha/2}\sigma/\sqrt{n}$ is $1 - \alpha$. It is important to note that there is some arbitrariness in choosing a confidence interval for a given problem. There may be several pivotals for $\hat{\theta}$ that could be used. Also, it is not necessary to allocate equal probability to the two tails of the distribution; however, doing so may result in the shortest length confidence interval for a given confidence coefficient.

When we make the statement of the form

$$P\left(\overline{X} - \frac{1.96}{\sqrt{n}} \le \mu \le \overline{X} + \frac{1.96}{\sqrt{n}}\right) = 0.95,$$

we mean that, in an infinite series of trials in which repeated samples of size n are drawn from the same population and 95% confidence intervals for μ are calculated by the same method for each of the samples, the proportion of intervals that actually include μ will be 0.95. Fig. 5.4 illustrates this idea, where the vertical line represents the position of true mean μ and each of the horizontal lines represents a 95% confidence interval of the sample, 20 samples of size n are taken.

A statement of the type $P(\bar{x} - (1.96/\sqrt{n}) \le \mu \le \bar{x} + (1.96/\sqrt{n})) = 0.95$, where $\bar{x}$ is the observed sample mean, is misleading. Once we calculate this interval using a particular sample, then either this interval contains the true mean μ or not, and hence the probability will be either 0 or 1. Thus the correct interpretation of confidence interval for the population mean is that if samples of the same size, n, are drawn repeatedly from a population, and a confidence interval is calculated from each sample, then 95% of these intervals should contain the population mean. This is often stated as "We are 95% confident that the true mean is in the interval $(\overline{X} - z_{\alpha/2}(\sigma/\sqrt{n}), \overline{X} + z_{\alpha/2}(\sigma/\sqrt{n}))$." Thus the correct interpretation requires the confidence limits to be variables. This concept of confidence interval is attributed to Neyman, where the vertical line representing mean μ only passes through about 95% of the horizontal lines representing the confidence intervals.

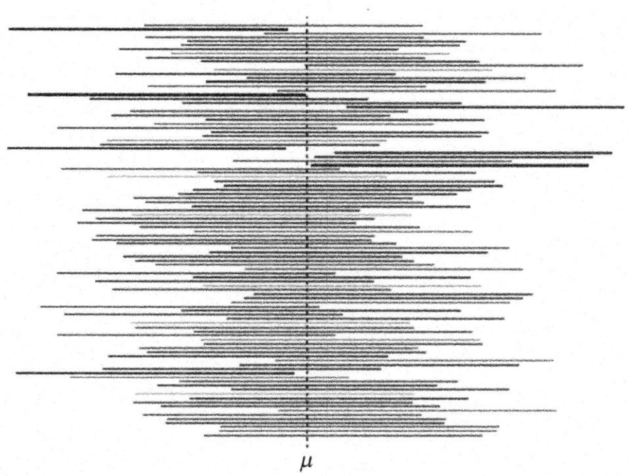

FIGURE 5.4 95% confidence intervals for μ. The plot consists of a central vertical line indicating the true mean mu. Surrounding this line are 20 horizontal lines, each representing a 95 percent confidence interval for a sample. These intervals vary in length and position, reflecting the variability in sample estimates of the mean. The intervals are distributed around the true mean, with some intervals overlapping the vertical line, indicating that the true mean falls within those intervals.

We can follow the accompanying procedure to find a confidence interval for the parameter θ.

Procedure to Find a Confidence Interval for θ Using the Pivot

1. Find an estimator $\hat{\theta}$ of θ: usually MLE of θ works.
2. Find a function of θ and $\hat{\theta}$, $p(\theta, \hat{\theta})$ (pivot), such that the probability distribution of $p(.,.)$ does not depend on θ.
3. Find a and b such that $P(a \le p(\theta, \hat{\theta}) \le b) = 1 - \alpha$.. Choose a and b such that $P(p(\theta, \hat{\theta}) \le a) = \alpha/2$ and $P(p(\theta, \hat{\theta}) \ge b) = \alpha/2$ (see Fig. 5.5 where the shaded area in each side is $\alpha/2$).
4. Now, transform the pivot confidence interval to a confidence interval for the parameter θ. That is, work with the inequality in step 3, and rewrite it as $P(L \le \theta \le U) = 1 - \alpha$, where L is the lower confidence limit and U is the upper confidence limit.

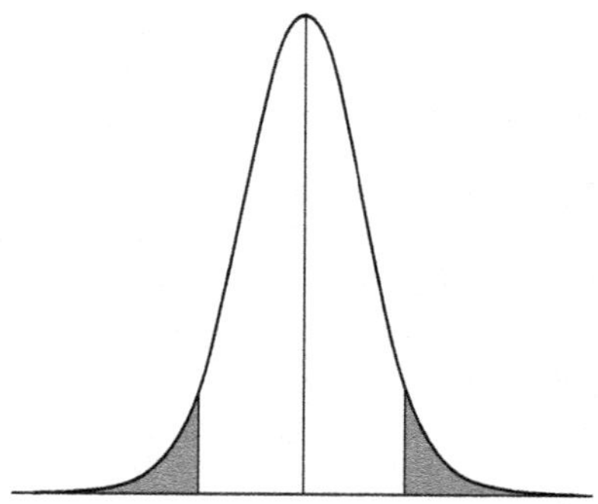

FIGURE 5.5 Probability density of the pivot.

The following example is given to show that the success of finding a pivotal quantity depends on our ability to find the right transformation of the statistic and its distribution so that the transformed variable is a pivot.

Example 5.4.3

Suppose the random sample $X_1,\ldots,X_n$ has $U(0,\theta)$ distribution. Construct a 90% confidence interval for θ and interpret. Identify the upper and lower confidence limits.

Solution

From Example 5.3.4, we know that

$$U = \max_{1 \le i \le n} X_i$$

is the MLE of θ. The random variable U has the pdf

$$f_U(u) = nu^{n-1}/\theta^n, \ 0 \le u \le \theta.$$

This is not independent of the parameter θ. Let $Y = U/\theta$; then (using the Jacobians described in Chapter 3) the pdf of Y is given by

$$f_Y(y) = ny^{n-1}, \ 0 \le y \le 1.$$

Hence Y satisfies the two characteristics of the pivotal quantity. Thus $Y = U/\theta$ is a pivot. Now, we have to find a and b such that

$$p\left(a \le \frac{U}{\theta} \le b\right) = 0.90.$$

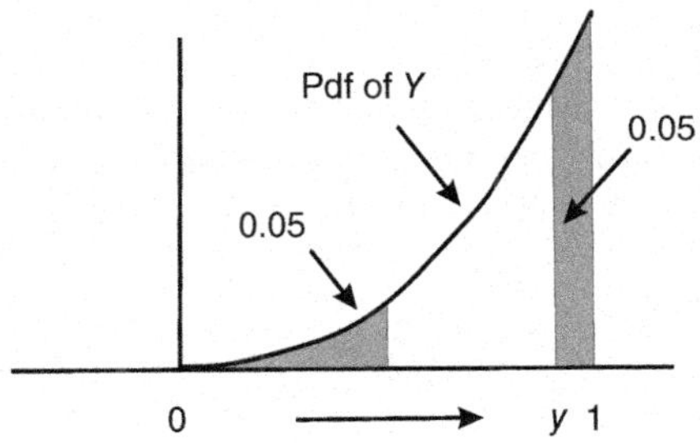

To find a and b, we use the cdf of Y, fy(y) = yn, 0 ≤ y ≤ 1, as follows:

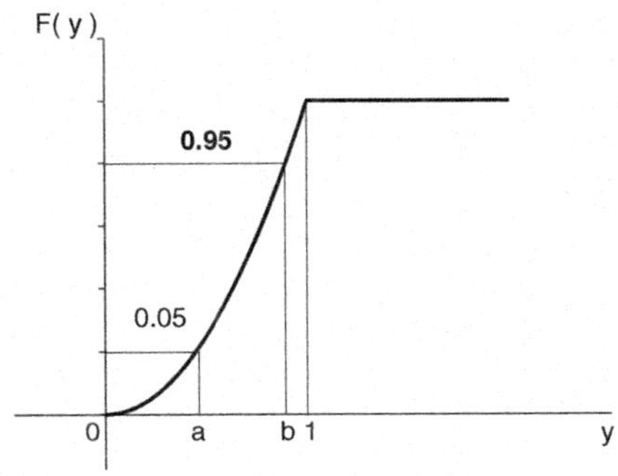

$$F_Y(a) = 0.05 \quad and \quad F_Y(b) = 0.95$$

which implies that

$$a^n = 0.05 \quad and \quad b^n = 0.95,$$

resulting in

$$a = \sqrt[n]{0.05} \quad and \quad b = \sqrt[n]{0.95}.$$

Write

$$P\left(\sqrt[n]{0.05} < \frac{U}{\theta} < \sqrt[n]{0.95}\right) = 0.90.$$

Solving, the 90% confidence interval for θ is

$$\left(\frac{U}{\sqrt[n]{0.95}}, \frac{U}{\sqrt[n]{0.05}}\right)$$

or

$$P\left(\frac{U}{\sqrt[n]{0.95}} \le \theta \le \frac{U}{\sqrt[n]{0.05}}\right) = 0.90.$$

Thus the lower confidence limit is $U/\sqrt[n]{0.95}$, *and the upper confidence limit is* $U/\sqrt[n]{0.05}$, *and the 90% confidence interval is* $(U/\sqrt[n]{0.95}, U/\sqrt[n]{0.05})$.

We can interpret this in the following manner. In a large number of trials in which repeated samples are taken from a population with uniform pdf with parameter θ, approximately 90% of the intervals will contain θ. For instance, if we observed $n = 20$ values from a uniform distribution with the maximum observed value being 15, then a 90% confidence interval for θ is (15.04, 17.42). Thus, we are 90% confident that these data came from a uniform distribution upper limit falling somewhere in this interval.

It is important to note that the pivotal method may not be applicable in all situations. For example, in the binomial case, to find a confidence interval for p, there is no quantity that satisfies the two conditions of a pivot. However, if sample size is large, then the z-score of sample proportion can be used as a pivot with approximate standard normal distribution. For the pivotal method to work, there is the practical necessity that the distribution of the pivotal quantity makes it easy to compute the probabilities. In cases where the pivotal method does not work, we may need to use other techniques such as the method based on sampling distributions (see Project 4A). A proper discussion of these methods is beyond the level of this book.

EXERCISES 5.4

5.4.1 **(a)** Suppose we construct a 99% confidence interval. What are we 99% confident about?

(b) Which of the confidence intervals is wider, 90% or 99%?

(c) In computing a confidence interval, when do you use the t-distribution, and when do you use z, with normal approximation?

(d) How does the sample size affect the width of a confidence interval?

5.4.2 Suppose X is a random sample of size $n = 1$ from a uniform distribution defined on the interval $(0, \theta)$. Construct a 98% confidence interval for θ, and interpret.

5.4.3 Consider the probability statement

$$P\left(-2.81 \le Z = \frac{\overline{X} - \mu}{\sigma/\sqrt{n}} \le 2.75\right) = \kappa,$$

where $\overline{X}$ is the mean of a random sample of size n from $N(\mu, \sigma^2)$ distribution with known σ^2.

(a) Find κ.

(b) Use this statement to find a confidence interval for μ.

(c) What is the confidence level of this confidence interval?

(d) Find a symmetric confidence interval for μ.

5.4.4 A random sample of size 50 from a particular brand of 16-ounce tea packets produced a mean weight of 15.65 ounces. Assume that the weights of these brands of tea packets are normally distributed with standard deviation of 0.59 ounce. Find a 95% confidence interval for the true mean μ.

5.4.5 Let $X_1,\ldots,X_n$ be a random sample from an $N(\mu, \sigma^2)$, where the value of σ^2 is unknown.

(a) Construct a $(1 - \alpha)$ 100% confidence interval for σ^2, choosing an appropriate pivot. Interpret its meaning.

(b) Suppose a random sample from a normal distribution gives the following summary statistics: $n = 21$, $\bar{x} = 44.3$, and $s = 3.96$. Using part (a), find a 90% confidence interval for σ^2. Interpret its meaning.

5.4.6 Let $X_1,\ldots, X_n$ be a random sample from a gamma distribution with $\alpha = 2$ and unknown β. Construct a 95% confidence interval for β.

5.4.7 Let $X_1,\ldots, X_n$ be a random sample from an exponential distribution with pdf $f(x) = (1/\theta)e^{-x/\theta}$, $\theta > 0$, $x > 0$. Construct a 95% confidence interval for θ and interpret. [Hint: Recall that $\sum_{i=1}^{n} X_i$ has a gamma distribution with $\alpha = n, \beta = \theta$.]

5.4.8 Let $X_1,\ldots, X_n$ be a random sample from a Poisson distribution with parameter λ.

(a) Construct a 90% confidence interval for λ.

(b) Suppose that the number of raisins in a bowl of a particular brand of cereal is observed to be 25. Assuming that the number of raisins in a bowl is Poisson distributed, estimate the expected number of raisins per bowl with a 90% confidence interval.

(c) How many bowls of cereal need to be sampled in order to estimate the expected number of raisins per bowl with a standard error of less than 0.2?

5.4.9 Let $X_1,\ldots, X_n$ be a random sample from an $N(\mu, \sigma^2)$.

(a) Construct a $(1 - \alpha)$ 100% confidence interval for μ when the value of σ^2 is known.

(b) Construct a $(1 - \alpha)$ 100% confidence interval for μ when the value of σ^2 is unknown.

5.4.10 Let $X_1,\ldots, X_n$ be a random sample from an $N(\mu_1, \sigma^2)$ population and $Y_1,\ldots, Y_n$ be an independent random sample from an $N(\mu_2, \sigma^2)$ distribution where σ^2 is assumed to be known. Construct a $(1 - \alpha)$ 100% interval for $(\mu_1 - \mu_2)$. Interpret its meaning.

5.4.11 Let $X_1,\ldots, X_n$ be a random sample from a uniform distribution on $[\theta, \theta + 1]$. Find a 99% confidence interval for θ, using an appropriate pivot.

5.5 ONE-SAMPLE CONFIDENCE INTERVALS

In this section, we will find confidence intervals for one-sample case for both large and small sample situations.

5.5.1 Large Sample Confidence Intervals

If the sample size is large, then by the Central Limit Theorem, certain sampling distributions can be assumed to be approximately normal. That is, if θ is an unknown parameter (such as μ, p, $(\mu_1 - \mu_2)$, $(p_1 - p_2)$), then for large samples, by the Central Limit Theorem, the z-transform

$$z = \frac{\hat{\theta} - \theta}{\sigma_{\hat{\theta}}}$$

possesses an approximately standard normal distribution, where $\hat{\theta}$ is the MLE of θ and $\sigma_{\hat{\theta}}$ is its standard deviation. Then, as in Example 5.4.1, the pivotal method can be used to obtain the confidence interval for the parameter θ. For $\theta = \mu$, $n \geq 30$ will be considered large; for the binomial parameter p, n is considered large if np and $n(1 - p)$ are both greater than 5. Note that these numbers are only a rule of thumb.

Procedure to Calculate Large Sample Confidence Interval for θ

1. Find an estimator (such as the MLE) of θ, say $\hat{\theta}$.
2. Obtain the standard error, $\sigma_{\hat{\theta}}$ of $\hat{\theta}$.
3. Find the z-transform $z = (\hat{\theta} - \theta)/\sigma_{\hat{\theta}}$. Then z has an approximately standard normal distribution.
4. Using the normal table, find two tail values $-z_{\alpha/2}$ and $z_{\alpha/2}$.
5. An approximate $(1 - \alpha)$ 100% confidence interval for θ is $(\hat{\theta} - z_{\alpha/2}\sigma_{\hat{\theta}}, \hat{\theta} + z_{\alpha/2}\sigma_{\hat{\theta}})$, that is,

$$P(\hat{\theta} - z_{\alpha/2}\sigma_{\hat{\theta}} \leq \theta \leq \hat{\theta} + z_{\alpha/2}\sigma_{\hat{\theta}}) = 1 - \alpha.$$

6. *Conclusion:* We are $(1 - \alpha)$ 100% confident that the true parameter θ lies in the interval $(\hat{\theta} - z_{\alpha/2}\sigma_{\hat{\theta}}, \hat{\theta} + z_{\alpha/2}\sigma_{\hat{\theta}})$.

Example 5.5.1

Let $\hat{\theta}$ be a statistic that is normally distributed with mean θ and standard deviation $\sigma_{\hat{\theta}}$, where σ is assumed to be known. Find a confidence interval for θ that possesses a confidence coefficient equal to $1 - \alpha$.

Solution

The z-transform of $\hat{\theta}$ *is*

$$Z = \frac{\hat{\theta} - \theta}{\sigma_{\hat{\theta}}}$$

and has a standard normal distribution. Select two tail values $-z_{\alpha/2}$ and $z_{\alpha/2}$ *such that*

$$P(-z_{\alpha/2} \leq Z \leq z_{\alpha/2}) = 1 - \alpha.$$

Because of symmetry, this is the shortest interval that contains the area $1 - \alpha$. Then,

$$P(\hat{\theta} - z_{\alpha/2}\sigma_{\hat{\theta}} \leq \theta \leq \hat{\theta} + z_{\alpha/2}\sigma_{\hat{\theta}}) = 1 - \alpha.$$

Therefore, the confidence limits of θ are $\hat{\theta} - z_{\alpha/2}\sigma_{\hat{\theta}}$ and $\hat{\theta} + z_{\alpha/2}\sigma_{\hat{\theta}}$. Hence, $(1 - \alpha)100\%$ confidence interval for θ is given by $\hat{\theta} \pm z_{\alpha/2}\sigma_{\hat{\theta}}$.

If in particular for a large sample of size n, let $\hat{\theta} = \overline{X}$ be the sample mean. Then the large sample $(1 - \alpha)$ 100% confidence interval for the population mean μ is

$$\overline{X} \pm z_{\alpha/2}\frac{\sigma}{\sqrt{n}} \simeq \overline{X} \pm z_{\alpha/2}\frac{S}{\sqrt{n}}$$

where S is a point estimate of σ. That is,

$$P\left(\overline{X} - z_{\alpha/2}\frac{S}{\sqrt{n}} \leq \mu \leq \overline{X} \pm z_{\alpha/2}\frac{S}{\sqrt{n}}\right) = 1 - \alpha.$$

As we have seen in Section 5.4, the correct interpretation of this confidence interval is that in a repeated sampling, approximately $(1 - \alpha)$ 100% of all intervals of the form $\overline{X} \pm z_{\alpha/2}(S/\sqrt{n})$ include μ, the true mean. Suppose $\overline{x}$ and s are the sample mean and the sample standard deviation, respectively, for a particular set of n observed sample values $x_1,\ldots,x_n$. Then we do not know whether the particular interval $(\overline{x} - z_{\alpha/2}(s/\sqrt{n}), \overline{x} - z_{\alpha/2}(s/\sqrt{n}))$ contains μ. However, the procedure that produced this interval does capture the true mean in approximately $(1 - \alpha)$ 100% of cases. This interpretation will be assumed hereafter, when we make a statement such as "We are 95% confident that the true mean will lie in the interval (74.1, 79.8)."

Example 5.5.2

Two statistics professors want to estimate average scores for an elementary statistics course that has two sections. Each professor teaches one section, and each section has a large number of students. A random sample of 50 scores from each section produced the following results:

(a) Section I: $\overline{x}_1 = 77.01, s_1 = 10.32$
(b) Section II: $\overline{x}_2 = 72.22, s_2 = 11.02$

Calculate 95% confidence intervals for each of these three samples.

Solution

Because $n = 50$ is large, we could use normal approximation. For $\alpha = 0.05$, from the normal table: $z_{\alpha/2} = z_{0.025} = 1.96$. The confidence intervals are:

(a) *We have*

$$\overline{x}_1 \pm z_{\alpha/2}\frac{s_1}{\sqrt{n}} = 77.01 \pm 1.96\left(\frac{10.32}{\sqrt{50}}\right),$$

which gives a 95% confidence interval (74.149, 79.871).

(b) *We can compute*

$$\bar{x}_2 \pm z_{\alpha/2}\frac{s_2}{\sqrt{n}} = 72.22 \pm 1.96\left(\frac{11.02}{\sqrt{50}}\right),$$

which gives the interval (69.165, 75.275).

It may be noted that if the population is normal with a known variance σ^2, we can use $\overline{X} \pm z_{\alpha/2}(\sigma/\sqrt{n})$ as the confidence interval for the population mean μ, irrespective of the sample size. However, if σ^2 is unknown, in order to use $\overline{X} \pm z_{\alpha/2}(s/\sqrt{n})$ as an approximate confidence interval for μ, the sample size has to be large for the Central Limit Theorem to hold. However, to use this approximate procedure, we do not need the condition that samples arise from a normal distribution. We will consider sample size to be large if $n \geq 30$ (applicable to estimators of the mean). If not, we shall use the small sample procedure discussed in the next section.

Example 5.5.3

Fifteen vehicles were observed at random for their speeds (in mph) on a highway with speed limit posted as 70 mph, and it was found that their average speed was 73.3 mph. Suppose that from past experience we can assume that vehicle speeds are normally distributed with $\sigma = 3.2$. Construct a 90% confidence interval for the true mean speed μ of the vehicles on this highway. Interpret the result.

Solution

Because the population is given to be normal with standard deviation $\sigma = 3.2$, *sample size need not be large given* $\bar{x} = 73.3$ *and* $\sigma = 3.2$. *Here,* $n = 15$, *and* $\alpha = 0.10$. *Thus,* $z_{\alpha/2} = z_{0.05} = 1.645$. *Hence a 90% confidence interval for* μ *is given by*

$$73.3 - 1.645\frac{3.2}{\sqrt{15}} < \mu < 73.3 + 1.645\frac{3.2}{\sqrt{15}}$$

or

$$71.681 < \mu < 74.919.$$

Interpretation: We are 90% confident that the true mean speed μ *of the vehicles on this highway is between 71.681 and 74.919.*

5.5.2 Confidence Interval for Proportion, p

Consider a binomial distribution with parameter p. Let X be the number of successes in n trials. Then, the maximum likelihood estimator $\hat{p}$ of p is $\hat{p} = X/n$. It can be shown, using the procedure outlined at the beginning of this section, that an approximate large sample $(1 - \alpha)$ 100% confidence interval for p is

$$\left(\hat{p} - z_{\alpha/2}\sqrt{\frac{\hat{p}(1-\hat{p})}{n}},\ \hat{p} + z_{\alpha/2}\sqrt{\frac{\hat{p}(1-\hat{p})}{n}}\right).$$

That is,

$$P\left(\hat{p} - z_{\alpha/2}\sqrt{\frac{\hat{p}(1-\hat{p})}{n}} < p < \hat{p} + z_{\alpha/2}\sqrt{\frac{\hat{p}(1-\hat{p})}{n}}\right) = 1 - \alpha.$$

A natural question is: "How do we determine the sample size that we have is sufficient for the normal approximation that is used in the foregoing formula?" There are various rules of thumb that are used to determine the adequacy of the sample size for normal approximation. Some of the popular rules are that np and $n(1 - p)$ should be greater than 10 or that $\hat{p} \pm 2\sqrt{\hat{p}(1-\hat{p})/n}$ should be contained in the interval (0,1), or $np(1 - p) \geq 10$, etc. All of these rules perform poorly when p is nearer to 0 or 1. Recently, there have been many works on coverage analysis for confidence intervals. We refer to a survey article by Lee et al. for more details on this topic. For simplicity of calculations, we will use the rule that np and $n(1 - p)$ are both greater than 5.

Example 5.5.4

An auto manufacturer gives a bumper-to-bumper warranty for 3 years or 36,000 miles for its new vehicles. In a random sample of 60 of its vehicles, 20 of them needed five or more major warranty repairs within the warranty period. Estimate the true proportion of vehicles from this manufacturer that need five or more major repairs during the warranty period, with confidence coefficient 0.95. Interpret.

Solution

Here we need to find a 95% confidence interval for the true proportion, p. Here, $\hat{p} = 20/60 = 1/3$. *For* $\alpha = 0.05$, $z_{\alpha/2} = z_{0.025} = 1.96$. *Hence a 95% confidence interval for p is*

$$\hat{p} \pm z_{\alpha/2}\sqrt{\frac{\hat{p}(1-\hat{p})}{n}} = \frac{1}{3} \pm 1.96\sqrt{\frac{\left(\frac{1}{3}\right)\left(\frac{2}{3}\right)}{60}},$$

which gives the confidence interval as (0.21405, 0.45262). *That is, we are 95% confident that the true proportion of vehicles from this manufacturer that need five or more major repairs during the warranty period will lie in the interval* (0.21405, 0.45262).

5.5.3 Margin of Error and Sample Size

In real-world problems, the estimates of the proportion p are usually accompanied by a margin of error, rather than a confidence interval. For example, in the news media, especially leading up to election time, we hear statements such as "The CNN/USA Today/Gallup poll of 818 registered voters taken on June 27–30 showed that if the election were held now, the president would beat his challenger 52% to 40%, with 8% undecided. The poll had a margin of error of plus or minus four percentage points." What is this "margin of error"? According to the American Statistical Association, the margin of error is a common summary of sampling error that quantifies uncertainty about a survey result. Thus, the margin of error is nothing but a confidence interval. The number quoted in the foregoing statement is half the maximum width of a 95% confidence interval, expressed as a percentage.

Let b be the width of a 95% confidence interval for the true proportion, p. Let $\hat{p} = x/n$ be an estimate for p where x is the number of successes in n trials. Then,

$$b = \frac{x}{n} + 1.96\sqrt{\frac{(x/n)(1-(x/n))}{n}} - \left(\frac{x}{n} - 1.96\sqrt{\frac{(x/n)(1-(x/n))}{n}}\right)$$
$$= 3.92\sqrt{\frac{(x/n)(1-(x/n))}{n}} \le 3.92\sqrt{\frac{1}{4n}},$$

because $(x/n)(1-(x/n)) = \hat{p}(1-\hat{p}) \le \frac{1}{4}$.

Thus the margin of error associated with $\hat{p} = (x/n)$ is $100d\%$, where

$$d = \frac{\max b}{2} = \frac{3.92\sqrt{\frac{1}{4n}}}{2} = \frac{1.96}{2\sqrt{n}}.$$

From the foregoing derivation, it is clear that we can compute the margin of error for other values of α by replacing 1.96 by the corresponding value of $z_{\alpha/2}$.

A quick look at the formula for the confidence interval for proportions reveals that a larger sample would yield a shorter interval (assuming other things being equal) and hence a more precise estimate of p. The larger sample is costlier in terms of time, resources, and money, whereas samples that are too small may result in inaccurate inferences. Then, it becomes beneficial for finding out the minimum sample size required (thus less costly) to achieve a prescribed degree of precision (usually, the minimum degree of precision acceptable). We have seen that the large sample $(1-\alpha)$ 100% confidence interval for p is

$$\hat{p} - z_{\alpha/2}\sqrt{\frac{\hat{p}(1-\hat{p})}{n}} < p < \hat{p} + z_{\alpha/2}\sqrt{\frac{\hat{p}(1-\hat{p})}{n}}.$$

Rewriting it, we have

$$|\hat{p} - p| \le z_{\alpha/2}\sqrt{\frac{\hat{p}(1-\hat{p})}{n}} = \frac{z_{\alpha/2}}{\sqrt{n}}\sqrt{\hat{p}(1-\hat{p})},$$

which shows that, with probability $(1-\alpha)$, the estimate $\hat{p}$ is within $z_{\alpha/2}\sqrt{\hat{p}(1-\hat{p})/n}$ units of p. Because $\hat{p}(1-\hat{p}) \le 1/4$, for all values of $\hat{p}$, we can write the foregoing inequality as

$$|\hat{p} - p| \le \frac{z_{\alpha/2}}{\sqrt{n}}\sqrt{\frac{1}{4}} = \frac{z_{\alpha/2}}{2\sqrt{n}}.$$

If we wish to estimate p at level $(1-\alpha)$ to within d units of its true value, that is, $|\hat{p} - p| \le d$, the sample size must satisfy the condition $(z_{\alpha/2}/(2\sqrt{n})) \le d$, or

$$n \ge \frac{z^2_{\alpha/2}}{4d^2}.$$

Thus to estimate p at level $(1-\alpha)$ to within d units of its true value, take the minimal sample size as $n = z^2_{\alpha/2}/(4d^2)$, and if this is not an integer, round up to the next integer.

Sometimes, we may have an initial estimate $\tilde{p}$ of the parameter p from a similar process or from a pilot study or simulation. In this case, we can use the following formula to compute the minimum required size of the sample to estimate p, at level $(1-\alpha)$, to within d units by using the formula

$$n = \frac{z^2_{\alpha/2}\tilde{p}(1-\tilde{p})}{d^2}$$

and, if this is not an integer, round up to the next integer.

A similar derivation for the calculation of sample size for estimation of the population mean μ at level $(1-\alpha)$ with margin of error E is given by

$$n = \frac{z^2_{\alpha/2}\sigma^2}{E^2}$$

and, if this is not an integer, rounding up to the next integer. This formula can be used only if we know the population standard deviation, σ. Although it is unlikely to know σ when the population mean itself is not known, we may be able to determine σ from an earlier similar study or from a pilot study/simulation.

Example 5.5.5

A dendritic tree is a branched formation that originates from a nerve cell. In order to study brain development, researchers want to examine the brain tissues from adult guinea pigs. How many cells must the researchers select (randomly) so as to be 95% sure that the sample mean is within 3.4 cells of the population mean? Assume that a previous study has shown $\sigma = 10$ cells.

Solution

A 95% confidence corresponds to $\alpha = 0.05$. Thus, from the normal table, $z_{\alpha/2} = z_{0.025} = 1.96$. Given that $E = 3.4$ and $\sigma = 10$, and using the sample size formula, the required sample size n is

$$n = \frac{z_{\alpha/2}^2 \sigma^2}{E^2} = \frac{(1.96)^2(10)^2}{(3.4)^2} = 33.232.$$

Thus, take n = 34.

Example 5.5.6

Suppose that a local TV station in a city wants to conduct a survey to estimate support for the president's policies on the economy within 3% error with 95% confidence.

(a) How many people should the station survey if they have no information on the support level?
(b) Suppose they have an initial estimate that 70% of the people in the city support the economic policies of the president. How many people should the station survey?

Solution
Here, $\alpha = 0.05$, *and thus* $z_{\alpha/2} = 1.96$. *Also,* $d = 0.03$.

(a) *With no information on p, we use the sample size formula:*

$$n = \frac{z_{\alpha/2}^2}{4d^2} = \frac{(1.96)^2}{4(0.03)^2} = 1067.1.$$

Hence the TV station must survey 1068 people.
(b) *Because* $\tilde{p} = 0.7$, *the required sample size is calculated from*

$$n = \frac{z_{\alpha/2}^2 \tilde{p}(1-\tilde{p})}{d^2} = \frac{(1.96)^2(0.70)(0.30)}{(0.03)^2} = 896.37.$$

Thus the TV station must survey at least 897 people.

In practice, we should realize that one of the key factors of a good design is not sample size by itself; it is getting representative samples. Even if we have a very large sample size, if the sample is not representative of our target population, then sample size means nothing. Therefore, whenever possible, we should use random sampling procedures (or other appropriate sampling procedures) to ensure that our target population is properly represented.

5.5.4 Small Sample Confidence Intervals For μ

Now we will consider the problem of finding a confidence interval for the true mean μ of a normal population when the variance σ^2 is unknown and obtaining a large sample is either impossible or impractical. Let $X_1,\ldots,X_n$ be a random sample from a normal population. We know that

$$T = \frac{\sqrt{n}\frac{\overline{X}-\mu}{\sigma}}{\sqrt{(n-1)S^2/[\sigma^2(n-1)]}} = \frac{\overline{X}-\mu}{S/\sqrt{n}}$$

has a t-distribution with $(n - 1)$ degrees of freedom, irrespective of the value of σ^2. Thus, $(\overline{X} - \mu)/(S/\sqrt{n})$ can be used as a pivot. Hence for n small ($n < 30$) and σ^2 unknown, we have the following result.

Theorem 5.5.1. *If* $\overline{X}$ *and S are the sample mean and the sample standard deviation of a random sample of size n from a normal population, then*

$$\bar{X} - t_{\alpha/2,n-1}\frac{S}{\sqrt{n}} < \mu < \bar{X} + t_{\alpha/2,n-1}\frac{S}{\sqrt{n}}$$

is a $(1 - \alpha)$ 100% *confidence interval for the population mean* μ.

Note that if the confidence coefficient, $1 - \alpha$, $\bar{X}$, and S remain the same, the confidence range $CR = \hat{\theta}_U - \hat{\theta}_L$ decreases as the sample size n increases, which means that we are closing in on the true parameter value of θ.

One can use the following procedure to find the confidence interval for the mean when a small sample is from an approximately normal distribution.

Procedure to Find Small Sample Confidence Interval For μ

1. Calculate the values of $\bar{X}$ and S, from the sample $X_1, \ldots, X_n$.
2. Using the t-table, select two tail values $-t_{\alpha/2}$ and $t_{\alpha/2}$.
3. The $(1 - \alpha)$ 100% confidence interval for μ is

$$\left(\bar{X} - t_{\alpha/2,n-1}\frac{S}{\sqrt{n}}, \bar{X} + t_{\alpha/2,n-1}\frac{S}{\sqrt{n}}\right),$$

that is, $P\left(\bar{X} - t_{\alpha/2,n-1}\frac{S}{\sqrt{n}} \le \mu \le \bar{X} + t_{\alpha/2,n-1}\frac{S}{\sqrt{n}}\right) = 1 - \alpha$.

4. *Conclusion*: We are $(1 - \alpha)$ 100% confident that the true parameter μ lies in the interval $\left(\bar{X} - t_{\alpha/2,n-1}(S/\sqrt{n}), \bar{X} + t_{\alpha/2,n-1}(S/\sqrt{n})\right)$.
5. **Assumption**: The population is normal.

In practice, the first step in the previous procedure should include a test of normality (see Project 4C). A built-in test of normality is available in most of the statistical software packages. In Example 5.5.9, we show how this test is utilized. Even when the data fail the normality test, most statistical software will produce a confidence interval based on normality or give an error report. We should understand that generally such answers are meaningless. In those cases, nonparametric methods (Chapter 12) such as the Wilcoxon rank sum method or bootstrap methods (Chapter 13) will be more appropriate. For more discussion, refer to Section 14.4.1.

Example 5.5.7

The following is a random data from a normal population.

7.2 5.7 4.9 6.2 8.5 2.8

Construct a 95% confidence interval for the population mean μ. Interpret.

Solution

The first step is to calculate the mean and standard deviation of the sample. We compute as the mean $\bar{x} = 5.883$ *and standard deviation,* $s = 1.959$. *For 5 degrees of freedom, and for* $\alpha = 0.05$, *from the t-table,* $t_{0.025} = 2.571$. *Hence a* 95% *confidence interval for* μ *is*

$$\left(\bar{x} - t_{\alpha/2,n-1}\frac{2}{\sqrt{n}}, \bar{x} + t_{\alpha/2,n-1}\frac{2}{\sqrt{n}}\right) = \left(5.883 - 2.571\left(\frac{1.959}{\sqrt{6}}\right), 5.5883 + 2.571\left(\frac{1.959}{\sqrt{6}}\right)\right) = (3.827, 7.939).$$

This can be interpreted as that we are 95% *confident that the true mean* μ *will be between* 3.827 *and* 7.939.

Example 5.5.8

The scores of a random sample of 16 people who took the TOEFL (Test of English as a Foreign Language) had a mean of 540 and a standard deviation of 50. Construct a 95% confidence interval for the population mean μ of the TOEFL score, assuming that the scores are normally distributed.

Solution

Because n=16 is small, using Theorem 6.3.1 with degrees of freedom 15, a 95% confidence interval for μ is

$$\bar{x} \pm t_{\alpha/2,n-1}\frac{s}{\sqrt{n}} = 540 \pm 2.131\left(\frac{50}{\sqrt{16}}\right).$$

So, the 95% confidence interval for the population mean μ of the TOEFL scores is (513.36, 566.64).

A Dobson unit is the most basic measure used in ozone research. The unit is named after G.M.B. Dobson, one of the first scientists to investigate atmospheric ozone (between 1920 and 1960). He designed the Dobson spectrometer–the standard instrument used to measure ozone from the ground. The data in Example 5.5.9 represent the total ozone levels at randomly selected points on the earth (represented by the pair [latitude, longitude]) on a particular day.

Example 5.5.9

The following data represent the total ozone levels measured in Dobson units at randomly selected locations of earth on a particular day.

269	246	388	354	266	303
295	259	274	249	271	254

Can we say that the data are approximately normally distributed? Construct a 95% confidence interval for the population mean μ of ozone levels on this day.

Solution

The following is the probability plot of these data created using software.

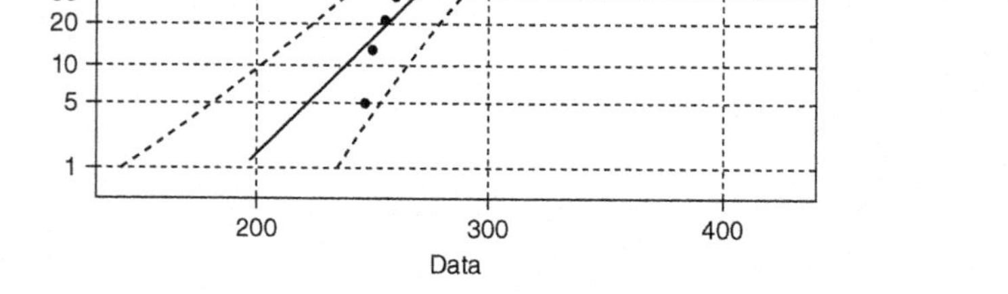

Because all the data values lie within the bounds on the normal probability plot (see the discussion in Section 3.2.4), we can assume that the data have approximate normality. We have $\bar{x}$ = 285.7 and s = 43.9. Also, n = 12. For α = 0.05, $t_{0.025,\,11}$ = 2.201. A *95% confidence* interval for μ is

$$\bar{X} \pm t_{\alpha/2,(n-1)}\frac{s}{\sqrt{n}} = 285.7 \pm 2.201\left(\frac{43.9}{\sqrt{12}}\right).$$

Hence a 95% confidence interval for μ, the average ozone level over the earth, lies in (257.81, 313.59).

EXERCISES 5.5

5.5.1 A survey indicates that it is important to pay attention to truth in political advertising. Based on a survey of 1200 people, 35% indicated that they found political advertisements to be untrue; 60% say that they will not vote for candidates whose advertisements are judged to be untrue; and of this latter group, only 15% ever complained to the media or to the candidate about their dissatisfaction.

(a) Find a 95% confidence interval for the percentage of people who find political advertising to be untrue.

(b) Find a 95% confidence interval for the percentage of voters who will not vote for candidates whose advertisements are considered to be untrue.

(c) Find a 95% confidence interval for the percentage of those who avoid voting for candidates whose advertisements are considered untrue and who have complained to the media or to the candidate about the falsehood in commercials.

(d) For each case above, interpret the results, and state any assumptions you have made.

5.5.2 Many mutual funds use an investment approach involving owning stocks whose price/earnings multiples (P/Es) are less than the P/E of the S&P 500. The following data give P/Es of 49 companies a randomly selected mutual fund owns in a particular year.

6.8	5.6	8.5	8.5	8.4	7.5	9.3	9.4	7.8	7.1
9.9	9.6	9.0	9.4	13.7	16.6	9.1	10.1	10.6	11.1
8.9	11.7	12.8	11.5	12.0	10.6	11.1	6.4	12.3	12.3
11.4	9.9	14.3	11.5	11.8	13.3	12.8	13.7	13.9	12.9
14.2	14.0	15.5	16.9	18.0	17.9	21.8	18.4	34.3	

Find a 98% confidence interval for the mean P/E multiples. Interpret the result, and state any assumptions you have made.

5.5.3 Let $X_1,\ldots, X_n$ be a random sample from $N(\mu, \sigma^2)$ distribution, σ^2 known.

(a) Show that $\hat{\mu} = \bar{X}$ is a maximum likelihood estimator of the population mean μ.

(b) Show that

$$P\left(\bar{X} - \frac{2\sigma}{\sqrt{n}} < \mu < \bar{X} + \frac{2\sigma}{\sqrt{n}}\right) = 0.954.$$

(c) Let

$$P\left(\bar{X} - \frac{k\sigma}{\sqrt{n}} < \mu < \bar{X} + \frac{k\sigma}{\sqrt{n}}\right) = 0.90.$$

Find k.

5.5.4 Let the observed mean of a sample of size 45 be $\bar{x} = 68.51$ from a distribution having variance 110. Find a 95% confidence interval for the true mean μ, and interpret the result, and state any assumptions you have made.

5.5.5 In a random sample of 50 college seniors, 18 indicated that they were planning to pursue a graduate degree. Find a 98% confidence interval for the true proportion of all college seniors planning to pursue a graduate degree, interpret the result, and state any assumptions you have made.

5.5.6 DVD players coming off an assembly line are automatically checked to make sure they are not defective. The manufacturer wants an interval estimate of the percentage of DVD players that fail the testing procedure. Compute a 90% confidence interval, based on a random sample of size 105 in which 17 DVD players failed the testing procedure. Also, interpret the result, and state any assumptions you have made.

5.5.7 Studies have shown that the risk of developing coronary disease increases with the level of obesity, or accumulation of body fat. A study was conducted on the effect of exercise on losing weight. Fifty men who exercised lost an average of 11.4 lb, with a standard deviation of 4.5 lb. Construct a 95% confidence interval for the mean weight loss through exercise. Interpret the result, and state any assumptions you have made.

5.5.8 Basing findings on 60 successful pregnancies involving natural birth, an experimenter found that the mean pregnancy term was 274 days, with a standard deviation of 14 days. Construct a 99% confidence interval for the true mean pregnancy term μ.

5.5.9 Let Y be the binomial random variable with parameter p and $n = 400$. If the observed value of Y is $y = 120$, find a 95% confidence interval for p.

5.5.10 For a health screening in a large company, the diastolic and systolic blood pressures of all the employees were recorded. In a random sample of 150 employees, 12 were found to suffer from hypertension. Find 95% and 98% confidence intervals for the proportion of the employees of this company with hypertension.

5.5.11 In a random sample of 500 items from a large lot of manufactured items, there were 40 defectives.

(a) Find a 90% confidence interval for the true proportion of defectives in the lot.

(b) Is the assumption of normal approximation valid?

(c) Suppose we suspect that another lot has the same proportion of defectives as in the first lot. What should be the sample size if we want to estimate the true proportion within 0.01 with 90% confidence?

5.5.12 Pesticide concentrations in sediment from irrigation areas can provide information required to assess exposure and fate of these chemicals in freshwater ecosystems and their likely impacts on the marine environment. In a study (Jochen F. Muller et al., "Pesticides in sediments from Queensland Irrigation channels and drains," *Marine Pollution Bulletin 41* (7–12), 294–301, 2000), 103 sediment samples were collected from irrigation channels and drains in 11 agricultural areas of Queensland. In 74 of these samples, they detected DDTs with concentration levels up to 840 ngg^{-1} dw. Obtain a 95% confidence interval for the proportion of the total number of sediments with detectable DDTs.

5.5.13 Let $\overline{X}$ be the mean of a random sample of size n from an $N(\mu,16)$ distribution. Find n such that $p(\overline{X} - 2 < \mu < \overline{X} + 2) = 0.95$.

5.5.14 Let X be a Poisson random variable with parameter λ. A sample of 150 observations from this population has a mean equal to 2.5. Construct a 98% confidence interval for λ.

5.5.15 An opinion poll conducted in March of 1996 by a newspaper *(Tampa Tribune)* among eligible voters with a sample size 425 showed that the president, who was seeking reelection, had 45% support. Give a 95% and a 98% confidence interval for the proportion of support for the president.

5.5.16 A random sample of 100 households located in a large city recorded the number of people living in the household, Y, and the monthly expenditure for food, X. The following summary statistics are given:

$$\sum_{i=1}^{100} Y_i = 340$$
$$\sum_{i=1}^{100} Y_i^2 = 1650$$
$$\sum_{i=1}^{100} X_i = 40{,}000$$
$$\sum_{i=1}^{100} X_i^2 = 44{,}000{,}000$$

(a) Form a 95% confidence interval for the mean number of people living in a household in this city.

(b) Form a 95% confidence interval for the mean monthly food expenses.

(c) For each case just given, interpret the results, and state any assumptions you have made.

5.5.17 Let $X_1, \ldots, X_n$ be a random sample from an exponential distribution with parameter θ. A sample of 350 observations from this population has a mean equal to 3.75. Construct a 90% confidence interval for θ.

5.5.18 Suppose a coin is tossed 100 times in order to estimate $p = p$ *(Head)*. It is observed that heads appeared 60 times. Find a 95% confidence interval for p.

5.5.19 Suppose the population is women at least 35 years of age who are pregnant with a fetus affected by Down syndrome. We are interested in testing positive on a noninvasive screening test for fetuses affected by Down syndrome in women at least 35 years of age. In an experiment, suppose 52 of 60 women tested positive. Obtain a

95% confidence interval for the true proportion of women at least 35 years of age who are pregnant with a fetus affected by Down syndrome who will receive positive test results from this procedure.

5.5.20

(a) Let $X_1,\ldots, X_n$ be a random sample from a Poisson distribution with parameter λ. Derive a $(1-\alpha)$ 100% large sample confidence interval for λ.

(b) To date nodes in a phylogenetic tree, the mean path length (MPL) is used in estimating the relative age of a node. The following data represent the MPL for 39 nodes (source: Tom Britton, Bengt Oxelman, Annika Vinnersten, and Kåre Bremer, "Phylogenetic dating with confidence intervals using mean path-lengths"). Assume that the data (given in centimeters) follows a Poisson distribution with parameter λ.

65.2	47.0	38.2	13.5	18.0	25.6	16.3	14.0	23.2	18.8
7.5	13.3	11.0	54.9	22.0	50.1	32.6	26.0	13.0	9.0
7.2	4.7	4.5	41.1	45.8	37.0	8.5	30.5	29.3	13.8
7.7	5.5	24.1	12.5	22.3	19.0	9.5	4.7	3.0	

Obtain a 95% confidence interval for λ, and interpret.

5.5.21 A person plans to start an Internet service provider in a large city. The plan requires an estimate of the average number of minutes of Internet use of a household in a week. How many households must be (randomly) sampled to be 95% sure that the sample mean is within 15 minutes of the population mean? Assume that a pilot study estimated the value of $\sigma = 35$ minutes.

5.5.22 The fruit fly *Drosophila melanogaster* normally has a gray color. However, because of mutation, a good portion of them are black. A biologist eager to learn about the effect of mutation wants to collect a random sample to estimate the proportion of black fruit flies of this type within 1% error with 95% confidence.

(a) How many individual flies should the researcher capture if there is no information on the population proportion of black flies?

(b) Suppose the researcher has the initial estimate that 25% of the fruit fly *Drosophila melanogaster* have been affected by this mutation. What is the sample size?

5.5.23 In a pharmacological experiment, 35 lab rats were not given water for 11 hours and were then permitted access to water for 1 hour. The amounts of water consumed (mL/hour) are given in the following table:

10.6	13.3	15.5	10.7	9.6	12.1	11.8	10.9	9.9	13.2
9.3	11.7	9.9	13.0	12.3	11.0	13.1	11.0	12.5	13.9
14.1	14.8	15.1	12.8	14.0	7.1	14.1	12.7	9.6	12.5
9.0	12.7	13.6	12.5	12.6					

Obtain a 98% confidence interval for the mean amount of water consumed.

5.5.24 In sociology, a social network is defined as the people you make frequent contact with, say through Facebook. The personal network size for each adult in a random sample of 3,000 adults was calculated. The sample had a mean personal network size of 190 with a known population standard deviation of 25. Find a 95% confidence interval for the mean personal network size of all adults in order to see if we have a normal number of friends in our network.

5.5.25 **(a)** How does the t-distribution compare with the normal distribution?

(b) How does the difference affect the size of confidence intervals constructed using z (normal approximation) relative to those constructed using the t-distribution?

(c) Does sample size make a difference?

(d) What assumptions do we need to make in using the t-distribution for the construction of a confidence interval?

5.5.26 Use the t-table to determine the values of $t_{\alpha/2}$ that would be used in the construction of a confidence interval for a population mean in each of the following cases:

(a) $\alpha = 0.99$, $n = 20$

(b) $\alpha = 0.95$, $n = 18$

(c) $\alpha = 0.90$, $n = 25$

5.5.27 Let $X_1, \ldots, X_n$ be a random sample from a normal population. A particular realization resulted in a sample mean of 20 with the sample standard deviation 4. Construct a 95% confidence interval for μ when: **(a)** $n = 5$, **(b)** $n = 10$, and **(c)** $n = 25$. What happens to the length of the confidence interval as n changes?

5.5.28 In a large university, the following are the ages of 20 randomly chosen employees:

24	31	28	43	28	56	48	39	52	32
38	49	51	49	62	33	41	58	63	56

Assuming that the data came from a normal population, construct a 95% confidence interval for the population mean μ of the ages of the employees of this university. Interpret your answer.

5.5.29 A random sample of size 26 is drawn from a population having a normal distribution. The sample mean and the sample standard deviation from the data are given, respectively, as $\bar{x} = -2.22$ and $s = 1.67$. Construct a 98% confidence interval for the population mean μ and interpret.

5.5.30 A drug is suspected of causing an elevated heart rate in a certain group of high-risk patients. Twenty patients from the group were given the drug. The changes in heart rates were found to be as follows:

−1	8	5	10	2	12	7	9	1	3
4	6	4	12	11	2	−1	10	2	8

Construct a 98% confidence interval for the mean change in heart rate. Assume that the population has a normal distribution. Interpret your answer.

5.5.31 Ten bearings made by a certain process have a mean diameter of 0.905 cm with a standard deviation of 0.0050 cm. Assuming that the data may be viewed as a random sample from a normal population, construct a 95% confidence interval for the actual average diameter of bearings made by this process, and interpret.

5.5.32 Air pollution in large US cities is monitored to see whether it conforms to requirements set by the Environmental Protection Agency. The following data, expressed as an air pollution index, give the air quality of a city for 10 randomly selected days.

57.3	58.1	58.7	66.7	58.6	61.9	59.0	64.4	62.6	64.9

Assuming that the data may be looked upon as a random sample from a normal population, construct a 95% confidence interval for the actual average air pollution index for this city, and interpret.

5.5.33 In order to find out the average hemoglobin (Hb) level in children with chronic diarrhea, a random sample of 10 children with chronic diarrhea is selected from a city, and their Hb levels (g/dL) are obtained as follows:

12.3	11.4	14.2	15.3	14.8	13.8	11.1	15.1	15.8	13.2

Assuming that the data may be looked upon as a random sample from a normal population, construct a 99% confidence interval for the actual average Hb level in children with chronic diarrhea for this city, and interpret. Draw a box plot and normal plot for this data, and comment.

5.5.34 Suppose that you need to estimate the mean number of typographical errors per page in the rough draft of a 400-page book. A careful examination of 10 pages gives an average of 6 errors per page with a standard deviation of 2 errors. Assuming that the data may be looked upon as a random sample from a normal population, construct a 99% confidence interval for the actual average number of errors per page in this book, and interpret. In this problem, is the normal model appropriate?

5.5.35 Creatine kinase (CK) is found predominantly in muscle and is released into circulation during muscular lesions. Therefore, serum CK activity has been theoretically expected to be useful as a marker in exercise physiology and sports medicine for the detection of muscle injury and overwork. The following data represent the peak CK activity (measured in IU/L) after 90 minutes of exercise in 15 healthy young men. (Source: Manabu Totsuka, Shigeyuki Nakaji, Katsuhiko Suzuki, Kazuo Sugawara, and Koki Sato, Break point of serum creatine kinase release after endurance exercise, http://jap.physiology.org/cgi/content/full/93/4/1280.)

1112	722	689	251	196	185	128	102	166	178
775	694	514	244	208					

Construct a 95% confidence interval for the mean peak CK activity.

5.5.36 A random sample of 20 observations gave the following summary statistics: $\sum x_i = 234$ and $\sum x_i^2 = 3048$. Assuming that the data may be looked upon as a random sample from a normal population, construct a 95% confidence interval for the actual average, μ.

5.5.37 Let a random sample of size 17 from a normal population for which both mean μ and variance σ^2 are unknown yield $\bar{x} = 3.12$ and $s^2 = 1.04$. Determine a 99% confidence interval for μ.

5.5.38 A random sample from a normal population yields the following 25 values:

90	87	121	96	106	107	89	107	83	92
117	93	98	120	97	109	78	87	99	79
104	85	91	107	89					

(a) Calculate an unbiased estimate $\hat{\theta}$ of the population mean.
(b) Give an approximate 99% confidence interval for the population mean.

5.5.39 The following are random data from a normal population:

3.3	3.3	4.7	2.6	6.4	4.7	1.7	4.5	5.0	3.0

Construct a 98% confidence interval for the population mean μ.

5.5.40 The following data represents the rates (micrometers per hour) at which a razor cut made in the skin of anesthetized newts is closed by new cells:

28	20	21	39	32	23	18	31	14	23
18	22	28	24	33	12	23	21	25	25

(a) Can we say that the data are approximately normally distributed?
(b) Find a 95% confidence interval for the population mean rate μ for the new cells to close a razor cut made in the skin of anesthetized newts.
(c) Find a 99% confidence interval for μ.
(d) Is the 95% CI wider or narrower than the 99% CI? Briefly explain why.

5.5.41 For a particular car, when the brake is applied at 62 mph, the following data give the stopping distance (in feet) for 10 random trials on a dry surface. (Source: http://www.nhtsa.dot.gov/cars/testing/brakes/b.pdf.)

146.9	148.4	149.4	148.6	150.3
147.5	147.5	149.3	148.4	145.5

(a) Can we say that the data are approximately normally distributed?
(b) Find a 95% confidence interval for the population mean stopping distance μ.

5.5.42 A pharmaceutical company tested a new drug to be marketed for the treatment of a particular type of virus. In order to obtain an estimate of the mean recovery time, this drug was tested on 15 volunteer patients, and the recovery time (in days) was recorded. The following data were obtained:

8	17	10	6	34	11	13	6	9	8
19	4	12	17	7					

(a) Obtain a 95% confidence interval estimate of the mean recovery.
(b) What assumptions do we need to make? Test for these assumptions.

5.6 A CONFIDENCE INTERVAL FOR THE POPULATION VARIANCE

In this section, we derive a confidence interval for the population variance σ^2 based on the chi-square distribution (χ^2-distribution). Recall that the χ^2-distribution, like the Student t-distribution, is indexed by a parameter called the degrees of freedom. However, the χ^2-distribution is not symmetric and covers positive values only, and hence it cannot be used to describe a random variable that assumes negative values. Let $X_1,\ldots, X_n$ be normally distributed with mean μ and variance σ^2, with both μ and σ unknown. We know that

$$\frac{\sum_{i=1}^{n}(X_i - \overline{X})^2}{\sigma^2} = \frac{(n-1)S^2}{\sigma^2}$$

has a χ^2-distribution with $(n - 1)$ degrees of freedom irrespective of σ^2. Hence, it can be used as a pivot. We now find two numbers χ_L^2 and χ_U^2 such that

$$P\left(\chi_L^2 \le \frac{(n-1)S^2}{\sigma^2} \le \chi_U^2\right) = 1 - \alpha.$$

The foregoing inequality can be rewritten as

$$P\left(\frac{(n-1)S^2}{\chi_U^2} \le \sigma^2 \le \frac{(n-1)S^2}{\chi_L^2}\right) = 1 - \alpha.$$

Hence a $(1 - \alpha)$ 100% confidence interval for σ^2 is given by $((n-1)S^2/\chi_U^2, (n-1)S^2/\chi_L^2)$. For convenience, we take the areas to the right of $\chi_U^2 = \chi_{\alpha/2}^2$ and to the left of $\chi_L^2 = \chi_{1-\alpha/2}^2$ to be both equal to $\alpha/2$; see Fig. 5.6. Using the chi-square table, we can find the values of $\chi_{\alpha/2}^2$ and $\chi_{1-\alpha/2}^2$. Then, we have the following result.

Theorem 5.6.1 *If $\overline{X}$ and S are the mean and standard deviation of a random sample of size n from a normal population, then*

$$P\left(\frac{(n-1)S^2}{\chi_{\alpha/2}^2} \le \sigma^2 \le \frac{(n-1)S^2}{\chi_{1-\alpha/2}^2}\right) = 1 - \alpha$$

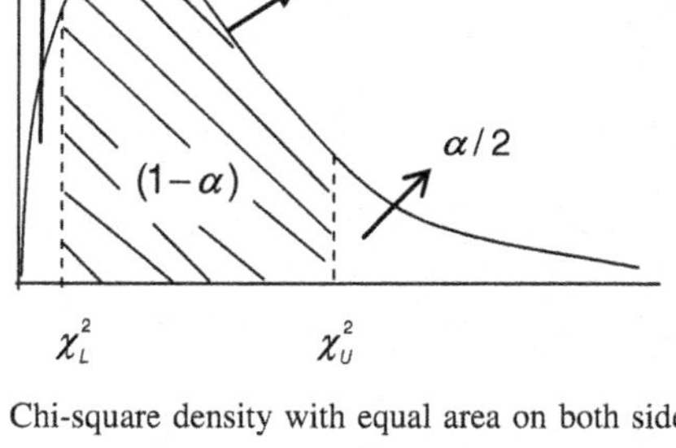

FIGURE 5.6 Chi-square density with equal area on both sides of the CI.

where the χ^2-distribution has $(n-1)$ degrees of freedom.

That is, we are $(1-\alpha)$ 100% confident that the population variance σ^2 falls in the interval $((n-1)S^2/\chi^2_{\alpha/2}, (n-1)S^2/\chi^2_{1-\alpha/2})$.

Example 5.6.1

A random sample of size 21 from a normal population gave a standard deviation of 9. Determine a 90% confidence interval for σ^2.

Solution

Here $n = 21$ and $s^2 = 81$. From the χ^2*-table with 20 degrees of freedom,* $\chi^2_{0.05} = 31.4104$, and $\chi^2_{0.95} = 10.8508$. *Therefore, a 90% confidence interval for σ^2 is obtained from*

$$\left(\frac{(n-1)S^2}{\chi^2_{\alpha/2}}, \frac{(n-1)S^2}{\chi^2_{1-\alpha/2}}\right).$$

Thus we get

$$\frac{(20)(81)}{31.4104} < \sigma^2 < \frac{(20)(81)}{10.8508},$$

or we are 90% confident that $51.575 < \sigma^2 < 149.298$.

We can summarize the steps for obtaining the confidence interval for the true variance as follows.

Procedure to Find Confidence Interval For σ^2

1. Calculate $\bar{x}$ and s^2 from the sample $x_1, \ldots, x_n$.
2. Find $\chi^2_U = \chi^2_{\alpha/2}$, and $\chi^2_L = \chi^2_{1-\alpha/2}$ using the χ^2-square table with $(n-1)$ degrees of freedom.
3. Compute the $(1-\alpha)$ 100% confidence interval for the population variance σ^2 as $\left((n-1)s^2/\chi^2_{\alpha/2}, (n-1)s^2/\chi^2_{1-\alpha/2}\right)$, where χ^2-values are with $(n-1)$ degrees of freedom.

Assumption: The population is normal.

Example 5.6.2

The following data represent cholesterol levels (in mg/dL) of 10 randomly selected patients from a large hospital on a particular day.

360 352 294 160 146 142 318 200 142 116

Determine a 95% confidence interval for σ^2.

Solution

From the data, we can get $\bar{x}$ = 223 and standard deviation s=96.9. The following probability graph is obtained by Software.

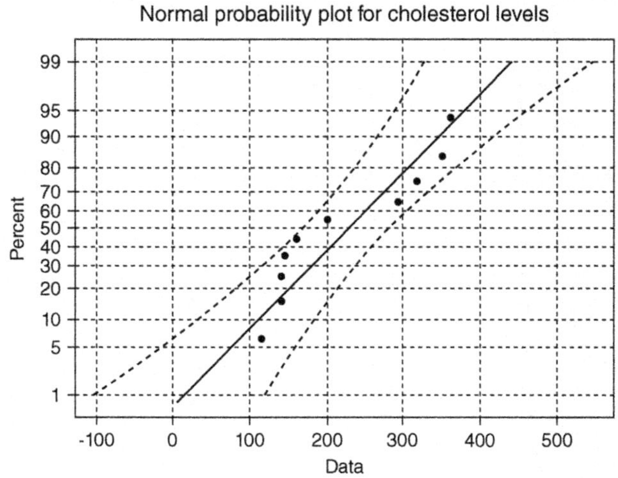

Even though the scattergram does not appear to follow a straight line, the data are still within the band, so we can assume approximate normality for the data. (In situations like this, it is more appropriate to use nonparametric tests explained in Chapter 12.) A box plot of the data shows that there are no outliers. From the χ^2-table, $\chi^2_{0.025}(9) = 19.023$, and $\chi^2_{0.975}(9) = 2.70$. Therefore a 90% confidence interval for σ^2 is obtained from

$$\left(\frac{(n-1)S^2}{\chi^2_{\alpha/2}(n-1)}, \frac{(n-1)S^2}{\chi^2_{1-\alpha/2}(n-1)} \right).$$

Thus we get

$$\frac{(9)(96.9)^2}{19.023} < \sigma^2 < \frac{(9)(96.9)^2}{2.70},$$

or we are 95% confident that $4442.3 < \sigma^2 < 31{,}299$. Note that the numbers look very large, but it is the value of variance. By taking the square root of the numbers on both sides, we can also get a confidence interval for the standard deviation σ.

As remarked in the previous exercise, in general, to find a $(1-\alpha)100\%$ confidence interval for the true population standard deviation, σ, take the square roots of the endpoints of the confidence interval of the variance.

EXERCISES 5.6

5.6.1 A random sample of size 20 is drawn from a population having a normal distribution. The sample mean and the sample standard deviation from the data are given, respectively, as $\bar{x} = -2.2$ and $s = 1.42$. Construct a 90% confidence interval for the population variance σ^2, and interpret.

5.6.2 A drug is suspected of causing an elevated heart rate in a certain group of high-risk patients. Twenty patients from the group were given the drug. The changes in heart rates were found to be as follows:

−1	8	5	10	2	12	7	9	1	3
4	6	4	12	11	2	−1	10	2	8

Construct a 95% confidence interval for the variance of change in heart rate. Assume that the population has a normal distribution, and interpret.

5.6.3 Air pollution in large US cities is monitored to see whether it conforms to requirements set by the Environmental Protection Agency. The following data, expressed as an air pollution index, give the air quality of a city for 10 randomly selected days.

56.23 57.12 57.7 65.80 59.40
62.90 58.00 64.56 63.92 63.45

Assuming that the data may be viewed as a random sample from a normal population, construct a 99% confidence interval for the actual variance of the air pollution index for this city, and interpret.

5.6.4 A random sample of 25 observations gave the following summary statistics: $\sum x_i = 234$ and $\sum x_i^2 = 3048$. Assuming that the data can be looked upon as a random sample from a normal population, construct a 95% confidence interval for the actual variance, σ^2.

5.6.5 Let a random sample of size 18 from a normal population with both mean μ and variance σ^2 unknown yield $\bar{x} = 2.27$ and $s^2 = 1.02$. Determine a 99% confidence interval for σ^2.

5.6.6 Suppose we want to study contaminated fish in a river. It is important for the study to know the size of the variance σ^2 in the fish weights. The 25 samples of fish in the study produced the following summary statistics, $\bar{x} = 1030.5g$, and the standard deviation $s = 200.6$ g. Construct a 95% confidence interval for the true variation in weights of contaminated fish in this river.

5.6.7 A random sample from a normal population yields the following 25 values:

90 87 121 96 106 107 89 107 83 92
117 93 98 120 97 109 78 87 99 79
104 85 91 107 89

(a) Calculate an unbiased estimate $\hat{\sigma}^2$ of the population variance.
(b) Give an approximate 99% confidence interval for the population variance.
(c) Interpret your results, and state any assumptions you made in order to solve the problem.

5.6.8 It is known that some brands of peanut butter contain impurities within an acceptable level. A test conducted on randomly selected 12 jars of a certain brand of peanut butter resulted in the following percentages of impurities:

1.9 2.7 2.1 2.8 2.3 3.6 1.4 1.8 2.1 3.2 2.0

Construct a 95% confidence interval for the average percentage of impurities in this brand of peanut butter.
Give an approximate 95% confidence interval for the population variance.
Interpret your results, and test for normality.

5.6.9 The following data represent the maximal head measurements (across the top of the skull) in millimeters of 15 Etruscans (inhabitants of ancient Etruria).

152 147 126 140 135 139 149 140
142 147 132 148 146 143 137

Calculate an unbiased estimate $\hat{\sigma}^2$ of the population variance.
Give an approximate 95% confidence interval for the population variance.
Interpret your results, and test for normality.

5.6.10 The rates of return (rounded to the nearest percentage) for 25 clients of a financial firm are given in the following table:

13 11 28 6 − 4 15 13 6 11 11
3 12 20 3 16 16 15 8 20 15
4 1 12 2 − 9

Find a 98% confidence interval for the variance σ^2 of rates of return. Use this to find the confidence interval for the population standard deviation, σ.

5.6.11 In order to test the precision of a new type of blood sugar monitor for diabetic patients, 20 randomly selected monitors of this type were used. A blood sample with 120 mg/dL was tested in each of these monitors, and the resulting readings are given in the following table:

117	116	121	120	122	117	120	120	118	119
118	123	119	123	119	122	118	122	121	120

(a) Obtain a 99% confidence interval for the variance σ^2.
(b) Is it reasonable to assume that the data follow a normal distribution?

5.7 CONFIDENCE INTERVAL CONCERNING TWO POPULATION PARAMETERS

In the earlier sections, we studied the confidence limits of true parameters from samples from single populations. Now, we consider the interval estimation based on samples from two populations. Our interest is to obtain a confidence interval for the parameters of interest based on two independent samples taken from these two populations.

Let $X_{11}, \ldots, X_{1n_1}$ be a random sample from a normal distribution with mean μ_1 and variance σ_1^2, and let $X_{21}, \ldots, X_{2n_2}$ be a random sample from a normal distribution with mean μ_2 and variance σ_2^2. Let $\overline{X}_1 = (1/n_1)\Sigma_{i=1}^{n1} X_{1i}$ and $\overline{X}_2 = (1/n_2)\Sigma_{i=1}^{n2} X_{2i}$. We will assume that the two samples are independent. Then, $\overline{X}_1$ and $\overline{X}_2$ are independent. The distribution of $\overline{X}_1 - \overline{X}_2$ is $N(\mu_1 - \mu_2, (1/n_1)\sigma_1^2 + (1/n_2)\sigma_2^2)$. Now, as in the one-sample case, the confidence interval for $\mu_1 - \mu_2$ is obtained as follows.

Large Sample Confidence Interval for the Difference of Two Means

(i). σ_1, σ_2 are known. The $(1 - \alpha)$ 100% large sample confidence interval for $\mu_1 - \mu_2$ is given by

$$(\overline{X}_1 - \overline{X}_2) \pm z_{\alpha/2}\sqrt{\left(\frac{\sigma_1^2}{n_1} + \frac{\sigma_2^2}{n_2}\right)}.$$

(ii). If σ_1 and σ_2 are not known, σ_1 and σ_2 can be replaced by the respective sample standard deviations S_1 and S_2 when $n_i \geq 30$, $i = 1, 2$. Thus, we can write

$$p\left((\overline{X}_1 - \overline{X}_2) - z_{\alpha/2}\sqrt{\left(\frac{S_1^2}{n_1} + \frac{S_2^2}{n_2}\right)} \leq \mu_1 - \mu_2 \right.$$
$$\left. \leq (\overline{X}_1 - \overline{X}_2) + z_{\alpha/2}\sqrt{\left(\frac{S_1^2}{n_1} + \frac{S_2^2}{n_2}\right)}\right) = 1 - \alpha.$$

Assumption: The population is normal, and the samples are independent.

Example 5.7.1

A study of two kinds of machine failures shows that 58 failures of the first kind took on average 79.7 minutes to repair with a standard deviation of 18.4 minutes, whereas 71 failures of the second kind took on average 87.3 minutes to repair with a standard deviation of 19.5 minutes. Find a 99% confidence interval for the difference between the true average amounts of time it takes to repair failures of the two kinds of machines.

Solution

Here, $n_1 = 58$, $n_2 = 71$, $\bar{x}_1 = 79.7$, $s_1 = 18.4$, $\bar{x}_2 = 87.3$, and $s_2 = 19.5$. Then, the 99% confidence interval for $\mu_1 - \mu_2$ is given by

$$(79.7 - 87.3) \pm 2.575\sqrt{\frac{(18.4)^2}{58} + \frac{(19.5)^2}{71}}.$$

That is, we are 99% certain that $\mu_1 - \mu_2$ is located in the interval (–16.215, 1.0149). Note that $-16.215 < \mu_1 - \mu_2 < 1.0149$ means that more than 90% of the length of this interval is negative. Thus, we can conclude that μ_2 dominates μ_1, that is, $\mu_2 > \mu_1$ more than 90% of the time.

In the small sample case, the problem of constructing confidence intervals for the difference of the means from the two normal populations with unknown variances can be a difficult one. However, if we assume that the two populations have a common but unknown variance, say $\sigma_1^2 = \sigma_2^2 = \sigma^2$, we can obtain an estimate of the variance by pooling the two-sample data sets. Define the pooled sample variance S_p^2 as

$$S_p^2 = \frac{\sum_{i=1}^{n_1}(X_{1i} - \bar{X}_1)^2 + \sum_{i=1}^{n_2}(X_{2i} - \bar{X}_2)^2}{n_1 + n_2 - 2}$$
$$= \frac{(n_1 - 1)S_1^2 + (n_2 - 1)S_2^2}{n_1 + n_2 - 2}.$$

Now, when the two samples are independent,

$$T = \frac{(\bar{X}_1 - \bar{X}_2) - (\mu_1 - \mu_2)}{S_p\sqrt{\frac{1}{n_1} + \frac{1}{n_2}}}$$

has a t-distribution with $n_1 + n_2 - 2$ degrees of freedom. We summarize the CI for $\mu_1 - \mu_2$ below.

Small Sample Confidence Interval for the Difference of Two Means ($\sigma_1^2 = \sigma_2^2$)

The small sample $(1 - \alpha)100\%$ confidence interval for $\mu_1 - \mu_2$ is

$$(\bar{X}_1 - \bar{X}_2) \pm t_{\alpha/2,(n_1+n_2-2)}S_p\sqrt{\frac{1}{n_1} + \frac{1}{n_2}}.$$

Assumption: The samples are independent from two normal populations with equal variances.

Example 5.7.2

Independent random samples from two normal populations with equal variances produced the following data:

Sample1: 1.2 3.1 1.7 2.8 3

Sample2: 4.2 2.7 3.6 3.9

(a) Calculate the pooled estimate of σ^2.

(b) Obtain a 90% confidence interval for $\mu_1 - \mu_2$.

Solution

(a) We have $n_1 = 5$ and $n_2 = 4$. Also

$$\bar{x}_1 = 2.36,\ s_1^2 = 0.733$$
$$\bar{x}_2 = 3.6,\ s_2^2 = 0.42.$$

Hence

$$s_p^2 = \frac{(n_1-1)s_1^2 + (n_2-1)s_2^2}{n_1+n_2-2} = 0.5989.$$

(b) For the confidence coefficient 0.90, $\alpha = 0.10$, and from the t-table, $t_{0.05,7} = 1.895$. Thus a 90% confidence interval for $\mu_1 - \mu_2$ is

$$\begin{aligned}(\bar{X}_1 - \bar{X}_2) &\pm t_{\alpha/2,(n_1+n_2-2)} s_p\sqrt{\frac{1}{n_1} + \frac{1}{n_2}} \\ &= (2.36 - 3.6) \pm 1.895\sqrt{0.5989\left(\frac{1}{5} + \frac{1}{4}\right)} \\ &= -1.24 \pm 0.98 = (-2.22, -0.26).\end{aligned}$$

Here μ_2 dominates μ_1 uniformly. Note that we can decrease the confidence range −2.22 to 0.26, by increasing n_1 and n_2 with $1 - \alpha = 0.90$ to remain the same. This means that we are closing on the unknown true value of $\mu_1 - \mu_2$.

In the small sample case, if the equality of the variances cannot be reasonably assumed, that is, $\sigma_1^2 \neq \sigma_2^2$, we can still use the previous procedure, except that we use the following degrees of freedom in obtaining the t-value from the table. Let

$$v = \frac{\left(\frac{s_1^2}{n_1} + \frac{s_2^2}{n_2}\right)^2}{\frac{\left(\frac{s_1^2}{n_1}\right)^2}{n_1-1} + \frac{\left(\frac{s_2^2}{n_2}\right)^2}{n_2-1}}.$$

The number given in this formula is always rounded down for the degrees of freedom. Hence, in this case, a small sample $(1-\alpha)$ 100% confidence interval for $\mu_1 - \mu_2$ is given by

$$(\bar{X}_1 - \bar{X}_2) \pm t_{\alpha/2,v}\sqrt{\frac{s_1^2}{n_1} + \frac{s_2^2}{n_2}},$$

where the t-distribution has v degrees of freedom as given previously.

Example 5.7.3

Assume that two populations are normally distributed with unknown and unequal variances. Two independent samples are taken with the following summary statistics:

$$n_1 = 16 \quad \bar{x}_1 = 20.17 s_1 = 4.3$$
$$n_2 = 11 \quad \bar{x}_2 = 19.23 s_2 = 3.8$$

Construct a 95% confidence interval for $\mu_1 - \mu_2$.

Solution

First let us compute the degrees of freedom

$$v = \frac{\left(\frac{s_1^2}{n_1} + \frac{s_2^2}{n_2}\right)^2}{\frac{\left(\frac{s_1^2}{n_1}\right)^2}{n_1 - 1} + \frac{\left(\frac{s_2^2}{n_1}\right)^2}{n_1 - 1}} = \frac{\left(\frac{(4.3)^2}{16} + \frac{(3.8)^2}{11}\right)}{\frac{\left(\frac{(4.3)^2}{16}\right)^2}{15} + \frac{\left(\frac{(3.8)^2}{11}\right)^2}{110}} = 23.312.$$

Hence $v = 23$, and $t_{0.025,23} = 2.069$.

Now, a 95% confidence interval for $\mu_1 - \mu_2$ is

$$(\bar{x}_1 - \bar{x}_2) \pm t_{\alpha/2,v}\sqrt{\frac{s_1^2}{n_1} + \frac{s_2^2}{n_2}} = (20.17 - 19.23)$$

$$\pm (2.069)\sqrt{\frac{(4.3)^2}{16} + \frac{(3.8)^2}{11}},$$

which gives the 95% confidence interval as

$$-2.3106 < \mu_1 - \mu_2 < 4.1906.$$

In a real-world problem, how do we determine if $\sigma_1^2 = \sigma_2^2$, or $\sigma_1^2 \neq \sigma_2^2$, so that we can select one of the two methods just given? In Chapter 14, we discuss a procedure that determines the homogeneity of the variances (i.e., whether $\sigma_1^2 = \sigma_2^2$). For the time being, a good indication is to look at the point estimators of σ_1^2 and σ_2^2, namely, S_1^2 and S_2^2. If the point estimators are fairly close to each other, then we can select $\sigma_1^2 = \sigma_2^2$. Otherwise, $\sigma_1^2 \neq \sigma_2^2$. For a more general method of testing for equality of variances, we refer to Section 14.4.3.

We now give a procedure for a large sample confidence interval for the difference of the true proportions, $p_1 - p_2$, in two binomially distributed populations.

Large Sample Confidence Interval For p_1–p_2

The $(1 - \alpha)100\%$ large sample confidence interval for p_1–p_2 is given by

$$(\hat{p}_1 - \hat{p}_2) \pm z_{\alpha/2}\sqrt{\left(\frac{\hat{p}_1(1 - \hat{p}_1)}{n_1} + \frac{\hat{p}_2(1 - \hat{p}_2)}{n_2}\right)},$$

where $\hat{p}_1$ and $\hat{p}_2$ are the points estimators of p_1 and p_2. This approximation is applicable if $\hat{p}_i n_i \geq 5$, $i = 1,2$ and$(1 - \hat{p}_i)n_i \geq 5$, $i = 1,2$.. The two samples are independent.

Example 5.7.4

Iron deficiency, the most common nutritional deficiency worldwide, has negative effects on work capacity and on motor and mental development. In a 1999–2000 survey by the National Health and Nutrition Examination Survey (NHANES), iron deficiency was detected in 58 of 573 white, non-Hispanic females (10% rounded to whole number) and 95 of 498 (19% rounded to whole number) black, non-Hispanic females (source: http://www.cdc.gov/mmwr/preview/mmwrhtml/mm5140a1.htm). Let p_1 be the proportion of black, non-Hispanic females with iron deficiency, and let p_2 be the proportion of white, non-Hispanic females with iron deficiency. Obtain a 95% confidence interval for $p_1 - p_2$.

Solution

Here, $n_1 = 573$ and $n_2 = 498$. Also, $\hat{p}_1 = \frac{58}{573} = 0.10122 \approx 0.1$, and $\hat{p}_2 = \frac{95}{498} = 0.1907 \approx 0.19$. For $\alpha = 0.05$, $z_{0.015} = 1.96$. Hence, a 95% confidence interval for $p_1 - p_2$ is

$$
\begin{aligned}
(\hat{p}_1 - \hat{p}_2) &\pm z_{\alpha/2}\sqrt{\left(\frac{\hat{p}_1(1-\hat{p}_1)}{n_1} + \frac{\hat{p}_2(1-\hat{p}_2)}{n_2}\right)} \\
&= (0.1 - 0.19) \pm (1.96)\sqrt{\frac{(0.1)(0.9)}{573} + \frac{(0.19)(0.81)}{498}} \\
&= (-0.13232, -0.047685).
\end{aligned}
$$

Here, the true difference of $p_1 - p_2$ is located in the negative portion of the real line, which tells us that the true proportion of black, non-Hispanic females with iron deficiency is larger than the proportion of white, non-Hispanic females with iron deficiency.

There are situations in applied problems that make it necessary to study and compare the true variances of two independent normal distributions. For this purpose, we will find a confidence interval for the ratio σ_1^2/σ_2^2 using the F-distribution. Let $X_1, \ldots, X_{n_1}$ and $Y_1, \ldots, Y_{n_2}$ *be* independent samples of size n_1 and n_2 from two normal distributions $N(\mu_1, \sigma_1^2)$ and $N(\mu_2, \sigma_2^2)$, respectively. Let S_1^2 and S_2^2 be the variances of the two random samples. The confidence interval for the ratio σ_1^2/σ_2^2 is given as follows.

A (1–α)100% confidence interval for $\frac{\sigma_1^2}{\sigma_2^2}$

A $(1-\alpha)100\%$ confidence interval for σ_1^2/σ_2^2 is given by

$$\left(\left(\frac{S_1^2}{S_2^2}\right)\left(\frac{1}{F_{n_1-1,n_2-1,1-\alpha/2}}\right), \left(\frac{S_1^2}{S_2^2}\right)\left(\frac{1}{F_{n_1-1,n_2-1,(\alpha/2)}}\right)\right).$$

That is,

$$P\left(\left(\frac{S_1^2}{S_2^2}\right)\left(\frac{1}{F_{n_1-1,n_2-1,1-\alpha/2}}\right) \le \frac{\sigma_1^2}{\sigma_2^2} \le \left(\frac{S_1^2}{S_2^2}\right)\left(\frac{1}{F_{n_1-1,n_2-1,(\alpha/2)}}\right)\right)$$

$$=1 - \alpha.$$

Assumptions: These two populations are normal, and the samples are independent.

Note that we can also write a $(1 - \alpha)100\%$ confidence interval for σ_1^2/σ_2^2 in the form

$$\left(\left(\frac{S_1^2}{S_2^2}\right)\left(\frac{1}{F_{n_1-1,n_2-1,1-\alpha/2}}\right), \left(\frac{S_1^2}{S_2^2}\right)F_{n_2-1,n_1-1,1-\alpha/2}\right).$$

The following example illustrates how to find the confidence interval for σ_1^2/σ_2^2.

Example 5.7.5

Assuming that two populations are normally distributed, two independent random samples are taken with the following summary statistics:

$$n_1 = 21 \quad \bar{x}_1 = 20.17 s_1 = 4.3$$
$$n_2 = 16\ \bar{x}_2 = 19.23 s_2 = 3.8$$

Construct a 95% confidence interval for σ_1^2/σ_2^2.

Solution

Here, $n_1 = 21$, $n_2 = 16$, and $\alpha = 0.05$. Using the F-table, we have

$$F_{n_1-1,n_2-1,1-\alpha/2} = F(20,15,0.975) = 2.76$$

and

$$F_{n_2-1,n_1-1,1-\alpha/2} = F(15,20,0.975) = 2.57.$$

A 95% confidence interval for σ_1^2/σ_2^2 is

$$\left(\left(\frac{S_1^2}{S_2^2}\right)\left(\frac{1}{F_{n_1-1,n_2-1,1-\alpha/2}}\right), \left(\frac{S_1^2}{S_2^2}\right)F_{n_2-1,n_1-1,1-\alpha/2}\right)$$

$$= \left(\left(\frac{(4.3)^2}{(3.8)^2}\right)\left(\frac{1}{2.76}\right), \left(\frac{(4.3)^2}{(3.8)^2}\right)(2.57)\right) = (0.46394, 3.2908).$$

That is, we are 95% confident that the ratio of true variance, σ_1^2/σ_2^2, is located in the interval that implies a 95% confidence interval (0.46394, 3.2908).

EXERCISES 5.7

5.7.1 A study was conducted to compare two different procedures for assembling components. Both procedures were implemented and run for a month to allow employees to learn each procedure. Then, each was observed for 10 days with the following results. Values are the number of components assembled per day.

Procedure I	115	101	113	64	104	97	114	96	87	93
Procedure II	86	99	100	78	97	111	102	94	88	99

Construct a 98% confidence interval for the difference in the mean number of components assembled by the two methods. Assume that the data for each procedure are from approximately normal populations with a common variance. Interpret the result.

5.7.2 A study was conducted to see the differences between oxygen consumption rates for male runners from a college who had been trained by two different methods, one involving continuous training for a period of time each day and the other involving intermittent training of about the same overall duration. The means, standard deviations, and sample sizes are shown in the following table.

Continuoustraining	$n_1 = 15$	$\bar{x}_1 = 46.28$	$s_1 = 6.3$
Intermittenttraining	$n_2 = 7$	$\bar{x}_2 = 42.34$	$s_2 = 7.8$

If the measurements are assumed to come from normally distributed populations with equal variances, estimate the difference between the population means, with a confidence coefficient 0.95, and interpret.

5.7.3 Studies have shown that the risk of developing coronary disease increases with the level of obesity. A study comparing two methods of losing weight: diet alone and exercise alone were conducted on 87 men over a 1-year period. Forty-two men dieted and lost an average of 16.0 lb. over the year, with a standard deviation of 5.6 lb. Forty-five men who exercised lost an average of 10.6 lb, with a standard deviation of 7.9 lb. Construct a 99% confidence interval for the difference in the mean weight loss by these two methods. State any assumptions you made, and interpret the result you obtained.

5.7.4 The following information was obtained from two independent samples selected from two normally distributed populations with unknown but equal variances.

Sample 1	14	15	12	13	6	14	11	12	17	19	23		
Sample 2	16	18	12	20	15	19	15	22	20	18	23	12	20

Construct a 95% confidence interval for the difference between the population means, and interpret.

5.7.5 In the academic year 2001–2002, two random samples of 25 male professors and 23 female professors from a large university produced a mean salary for male professors of \$58,550 with a standard deviation of \$4000; the mean for female professors was \$53,700 with a standard deviation of 3200. Construct a 90% confidence interval for the difference between the population mean salaries. Assume that the salaries of male and female professors are both normally distributed with equal standard deviations. Interpret the result.

5.7.6 Let the random variables X_1 and X_2 follow binomial distributions that have parameters $n_1 = 100$, $n_2 = 75$. Let $x_1 = 35$ and $x_2 = 27$ be observed values of X_1 and X_2. Let p_1 and p_2 be the true proportions. Determine an appropriate 95% confidence interval for $p_1 - p_2$.

5.7.7 The following information is obtained from two independent samples selected from two populations:

$$n_1 = 40 \quad \bar{x}_1 = 28.4 \quad s_1 = 4.1$$
$$n_2 = 32 \quad \bar{x}_2 = 25.6 \quad s_2 = 4.5$$

(a) What is the maximum likelihood estimator of $\mu_1 - \mu_2$?
(b) Construct a 99% confidence interval for $\mu_1 - \mu_2$.

5.7.8 In order to compare the mean hemoglobin (Hb) levels of well-nourished and undernourished groups of children, random samples from each of these groups yielded the following summary.

	Number of Children	Sample Mean	Sample Standard Deviation
Well-nourished	95	11.2	0.9
Undernourished	75	9.8	1.2

Construct a 95% confidence interval for the true difference of means, $\mu_1 - \mu_2$.

5.7.9 In a certain part of a city, the average price of homes in 2000 was \$148,822, and in 2001 it was \$155,908. Suppose these means were based on a random sample of 100 homes in 1997 and 150 homes in 1998 and that the sample standard deviations of sale prices were \$21,000 for 2000 and \$23,000 for 2001. Find a 98% confidence interval for the difference in the two population means.

5.7.10 Two independent samples from a normal population are taken with the following summary statistics:

$$n_1 = 16 \quad \bar{x}_1 = 2.4 \quad s_1 = 0.1$$
$$n_2 = 11 \quad \bar{x}_2 = 2.6 \quad s_2 = 0.5$$

Construct a 95% confidence interval for σ_1^2/σ_2^2.

5.7.11 The following information was obtained from two independent samples selected from two normally distributed populations:

Sample 1	35	36	33	34	27	35	32	33	38	40	44		
Sample 2	37	39	33	41	36	40	36	43	41	39	44	33	41

Construct a 90% confidence interval for σ_1^2/σ_2^2.

5.7.12 The management of a supermarket wanted to study the spending habits of its male and female customers. A random sample of 16 male customers who shopped at this supermarket showed that they spent an average of \$55 with a standard deviation of \$12. Another random sample of 25 female customers showed that they spent \$85 with a standard deviation of \$20.50. Assuming that the amounts spent at this supermarket by all its male and female customers were approximately normally distributed, construct a 90% confidence interval for the ratio of variance in spending for males and females, σ_1^2/σ_2^2.

5.7.13 An experiment is conducted comparing the effectiveness of a new method of teaching algebra for eighth-grade students. Twelve gifted and 12 regular students are taught using this method. Their scores on a final exam are shown in the following table:

Average	58	69	55	65	88	52	99	76	45	86	55	79
Gifted	77	86	84	93	77	91	87	95	68	78	74	58

(a) Compute the 95% confidence interval on the difference between the mean of the students being taught by this new method.

(b) Construct a 95% confidence interval for the ratio of variance in test scores for regular and gifted students, σ_1^2/σ_2^2.

(c) What are the assumptions you made in parts (a) and (b)? Are these assumptions justified?

5.7.14 Assume that two populations have the same variance σ^2. If a sample of size n_1 produced a variance S_1^2 from population I and a sample of size n_2 produced a variance S_2^2 from population II, show that the pooled variance

$$S_p^2 = \frac{(n_1 - 1)S_1^2 + (n_2 - 1)S_2^2}{n_1 + n_2 - 2},$$

is an unbiased estimator of σ^2. Show that $(S_1^2 + S_2^2)/2$ is also an unbiased estimator of σ^2. Which of the two estimators would you prefer? Give reasons for your choice.

5.8 CHAPTER SUMMARY

In this chapter, we have discussed the basic concepts of estimation, both point estimation and interval estimation. Two methods of finding point estimators were described—the method of moments and the method of maximum likelihood. Some desirable properties of the point estimators that we have discussed are unbiasedness and sufficiency. Unbiasedness guards against consistently producing under- or overestimates of the parameter in repeated sampling. A sufficient estimator is a "good" estimator of the population parameter θ in the sense that it depends on fewer data values. Later, this chapter discusses the concept of interval estimation. A $(1 - \alpha)$ 100% confidence interval (CI) for an unknown parameter θ is computed from sample data. The so-called pivotal method is introduced for deriving confidence intervals. Large sample and small sample confidence intervals are derived for the population mean μ. Confidence intervals in the case of two samples are also discussed. Additionally, confidence intervals for variance and the ratio of variances are derived.

We will now list some of the key definitions introduced in this chapter.

- Method of moments
- Likelihood function
- Maximum likelihood equations
- Unbiased estimator
- Mean square error
- Minimum variance unbiased estimator
- Sufficient estimator
- Jointly sufficient
- Upper and lower confidence limits
- Confidence coefficient
- $100\,(1 - \alpha)\%$ confidence interval for θ
- Interval estimation

- Confidence interval

In this chapter, we have also learned the following important concepts and procedures:

- The method of moments procedure
- Procedure to find the MLE
- Procedure to verify
- Pivotal method
  - Procedure to find a confidence interval for θ using the pivot
  - Procedure to find a large sample confidence interval for θ
  - Procedure to find a small sample confidence interval for μ
  - Procedure to find a confidence interval for the population variance σ^2
  - Large sample confidence interval for the difference of the means
  - Small sample confidence interval for the difference of two means ($\sigma_1^2 = \sigma_2^2$)
  - Small sample confidence interval for the difference of two means ($\sigma_1^2 \neq \sigma_2^2$)
  - Large sample confidence interval for $p_1 - p_2$
  - A $(1 - \alpha)$ 100% confidence interval for σ_1^2/σ_2^2

5.9 COMPUTER EXAMPLES

5.9.1 Examples Using R

It should be noted that for the problems where you're generating random samples, your answers will vary!

EXAMPLE 5.9.1

Descriptive Point Estimates

Generate 50 sample points from an $N(4,4)$ distribution and find the descriptive statistics. Obtain an unbiased and sufficient estimate of μ.

R-code

```
sample=rnorm(50,4,4);
summary(sample);
sd(sample);
sd(sample)/sqrt(length(sample));   <- Standard error of the mean
```

Output

Your output will be unique since the samples are generated randomly; take notice of standard error.

```
Min.     1st Qu.   Median   Mean 3rd   Qu.     Max.
-4.292   1.105     4.012    3.865      6.478   14.790
```

Notice this is an estimate;
we know that the population
mean is 4 as we defined it.

Notice this is an estimate;
we know that the population
mean is 4 as we defined it.

```
4.288085    <- Standard deviation
0.6064268   <- Standard error of the mean
```

EXAMPLE 5.9.2 Uniform Maximum likelihood

Generate 35 samples from a $U(0,5)$ distribution, and using the descriptive statistics command, find the maximum likelihood estimate for this data.

Solution

We know that for a random sample $X_1, \ldots, X_n$ from $U(0, \theta)$ the MLE, $\hat{\theta} = \max(X_i) = X_{(n)}$, the nth order statistic. We can use the following steps to obtain the estimate.

R-code

```
sample=runif(35,0,5);
summary(sample);
```

Output

Your output will be unique since the samples are generated randomly.

```
Min.   1st Qu.  Median  Mean    3rd Qu.  Max.
0.1155 1.5710   2.9520  2.7620  4.0920   4.9900
```

The *MLE* of the data 4.99

EXAMPLE 5.9.3 Confidence Interval

Obtain a 95% confidence interval for μ using the following data:

Sample (x): 7.227 5.7383 4.9369 6.238 8.4876 2.7618

This example assumes you've stored your data into variable x. Please modify code appropriately.

R-code

```
t.test(x,conf.level=0.95);
```

Output

```
One Sample t-test.
data: x
t = 7.3399, df = 5, p-value = 0.0007365
alternative hypothesis: true mean is not equal to 0
95 percent confidence interval:
3.832566 7.963967
sample estimates:
mean of x
5.898267
```

Notice the interval rather than point estimate.

Example 5.9.4 Confidence Interval

For the following data, obtain a 98% confidence interval for μ.

Sample (x): 6.8 5.6 8.5 8.5 8.4 7.5 9.3 9.4 7.8 7.1 9.9 9.6 9.0 13.7 9.4 16.6 9.1 10.1 10.6 11.1 8.9 11.7 12.8 11.5 10.6 12.0 11.1 6.4 12.3 12.3 11.4 9.9 15.5 14.3 11.5 13.3 11.8 12.8 13.7 13.9 12.9 14.2 14.0

This example assumes you've stored the data into variable x. Please modify your code appropriately.

R-code

```
t.test(x,conf.level=0.98);
```

Output

```
One Sample t-test
data: x
t = 27.7762, df = 42, p-value < 2.2e-16
alternative hypothesis: true mean is not equal to 0
98 percent confidence interval:
9.910598 11.801030
sample estimates:
mean of x
10.85581
```

Notice the interval rather than point estimate.

EXAMPLE 5.9.5 Confidence Interval

For the following data, find a 90% confidence interval for $\mu_1 - \mu_2$, using the following data.

Sample (x): 1.2 3.1 1.7 2.8 3.0

Sample (y): 4.2 2.7 3.6 3.9

This example assumes you've stored your data into variables x and y. Please modify your code appropriately.

R-code

```
t.test(x,y,conf.level=0.90);
```

Output

```
Welch Two Sample t-test
data: x and y
t = −2.4721, df = 6.996, p-value = 0.04272
alternative hypothesis: true difference in means is not equal to 0
90 percent confidence interval:
−2.1903896 −0.2896104
sample estimates:
mean of x mean of y
     2.36      3.60
```

90% Confidence Interval

5.9.2 SPSS Examples

EXAMPLE 5.9.6

Consider the data

66 74 79 80 77 78 65 79 81 69

Using SPSS, obtain a 99% confidence interval for μ.

Solution

*One easy way to obtain the confidence interval in SPSS is to use the hypothesis testing procedure. The procedure is as follows: First, enter the data in **C1**. Then click*

Analyze> Compare Means> One-sample t-Test...,> Move ***var00001*** to ***Test Variable(s)***, and Click ***Options...***, and enter **99** in ***Confidence interval:***, click ***Continue***, and **OK**.

Note that the default value is 95%.

5.9.3 SAS Examples

We will not give the output in this section.

EXAMPLE 5.9.7

The following data give P/E for a particular year of 49 mutual fund companies owned by a randomly selected mutual fund:

6.8	5.6	8.5	8.5	8.4	7.5	9.3	9.4	7.8	7.1
9.9	9.6	9.0	16.6	9.1	10.1	10.6	11.1	8.9	11.7
12.8	11.5	12.0	10.6	11.1	6.4	11.4	9.9	14.3	11.5
11.8	13.3	13.9	12.9	14.2	14.0	15.5	17.9	21.8	18.4
34.3	13.7	12.3	18.0	9.4	12.3	16.9	12.8	13.7	

Find a 98% confidence interval for the mean P/E multiples. Use SAS procedures.

Solution

We could use the following procedure:

```
DATA peratio;
INPUT patio @@;
DATALINES;
6.85.68.58.58.47.59.39.47.8
7.19.99.69.09.413.716.69.110.110.6
11.18.911.712.811.512.010.611.16.412.3
12.311.49.914.311.511.813.312.813.713.912.9
14.214.015.516.918.017.921.818.434.3
;
PROC MEANS data = peratio lclm uclm alpha = 0.02;
var peratio;
RUN;
```

EXERCISES 5.9

5.9.1 Using any of the software packages (R, SPSS, or SAS), obtain confidence intervals for at least one data set taken from each section of this chapter.

5.10 PROJECTS FOR CHAPTER 5

5.10.1 Asymptotic Properties

In general, we do not have a single sample with one estimator of the unknown parameter θ. Rather, we will have a general formula that defines an estimator for any sample size. This gives a sequence of estimators of θ:

$$\hat{\theta} = h_n(X_1, \ldots, X_n),\ n = 1,2,\ \ldots..$$

In this case, we can define the following asymptotic properties:

(i) The sequence of estimators $\widehat{\theta}_n$ is said to be *asymptotically unbiased* for θ if *bias* $(\widehat{\theta}_n) \to 0$ as n → ∞.

(ii) Suppose $(\widehat{\theta}_n)$ and $\widehat{y}_n$ are two sequences of estimators that are asymptotically unbiased for θ. The *asymptotic relative efficiency* of $\widehat{\theta}_n$ to $\widehat{y}_n$ is defined by

(a) Show that $\widehat{\theta}_n$ is asymptotically unbiased if and only if

$$E(\hat{\theta}_n) \to \theta \text{ as } n \to \infty.$$

(b) Let $X_1, \ldots, X_n$ be a random sample from a distribution with unknown mean μ and variance σ^2. It is known that the method of moments estimators for μ and σ^2 are, respectively, the sample mean $\overline{X}$ and$S'^2_n = (1/n)\Sigma_{i=1}^n (X_i - \overline{X})^2 = ((n-1)/n)S_n^2$, where S_n^2 is the sample variance.

(i) Show that S'^2_n is an asymptotically unbiased estimator of σ^2.

(ii) Show that the asymptotic relative efficiency of S'^2_n to S_n^2 is 1.

(iii) Show that $MSE(S'^2_n) < MSE(S_n^2)$. Thus, (S_n^2) is unbiased but (S'^2_n) has a smaller mean square error. However, it should be noted that the difference is very small and approaches zero as n becomes large.

$$\lim_{n} \frac{Var(\hat{\theta}_n)}{Var\left(\hat{y}_n\right)}.$$

5.10.2 Robust Estimation

The estimators derived in this chapter are for particular parameters of a presumed underlying family of distributions. However, if the choice of the underlying family of distributions is based on past experience, there is a possibility that the true population will be slightly different from the model used to derive the estimators. Formally, a statistical procedure is *robust* if its behavior is relatively insensitive to deviations from the assumptions on which it is based. If the behavior of an estimator is taken as its variance, a given estimator may have minimum variance for the distribution used, but it may not be very good for the actual distribution. Hence, it is desirable for the derived estimators to have small variance over a range of distributions. We call such estimators *robust estimators.* The following illustrates how the variance of an estimator can be affected by deviations from the presumed underlying population model.

Consider estimating the mean of a standard normal distribution. Let $X_1,\ldots,X_n$ be a random sample from a standard normal distribution. Suppose the population actually follows a contaminated normal distribution. That is, for $0 \le \delta \le 1$, 100 $(1-\delta)$ % of the observations come from an $N(0, 1)$ distribution, and the remaining 100δ% of observations come from an $N(0, 5)$ distribution. We already know that the MVUE of the mean μ of an uncontaminated normal distribution is the sample mean. A less effective alternative would be the sample median.

(a) Conduct a simulation study with sample size n that takes, say, 5000 random samples of 100 observations each. Find the mean and median. Also, find the sample variance of each. For various values of δ, say 0.0, 0.01, 0.05, 0.1, 0.2, 0.3, and 0.4, create a table of variances of sample mean and sample variance. Compare the variances as the value of δ increases.

(b) The aim of robust estimation is to derive estimators with variance near that of the sample mean when the distribution is standard normal while having the variance remain relatively stable as δ increases. One such estimator is the $\alpha -$ *trimmed mean.* Let $0 \le \alpha \le 0.5$, and define $k = [n\alpha]$, where $[x]$ is the greatest integer that is less than or equal to x. For the ordered sample, discard the k highest and lowest observations, and find the mean of the remaining $n - k$ observations. That is, let $X_{(1)} \le X_{(2)} \le \ldots < x_{(n)}$ be the ordered sample, and define

$$\overline{X_\alpha} = \frac{X_{(1+k)} + X_{(2+k)} + \ldots + X_{(n+k)}}{n - 2k}.$$

For the values of δ and the samples in part (a), compute the mean and the 0.05-, 0.1-, 0.25-, and 0.5-trimmed means. Discuss the robustness.

5.10.3 Numerical Unbiasedness and Consistency

(a) Run the simulation of a normal experiment with increasing sample size. Numerically show the unbiased and consistent properties of the sample mean. Run the experiment at least up to $n = 1000$.

(b) Repeat the experiment of part (a), now with an exponential distribution.

5.10.4 Averaged Squared Errors

Generate 25 samples of size 40 from a normal population with $\mu = 10$, and $\sigma^2 = 4$. For each of the 25 samples:

(a) Compute: $\overline{x}$, $s^2 = \frac{\sum_{i=1}^{40}(x_i - \overline{x})^2}{39}$, $s_1^2 = \frac{\sum_{i=1}^{40}(x_i - \overline{x})^2}{40}$, and $s_2^2 = \frac{\sum_{i=1}^{40}(x_i - \overline{x})^2}{41}$.

(b) Compute the average squared error (ASE) for each of the estimates s^2, s_1^2, s_2^2 as follows.

Let $K^{s^2} = \left[\left[\sum_{i=1}^{K}(x_i - \overline{x})^2\right]/39\right]$ for $K = 1, 2, \ldots, 25$; and K^{s^2} be the sample variance for the Kth sample. Then, the average squared error is

$$\text{ASE} = \frac{\sum_{i=1}^{25}\left(K^{s^2} - \sigma^2\right)^2}{25}.$$

Repeat this procedure for the other two estimators. Compare the three ASEs and check which has the least ASE.

(c) Repeat (a) and (b) with a sample size of 15.

5.10.5 Alternate Method of Estimating the Mean and Variance

(a) Consider the following alternative method of estimating μ and σ^2. We sample sequentially, and at each stage we compute the estimates of μ and σ^2 as follows.
Let $x_1, \ldots, x_n, x_{n+1}$ be the sample values.
Compute

$$\overline{X}_n = \frac{\sum_{i=1}^{n} X_i}{n}, \ \overline{X}_{n+1} = \frac{\sum_{i=1}^{n+1} X_i}{n+1}, \ S_n^2 = \frac{\sum_{i=1}^{n}(X_i - \overline{X}_n)^2}{n-1}, \text{ and}$$
$$S_{n+1}^2 = \frac{\sum_{i=1}^{n+1}(X_i - \overline{X}_n)^2}{n}.$$

The sequential procedure is stopped when

$$|S_n^2 - S_{n+1}^2| \le 0.01.$$

This will also determine the sample size.
(b) Compare the sample sizes and estimates in 5D and 5E.

5.10.6 Newton–Raphson in One Dimension

For a given function $g(x)$, suppose we need to solve $g(\theta) = 0$. Using the first-order Taylor expansion, $g(\theta) \approx g(x) + (\theta - x)g'(x)$, where $g'(x) = \frac{dg}{dx}$, and setting$g(\theta) = 0$, we get$\theta \approx x - \frac{g(x)}{g'(x)}$. Thus, starting with an initial guess solution x, the guess is updated by θ using the previous formula. This derivation is the basis for the Newton–Raphson iterative method for obtaining the solution of $g(\theta) = 0$. This is given by

$$\theta_{(n+1)} = \theta_n - \frac{g(\theta_n)}{g'(\theta_n)}, \ n \ge 0,$$

where θ_n is the value of θ at the nth iteration, starting with the initial guess, θ_0. For a good approximation of the solution, the choice of θ_0 is important. The convergence of this algorithm cannot be guaranteed.

For the MLE, we want to find a solution of

$$g(\theta) = \frac{dL}{d\theta} = 0,$$

where $L = L(\theta)$ is the likelihood function of the random sample $X_1, \ldots, X_n$. An iterative algorithm for finding the MLE can be given by

$$\theta_{(n+1)} = \theta_n - \frac{\frac{dL}{d\theta}(\theta_n)}{\frac{d^2L}{d\theta^2}(\theta_n)}, \ n \ge 0.$$

Write a computer program to find the MLE of a gamma distribution with parameters α and β.

5.10.7 The Empirical Distribution Function

The estimators in this chapter yield a single real value (point estimate) for each parameter. In Chapter 6, we will learn about so-called interval estimates. In this project, we use an estimation procedure that estimates the whole distribution function, F, of a random variable X. We now define the empirical distribution.

The *empirical distribution function* for a random sample $X_1, \ldots, X_n$ from a distribution F is the function defined by

$$F_n(x) = \frac{1}{n}\#\{i,\ 1 \le i \le n:\ X_i \le x\}.$$

It can be shown that $nF_n(x)$ is a binomial random variable with

$$E[F_n(x)] = F(x) \text{and} Var[F_n(x)] = \frac{1}{n}F(x)[1 - F(x)].$$

Also, by the strong law of large numbers, for each real number x,

$$\lim_{n\to\infty} F_n(x) = F(x) \text{with probability } 1.$$

One of the tests to determine whether a random sample comes from a specific distribution is the Kolmogorov–Smirnov (K-S) test. The K-S test is based on the maximum distance between the empirical distribution function and the actual cumulative distribution function of this specific distribution (such as, say, the normal distribution).

Using the method of Project 4A (or using any statistical software), generate 100 sample points from a normal distribution with mean 2 and variance 9. Graph the empirical distribution function for this sample. Compare this graph with the graph of the $N(2, 9)$ distribution.

5.10.8 Simulation of Coverage of the Small Confidence Intervals for μ

(a) Generate 25 samples of size 15 from a normal population with $\mu = 10$ and $\sigma^2 = 4$. Using a statistical package (such as Software), compute the 95% confidence intervals for each of the samples using the small sample formula. From your output, determine the proportion of the 25 intervals that cover the true mean $\mu = 10$.

(b) What would you expect if the sample size increased to 100? Would the width of the interval increase or decrease? Would you expect more or fewer of these intervals to contain the true mean 10? Check your answers with actual computation.

(c) Repeat with 20 samples of size 10.

5.10.9 Confidence Intervals Based on Sampling Distributions

If we want to obtain a $(1 - \alpha)$ 100% confidence interval for θ, begin with an estimator $\hat{\theta}$ of θ, and determine its sampling distribution. Now, select two probability levels, α_1 and α_2, so that $\alpha = \alpha_1 + \alpha_2$. Generally, we let $\alpha_1 = \alpha_2$. Take a sample, and calculate the value of$\hat{\theta}$, say $\hat{\theta} = k$. Now we need to determine the values of the upper and lower confidence limits. Find a value θ_L such that

$$p(\hat{\theta} \ge k) = \alpha_1$$

and θ_U such that

$$p(\hat{\theta} \le k) = \alpha_2.$$

Then a $(1 - \alpha)$ 100% confidence interval for θ will be

(a) Let $X_1, \ldots, X_n$ be a random sample from $U(0,\theta)$ distribution. Obtain a $(1 - \alpha)$ 100% confidence interval for θ, using the method of sampling distribution.

(b) Let X have a binomial distribution with parameters n and p. First, show that there is no quantity that satisfies the conditions of a pivotal quantity. Then, using the method of sampling distributions, obtain a $(1 - \alpha)$ 100% confidence interval for p.

$$\theta_L < \theta < \theta_U.$$

5.10.10 Large Sample Confidence Intervals: General Case

The method of finding a confidence interval for a parameter θ that we described in this chapter depends on our ability to find the pivotal quantity. We have seen that such a quantity may not exist. In those cases, the method of sampling distribution described in the previous project could be used. However, this method can involve some difficult calculations. For large samples, we can utilize the following procedure, which is based on the asymptotic distribution of MLEs. Under fairly general conditions, the MLEs have a limiting distribution that is normal. Also, MLEs are asymptotically efficient. Hence, for a large sample, the maximum likelihood estimator $\hat{\theta}$ of θ will have approximately normal distribution with mean θ. Also, if the Cramér–Rao lower bound exists, the limiting variance of $\hat{\theta}$ will be

$$\sigma_{\hat{\theta}}^2 = \frac{1}{E\left[\left(\frac{\partial \ln L}{\partial \theta}\right)^2\right]}.$$

Hence,

$$Z = \frac{\hat{\theta} - \theta}{\sigma_{\hat{\theta}}} \sim N(0,1).$$

Then a large sample $(1 - \alpha)$ 100% confidence interval is obtained from the probability statement

$$P\left(-z_{\alpha/2} < \frac{\hat{\theta} - \theta}{\sigma_{\hat{\theta}}} < z_{\alpha/2}\right) \approx 1 - \alpha.$$

We summarize the procedure to construct large sample confidence intervals.

1. Determine the maximum likelihood estimator, $\hat{\theta}$, of θ. Also, find the MLEs of all other unknown parameters.
2. Obtain the variance $\sigma_{\hat{\theta}}$ (if possible, directly, otherwise by using the Cramér–Rao lower bound).
3. In the expression for $\sigma_{\hat{\theta}}$, substitute $\hat{\theta}$ for θ. Replace all other unknown parameters by its MLEs. Let the resulting quantity be denoted by $s_{\hat{\theta}}$.
4. Now construct a $(1 - \alpha)$ 100% confidence interval for θ from

(a) Using the foregoing procedure, show that a large sample $(1 - \alpha)$ 100% confidence interval for the parameter p in a binomial distribution based on n trials is

$$\hat{p} - z_{\alpha/2}\sqrt{\frac{\hat{p}(1 - \hat{p})}{n}} < p < \hat{p} + z_{\alpha/2}\sqrt{\frac{\hat{p}(1 - \hat{p})}{n}}.$$

(b) Let $X_1, \ldots, X_n$ be a random sample from a normal population with parameters μ and σ^2. Derive a large sample confidence interval for σ^2 using the above procedure.

(c) Let $X_1, \ldots, X_n$ be a random sample from a population with a pdf given by

$$\hat{\theta} - z_{\alpha/2}s_{\hat{q}} < \theta < \hat{\theta} + z_{\alpha/2}s_{\hat{q}}.$$

$$f(x) = \begin{cases} \frac{1}{\theta}e^{-x/\theta}, & x > 0 \\ 0, & \text{otherwise.} \end{cases}$$

Derive a large sample confidence interval for θ.

5.10.11 Prediction Interval for an Observation from a Normal Population

In many cases, we may be interested in predicting future observations from a population, rather than making an inference. A $(1 - \alpha)$ 100% *prediction interval* for a future observation X is an interval of the form (X_L, x_u) such that $P(X_L < X < X_U) = 1 - \alpha$. Similarly to confidence intervals, we can also define one-sided prediction intervals. Assume that the population is normal with known variance σ^2. Let $X_1,\ldots,X_n$ be a random sample from this population. Then, the sampling distribution of the difference $X - \overline{X}$ (we use $\overline{X}$ to denote $\overline{X}_n$) is normal with mean zero and variance $\sigma^2 + \sigma^2_{\overline{X}} = (1 + (1/n))\sigma^2$. Then, a $(1 - \alpha)$ 100% prediction interval for X is given by

$$\left(\overline{X} - z_{\alpha/2}\sqrt{\left(1 + \frac{1}{n}\right)\sigma^2}, \overline{X} + z_{\alpha/2}\sqrt{\left(1 + \frac{1}{n}\right)\sigma^2}\right).$$

Thus we are $(1 - \alpha)$ 100% confident that the next observation, X_{n+1}, will lie in this interval. As in confidence intervals, if the sample size is large, replace σ by sample standard deviation s.

In the case where both μ and σ are not known, and the sample size is small (so that the Central Limit Theorem cannot be applied), it can be shown that $\left[(X_{n+1} - \overline{X}_n)/(S_n\sqrt{1 + (1/n)}) \right]$ has a t-distribution with $(n - 1)$ degrees of freedom. Thus a $(1 - \alpha)$ 100% prediction interval for X_{n+1} is given by

$$\left(\overline{X} - t_{\alpha/2,n-1}\sqrt{(1 + (1/n))S^2}, \overline{X} + t_{\alpha/2,n-1}\sqrt{(1 + (1/n))S^2}\right).$$

A standard measure of the capacity of lungs to expel air in breathing is called forced expiratory volume (FEV). The FEV1 is the volume exhaled during the first second of a forced expiratory maneuver started from the level of total lung capacity. The following data (source: M. Bland, *An Introduction to Medical Statistics*, Oxford University Press, 1995) represents FEV measurements (in liters) from 57 male medical students:

4.47	3.10	4.50	4.90	3.50	4.14	4.32	4.80	3.10	4.68
4.47	3.57	2.85	5.10	5.20	4.80	5.10	4.30	4.70	4.08
3.48	4.20	3.70	5.30	4.71	4.10	4.30	3.39	3.69	4.44
5.00	4.50	4.20	4.16	3.70	3.83	3.90	4.47	3.30	5.43
3.42	3.60	3.20	4.56	4.78	3.60	3.96	3.19	2.85	3.04
3.78	3.75	4.05	3.54	4.14	2.98	3.54			

Obtain a 95% prediction interval for a future observation X_{n+1}.

5.10.12 Empirical Distribution function (EDF) as Estimator for Cumulative Distribution Function (CDF)

In Chapter 3, we saw that probabilistically the cumulative distribution function defined as $F(x) = P(X \le x), \textit{for all } x \in (-\infty, \infty)$. The question is, given a random sample $X_1, \ldots, X_n$ with common CDF $F(x)$, how do we estimate the CDF, $F(x)$? The EDF is the "data analogue" of CDF of a random variable. The EDF is defined as

$$\widehat{F}_n(x) = \frac{\text{number of elements in the sample} \le \text{x}}{n} = \frac{1}{n}\sum_{i=1}^{n} I_{(X_i \le x)},$$

where I_A is the indicator of event A. Note that EDF is a step function that jumps $\frac{1}{n}$ at each $X_i = x_i$. We can see that EDF describes the data in more detail than the histogram. Using the strong law of large numbers, it can be shown that $\widehat{F}_n(x) \to F(x)$, with probability 1. We refer readers to look for other interesting properties of EDF.

Using R, create 100 data values from each of uniform and normal distributions. Draw the theoretical CDF and EDF on the same graph for each of the distributions, respectively.

Chapter 6

Hypothesis Testing

Chapter outline

OBJECTIVE

In this chapter, various methods of testing hypotheses will be discussed.

Jerzy Neyman
(Source: *http://sciencematters.berkeley.edu/archives/volume2/issue12/legacy.php*).

Jerzy Neyman (1894–1981) was a Polish statistician and mathematician who, after spending time in various institutions in Warsaw, Poland, came to the University of California, Berkeley. He had made far-reaching contributions in hypothesis testing, confidence intervals, probability theory, and other areas of mathematical statistics. His work with Egon Pearson gave a logical foundation and mathematical rigor to the theory of hypothesis testing. Neyman made a broader impact in statistics throughout his lifetime.

6.1 INTRODUCTION

Statistics play an important role in decision-making. In statistics, one utilizes random samples to make inferences about the population from which the samples were obtained. Statistical inference regarding population parameters takes two

Mathematical Statistics with Applications in R. DOI: https://doi.org/10.1016/B978-0-443-27548-7.00006-2

forms, estimation and hypothesis testing, although both may be viewed as different aspects of the same general problem of arriving at decisions on the basis of observed data. We already saw several estimation procedures in earlier chapters. Hypothesis testing is the subject of this chapter. This has an important role in the application of statistics to real-life problems. Here we utilize the sampled data to make decisions concerning the unknown distribution of a population or its parameters. Pioneering work on the explicit formulation as well as the fundamental concepts of the theory of hypothesis testing is due to J. Neyman and E. S. Pearson.

A statistical hypothesis is a statement concerning the probability distribution of a random variable or population parameters that are inherent in a probability distribution. The following example illustrates the concept of hypothesis testing. An important industrial problem is that of accepting or rejecting lots of manufactured products. Before releasing each lot for the consumer, the manufacturer usually performs some tests to determine whether the lot conforms to acceptable standards. Let us say that both the manufacturer and the consumer agree that if the proportion of defectives in a lot is less than or equal to a certain number p, the lot will be released. Very often, instead of testing every item in the lot, we may test only a few at random from the lot and make decisions about the proportion of defective items in the lot; that is, we make the decisions about the population on the basis of sample information. Such decisions are called *statistical decisions*. In attempting to reach decisions, it is useful to make some initial conjectures about the population involved. Such conjectures are called statistical hypotheses. Sometimes the results from the sample may be markedly different from those expected under the hypothesis. Then we can say that the observed differences are significant, and we would be inclined to reject the initial hypothesis. The procedures that enable us to decide whether to reject hypotheses or to determine whether observed samples differ significantly from expected results are called *tests of hypotheses, tests of significance, or rules of decision*.

In any hypothesis testing problem, we formulate a *null hypothesis* and an *alternative hypothesis* such that if we reject the null, then we have to accept the alternative. The null hypothesis usually is a statement of either the "status quo" or "no effect" or a "belief." A guideline for selecting a null hypothesis is that when the objective of an experiment is to establish a claim, the nullification of the claim should be taken as the null hypothesis. The experiment is often performed to determine whether the null hypothesis is false. For example, suppose the prosecution wants to establish that a certain person is guilty. The null hypothesis would be that the person is innocent, and the alternative would be that the person is guilty. Thus the claim itself becomes the alternative hypothesis. Customarily, the alternative hypothesis is the statement that the experimenter believes to be true. For example, the alternative hypothesis is the reason a person is arrested (police suspect the person is not innocent). Once the hypotheses have been stated, appropriate statistical procedures are used to determine whether to reject the null hypothesis. For the testing procedure, one begins with the assumption that the null hypothesis is true. If the information furnished by the sampled data strongly contradicts (beyond a reasonable doubt) the null hypothesis, then we reject it in favor of the alternative hypothesis. If we do not reject the null, then we automatically reject the alternative. Note that we always make a decision with respect to the null hypothesis. Failure to reject the null hypothesis does not necessarily mean that the null hypothesis is true. For example, a person being judged "not guilty" does not mean the person is innocent. This basically means that there is not enough evidence to reject the null hypothesis (presumption of innocence) beyond "a reasonable doubt."

We summarize the elements of a statistical hypothesis in the following.

The elements of a statistical hypothesis

1. The *null hypothesis*, denoted by H_0, is usually the nullification of a claim. Unless evidence from the data indicates otherwise, the null hypothesis is assumed to be true.
2. The *alternate hypothesis*, denoted by H_a (or sometimes denoted by H_1), is customarily the claim itself.
3. The *test statistic*, denoted by *TS*, is a function of the sample measurements upon which the statistical decision, to reject or not to reject the null hypothesis, will be based.
4. A *rejection region* (or a *critical region*) is the region (denoted by RR) that specifies the values of the observed test statistic for which the null hypothesis will be rejected. This is the range of values of the test statistic that corresponds to the rejection of H_0 at some fixed level of significance, α, which will be explained later.
5. **Conclusion**: If the value of the observed test statistic falls in the rejection region, the null hypothesis is rejected, and we will conclude that there is enough evidence to decide that the alternative hypothesis is true. If the TS does not fall in the rejection region, we conclude that we cannot reject the null hypothesis.

In practice, one may have hypotheses such as H_0: $\mu = \mu_0$ against one of the following alternatives:

$$\begin{cases} H_a: \mu \neq \mu_0, \text{ called a two-tailed alternative.} \\ \text{or} \\ H_a: \mu < \mu_0, \text{ called a lower(or left)tailed alternative.} \\ \text{or} \\ H_a: \mu > \mu_0, \text{ called an upper(or right)tailed alternative.} \end{cases}$$

A test with a lower or upper tailed alternative is called a *one-tailed test.* One of the issues in hypothesis testing is the choice of the form of alternative hypothesis. Note that, as we discussed earlier, the null hypothesis is always concerned with the question posed—the claim. The alternative hypothesis must reflect the aim of the claim when, in fact, we reject the claim; we want to know why we rejected it. For example, suppose that a pharmaceutical company claims that drug A is 80% effective (i.e., $p = 0.8$). We conduct an experiment—clinical trials—to test this claim. Thus the null hypothesis is that the claim is true. Now, if we don't want to reject the null hypothesis, no problem, but if we reject the null hypothesis, we want to know why. Thus the alternative must be a one-tailed test, $p < 0.8$; that is, the claim is not true. If we were to use a two-tailed test, we wouldn't know whether the rejection was due because $p < 0.8$ or $p > 0.8$. $p > 0.8$ in this case is actually part of the null hypothesis. It is important to note that when using a one-sided test in a certain direction, if the consequence of missing an effect in the other direction is not negligible, it is better to use a two-sided test. Also, choosing a one-tailed test after doing a two-tailed test that failed to reject the null hypothesis is not appropriate. Therefore the choice of alternative is based on what happens if we reject the null hypothesis. In an applied hypothesis testing problem, we can use the following general steps.

General method for hypothesis testing

1. From the (word) problem, determine the appropriate null hypothesis, H_0, and the alternative, H_a.
2. Identify the appropriate test statistics and calculate the observed test statistic from the data.
3. Find the rejection region by looking up the *critical value* in the appropriate table.
4. Draw the conclusion: Reject or fail to reject the null hypothesis, H_0, based on a given level of significance α.
5. Interpret the results: State in words what the conclusion means to the problem we started with.

It is always necessary to state a null and an alternate hypothesis for every statistical test performed. All possible outcomes should be accounted for by the two hypotheses. Note that a critical value is the value that a test statistic must surpass for the null hypothesis to be rejected and is derived from the level of significance α of the test. Thus the critical values are the boundaries of the rejection region. It is important to observe that both null and alternate hypotheses are stated in terms of parameters, not in terms of statistics.

EXAMPLE 6.1.1

In a coin-tossing experiment, let p be the probability of heads. We start with the claim that the coin is fair, that is, H_0: $p = 1/2$. We test this against one of the following alternatives:

(a) H_a: The coin is not fair ($p \neq 1/2$). This is a two-tailed alternative.
(b) H_a: The coin is biased in favor of heads ($p > 1/2$). This is an upper tailed alternative.
(c) H_a: The coin is biased in favor of tails ($p < 1/2$). This is a lower tailed alternative.

It is important to observe that the test statistic is a function of a random sample. Thus the test statistic itself is a random variable whose distribution is known under the null hypothesis. The value of a test statistic when specific sample values are substituted is called the observed test statistic or simply the test statistic.

For example, consider the hypothesis H_0: $\mu = \mu_0$ versus H_a: $\mu \neq \mu_0$, where μ_0 is known. Assume that the population is normal with a known variance σ^2. Consider $\overline{X}$, an unbiased estimator of μ based on the random sample, $X_1, \ldots, X_n$. Then $Z = (\overline{X} - \mu_0)/(\sigma/\sqrt{n})$ is a function of the random sample $X_1, \ldots, X_n$, and has a known distribution, say a standard normal, under H_0. If $x_1, x_2, \ldots, x_n$ are specific sample values, then $z = (\overline{x} - \mu_0)/(\sigma/\sqrt{n})$ is called the *observed sample statistic* or simply *the sample statistic*.

Definition 6.1.1. *A hypothesis is said to be a* **simple hypothesis** *if that hypothesis uniquely specifies the distribution from which the sample is taken. Any hypothesis that is not simple is called a* **composite hypothesis**.

Example 6.1.2

Refer to Example 6.1.1. The null hypothesis $p = 1/2$ is simple, because the hypothesis completely specifies the distribution, which in this case will be a binomial with $p = 1/2$ and with n being the number of tosses. The alternative hypothesis $p \neq 1/2$ is composite because the distribution now is not completely specified (we do not know the exact value of p).

Because the decision is based on the sample information, we are prone to commit errors. In a statistical test, it is impossible to establish the truth of a hypothesis with 100% certainty. There are two possible types of errors. On the one hand, one can make an error by rejecting H_0 when in fact it is true. On the other hand, one can also make an error by failing to reject the null hypothesis when in fact it is false. Because the errors arise as a result of wrong decisions, and the decisions themselves are based on random samples, it follows that the errors have probabilities associated with them. We now have the following definitions.

The decision and the errors are represented in Table 6.1.

TABLE 6.1 Statistical Decision and Error Probabilities.

Statistical Decision	True State of Null Hypothesis	
	H_0 True	**H_0 False**
Do not reject H_0	Correct decision	Type II error (β)
Reject H_0	Type I error (α)	Correct decision

Definition 6.1.2.

(a) A **type I error** *is made if H_0 is rejected when in fact H_0 is true. The probability of a type I error is denoted by α. That is,*

$$P(\textit{rejecting } H_0 | H_0 \textit{ is true}) = \alpha.$$

The probability of a type I error, α, is called **the level of significance**.

(b) A **type II error** *is made if H_0 is accepted when in fact H_a is true. The probability of a type II error is denoted by β. That is,*

$$P(\textit{not rejecting } H_0 | H_0 \textit{ is false}) = \beta.$$

It is desirable that a test should have $a = \beta = 0$ (this can be achieved only in trivial cases), or at least we prefer to use a test that minimizes both types of errors. Unfortunately, it so happens that for a fixed sample size, as α decreases, β tends to increase and vice versa. There are no hard and fast rules that can be used to make the choice of α and β. This decision must be made about each problem based on quality and economic considerations. However, in many situations, it is possible to determine which of the two errors is more serious. It should be noted that a type II error is only an error in the sense that a chance to correctly reject the null hypothesis was lost. It is not an error in the sense that an incorrect conclusion was drawn, because no conclusion is made when the null hypothesis is not rejected. In the case of a type I error, a conclusion is drawn that the null hypothesis is false when, in fact, it is true. Therefore type I errors are generally considered more serious than type II errors. For example, it is mostly agreed that finding an innocent person guilty is a more serious error than finding a guilty person innocent. Here, the null hypothesis is that the person is innocent, and the alternate hypothesis is that the person is guilty. "Not rejecting the null hypothesis" is equivalent to acquitting a defendant. It does not prove that the null hypothesis is true or that the defendant is innocent. In statistical testing, the significance level α is the probability of wrongly rejecting the null hypothesis when it is true (i.e., the risk of finding an innocent person guilty). Here, the type II risk is acquitting a guilty defendant. The usual approach to hypothesis testing is to find a test procedure that limits α, the probability of type I error, to an acceptable level while trying to lower β as much as possible. In hypothesis testing, a *false positive* (type I error) occurs when we incorrectly reject a true null hypothesis. A false negative (type II error) is when we fail to reject a false null hypothesis, thus, missing a real effect that does exist. For instance, in a blood test for a certain disease, a false positive indicates a disease is present when it is not, while a false negative indicates a disease is absent when it is truly present.

The consequences of different types of errors are, in general, very different. For example, if a doctor tests for the presence of a certain illness, incorrectly diagnosing the presence of the disease (type I error) will cause a waste of resources, not to mention mental agony to the patient. On the other hand, failure to determine the presence of the disease (type II error) can lead to a serious health risk.

To formulate a hypothesis testing problem, consider the following situation. Suppose a toy store chain claims that at least 80% of girls under 8 years old prefer dolls over other types of toys. We feel that this claim is inflated. In an attempt to dispose of this claim, we observe the buying pattern of 20 randomly selected girls under 8 years old. and we observe X, the number of girls under 8 years old who buy stuffed toys or dolls. Now the question is, how can we use X to confirm or reject the store's claim? Let p be the probability that a girl under 8 chosen at random prefers stuffed toys or dolls. The question now can be reformulated as a hypothesis testing problem. Is $p \geq 0.8$ or $p < 0.8$? Because we would like to reject the store's claim only if we are highly certain of our decision, we should choose the null hypothesis to be H_0: $p \geq 0.8$, the rejection of which is considered to be more serious. The null hypothesis should be H_0: $p \geq 0.8$, and the alternative H_a: $p < 0.8$. To make the null hypothesis simple, we will use H_0: $p = 0.8$, which is the boundary value with the understanding that it really represents H_0: $p \geq 0.8$. We note that X, the number of girls under 8 years old who prefer stuffed toys or dolls, is a binomial random variable. Clearly, a large sample value of X would favor H_0. Suppose we arbitrarily choose to accept the null hypothesis if $X > 12$. Because our decision is based on only a sample of 20 girls under 8, there is always a possibility of making errors, whether we accept or reject the store chain's claim. In the following example, we will now formally state this problem and calculate the error probabilities based on our decision rule.

EXAMPLE 6.1.3

A toy store chain claims that at least 80% of girls under 8 years old prefer dolls over other types of toys. After observing the buying pattern of many girls under 8 years old, we feel that this claim is inflated. In an attempt to dispose of this claim, we observe the buying pattern of 20 randomly selected girls under 8 years old, and we observe x, the number of girls who buy stuffed toys or dolls. We wish to test the hypothesis H_0: $p = 0.8$ against H_a: $p < 0.8$. Suppose we decide to accept H_0 if $X > 12$ (i.e., $X \geq 13$). This means that if $\{X \leq 12\}$ (i.e., $X < 13$), we will reject H_0.

(a) Find α.
(b) Find β for $p = 0.6$.
(c) Find β for $p = 0.4$.
(d) Find the rejection region of the form $\{X \leq K\}$ so that (i) $\alpha = 0.01$; (ii) $\alpha = 0.05$.
(e) For the alternative H_a: $p = 0.6$, find β for the values of α in part (d).

Solution

The TS X is the number of girls under 8 years old who buy dolls. X follows the binomial distribution with n=20 and p, the unknown population proportion of girls under 8 who prefer dolls. We now calculate α and β.

(a) *For p=0.8, the probability of type I error is*

$$\begin{aligned}\alpha &= P\{\text{reject}\,H_0|H_0\,\text{is true}\}\\ &=P\{X \le 12|p = 0.8\}\\ &= \sum_{x=0}^{12}\binom{20}{x}(0.8)^x(0.2)^{20-x}\\ &=0.0321.\end{aligned}$$

If we calculate α for any other value of $p > 0.8$, then we will find that it is smaller than 0.0321. Hence, there is at most a 3.21% chance of rejecting a true null hypothesis. That is, if the store's claim is in fact true, then the chance that our test will erroneously reject that claim is at most 3.21%.

(b) *Here, p=0.6. The probability of type II error is*

$$\begin{aligned}\beta &= P\{\text{accept}\,H_0|H_0\,\text{false}\}\\ &=P\{X > 12|p = 0.6\}\\ &=1 - P\{X \le 12|p = 0.6\}\\ &=1 - 0.584\\ &=0.416\end{aligned}$$

that is, there is a 41.6% chance of accepting a false null hypothesis. Thus, in case the store's claim is not true, and the truth is that only 60% of the girls under 8 years old prefer dolls over other types of toys, then there is a 41.6% chance that our test will erroneously conclude that the store's claim is true.

(c) *If p=0.4, then*

$$\begin{aligned}\beta &= P\{\text{accept}\,H_0|H_0\,\text{false}\}\\ &=P\{X > 12|p = 0.4\}\\ &=1 - P\{X \le 12|p = 0.4\}\\ &=1 - 0.979\\ &=0.021.\end{aligned}$$

That is, there is a 2.1% chance of not rejecting a false null hypothesis.

(d)

(i). *To find K such that*

$$\alpha = P\{X \le K|p = 0.8\} = 0.01,$$

from the binomial table, K=11. Hence, the rejection region is: Reject H_0 if $\{X \le 11\}$.

(ii). *To find K such that*

$$\alpha = P\{X \le K|p = 0.8\} = 0.05.$$

From the binomial table, α=0.05 falls between K=12 and K=13. However, for K=13, the value for α is 0.087, exceeding 0.05. If we want to limit α to be no more than 0.05, we will have to take K=12. That is, we reject the null hypothesis if $X \le 12$, yielding an α=0.0321 as shown in (a).

(e)

(i) *When a=0.01, from (d), the rejection region is of the form {X ≤ 11}. For p=0.6,*

$$\begin{aligned}\beta &= P\{\text{accept} H_0|H_0 \text{ false}\}\\ &=P\{Y > 11|p = 0.6\}\\ &=1 - P\{Y \le 11|p = 0.6\}\\ &=1 - 0.404\\ &=0.596.\end{aligned}$$

(ii) *From (a) and (b) for testing the hypothesis H_0: p=0.8 against H_a: p < 0.8 with n=20. We see that when α is 0.0321, β is 0.416. From (d)(i) and (e)(i) for the same hypothesis, we see that when α is 0.01, β is 0.596. This holds in general. Thus, we observe that for fixed n as α decreases, β increases and vice versa.*

In the next example, we explore what happens to β as the sample size increases, with α fixed.

EXAMPLE 6.1.4

Let X be a binomial random variable. We wish to test the hypothesis H_0: p =0.8 against H_a: p =0.6. Let α =0.03 be fixed. Find β for n =10 and n =20.

Solution

For n =10, using the binomial tables, we obtain $P\{X \le 5|p = 0.8\} \cong 0.03$. Hence, the rejection region for the hypothesis H_0: p = 0.8 versus H_a: p = 0.6 is given by reject H_0 if $X \le 5$. The probability of type II error is

$$\beta = P\{\text{accept } H_0|p = 0.6\}$$
$$\beta = P\{X > 5|p = 0.6\} = 1 - P\{X \le 5|p = 0.6\} = 0.633.$$

For n =20, as shown in Example 6.1.3, if we reject H_0 for $X \le 12$, we obtain

$$P(X \le 12|p = 0.8) \cong 0.03$$

and

$$\beta = P(X > 12|p = 0.6) = 1 - P\{X \le 12|p = 0.6\} = 0.416.$$

We see that for a fixed α, as n increases, β decreases and vice versa. It can be shown that this result holds in general.

For us to compute the value of β, it is necessary that the alternate hypothesis is simple. Now we will discuss a three-step procedure to calculate β.

Steps to calculate β

1. Decide an appropriate test statistic (usually this is a sufficient statistic or an estimator for the unknown parameter, whose distribution is known under H_0).
2. Determine the rejection region using a given α and the distribution of the TS.
3. Find the probability that the observed test statistic does not fall in the rejection region, assuming H_a is true. This gives β. That is,

β=P(*T.S.* falls in the complement of the rejection region | H_a is true).

EXAMPLE 6.1.5

A random sample of size 36 from a population with known variance, σ^2=9, yields a sample mean of $\bar{x}$ = 17. For the hypothesis H_0: μ =15 versus H_a: μ > 15, find β when μ =16. Assume α =0.05.

Solution

Here $n = 36$, $\bar{x} = 17$, and $\sigma^2 = 9$. In general, to test $H_0: \mu = \mu_0$ versus $H_a: \mu > \mu_0$, we proceed as follows. An unbiased estimator of μ is $\bar{X}$. Intuitively, we would reject H_0 if $\bar{X}$ is large, say $\bar{X} > c$. Now using $\alpha = 0.05$, we will determine the rejection region. By the definition of α, we have

$$P(\bar{X} > c | \mu = \mu_0) = 0.05$$

or

$$P\left(\frac{\bar{X} - \mu_0}{\sigma/\sqrt{n}} > \frac{c - \mu_0}{\sigma/\sqrt{n}} | \mu = \mu_0\right) = 0.05$$

But if $\mu = \mu_0$, because the sample size $n \geq 30$, $[(\bar{X} - \mu_0)/(\sigma/\sqrt{n})] \sim N(0,1)$. Therefore $P\left(\frac{\bar{X} - \mu_0}{(\sigma/\sqrt{n})} > \frac{c - \mu_0}{(\sigma/\sqrt{n})}\right) = 0.05$ is equivalent to $P\left(Z > \frac{c - \mu_0}{(\sigma/\sqrt{n})}\right) = 0.05$. From standard normal tables, we obtain $P(Z > 1.645) = 0.05$. Hence $\frac{c - \mu_0}{(\sigma/\sqrt{n})} = 1.645$ or $c = \mu_0 + 1.645(\sigma/\sqrt{n})$. Therefore the rejection region is the set of all sample means $\bar{x}$ such that

$$\bar{x} > \mu_0 + 1.645\left(\frac{\sigma}{\sqrt{n}}\right).$$

Substituting $\mu_0 = 15$, and $\sigma = 3$, we obtain

$$\mu_0 + 1.645(\sigma/\sqrt{n}) = 15 + 1.645\left(\frac{3}{\sqrt{36}}\right) = 15.8225.$$

The rejection region is the set of $\bar{x}$ such that $\bar{x} \geq 15.8225$.
Then by definition,

$$\beta = P(\bar{X} \leq 15.8225 \text{ when } \mu = 16).$$

Consequently, for $\mu = 16$,

$$\begin{aligned}\beta &= P\left(\frac{\bar{X} - 16}{\sigma/\sqrt{n}} \leq \frac{15.8225 - 16}{3/\sqrt{36}}\right)\\ &= P(Z \leq -0.36)\\ &= 0.3594.\end{aligned}$$

That is, under the given information, there is a 35.94% chance of not rejecting a false null hypothesis.

6.1.1 Sample Size

It is clear from the preceding example that once we are given the sample size n, an α, a simple alternative H_a, and a test statistic, we have no control over β. Hence, for a given sample size and the test statistic, any effort to lower β will lead to an increase in α and vice versa. This means that, for a test with a fixed sample size, it is not possible to simultaneously reduce both α and β. We also notice from Example 6.1.4 that by increasing the sample size n, we can decrease β (for the same α) to an acceptable level. The following discussion illustrates that it may be possible to determine the sample size for a given α and β.

Suppose we want to test $H_0: \mu = \mu_0$ versus $H_a: \mu > \mu_0$. Given α and β, we want to find n, the sample size, and K, the point at which the rejection begins. We know that

$$\begin{aligned}\alpha &= P(\bar{X} > K, \text{ when } \mu = \mu_0)\\ &= P\left(\frac{\bar{X} - \mu_0}{\sigma/\sqrt{n}} > \frac{K - \mu_0}{\sigma/\sqrt{n}}, \text{ when } \mu = \mu_0\right)\\ &= P(Z > z_\alpha)\end{aligned} \tag{6.1}$$

and for some particular value $\mu = \mu_a > \mu_0$,

$$\begin{aligned}\beta &= P(\overline{X} \le K, \text{ when } \mu = \mu_a)\\ &= P\left(\frac{\overline{X}-\mu_a}{\sigma/\sqrt{n}} \le \frac{K-\mu_a}{\sigma/\sqrt{n}}, \text{ when } \mu = \mu_a\right)\\ &= P(z \le z_\beta).\end{aligned} \tag{6.2}$$

From Eqs. (6.1) and (6.2),

$$z_\alpha = \frac{K-\mu_0}{\sigma/\sqrt{n}}$$

and

$$-z_\beta = \frac{K-\mu_a}{\sigma/\sqrt{n}}.$$

This gives us two equations with two unknowns (K and n), and we can proceed to solve them. Eliminating K, we get

$$\mu_0 + z_\alpha\left(\frac{\sigma}{\sqrt{n}}\right) = \mu_a - z_\beta\left(\frac{\sigma}{\sqrt{n}}\right).$$

From this, we can derive

$$\sqrt{n} = \frac{(z_\alpha + z_\beta)\sigma}{\mu_a - \mu_0}.$$

Thus the sample size for an upper tailed alternative hypothesis is

$$n = \frac{(z_\alpha + z_\beta)^2\sigma^2}{(\mu_a - \mu_0)^2}.$$

The sample size increases with the square of the standard deviation and decreases with the square of the difference between the mean value of the alternative hypothesis and the mean value under the null hypothesis. Note that in real-world problems, care should be taken in the choice of the value of μ_a for the alternative hypothesis. It may be tempting for a researcher to take a large value of μ_a to reduce the required sample size. This will seriously affect the accuracy (power) of the test. This alternative value must be realistic within the experiment under study. Care should also be taken in the choice of the standard deviation σ. Using an underestimated value of the standard deviation to reduce the sample size will result in inaccurate conclusions similar to overestimating the difference of means. Usually, the value of σ is estimated using a similar study conducted earlier. The problem could be that the previous study may be old and may not represent the new reality. When accuracy is important, it may be necessary to conduct a pilot study only to get some idea of the estimate of σ. Once we determine the necessary sample size, we must devise a procedure by which the appropriate data can be randomly obtained. This aspect of the design of experiments is discussed in Chapter 9.

EXAMPLE 6.1.6

Let $\sigma = 3.1$ be the true standard deviation of the population from which a random sample is chosen. How large should the sample size be for testing H_0: $\mu = 5$ versus H_a: $\mu = 5.5$, in order that $\alpha = 0.01$ and $\beta = 0.05$?

Solution

We are given $\mu_0 = 5$ and $\mu_a = 5.5$. *Also,* $z_\alpha = z_{0.01} = 2.33$ *and* $z_\beta = z_{0.05} = 1.645$. *Hence, the sample size*

$$n = \frac{(z_\alpha + z_\beta)^2\sigma^2}{(\mu_a - \mu_0)^2} = \frac{(2.33 + 1.645)^2(3.1)^2}{(0.5)^2} = 607.37.$$

So, $n = 608$ *will provide the desired levels. That is, for us to test the foregoing hypothesis, we must randomly select 608 observations from the given population.*

From a practical standpoint, the researcher typically chooses α, and the sample size, β, is ignored. Because a trade-off exists between α and β, choosing a very small value of α will tend to increase β in a serious way. A general rule of thumb is to pick reasonable values of α, possibly in the 0.05 to 0.10 range, so that β will remain reasonably small.

EXERCISES 6.1

6.1.1 An appliance manufacturer is considering the purchase of a new machine for stamping out sheet metal parts. If μ_0 (given) is the true average of the number of good parts stamped out per hour by their old machine and μ is the corresponding true unknown average for the new machine, the manufacturer wants to test the null hypothesis $\mu = \mu_0$ versus a suitable alternative. What should the alternative be if he does not want to buy the new machine unless it is (a) more productive than the old one or (b) at least 20% more productive than the old one?

6.1.2 Formulate an alternative hypothesis for each of the following null hypotheses.

(a) H_0: Support for a presidential candidate is unchanged after starting the use of TV commercials.

(b) H_0: The proportion of viewers watching a particular local news channel is less than 30%.

(c) H_0: The median grade point average of undergraduate mathematics majors is 2.9.

6.1.3 It is suspected that a coin is not balanced (not fair). Let p be the probability of tossing a head. To test H_0: $p=0.5$ against the alternative hypothesis H_a: $p > 0.5$, a coin is tossed 15 times. Let Y equal the number of times a head is observed in the 15 tosses of this coin. Assume the rejection region to be $\{Y \geq 10\}$.

(a) Find α.

(b) Find β for $p = 0.7$.

(c) Find β for $p = 0.6$.

(d) Find the rejection region for $\{Y \geq K\}$ for $\alpha = 0.01$ and $\alpha = 0.03$.

(e) For the alternative H_a: $p = 0.7$, find β for the values of α given in (d).

6.1.4 In Exercise 6.1.3:

(a) Assume that the rejection region is $\{Y \geq 8\}$. Calculate α and β if $p = 0.6$. Compare the results with the corresponding values obtained in Exercise 6.1.3. (This gives the effect of enlarging the rejection region on α and β.)

(b) Assume that the rejection region is $\{Y \geq 8\}$. Calculate α and β if $p = 0.6$, and (i) the coin is tossed 20 times, or (ii) the coin is tossed 25 times. (This shows the effect of increasing the sample size on α and β for a fixed rejection region.)

6.1.5 Suppose we have a random sample of size 25 from a normal population with an unknown mean μ and a standard deviation of 4. We wish to test the hypothesis H_0: $\mu=10$ versus H_a: $\mu > 10$. Let the rejection region be defined by rejecting H_0 if the sample mean $\overline{X} > 11.2$.

(a) Find α.

(b) Find β for H_a: $\mu=11$.

(c) What should the sample size be if $\alpha = 0.01$ and $\beta = 0.2$?

6.1.6 A process for making steel pipe is under control if the diameter of the pipe has a mean of 3.0 in. with a standard deviation of no more than 0.0250 in. To check whether the process is under control, a random sample of size $n = 30$ is taken each day, and the null hypothesis $\mu = 3.0$ is rejected if $\overline{X}$ is less than 2.9960 or greater than 3.0040. Find (a) the probability of type I error and (b) the probability of type II error when $\mu = 3.0050$ in. and $\sigma = 0.0250$ in.

6.1.7 A bowl contains 20 balls, of which x are green and the remainder red. To test H_0: $x=10$ versus H_a: $x=15$, three balls are selected at random without replacement, and H_0 is rejected if all three balls are green. Calculate α and β for this test.

6.1.8 Suppose we have a sample of size 6 from a population with pdf $f(x) = (1/\theta)e^{-x/\theta}$, $x > 0$, $\theta > 0$. We wish to test H_0: $\theta=1$ versus H_a: $\theta > 1$. Let the rejection region be defined by rejecting H_0 if $\Sigma_{i=1}^{6} X_i > 8$. (a) Find α. (b) Find β for H_a: $\theta = 2$.

6.1.9 Let $\sigma^2 = 16$ be the variance of a normal population from which a random sample is chosen. How large should the sample size be for testing H_0: $\mu = 25$ versus H_a: $\mu = 24$, in order that $\alpha = 0.05$ and $\beta = 0.05$?

6.2 THE NEYMAN–PEARSON LEMMA

In practical hypothesis testing situations, there are typically many tests possible with significance level α (which is also called the *size of the test*) for a null hypothesis versus an alternative hypothesis (see Project 6A). This leads to some important questions, such as (1) how to decide on the test statistic and (2) how to know that we selected the best rejection region. In this section, we study the answer to these questions using the Neyman–Pearson approach introduced by Jerzy Neyman and Egon Pearson in a paper published in 1933.

Definition 6.2.1. *Suppose that W is the test statistic and RR is the rejection region for a test of the hypothesis concerning the value of a parameter θ. Then the power of the test is the probability that the test rejects H_0 when the alternative is true. That is,*

$$\pi = Power(\theta)$$
$$= P(W \text{ in } RR \text{ when the parameter value is an alternative } \theta).$$

If H_0: $\theta=\theta_0$ and H_a: $\theta \neq \theta_0$, then the power of the test for some $\theta=\theta_1 \neq \theta_0$ is

$$Power(\theta_1) = P(\text{reject } H_0|\theta = \theta_1).$$

But, $\beta(\theta_1)=P(\text{accept } H_0|\theta=\theta_1)$. Therefore,

$$Power(\theta_1) = 1 - \beta(\theta_1).$$

In other words, *power* refers to the probability that the test will find a statistically significant difference when such a difference actually exists. A good test will have high power. In statistical tests, it is generally accepted that the power should be 8 or greater. For instance, in an AI algorithm, the power of a test will give the probability that a statistical test will correctly detect a real bias in the algorithm.

Note that the power of a test H_0 cannot be found until some true situation H_a is specified. That is, the sampling distribution of the test statistic when H_a is true must be known or assumed. Because β depends on the alternative hypothesis, which, being composite most of the time, does not specify the distribution of the test statistic, it is important to observe that the experimenter cannot control β. For example, the alternative H_a: $\mu < \mu_0$ does not specify the value of μ, as in the case of the null hypothesis, H_0: $\mu = \mu_0$.

EXAMPLE 6.2.1

Let $X_1, \ldots, X_n$ be a random sample from a Poisson distribution with parameter λ, that is, the pdf is given by $f(x)=e^{-\lambda}\lambda^x/(x!)$. Then the hypothesis H_0: $\lambda=1$ uniquely specifies the distribution, because $f(x)=e^{-1}/(x!)$ and hence is a simple hypothesis. The hypothesis H_a: $\lambda > 1$ is composite, because $f(x)$ is not uniquely determined.

Definition 6.2.2. *A test at a given α of a simple hypothesis H_0 versus the simple alternative H_a that has the largest power among the tests with the probability of type I error is no larger than the given α is called a* **most powerful** *test.*

Consider the test of hypothesis H_0: $\theta=\theta_0$ versus H_a: $\theta=\theta_1$. If α is fixed, then our interest is to make β as small as possible. Because $\beta=1 - Power(\theta_1)$, by minimizing β, we would obtain the most powerful test. The following result says that among all tests with a given probability of type I error, the likelihood ratio test given later minimizes the probability of a type II error. In other words, it is the most powerful.

Theorem 6.2.1. (Neyman–Pearson Lemma) *Suppose that one wants to test a simple hypothesis H_0: $\theta=\theta_0$ versus the simple alternative hypothesis H_a: $\theta=\theta_1$ based on a random sample $X_1, \ldots, X_n$ from a distribution with parameter θ. Let $L(\theta) \equiv L(\theta; X_1, \ldots, X_n) > 0$ denote the likelihood of the sample when the value of the parameter is θ. If there exists a positive constant K and a subset C of the sample space $\mathbb{R}^n$ (the Euclidean n-space) such that*

1. $\frac{L(\theta_0)}{L(\theta_1)} \leq K$ *for* $(x_1, x_2, \ldots, x_n) \in C$
2. $\frac{L(\theta_0)}{L(\theta_1)} \geq K$ *for* $(x_1, x_2, \ldots, x_n) \in C'$, *where C' is the complement of C*
3. $P[(X_1, \ldots, X_n) \in C; \theta_0] = \alpha$

Then the test with critical region C will be the most powerful test for H_0 versus H_a. We call α the size of the test and C the best critical region of size α.

Proof. We prove this theorem for continuous random variables. For discrete random variables, the proof is identical with sums replacing the integral. Let S be some region in $\mathbb{R}^n$, an n-dimensional Euclidean space. For simplicity, we will use the following notation:

$$\int_S L(\theta) = \int_S \cdots \int_S L(\theta; x_1, x_2, \ldots, x_n) dx_1 dx_2, \ldots, dx_n.$$

Note that

$$\begin{aligned} P((X_1, \ldots, X_n) \in C; \theta_0) &= \int_C f(x_1, \ldots, x_n; \theta_0) dx_1, \ldots, dx_n \\ &= \int_C L(\theta_0; x_1, \ldots, x_n) dx_1, \ldots, dx_n. \end{aligned}$$

Suppose that there is another critical region, say B, of size less than or equal to α, that is, $\int_B L(\theta_0) \le \alpha$. Then

$$0 \le \int_C L(\theta_0) - \int_B L(\theta_0), \text{ because } \int_C L(\theta_0) = \alpha \text{ by assumption 3.}$$

Therefore,

$$\begin{aligned} 0 &\le \int_C L(\theta_0) - \int_B L(\theta_0) \\ &= \int_{C \cap B} L(\theta_0) + \int_{C \cap B'} L(\theta_0) - \int_{C \cap B} L(\theta_0) - \int_{C' \cap B} L(\theta_0) \\ &= \int_{C \cap B'} L(\theta_0) - \int_{C' \cap B} L(\theta_0). \end{aligned}$$

Using assumption 1 of Theorem 6.2.1, $KL(\theta_1) \ge L(\theta_0)$ at each point in the region C and hence in $C \cap B'$. Thus,

$$\int_{C \cap B'} L(\theta_0) \le K \int_{C \cap B'} L(\theta_1).$$

By assumption 2 of the theorem, $KL(\theta_1) \le L(\theta_0)$ at each point in C' and hence in $C' \cap B$. Thus,

$$\int_{C' \cap B} L(\theta_0) \ge K \int_{C' \cap B} L(\theta_1).$$

Therefore,

$$\begin{aligned} 0 &\le \int_{C \cap B'} L(\theta_0) - \int_{C' \cap B} L(\theta_0) \\ &\le K \left\{ \int_{C \cap B'} L(\theta_1) - \int_{C' \cap B} L(\theta_1) \right\}. \end{aligned}$$

That is,

$$\begin{aligned} 0 &\le K \left\{ \int_{C \cap B} L(\theta_1) + \int_{C \cap B'} L(\theta_1) - \int_{C \cap B} L(\theta_1) - \int_{C \cap B} L(\theta_1) \right\} \\ &= K \left\{ \int_C L(\theta_1) - \int_B L(\theta_1) \right\}. \end{aligned}$$

As a result,

$$\int_C L(\theta_1) \ge \int_B L(\theta_1).$$

Because this is true for every critical region B of size $\le \alpha$, C is the best critical region of size α, and the test with critical region C is the **most powerful test** of size α.

When testing two simple hypotheses, the existence of a best critical region is guaranteed by the Neyman–Pearson lemma. In addition, the foregoing theorem provides a means for determining what the best critical region is. In this case, given a choice of α, we will get a test that is the one with the greatest statistical power in terms of the choice of a critical region. In addition, this lemma tells us that good hypothesis tests are, in fact, the likelihood ratio tests. However, it is important to note that Theorem 6.2.1 gives only the form of the rejection region; the actual rejection region depends on the specific value of α.

In real-world situations, we are seldom presented with the problem of testing two simple hypotheses. There is no general result in the form of Theorem 6.4.1 for composite hypotheses. However, for hypotheses of the form H_0: $\theta = \theta_0$ versus H_a: $\theta > \theta_0$, we can take a particular value $\theta_1 > \theta_0$ and then find a most powerful test for H_0: $\theta = \theta_0$ versus H_a: $\theta > \theta_1$. If this test (i.e., the rejection region of the test) does not depend on the particular value θ_1, then this test is said to be the *uniformly most powerful test* for H_0: $\theta = \theta_0$ versus H_a: $\theta > \theta_0$.

The following example illustrates the use of the Neyman–Pearson lemma.

EXAMPLE 6.2.2

Let $X_1, \ldots, X_n$ denote an independent random sample from a population with a Poisson distribution with mean λ. Derive the most powerful test for testing H_0: $\lambda = 2$ versus H_a: $\lambda = 1/2$.

Solution

Recall that the pdf of a Poisson variable is

$$p(x) = \begin{cases} \frac{e^{-\lambda}\lambda^x}{x!} & \lambda > 0,\ x = 0,1,2,\ \ldots \\ 0, & \text{otherwise.} \end{cases}$$

Thus the likelihood function is

$$L = \frac{[\lambda^{(\sum_{i=1}^{n} x_i)} e^{-\lambda n}]}{\prod_{i=1}^{n} (x_i!)}.$$

For λ=2,

$$L(\theta_0) = L(\lambda = 2) = \frac{\left[2^{\left(\sum_{i=1}^{n} x_i\right)} e^{-2n}\right]}{\prod_{i=1}^{n} (x_i!)},$$

and for λ=1/2,

$$L(\theta_1) = L(\lambda = 1/2) = \frac{\left[(1/2)^{\left(\sum_{i=1}^{n} x_i\right)} e^{-(1/2)n}\right]}{\prod_{i=1}^{n} (x_i!)}.$$

Thus,

$$\frac{L(\theta_0)}{L(\theta_1)} = \frac{(2^{(\sum x_i)} e^{-2n})}{\left(\frac{1}{2}\right)^{\sum x_i} e^{-\frac{n}{2}}} < K,$$

which implies

$$(4)^{\sum x_i}\left(e^{-\frac{3n}{2}}\right) < K$$

or, taking natural logarithms,

$$\left(\sum x_i\right)\ln 4 - \frac{3n}{2} < \ln K.$$

Solving for $(\sum x_i)$ *and letting* $\{[\ln K\,(3n/2)]/\ln 4\} = K'$, *we will reject* H_0 *whenever* $(\sum x_i) < K'$.

A step-by-step procedure for applying the Neyman–Pearson lemma is now given.

Procedure for applying the neyman–pearson lemma

1. Determine the likelihood functions under both null and alternative hypotheses.
2. Take the ratio of the two likelihood functions to be less than a constant K.
3. Simplify the inequality in step 2 to obtain a rejection region.

EXAMPLE 6.2.3

Suppose $X_1, \ldots, X_n$ is a random sample from a normal distribution with a known mean of μ and an unknown variance of σ^2. Find the most powerful α-level test for testing H_0: $\sigma^2 = \sigma_0^2$ versus H_a: $\sigma^2 = \sigma_1^2$, $(\sigma_1^2 > \sigma_0^2)$. Show that this test is equivalent to the χ^2-test. Is the test uniformly most powerful for H_a: $\sigma^2 > \sigma_0^2$?

Solution

To test H_0: $\sigma^2 = \sigma_0^2$ versus H_a: $\sigma^2 > \sigma_1^2$. We have,

$$L(\sigma_0^2) = \Pi_{i=1}^{n} \frac{1}{\sqrt{2\pi}\sigma_0^n} e^{-\frac{(x_i-\mu)^2}{2\sigma_0^2}}$$

$$= \frac{1}{(\sqrt{2\pi})^n \sigma_0^n} e^{-\frac{\Sigma(x_i-\mu)^2}{2\sigma_0^2}}.$$

Similarly,

$$L(\sigma_1^2) = \frac{1}{(\sqrt{2\pi})^n \sigma_1^n} e^{-\frac{\Sigma(x_i-\mu)^2}{2\sigma_1^2}}.$$

Therefore the most powerful test is, reject H_0 if,

$$\frac{L(\sigma_0^2)}{L(\sigma_1^2)} = \left(\frac{\sigma_1^2}{\sigma_0^2}\right)^n e^{\left[-\frac{(\sigma_1^2-\sigma_0^2)^2}{2\sigma_1^2\sigma_0^2}\Sigma(x_i-\mu)^2\right]} \le K$$

for some K.
Taking the natural logarithms, we have

$$n \ln\left(\frac{\sigma_1}{\sigma_0}\right) - \frac{(\sigma_1^2 - \sigma_0^2)}{2\sigma_1^2\sigma_0^2} \sum (x_i - \mu)^2 \le \ln K,$$

or

$$\sum (x_i - \mu)^2 \ge \left[n \ln\left(\frac{\sigma_1}{\sigma_0}\right) - \ln K\right]\left(\frac{2\sigma_1^2\sigma_0^2}{\sigma_1^2 - \sigma_0^2}\right) - C.$$

To find the rejection region for a fixed value of α, we write the region as

$$\frac{\Sigma(x_i - \mu)^2}{\sigma_0^2} \ge \frac{C}{\sigma_0^2} = C'.$$

Note that by Theorem 4.2.7, $\Sigma(x_i - \mu)^2/\sigma_0^2$ has a χ^2-distribution with n degrees of freedom. Thus this test is equivalent to the χ^2-test. Under the H_0, because the same rejection region (does not depend upon the specific value of σ_1^2 in the alternative) would be used for any $\sigma_1^2 > \sigma_0^2$, the test is uniformly most powerful.

The foregoing example shows that, to test for variance using a sample from a normal distribution, we could use the chi-square table to obtain the critical value for the rejection region given α.

EXAMPLE 6.2.4

Suppose X is a single observation from a pdf $f(x) = \lambda x^{\lambda-1}$ for $0 < x < 1$. With $\alpha = 0.05$, find the most powerful test for H_0: $\lambda = 3$ against H_a: $\lambda = 2$.

Solution

Here we want to test H_0: $\lambda = 3$ against H_a: $\lambda = 2$.
Therefore the most powerful test is, reject H_0 if,

$$\frac{L(\lambda_0)}{L(\lambda_1)} = \frac{3x^2}{2x} = \frac{3}{2}x \leq C$$

Thus $x \leq C^*$. Now, $\alpha = 0.05 = P(X < C^*$ when $\lambda = 3) = \int_0^{C_*} 3x^2 dx$, we get $C_* = (0.05)^{1/3} = 0.368$. Thus the rejection region of the most powerful test in this case is $x < 0.368$.

In the following example, we obtain the most powerful test and power. Also, we give the R-code to run the data.

EXAMPLE 6.2.5

We are interested in seed germination. Suppose we plant $n = 10$ seeds. Each germinates independently of the others.

(a) Obtain the most powerful test for H_0: $p = 0.3$ versus H_a: $p = 0.6$ at level $\alpha = 0.05$.
(b) Find the power of the test.

Solution

(a) *We want to test H_0: $p = 0.3$ versus H_a: $p = 0.6$.*
Therefore the most powerful test is, reject H_0 if,

$$\frac{L(p = 0.3)}{L(p = 0.6)} = \frac{\binom{10}{x}(0.3)^x(0.7)^{10-x}}{\binom{10}{x}(0.6)^x(0.4)^{10-x}} = \left(\frac{0.3}{0.6}\right)^x\left(\frac{0.7}{0.4}\right)^{10-x} \leq C$$

Taking the natural logarithms, we have,

$$x \ln\left(\frac{0.3}{0.6}\right) + (10 - x)\ln\left(\frac{0.7}{0.4}\right) \leq \ln C.$$

Simplifying, we will get,

$$-1.25x \leq \ln C - 5.6.$$

That is, $x \geq \frac{1}{1.25}(\ln C - 5.6) = C^*$.
Thus the most powerful test is to reject H_0 for larger values of x.
For example, using a binomial table, we get,
$P_{H_0}(X \geq 6) = 1 - P(X \leq 5) = 0.047$, and $P_{H_0}(X \geq 5) = 1 - P(X \leq 4) = 0.15$. Thus, for the most powerful test, we will reject when $X \geq 6$. Thus, at 5% level, we will reject the null hypothesis if six or more seeds germinate.

(b) *The power of the test at $p = 0.6$,*

$$\text{Power} = P_{H_1}(X \geq 6) = 0.6336.$$

Thus this test has approximately 63.4% power.
R-code **(a)**

```
n<-10; p0<-0.3; p1<-0.6
#tailed probabilities under H0
p_ge_5_H0<-1-pbinom(4, size=n, prob=p0)
```

```
p_ge_6_H0<-1-pbinom(5, size=n, prob=p0)
p_ge_5_H0; p_ge_6_H0
(b) #power at H1 for critical region X>=6
power<-1-pbinom(5, size=n, prob=p1)
power.
```

EXERCISES 6.2

6.2.1 Suppose $X_1,\ldots,X_n$ is a random sample from a normal distribution with a known variance of σ^2 and an unknown mean of μ. Find the most powerful α-level test of H_0: $\mu = \mu_0$ versus H_a: $\mu = \mu_a$ if (a) $\mu_0 > \mu_a$, and (b) $\mu_a > \mu_0$.

6.2.2 Show that the most powerful test obtained in Example 6.2.1 is uniformly most powerful for testing H_0: $\mu \le \mu_0$ versus H_a: $\mu > \mu_a$, but not uniformly most powerful for testing H_0: $\mu = \mu_0$ versus H_a: $\mu \ne \mu_0$.

6.2.3 Suppose $X_1,\ldots,X_n$ is a random sample from a $U(0, \theta)$ distribution. Find the most powerful α-level test for testing H_0: $\theta = \theta_0$ versus H_a: $\theta = \theta_1$, where $\theta_0 < \theta_1$.

6.2.4 Let $X_1,\ldots,X_n$ be a random sample from a geometric distribution with parameter p. Find the most powerful test of H_0: $p = p_0$ versus H_a: $p = p_a (> p_0)$. Is this uniformly most powerful test for H_0: $p = p_0$ versus H_a: $p > p_0$?

6.2.5 Let $X_1,\ldots, X_n$ be a random sample from a distribution having a pdf of

$$f(y) = \begin{cases} \frac{2y}{n^2} e^{-\frac{y^2}{n^2}}, & \text{if } x > 0 \\ 0, & \text{otherwise.} \end{cases}$$

Find a uniformly most powerful test for testing H_0: $\eta = \eta_0$ versus H_a: $\eta < \eta_0$.

6.2.6 Let X be a single observation from the pdf

$$f(x) = \begin{cases} \theta x^{\theta-1}, & 0 < x < 1 \\ 0, & \text{otherwise.} \end{cases}$$

Find the most powerful test with a level of significance $\alpha = 0.01$ to test H_0: $\theta = 3$ versus H_a: $\theta = 4$.

6.2.7 Let $X_1,\ldots,X_n$ be a random sample from a Bernoulli distribution with parameter p. Find the most powerful test of H_0: $p = p_0$ versus H_a: $p = p_a$, where $p_a > p_0$.

6.2.8 Let $X_1,\ldots, X_n$ be a random sample from a Poisson distribution with mean λ. Find the best critical region for testing H_0: $\lambda = 3$ against H_a: $\lambda = 6$.

6.2.9 Let $X_1,\ldots, X_n$ be a random sample from a population with pdf

$$f(x) = \begin{cases} \frac{\lambda}{x^2}, & \text{if } 0 < \lambda \le x < \infty \\ 0, & \text{otherwise.} \end{cases}$$

(a) Find the most powerful test to test $\lambda = \lambda_0$ against $\lambda = \lambda_1 (\ne \lambda_0)$.

(b) Suppose sample size 1 is taken from this pdf, what is the most powerful test for $\lambda = 4$ against $\lambda = 3$, with $\alpha = 0.05$?

6.2.10 Let $X_1,\ldots, X_n$ be a random sample from a normal population with mean μ and variance 25. Find the most powerful test, with a sample size of 20, and the size of the test $\alpha = 0.05$ to test H_0: $\mu = 5$ against H_a: $\mu = 10$.

6.3 LIKELIHOOD RATIO TESTS

The Neyman–Pearson lemma provides a method for constructing the most powerful tests for simple hypotheses. We have also seen that in some instances, when a hypothesis is not simple, it is also possible to find uniformly most powerful tests. In general, uniformly most powerful (UMP) tests do not exist for composite hypotheses. As an example, consider the two-sided hypothesis, at level α, given by

$$H_0\text{: } \mu = \mu_0 \text{ vs. } H_a\text{: } \mu \neq \mu_0$$

where μ is the mean of a normal population with known variance σ^2. If $\overline{X}$ is the sample mean of a random sample of size n, then, as shown earlier, we can use the test statistic

$$Z = \frac{\overline{X} - \mu_0}{\sigma / \sqrt{n}}.$$

For H_a: $\mu = \mu_1 > \mu_0$, the rejection region for the most powerful test would be

$$\text{Reject } H_0 \text{ if} z > z_\alpha.$$

On the other hand, for H_a: $\mu = \mu_2 < \mu_0$, the rejection region for the most powerful test would be

$$\text{Reject } H_0 \text{ if} z < -z_\alpha.$$

Thus the rejection region depends on the specific alternative. Consequently, the two-sided hypothesis just given has no UMP test.

In this section, we shall study a general procedure that is applicable when one or both H_0 and H_a are composite. In fact, this procedure works for simple hypotheses as well. This method is based on the maximum likelihood estimation and the ratio of likelihood functions used in the Neyman–Pearson lemma. We assume that the pdf or pmf of the random variable X is $f(x, \theta)$, where θ can be one or more unknown parameters. Let Θ represent the total parameter space, that is, the set of all possible values of the parameter θ given by either H_0 or H_a.

Consider the hypotheses

$$H_0\text{: } \theta \in \Theta_0 \text{ versus } H_a\text{: } \theta \in \Theta_a = \Theta - \Theta_0,$$

where θ is the unknown population parameter (or parameters) with values in Θ and Θ_0 is a subset of Θ.

Let $L(\theta)$ be the likelihood function based on the sample $X_1,\ldots,X_n$. Now we define the likelihood ratio corresponding to the hypotheses H_0 and H_a. This ratio will be used as a test statistic for the testing procedure that we develop in this section. This is a natural generalization of the ratio test used in the Neyman–Pearson lemma when both hypotheses were simple.

Definition 6.3.1. *The* **likelihood ratio** λ *is the ratio*

$$\lambda = \frac{\max_{\theta \in \Theta_0} L(\theta; x_1,\ldots,x_n)}{\max_{\theta \in \Theta} L(\theta; x_1,\ldots,x_n)} = \frac{L_0^*}{L_*}.$$

We note that $0 \le \lambda \le 1$. Because λ is the ratio of nonnegative functions, we have $\lambda \ge 0$. Because Θ_0 is a subset of Θ, we know that $\max_{\theta \in \Theta_0} L(\theta) \le \max_{\theta \in \Theta} L(\theta)$. Hence, $\lambda \le 1$.

If the maximum of L in Θ_0 is much smaller as compared with the maximum of L in Θ, that is, if λ is small, it would appear that the data $X_1,\ldots,X_n$ do not support the null hypothesis $\theta \in \Theta_0$. Thus there are some parameter values in H_a from which the observed sample more likely came from than for any parameter values in H_0. On the other hand, if λ is close to 1, one could conclude that the data support the null hypothesis, H_0. Therefore small values of λ would result in rejection of the null hypothesis, and large values nearer to 1 would result in a decision in support of the null hypothesis.

For the evaluation of λ, it is important to note that $\max_{\theta \in \Theta} L(\theta) = L(\hat{\theta}_{ml.})$, where $\hat{\theta}_{ml.}$ is the maximum likelihood estimator of $\theta \in \Theta$ and $\max_{\theta \in \Theta_0} L(\theta)$ is the likelihood function with unknown parameters replaced by their maximum likelihood estimators, subject to the condition that $\theta \in \Theta_0$. We can summarize the likelihood ratio test as follows.

Likelihood ratio tests (lrt)

To test

$$H_0\text{: } \theta \in \Theta_0 \text{ versus } H_a\text{: } \theta \in \Theta_a,$$

$$\lambda = \frac{\max_{\theta \in \Theta_0} L(\theta; x_1, \ldots, x_n)}{\max_{\theta \in \Theta} L(\theta; x_1, \ldots, x_n)} = \frac{L_0^*}{L_*},$$

will be used as the test statistic.

The rejection region for the LRT is given by

$$\text{Reject } H_0, \text{ if } \lambda \le K.$$

K is selected such that the test has the given significance level α.

Note that different choices of $K \in [0,1]$ will give different tests and rejection regions. Smaller values of K will result in smaller values of type I error probabilities, and the larger values of K will result in smaller type II error probabilities.

EXAMPLE 6.3.1

Let $X_1,\ldots,X_n$ be a random sample from an $N(\mu, \sigma^2)$. Assume that σ^2 is known. At level α, we wish to test H_0: $\mu=\mu_0$ versus H_a: $\mu \neq \mu_0$. Find an appropriate LRT.

Solution

We have seen that to test

$$H_0: \mu = \mu_0 \text{ versus } H_a: \mu \neq \mu_0$$

there is no uniformly most powerful test. The likelihood function is

$$L(\mu) = \left(\frac{1}{\sqrt{2\pi}\sigma}\right)^n e^{-\frac{\sum_{i=1}^{n}(x_i-\mu)^2}{2\sigma^2}}.$$

Here, $\Theta_0=\{\mu_0\}$ and $\Theta_a=\mathbb{R} - \{\mu_0\}$.
Hence

$$L_0^* = \max_{\mu=\mu_0}\left(\frac{1}{\sigma\sqrt{2\pi}}\right)^n e^{-\frac{\sum_{i=1}^{n}(x_i-\mu)^2}{2\sigma^2}}$$
$$=\left(\frac{1}{\sqrt{2\pi}\sigma}\right)^n e^{-\frac{\sum_{i=1}^{n}(x_i-\mu_0)^2}{2\sigma^2}}.$$

Similarly,

$$L^* = \max_{-\infty<\mu<\infty}\left(\frac{1}{\sqrt{2\pi}\sigma}\right)^n e^{-\frac{\sum_{i=1}^{n}(x_i-\mu)^2}{2\sigma^2}}.$$

Because the only unknown parameter in the parameter space Θ is μ, $-\infty < \mu < \infty$, the maximum of the likelihood function is achieved when μ equals its maximum likelihood estimator, that is,

$$\hat{\mu}_{ml.} = \overline{X}.$$

Therefore, with simple calculation, we have

$$\lambda = \frac{e^{-\left(\sum_{i=1}^{n}(x_i-\mu_0)^2\right)/2\sigma^2}}{e^{-\left(\sum_{i=1}^{n}(x_i-\overline{x})^2\right)/2\sigma^2}} = e^{-n(\overline{x}-\mu_0)^2/2\sigma^2}.$$

Thus the LRT has the rejection region

$$\text{Reject } H_0 \text{ if } \lambda \leq K,$$

which is equivalent to

$$-\frac{n}{2\sigma^2}(\overline{X} - \mu_0)^2 \leq \ln K \Leftrightarrow$$
$$\frac{(\overline{X} - \mu_0)^2}{\sigma^2/n} \geq 2 \ln K \Leftrightarrow$$
$$\left|\frac{\overline{X} - \mu_0}{\sigma/\sqrt{n}}\right| \geq 2 \ln K = c_1.$$

Note that we use the symbol $\Leftrightarrow$ to mean "if and only if." We now compute c_1. Under H_0, $[(\overline{X} - \mu_0)/(\sigma/\sqrt{n})] \sim N(0,1)$. Observe that

---

$$\alpha = P\left\{\left|\frac{\overline{X} - \mu_0}{\sigma/\sqrt{n}}\right| \ge c_1\right\}$$

This gives a possible value of c_1 as $c_1 = z_{\alpha/2}$. Hence LRT for the given hypothesis is

$$\text{Reject } H_0, \text{ if} \left|\frac{\overline{X} - \mu_0}{\sigma/\sqrt{n}}\right| \ge z_{a/2}.$$

Thus, in this case, the LRT is equivalent to the z-test for large random samples.

In fact, when both the hypotheses are simple, the LRT is identical to the Neyman–Pearson test. We can now summarize the procedure for the likelihood ratio test, LRT.

Procedure for the likelihood ratio test (LRT)

1. Find the largest value of the likelihood $L(\theta)$ for any $\theta_0 \in \Theta_0$ by finding the maximum likelihood estimate within Θ_0 and substituting back into the likelihood function.
2. Find the largest value of the likelihood $L(\theta)$ for any $\theta \in \Theta$ by finding the maximum likelihood estimate within Θ and substituting back into the likelihood function.
3. Form the ratio

$$\lambda = \lambda(x_1, x_2, \ldots, x_n) = \frac{L(\theta)\,\text{in}\,\Theta_0}{L(\theta)\,\text{in}\,\Theta}.$$

4. Determine a K so that the test has the desired probability of type I error, α.
5. Reject H_0 if $\lambda \le K$.

In the next example, we find an LRT for a testing problem when both H_0 and H_a are simple

EXAMPLE 6.3.2

Machine I produces 5% defective products. Machine 2 produces 10% defectives. Ten items produced by each of the machines are sampled randomly: X = the number of defectives. Let θ be the true proportion of defectives. Test H_0: $\theta = 0.05$ versus H_a: $\theta = 0.1$. Use $\alpha = 0.05$.

Solution

We need to test H_0: $\theta = 0.05$ versus H_a: $\theta = 0.1$. Let

$$L(\theta) = \begin{cases} \binom{10}{x}(0.05)^x(0.95)^{10-x}, & \text{if}\,\theta = 0.05 \\ \binom{10}{x}(0.1)^x(0.90)^{10-x}, & \text{if } \theta = 0.10, \end{cases}$$

and

$$L_1 = L(0.05) = \binom{10}{x}(0.05)^x(0.95)^{10-x},$$

and

$$L_2 = L(0.1) = \binom{10}{x}(0.1)^x(0.90)^{10-x}.$$

Thus we have

$$\frac{L_1}{L_2} = \frac{0.05^x}{0.1^x}\frac{(0.95)^{10-x}}{(0.9)^{10-x}} = \left(\frac{1}{2}\right)^x\left(\frac{19}{18}\right)^{10-x}.$$

The LRT ratio is

$$\lambda = \frac{L_1}{\max(L_1, L_2)}.$$

Note that if $\max(L_1, L_2) = L_1$, *then,* $\lambda = 1$. *Because we want to reject for small values of* λ, $\max(L_1, L_2) = L_2$, *and we reject* H_0 *if* $(L_1/L_2) \leq K$ *or* $(L_2/L_1) > K$ *(note that* $\frac{L_2}{L_1} = 2^x\left(\frac{18}{19}\right)^{10-x}$*).*
That is, reject H_0 *if*

$$2^x\left(\frac{18}{19}\right)^{10-x} > K$$
$$\Leftrightarrow \left(\frac{2}{\frac{18}{19}}\right)^x > K_1$$
$$\Leftrightarrow \left(\frac{19}{9}\right)^x > K_1.$$

Hence, reject H_0 if $X > C$; $P(X > C|H_0\text{: } \theta=0.05) \leq 0.05$.
Using the binomial tables, we have

$$P(X > 2|\theta = 0.05) = 0.0116$$

and

$$P(X \geq 2|\theta = 0.05) = 0.0862.$$

Reject H_0 if $X > 2$. If we want α to be exactly 0.05, we have to use a randomized test. Reject with probability $\frac{0.0384}{0.0762} = 0.5039$ if $X=2$.

The LRTs do not always produce a test statistic with a known probability distribution, such as the z-statistic of Example 6.3.1. If we have a large sample size, then we can obtain an approximation to the distribution of the statistic λ, which is beyond the level of this book.

EXERCISES 6.3

6.3.1 Let $X_1,\ldots,X_n$ be a random sample from an $N(\mu,\sigma^2)$. Assume that σ^2 is unknown. We wish to test, at level α, H_0: $\mu = \mu_0$ versus H_a: $\mu < \mu_0$. Find an appropriate LRT.

6.3.2 Let $X_1,\ldots,X_n$ be a random sample from an $N(\mu,\sigma^2)$. Assume that both μ and σ^2 are unknown. We wish to test, at level α, H_0: $\sigma^2 = \sigma_0^2$ versus H_a: $\sigma^2 > \sigma_0^2$. Find an appropriate LRT.

6.3.3 Let $X_1,\ldots, X_n$ be a random sample from an $N(\mu_1, \sigma^2)$, and let $Y_1,Y_2,\ldots, Y_n$ be an independent sample from an $N(\mu_2, \sigma^2)$, where σ^2 is unknown. We wish to test, at level α, H_0: $\mu_1 = \mu_2$ versus H_a: $\mu_1 \neq \mu_2$. Find an appropriate LRT.

6.3.4 Let $X_1,\ldots, X_n$ be a sample from a Poisson distribution with parameter λ. Show that a LRT of H_0: $\lambda = \lambda_0$ versus H_a: $\lambda \neq \lambda_0$ rejects the null hypothesis if $\overline{X} \geq m_1$ or $\overline{X} \leq m_2$.

6.3.5 Let $X_1, \ldots, X_n$ be a sample from an exponential distribution with parameter θ. Show that a LRT of H_0: $\theta = \theta_0$ versus H_a: $\theta \neq \theta_0$ rejects the null hypothesis if $\sum_{i=1}^n X_i \geq m_1$ or $\sum_{i=1}^n X_i \leq m_2$.

6.3.6 A clinical oncology program developed a set of guidelines for their cancer patients to follow. It is believed that the proportion of patients who are still living after 24 months is greater for those who follow the guidelines. Of the 40 patients who followed the guidelines, 30 are still living after 24 months, whereas of 32 patients who did not follow the guidelines, 21 are living after 24 months. Find a likelihood ratio test at α=0.01 to decide whether the program is effective.

6.4 HYPOTHESES FOR A SINGLE PARAMETER

In this section, we first introduce the concept of P-value. After that, we study hypothesis testing concerning a single parameter.

6.4.1 The *P*-Value

In hypothesis testing, the choice of the value of α is somewhat arbitrary. For the same data, if the test is based on two different values of α, the conclusions could be different. Many statisticians prefer to compute the so-called P-value, which is calculated based on the observed test statistic. For computing the P-value, it is not necessary to specify the value of α. We can use the given data to obtain the P-value.

Definition 6.4.1. *Corresponding to an observed value of a test statistic, the* **P-value** (*or* **attained significance level**) *is the lowest level of significance at which the null hypothesis would have been rejected.*

For example, if we are testing a given hypothesis with α =0.05 and we make a decision to reject H_0 and we proceed to calculate the P-value equal to 0.03, this means that we could have used an α as low as 0.03 and still maintain the same decision, rejecting H_0.

Based on the alternative hypothesis, one can use the following steps to compute the P-value.

Steps to find the *P*-value

1. Let TS be the test statistic.
2. Compute the value of TS using the sample $X_1, \ldots, X_n$. Say the computed value of TS is a.
3. The P-value is given by

$$P\text{-value} = \begin{cases} P(TS < a|H_0), & \text{if lower tail test} \\ P(TS > a|H_0), & \text{if upper tail test} \\ P(|TS| > |a||H_0), & \text{if two tail test.} \end{cases}$$

EXAMPLE 6.4.1

To test H_0: μ =0 versus H_a: $\mu \neq 0$, suppose that the test statistic Z results in a computed value of 1.58.

Then, the P-value =$P(|Z| > 1.58)$ =2(0.0571)=0.1142. That is, we must have a type I error of 0.1142 to reject H_0. Also, if H_a: $\mu >$ 0, then the P-value would be $P(Z > 1.58)$ =0.0582. In this case, we must have an α of 0.0582 to reject H_0.

The P-value can be thought of as a measure of support for the null hypothesis: The lower its value, the lower the support. Typically, one decides that the support for H_0 is insufficient when the P-value drops below a particular threshold, which is the significance level of the test, α.

Reporting test result as *P*-values

1. Choose the maximum value of α that you are willing to tolerate.
2. If the P-value of the test is less than the maximum value of α, reject H_0.

If the exact P-value cannot be found, one can give an interval in which the P-value can lie. For example, if the test is significant at α =0.05 but not significant for α = 0.025, report that $0.025 \le P\text{-value} \le 0.05$. So for $\alpha > 0.05$, reject H_0, and for $\alpha < 0.025$, do not reject H_0.

In another interpretation, 1 − (P-value) is considered as an index of the strength of the evidence against the null hypothesis provided by the data. It is clear that the value of this index lies in the interval [0, 1]. If the P-value is 0.02, the value of the index is 0.98, supporting the rejection of the null hypothesis. Not only do P-values provide us with a yes or no answer, but they also provide a sense of the strength of the evidence against the null hypothesis. The lower the P-value, the stronger the evidence. Thus, in any test, reporting the P-value of the test is good practice.

Because most of the outputs from statistical software used for hypothesis testing include the P-value, the P-value approach to hypothesis testing is becoming more and more popular. In this approach, the decision of the test is made in the following way. If the value of α is given, and if the P-value of the test is less than the value of α, we will reject H_0. If the value of α is not given and the P-value associated with the test is small (usually set at P-value < 0.05), there is evidence to reject the null hypothesis in favor of the alternative. In other words, there is evidence that the value of the true parameter (such as the population mean) is significantly different (greater or lesser) than the hypothesized value. If the *P*-value associated with the test is not small ($p > 0.05$), we conclude that there is not enough evidence to reject the null hypothesis. In most of the examples in this chapter, we give both the rejection region and *P*-value approaches.

EXAMPLE 6.4.2

The management of a local health club claims that its members lose on average 15 pounds or more within the first 3 months after joining the club. To check this claim, a consumer agency took a random sample of 45 members of this health club and found that they lost an average of 13.8 pounds within the first 3 months of participation, with a sample standard deviation of 4.2 pounds.

(a) Find the *P*-value for this test.
(b) Based on the *P*-value in (a), would you reject the null hypothesis at α=0.01?

Solution

(a) *Let μ be the true mean weight loss in pounds within the first 3 months of participation in this club. Then we have to test the hypothesis*

$$H_0: \mu = 15 \text{ versus } H_a: \mu < 15$$

Here, n=45, $\bar{x} = 13.8$, and s=4.2. Because n=45 > 30, we can use a normal approximation. Hence, the test statistic is,

$$z = \frac{13.8 - 15}{4.2/\sqrt{45}} = -1.9166,$$

and

$$P\text{-value} = P(Z < -1.9166) \simeq P(Z < -1.92) = 0.0274.$$

Thus we can use α as small as 0.0274 and still reject H_0.

(b) *No. Because the P-value=0.0274 is greater than α=0.01, one cannot reject H_0.*

In any hypothesis testing, after an experimenter determines the objective of an experiment and decides on the type of data to be collected, we recommend the following step-by-step procedure for hypothesis testing.

Steps in any hypothesis testing problem

1. State the alternative hypothesis, H_a (what is believed to be true).
2. State the null hypothesis, H_0 (what is doubted to be true).
3. Decide on a level of significance α.
4. Choose appropriate TS, and compute the observed test statistic.
5. Using the distribution of TS and α, determine the rejection region(s) (RR).
6. Conclusion: If the observed test statistic falls in the RR, reject H_0, and conclude that based on the sampled information, we are $(1 - \alpha)100\%$ confident that H_a is true. Otherwise, conclude that there is not sufficient evidence to reject H_0. In all the applied problems, interpret the meaning of your decision.
7. State any assumptions you made in testing the given hypothesis.
8. Compute the *P*-value from the null distribution of the test statistic, and interpret it.

6.4.2 Hypothesis Testing for a Single Parameter

Now we study the testing of a hypothesis concerning a single parameter, θ, based on a random sample $X_1,\ldots, X_n$. Let $\hat{\theta}$ be the sample statistic. First, we deal with tests for the population mean μ for large and small samples. Next, we study procedures for testing the population variance σ^2. We conclude the section by studying a test procedure for the true proportion p.

To test the hypothesis H_0: $\mu = \mu_0$ concerning the true population mean μ, when we have a large sample ($n \geq 30$), we use the test statistic Z given by

$$Z = \frac{\overline{X} - \mu_0}{S/\sqrt{n}}$$

where S is the sample standard deviation and μ_0 is the claimed mean under H_0 (if the population variance is known, we replace S with σ).

For a small random sample ($n < 30$), the test statistic is

$$T = \frac{\overline{X} - \mu_0}{S/\sqrt{n}}$$

where μ_0 is the claimed value of the true mean and $\overline{X}$ and S are the sample mean and standard deviation, respectively. Note that we are using the lowercase letters, such as z and t, to represent the observed values of the test statistics Z and T, respectively.

In practice, with raw data, it is important to verify the assumptions. For example, in the small sample case, it is important to check for normality by using normal plots. If this assumption is not satisfied, the nonparametric methods described in Chapter 12 may be more appropriate. In addition, because the sample statistic such as $\overline{X}$ and S will be greatly affected by the presence of outliers, drawing a box plot to check for outliers is a basic practice we should incorporate in our analysis.

We now summarize the typical test of hypothesis for tests concerning the population (true) mean.

To compute the observed test statistic, z in the large sample case and t in the small sample case, calculate the values of $z = (\bar{x} - \mu_0)/(s/\sqrt{n})$ and $t = [(\bar{x} - \mu_0)/(s/\sqrt{n})]$, respectively.

Summary of hypothesis tests for μ

Large Sample ($n \geq 30$)
To test
$H_0{:}\mu=\mu_0$
versus

$$H_a: \begin{matrix} \mu > \mu_0 \text{ upper tailed test} \\ \mu < \mu_0 \text{ lower tailed test} \\ \mu \neq \mu_0, \text{ two-tailed test} \end{matrix}$$

Test statistic:

$$Z = \frac{\overline{X} - \mu_0}{\sigma/\sqrt{n}}$$

Replace σ by S, if σ is unknown.

$$\text{Rejection Region: } \begin{cases} z > z_\alpha, & \text{upper tailed RR} \\ z < -z_\alpha, & \text{lower tailed RR} \\ |z| > z_{\alpha/2}, & \text{two tailed RR} \end{cases}$$

Assumption: $n \geq 30$ and $\sigma^2 < \infty$.

Small Sample ($n < 30$)

To test
$H_0{:}\mu=\mu_0$
versus

$$H_a: \begin{matrix} \mu > \mu_0 \text{ upper tailed test} \\ \mu < \mu_0 \text{ lower tailed test} \\ \mu \neq \mu_0, \text{ two-tailed test} \end{matrix}$$

Test statistic:

$$T = \frac{\overline{X} - \mu_0}{S/\sqrt{n}}$$

$$\text{RR: } \begin{cases} t > t_{\alpha,n-1}, & \text{upper tailed RR} \\ t < -t_{\alpha,n-1}, & \text{lower tailed RR} \\ |t| > t_{\alpha/2,(n-1)}, & \text{two tailed RR} \end{cases}$$

Assumption: Random sample comes from a normal population

Decision: Reject H_0 if the observed test statistic falls in the RR, and conclude that H_a is true with $(1 - \alpha)100\%$ confidence. Otherwise, keep H_0 so that there s not enough evidence to conclude that H_a is true for the given α, and more data may be needed.

EXAMPLE 6.4.3

It is claimed that sports-car owners drive on average 18,000 miles per year. A consumer firm believes that the average mileage is probably lower. To check, the consumer firm obtained information from 40 randomly selected sports-car owners that resulted in a sample mean of 17,463 miles with a sample standard deviation of 1348 miles. What can we conclude about this claim? Use α=0.01. What is the *P*-value?

Solution

Let μ be the true population mean. We can formulate the hypotheses as H_0: μ=18,000 versus H_a: μ < 18,000. The observed test statistic (for $n \geq 30$) is

$$z = \frac{\bar{x}-\mu_0}{s/\sqrt{n}} \cong \frac{17{,}463-18{,}000}{1348/\sqrt{40}}$$
$$= -\ 2.52.$$

Rejection region is $\{z < -\ z_{0\cdot 01}\}=\{z < -2.33\}$.

Decision: *Because z=–2.52 is less than –2.33, the null hypothesis is rejected at α =0.01. There is sufficient evidence to conclude that the mean mileage on sports cars is less than 18,000 miles per year.*

P-value=$P(z < -2.52)$ = 0.0059. *This P-value is less than 0.01 also supports rejection of the null hypothesis.*

EXAMPLE 6.4.4

In a frequently traveled stretch of the I-75 highway, where the posted speed is 70 mph, it is thought that people travel at an average of at least 70 mph. To check this claim, the following radar measurements of the speeds (in mph) are obtained for 10 vehicles traveling on this stretch of the interstate highway.

66 74 79 80 69 77 78 65 79 81.

Do the data provide sufficient evidence to indicate that the mean speed at which people travel on this stretch of highway is at least 70 mph (the posted speed limit)? Test the appropriate hypothesis using α=0.01. Draw a box plot and normal plot for this data and comment.

Solution

We need to test

$$H_0: \mu = 70 \text{ versus } H_a: \mu > 70.$$

Here $n < 30$. For this sample, the sample mean is $\bar{x}$ = 74.8 mph, and the sample standard deviation is s = 5.9963 mph. Hence, the observed test statistic is

$$t = \frac{\bar{x}-\mu_0}{s/\sqrt{n}} = \frac{74.8-70}{5.9963/\sqrt{10}}$$
$$=2.5314.$$

From the t-table, $t_{0.01,9}$=2.821. Hence, the rejection region is $\{t > 2.821\}$.

Because t = 2.5314 *does not fall in the rejection region, we do not reject the null hypothesis at α.* This can also be verified by the fact that the *P*-value of 0.01608 is larger than α=0.01. This *P*-value is obtained from R (if we use the *t*-table, we will see that $0.01 < P - value < 0.025$). *Note that we assumed that the vehicles were randomly selected and that collected data follow the normal distribution. Because of the small sample size, n < 30, we use the t-test.*

Figs 6.1 and 6.2 are the box plot and the normal plot of the data, respectively.

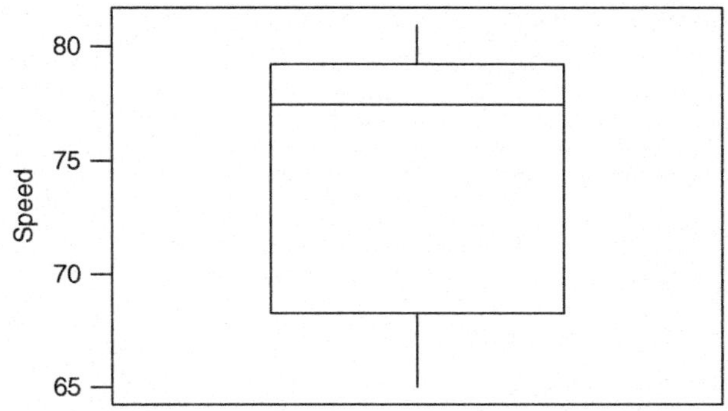

FIGURE 6.1 Box plot of speed data. A simple box plot showing the distribution of a single variable. The plot displays the median, quartiles and potential outliers.

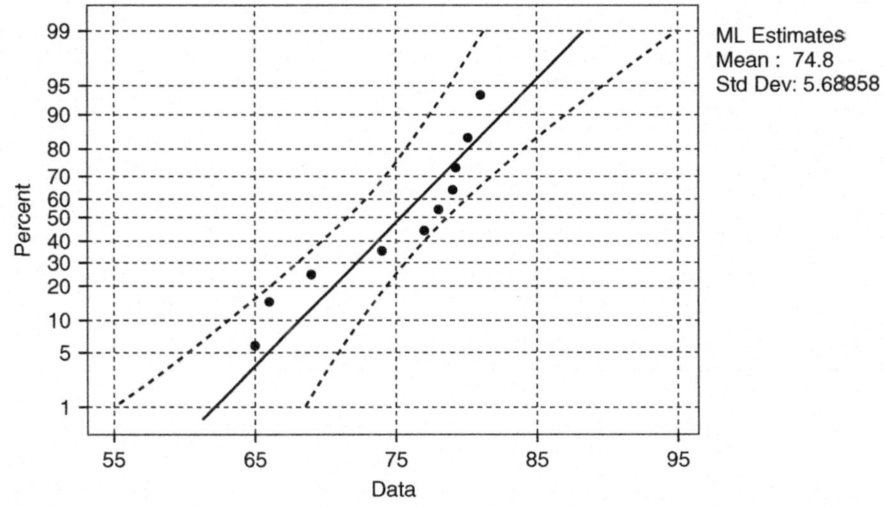

FIGURE 6.2 Normal probability plot for speed.

The box plot suggests that there are no outliers present. However, the normal plot indicates that the normality assumption for this data set is not justified. Hence, it may be more appropriate to do a nonparametric test or obtain more data.

EXAMPLE 6.4.5

In attempting to control the strength of the waste discharged into a nearby river, an industrial firm has taken a number of restorative measures. The firm believes that they have lowered the oxygen-consuming power of their wastes from a previous mean of 450 manganate in parts per million. To test this belief, readings are taken on n=20 successive days. A sample mean of 312.5 and the sample standard deviation 106.23 are obtained. Assume that these 20 values can be treated as a random sample from a normal population. Test the appropriate hypothesis. Use α =0.05.

Solution

Here we need to test the following hypothesis:

$$H_0\text{: } \mu = 450 \text{ versus } H_a\text{: } \mu < 450$$

Given n=20, $\bar{x} = 312.5$, *and* s=106.23. *The observed test statistic is*

$$t = \frac{312.5 - 450}{106.23/\sqrt{20}} = -5.79.$$

The rejection region for α *=0.05 and with 19 degrees of freedom is the set of t-values such that*

$$\{t < -t_{0.05,19}\} = \{t < -1.729\}.$$

Decision: *Because* $t = -5.79$ *is less than* -1.729*, reject* H_0*. There is sufficient evidence to confirm the firm's belief.*

For large random samples, the following procedure is used to perform tests of hypotheses about the population proportion, p.

EXAMPLE 6.4.6

A machine is considered to be unsatisfactory if it produces more than 8% defectives. It is suspected that the machine is unsatisfactory. A random sample of 120 items produced by the machine contains 14 defectives. Does the sample evidence support the claim that the machine is unsatisfactory? Use $\alpha = 0.01$.

Solution

Let Y be the number of observed defectives. This follows a binomial distribution. However, because np_0 *and* nq_0 *are greater than 5, we can use a normal approximation to the binomial to test the hypothesis. So we need to test* H_0: $p=0.08$ *versus* H_a: $p > 0.08$*. Let the point estimate of p be* $\hat{p} = (Y/n) = 0.117$*, the sample proportion. Then the value of the TS is*

$$z = \frac{\hat{p} - p_0}{\sqrt{\frac{p_0 q_0}{n}}} = \frac{0.117 - 0.08}{\sqrt{\frac{(0.08)(0.92)}{120}}} = 0.137.$$

For $\alpha = 0.01$, $z_{0.01} = 2.33$. *Hence the rejection region is* $\{z > 2.33\}$.

Decision: *Because 0.137 is not greater than 2.33, we do not reject* H_0*. We conclude that the evidence does not support the claim that the machine is unsatisfactory.*

Summary of large sample hypothesis test for p

To test

$$H_0: p = p_0$$

versus

$$H_a: \begin{matrix} p > p_0, & \text{upper tailed test} \\ p < p_0, & \text{lower tailed test} \\ p \neq p_0, & \text{two-tailed test.} \end{matrix}$$

Test statistic

$$Z = \frac{\hat{p} - p_0}{\sigma_{\hat{p}}}, \text{ where } \sigma_{\hat{p}} = \sqrt{\frac{p_0 q_0}{n}}, \text{ where } q_0 = 1 - p_0.$$

$$\text{Rejection region: } \begin{cases} z > z_\alpha, & \text{upper tailed RR} \\ z < -z_\alpha, & \text{lower tailed RR} \\ |z| > z_{\alpha/2}, & \text{two-tailed RR,} \end{cases}$$

where z is the observed test statistic.

Assumption: n is large. A good rule of thumb is to use the normal approximation to the binomial distribution only when np_0 and $n(1 - p_0)$ are both greater than 5.

Decision: Reject H_0, if the observed test statistic falls in the RR, and conclude that H_a is true with $(1 - \alpha)100\%$ confidence. Otherwise, do not reject H_0 because there is not enough evidence to conclude that H_a is true for given α, and more data is needed.

Note that this is an approximate test, and the test can be improved by increasing the sample size.
Now we give the procedure for testing the population variance when the samples come from a normal population.

Summary of hypothesis test for the variance σ^2

To test

H_0: $\sigma^2 = \sigma_0^2$

versus

$\sigma^2 > \sigma_0^2$, upper tailed test
H_a: $\sigma^2 < \sigma_0^2$, lower tailed test
$\sigma^2 \neq \sigma_0^2$, two-tailed test.

Test statistic:

$$\chi^2 = \frac{(n-1)S^2}{\sigma_0^2}$$

where S^2 is the sample variance.
Observed value of the test statistic: $\frac{(n-1)s^2}{\sigma_0^2}$

$$\text{Rejection region: } \begin{cases} \chi^2 > \chi^2_{\alpha,n-1}, \text{ upper tailed RR} \\ \chi^2 < \chi^2_{1-\alpha,n-1}, \text{ lower tailed RR} \\ \chi^2 > \chi^2_{\alpha/2,n-1} \text{ or } \chi^2 < \chi^2_{1-\alpha/2,n-1}, \text{ two–tailed RR} \end{cases}$$

where $\chi^2_{\alpha,n-1}$ is such that the area under the chi-square distribution with $(n-1)$ degrees of freedom to its right is equal to α.

Assumption: Sample comes from a normal population.

Decision: Reject H_0 if the observed test statistic falls in the RR, and conclude that H_a is true with $(1-\alpha)100\%$ confidence. Otherwise, do not reject H_0 because there is not enough evidence to conclude that H_a is true for given α, and more data is needed.

Because the chi-square distribution is not symmetric, the "equal tailed" used for the two-sided alternative may not be the best procedure. However, in real-world problems, we seldom use a two-tailed test for the population variance.

EXAMPLE 6.4.7

A physician claims that the variance in cholesterol levels of adult men in a certain laboratory is at least 100 mg/dL. A random sample of 25 adult males from this laboratory produced a sample standard deviation of cholesterol levels of 12 mg/dL. Test the physician's claim at 5% level of significance.

Solution

To test

$$H_0: \sigma^2 = 100 \text{ versus } H_a: \sigma^2 < 100$$

for α=0.05, and 24 degrees of freedom, the rejection region is

$$RR = \{\chi^2 < \chi^2_{1-\alpha,n-1}\} = \{\chi^2 < 13.484\}.$$

The observed value of the TS is

$$\chi^2 = \frac{(n-1)s^2}{\sigma_0^2} = \frac{(24)(144)}{100} = 34.56.$$

Because the value of the test statistic does not fall in the rejection region, we cannot reject H_0 at 5% level of significance. Here, we assumed that the 25 cholesterol measurements follow the normal distribution.

EXAMPLE 6.4.8

An insurer notes that historically, the mean claim amount for a policy type is μ_0 = \$1200 and the standard deviation is around σ_0 = \$500. After a loss-control program (a proactive risk management strategy), the insurer randomly samples 40 claims for which the sample mean was \$1080 and the standard deviation was \$520.

(a). Has the mean sample size decreased after the new program?
(b). Has the variance increased?
(c). Obtain a 95% confidence interval for the variance, σ^2.

Solution

Given the sample mean is $\bar{x} = 1080$, *and the sample deviation is* $s = 520$. *Sample size,* $n = 40$.

(a). *To test* H_0: $\mu = 1200$ *versus* H_a: $\mu < 1200$.

The test statistic

$$Z = \frac{1080 - 1200}{500/\sqrt{40}} \approx -1.52$$

With $\alpha = 0.05$, *the rejection region is* $\{Z < -1.645\}$. *Since the test statistic does not fall in the rejection region, we fail to reject* H_0*at a 95% confidence level and conclude that there is not enough evidence to show that the mean claim decreased.*

EXERCISES 6.4

6.4.1 A random sample of 50 measurements resulted in a sample mean of 62 with a sample standard deviation of 8. It is claimed that the true population mean is at least 64.

(a) Is there sufficient evidence to refute the claim at the 2% level of significance?

(b) What is the P-value?

(c) What is the smallest value of α for which the claim will be rejected?

6.4.2 A machine in a certain factory must be repaired if it produces more than 12% defectives among the large lot of items it produces in a week. A random sample of 175 items from a week's production contains 35 defectives, and it is decided that the machine must be repaired.

(a) Does the sample evidence support this decision? Use α =0.02.

(b) Compute the P-value.

6.4.3 A random sample of 78 observations produced the following sums:

$$\sum_{i=1}^{78} x_i = 22.8, \ \sum_{i=1}^{78} (x_i - \bar{x})^2 = 2.05.$$

(a) Test the null hypothesis that μ=0.45 against the alternative hypothesis that $\mu < 0.45$ using $\alpha = 0.01$. Also, find the P-value.

(b) Test the null hypothesis that μ=0.45 against the alternative hypothesis that $\mu \neq 0.45$ using $\alpha = 0.01$. Also, find the P-value.

(c) What assumptions did you make for solving (a) and (b)?

6.4.4 Consider the test H_0: $\mu = 35$ versus H_a: $\mu > 35$ for a population that is normally distributed.

(a) A random sample of 18 observations taken from this population produced a sample mean of 40 and a sample standard deviation of 5. Using α =0.025, would you reject the null hypothesis?

(b) Another random sample of 18 observations produced a sample mean of 36.8 and a sample standard deviation of 6.9. Using α =0.025, would you reject the null hypothesis?

(c) Compare and discuss the decisions of parts (a) and (b).

6.4.5 According to the information obtained from a large university, professors there earned an average annual salary of $55,648 in 1998. A recent random sample of 15 professors from this university showed that they earn an average annual salary of $58,800 with a sample standard deviation of $8300. Assume that the annual salaries of all the professors in this university are normally distributed.

(a) Suppose the probability of making a type I error is chosen to be zero. Without performing all the steps of the test of hypothesis, would you accept or reject the null hypothesis that the current mean annual salary of all professors at this university is $55,648?

(b) Using the 1% significance level, can you conclude that the current mean annual salary of professors at this university is more than $55,648?

6.4.6 A check-cashing service company found that approximately 7% of all checks submitted to the service were without sufficient funds. After instituting a random check verification system to reduce its losses, the service company found that only 70 were rejected in a random sample of 1125 that were cashed. Is there sufficient evidence that the check verification system reduced the proportion of bad checks at α =0.01? What is the P-value associated with the test? What would you conclude at the α =0.05 level?

6.4.7 Preliminary results of a study (the journal Environmental News reported in April 1975 that "The continuing analysis of lead levels in the drinking water of several Boston communities has verified elevated lead concentrations in the water supplies of Somerville, Brighton, and Beacon Hill") found that "20% of the 248 randomly chosen households tested in these communities showed lead levels exceeding the U.S. Public Health Service standard of 50 parts per million." In contrast, in Cambridge, which adds anticorrosive to its water in an attempt to keep the lead from leaching out of the pipes, "only 5% of the 100 randomly sampled households showed lead levels exceeding the standard." Find a 95% confidence interval for the difference in the proportions of households in Somerville, Brighton, and Beacon Hill on the one hand and Cambridge on the other that had lead levels exceeding the government standard, and carry out a test of the hypothesis of no difference at $\alpha = 0.05$.

6.4.8 A manufacturer of washers provides a particular model in one of three colors: white, black, or ivory. Of the first 1500 washers sold, it is noticed that 550 were of ivory color. Would you conclude that customers have a preference for the ivory color? Justify your answer. Use α =0.01.

6.4.9 A test of the breaking strength of six ropes manufactured by a company showed a mean breaking strength of 7225 lb and a standard deviation of 120 lb. However, the manufacturer claimed a mean breaking strength of 7500 lb.

(a) Can we support the manufacturer's claim at a level of significance of 0.10?

(b) Compute the *P*-value. What assumptions did you make about this problem?

6.4.10 A sample of 10 observations taken from a normally distributed population produced the following data:

44 31 52 48 46 39 43 36 41 49

(a) Test the hypothesis that H_0: μ=44 versus H_a: $\mu \neq 44$ using α =0.10. Draw a box plot and normal plot for this data and comment.

(b) Find a 90% confidence interval for the population mean μ.

(c) Discuss the meanings of (a) and (b). What can we conclude?

6.4.11 The principal of a charter school in Tampa believes that the intelligence quotients (IQs) of its students are above the national average of 100. From past experience, IQ is normally distributed with a standard deviation of 10. A random sample of 20 students is selected from this school, and their IQs are observed. The following are the observed values:

95 91 110 93 133 119 113 107 110 89
113 100 100 124 116 113 110 106 115 113

(a) Test for the normality of the data.

(b) Do the IQs of students at the school run above the national average at α =0.01?

6.4.12 To find out whether children with chronic diarrhea have the same average hemoglobin level (Hb) that is normally seen in healthy children in the same area, a random sample of 10 children with chronic diarrhea is selected, and their Hb levels (g/dL) are obtained as follows:

12.3 11.4 14.2 15.3 14.8 13.8 11.1 15.1 15.8 13.2

Do the data provide sufficient evidence to indicate that the mean Hb level for children with chronic diarrhea is less than that of the normal value of 14.6 g/dL? Test the appropriate hypothesis using α=0.01. Draw a box plot and a normal plot for this data, and comment.

6.4.13 A company that manufactures precision special-alloy steel shafts claims that the variance in the diameters of shafts is no more than 0.0003. A random sample of 10 shafts gave a sample variance of 0.00027. At the 5% level of significance, test whether the company's claim can be substantiated.

6.4.14 It was claimed that the average annual expenditure per consumer unit had continued to rise, as measured by the Consumer Price Index annual averages (Bureau of Labor Statistics report, 1995). To test this claim, 100 consumer units were randomly selected in 1995 and found to have an average annual expenditure of $32,277 with a standard deviation of $1200. Assuming that the average annual expenditure of all consumer units was $30,692 in 1994, test at the 5% significance level whether the annual expenditure per consumer unit had really increased from 1994 to 1995.

6.4.15 It is claimed that two of three Americans say that the chances of world peace are seriously threatened by the nuclear capabilities of other countries. If in a random sample of 400 Americans, it is found that only 252 hold this view. Do you think the claim is correct? Use α =0.05. State any assumptions you make in solving this problem.

6.4.16 According to the Bureau of Labor Statistics (1996), the average price of a gallon of gasoline in all US cities in the United States in January 1996 was $1.129. A later random sample in 24 cities found the mean price to be $1.14 with a standard deviation of 0.01. Test at α =0.05 to see whether the average price of a gallon of gas in the cities had recently changed.

6.4.17 A manufacturer claims that the mean life of batteries manufactured by his company is at least 44 months. A random sample of 40 of these batteries was tested, resulting in a sample mean life of 41 months with a sample standard deviation of 16 months. Test at α=0.01 whether the manufacturer's claim is correct.

6.5 TESTING OF HYPOTHESES FOR TWO SAMPLES

In this section, we study the hypothesis testing procedures for comparing the means and variances of two populations. For example, suppose that we want to determine whether a particular drug is effective for a certain illness. The sample subjects will be randomly selected from a large pool of people with that particular illness and will be assigned randomly to the two groups. To one group we will administer a placebo; to the other we will administer the drug of interest. After a period of time, we measure a physical characteristic—say the blood pressure—of each subject that is an indicator of the severity of the illness. The question is whether the drug can be considered effective on the population from which our samples have been selected. We will consider the cases of independent and dependent samples.

Independent Samples

Two random samples are drawn independently of each other from two populations, and the sample information is obtained. We are interested in testing a hypothesis about the difference in the true means. Let $X_{11}, \ldots, X_{1n_1}$ be a random sample from population 1 with mean μ_1 and variance σ_1^2 and $X_{11}, \ldots, X_{1n_2}$ be a random sample from population 2 with mean μ_2 and variance σ_2^2. Let $\overline{X}_i$, i=1, 2, represent the respective sample means and S_i^2,i=1, 2, represent the sample variances. In this case, we shall consider the following three cases in testing hypotheses about μ_1 and μ_2: (i) when σ_1^2 and σ_2^2 are known, (ii) when σ_1^2 and σ_2^2 are unknown and $n_1 \geq 30$ and $n_2 \geq 30$, and (iii) when σ_1^2 and σ_2^2 are unknown and $n_1 < 30$ and $n_2 < 30$. In case (iii) we have the following two possibilities: (a) $\sigma_1^2 = \sigma_2^2$, and (b)$\sigma_1^2 \neq \sigma_2^2$.

In the large sample case, knowledge of population variances σ_1^2 and σ_2^2 does not make much difference. If the population variances are unknown, we could replace them with sample variances as an approximation. If both $n_1 \geq 30$ and $n_2 \geq 30$ (large sample case), we can use a normal approximation. The following box sums up a large sample hypothesis testing procedure for the difference of means for the large sample case.

Summary of hypothesis test for $\mu_1 - \mu_2$ for large samples (n_1 and $n_2 \geq 30$)

To test

$$H_0: \mu_1 - \mu_2 = D_0$$

versus

$$H_a: \begin{cases} \mu_1 - \mu_2 > D_0, & \text{upper tailed test} \\ \mu_1 - \mu_2 < D_0, & \text{lower tailed test} \\ \mu_1 - \mu_2 \neq D_0, & \text{two-tailed test.} \end{cases}$$

The test statistic is

$$Z = \frac{\overline{X}_1 - \overline{X}_2 - D_0}{\sqrt{\frac{\sigma_1^2}{n_1} + \frac{\sigma_2^2}{n_2}}}.$$

Replace σ_i by s_i, if σ_i, i=1,2 are not known.
Rejection region is

$$RR: \begin{cases} z > z_\alpha, & \text{upper tailed RR} \\ z < -z_\alpha, & \text{lower tailed RR} \\ |z| > z_{\alpha/2}, & \text{two–tailedRR,} \end{cases}$$

where z is the observed test statistic given by

$$z = \frac{\overline{x}_1 - \overline{x}_2 - D_0}{\sqrt{\frac{\sigma_1^2}{n_1} + \frac{\sigma_2^2}{n_2}}}.$$

Assumption: The samples are independent and n_1 and n_2 $\geq$ 30.

Decision: Reject H_0, if the test statistic falls in the RR, and conclude that H_a is true with $(1 - \alpha)100\%$ confidence. Otherwise, do not reject H_0 because there is not enough evidence to conclude that H_a is true for given α and more data is needed.

EXAMPLE 6.5.1

In a salary equity study of faculty at a certain university, sample salaries of 50 male assistant professors and 50 female assistant professors yielded the following basic statistics.

	Sample Mean Salary	Sample Standard Deviation
Male assistant professor	$46,400	360
Female assistant professor	$46,000	220

Test the hypothesis that the mean salary of male assistant professors is more than the mean salary of female assistant professors at this university. Use $\alpha = 0.05$.

Solution

Let μ_1 be the true mean salary for male assistant professors and μ_2 be the true mean salary for female assistant professors at this university. To test

$$H_0: \mu_1 - \mu_2 = 0 \text{ versus } H_a: \mu_1 - \mu_2 > 0$$

the test statistic is

$$z = \frac{\bar{x}_1 - \bar{x}_2 - D_0}{\sqrt{\frac{s_1^2}{n_1} + \frac{s_2^2}{n_2}}} = \frac{46{,}400 - 46{,}000}{\sqrt{\frac{(360)^2}{50} + \frac{(220)^2}{50}}} = 6.704.$$

The rejection region for α=0.05 is $\{z > 1.645\}$.

Because $z = 6.704 > 1.645$, we reject the null hypothesis at $\alpha = 0.05$. We conclude that the salary of male assistant professors at this university is higher than that of female assistant professors for $\alpha = 0.05$. Note that even though σ_1^2 and σ_2^2 are unknown, because $n_1 \geq 30$ and $n_2 \geq 30$, we could replace σ_1^2 and σ_2^2 by the respective sample variances. We are assuming that the salaries of male and female are sampled independently of each other.

Equal Variances

Given next is the procedure we follow to compare the true means from two independent normal populations when n_1 and n_2 are small ($n_1 < 30$ or $n_2 < 30$), and we can assume homogeneity in the population variances, that is, $\sigma_1^2 = \sigma_2^2$. In this case, we pool the sample variances to obtain a point estimate of the common variance.

Comparison of two population means, small sample case (pooled *t*-test)

To test

$$H_0: \mu_1 - \mu_2 = D_0$$

versus

$$H_a: \begin{cases} \mu_1 - \mu_2 > D_0, & \text{upper tailed test} \\ \mu_1 - \mu_2 < D_0, & \text{lower tailed test} \\ \mu_1 - \mu_2 \neq D_0, & \text{two-tailed test.} \end{cases}$$

The test statistic is

$$T = \frac{\bar{X}_1 - \bar{X}_2 - D_0}{S_p\sqrt{\frac{1}{n_1} + \frac{1}{n_2}}}.$$

Here, the pooled sample variance is

$$S_p^2 = \frac{(n_1 - 1)S_1^2 + (n_2 - 1)S_2^2}{n_1 + n_2 - 2}.$$

Then the rejection region is

$$RR: \begin{cases} t > t_\alpha, & \text{upper tailed test} \\ t < -t_\alpha, & \text{lower tailed test} \\ |t| > t_{\alpha/2}, & \text{two-tailed test} \end{cases}$$

where t is the observed test statistic and t_α is based on $(n_1\ n_2 - 2)$ degrees of freedom such that $P(T > t_\alpha) = \alpha$.

Decision: Reject H_0 if the test statistic falls in the RR, and conclude that H_a is true with $(1-\alpha)100\%$ confidence. Otherwise, do not reject H_0 because there is not enough evidence to conclude that H_a is true for given α.

Assumptions: The samples are independent and come from normal populations with means μ_1 and μ_2, and with equal (but unknown) variances, that is, $\sigma_1^2 = \sigma_2^2$.

Unequal Variances: Welch's *t*-Test ($\sigma_1^2 \neq \sigma_2^2$)

Now we shall consider the case where σ_1^2 and σ_2^2 are unknown and cannot be assumed to be equal. Welch's t-test is designed for this case; however, it is still necessary to assume the samples are coming from normal distributions. In such a case, the following test is often used. For the hypothesis

$$H_0: \mu_1 - \mu_2 = D_0 \text{ versus } H_a: \begin{cases} \mu_1 - \mu_2 > D_0 \\ \mu_1 - \mu_2 < D_0 \\ \mu_1 - \mu_2 \neq D_0 \end{cases}$$

define the test statistic T_v as

$$T_v = \frac{\overline{X}_1 - \overline{X}_2 - D_0}{\sqrt{\frac{S_1^2}{n_1} + \frac{S_2^2}{n_2}}}$$

where T_v has a t-distribution with v degrees of freedom, and for a particular sample with $S_1^2 = s_1^2$ and $S_2^2 = s_2^2$,

$$v = \frac{[(s_1^2/n_1) + (s_2^2/n_2)]^2}{\frac{(s_1^2/n_1)^2}{n_1 - 1} + \frac{(s_2^2/n_2)^2}{n_2 - 1}} = \frac{[(s_1^2/n_1) + (s_2^2/n_2)]^2}{\frac{s_1^4}{n_1^2(n_1 - 1)} + \frac{s_2^4}{n_2^2(n_2 - 1)}}.$$

The value of v will not necessarily be an integer. In that case, we will round it down to the nearest integer. This method of hypothesis testing with unequal variances is called the **Smith–Satterthwaite** or also called **Welch's** procedure. Even though this procedure is not widely used, some simulation studies have shown that the Smith–Satterthwaite procedure performs well when variances are unequal, and it gives results that are more or less equivalent to those obtained with the pooled t-test when the variances are equal. However, when the sample sizes are approximately equal, the pooled t-test may still be used. Note that in addressing the question, which of the cases (iii)(a) or (iii)(b) to use in a given problem, we suggest that if the point estimates S_1^2 ofσ_1^2 and S_2^2 of σ_2^2 are approximately the same, then it is logical to assume homogeneity, $\sigma_1^2 = \sigma_2^2$, and use (iii)(a), whereas if S_1^2 and S_2^2 are significantly different, we use (iii)(b). More appropriately, we have tests that can be used to test hypotheses concerning $\sigma_1^2 = \sigma_2^2$ or $\sigma_1^2 \neq \sigma_2^2$, known as the F-test, which we discuss at the end of this subsection. Some authors do suggest doing Welch's t-test all the time, to avoid the test of equality of variances. It should be noted that the assumption of normality is crucial.

EXAMPLE 6.5.2

The IQs of 17 students from one area of a city showed a sample mean of 106 with a sample standard deviation of 10, whereas the IQs of 14 students from another area chosen independently showed a sample mean of 109 with a sample standard deviation of 7. Is there a significant difference between the IQs of the two groups at α=0.01? Assume that the population variances are equal.

Solution

We need to test

$$H_0: \mu_1 - \mu_2 = 0 \text{ versus } H_a: \mu_1 - \mu_2 \neq 0$$

Here n_1=17, $\overline{x}_1 = 106$, *and* s_1=10. *Also,* n_2=14, $\overline{x}_2 = 109$, *and* s_2=7.
We have

$$s_p^2 = \frac{(n_1 - 1)s_1^2 + (n_2 - 1)s_2^2}{n_1 + n_2 - 2} = \frac{(16)(10)^2 + (13)(7)^2}{29} = 77.138.$$

The test statistic is

$$t = \frac{\bar{x}_1 - x_2 - D_0}{s_p\sqrt{\frac{1}{n_1} + \frac{1}{n_2}}} = \frac{106 - 109}{(\sqrt{77.138})\sqrt{\frac{1}{17} + \frac{1}{14}}} = -0.94644.$$

For α=0.01, $t_{0.01,29}$ =2.462. Hence, the rejection region is t < − 2.462 or t > 2.462.

Because the observed value of the test statistic, T =−0.94644, does not fall in the rejection region, there is not enough evidence to conclude that the mean IQs are different for the two groups. Here, we assume that the two samples are independent and taken from normal populations.

EXAMPLE 6.5.3

Assume that two populations are normally distributed with unknown and unequal variances. Two independent samples were drawn from these populations, and the data obtained resulted in the following basic statistics:

$$n_1 = 18 \quad \bar{x}_1 = 20.17 \quad s_1 = 4.3$$
$$n_2 = 12 \quad \bar{x}_2 = 19.23 \quad s_2 = 3.8.$$

Test at the 5% level of significance whether the two population means are different.

Solution

We need to test the hypothesis

$$H_0: \mu_1 - \mu_2 = 0 \text{ versus } H_a: \mu_1 - \mu_2 \neq 0.$$

Here n_1=18, $\bar{x}_1$ = 20.17, and s_1=4.3. Also, n_2=12, $\bar{x}_2$ = 19.23, and s_2=3.8.

The degrees of freedom for the t-distribution are given by

$$v = \frac{\left(s_1^2/n_1 + s_2^2/n_2\right)^2}{\frac{\left(s_1^2/n_1\right)^2}{n_1 - 1} + \frac{\left(s_2^2/n_2\right)^2}{n_2 - 1}}$$
$$= \frac{\left(\frac{(4.3)^2}{18} + \frac{(3.8)^2}{12}\right)^2}{\frac{\left(\frac{(4.3)^2}{18}\right)^2}{17} + \frac{\left(\frac{(3.8)^2}{12}\right)^2}{11}} = 25.685.$$

Hence, rounding down, we have v=25 degrees of freedom. For α=0.05, $t_{0.025,25}$ =2.060. Thus the rejection region is t < −2.060 or t > 2.060.

The test statistic is given by

$$t_v = \frac{\bar{x}_1 - \bar{x}_2 - D_0}{\sqrt{\frac{s_1^2}{n_1} + \frac{s_2^2}{n_2}}}$$
$$= \frac{20.17 - 19.23}{\sqrt{\frac{(4.3)^2}{18} + \frac{(3.8)^2}{12}}} = 0.62939.$$

Because the observed value of the test statistic, t_υ = 0.62939, does not fall in the rejection region, we do not reject the null hypothesis. At α = 0.05, there is not enough evidence to conclude that the means of populations are different. Note that the assumptions we made are that the samples are independent and come from two normal populations. No homogeneity assumption of the variance is made.

EXAMPLE 6.5.4

Infrequent or suspended menstruation can be a symptom of serious metabolic disorders in women. In a study to compare the effect of jogging and running on the number of menses, two independent subgroups were chosen from a large group of women who were similar in physical activity (aside from running), height, occupation, distribution of age, and type of birth control method being used. The first group consisted of a random sample of 26 women joggers who jogged "slow and easy" 5–30 miles per week, and the second group consisted of a random sample of 26 women runners who ran more than 30 miles per week and combined long, slow distance with speed work. The following summary statistics were obtained (E. Dale, D. H. Gerlach, and A. L. Wilhite, "Menstrual Dysfunction in Distance Runners," *Obstet. Gynecol.* **54**, 47–53, 1979):

Joggers $\bar{x}_1 = 10.1,\ s_1 = 2.1$
Runners $\bar{x}_2 = 9.1,\ s_2 = 2.4$

Using α=0.05, (a) test for differences in mean number of menses for each group assuming equality of population variances, and (b) test for differences in mean number of menses for each group assuming inequality of population variances.

Solution

Here we need to test

$$H_0: \mu_1 - \mu_2 = 0 \text{ versus } H_a: \mu_1 - \mu_2 \neq 0.$$

We are given, $n_1 = 26$, $\bar{x}_1 = 10.1$, *and* $s_1 = 2.1$. *Also,* $n_2 = 26$, $\bar{x}_2 = 9.1$, *and* $s_2 = 2.4$.

(a) *Under the assumption* $\sigma_1^2 = \sigma_2^2$, *we have*

$$s_p^2 = \frac{(n_1-1)s_1^2 + (n_2-1)s_2^2}{n_1+n_2-2} = \frac{(25)(2.1)^2 + (25)(2.4)^2}{50} = 5.085.$$

The test statistic is

$$t = \frac{\bar{x}_1 - \bar{x}_2 - D_0}{s_p\sqrt{\frac{1}{n_1}+\frac{1}{n_2}}} = \frac{10.1 - 9.1}{(\sqrt{5.085})\sqrt{\frac{1}{26}+\frac{1}{26}}} = 1.5989.$$

For α=0.05, $t_{0.025,50} \approx 1.96$. *Hence, the rejection region is* $t < -1.96$ *and* $t > 1.96$. *Because* t=1.589 *does not fall in the rejection region, we do not reject the null hypothesis. At* $\alpha = 0.05$, *there is not enough evidence to conclude that the population mean number of menses for joggers and runners is different.*

(b) *Under the assumption* $\sigma_1^2 \neq \sigma_2^2$, *we have*

$$v = \frac{\left(s_1^2/n_1 + s_2^2/n_2\right)^2}{\frac{(s_1^2/n_1)^2}{n_1-1} + \frac{(s_2^2/n_2)^2}{n_2-1}} = \frac{\left(\frac{(2.1)^2}{26} + \frac{(2.4)^2}{26}\right)^2}{\frac{\left(\frac{(2.1)^2}{26}\right)^2}{25} + \frac{\left(\frac{(2.4)^2}{26}\right)^2}{25}} = 49.134.$$

Hence, we have v=49 *degrees of freedom. Because this value is large, the rejection region is still approximately* $t < -1.96$ *and* $t > 1.96$. *Hence, the conclusion is the same as that of part (a). In both parts (a) and (b), we assumed that the samples are independent and come from two normal populations.*

Now we present the summary of the test procedure for testing the difference of two proportions, inherent in two binomial populations. Here, again, we assume that the binomial distribution is approximated by the normal distribution, and thus it is an approximate test.

Summary of hypothesis test for $(p_1 - p_2)$ for large samples ($n_i p_i$ >5 and $n_i q_i$ >5, for i=1, 2)

To test

$$H_0: p_1 - p_2 = D_0$$

versus

$$H_a: \begin{array}{ll} p_1 - p_2 < D_0, & \text{upper tailed test} \\ p_1 - p_2 > D_0, & \text{lower tailed test} \\ p_1 - p_2 \neq D_0, & \text{two-tailed test} \end{array}$$

at the level of significance α, the test statistic is

$$Z = \frac{\hat{p}_1 - \hat{p}_2 - D_0}{\sqrt{\frac{\hat{p}_1 \hat{q}_1}{n_1} + \frac{\hat{p}_2 \hat{q}_2}{n_2}}}$$

where z is the observed value of Z.

The rejection region is

$$RR: \begin{cases} z > z_\alpha, & \text{upper tailed RR} \\ z < -z_\alpha, & \text{lower tailed RR} \\ |z| > z_{\alpha/2}, & \text{two-tailed RR} \end{cases}$$

Assumption: The samples are independent and

$$n_i p_i > 5 \text{ and } n_i q_i > 5, \text{ for } i = 1,2.$$

Decision: Reject H_0 if the test statistic falls in the RR, and conclude that H_a is true with $(1 - \alpha)100\%$ confidence. Otherwise, do not reject H_0, because there is not enough evidence to conclude that H_a is true for given α, and more data is needed.

Example 6.5.5

Because of the impact of the global economy on a high-wage country such as the United States, it is claimed that the domestic content in manufacturing industries fell between 1977 and 1997. A survey of 36 randomly picked US companies gave the proportion of domestic content in total manufacturing in 1977 as 0.37 and in 1997 as 0.36. At the 1% level of significance, test the claim that the domestic content really fell during the period 1977–97.

Solution

Let p_1 be the domestic content in 1977 and p_2 be the domestic content in 1997.
Given $n_1 = n_2 = 36$, $\hat{p}_1 = 0.37$, and $\hat{p}_2 = 0.36$. We need to test

$$H_0: p_1 - p_2 = 0 \text{ versus } H_a: p_1 - p_2 > 0.$$

The test statistic is

$$z = \frac{\hat{p}_1 - \hat{p}_2}{\sqrt{\frac{\hat{p}_1 \hat{q}_2}{n_1} + \frac{\hat{p}_1 \hat{q}_2}{n_2}}}$$

$$= \frac{0.37 - 0.36}{\sqrt{\frac{(0.37)(0.63)}{36} + \frac{(0.36)(0.64)}{36}}} = 0.08813.$$

For $\alpha = 0.01$, $z_{0.01} = 2.325$. *Hence, the rejection region is $z > 2.325$.*
Because the observed value of the test statistic does not fall in the rejection region, at $\alpha = 0.01$, there is not enough evidence to conclude that the domestic content in manufacturing industries fell between 1977 and 1997.

Let $X_1,\ldots,X_n$ and $Y_1,\ldots,Y_n$ be two independent random samples from two normal populations with sample variances S_1^2 and S_2^2, respectively. The problem here is of testing for the equality of the variances, H_0: $\sigma_1^2 = \sigma_2^2$. We have already seen in Chapter 4 that

$$F = \frac{S_1^2/\sigma_1^2}{S_2^2/\sigma_2^2}$$

follows the F-distribution with $v_1 = n_1 - 1$ numerator and $v_2 = n_2 - 1$ denominator degrees of freedom. Under the assumption H_0: $\sigma_1^2 = \sigma_2^2$, we have

$$F = \frac{S_1^2}{S_2^2},$$

which has an F-distribution with (ν_1, ν_2) degrees of freedom. We summarize the test procedure for the equality of variances.

Testing for the equality of variances

To test

$$H_0: \sigma_1^2 = \sigma_2^2$$

versus

$$H_a: \begin{cases} \sigma_1^2 > \sigma_2^2, \text{ lower tailed test} \\ \sigma_1^2 < \sigma_2^2, \text{ upper tailed test} \\ \sigma_1^2 \neq \sigma_2^2, \text{ two-tailed test} \end{cases}$$

at significance level α, the test statistic is

$$F = \frac{S_1^2}{S_2^2}.$$

The rejection region is

$$RR: \begin{cases} f > F_\alpha(\nu_1, \nu_2), \text{ upper tailed RR} \\ f < F_{1-\alpha}(\nu_1, \nu_2), \text{ lower tailed RR} \\ f > F_{\alpha/2}(\nu_1, \nu_2) \text{or} f < F_{1-\alpha/2}(\nu_1, \nu_2), \text{ two-tailed RR} \end{cases}$$

where f is the observed test statistic given by $f = \frac{s_1^2}{s_2^2}$.

Decision: Reject H_0 if the test statistic falls in the RR, and conclude that H_a is true with $(1 - \alpha)100\%$ confidence. Otherwise, keep H_0, because there is not enough evidence to conclude that H_a is true for a given α, and more data is needed.

Assumption:

(i) The two random samples are independent.
(ii) Both populations are normal.

Recall from the section Testing for the equality of variances that to find $F_{1-\alpha}(\nu_1, \nu_2)$, we use the identity $F_{1-\alpha}(\nu_1, \nu_2) = (1/F_\alpha(\nu_2, \nu_1))$.

EXAMPLE 6.5.6

Consider two independent random samples $X_1,\ldots, X_n$ from an $N(\mu_1, \sigma_1^2)$ distribution and $Y_1, \ldots, Y_n$ from an $N(\mu_2, \sigma_2^2)$ distribution. Test $H_0: \sigma_1^2 = \sigma_2^2$ versus $H_a: \sigma_1^2 \neq \sigma_2^2$ for the following basic statistics:

$$n_1 = 25, \bar{x}_1 = 410, s_1^2 = 95, \text{ and } n_2 = 16, \bar{x}_2 = 390, s_2^2 = 300.$$

Use α =0.20.

Solution

Test $H_0: \sigma_1^2 = \sigma_2^2$ versus $H_a: \sigma_1^2 \neq \sigma_2^2$. This is a two-tailed test.
Here the degrees of freedom are ν_1=24 and ν_2=15. The test statistic is

$$F = \frac{s_1^2}{s_2^2} = \frac{95}{300} = 0.317.$$

From the F-table with $\alpha/2$ = 0.10, $F_{0.10}(24, 15)$=1.90 and $F_{0.90}(24, 15)$ =(1/$F_{0.10}(15, 24)$) =1/1.78= 0.56.
Hence, the rejection region is F > 1.90 or F < 0.56. Because the observed value of the test statistic, 0.317, is less than 0.56, we reject the null hypothesis. There is evidence that the population variances are not equal.

6.5.1 Dependent Samples

We now consider the case where the two random samples are not independent. When two samples are dependent (the samples are dependent if one sample is related to the other), then each data point in one sample can be coupled in some natural, nonrandom fashion with each data point in the second sample. This situation occurs when each individual data point within a sample is paired (matched) to an individual data point in the second sample. The pairing may be the result of the individual observations in the two samples: (1) representing before and after a program (such as weight before and after following a certain diet program), (2) sharing the same characteristic, (3) being matched by location, (4) being matched by time, (5) control and experimental, and so forth. Let (X_{1i}, X_{2i}), for i=1, 2, …, n, be a random sample. X_{1i} and X_{2j} $(i \neq j)$ are independent. To test the significance of the difference between two population means when the samples are dependent, we first calculate for each pair of scores the difference, $D_i = X_{1i} - X_{2i}$, i=1, 2, …, n, between the two scores. Let $\mu_D = E(D_i)$, the expected value of D_i. Because pairs of observations form a random sample $D_1, \ldots, D_n$ are independent and identically distributed random variables, if $d_1, \ldots, d_n$ are the observed values of $D_1, \ldots, D_n$, then we define

$$\overline{d} = \frac{1}{n}\sum_{i=1}^{n} d_i \text{ and } s_d^2 = \frac{1}{n-1}\sum_{i=1}^{n}(d_i - \overline{d})^2 = \frac{\sum_{i=1}^{n} d_i^2 - \frac{1}{n}\left(\sum_{i=1}^{n} d_i\right)^2}{n-1}.$$

Now the testing for these n observed differences will proceed as in the case of a single sample. If the number of differences is large ($n \geq 30$), large sample inferential methods for one sample case can be used for the paired differences. We now summarize the hypothesis testing procedure for small samples.

Summary of testing for matched pairs experiment

To test

$$H_0: \mu_D = d_0 \text{ versus } H_a: \begin{array}{l} \mu_D > d_0, \text{ upper tailed test} \\ \mu_D < d_0, \text{ lower tailed test} \\ \mu_D \neq d_0, \text{ two-tailed test} \end{array}$$

the test statistic: $T = \frac{\overline{D} - D_0}{S_D/\sqrt{n}}$ (this approximately follows a Student t-distribution with $(n-1)$ degrees of freedom). The rejection region is

$$\begin{cases} t > t_{\alpha,n-1}, \text{ upper tailed RR} \\ t < -t_{\alpha,n-1}, \text{ lower tailed RR} \\ |t| > t_{\alpha/2,n-1}, \text{ two-tailed RR} \end{cases}$$

where t is the observed test statistic.

Assumptions: The differences are approximately normally distributed.

Decision: Reject H_0 if the test statistic falls in the RR, and conclude that H_a is true with $(1-\alpha)100\%$ confidence. Otherwise, do not reject H_0, because there is not enough evidence to conclude that H_a is true for a given α, and more data is needed.

EXAMPLE 6.5.7

A new diet and exercise program has been advertised as a remarkable way to reduce blood glucose levels in diabetic patients. Ten randomly selected diabetic patients are put on the program, and the results after 1 month are given by the following table:

Before	268	225	252	192	307	228	246	298	231	185
After	106	186	223	110	203	101	211	176	194	203

Does the data provide sufficient evidence to support the claim that the new program reduces blood glucose level in diabetic patients? Use α=0.05.

Solution

We need to test the hypothesis

$$H_0: \mu_D = 0 \text{ versus } H_a: \mu_D < 0.$$

First, we calculate the difference of each pair given in the following table:

Before	268	225	252	192	307	228	246	298	231	185
After	106	186	223	110	203	101	211	176	194	203
Difference (after–before)	−162	−39	−29	−82	−104	−127	−35	−122	−37	18

From the table, the mean of the differences is $\bar{d} = -71.9$, *and the standard deviation* $s_d = 56.2$. *The test statistic is*

$$t = \frac{\bar{d} - d_0}{s_d/\sqrt{n}} = \frac{-71.9}{56.2/\sqrt{10}} = -4.0457 \approx -4.05.$$

From the t-table, $t_{0,05,9} = 1.833$. *Because the observed value of* $t = -4.05 < -t_{0,05,9} = -1.833$, *we reject the null hypothesis and conclude that the sample evidence suggests that the new diet and exercise program is effective. Here we assume the differences follow the normal distribution.*

We can also obtain a $(1 - \alpha)$ 100% confidence interval for μ_D using the formula

$$\left(\bar{D} - t_{\alpha/2}\frac{S_d}{\sqrt{n}}, \bar{D} + t_{\alpha/2}\frac{S_d}{\sqrt{n}}\right)$$

where $t_{\alpha/2}$ is obtained from the t-table with $(n - 1)$ degrees of freedom. The interpretation of the confidence interval is identical to the earlier interpretation.

EXAMPLE 6.5.8

For the data in Example 6.5.7, obtain a 95% confidence interval for μ_D, and interpret its meaning.

Solution

We have already calculated $\bar{d} = -71.9$ *and* $s_d = 56.2$. *From the t-table,* $t_{0.025,9} = 2.262$. *Hence a 95% confidence interval for* μ_D *is* $(-112.1, -31.7)$. *That is,* $P(-112.1 \le \mu_D \le -31.7) \ge 0.95$. *Note that* $\mu_D = \mu_1 - \mu_2$, *and from the confidence limits, we can conclude with at least 95% confidence that* μ_2 *is always greater than* μ_1, *that is,* $\mu_2 > \mu_1$.

It is interesting to compare the matched pairs test with the corresponding two independent sample tests. One of the natural questions is, why must we take paired differences and then calculate the mean and standard deviation for the differences—why can't we just take the difference of means of each sample, as we did for independent samples? The answer lies in the fact that $\sigma_{\bar{D}}^2$ need not be equal to $\sigma_{(\bar{X}_1 - \bar{X}_2)}^2$. Assume that

$$E(X_{ji}) = \mu_j,\ Var(X_{ji}) = \sigma_j^2, \text{ for } j = 1,2,$$

and

$$Cov(X_{1i}, X_{2i}) = \rho\sigma_1\sigma_2$$

where ρ denotes the assumed common correlation coefficient of the pair (X_{1i}, X_{2i}) for $i = 1, 2, \ldots, n$. Because the values of D_i, $i = 1, 2, \ldots, n$, are independent and identically distributed,

$$\mu_D = E(D_i) = E(X_{1i}) - E(X_{2i}) = \mu_1 - \mu_2$$

and

$$\sigma_D^2 = Var(D_i) = Var(X_{1i}) + Var(X_{2i}) - 2Cov(X_{1i}, X_{2i})$$
$$= \sigma_1^2 + \sigma_2^2 - 2\rho\sigma_1\sigma_2.$$

From these calculations,

$$E(\overline{D}) = \mu_D = \mu_1 - \mu_2$$

and

$$\sigma_{\overline{D}}^2 = Var(\overline{D}) = \frac{\sigma_D^2}{n} = \frac{1}{n}(\sigma_1^2 + \sigma_2^2 - 2\rho\sigma_1\sigma_2).$$

Now, if the samples were independent with $n_1 = n_2 = n$, we have

$$E(\overline{X}_1 - \overline{X}_2) = \mu_1 - \mu_2$$

and

$$\sigma_{(\overline{X}_1 - \overline{X}_2)}^2 = \frac{1}{n}(\sigma_1^2 + \sigma_2^2).$$

Hence, if $\rho > 0$, then $\sigma_{\overline{D}}^2 < \sigma_{(\overline{X}_1 - \overline{X}_2)}^2$. As a result, we can see that the matched pairs test reduces any variability introduced by differences in physical factors in comparison to the independent samples test when $\rho > 0$. It is also important to observe that the normality assumption for the difference does not imply that the individual samples themselves are normal. Also, in a matched pairs experiment, there is no need to assume equality of variances for the two populations. Matching also reduces degrees of freedom, because in the case of two independent samples, the degrees of freedom are $(n_1\ n_2 - 2)$, whereas for the case of two dependent samples, it is only $(n - 1)$.

EXERCISES 6.5

6.5.1 Two sets of elementary school children were taught to read by different methods, 50 by each method. At the conclusion of the instructional period, a reading test gave the results $\overline{y_1} = 74$, $\overline{y_2} = 71$, $s_1 = 9$, and $s_2 = 10$. What is the attained significance level if you wish to see if there is evidence of a real difference between the two population means? What would you conclude if you desired an α-value of 0.05?

6.5.2 The following information was obtained from two independent samples selected from two normally distributed populations with unknown but equal variances.

Sample 1	14	15	11	14	10	8	13	10	12	16	15		
Sample 2	17	16	21	12	20	18	16	14	21	20	13	20	13

Test whether μ_1 is lower than μ_2 at $\alpha = 0.02$.

6.5.3 In the academic year 1997–98, two random samples of 25 male professors and 23 female professors from a large university produced a mean salary for male professors of \$58,550 with a standard deviation of \$4000 and an average for female professors of \$53,700 with a standard deviation of \$3200. At the 5% significance level, can you conclude that the mean salary of all male professors for 1997–98 was higher than that of all female professors? Assume that the salaries of male and female professors are both normally distributed with equal standard deviations.

6.5.4 It is believed that the effect of smoking differs depending on race. The following table gives the results of a statistical study for this question.

	Number in the Study	Average Number of Cigarettes per day	Number of Lung Cancer Cases
Whites	400	15	78
African Americans	280	15	70

Does the data indicate that African Americans are more likely to develop lung cancer due to smoking? Use α=0.05.

6.5.5 A supermarket chain is considering two sources, A and B, for the purchase of 50-pound bags of onions. The following table gives the results of a study.

	Source A	Source B
Number of bags weighed	80	100
Mean weight	105.9	100.5
Sample variance	0.21	0.19

Test at α =0.05 whether there is a difference in the mean weights.

6.5.6 To compare the mean hemoglobin (Hb) levels of well-nourished and undernourished groups of children, random samples from each of these groups yielded the following summary.

	Number of Children	Sample Mean	Sample Standard Deviation
Well nourished	95	11.2	0.9
Undernourished	75	9.8	1.2

Test at α =0.01 whether the mean Hb levels of well-nourished children were higher than those of undernourished children.

6.5.7 An aquaculture farm takes water from a stream and returns it after it has circulated through the fish tanks. To find out how much organic matter is left in the wastewater after the circulation, some samples of the water are taken at the intake, and other samples are taken at the downstream outlet and tested for biochemical oxygen demand (BOD). BOD is a common environmental measure of the quantity of oxygen consumed by microorganisms during the decomposition of organic matter. If BOD increases, it can be said that the waste matter contains more organic matter than the stream can handle. The following table gives data for this problem.

Upstream	9.0	6.8	6.5	8.0	7.7	8.6	6.8	8.9	7.2	7.0
Downstream	10.2	10.2	9.9	11.1	9.6	8.7	9.6	9.7	10.4	8.1

Assuming that the samples come from a normal distribution,

(a) Test that the mean BOD for the downstream samples is more than for the samples upstream at α=0.05. Assume that the variances are equal.

(b) Test for the equality of the variances at α=0.05.

(c) In parts (a) and (b), we assumed samples are independent. Now, we feel this assumption is not reasonable. Assuming that the difference of each pair is approximately normal, test that the mean BOD for the downstream samples is more than for the upstream samples at α =0.05.

6.5.8 Suppose we want to know the effect on driving of a drug for cold and allergy, in a study in which the same people were tested twice, once after 1 hour of taking the drug and once when no drug is taken. Suppose we obtain the following data, which represent the number of cones (placed in a certain pattern) knocked down by each of the nine individuals before taking the drug and after an hour of taking the drug.

No drug	0	0	3	2	0	0	3	3	1
After drug	1	5	6	5	5	5	6	1	6

Assuming that the difference of each pair comes from an approximately normal distribution, test if there is any difference in the individuals' driving ability under the two conditions. Use α =0.05. What is the *P*-value?

6.5.9 Suppose that we want to evaluate the role of intravenous pulse cyclophosphamide (IVCP) infusion in the management of nephrotic syndrome in children with steroid resistance. Children were given a monthly infusion of IVCP in a dose of 500–750 mg/m^2. The following data (source: Gulati, S., Kher, V., 2000. Intravenous pulse cyclophosphamide—a new regime for steroid resistant focal segmental glomerulosclerosis. *Indian Pediatr.* 37) represent levels of serum albumin (g/dL) before and after IVCP in 14 randomly selected children with nephrotic syndrome:

Pre-IVCP	2.0	2.5	1.5	2.0	2.3	2.1	2.3	1.0	2.2	1.8	2.0	2.0	1.5	3.4
Post-IVCP	3.5	4.3	4.0	4.0	3.8	2.4	3.5	1.7	3.8	3.6	3.8	3.8	4.1	3.4

Assuming that the samples come from a normal distribution:

(a) Here, we cannot assume that samples are independent. Assuming that the difference of each pair is approximately normal, test that the mean Pre-IVCP is less than the Post-IVCP at α =0.05.

(b) Test for the equality of the variances at α =0.05.

6.5.10 Show that S_D^2 is an unbiased estimator of σ_D^2.

6.5.11 Test H_0: $\sigma_1^2 = \sigma_2^2$ versus H_a: $\sigma_1^2 \neq \sigma_2^2$ for the following data:

$$n_1 = 10,\ \bar{x}_1 = 71,\ s_1^2 = 64 \text{ and } n_2 = 25,\ \bar{x}_2 = 131,\ s_2^2 = 96.$$

Use α=0.10.

6.5.12 The IQs of 17 students from one area of a city showed a mean of 106 with a standard deviation of 10, whereas the IQs of 14 students from another area showed a mean of 109 with a standard deviation of 7. Test for equality of variances between the IQs of the two groups at α=0.02.

6.5.13 The following data give SAT mean scores for math by state for 1989 and 1999 for 20 randomly selected states (source: *The World Almanac and Book of Facts 2000*).

State	1989	1999
Arizona	523	525
Connecticut	498	509
Alabama	539	555
Indiana	487	498
Kansas	561	576
Oregon	509	525
Nebraska	560	571
New York	496	502
Virginia	507	499
Washington	515	526
Illinois	539	585
North Carolina	469	493
Georgia	475	482
Nevada	512	517
Ohio	520	568
New Hampshire	510	518

Assuming that the samples come from a normal distribution:

(a) Test that the mean SAT score for math in 1999 is greater than that of 1989 at α =0.05. Assume the variances are equal.

(b) Test for the equality of the variances at α =0.05.

6.6 CHAPTER SUMMARY

In this chapter, we have learned various aspects of hypothesis testing. First, we dealt with hypothesis testing for one sample where we used test procedures for testing hypotheses about true mean, true variance, and true proportion. Then we discussed the comparison of two populations through their true means, true variances, and true proportions. We also introduced the Neyman–Pearson lemma and discussed LRTs and chi-square tests for categorical data.

We now list some of the key definitions in this chapter.

- Statistical hypotheses
- Tests of hypotheses, tests of significance, or rules of decision
- Simple hypothesis
- Composite hypothesis
- Type I error
- Type II error
- The level of significance
- The *P*-value or attained significance level
- The Smith–Satterthwaite procedure
- Power of the test
- Most powerful test
- Likelihood ratio

In this chapter, we also learned the following important concepts and procedures:

- General method for hypothesis testing
- Steps to calculate β
- Steps to find the *P*-value
- Steps in any hypothesis testing problem
- Summary of hypothesis tests for μ
- Summary of large sample hypothesis tests for p
- Summary of hypothesis tests for the variance σ^2
- Summary of hypothesis tests for $\mu_1 - \mu_2$ for large samples (n_1 and $n_2 \geq 30$)
- Summary of hypothesis tests for $p_1 - p_2$ for large samples
- Testing for the equality of variances
- Summary of testing for a matched pairs experiment
- Procedure for applying the Neyman–Pearson lemma
- Procedure for the likelihood ratio test

6.7 COMPUTER EXAMPLES

In the following examples, if the value of α is not specified, we will always take it as 0.05.

6.7.1 R-Examples

EXAMPLE 6.7.1

One-sample *t*-test:

Using the following data:

Sample x: 66 74 79 80 69 77 78 65 79 81

Test H_0: $\mu = 70$ versus H_a: $\mu > 70$

This example assumes you've stored the data in variable x, please modify the code appropriately.

R-Code:

```
t.test( x, mu=70, alternative="greater");
```

Output:

```
        One-Sample t-test
data: x
t=2.5314, df=9, P-value=0.01608
alternative hypothesis: true mean is greater than 70
95 percent confidence interval:
```

```
71.32406      Inf
sample estimates:
mean of x
   74.8
```

Conclusion: since the P-value=0.01608>0.01, we will not reject H_0 at $\alpha = 0.01$. However, if α is greater than 0.01608, then we will reject the null hypothesis.

EXAMPLE 6.7.2

The management of a local health club claims that its members lose on the average 15 pounds or more w thin the first 3 months after joining the club. To check this claim, a consumer agency took a random sample of 45 members of this health club and found that they lost an average of 13.8 pounds within the first 3 months of membership, with a sample standard deviation of 4.2 pounds.

(a) Find the *P*-value for this test.
(b) Based on the *P*-value in (a), would you reject the null hypothesis at α =0.01?

R-code:

```
> xbar=13.8    #sample mean
> mu0=15         #hypothesized value
> sigma=4.2
> n=45
> z=(xbar-mu0)/(sigma/sqrt(n))
> z
[1] -1.91663
> alpha=.01
> z.alpha=qnorm(1-alpha)
> -z.alpha
[1] -2.326348
```

Since observed z- does not fall in the rejection region, we do not reject the null hypothesis at $\alpha = 0.01$.

If we need P-value approach, then:

```
> pval=pnorm(z)
> pval
```

Output:

```
[1] 0.0276425
```

Again, since *P*-value is larger that $\alpha = 0.01$, we do not reject the null hypothesis.

EXAMPLE 6.7.3 R-code for Exercise 6.4.9

```
> xbar=7225
> mu0=7500
> s=120
> n=6
> t=(xbar-mu0)/(s/sqrt(n))
> t
[1] -5.613414
> alpha=0.01
> t.alpha=qt(1-alpha, df=n-1)
> -t.alpha
[1] -3.36493
> pval=pt(t, df=n-1)
> pval
[1] 0.001240944
```

EXAMPLE 6.7.4 Two sample *t*-test:

Using the following data:

Sample x: 16 18 21 13 19 16 18 15 20 19 14 21 14
Sample y: 14 15 10 13 11 7 12 11 12 15 14

Test H_0: $\mu_x = \mu_y$ versus H_a: $\mu_x < \mu_y$ using $\alpha = 0.02$.
This example assumes you've stored the data in variables x and y. Please modify your code appropriately.

R-Code:

```
t.test(x, y, alternative="less");
```

Output:

```
        Welch Two Sample t-test
data: x and y
```

Since our *P*-value is greater than 0.02 we fail to reject the null.

```
t=4.8077, df=21.963, P-value=1
alternative hypothesis: true difference in means is less than 0
95 percent confidence interval:
    -Inf 6.852384
sample estimates:
mean of x mean of y
  17.23077 12.18182
```

EXAMPLE 6.7.5 One-sample *t*-test (two-taileded):

Using the following data:

Sample X: 6.8 5.6 8.5 8.5 8.4 7.5 9.3 9.4 7.8 7.1 9.9 9.6 9.0 9.4 13.7 16.6 9.1 10.1 10.6 11.1 8.9 11.7 12.8 11.5 12.0 10.6 11.1 6.4 12.3 12.3 11.4 9.9 14.3 11.5 11.8 13.3 12.8 13.7 13.9 12.9 14.2 14.0 15.5 16.9 18.0 17.9 21.8 18.4 34.3

Test H_0: $\mu_x = 12$ versus H_a: $\mu_x \neq 12$ using $\alpha = 0.05$.
This example assumes you've stored the data in variable x. Please modify your code appropriately.

R-Code:

```
t.test(x, mu=12);
```

Output:

```
      One-Sample t-test
data: x
```

Since the *P*-value is greater than 0.05 we fail to reject the null hypothesis true mean could be 12.

```
t=0.1854, df=48, P-value=0.8537
alternative hypothesis: true mean is not equal to 12
95 percent confidence interval:
  10.77437 13.47461
sample estimates:
mean of x
  12.12449
```

EXAMPLE 6.7.6 Paired samples *t*-test:

Using data:

Upstream (x)	9.0	6.8	6.5	8.0	7.7	8.6	6.8	8.9	7.2	7.0
Downstream (y)	10.2	10.2	9.9	11.1	9.6	8.7	9.6	9.7	10.4	8.1

Test H_a: $\mu_d = 0$ versus H_a: $\mu_d < 0$ using $\alpha = 0.05$.
This is a paired *t*-test and assumes you've stored the data in variables *x* and *y*. Please modify your code appropriately.
R-Code:

```
t.test(x, y, paired=TRUE, alternative="less");
```

Output:

```
    Paired t-test
data: x and y
t=-5.3982, df=9, P-value=0.000217
```

We reject the null hypothesis since our *P*-value is less than 0.05 suggesting than the mean difference is less than 0.

```
alternative hypothesis: true difference in means is less than 0
95 percent confidence interval:
    -Inf -1.38689
sample estimates:
mean of the differences
                    -2.1
```

6.7.2 SPSS Examples

EXAMPLE 6.7.7

Consider the data

66 74 79 80 69 77 78 65 79 81

Using SPSS, test H_0: μ=75 versus H_1: $\mu > 75$.

Solution

Use the following procedure:

1. Enter the data in column 1.
2. Click ***Analyze > Compare Means > One-sample t-Test...***, Move ***var00001*** to ***Test Variable(s)***, and change ***Test Value: 0*** to ***75***. Click ***OK***.

If we want the computer to calculate the *P*-value in the previous example, use the following procedure.

1. Enter the test statistic ***(−0.105)*** in the data editor using "***teststat***".
2. Click ***Transform > compute...***
3. Type "***P-value***" in the box called ***Tarobtain value***. In the box called ***Functions***: scroll and click on ***CDF.T(q,df)*** and move to Numeric Expressions.
4. The CDF(q,df) will appear as ***CDF(?,?)*** in the Numeric Expressions box. Replace teststat for ***q*** and **9** for ***df*** (the degree of freedom in this example is 9). Click ***OK***.

EXAMPLE 6.7.8

For the following data

Sample 1: 16 18 21 13 19 16 18 15 20 19 14 21 14

Sample 2: 14 15 10 13 11 7 12 11 12 15 14

Test H_0: $\mu_1=\mu_2$ versus H_1: $\mu_1 < \mu_2$. Use α=0.02.

Solution

In column 1, under the title "group" enter 1 s to identify the sample 1 data and 2 s to identify sample 2 data. In column ***C2****, under the title "data" enter the data corresponding to samples 1 and 2. Then:*

Analyze > Compare Means > Independent Samples t-test... > bring ***Data*** to ***Test Variable(s)***: and ***group*** to ***Grouping Variable***:, click ***Define Groups...***, and enter ***1*** for ***sample 1***, ***2*** for ***sample 2*** > click ***continue*** > click ***Options...*** Enter ***98*** in ***Confidence interval***: > click ***continue*** > ***OK***

EXAMPLE 6.7.9 Paired *t*-test

For the data of Example 6.5.7, use SPSS to test whether the data provide sufficient evidence for the claim that the new program reduces blood glucose level in diabetic patients. Use α =0.05.

Solution

Enter **after** *data in column* ***C1*** *and* **before** *the data in column* ***C2****. Then:*

Analyze > ***Compare Means*** > ***Paired-Sample T-Test*** > bring after and before to ***Paired Variables***: so that it will look **after-before** > click ***OK***

6.7.3 SAS Examples

To conduct a hypothesis test using SAS, we could use proc t-test, or proc means with option of computing the t-value and corresponding probability. However, to use this, we need a hypothesis of the form H_0: μ =0. For testing nonzero values, H_0: μ =μ_0, we must create a new variable by subtracting μ_0 from each observation and then use the test procedure for this new variable. The following example illustrates this concept.

EXAMPLE 6.7.10 *t*-Test

The following radar measurements of speed (in miles per hour) are obtained for 10 vehicles traveling on a stretch of interstate highway.

66 74 79 80 69 77 78 65 79 81.

Do the data provide sufficient evidence to indicate that the mean speed at which people travel on this stretch of highway is at least 75 mph? Test using α =0.01. Use an SAS procedure to do the analysis.

Solution

In the SAS editor, type in the following commands:

```
data speed;
    title 'Test on highway speed';
    input X @@;
    Y=X-75;
datalines;
66 74 79 80 69 77 78 65 79 81
;
PROC T-TEST data=speed;
run;
```

We obtain the following output.

Test on highway speed

The T-TEST Procedure Statistics

Statistics

	Lower CL		Upper	CL	Lower CL	Upper CL	
Variable N	Mean	Mean	Mean	Std Dev	Std Dev	Std Dev	Std Err
X	10 1.8962	70.511	74.8	79.089	4.1245	5.9963	10.947
Y	10	−4.489	−0.2	4.0895	4.1245	5.9963	10.947

T-Tests

Variable	DF	t-Value	Pr > |t|
X	9	39.45	<.0001
Y	9	−0.11	0.9183

To test H_0: μ=75, we need to look at the Y-values. The corresponding t-value is −0.11, and because this is a one-sided test, we need to divide 0.9183 by 2 to obtain the P-value as P =0.45915. Because the P-value is larger than 0.01 =α, we cannot reject the null hypothesis.

One of the easier ways to conduct large sample hypothesis testing using SAS procedures is through the computation of the P-value. The following example illustrates the procedure:

EXAMPLE 6.7.11 z-Test

It is claimed that the average miles driven per year for sports cars is at least 18,000 miles. To check the claim, a consumer firm tests 40 of these cars randomly and obtains a mean of 17,463 miles with standard deviation of 1348 miles. What can it conclude if $\alpha = 0.01$?

Solution

Here we will find the P-value and compare that with α to test the hypothesis. We use the following SAS procedure:

```
Data ex888;
z=(17463–18000)/(1348/(SQRT(40)));
                         pval=probnorm(z);
run;
proc print data=ex888;
title 'Test of mean, large sample';
run;
```

We obtain the following output:

Test of mean		large sample
Obs	z	pval
1	2.51950	.005876079

Because the P-value of 0.005876079 is less than $\alpha = 0.01$, we reject the null hypothesis. There is sufficient evidence to conclude that the mean miles driven per year for sport cars is less than 18,000.

Note that in the previous example, the value of z was negative. If the value of z is positive, use pval=probnorm($-z$);, also, if it is a two-sided hypothesis, we need to multiply by 2, so use pval=probnorm(z)*2; to obtain the P-value.

EXAMPLE 6.7.12 Paired t-test

For the data of Example 6.5.7, use SAS to test whether the data provide sufficient evidence for the claim that the new program reduces blood glucose level in diabetic patients. Use $\alpha = 0.05$.

Solution

We can use the following commands:

```
data dietexr;
input before after;
diff=after - before;
datalines;

268  106
225  186
252  223
192  110
307  203
228  101
246  211
298  176
231  194
185  203

;
run;
proc means data=dietexr t prt;
```

```
var diff;
title 'Test of mean, Paired difference';
run;
```

PROJECTS FOR CHAPTER 6

6A. Testing on Computer-Generated Samples

(a) Small sample test:
Generate a sample of size 20 from a normal population with μ =10, and σ^2 =4.
(i) Perform a t-test for the test H_0: μ =10 versus H_a: $\mu \neq 10$ at level α =0.05.
(ii) Perform the test H_0: σ^2 =4 versus H_a: $\sigma^2 \neq 4$ at level α =0.05.
Repeat the procedure 10 times, and comment on the results.

(b) Large sample test:
Generate a sample of size 50 from a normal population with μ =10, and σ^2 =4. Perform a z-test for the test H_0: μ =10 versus H_a: $\mu \neq 10$ at level α =0.05. Repeat the procedure 10 times, and comment on the results.

6B. Conducting a Statistical Test with Confidence Interval

Let θ be any population parameter. Consider the three tests of hypotheses

$$H_0\text{: } \theta = \theta_0 \text{ versus } H_a\text{: } \theta > \theta_0 \quad (1)$$

$$H_0\text{: } \theta = \theta_0 \text{ versus } H_a\text{: } \theta < \theta_0 \quad (2)$$

$$H_0\text{: } \theta = \theta_0 \text{ versus } H_a\text{: } \theta \neq \theta_0 \quad (3)$$

The following procedure can be exploited to test a statistical hypothesis utilizing the confidence intervals.

Procedure to Use Confidence Interval for Hypothesis Testing

Let θ be any population parameter.

(a) For test (1), that is,

$$H_0\text{: } \theta = \theta_0 \text{ versus } H_a\text{: } \theta > \theta_0$$

Choose a value for α. From a random sample, compute a confidence interval for θ using a confidence coefficient equal to $1 - 2\alpha$. Let L be the lower end point of this confidence interval.

$$\text{Reject } H_0 \text{ if } \theta_0 < L.$$

That is, we will reject the null hypothesis if the confidence interval is completely to the right of θ_0.

(b) For test (2), that is,

$$H_0\text{: } \theta = \theta_0 \text{ versus } H_a\text{: } \theta < \theta_0$$

Choose a value for α. From a random sample, compute a confidence interval for θ using a confidence coefficient equal to $1 - 2\alpha$. Let U be the upper end point of this confidence interval.

$$\text{Reject } H_0 \text{ if } U < \theta_0.$$

That is, we will reject the null hypothesis if the confidence interval is completely to the left of θ_0.

(c) For test (3), that is,

$$H_0\colon \theta = \theta_0 \text{ versus } H_a\colon \theta \neq \theta_0$$

Choose a value for α. From a random sample, compute a confidence interval for θ using a confidence coefficient equal to $1 - \alpha$. Let L be the lower end point and U be the upper end point of this confidence interval.

$$\text{Reject } H_0 \text{ if } \theta_0 < L \text{ or } U < \theta_0.$$

That is, we will reject the null hypothesis if the confidence interval does not contain θ_0.

(i) For any large data set, conduct all three of these hypothesis tests using a confidence interval for the population mean.

(ii) For any small data set, conduct all three of these hypothesis tests using a confidence interval for the population mean.

Chapter 7

Linear Regression Models

Chapter outline

OBJECTIVE

In this chapter we will study linear relationships in random sample data and use the method of least squares to estimate the necessary parameters.

Sir Francis Galton
(Source: *http://en.wikipedia.org/wiki/francis_galton*)

English scientist Sir Francis Galton (1822–1911), a cousin of Charles Darwin, made significant contributions to both genetics and psychology. He is the inventor of regression and a pioneer in applying statistics to biology. One of the data sets that he considered consisted of the heights of fathers and first sons. He was interested in predicting the height of the son based on the height of the father. Looking at the scatterplots of these heights, Galton saw that the trend was linear and increasing. After fitting a line to these data (using the techniques described in this chapter), he observed that for fathers

Mathematical Statistics with Applications in R. DOI: https://doi.org/10.1016/B978-0-443-27548-7.00007-4

whose heights were taller than the average, the regression line predicted that taller fathers tended to have shorter sons and shorter fathers tended to have taller sons. There is a regression toward the mean. That is how the method of this chapter got its name: regression.

7.1 INTRODUCTION

In earlier chapters, we were primarily concerned about inferences on population parameters. In this chapter, we examine the relationship between one or more variables and create a model that can be used for predictive purposes. For example, consider the question "Is there statistical evidence to conclude that the countries with the highest average blood-cholesterol levels have the greatest incidence of heart disease?" It is important to answer this if we want to make appropriate lifestyle and medical choices. We will study the relationship between variables using regression analysis. Our aim is to create a model and study inferential procedures when one dependent and several independent variables are present. We denote by Y the random variable to be predicted, also called the *dependent* variable (or response variable), and by X_i the *independent* (or predictor) variables used to model (or predict) Y. For example, let (x, y) denote the height and weight of an adult male. Our interest may be to find the relationship between height and weight from a sample measurement of n individuals. The process of finding a mathematical equation that best fits the noisy data is known as *regression analysis*. In his book *Natural Inheritance*, Sir Francis Galton introduced the word regression in 1889 to describe certain genetic relationships. The technique of regression is one of the most popular statistical tools to study the dependence of one variable with respect to another. There are different forms of regression: *simple linear*, *nonlinear*, *multiple*, and others. The primary use of a regression model is prediction. When using a model to predict Y for a particular set of values of $x_1, \ldots, x_k$, one may want to know how large the error of prediction might be. Regression analysis in general, after collecting the sample data, involves the following steps.

Procedure for regression modeling

1. Hypothesize the form of the model as $Y = f(x_1, \ldots, x_k; \beta_0, \beta_1, \ldots, \beta_k) + \varepsilon$. Here ε represents the random error term. We assume that $E(\varepsilon) = 0$ but $Var(\varepsilon) = \sigma^2$ is unknown. From this model we can obtain $E(Y) = f(x_1, \ldots, x_k; \beta_0, \beta_1, \ldots, \beta_k)$.
2. Use the sample data to estimate the unknown parameters in the model.
3. Check for goodness of fit of the proposed model.
4. Use the model for prediction.

The function $f(x_1, \ldots, x_k; \beta_0, \beta_1, \ldots, \beta_k)$ $(k \geq 1)$ contains the independent, explanatory, or predictor variables $x_1, \ldots, x_n$ (assumed to be nonrandom) and unknown parameters or weights $\beta_0, \beta_1, \ldots, \beta_k$ and ε representing the random error or error variable. The variable y is regarded as the response, outcome, or dependent variable. We now proceed to introduce the simplest form of a regression model, called simple linear regression.

7.2 THE SIMPLE LINEAR REGRESSION MODEL

Consider a random sample of n observations of the form $(x_1, y_1), (x_2, y_2), \ldots, (x_n, y_n)$, where X is the independent variable, and Y is the dependent variable, both being scalars. A preliminary descriptive technique for determining the form of the relationship between X and Y is the *scatter diagram* or the *scatter plot*. A scatter diagram is drawn by plotting the sample observations in Cartesian coordinates. The pattern of the points gives an indication of a linear, nonlinear, or no relationship between the variables. A no relationship may indicate that events are happening randomly, and any effort to predict based on these data will be futile. Thus, we can consider the scatter plots as a visualization and discovery tool. In practice with very large data sets, scatter plots may show trends, clusters, patterns, and relationships among the data points. In this chapter, we will use the scatter plots for identifying only possible linear or nonlinear relationships.

Fig. 7.1A displays the relationship between x and y, which is fairly linear, whereas the relationship is somewhat like a parabola in Fig. 7.1B, and in Fig. 7.1C there is no obvious relationship between the variables.

Once the scatter diagram reveals a linear relationship, the problem then is to find the linear model that "best" fits the given data. To this end, we will first give a general definition of a linear statistical model, called a multiple linear regression model.

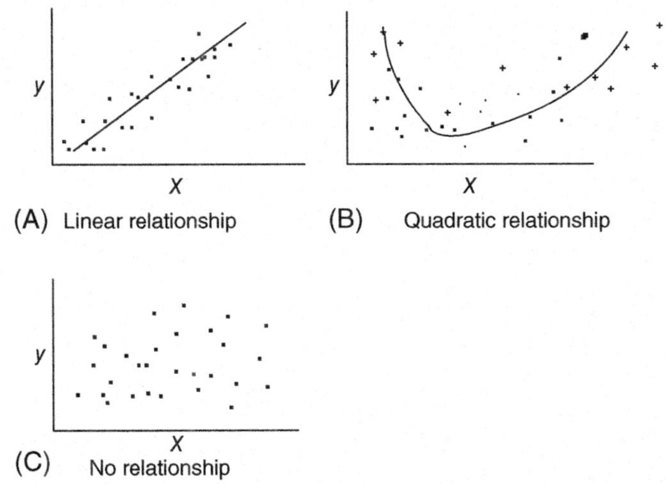

FIGURE 7.1 Scatter diagrams. Scatter plots comparing three types of relationships between an x axis and a y axis, arranged as separate panels. Panel A shows a scatter plot with the x axis as the horizontal axis and the y axis as the vertical axis, where square data points cluster around a straight solid line with a positive slope, indicating a linear relationship as x increases then y increases proportionally. Panel B shows a scatter plot with the same axes, where mixed square and cross shaped data points follow a curved solid line that decreases to a minimum and then increases, indicating a quadratic relationship with a single low point in the middle range of x values. Panel C shows a scatter plot where square data points are widely dispersed across the plotting area without any visible trend, indicating no relationship between x axis values and y axis values.

Definition 7.2.1. *A* **multiple linear regression model** *relating a random response Y to a set of predictor variables $x_1, \ldots, x_k$ is an equation of the form*

$$Y = \beta_0 + \beta_1 x_1 + \beta_2 x_2 + \ldots + \beta_k x_k + \varepsilon,$$

where $\beta_0, \ldots, \beta_k$ are unknown parameters, $x_1, \ldots, x_k$ are the independent nonrandom variables, and ε is a random variable representing an error term. We assume that $E(\varepsilon) = 0$, or equivalently,

$$E(Y) = \beta_0 + \beta_1 x_1 + \beta_2 x_2 + \ldots + \beta_k x_k.$$

To understand the basic concepts of regression analysis, we shall consider a single dependent variable Y and a single independent nonrandom variable x. We assume that there are no measurement errors in x_i. The possible measurement errors in Y and the uncertainties in the assumed model are expressed through the random error ε. Our inability to provide an exact model for a natural phenomenon is expressed through the random term ε, which will have a specified probability distribution (such as a normal) with mean zero. Thus, one can think of Y as having a deterministic component, $E(Y)$, and a random component, ε. If we take $k = 1$ in the multiple linear regression model, we have a simple linear regression model.

Definition 7.2.2. *If $Y = \beta_0 + \beta_1 x + \varepsilon$, this is called a* **simple linear regression model**. *Here, β_0 is the y-intercept of the line, and β_1 is the slope of the line. The term ε is the error component.*

This basic linear model assumes the existence of a linear relationship between the variables x and Y that is disturbed by a random error ε. The known data points are the pairs $(x_1, y_2), (x_2, y_2), \ldots, (x_n, y_n)$; the problem of simple linear regression is to fit a straight line optimal in some sense to the set of data, as shown in Fig. 7.2, below.

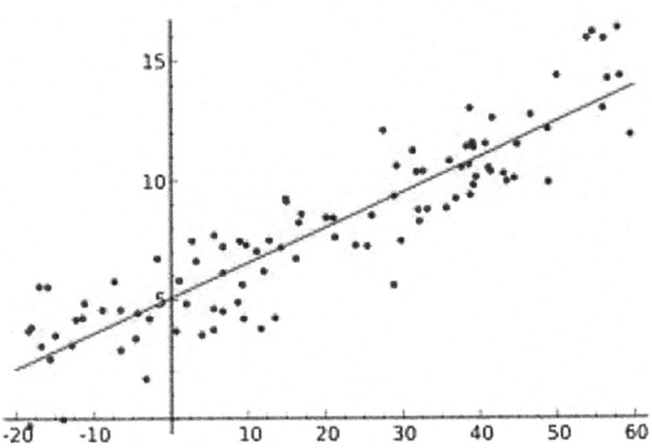

FIGURE 7.2 Scatterplot and least-squares regression line. Scatter plot depicting a positive linear relationship between x axis values and y axis values with clustered data points around a fitted solid line.

Now, the problem becomes one of finding estimators for β_0 and β_1. Once we obtain the "good" estimators $\hat{\beta}_0$ and $\hat{\beta}_1$, we can fit a line to the data given by the prediction equation $\hat{Y} = \hat{\beta}_0 + \hat{\beta}_1 x$. Unlike the single variable estimation problems, now the response variable is dependent on the independent variables, and thus estimators have to reflect this aspect. The question is whether this predicted line gives the "best" (in some sense) description of the data. This necessitates a new method of estimation. We now describe the most widely used technique, called the method of least squares, to obtain the estimators or weights of the parameters.

7.2.1 The Method of Least Squares

As stated $(x_1, y_1), (x_2, y_2), \ldots,(x_n, y_n)$ are the n observed data points, with corresponding errors ε_i, $i = 1, \ldots, n$. That is,

$$Y_i = \beta_0 + \beta_1 x_i + \varepsilon_i,\ i = 1,2, \ldots,n.$$

We assume that the errors ε_i, $i = 1, \ldots, n$ are independent and identically distributed with $E(\varepsilon_i) = 0$, $i = 1, \ldots, n$, and $Var(\varepsilon_i) = \sigma^2$, $i = 1, \ldots, n$. One of the ways to decide on how well a straight line fits the set of data is to determine the extent to which the data points deviate from the line. The straight line model for the response Y for a given x is

$$Y = \beta_0 + \beta_1 x + \varepsilon.$$

Because we assumed that $E(\varepsilon) = 0$, the expected value of Y is given by

$$E(Y) = \beta_0 + \beta_1 x.$$

The estimator of the $E(Y)$, denoted by $\hat{Y}$, can be obtained by using the estimators $\hat{\beta}_0$ and $\hat{\beta}_1$ of the parameters β_0 and β_1, respectively. Then, the fitted regression line we are looking for is given by

$$\hat{Y} = \hat{\beta}_0 + \hat{\beta}_1 x.$$

For observed values (x_i,y_i), we obtain the estimated value of y_i as

$$\hat{y}_i = \hat{\beta}_0 + \hat{\beta}_1 x_i.$$

The deviation of the observed y_i from its predicted value $\hat{y}_i$, called the *ith residual*, is defined by

$$e_i = (y_i - \hat{y}_i) = [y_i - (\hat{\beta}_0 + \hat{\beta}_1 x_i)].$$

The residuals, or errors, e_i, are the vertical distances between observed and predicted values of $y'_i s$ (see in Fig. 7.3 below).

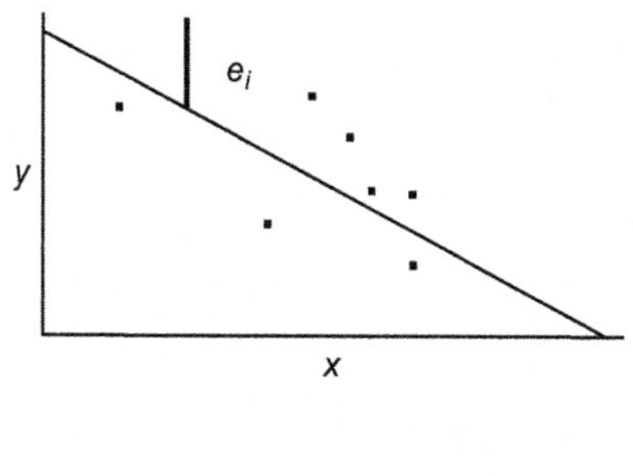

FIGURE 7.3 Illustration of error e_i.

Definition 7.2.3. *The sum of squares for errors (SSE) or sum of squares of the residuals for all of the n data points is*

$$SSE = \sum_{i=1}^{n} e_i^2 = \sum_{i=1}^{n}\left[y_i - \left(\hat{\beta}_0 + \hat{\beta}_1 x_i\right)\right]^2.$$

The least-squares approach to estimation is to find $\hat{\beta}_0$ and $\hat{\beta}_1$ that minimize the sum of squared residuals, SSE. Thus, in the method of least squares, we choose β_0 and β_1 so that SSE is a minimum. The quantities $\hat{\beta}_0$ and $\hat{\beta}_1$ that make the SSE a minimum are called the *least-squares estimates* of the parameters β_0 and β_1, and the corresponding line $\hat{y} = \hat{\beta}_0 + \hat{\beta}_1 x$ is called the *least-squares line.*

Definition 7.2.4. *The* **least-squares line** $\hat{y} = \hat{\beta}_0 + \hat{\beta}_1 x$ *is one that satisfies the following property:*

$$SSE = \sum_{i=1}^{n}\left(y_i - \hat{y}_i\right)^2,$$

is a minimum for any other straight line model with the sum of errors

$$SE = \sum_{i=1}^{n}\left(y_i - \hat{y}_i\right) = 0.$$

Thus, the least-squares line is a line of the form $y = b_0 + b_1 x$ for which the error sum of squares $\sum_{i=1}^{n}(y_i - b_0 - b_1 x)^2$ is a minimum. The minimum is taken over all values of b_0 and b_1, and $(x_1, y_1), (x_2, y_2), \ldots, (x_n, y_n)$ are observed data pairs.

The problem of fitting a least-squares line now reduces to finding the quantities $\hat{\beta}_0$ and $\hat{\beta}_1$ that minimize the error sum of squares.

7.2.2 Derivation of $\hat{\beta}_0$ and $\hat{\beta}_1$

Now we derive expressions for $\hat{\beta}_0$ and $\hat{\beta}_1$ using the methods of calculus. If *SSE* attains a minimum, then the partial derivatives of *SSE* with respect to β_0 and β_1 are zeros. That is,

$$\begin{aligned}\frac{\partial SSE}{\partial \beta_0} &= \frac{\partial\left\{\sum_{i=1}^{n}[y_i - (\beta_0 + \beta_1 x_i)]^2\right\}}{\partial \beta_0} \\ &= -\sum_{i=1}^{n} 2\left[y_i - \left(\beta_0 + \beta_1 x_i\right)\right] \\ &= 2\left(\sum_{i=1}^{n} y_i - n\beta_0 - \beta_1 \sum_{i=1}^{n} x_i\right) = 0,\end{aligned} \tag{7.1}$$

and

$$\begin{aligned}\frac{\partial SSE}{\partial \beta_1} &= \frac{\partial\left\{\sum_{i=1}^{n}\left[y_i - \left(\beta_0 + \beta_1 x_i\right)\right]^2\right\}}{\partial \beta_1} \\ &= -\sum_{i=1}^{n} 2\left[y_i - \left(\beta_0 + \beta_1 x_i\right)\right]x_i = -2\left(\sum_{i=1}^{n} x_i y_i - \beta_0 \sum_{i=1}^{n} x_i - \beta_1 \sum_{i=1}^{n} x_i^2\right) = 0.\end{aligned} \tag{7.2}$$

Eqs. (7.1) and (7.2) are called the *least-squares equations* for estimating the parameters of a line. From (7.1) and (7.2), we obtain a set of linear equations called the *normal equations,*

$$\sum_{i=1}^{n} y_i = n\beta_0 + \beta_1 \sum_{i=1}^{n} x_i, \tag{7.3}$$

and

$$\sum_{i=1}^{n} x_i y_i = \beta_0 \sum_{i=1}^{n} x_i + \beta_1 \sum_{i=1}^{n} x_i^2. \tag{7.4}$$

Solving for β_0 and β_1 from Eqs. (7.3) and (7.4), we obtain

$$\hat{\beta}_1 = \frac{\sum_{i=1}^{n}(x_i-\overline{x})(y_i-\overline{y})}{\sum_{i=1}^{n}(x_i-\overline{x})^2} = \frac{n\sum_{i=1}^{n}x_iy_i - \sum_{i=1}^{n}x_i\sum_{i=1}^{n}y_i}{n\sum_{i=1}^{n}x_i^2-(\sum_{i=1}^{n}x_i)^2} = \frac{\sum_{i=1}^{n}x_iy_i - \frac{\sum_{i=1}^{n}x_i\sum_{i=1}^{n}y_i}{n}}{\sum_{i=1}^{n}x_i^2 - \frac{(\sum_{i=1}^{n}x_i)^2}{n}}, \tag{7.5}$$

and

$$\hat{\beta}_0 = \overline{y} - \hat{\beta}_1\overline{x}. \tag{7.6}$$

To simplify the formula for $\hat{\beta}_1$, set

$$S_{xx} = \sum_{i=1}^{n} x_i^2 - \frac{(\sum_{i=1}^{n}x_i)^2}{n}, \quad S_{xy} = \sum_{i=1}^{n} x_iy_i - \frac{(\sum_{i=1}^{n}x_i)\left(\sum_{i=1}^{n}y_i\right)}{n},$$

Thus, we can rewrite Eq. (7.5) as

$$\hat{\beta}_1 = \frac{S_{xy}}{S_{xx}}.$$

It can be shown (by using the second derivatives) that Eqs. (7.5) and (7.6) do indeed minimize *SSE*. Now we will summarize the procedure for fitting a least-squares line.

Procedure for fitting a least-squares line

1. Form the n data points $(x_1, y_1), (x_2, y_2), \ldots, (x_n, y_n)$, compute the following quantities: $\sum_{i=1}^{n} x_i$, $\sum_{i=1}^{n} x_i^2$, $\sum_{i=1}^{n} y_i$, $\sum_{i=1}^{n} y_i^2$, and $\sum_{i=1}^{n} x_iy_i$. Also compute the sample means, $\overline{x} = (1/n)\sum_{i=1}^{n} x_i$ and $\overline{y} = (1/n)\sum_{i=1}^{n} y_i$.
2. Compute

$$S_{xx} = \sum_{i=1}^{n} x_i^2 - \frac{(\sum_{i=1}^{n}x_i)^2}{n} = \sum_{i=1}^{n}(x_i-\overline{x})^2$$

and

$$S_{xy} = \sum_{i=1}^{n} x_iy_i - \frac{(\sum_{i=1}^{n}x_i)(\sum_{i=1}^{n}y_i)}{n} = \sum_{i=1}^{n}(x_i-\overline{x})(y_i-\overline{y}).$$

3. Compute $\hat{\beta}_0$ and $\hat{\beta}_1$ by substituting the computed quantities from step 1 into the equations

$$\hat{\beta}_1 = \frac{S_{xy}}{S_{xx}}$$

and

$$\hat{\beta}_0 = \overline{y} - \hat{\beta}_1\overline{x}.$$

4. The fitted least-squares line is

$$\hat{y} = \hat{\beta}_0 + \hat{\beta}_1 x.$$

For a graphical representation, in the xy-plane, plot all the data points and draw the least-squares line obtained in step 4.

Once we have accomplished the best-fit combination of the two parameters β_0 and β_1, any deviation of either parameter away from its optimum value will cause the sum of squares error to increase. Thus, the optimum combination of the pairs $(\hat{\beta}_0, \hat{\beta}_1)$ forms a global minimum point of the error sum of squares among all possible values of β_0 and β_1 for the given data set.

EXAMPLE 7.2.1

Use the method of least squares to fit a straight line to the accompanying data points. Give the estimates of β_0 and β_1. Plot the points and sketch the fitted least-squares line. The observed data values are given in the following table.

x	-1	0	2	-2	5	6	8	11	12	-3
y	-5	-4	2	-7	6	9	13	21	20	-9

Solution

Form the data table to compute the necessary terms

x_i	y_i	x_iy_i	x_i^2
-1	-5	5	1
0	-4	0	0
2	2	4	4
-2	-7	14	4
5	6	30	25
6	9	54	36
8	13	104	64
11	21	231	121
12	20	240	144
-3	-9	27	9
$\Sigma x_i = 38$	$\Sigma y_i = 46$	$\Sigma x_iy_i = 709$	$\Sigma x_i^2 = 408$

$$S_{xx} = \sum_{i=1}^{n} x_i^2 - \frac{(\sum_{i=1}^{n} x_i)^2}{n} = 408 - \frac{(38)^2}{10} = 263.6,$$

and

$$S_{xy} = \sum_{i=1}^{n} x_i y_i - \frac{(\sum_{i=1}^{n} x_i)(\sum_{i=1}^{n} y_i)}{n} = 709 - \frac{(38)(46)}{10} = 534.2,$$

$$\bar{x} = 3.8 \text{ and } \bar{y} = 4.6.$$

Therefore,

$$\hat{\beta}_1 = \frac{S_{xy}}{S_{xx}} = \frac{534.2}{263.6} = 2.0266,$$

and

$$\hat{\beta}_0 = \bar{y} - \hat{\beta}_1 \bar{x}$$
$$= 4.6 - (2.0266)(3.8) = -3.1011.$$

Hence, the least-squares regression line for these data is

$$\hat{y} = \hat{\beta}_0 + \hat{\beta}_1 x = -3.1011 + 2.0266x$$

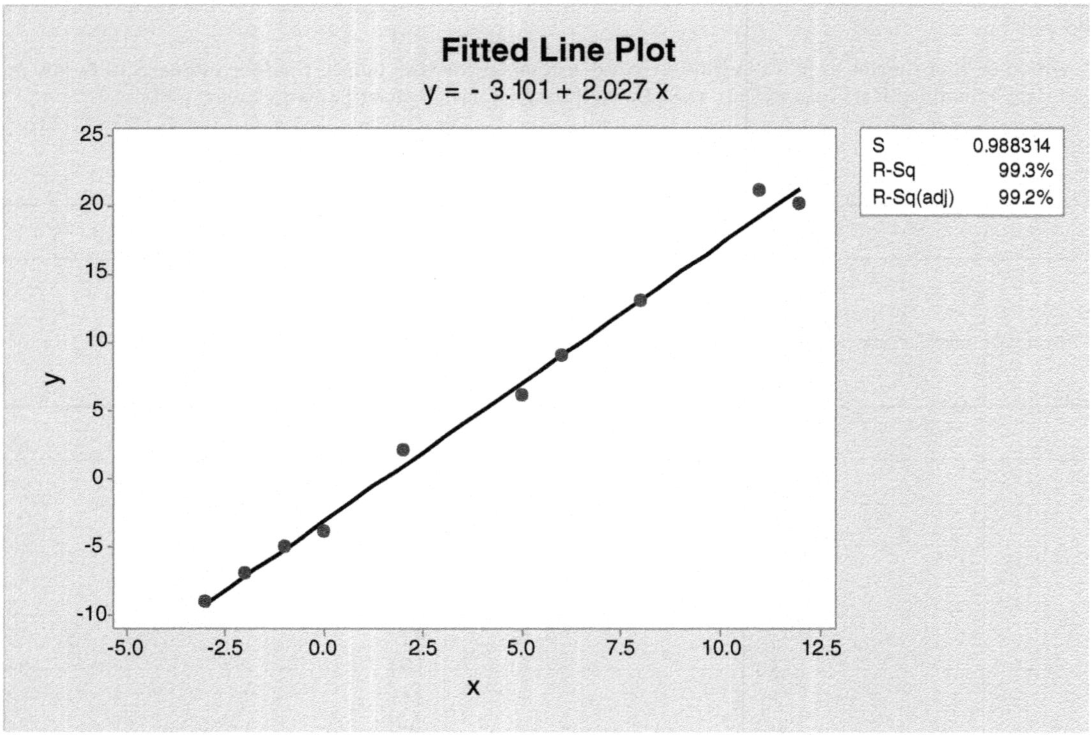

FIGURE 7.4 Simple regression line. Fitted line plot showing a strong positive linear relationship between x axis and y axis values with data points closely aligned to a solid fitted line.

and its plot is shown in Fig. 7.4.

```
# Data
x <- c(-1, 0, 2, -2, 5, 6, 8, 11, 12, -3)
y <- c(-5, -4, 2, -7, 6, 9, 13, 21, 20, -9)
# Linear regression
model <- lm(y ~ x)
# Summary of the model
summary(model)
# Plotting
plot(x, y, main = "Simple Linear Regression", xlab = "x", ylab = "y", pch = 19, col = "blue")
abline(model, col = "red", lwd = 2)
# Add regression equation to plot
coeff <- coef(model)
eq <- paste0("y = ", round(coeff[1], 2), " + ", round(coeff[2], 2), "x")
legend("topleft", legend = eq, col = "red", lwd = 2, bty = "n")
```

Running this code, we will obtain the line in Fig. 7.4.

Recalling that for the regression line $\hat{y} = \hat{\beta}_0 + \hat{\beta}_1 x$, we have defined SSE to be

$$SSE = \sum_{i=1}^{n} (y_i - \hat{y}_i)^2 = \sum_{i=1}^{n} (y_i - \hat{\beta}_0 - \hat{\beta}_1 x_i)^2.$$

We now show that

$$SSE = S_{yy} - \hat{\beta}_1 S_{xy}, \text{ where } S_{yy} = \sum_{i=1}^{n} y_i^2 - \frac{(\sum_{i=1}^{n} y_i)^2}{n} = \sum_{i=1}^{n} (y_i - \bar{y})^2.$$

We know that

$$
\begin{aligned}
SEE &= \sum_{i=1}^{n}\left(y_i-\hat{\beta}_0-\hat{\beta}_1 x_i\right)^2 \\
&= \sum_{i=1}^{n}\left(y_i-\overline{y}+\hat{\beta}_1\overline{x}-\hat{\beta}_1 x_i\right)^2 \\
&= \sum_{i=1}^{n}\left[(y_i-\overline{y})-\hat{\beta}_1(x_i-\overline{x})\right]^2 \\
&= \sum_{i=1}^{n}(y_i-\overline{y})^2+\hat{\beta}_1^2\sum_{i=1}^{n}(x_i-\overline{x})^2-2\hat{\beta}_1\sum_{i=1}^{n}(x_i-\overline{x})(y_i-\overline{y}) \\
&= S_{yy}+\hat{\beta}_1^2 S_{xx}-2\hat{\beta}_1 S_{xy}.
\end{aligned}
$$

Recall that $\hat{\beta}_1 = \frac{S_{xy}}{S_{xx}}$.

Substituting for $\hat{\beta}_1$, we obtain

$$
\begin{aligned}
SSE &= S_{yy}-\left(\frac{S_{xy}}{S_{xx}}\right)^2 S_{xx}-2\frac{S_{xy}}{S_{xx}}S_{xy} \\
&= S_{yy}-\frac{S_{xy}}{S_{xx}}S_{xy} \\
&= S_{yy}-\hat{\beta}_1 S_{xy}.
\end{aligned}
$$

7.2.3 Quality of the Regression

Once we obtain the linear model, the question is, how well does this line fit the data? We could make use of the residuals, that is,

$$e_i = y_i-\hat{\beta}_0-\hat{\beta}_1 x_i,$$

to answer the question and to assess the quality of the fit. If our model is good, then the residual e_i should be close to the random error ε with mean zero. Furthermore, the residuals should contain little or no information about the model, and there should be no recognizable pattern. If we plot the residuals versus the independent variables on the x-axis, ideally, the plot should look like a horizontal blur, the residuals showing no relationship to the x-values, as shown in Fig. 7.5. Otherwise, these plots reveal a not very good fit of the given data, as shown in Fig. 7.6, and we need to improve our model specifications. Thus, a symmetric trend in the plot of residuals e_i versus x_i or $\hat{y}_i (i = 1, \ldots, n)$ indicates that the assumed regression model is not correct.

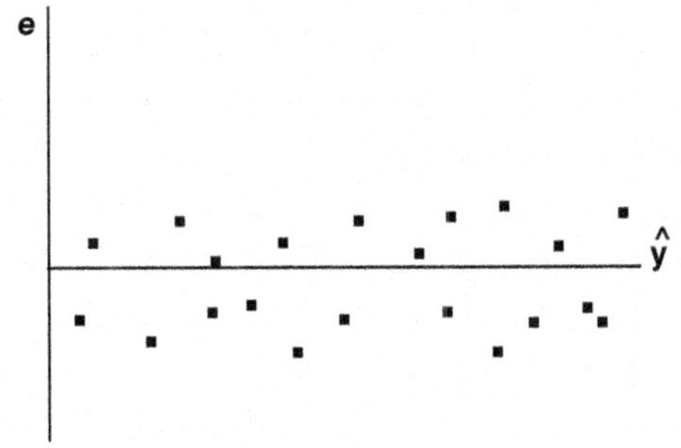

FIGURE 7.5 Good fit to the data.

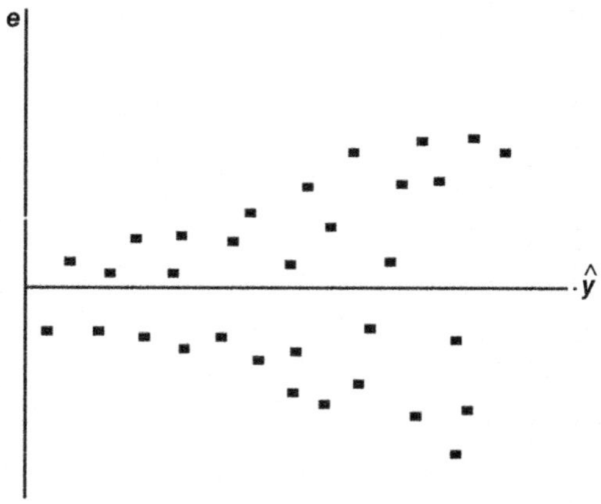

FIGURE 7.6 Scatter plot depicting a residual analysis with the horizontal axis representing predicted values labeled y hat and the vertical axis representing residuals labeled e. A solid horizontal reference line runs across the plot at zero residual. Square data points are scattered above and below this line across the full range of predicted values without a visible pattern, trend, or curvature. The roughly symmetrical dispersion around the zero line indicates constant variance and no systematic relationship between residuals and predicted values.

Whereas the residual plots give us a visual representation of the quality of the fit, a numerical measure of how well the regression explains the data is obtained by calculating the *coefficient of determination*, also called the R^2 of the regression. Particular (observed) value of realized R^2 is

$$r^2 = \frac{S_{yy}-SSE}{S_{yy}} = 1-\frac{\sum_{i=1}^{n}(y_i-\hat{y}_i)^2}{\sum_{i=1}^{n}(y_i-\bar{y})^2}.$$

Further discussion is given in Project 7B. Regression analysis with any of the standard statistical software packages will contain an output value of the R^2. This value will be between 0 and 1; closer to 1 means a better fit. For example, if the value of R^2 is 0.85, the regression captures 85% of the variation in the dependent variable. This is generally considered good regression. It should be noted that $R^2 = 0$ only implies that there is no linear relationship between x and y, and there could be nonlinear (or curvilinear) relationships between these variables. Also, a high value of R^2 doesn't mean that the derived regression line fits the data well. There is a possibility of another function fitting the data better. Thus, R^2 should only be interpreted in terms of a linear relationship. Also, R^2 value can be hugely affected by just one data value, such as an outlier. A large R^2 value does not necessarily mean that we will get a useful predicted response for a new value. Thus, we have to be careful while interpreting our model in terms of R^2.

7.2.4 Properties of the Least-Squares Estimators for the Model $Y = \beta_0 + \beta_1 x + \varepsilon$

We discussed in Chapter 4 the concept of sampling distribution of sample statistics, such as that of $\overline{X}$. Similarly, knowledge of the distributional properties of the least-squares estimators $\hat{\beta}_0$ and $\hat{\beta}_1$ is necessary to allow any statistical inferences to be made about them. The following result gives the sampling distribution of the least-squares estimators.

Theorem 7.2.1. *Let $Y = \beta_0 + \beta_1 x + \varepsilon$ be a simple linear regression model with $\varepsilon \sim N(0, \sigma^2)$, and let the errors ε_i associated with different observations $y_i(i=1, \ldots, n)$ be independent. Then*

(a) *$\hat{\beta}_0$ and $\hat{\beta}_1$ have normal distributions.*
(b) *The mean and variance are given by*

$$E(\hat{\beta}_0) = \beta_0,\ Var(\hat{\beta}_0) = \left(\frac{1}{n} + \frac{\bar{x}^2}{S_{xx}}\right)\sigma^2,$$

and

$$E(\hat{\beta}_1) = \beta_1,\ Var\left(\hat{\beta}_1\right) = \frac{\sigma^2}{S_{xx}},$$

where $S_{xx} = \sum_{i=1}^{n}x_i^2 - \frac{1}{n}(\sum_{i=1}^{n}x_i)^2$. In particular, the least-squares estimators $\hat{\beta}_0$ and $\hat{\beta}_1$ are unbiased estimators of β_0 and β_1, respectively

Proof. We know that

$$\begin{aligned}
\hat{\beta}_1 &= \frac{S_{xy}}{S_{xx}} \\
&= \frac{1}{S_{xx}} \sum_{i=1}^{n} (x_i - \overline{x})(Y_i - \overline{Y}) \\
&= \frac{1}{S_{xx}} \left[\sum_{i=1}^{n} (x_i - \overline{x})\, Y_i - \overline{Y} \sum_{i=1}^{n} (x_i - \overline{x}) \right] \\
&= \frac{1}{S_{xx}} \sum_{i=1}^{n} (x_i - \overline{x})\, Y_i,
\end{aligned}$$

where the last equality follows from the fact that $\sum_{i=1}^{n} (x_i - \overline{x}) = \sum_{i=1}^{n} x_i - n\overline{x} = 0$. Because Y_i is normally distributed, the sum $\frac{1}{S_{xx}} \sum_{i=1}^{n} (x_i - \overline{x}) Y_i$ is also normal. Furthermore,

$$\begin{aligned}
E[\hat{\beta}_1] &= \frac{1}{S_{xx}} \sum_{i=1}^{n} (x_i - \overline{x}) E[Y_i] \\
&= \frac{1}{S_{xx}} \sum_{i=1}^{n} (x_i - \overline{x})(\beta_0 + \beta_1 x_i) \\
&= \frac{\beta_0}{S_{xx}} \sum_{i=1}^{n} (x_i - \overline{x}) + \frac{\beta_1}{S_{xx}} \sum_{i=1}^{n} (x_i - \overline{x}) x_i \\
&= \beta_1 \frac{1}{S_{xx}} \sum_{i=1}^{n} (x_i - \overline{x}) x_i \\
&= \beta_1 \frac{1}{S_{xx}} \left[\sum_{i=1}^{n} x_i^2 - \overline{x} \sum_{i=1}^{n} x_i \right] \\
&= \beta_1 \frac{1}{S_{xx}} \left[\sum_{i=1}^{n} x_i^2 - \left(\sum_{i=1}^{n} x_i \right) \left(\frac{\sum_{i=1}^{n} x_i}{n} \right) \right] \\
&= \beta_1 \frac{1}{S_{xx}} \left[\sum_{i=1}^{n} x_i^2 - \frac{(\sum_{i=1}^{n} x_i)^2}{n} \right] \\
&= \beta_1 \frac{1}{S_{xx}} S_{xx} = \beta_1.
\end{aligned}$$

For the variance we have,

$$\begin{aligned}
Var[\hat{\beta}_1] &= Var\left[\frac{1}{S_{xx}} \sum_{i=1}^{n} (x_i - \overline{x}) Y_i \right] \\
&= \frac{1}{S_{xx}^2} \sum_{i=1}^{n} (x_i - \overline{x})^2\, Var[Y_i] (\text{since the } Y'_i \text{ s are independent}) \\
&= \sigma^2 \frac{1}{S_{xx}^2} \sum_{i=1}^{n} (x_i - \overline{x})^2 \left(Var(Y_i) = Var(\beta_0 + \beta_1 + \varepsilon_i) = Var(\varepsilon_i) = \sigma^2 \right) \\
&= \frac{\sigma^2}{S_{xx}}.
\end{aligned}$$

Note that both $\overline{Y}$ and $\hat{\beta}_1$ are normal random variables. It can be shown that they are also independent (see Exercise 7.3.3). Because $\hat{\beta}_0 = \overline{y} - \hat{\beta}_1 \overline{x}$ is a linear combination of $\overline{Y}$ and $\hat{\beta}_1$, it is also normal. Now,

$$E[\hat{\beta}_0] = E[\overline{Y}-\hat{\beta}_1\overline{x}] = E[\overline{Y}]-\overline{x}E[\hat{\beta}_1]$$
$$=E[\tfrac{1}{n}\textstyle\sum_{i=1}^{n} Y_i]-\overline{x}\beta_1 = \tfrac{1}{n}\textstyle\sum_{i=1}^{n}(\beta_0 + \beta_1 x)-\overline{x}\beta_1$$
$$=\beta_0 + \overline{x}\beta_1-\overline{x}\beta_1 = \beta_0.$$

The variance of $\hat{\beta}_0$ is given by

$$Var[\hat{\beta}_0] = Var[\overline{Y}-\hat{\beta}_1\overline{x}]$$
$$=Var[\overline{Y}] + \overline{x}^2\, Var[\hat{\beta}_1]\text{(since } \overline{Y} \text{ and } \hat{\beta}_1 \text{ are independent)}$$
$$=\frac{\sigma^2}{n} + \frac{\overline{x}^2\sigma^2}{S_{xx}} = \left(\frac{1}{n} + \frac{\overline{x}^2}{S_{xx}}\right)\sigma^2.$$

If an estimator $\hat{\theta}$ of θ is a linear combination of the sample observations and has a variance that is less than or equal to that of any other estimator that is also a linear combination of the sample observations, then $\hat{\theta}$ is said to be a *best linear unbiased estimator* (BLUE) for θ. The following result states that among all unbiased estimators for β_0 and β_1 which are linear in Y_i, the least-square estimators have the smallest variance.

Gauss–Markov Theorem

Theorem 7.2.2. *Let $Y = \beta_0 + \beta_1 x + \varepsilon$ be the simple regression model such that for each x_i fixed, each Y_i is an observable random variable and each $\varepsilon = \varepsilon_i$, $i = 1, 2, \ldots, n$ is an unobservable random variable. Also, let the random variable ε_i be such that $E[\varepsilon_i] = 0$, $Var(\varepsilon_i) = \sigma^2$ and $Cov(\varepsilon_i, \varepsilon_j) = 0$, if $i \neq j$. Then the least-squares estimators for β_0 and β_1 are best linear unbiased estimators.*

It is important to note that even when the error variances are not constant, there still can exist unbiased least-square estimators, but the least-squares estimators do not have a minimum variance.

7.2.5 Estimation of Error Variance σ^2

The greater the variance, σ^2, of the random error ε, the larger will be the errors in the estimation of model parameters β_0 and β_1. We can use already-calculated quantities to estimate this variability of errors. It can be shown that (see Exercise 7.2.1(b))

$$E(SSE) = (n-2)\sigma^2.$$

Thus, an unbiased estimator of the error variance, σ^2, is $\hat{\sigma}^2 = (SSE)/(n-2)$. We will denote $(SSE)/(n-2)$ by MSE (Mean Square Error).

EXERCISES 7.2

7.2.1 For a random sample of size n,

(a) Show that the error sum of squares can be expressed by

$$SSE = S_{yy}-\hat{\beta}_1 S_{xy}.$$

(b) Show that $E[SSE] = (n-2)\sigma^2$.

7.2.2 The following are midterm and final examination test scores for 10 students from a calculus class, where x denotes the midterm score, and y denotes the final score for each student.

x	68	87	75	91	82	77	86	82	75	79
y	74	79	80	93	88	79	97	95	89	92

(a) Calculate the least-squares regression line for these data.
(b) Plot the points and the least-squares regression line on the same graph.

7.2.3 The following data gives the annual incomes (in thousands of dollars) and amounts (in thousands of dollars) of life insurance policies for eight persons.

Annual income	42	58	27	36	70	24	53	37
Life insurance	150	175	25	75	250	50	250	100

(a) Calculate the least-squares regression line for these data.
(b) Plot the points and the least-squares regression line on the same graph.

7.2.4 Consider a simple linear model $Y = \beta_0 + \beta_1 x + \varepsilon$, with $\varepsilon \sim N(0, \sigma^2)$. Show that

$$\operatorname{cov}(\hat{\beta}_0, \hat{\beta}_1) = \frac{-\sigma^2 \sum_{i=1}^n x_i}{n\sum_{i=1}^n x_i^2 - (\sum_{i=1}^n x_i)^2}.$$

7.2.5 **(a)** Show that the least-squares estimates of β_0 and β_1 of a line can be expressed as

$$\hat{\beta}_0 = \bar{y} - \hat{\beta}_1 \bar{x}$$

and

$$\hat{\beta}_1 = \frac{\sum_{i=1}^n (x_i - \bar{x})(y_i - \bar{y})}{\sum_{i=1}^n (x_i - \bar{x})^2}.$$

(b) Using part (a), show that the line fitted by the method of least squares passes through the point $(\bar{x}, \bar{y})$.

7.2.6 Crickets make their chirping sounds by rapidly sliding one wing over the other. The faster they move their wings, the higher the number of chirping sounds that are produced. Scientists have noticed that crickets move their wings faster in warm temperatures than in cold temperatures (they also do this when they are threatened). Therefore, by listening to the pitch of the chirp of crickets, it is possible to tell the temperature of the air. The following table gives the number of cricket chirps per 13 seconds recorded at 10 different temperatures. Assume that the crickets are not threatened.

Temperature	60	66	70	73	78	80	82	87	90	92
Number of chirps	20	25	31	33	36	39	42	48	49	52

Calculate the least-squares regression line for this data and discuss its usefulness.

7.2.7 Consider the regression model

$$Y = \beta_1 x + \varepsilon$$

where $\varepsilon \sim N(0, \sigma^2)$. Show that

$$\hat{\beta}_1 = \frac{\sum_{i=1}^n x_i y_i}{\sum_{i=1}^n x_i^2}.$$

7.2.8 A farmer collected the following data, which show crop yields for various amounts of fertilizer used.

Fertilizer (pounds/100 sq. ft)	0	4	8	10	15	18	20	25
Yield (bushels)	6	7	10	13	17	18	22	23

(a) Calculate the least-squares regression line for this data.
(b) Plot the points and the least-squares regression line on the same graph.

7.2.9 An economist desires to estimate a line that relates personal disposable income (DI) to consumption expenditures (CE). Both DI and CE are in thousands of dollars. The following gives the data for a random sample of nine households of size four.

DI	25	22	19	36	40	47	28	52	60
CE	21	20	17	28	34	41	25	45	51

(a) Calculate the least-squares regression line for this data.
(b) Plot the points and the least-squares regression line on the same graph.

7.2.10 The following data represent systolic blood pressure readings on 10 randomly selected females between the ages of 40 and 82.

Age (x)	63	70	74	82	60	44	80	71	71	41
Systolic (y)	151	149	164	157	144	130	157	160	121	125

(a) Calculate the least-squares regression line for these data.
(b) Plot the points and the least-squares regression line on the same graph.

7.2.11 It is believed that exposure to solar radiation increases the pathogenesis of melanoma. Suppose that the following data gives sunspot relative number and age-adjusted total incidence (incidence is the number of cases per 100,000 population) for 8 different years in a certain region.

Sunspot relative number	104	12	40	75	110	180	175	30
Incidence total	4.7	1.9	3.8	2.9	0.9	2.7	3.9	1.6

(a) Calculate the least-squares regression line for this data.
(b) Plot the points and the least-squares regression line on the same graph.

7.2.12 It is believed that the average size of a mammal species is a major factor in the period of gestation (the period of development in the uterus from conception until birth). In general, it is observed that the bigger the mammal is, the longer the gestation period. Table 7.1 gives the adult mass in kilograms and gestation period in weeks of some species (source: http://www.saburchill.com/chapters/chap0037.html).
(a) Calculate the least-squares regression line for this data with adult mass as the independent variable.
(b) Plot the points and the least-squares regression line on the same graph.
(c) Calculate the least-squares regression line for this data with gestation period as the independent variable.

TABLE 7.1 Adult Mass and Gestation Period Data.

Species	Adult Mass (kg)	Gestation Period (Weeks)
African elephant	6000	88
Horse	400	48
Grizzly bear	400	30
Lion	200	17
Wolf	34	9
Badger	12	8
Rabbit	2	4.5
Squirrel	0.5	3.5

(d) Assuming that the regression model of part (c) holds for all mammals, estimate the adult mass in kilograms for the mammals given in Table 7.2.

TABLE 7.2 Mammals and Gestation Period Data.

Species	Gestation Period (Weeks)
Indian elephant	89.0
Camel	57.0
Sea lion	51.4
Dog	8.7
Rat	3.0
Hamster	2.3

7.2.13 Using the internet, obtain home sales data relating square footage to sale price for randomly selected ten homes for your area of interest and obtain a least-squares regression line for these data. Test all the assumptions for this analysis and see if your data satisfies these assumptions.

7.2.14 The following data represents sales volume as a fraction of number of visits to a company website.

Average number of visits per month x	100	150	175	200	240	464	530	480	598	650
Sales volume (\$1000)y	16	25	27	31	34	88	108	95	132	165

(a) Calculate the least-squares regression line for this data with adult mass as the independent variable.
(b) Plot the points and the least-squares regression line on the same graph.
(c) Calculate the least-squares regression line for this data with average number of visits as the independent variable.
(d) Predict sales volume if the number of visits is x=490.

7.3 INFERENCES ON THE LEAST-SQUARES ESTIMATORS

Once we obtain the estimators of the slope β_1 and intercept β_0 of the model regression line, we are in a position to use Theorem 7.2.1 to make inferences regarding these model parameters. Using the properties of $\hat{\beta}_0$ and $\hat{\beta}_1$, in this section we study the confidence intervals and hypothesis tests concerning these parameters.

From Theorem 7.2.1, we can write

$$Z_1 = \frac{\hat{\beta}_1 - \beta_1}{\frac{\sigma}{\sqrt{S_{xx}}}} \sim N(0,1).$$

Also, it can be shown that SSE/σ^2 is independent of $\hat{\beta}_1$ and has a chi-square distribution with $n-2$ degrees of freedom. Let the MSE be defined by

$$MSE = \frac{SSE}{n-2} = \frac{1}{n-2} = \sum_{i=1}^{n} [y_i - (\hat{\beta}_0 + \hat{\beta}_1 x_i)]^2.$$

Then, using Definition 4.2.2, we have

$$t_{\beta_1} = \frac{Z}{\sqrt{\frac{\left(\frac{SSE}{\sigma^2}\right)}{n-2}}} = \frac{\hat{\beta}_1 - \beta_1}{\sqrt{\frac{MSE}{S_{xx}}}},$$

which follows the t-distribution with $n - 2$ degrees of freedom.
Similarly, let

$$Z_0 = \frac{\hat{\beta}_0 - \beta_0}{\sigma\left(\frac{1}{n} + \frac{\bar{x}^2}{S_{yy}}\right)} \sim N(0,1).$$

Also, it can be shown that $\hat{\beta}_0$ and SSE are independent. Hence,

$$t_{\beta_0} = \frac{z_0}{\sqrt{\frac{\frac{SSE}{\sigma^2}}{n-2}}} = \frac{\hat{\beta}_0 - \beta_0}{\left[MSE\left(\frac{1}{n} + \frac{\bar{x}^2}{S_{xx}}\right)\right]^{1/2}},$$

follows the t-distribution with $n - 2$ degrees of freedom.

From these derivations, we can obtain the following procedure about the confidence intervals for the slopes β_1 and for the intercept β_0.

Procedure for obtaining confidence intervals for β_0 and β_1

1. Compute S_{xx}, S_{xy}, S_{yy}, $\bar{y}$, and $\bar{x}$ as in the procedure for fitting a least-squares line.
2. Compute $\hat{\beta}_1$, $\hat{\beta}_0$ using equations $\hat{\beta}_1 = (S_{xy})/(S_{xx})$ and $\hat{\beta}_0 = \bar{y} - \hat{\beta}_1\bar{x}$, respectively.
3. Compute SSE by $SSE = S_{yy} - \hat{\beta}_1 S_{xy}$.
4. Define MSE (mean square error) to be $MSE = \frac{SSE}{n-2}$, where n =number of pairs of observations $(x_1,y_1),\ldots,(x_n,y_n)$
5. A$(1-\alpha)100\%$ confidence interval for β_1 is given by $\left(\hat{\beta}_1 - t_{\alpha/2,n-2}\sqrt{\frac{MSE}{S_{xx}}}, \hat{\beta}_1 + t_{\alpha/2,n-2}\sqrt{\frac{MSE}{S_{xx}}}\right)$, where $t_{\alpha/2}$ is the upper tail $\alpha/2$-point based on a t-distribution with $(n-2)$ degrees of freedom.
6. A$(1-\alpha)100\%$ confidence interval for β_0 is given by $\left(\hat{\beta}_0 - t_{\alpha/2,n-2}\left[MSE\left(\frac{1}{n} + \frac{\bar{x}^2}{S_{xx}}\right)\right]^{1/2}, \hat{\beta}_0 + t_{\alpha/2,n-2}\left[MSE\left(\frac{1}{n} + \frac{\bar{x}^2}{S_{xx}}\right)\right]^{1/2}\right)$.

We illustrate this procedure for obtaining confidence limits for the following example.

EXAMPLE 7.3.1

For the data of Example 7.2.1:

(a) Construct a 95% confidence interval for β_0 and interpret.
(b) Construct a 95% confidence interval for β_1 and interpret.

Solution

The following calculations were obtained in Example 7.2.1:

$$S_{xx} = 263.6,\ S_{xy} = 534.2,\ \bar{y} = 4.6 \text{ and } \bar{x} = 3.8.$$

Also,

$$\hat{\beta}_1 = 2.0266,\ \hat{\beta}_0 = -3.1011.$$

In addition to those calculations, we can compute

$$\sum_{i=1}^{n} y_i^2 = 1302 \text{ and } S_{yy} = \sum_{i=1}^{n} y_i^2 - \frac{(\sum_{i=1}^{n} y_i)^2}{n} = 1302 - \frac{(46)^2}{10} = 1090.4.$$

Now,

$$\begin{aligned} SSE &= S_{yy} - \hat{\beta}_1 S_{xy} \\ &= 1090.4 - (2.0266)(534.2) \\ &= 7.79028. \end{aligned}$$

Hence,

$$MSE = \frac{SSE}{n-2} = \frac{7.79028}{8} = 0.973785.$$

Now from the t-table, we have $t_{0.025,8} = 2.306$.

(a) *A 95% confidence interval for* β_0 *is given by*

$$\left(\hat{\beta}_0 - t_{\alpha/2,n-2}\left[MSE\left(\frac{1}{n}+\frac{\bar{x}^2}{S_{xx}}\right)\right]^{1/2}, \hat{\beta}_0 + t_{\alpha/2,n-2}\left[MSE\left(\frac{1}{n}+\frac{\bar{x}^2}{S_{xx}}\right)\right]^{1/2}\right)$$

$$= \left(-3.1011-(2.306)\left[(0.973785)\left(\frac{1}{10}+\frac{(3.8)^2}{263.6}\right)\right]^{1/2}\right.$$

$$\left. - 3.1011 + (2.306)\left[(0.973785)\left(\frac{1}{10}+\frac{(3.8)^2}{263.6}\right)\right]^{1/2}\right).$$

From which we obtain a 95% confidence interval for β_0 *as (−3.9846, −2.2176). Thus, we can conclude with at least 95% confidence that the true value of the intercept,* β_0*, is between −3.9846 and −2.2176.*

(b) A 95% confidence interval for β_1 is given by

$$\left(\hat{\beta}_1 - t_{\alpha/2,n-2}\sqrt{\frac{MSE}{S_{xx}}}, \hat{\beta}_1 + t_{\alpha/2,n-2}\sqrt{\frac{MSE}{S_{xx}}}\right)$$

$$= \left(2.0266-(2.306)\sqrt{\frac{0.973785}{236.6}}, 2.0266 + (2.306)\sqrt{\frac{0.973785}{236.6}}\right)$$

from which we obtain a 95% confidence interval for β_1 *as (1.8864, 2.1668). Thus, we can conclude with at least 95% confidence that the true value of the slope of the linear regression model is between 1.8864 and 2.1663.*

One of the assumptions for the linear regression model that we have made is that the variance of errors is a constant and independent of x. Errors with this property are called *homoscedastic*. If the variance of the errors is not constant, the errors are called *heteroscedastic*. In the heteroscedastic case, standard errors and confidence intervals based on the assumption that s^2 is an estimate of σ^2 may be somewhat deceptive.

Now, we introduce hypothesis testing concerning the slope and intercept of the fitted least-squares line. We use t_{β_0} and t_{β_1} defined earlier as the test statistic for testing hypotheses concerning β_0 and β_1, respectively. The usual one- and two-sided alternatives apply. We proceed to summarize these test procedures.

Hypothesis test for β_0

One-sided test	Two-sided test
H_0: $\beta_0 = \beta_{00}$ (β_{00} is a specific value of β_0) H_a: $\beta_0 > \beta_{00}$ or $\beta_0 < \beta_{00}$ Test statistic: $t_{\beta_0} = \dfrac{\hat{\beta}_0 - \beta_{00}}{\left[MSE\left(\frac{1}{n}+\frac{\bar{x}}{S_{xx}}\right)\right]^{1/2}}$ Rejection region: $t > t_{\alpha,(n-2)}$, (upper tail region) $t < -t_{\alpha,(n-2)}$, (lower tail region)	H_0: $\beta_0 = \beta_{00}$ H_a: $\beta_0 \neq \beta_{00}$ Test statistic: $t_{\beta_0} = \dfrac{\hat{\beta}_0 - \beta_{00}}{\left[MSE\left(\frac{1}{n}+\frac{\bar{x}}{S_{xx}}\right)\right]^{1/2}}$ Rejection region: $\lvert t\rvert > t_{\alpha/2,(n-2)}$

Decision: If t_{β_0} falls in the rejection region, reject the null hypothesis at the level of significance α.
Assumptions: Assume that the errors ε_i, $i=1,\ldots, n$ are independent and normally distributed with $E(\varepsilon_i)=0$, $i=1,\ldots, n$, and $Var(\varepsilon_i) = \sigma^2$, $i=1,\ldots, n$.

We now illustrate this procedure with the following example.

EXAMPLE 7.3.2

Using the data given in Example 7.2.1, test the hypothesis H_0: β_0 =–3 versus H_a: $\beta_0 \neq -3$ using the 0.05 level of significance.

Solution

We test H_0: β_0=-3 versus H_a: $\beta_0 \neq$ -3.
Here, β_{00} = -3. The rejection region is $t < -2.306$ or $t > 2.306$.
From the calculations of the previous example, we have

$$t_{\beta_0} = \frac{\hat{\beta}_0 - \beta_{00}}{\left[MSE\left(\frac{1}{n} + \frac{\bar{x}^2}{S_{xx}}\right)\right]^{1/2}}$$
$$= \frac{-3.1011 - (-3)}{\left[(0.973785)\left(\frac{1}{10} + \frac{(3.8)^2}{263.2}\right)\right]^{1/2}}$$
$$= -0.26041.$$

Because the test statistics do not fall in the rejection region, at α = 0.05, we do not reject H_0.

Hypothesis test for β_1

One-sided test	**Two-sided test**
H_0: β_1=β_{10} (β_{10} is a specific value of β_1)	H_0: $\beta_1 = \beta_{10}$
H_a: β_1>β_{10} or β_1<β_{10}	H_a: $\beta_1 \neq \beta_{10}$
Test statistic:	Test statistic:
$t_{\beta_1} = \frac{\hat{\beta}_1 - \beta_{10}}{\sqrt{\frac{MSE}{S_{xx}}}}$	$t_{\beta_1} = \frac{\hat{\beta}_1 - \beta_{10}}{\sqrt{\frac{MSE}{S_{xx}}}}$
Rejection region:	Rejection region:
$t > t_{\alpha,(n-2)}$ (upper tail region)	$\lvert t\rvert > t_{\alpha/2,\,(n-2)}$
$t < -t_{\alpha,(n-2)}$ (lower tail region)	

Decision: If t_{β_1} falls in the rejection region, reject the null hypothesis at the level of significance α.
Assumptions: Assume that the errors ε_i, $i = 1, \ldots, n$ are independent and normally distributed with $E(\varepsilon_i) = 0$, $i = 1, \ldots, n$, and $Var(\varepsilon_i) = \sigma^2$, $i = 1, \ldots, n$.

The test of hypothesis H_0: $\beta_1 = 0$ answers the question, Is the regression significant? If $\beta_1 = 0$, we conclude that there is no significant linear relationship between X and Y, and hence, the independent variable X is not important in predicting the values of Y if the relationship of Y and X is not linear. Note that if $\beta_1 = 0$, then the model becomes $y = \beta_0 + \varepsilon$. Thus, the question of the importance of the independent variable in the regression model translates into a narrower question of the test of hypothesis H_0: $\beta_1 = 0$. That is, the regression line is actually a horizontal line through the intercept, β_0.

EXAMPLE 7.3.3

Using the data given in Example 7.2.1, test the hypothesis H_0: $\beta_1 = 2$ versus H_a: $\beta_1 \neq 2$ using the 0.05 level of significance.

Solution

We test

$$H_0: \beta_1 = 2 \text{ versus } H_a: \beta_1 \neq 2.$$

We know that $\hat{\beta}_1 = 2.0266$.
For α = 0.05 and n = 10, the rejection region is $t_{\beta_1} < -2.306$ or $t_{\beta_1} > -2.306$. The test statistic is

$$t_{\beta_1} = \frac{\hat{\beta}_1 - \beta_{10}}{\sqrt{\frac{MSE}{S_{xx}}}}$$
$$= \frac{2.0266 - 2}{\sqrt{\frac{2.0266 - 2}{263.6}}} = 0.4376.$$

Because the test statistic does not fall in the rejection region, at $\alpha = 0.05$, we do not reject H_0. Thus, for $\alpha = 0.05$, the given data support the null hypothesis that the true value of the slope, β_1, of the regression line is equal to 2.

As we already know, estimates of the regression coefficients β_0 and β_1 are subject to sampling uncertainty. Therefore, we will never estimate the true value accurately of these parameters from sample data. However, we may construct confidence intervals for the intercept and the slope parameters. Thus, a problem closely related to the problem of estimating the regression coefficients β_0 and β_1 is that of estimating the mean of the distribution of Y for a given value of x, that is, estimating $\beta_0 + \beta_1 x$. For a fixed value of x, say x_0, we have the following confidence limits.

A $(1 - \alpha)100\%$ confidence interval for $\beta_0 + \beta_1 x$ is given by

$$\left(\hat{\beta}_0 + \hat{\beta}_1 x\right) \pm t_{\alpha/2} s_e \sqrt{\frac{1}{n} + \frac{(x_0 - \bar{x})^2}{S_{xx}}}$$

where

$$s_e = \sqrt{\frac{S_{yy} - (S_{xy})^2}{(n-2)S_{xx}}}$$

We could use the data from the previous example to easily calculate a confidence interval for $\beta_0 + \beta_1 x$.

7.3.1 Analysis of Variance (ANOVA) Approach to Regression

Another approach to hypothesis testing is based on ANOVA. A detailed explanation of this approach is given in Chapter 10. Here we present the necessary steps for regression. The main reason for this presentation is the fact that most of the major statistical software outputs for regression analysis (see Section 7.9) are given in the form of ANOVA tables.

It can be verified that (see Exercise 7.3.7),

$$\sum_{i=1}^{n} (y_i - \bar{y})^2 = \sum_{i=1}^{n} (y_i - \hat{y}_i)^2 + \sum_{i=1}^{n} (\hat{y}_i - \bar{y})^2.$$

Denoting

$$SST = \sum_{i=1}^{n} (y_i - \bar{y})^2,\ SSE = \sum_{i=1}^{n} (y_i - \hat{y}_i)^2, \text{ and } SSR = \sum_{i=1}^{n} (\hat{y}_i - \bar{y})^2,$$

the foregoing equation can be written as

$$SST = SSR + SSE.$$

Note that the total sum of squares (SST) is a measure of the variation of y_i's around the mean $\bar{y}$, and SSE is the residual or error sum of squares that measures the lack of fit of the regression model. Hence, SSR (sum of squares of regression or model) measures the variation that can be explained by the regression model.

We saw that to test the hypothesis H_0: $\beta_1 = 0$ versus H_a: $\beta_1 \neq 0$, the statistic

$$t_{\beta_1} = \frac{\hat{\beta}_1}{\sqrt{\frac{MSE}{S_{xx}}}},$$

was used, where t_{β_1} follows a t-distribution with $(n - 2)$ degrees of freedom. From Exercise 4.2.18, we know that

$$t_{\hat{\beta}_1}^2 = \frac{\hat{\beta}_1^2}{\left(\frac{MSS}{S_{xx}}\right)},$$

follows an F-distribution with numerator degrees of freedom 1 and denominator degrees of freedom $(n-2)$. We can also verify that

$$t_{\hat{\beta}_1}^2 = \frac{MSR}{MSE}.$$

Thus, to test H_0: $\beta_1 = 0$ versus H_a: $\beta_1 \neq 0$, we could use the statistic

$$\frac{MSR}{MSE} \sim F(1, n-2),$$

and reject H_0 if

$$\frac{MSR}{MSE} \geq F_\alpha(1, n-2).$$

The procedure is summarized in Table 7.3, known as the ANOVA table.

The last column in the ANOVA table gives the statistic $(MSR)/(MSE)$. It is also customary to give another column with the P-value of the test.

TABLE 7.3 ANOVA Table for Simple Regression.

Source of Variation	Degrees of Freedom	Sum of Squares	Mean Sum of Squares	F-Ratio
Regression (model)	1	SSR	$MSR = \frac{SSR}{d.f.}$	$\frac{MSR}{MSE}$
Error (residuals)	$n-2$	SSE	$\frac{SSE}{d.f.}$	
Total	$n-1$	SST		

Example 7.3.4

In a study of baseline characteristics of 20 patients with foot ulcers, we want to see the relationship between the stage of ulcer that is determined using the Yarkony–Kirk scale, a higher number indicating a more severe stage, with range 1–6, and duration of the ulcer in days. Suppose we have the data shown in Table 7.4.

TABLE 7.4 ANOVA Table for Multiple Regression.

Stage of Ulcer (x)	**4**	**3**	**5**	**4**	**4**	**3**	**3**	**4**	**6**	**3**
Duration (d)	18	6	20	15	16	15	10	18	26	15
Stage of ulcer (x)	3	4	3	2	3	2	2	3	5	6
Duration (d)	8	16	17	6	7	7	8	11	21	24

(a) Develop an ANOVA table to test H_0: $\beta_1 = 0$ versus H_a: $\beta_1 \neq 0$. What is the conclusion of the test based on $\alpha = 0.05$?
(b) Write down the expression for the least-squares line.

Solution

(a) *We test H_0: $\beta_1 = 0$ versus H_a: $\beta_1 \neq 0$. We will use Minitab to generate the ANOVA table* (Table 7.5). *Because the P-value is less than 0.001, for $\alpha = 0.05$, we reject the null hypothesis that $\beta_1 = 0$ and conclude that there is a relationship between the stage of ulcer and its duration.*

TABLE 7.5 ANOVA Table for the Data.

Source of Variation	Degrees of Freedom	Sum of Squares	Mean Sum of Squares	*F*-Ratio	*P*-Value
Regression (model)	1	570.04	570.04	77.05	0.000
Error (residuals)	18	133.16	7.40		
Total	19	703.20			

(b) *Again, using the Minitab output, we obtain the least-squares line as*

$$d = 4.61x - 2.40.$$

EXERCISES 7.3

7.3.1 An experiment was conducted to observe the effect of an increase in temperature on the potency of an antibiotic. Three one-ounce portions of the antibiotic were stored for equal lengths of time at each of the following Fahrenheit temperatures: 40, 55, 70, and 90 degrees. The potency readings observed at the end of the experimental period were

Potency reading, y	49	38	27	24	38	33	19	28	16	18	23
Temperature (degrees), x	40			55			70			90	

(a) Find the least-squares line appropriate for this data.
(b) Plot the points and graph the line as a check for your calculations.
(c) Calculate the 95% confidence intervals for β_0 and β_1, respectively.

7.3.2 Consider the data

x	38	26	48	22	40	15	30	33
y	10	11	16	8	12	5	10	11

(a) Find the least-squares line appropriate for these data.
(b) Plot the points and graph the line as a check for your calculations.
(c) Calculate the 95% confidence intervals for β_0 and β_1, respectively.

7.3.3 Show that $\overline{Y}$ and $\hat{\beta}_1$ are independent, under the usual assumptions of a simple linear regression model.
7.3.4 Using the data of Exercise 7.2.10, calculate the 95% confidence intervals for β_0 and β_1, respectively.
7.3.5 The following data represent survival time in days after a heart transplant and patient age in years at the time of transplant for 10 randomly selected patients.

Age at transplant	28	41	46	53	39	36	47	29	48	44
Survival time, in days	7	278	44	48	406	382	1995	176	323	1846

(a) Find the least-squares line appropriate for this data.

(b) Plot the points and graph the line.
(c) Calculate the 95% confidence intervals for β_0 and β_1, respectively.

7.3.6 The following data represent the weights of cigarettes (g) from different manufacturers and their nicotine contents (mg).

Weight	15.8	14.9	9.0	4.5	15.0	17.0	8.6	12.0	4.1	16.0
Nicotine	0.957	0.886	0.852	0.911	0.889	0.919	0.969	1.118	0.946	1.094

(a) Find the least-squares line appropriate for this data.
(b) Plot the points and graph the line. Do you think linear regression is appropriate?
(c) Calculate the 95% confidence intervals for β_0 and β_1, respectively.

7.3.7 The following data represents total CO_2 emissions per vehicle (in Metric tons per vehicle) (http://corporate.ford.com/microsites/sustainability-report-2012–13/environment-data-energy).

Year	2007	2008	2009	2010	2011	2012
Total	1.01	1.09	1.07	1.01	0.91	0.90

(a) Find the least-squares line appropriate for this data.
(b) Plot the points and graph the line.
(c) Calculate the 95% confidence intervals for β_0 and β_1, respectively.

7.3.8 Show that

$$\sum_{i=1}^{n}(y_i-\bar{y})^2=\sum_{i=1}^{n}(y_i-\hat{y}_i)^2+\sum_{i=1}^{n}(\hat{y}_i-\bar{y})^2.$$

7.4 PREDICTING A PARTICULAR VALUE OF *Y*

In the earlier sections, we have seen how to fit a least-squares line for a given set of data. Also using this line, we could find $E(Y)$, for any given value of x. Instead of obtaining this mean value, we may be interested in predicting the particular value of Y for a given x. In fact, one of the primary uses of the estimated regression line is to predict the response value of Y for a given predictor value of x. Prediction problems are very important in several real-world problems; for example, in economics one may be interested in predicting a particular gain associated with an investment.

Let $\hat{Y}_0$ denote a predictor of a particular value of $Y = Y_0$ and let the corresponding values of x be x_0. We shall choose $\hat{Y}_0$ to be $E(\hat{Y}|x_0)$. Let $\hat{Y}$ denote a predictor of a particular value of Y. Then the error η of the predictor in comparison to a particular value of Y is

$$\eta = Y-\hat{Y}_0.$$

Both Y and $\hat{Y}$ are normal random variables, and the error is a linear function of Y and $\hat{Y}$. This means that η itself is normally distributed. Also, because $E(\hat{Y}) = E(Y)$, we have

$$E(\eta) = E(Y|x_0)-E(\hat{Y}) = 0.$$

Furthermore,

$$Var(\eta) = Var(Y-\hat{Y}) = Var(Y) + Var(\hat{Y})-2Cov(Y, \hat{Y}).$$

We can consider Y and $\hat{Y}$ as independent, because we are predicting a different value of Y, not used in the calculation of $\hat{Y}$. Therefore, $Cov(Y, \hat{Y}) = 0$. In that case

$$Var(\eta) = Var(Y_0) + Var(\hat{Y}_0)$$
$$=\sigma^2 + \sigma^2\left[\frac{1}{n} + \frac{(x-\overline{x})^2}{S_{xx}}\right]$$
$$=\left[1 + \frac{1}{n} + \frac{(x-\overline{x})^2}{S_{xx}}\right]\sigma^2.$$

Hence, the error of predicting a particular value of Y, given x, is normally distributed with mean zero and variance $\left[1 + \frac{1}{n} + \frac{(x-\overline{x})^2}{S_{xx}}\right]\sigma^2$.

That is,

$$\eta \sim N\left(0, \left[1 + \frac{1}{n} + \frac{(x-\overline{x})^2}{S_{xx}}\right]\sigma^2\right),$$

and

$$Z = \frac{Y-\hat{Y}}{\sigma\sqrt{\left[1 + \frac{1}{n} + \frac{(x-\overline{x})^2}{S_{xx}}\right]}} \sim N(0,1).$$

If we substitute the sample standard deviation S for σ, then we can show that

$$T = \frac{Y-\hat{Y}}{S\sqrt{\left[1 + \frac{1}{n} + \frac{(x-\overline{x})^2}{S_{xx}}\right]}},$$

follows the t-distribution with $[n-(k+1)]$ degrees of freedom. Using this fact, we now give a prediction interval for the random variable Y, the response of a given situation.

We know that

$$P(-t_{\alpha/2} < T < t_{\alpha/2}) = 1-\alpha.$$

Substituting for T, we have

$$P\left(-t_{\alpha/2} < \frac{Y-\hat{Y}}{S\sqrt{\left[1 + \frac{1}{n} + \frac{(x-\overline{x})^2}{S_{xx}}\right]}} < t_{\alpha/2}\right) = 1-\alpha,$$

which implies that

$$P\left[\hat{Y}-t_{\alpha/2}S\sqrt{\left[1 + \frac{1}{n} + \frac{(x-\overline{x})^2}{S_{xx}}\right]} < Y < \hat{Y} + t_{\alpha/2}S\sqrt{\left[1 + \frac{1}{n} + \frac{(x-\overline{x})^2}{S_{xx}}\right]}\right] = 1-\alpha.$$

Hence, we have the following.

A (1 - α)100% prediction interval for Y is

$$\hat{Y} \pm t_{\alpha/2}S\sqrt{\left[1 + \frac{1}{n} + \frac{(x-\overline{x})^2}{S_{xx}}\right]}$$

where $t_{\alpha/2}$ is based on (n - 2) degrees of freedom and $S^2 = \frac{SSE}{n-2} = \sqrt{MSE}$.

We illustrate this statistical procedure with the following example.

EXAMPLE 7.4.1

Using the data given in Example 7.2.1, obtain a 95% prediction interval at $x = 5$.

Solution

We have shown that $\hat{y} = -3.1011 + 2.0266x$. *Hence, at* $x = 5$, $\hat{y} = 7.0319$.

Also, $\bar{x} = 3.8$, $S_{xx} = 263.6$, $SSE = 7.79028$, *and* $S = \sqrt{\frac{7.79028}{8}} = 2.306$.
From the t-table, $t_{0.025,8} = 2.306$.
Thus, we have

$$7.0319 \pm (2.306)(0.98681)\sqrt{\left[1 + \frac{1}{10} + \frac{(5-3.8)^2}{263.6}\right]},$$

which gives the 95% prediction interval as (4.6393, 9.4245).

We can conclude with at least 95% confidence that the true value of Y at point x = 5 will be somewhere between 4.6393 and 9.4245.

EXERCISES 7.4

7.4.1 The following are midterm and final examination test scores for 10 calculus students, where x denotes the midterm score, and y denotes the final score for each student.

x	68	87	75	91	82	77	86	82	75	79
y	74	89	80	93	88	79	97	95	89	92

Obtain a 95% prediction interval for $x = 92$ and interpret its meaning.

7.4.2 The following data give the annual income (in thousands of dollars) and amounts (in thousands of dollars) of life insurance policies for eight persons.

Annual income	42	58	27	36	70	24	53	37
Life insurance	150	175	25	75	250	50	250	100

Obtain a 90% prediction interval for $x = 59$ and interpret its meaning.

7.4.3 For the following data, construct a 95% prediction interval for $x = 12$.

x	1	3	5	7	9	11
y	16	36	43	65	80	88

7.4.4 The data given below are from a random sample of height (in inches) and weight (in pounds) of seven basketball players.

Height	73	83	77	80	85	71	80
Weight	186	234	208	237	265	190	220

Construct a 99% prediction interval for height equal to 90. Interpret the result and state any assumptions.

7.4.5 For the data in Exercise 7.2.10, obtain a 95% prediction interval for the age, $x = 85$, interpret and state any assumptions.

7.4.6 For the CO_2 emission data of Exercise 7.3.7, construct a 95% prediction interval for the year 2013 emission.

7.5 CORRELATION ANALYSIS

Using the regression model, we can evaluate the magnitude of change in the dependent variable due to certain changes in the independent variables. One of the main assumptions we have used is that the independent variables are known. However, there are problems where the x-values as well as the y-values are assumed to be random variables. This would be the case, for example, if we studied the relationship between secondhand smoking and the incidence of a certain disease. Here, basically, one treats X as random, and hence the simple linear regression model is

$$Y = \beta_0 + \beta_1 X + \varepsilon.$$

This implies that

$$E(Y|X = x) = \beta_0 + \beta_1 x,$$

and one looks for dependence of X and Y. Once we have determined that there is a relationship between the variables, the next question that arises is how closely the variables are associated. A measure of the amount of linear dependency of the two random variables is the *correlation*. The correlation coefficient tells us how strongly two variables are linearly related. The statistical method used to measure the degree of correlation is referred to as the *correlation analysis*. We will assume that the vector random variable (X, Y) has a bivariate normal distribution. In this case, it can be shown that

$$E(Y|X = x) = \beta_0 + \beta_1 x.$$

At times, our interest may not be in the linear relationship; rather, we may merely want to know whether X and Y are independent random variables. If (X, Y) has a bivariate normal distribution, then testing for independence is equivalent to testing that the correlation coefficient, $\rho = \sigma_{xy}/(\sigma_x\sigma_y)$, is equal to zero. Note that ρ is positive if X and Y increase together and ρ is negative if Y decreases as X increases. If $\rho = 0$, there is no relation between X and Y; if $\rho > 0$, there is a positive relation between X and Y (increasing slope); and when $\rho < 0$, we have a negative relationship (decreasing slope). Thus, the correlation coefficient can be used to measure how well the linear regression model fits the data.

Let $(X_1, Y_1), (X_2, Y_2), \ldots,(X_n, Y_n)$ be a random sample from a bivariate normal distribution. The maximum likelihood estimator of ρ is the sample correlation coefficient defined by $\hat{\rho}$ or r,

$$\begin{aligned} r &= \frac{\sum_{i=1}^{n}(X_i - \overline{X})(Y_i - \overline{Y})}{\sqrt{\sum_{i=1}^{n}(X_i - \overline{X})^2 \sum_{i=1}^{n}(Y_i - \overline{Y})^2}} \\ &= \frac{S_{xy}}{\sqrt{S_{xx}S_{yy}}}. \end{aligned} \tag{7.7}$$

Equivalently, we can rewrite Eq. (7.7) by

$$r = \frac{n\sum_{i=1}^{n} X_i Y_i - \sum_{i=1}^{n} X_i \sum_{i=1}^{n} Y_i}{\sqrt{[n\sum_{i=1}^{n} X_i^2 - (\sum_{i=1}^{n} X_i)^2][n\sum_{i=1}^{n} Y_i^2 - (\sum_{i=1}^{n} Y_i)^2]}}.$$

We can see that $-1 \le r \le 1$. The value of r could readily be obtained by the calculations one has already performed for the regression analysis. Observe that the numerator of r is exactly the same as the numerator of $\hat{\beta}_1$ derived in Section 7.2. Because the denominators of both $\hat{\beta}_1$ and r are nonnegative, they have the same sign. It can be shown that this estimator is not unbiased. If the value of r is near or equal to zero, this implies little or no linear relationship between x and y. On the other hand, the closer r is to 1 or -1, the stronger the linear relationship between x and y. When $r > 0$, values of y increase as the values of x increase, and the data set is said to be *positively correlated*. When $r < 0$, values of y decrease as the values of x increase, and the data set is said to be *negatively correlated*. $r = 0$ indicates no linear relationship between x and y, however, there can be a nonlinear relationship in this case. In this book, we use the term *correlation* only when referring to linear relationships. In actual practice, we can use the value of r to decide whether it is appropriate to develop linear regression models in a given situation. As a rule of thumb, if $r > 0.30$ or $r < -0.30$, we proceed with developing a linear regression model. However, a much higher or lower value is desirable. For example, if in a given problem where $r = 0.77$, it conveys to us that approximately 77% of the data we have are linearly related.

The probability distribution for r is difficult to obtain. For large samples, this difficulty could be overcome by using the fact that the Fisher z-transform, given by

$$z = (1/2)\ln[(1 + r)/(1-r)],$$

is approximately normally distributed with mean $\mu_z = (1/2)\ \ln[(1 + \rho)(1 - \rho)]$ and variance $\sigma_z = 1/(n - 3)$. Thus, for large random samples, we can test hypotheses about ρ using the approximate test statistic:

$$Z = \frac{z - \mu_z}{\sigma_z}$$

$$= \frac{(1/2)\ln\left(\frac{1+r}{1-r}\right) - (1/2)\ln\left(\frac{1+\rho}{1-\rho}\right)}{\frac{1}{\sqrt{n-3}}}.$$

For example, suppose we are interested in testing the hypothesis that the true value of ρ is a specific number, say, ρ_0, with a certain value of α. We can proceed to make a decision by following the procedure given below.

HYPOTHESIS TEST FOR ρ

One-sided test	Two-sided test
H_0: $\rho = \rho_0$ H_a: $\rho > \rho_0$ or H_a: $\rho < \rho_0$ Test statistic: $Z = \frac{(1/2)\ln\left(\frac{1+r}{1-r}\right) - (1/2)\ln\left(\frac{1+\rho_0}{1-\rho_0}\right)}{\frac{1}{\sqrt{n-3}}}$ Rejection region: $z > z_\alpha$ (upper tail region) $z < -z_\alpha$ (lower tail region)	H_0: $\rho = \rho_0$ H_a: $\rho = \rho_0$ Test statistic: $Z = \frac{(1/2)\ln\left(\frac{1+r}{1-r}\right) - (1/2)\ln\left(\frac{1+\rho_0}{1-\rho_0}\right)}{\frac{1}{\sqrt{n-3}}}$ Rejection region: $\lvert z\rvert > z_{a/2}$

Decision: If z falls in the rejection region, reject the null hypothesis at the level of significance α.
Assumption: (X, Y) follow the bivariate normal, and this test procedure is approximate.

EXAMPLE 7.5.1

For the data given in Example 7.2.1, would you say that the variables X and Y are independent? Use $\alpha = 0.05$.

Solution

We test

$$H_0\text{: } \rho = 0 \text{ versus } H_a\text{: } \rho \neq 0.$$

From Example 7.2.1, for n=10, we have the following summary:,

$$\sum_{i=1}^{10} x_i = 38;\ \sum_{i=1}^{10} y_i = 46;\ \sum_{i=1}^{10} x_i y_i = 709,$$

and

$$\sum_{i=1}^{10} x_i^2 = 408;\ \sum_{i=1}^{10} y_i^2 = 1302.$$

Hence

$$r = \frac{n\sum_{i=1}^{n}X_iY_i - \sum_{i=1}^{n}X_i\sum_{i=1}^{n}Y_i}{\sqrt{\left[n\sum_{i=1}^{n}X_i^2 - \left(\sum_{i=1}^{n}X_i\right)^2\right]\left[n\sum_{i=1}^{n}Y_i^2 - \left(\sum_{i=1}^{n}Y_i\right)^2\right]}}$$
$$= \frac{(10)(709) - (38)(46)}{\sqrt{[(10)(408) - (38)^2][(10)(1302) - (46)^2]}}$$
$$= 0.99641.$$

The test statistic is

$$z = \frac{(1/2)\ln\left(\frac{1+r}{1-r}\right) - (1/2)\ln\left(\frac{1+\rho_0}{1-\rho_0}\right)}{\frac{1}{\sqrt{n-3}}}$$
$$= \frac{(1/2)\ln\left(\frac{1+0.99641}{1-0.99641}\right) - (1/2)\ln\left(\frac{1+0}{1-0}\right)}{\frac{1}{\sqrt{7}}}$$
$$= 8.3618.$$

For $z_{\alpha/2} = z_{0.025} = 1.96$, the rejection region is $|z| > 1.96$. Because the observed value of the test statistic falls in the rejection region, we reject the null hypothesis and conclude that at $\alpha = 0.05$, the variables X and Y are dependent

EXERCISES 7.5

7.5.1 The table shows the midterm and final examination test scores for 10 students from a differential equations class, where x denotes the midterm scores and y denotes the final scores.

x	68	87	75	91	82	77	86	82	75	79
y	74	89	80	93	88	79	97	95	89	92

(a) At 95% confidence level, test whether X and Y are independent.
(b) Find the P-value.
(c) State any assumptions you have made in solving the problem.

7.5.2 The following table gives the annual income (in thousands of dollars) and amounts (in thousands of dollars) of life insurance policies for eight persons.

Annual income	42	58	27	36	70	24	53	37
Life insurance	150	175	25	75	250	50	250	100

(a) At the 98% confidence level, test whether annual income and the amount of life insurance policies are independent.
(b) Find the attained significance level.
(c) State any assumptions you have made in solving the problem.

7.5.3 Show that

$$r = \frac{n\sum_{i=1}^{n}X_iY_i - \sum_{i=1}^{n}X_i\sum_{i=1}^{n}Y_i}{\sqrt{[n\sum_{i=1}^{n}X_i^2 - (\sum_{i=1}^{n}X_i)^2][n\sum_{i=1}^{n}Y_i^2 - (\sum_{i=1}^{n}Y_i)^2]}},$$

is not an unbiased estimator of the population coefficient, ρ.

7.5.4 Using the data in Example 7.2.1:
(a) Compute r, the coefficient of correlation.
(b) Would you say that the variables X and Y are independent? Use $\alpha = 0.05$.
(c) State any assumptions you have made in solving the problem.

7.5.5 A new drug is tested for serum cholesterol-lowering properties on six randomly selected volunteers. The serum cholesterol values are given in the following table.

Before treatment	232	254	220	200	213	222
After treatment	212	240	225	205	204	218

(a) At 95% confidence level, test whether X and Y are independent.
(b) Find the P-value.
(c) Calculate the least-squares regression line for this data.
(d) Interpret the usefulness of the model.
(e) State any assumptions you have made in solving the problem.

7.5.6 Find the relationship between engine size in cars and CO_2 emissions. The data is given in the following table.

Engine size (L)	1.2	1.6	2.0	2.4	3.0
CO_2 emissions (g/km)	120	150	180	200	250

(a) Obtain a least-squares regression line.
(b) Obtain R^2 and interpret.
(c) Obtain a 95% confidence interval for the mean CO_2 at $x_0 = 2.0$.
(d) Obtain a 95% prediction interval for the mean CO_2 at $x_0 = 2.0$.
(e) Test H_0: $\beta_1 = 0$ versus H_a: $\beta_1 \neq 0$, at $\alpha = 0.05$.
(f) Write and run R-code for (a) through (e).

7.6 MATRIX NOTATION FOR LINEAR REGRESSION

Most real-life applications of regression analysis use models that are more complex than the simple straight line model. For example, a person's body weight may depend not just on the person's eating habits; it may depend on additional factors such as heredity, exercise, and type of work. Hence, we may want to incorporate other potential independent variables in modeling. We now study the situation where k (>1) independent variables are used to predict the dependent variable. The model to be studied is of the form

$$Y = \beta_0 + \beta_1 x_1 + \beta_1 x_2 + \cdots + \beta_k x_k + \varepsilon.$$

Here, $\varepsilon \sim N(0, \sigma^2)$. This model is called a **multiple regression model**.

Let $y_1, y_2, \ldots, y_n$ be n independent observations on Y. Then each observation y_i can be written as

$$y_i = \beta_0 + \beta_1 x_{i1} + \beta_2 x_{i2} + \cdots + \beta_k x_{ik} + \varepsilon,$$

where x_{ij} is the jth independent variable for the ith observation, $i = 1, 2, \ldots, n$, and $\varepsilon'_i s$ are independent as in the simple linear regression case. It is sometimes advantageous to introduce matrices to study the linear equations. Let $x_0 = 1$. Define the following matrices:

$$X = \begin{bmatrix} x_0 & x_{11} & x_{12}\cdot\cdot & x_{1k} \\ x_0 & x_{21} & x_{22}\cdot\cdot & x_{2k} \\ \vdots & \vdots & \vdots\ \vdots & \vdots \\ x_0 & x_{n1} & x_{n2}\cdot\cdot & x_{nk} \end{bmatrix}, \quad Y = \begin{bmatrix} y_1 \\ y_2 \\ \vdots \\ y_n \end{bmatrix},$$
$$\beta = \begin{bmatrix} \beta_0 \\ \beta_1 \\ \vdots \\ \beta_n \end{bmatrix} \quad \text{and} \quad \varepsilon = \begin{bmatrix} \varepsilon_1 \\ \varepsilon_2 \\ \vdots \\ \varepsilon_n \end{bmatrix}. \tag{8.8}$$

Thus, the n equations representing the linear equations can be rewritten in the matrix form as

$$Y = X\beta + \varepsilon.$$

In particular, for the n observations from the simple linear model of the form

$$Y = \beta_0 + \beta_1 x + \varepsilon$$

we can write

$$Y = X\beta + \varepsilon,$$

where

$$Y = \begin{bmatrix} y_1 \\ y_2 \\ \vdots \\ y_n \end{bmatrix}, X = \begin{bmatrix} 1 & x_1 \\ 1 & x_2 \\ 1 & . \\ 1 & . \\ 1 & . \\ 1 & x_n \end{bmatrix}, \varepsilon = \begin{bmatrix} \varepsilon_1 \\ \varepsilon_2 \\ \vdots \\ \varepsilon_n \end{bmatrix}, \text{ and } \beta = \begin{bmatrix} \beta_0 \\ \beta_1 \end{bmatrix}.$$

We can see that

$$X'X = \begin{bmatrix} 1 & 1 & . & . & . & 1 \\ x_1 & x_2 & . & . & . & x_n \end{bmatrix} \begin{bmatrix} 1 & x_1 \\ 1 & x_2 \\ \vdots & \vdots \\ 1 & x_n \end{bmatrix} = \begin{bmatrix} n & \sum_{i=1}^{n} x_i \\ \sum_{i=1}^{n} x_i & \sum_{i=1}^{n} x_i^2 \end{bmatrix},$$

where $'$ denotes the transpose of a matrix.
Also,

$$X'Y = \begin{bmatrix} \sum_{i=1}^{n} y_i \\ \sum_{i=1}^{n} x_i y_i \end{bmatrix}.$$

Let us now go back to the multiple regression model

$$Y = \beta_0 + \beta_1 x_1 + \beta_1 x_2 + \cdots + \beta_k x_k + \varepsilon.$$

The least-squares estimators $\hat{\beta}_i$ of β_i for $i = 0, 1, 2,\ldots,k$ are the ones that minimize the sum of squares

$$\begin{aligned} SSE &= \sum_{i=1}^{n} e_i^2 = \sum_{i=1}^{n} [y_i - (\hat{\beta}_0 + \hat{\beta}_1 x_1 + \hat{\beta}_2 x_2 + \cdots + \hat{\beta}_k x_k)]^2 \\ &= (y - X\hat{\beta})'(y - X\hat{\beta}) \\ &= y'y - y'X\hat{\beta} - (X\hat{\beta})'y + (\hat{\beta}X)'X\hat{\beta}. \end{aligned}$$

To minimize SSE with respect to β, we differentiate SSE with respect to β and equate it to zero. Thus,

$$\frac{\partial}{\partial \beta}(y'y - y'X'\beta - \beta'X'y + X'\beta'X\beta) = 0,$$

yielding

$$(X'X)\hat{\beta} = X'Y.$$

Assuming the matrix $(X'X)$ is invertible, we obtain

$$\hat{\beta} = (X'X)^{-1}X'Y.$$

Now we summarize the procedure to obtain a multiple linear regression equation.

Procedure to obtain a multiple linear regression equation

1 Rewrite the n observations as

$$Y_i = \beta_0 + \beta_1 x_{1i} + \beta_1 x_{2i} + \cdots + \beta_k x_{ki},\ i = 1,2,\ \ldots,n$$

in the matrix notation as

$$Y = X\beta + \varepsilon$$

where X, Y, and β are defined in (1).

2 Compute $(X'X)^{-1}$ and obtain the estimators of β as

$$\hat{\beta} = (X'X)^{-1}X'Y.$$

3 Then the regression equation is

$$\hat{Y} = X\hat{\beta}.$$

EXAMPLE 7.6.1

Using the data given in Example 7.2.1, use the matrix approach to solve the problem of operations.

Solution

From the data in Example 7.2.1, we have,

$$Y = \begin{bmatrix} -9 \\ -7 \\ -5 \\ -4 \\ 2 \\ 6 \\ 9 \\ 13 \\ 21 \\ 20 \end{bmatrix} \quad \text{and } X = \begin{bmatrix} 1 & -3 \\ 1 & -2 \\ 1 & -1 \\ 1 & 0 \\ 1 & 2 \\ 1 & 5 \\ 1 & 6 \\ 1 & 8 \\ 1 & 11 \\ 1 & 12 \end{bmatrix}.$$

Thus, we can write

$$X'X = \begin{bmatrix} 10 & 38 \\ 38 & 408 \end{bmatrix},\ X'Y = \begin{bmatrix} 46 \\ 709 \end{bmatrix},\ (X'X)^{-1} = \begin{bmatrix} 0.1548 & -0.0144 \\ -0.0144 & 0.0038 \end{bmatrix}.$$

Hence,

$$\hat{\beta} = (X'X)^{-1}(X'Y) = \begin{bmatrix} 0.1548 & -0.0144 \\ -0.0144 & 0.0038 \end{bmatrix}\begin{bmatrix} 46 \\ 709 \end{bmatrix}$$

$$= \begin{bmatrix} -3.1009 \\ 2.0266 \end{bmatrix} = \begin{bmatrix} \hat{\beta}_0 \\ \hat{\beta}_1 \end{bmatrix}.$$

Thus, the least-squares line is given by

$$\hat{y} = -3.1009 + 2.0266X,$$

which is identical to the regression line we obtained in Example 8.2.1.

EXAMPLE 7.6.2

The following data relate to the prices (Y) of five randomly chosen houses in a certain neighborhood, the corresponding ages of the houses (x_1), and the square footage (x_2).

Price y in Thousands of Dollars	Age x_1 in Years	Square Footage x_2 in Thousands of Square Feet
100	1	1
80	5	1
104	5	2
94	10	2
130	20	3

Fit a multiple linear regression model

$$Y = \beta_0 + \beta_1 x_1 + \beta_2 x_2 + \varepsilon$$

to the foregoing data.

Solution

We have,

$$Y = \begin{bmatrix} 100 \\ 80 \\ 104 \\ 94 \\ 130 \end{bmatrix}; X = \begin{bmatrix} 1 & 1 & 1 \\ 1 & 5 & 1 \\ 1 & 5 & 2 \\ 1 & 0 & 2 \\ 1 & 20 & 3 \end{bmatrix}; X'X = \begin{bmatrix} 5 & 41 & 9 \\ 41 & 551 & 96 \\ 9 & 96 & 19 \end{bmatrix};$$

$$X'Y = \begin{bmatrix} 508 \\ 4560 \\ 966 \end{bmatrix}$$

and

$$(X'X)^{-1} = \begin{bmatrix} 2.3076 & 0.1565 & -1.8840 \\ 0.1565 & 0.0258 & -0.2044 \\ -1.8840 & -0.2044 & 1.9779 \end{bmatrix}.$$

Hence,

$$(X'X)^{-1}(X'Y) = \begin{bmatrix} 66.1252 \\ -0.3794 \\ 21.4365 \end{bmatrix}.$$

Thus, the regression model is

$$y = 66.12 - 0.3794x_1 + 21.4365x_2.$$

Thus, for a given x_1 and x_2 *we can estimate (predict) the value of the house.*

7.6.1 ANOVA for Multiple Regression

As in Section 7.3, we can obtain an ANOVA table for multilinear regression (with k independent or explanatory variables) to test the hypothesis

$$H_0\colon \beta_1 = \beta_2 = \cdots = \beta_k = 0$$

versus,

$$H_a\colon \text{At least one of the parameters } \beta_j \neq 0, j = 1, \ldots, k.$$

The calculations for multiple regression are almost identical to those for simple linear regression, except that the test statistic $(MSR)/(MSE)$ has an $F(k, n-k-1)$ distribution. Note that the F-test does not indicate which of the parameters $\beta_j \neq 0$, except to say that at least one of them is not zero. The ANOVA table for multiple regression is shown in Table 7.6

TABLE 7.6 ANOVA Table for Multiple Regression.

Source of Variation	Degrees of Freedom	Sum of Squares	Mean Sum of Squares	F-Ratio
Regression (model)	k	SSR	$MSR = \frac{SSR}{d.f.}$	$\frac{MSR}{MSE}$
Error (residuals)	$n-k-1$	SSE	$\frac{SSE}{d.f.}$	
Total	$n-1$	SST		

EXAMPLE 7.6.3

For the data in Example 7.6.2, obtain an ANOVA table and test the hypothesis

$$H_0: \beta_1 = \beta_2 = 0 \text{ versus } H_a: \text{at least one of the } \beta_i \neq 0, i = 1,2.$$

Use $\alpha = 0.05$.

Solution

We test H_0: $\beta_1 = \beta_2 = 0$ versus H_a: At least one of the $\beta_i \neq 0$, $i = 1, 2$. Here $n = 5$, $k = 2$. Using Minitab, we obtain the ANOVA table (Table 7.7). *Based on the P-value, we cannot reject the null hypothesis at $\alpha = 0.05$.*

TABLE 7.7 ANOVA Table for the Data.

Source of Variation	Degrees of Freedom	Sum of Squares	Mean Sum of Squares	F-Ratio	P-Value
Regression (model)	2	956.5	478.2	2.50	0.286
Error (residuals)	2	382.7	191.4		
Total	4	1339.2			

EXERCISES 7.6

7.6.1 Given the data

X_1	X_2	y
3	1	4
2	5	3
3	3	6
1	2	5

(a) Write the multiple regression model in matrix form.
(b) Find $X'X$, $(X'X)^{-1}$, and $X'y$.
(c) Estimate β.

7.6.2 A study is conducted to estimate the demand for housing (y) based on the current interest rate X_1 and the rate of unemployment. The data in Table 7.8 is obtained.
(a) Fit the multiple regression model

TABLE 7.8 Units Sold, Interest Rate and Unemployment Rate Data.

Units Sold	Interest Rate (%)	Unemployment Rate (%)
65	9.0	10.0
59	9.3	8.0
80	8.9	8.2
90	9.1	7.7
100	9.0	7.1
105	8.7	7.2

$$y = \beta_0 + \beta_1 x_1 + \beta_1 x_2 + \varepsilon.$$

(b) Test whether the model is significant.

7.6.3 The following data give the annual incomes (in thousands of dollars) and amounts (in thousands of dollars) of life insurance policies for eight persons.

Annual income	42	58	27	36	70	24	53	37
Life insurance	150	175	25	75	250	50	250	100

Calculate the least-squares regression line for this data using matrix operations.

7.6.4 The following is a random sample of height (in inches) and weight (in pounds) of seven basketball players.

Height	73	83	77	80	85	71	80
Weight	186	234	208	237	265	190	220

Calculate the least-squares regression line for this data using matrix operations.

7.7 REGRESSION DIAGNOSTICS

In the previous sections, we derived least-squares estimators for the parameters in the linear regression model. These estimators are useful as long as we can determine (1) how well the model fits the data and (2) how good our estimates are in providing possible relationships between variables of interest. Some of these problems are discussed in Chapter 15 in a unified manner. We now briefly discuss some aspects of the adequacy of the simple linear regression model. In multiple regression, in addition to the problems discussed here, there are other problems, such as collinearity and model specification (inclusion of all relevant variables, as well as exclusion of irrelevant variables), that need to be examined. They are beyond the level of this text. Many graphical methods and numerical tests dealing with these problems are available in the literature and are often called regression diagnostics. Most of the major statistical software packages incorporate these tests, making it easier to perform regression diagnostics so as to detect potential problems.

We have seen that the (ordinary) least-squares regression model must meet the following assumptions.

1. **Linearity**. The existence of a linear relationship between x and y is the basis of the simple linear regression model. A simple method to test for linearity is to draw a scatterplot of data points. As we explained in Section 7.2, we could also plot residual e_i versus x_i or $\hat{Y}_i$. A symmetric trend in the plot of the residuals versus the explanatory variable or the fitted values indicates there is a problem with the obtained regression model. For a correct model, the residuals should center around zero across the explanatory variables and the fitted values. The degree of linear relationship can be ascertained

by the correlation coefficient, r, given in Section 7.5 or by using the value of the coefficient of determination r^2, explained in Project 8B. Most statistical software packages give the value of r^2 (refer to outputs given in Section 8.9). The closer the value of r^2 is to 1, the better the least-squares equation $\hat{y} = \hat{\beta}_1 x + \hat{\beta}_0$ performs as a predictor of y.

2. **Homoscedasticity** (homogeneity of variance). This assumption says that the variance of the error term remains constant across all values of x. In this case, we know by the Gauss–Markov theorem that the least-squares estimators $\hat{\beta}_0$ and $\hat{\beta}_1$ are the best linear unbiased estimators of β_0 and β_1. A frequently used graphical method is to draw the residuals versus a fitted plot. This can be easily done using statistical software packages. The graph of residuals e_i versus fitted values $\hat{Y}_i$ or explanatory variable x_i indicates a change in the spread of residuals as $\hat{Y}$ or x changes. It may look like Fig. 7.7.

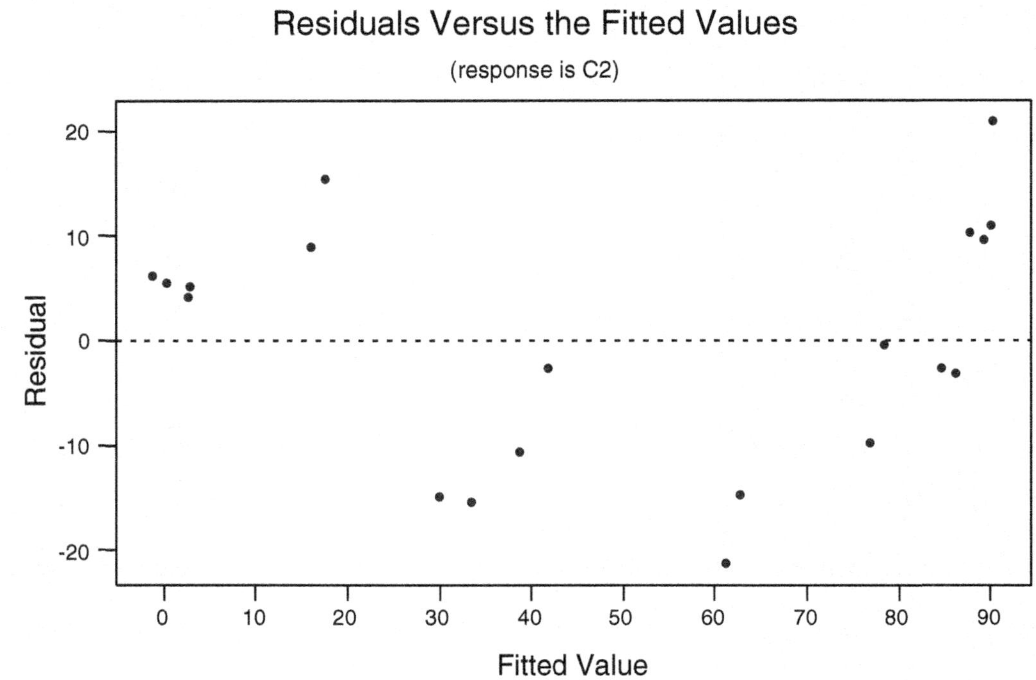

FIGURE 7.7 Scatterplot of fitted values versus residuals.

If the variances of y_i values are not constant, the inferences we made, such as confidence intervals on means, prediction, and so forth, are off. The severity of this discrepancy depends on the degree of the assumption violation. If we see that the pattern of data points only changes slightly, that will indicate a mild heteroscedasticity. Two numerical tests for heteroscedasticity are explained in Section 14.4.3.

3. **Independence of ε_i and ε_j, for $i \neq j$.** This assumption specifies that the errors associated with one observation should not be correlated with the errors of any other observation. In general, whether the two samples are independent of each other is decided by the structure of the experiment from which they arise. Violation of the independence assumption can occur in a variety of situations. For example, if we take a survey on a certain issue on children's education from one particular school, these observations may reflect some pattern, thus violating the independence assumption. If data is collected on the same variable over time, then the assumption of independence will be violated. Project 12B explains a run test for the checking of this assumption. Also, see Section 14.4.4.
4. **Normality of the errors**. This assumption specifies that the distribution of the ε_i values should be normal. This assumption is crucial when the sample size is small, if the P-value for the test is to be valid. For large samples, by the Central Limit Theorem, this assumption becomes less important unless the prediction of a single value of y is involved. Thus, a test of normality is necessary mainly when the t-test is used. Section 14.4.1 explains some of the tests for normality. A simple way is to draw a probability plot for the errors to conform to the assumption of normality. If we observe nonnormality, one of the ways to overcome the problem is to use data transformation, such as logarithmic

transformation as explained in Section 14.4.2, and perform the regression analysis on the transformed data. Sometimes non-parametric methods may be more appropriate, but we will not deal with this topic in this book.

Another important issue is the existence of *influential observations*, individual observations that have a strong influence on estimated coefficients. If a single observation substantially changes our results, we need to do further investigation. The ordinary least-squares method is quite sensitive for outlying observations, both for independent variables and for dependent variables, and can have an adverse effect on the estimate. In higher-dimensional data, these outlying observations can remain unnoticed. This aspect in one explanatory variable case is discussed in Project 8C. One of the simple ways to identify such observations is to draw a scatterplot. In the scatterplot, if we see a data point that is farther away from the rest of the data points, that is an indication of a possible influential point or an outlier.

The natural question is, if we find that the data violates one or more of the assumptions, what can we do about it? We have already explained that violation of the normality assumption in large samples is not an issue unless prediction is involved, because prediction depends on the normality of an individual observation. Thus, if the inferences are based on the t- or F-tests or prediction is involved, we may be able to transform Y to Y' to achieve normality. If we have predicted Y', then back-transform to predict Y. If we observe nonlinearity of data, we may be able to transform x to $x' = h(x)$ such that Y is linear in x', or consider a polynomial model in x, in which case the ideas of multiple linear regression may be utilized. Robust estimates of variances of β_0 and β_1, or the method of weighted least squares, may be used to deal with the case of nonconstant variance. Often, careful experimental design could be done to remove possible correlation in errors. There are also robust methods available for correlation analysis. We refer to specialized books on regression methods for further details on these issues. If we detect influential observations, there are statistical techniques available, such as least trimmed squares estimators, to deal with outlying observations.

7.8 CHAPTER SUMMARY

In this chapter, we first derived the least-squares line and its properties. Then we learned about the confidence intervals for the coefficients in the regression model and did hypothesis tests on the values of the coefficients. We introduced the matrix notation for linear regression as well as for multiple regression. We discussed how to predict a particular value of Y for a given value of X. In order to study the dependence of X and Y, we presented correlation analysis.

The following are some of the key definitions we have used in this chapter:

- Predictors
- Response variable
- Regression analysis
- Multiple linear regression model
- Simple linear regression model
- Sum of squares for errors (SSE)
- Sum of squares of the residuals
- Least-squares line
- Least-squares equations
- Normal equations
- Best linear unbiased estimator (BLUE)
- Correlation analysis

The following important concepts and procedures were discussed in this chapter:

- Procedure for regression modeling
- Procedure for fitting a least-squares line
- Properties of the least-squares estimators for the model $Y = \beta_0 + \beta_1 x + \varepsilon$
- The Gauss–Markov theorem
- Procedure for obtaining confidence intervals of β_0 and β_1
- Procedure to obtain a multiple linear regression equation
- Prediction interval for the response variable Y
- Hypothesis testing for correlation, ρ
- Linearity
- Homoscedasticity
- Independence of ε_i and ε_j, for $i \neq j$

- Normality of the errors
- Influential observations

7.9 COMPUTER EXAMPLES

7.9.1 Examples Using R

EXAMPLE 7.9.1

For the following data, use the method of least-squares regression to fit a straight line to the accompanying data points. Give the estimates of β_0 and β_1. Plot the points and sketch the fitted least-squares line.

Sample (x):	–1	0	2	–2	5	6	8	11	12	-3
Sample (y):	–5	–4	2	–7	6	9	13	21	20	-9

This example assumes you put the data into variables *x* and *y*. Please modify your code appropriately.

R-Code:

```
model=lm(y~x);
summary(model);
```

Solution

From the output below, the estimate of β_0 is –3.10091, and the estimate of β_1 is 2.02656. Hence, the regression line is $\hat{y} = -3.10091 + 2.02656x$.

Output:

```
Residuals:
Min        1Q       Median      3Q       Max
-1.21775-0.70220 0.03452 0.17394 1.80880
Coefficients:
Estimate Std     Error t value Pr(>ltl)
(Intercept) -3.10091      0.38882-7.975 4.47e-05 ***
x           2.02656        0.06087     33.292 7.23e-10 ***
--
Signif. codes: 0 "***" 0.001 "**" 0.01 "*" 0.05 "." 0.1 " " 1
Residual standard error: 0.9883 on 8 degrees of freedom
Multiple R-squared: 0.9928, Adjusted R-squared: 0.9919
F-statistic: 1108 on 1 and 8 DF, P-value: 7.232e-10
```

EXAMPLE 7.9.2

Now obtain the fitted regression line, using results from the previous example.

This example assumes you have your linear model stored in the variable model from the previous example. This example also assumes you have the data from the previous example stored in *x* and *y*. Please modify your code appropriately.

R-Code:

```
yhat=predict(model,data=x);
plot(x,y);
lines(x,yhat);
c=confint(model);
m=model;
m$coefficients[1]=c[1];
lines(x,predict(m,data=x),col="blue");
m$coefficients[1]=c[3];
```

```
lines(x,predict(m,data=x),col="blue");
m=model;
m$coefficients[2]=c[2];
lines(x,predict(m,data=x),col="red");
m$coefficients[2]=c[4];
lines(x,predict(m,data=x),col="red");
```

New Command for confidence interval of model estimates

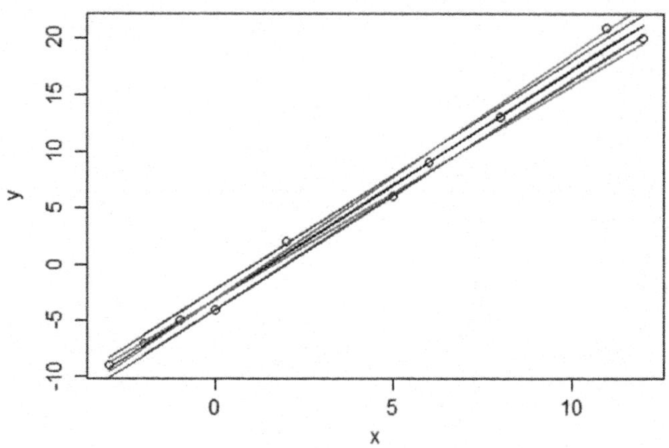

Output:

We obtain a graph with confidence intervals for the intercept in blue and confidence intervals for the slope in red. The coefficient of determination r^2 *is 0.9928, and the P-value is small, suggesting the model fits pretty well.*

EXAMPLE 7.9.3

In this example, we'll be using matrix multiplication to perform linear regression. The following is a random sample of height (in inches) and weight (in pounds) of several basketball players.

Sample (x):	73	83	77	80	85	71	80
Sample (y):	186	234	208	237	265	190	220

Calculate the least-squares regression line for this data. This example assumes you've placed the data into variables x and y. Please modify your code appropriately.

R-Code:

```
library('MASS');
```
Required for ginv() function

```
x=cbind(c(1:length(x))*0+1,x);
```
Creates a matrix with a column of 1's for the intercept

```
b=ginv(t(x)%*%x)%*%t(x)%*%y;
```
Store coefficients into b

```
yhat=x%*%b;
```
Calculate yhat using the regression equation

```
plot(x[,2],y);
lines(x[,2],yhat);
```

Looking at the coefficients, we see that $\hat{\beta}_0 = -188.476$ *and* $\hat{\beta}_1 = 5.208$. *Hence, the regression line is given by* $\hat{y} = -188.476 + 5.208x$. *It is more difficult to perform confidence intervals and other tasks since this is done using matrices instead of model objects.*

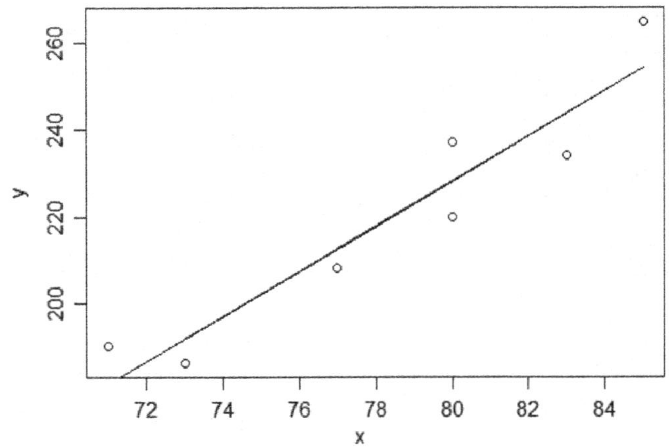

EXAMPLE 7.9.4

Consider the following advertisement expense versus total sales data.

Year	Advertising Cost ($)	Yearly Sales Volume (Units)
1999	20210	112485
2000	22469	118332
2001	23982	122435
2002	24645	125569
2003	24988	125880
2004	25250	127362
2005	25978	125967
2006	26556	127252
2007	26978	127456
2008	27125	127789
2009	27461	128313
2010	28120	128662
2011	28888	128879
2012	29200	129290

Use the method of least-squares regression to fit a straight line to the accompanying data points. Plot the points and sketch the fitted least-squares line. Interpret the output.

R-code

```
> x<-c(22469, 23982,24645,24988,25250,25978,26556,26978,27125,27461,28120,28888,29200)
> y<-c(118332,122435,125569,125880,127362,125967,127252,127456,127789,128313,128662,
128879,129290)
> model=lm(y~x)
> summary(model)
```

7.9.2 SPSS Examples

A detailed explanation of regression methods, including diagnostics using SPSS, can be obtained at the site: http://www.ats.ucla.edu/stat/spss/webbooks/reg/. We will just demonstrate a simple case with an example.

EXAMPLE 7.9.5

The following is a random sample of height (in inches) and weight (in pounds) of seven basketball players.

Height 73 83 77 80 85 71 80

Weight 186 234 208 237 265 190 220

Calculate the least-squares regression line for this data using SPSS.

Solution

Enter height in column 1 and weight in column 2. Then

Analyze>Regression>Linear>move ***var00002*** to ***dependent***, and ***var00001*** to ***Independent(s)>***click ***OK***.

7.9.3 SAS Examples

For regression analysis, we can use the SAS procs called GLM, which stands for General Linear Model, and REG, which stands for regression. In the following example, we will give a simplified version of the foregoing procedure. A good explanation of regression methods, including diagnostics using SAS, can be obtained at http://www.ats.ucla.edu/stat/sas/webbooks/reg/.

EXAMPLE 7.9.6

Using the SAS commands, redo Example 7.9.1.

Solution

We can use the following commands.

```
options nodate nonumber;
data exreg;
INPUT x y @@;
datalines;
-1 -5
0 -4
22
-2 -7
56
69
813
11 21
12 20
-3 -9
;
proc reg data=exreg;
title 'Regression of Y on X';
model y=x / p clm;
run;
```

We obtain the following output.

```
Regression of Y on X
The REG Procedure
Model: MODEL1
Dependent Variable: y
Analysis of Variance
```

```
  Source          DF  Sum       of       Squares        Mean       Square       F       Value        Pr       >       F
Model              1             1082.58589                      1082.58589            1108.34        <.0001
  Error                  8         7.81411                         0.97676
  Corrected Total   9            1090.40000
  Root MSE                       0.98831                          R-Square                          0.9928
  Dependent Mean                  4.60000                          Adj R-Sq                          0.9919
  Coeff Var                       21.48508
                                                                  ParameterEstimates
                                        Parameter    Standard
  Variable         DF        Estimate        Error         t Value         Pr > |t|
  Intercept         1     -3.10091                 0.38882               -7.98            <.0001
  x        1           2.02656              0.06087               33.29          <.0001
                                        Regression of Y on X
                                          The REG Procedure
                                               Model: MODEL1
                                        Dependent Variable: y
                            Output Statistics
                        Dep Var Predicted StdError
```

Obs	y	Value	Mean	Predict	95% CL Mean	Residual
1	-5.0000	-5.1275	0.4278	-6.1141	-4.1409	0.1275
2	-4.0000	-3.1009	0.3888	-3.9975	-2.2043	0.8991
3	2.0000	0.9522	0.3312	0.1885	1.7159	1.0478
4	-7.0000	-7.1540	0.4715	-8.2413	-6.0667	0.1540
5	6.0000	7.0319	0.3210	6.2917	7.7720	-1.0319
6	9.0000	9.0584	0.3400	8.2743	9.8425	-0.0584
7	13.0000	13.1115	0.4038	12.1804	14.0427	-0.1115
8	21.0000	19.1912	0.5383	17.9499	20.4325	1.8088
9	20.0000	21.2178	0.5889	19.8597	22.5758	-1.2178
10	-9.0000	-9.1806	0.5187	-10.3766	-7.9845	0.1806

```
Sum of Residuals                                     0
Sum of Squared Residuals                    7.81411
Predicted Residual SS (PRESS)                   14.18340
```

By looking at the parameter estimates in the foregoing output, we see that an intercept value of -3.10091 is the estimate of β_0, and the estimate of β_1 is 2.02656, corresponding to the variable x. For each value of x, the actual value and predicted value of y are given as the output statistics.

It is important to note that the presentation of results of analysis in a simple way is as important as the analysis itself. For example, if one is interested only in a simple linear regression, most of the output values in the foregoing output may not be necessary. All the values until the parameter estimates are giving us the analysis of variance results, and all the values in the REG procedure are dealing with prediction and confidence intervals. For clarity and simplicity of report, we may only need to report the regression line, and perhaps the graph of the line.

If we need the plot of the points (x, y), add the following commands to the previous program. We will not give the corresponding graph.

```
proc plot data=exreg;
title 'Plot of Y Vs. X';
plot y*x;
run;
```

If we need the graph of the regression line along with, say, 95% prediction and confidence intervals, we add the following.

```
proc gplot data=exreg;
plot y*x
y*x
y*x / overlay frame vaxis =axis1 haxis=axis2;
symbol1 v = - h = 1.5 i=none c=black;
symbol2 v=none i=rlclm95 c=red;
symbol3 v=none i=rlcli95 c=blue;
axis1 order = (-5 to 14 by 1)
offset = (1)
label = (h =1.5 f=duplex);
axis2 order = (-10 to 20 by 1)
offset = (1)
label = (h =1.5 f = duplex);
title h = 1.5
'Effect of X on Y';
title2 h = 1.2 f = duplex
'Common regression line with 95% confidence
interval';
title3 h=1.5 f = duplex
'Regression line is predicted Y=-3.1011
+2.0266X';
run;
```

PROJECTS FOR CHAPTER 7

7A. Checking the Adequacy of the Model by Scatterplots

If the regression model is adequate, then the fitted equation can be used to make inferences. Otherwise, the inferences made will be practically useless. Note that the residuals give all the information on the lack of fit. Figs. 7.5 and 7.6 give an indication of good fit and misfit.

(i) Collect a couple of real-life data points, and find a regression line for each.
(ii) Draw the scatterplot for the residuals e_i versus x and determine whether the regression lines obtained in (i) are a good fit or not.

7B. The Coefficient of Determination

One of the ways to measure the contribution of x in predicting y is to consider how much the prediction errors were reduced by using the information provided by the variable x. The quantity called the coefficient of determination measures how well the least-squares equation $\hat{y} = \hat{\beta}_1 x + \hat{\beta}_0$ performs as a predictor of y. If x contributes no information for predicting y, then the best prediction for values of y is simply the sample mean $\bar{y}$. The resulting sum of squares of deviation for this model $\hat{y} = \bar{y}$ is $S_{yy} = \Sigma_{i=1}^{n} (y_i - \bar{y})^2$. In the case where x contributes information for predicting y, then we have seen that the sum of squares of deviation for the model $\hat{y} = \hat{\beta}_1 x + \hat{\beta}_0$ is $S_{yy} = \Sigma_{i=1}^{n} (y_i - \hat{y}_i)^2$. It can be shown that $\Sigma_{i=1}^{n} (y_i - \hat{y}_i)^2 \le \Sigma_{i=1}^{n} (y_i - \bar{y})^2$.

The *coefficient of determination* is the proportion of the sum of squares of deviations of the y values that can be credited to a linear relationship between x and y. This is defined by

$$r^2 = \frac{S_{yy} - SSE}{S_{yy}}$$
$$= 1 - \frac{SSE}{S_{yy}}$$
$$= 1 - \frac{\Sigma_{i=1}^{n} (y_i - \hat{y}_i)^2}{\Sigma_{i=1}^{n} (y_i - \bar{y})^2}.$$

We can see that $0 \le r^2 \le 1$. We can interpret r^2 to be the proportion of variability explained by the regression line. When x contributes no information for predicting y, S_{yy} and SSE will be nearly equal, and hence r^2 will be near to zero. If x

contributes information for predicting y, S_{yy} will be larger than SSE, and hence r^2 will be greater than zero. Thus, $r^2 = 0.75$ means that the use of $\hat{y}$ instead of $\bar{y}$ to predict y reduced the sum of squares of deviations of the y values about their predicted values $\hat{y}$ by 75%. This can also be interpreted as meaning that nearly 75% of the variation is explained by the independent variable x. In general, about $(r^2 \times 100)\%$ of the sample variation in y can be attributed to using x to predict y in the linear model. The *coefficient of nondetermination* is the percent of variation that is unexplained by the regression equation and is given by $1 - r^2$.

(i) For Exercises 7.2.2 and 7.2.3, find the coefficient of determination, and discuss the information contributed by x in predicting y.

(ii) Collect a couple of real-life data points, and find the corresponding regression lines. Also, draw the scatterplot for e_i versus $\hat{y}$ and determine whether the regression line obtained is a good fit or not based on the coefficient of determination.

7C. Outliers and High Leverage Points

One of the important aspects of residual analysis is to identify the existence of unusual observations in a data set. There are two possibilities for a data point to be unusual. It could be in the response variable (i.e., in the horizontal direction) representing model failure, or in the predictor variable (i.e., in the vertical direction). It should be noted that unusual observations in the horizontal direction occur when we assume that the independent variable X in the linear model is random. An observation that is unusual in the vertical direction is called an *outlier*. An observation that is unusual in the horizontal direction is called a *high leverage point* (or just *leverage point*).

Consider the following 10 points, which we will call base points, and three additional points representing an outlier (O), a high leverage point (H), and both (OH), respectively.

	10 Base points										O	H	OH
x	−1	0	2	−2	5	6	8	11	12	−3	6	19	19
y	−5	−4	2	−7	6	9	13	21	20	−9	30	13	30

Investigate the effect of adding a single aberrant point by running four separate regressions: (i) regression for 10 base points; (ii) regression for 10 base points plus O; (iii) regression for 10 base points plus H; and (iv) regression for 10 base points plus OH. For each of them, find $\hat{\beta}_0$ and $\hat{\beta}_1$ as well as the coefficient of determination. Discuss the effects of each type of outlier on the regression line.

Chapter 8

Design of Experiments

Chapter outline

OBJECTIVE

To study the basic design concepts for experiments through which we can make comparisons of treatments with respect to the observed responses.

Genichi Taguchi
(Source: *http://www.amsup.com/bios/g_taguchi.html*)

Genichi Taguchi (1924–2012) acquired his statistical skills under the guidance of Prof. Motosaburo Masuyama, one of the best statisticians of his time. After World War II, Japanese manufacturers were struggling to survive with very limited resources. Taguchi revolutionized the manufacturing process in Japan through cost savings. He understood that all manufacturing processes are affected by outside influences—noise. However, Taguchi found methods of identifying those noise sources that have the greatest effects on product variability. Isolating these factors to determine their individual effects can be a very costly and time-consuming process. Taguchi devised a way to use the so-called orthogonal arrays to isolate these noise factors from all others in a cost-effective manner. He introduced the loss function to quantify the decline of a customer's perceived value of a product as its quality declines. Taguchi referred to the ability of a process or

Mathematical Statistics with Applications in R. DOI: https://doi.org/10.1016/B978-0-443-27548-7.00008-6

product to work as intended, regardless of uncontrollable outside influences, as robustness. This was a novel concept in the design of experiments with profound influence on manufacturing. His ideas have been adopted by successful manufacturers around the globe because of their results in creating superior production processes at much lower costs.

8.1 INTRODUCTION

In statistics, we are concerned with the analysis of data generated from an experiment. How do we collect data to answer our research questions? What should our design be? It is desirable to take the necessary time and effort to organize the experiment appropriately so that we have the right type and a sufficient amount of data to answer the questions of interest as clearly and efficiently as possible. This systematic procedure determines the relationship between factors affecting a process, and the output of that process is called *experimental design* (or *design of experiment [DOE]*). In any experiment, there are uncontrollable and controllable factors. The idea of DOE is to modify controllable factors given the uncontrollable factors for an "optimal" output. We can trace the roots of modern experimental design to the 1935 publication of the book *The Design of Experiments*, written by Sir Ronald A. Fisher. He showed how one could conduct credible experiments in the presence of many naturally fluctuating conditions, such as soil condition, temperature, and rainfall, in an agricultural experiment. The design principles that were developed for agricultural experiments were successfully modified and adapted to industrial, military, and other applications. In modern industry, it is essential to manufacture parts efficiently and with practically no defects. As a result, variation reduction in quality characteristics of these parts has become a major focus of quality and productivity improvement. Dr. Genichi Taguchi pioneered the use of DOE in designing robust products—those relatively insensitive to changes in design parameters. Presently, DOE is used as an essential tool for improving the quality of goods and services. It is important to note that, unless a sound design is employed, it may be very difficult or even impossible to obtain valid conclusions from the resulting data. Also, properly designed experiments will generate more precise data while using substantially fewer experimental runs than ad hoc approaches. In industrial manufacturing, some of the major benefits of DOE are lower costs, simultaneous optimization of several factors, fast generation and organization of quantitative information, and overall quality improvement.

It is important to clearly identify the particular questions that an experiment is intended to answer—the major objective of the experiments—before conducting the experiment. These objectives may be to estimate or predict some unknown parameters, to explore relationships among various factors, to compare a collection of effects or parameters, or any combination of these factors. When the intention is to compare parameters, the objective may be to corroborate a hypothesis, or to explore some simple relationships. In any design, it is necessary to identify the populations that need to be studied and the type of information about these populations that will be needed to answer the desired questions. While planning an experiment to investigate the primary objectives of the investigation, we need to ensure that the measurement process is simple, the cost of the study is reasonable, the study can be concluded in a reasonable time frame, and the study produces reliable data. Because of the complex nature of real-world problems, planning an effective experiment is not an easy task. The important issues confronting one area, say engineering, will be different from those for another area, such as biology or medicine. As a result, DOEs can take several forms. In this chapter, we will follow a general framework. Two of the major distinguishing elements of DOE are (1) simultaneous variation and evaluation of various factors, and (2) systematic removal of some of the possible test combinations to cut back experimental time and cost. Thus, a researcher should ensure that the statistical design is as simple as possible, given the objectives of the experiment and within the practical constraints such as material, labor, and cost. Some other desirable criteria of a good design are that it provides unbiased estimates of treatment effects and the experimental error. In addition, it should be able to detect important small differences with sufficient precision, and it should provide an estimation of uncertainty in the conclusions and the confidence with which the result can be extended to other analogous situations. The experimental design determines the basic characteristics of the data collected. These data are then processed using statistical analysis methods, with the goals of these analyses being determined by the experimental objectives. Conclusions are obtained by looking at the results of the statistical analyses.

8.2 CONCEPTS FROM EXPERIMENTAL DESIGN

In this section, we introduce some of the basic definitions, methods, and procedures used in experimental design. Many of the terms used have an agricultural basis, because the early development and applications of DOE were in the field of agriculture.

8.2.1 Basic Terminology

The first step in planning an experiment is to formulate a clear statement of the objectives of the test program. The purpose of most statistical experiments is to determine the effect of one or more independent variables on the response variable. The main variable of interest in a study is the *response variable*, also called an *output variable*. These are the dependent variables (also referred to as criteria, effect, or predicted variable) in an experiment that describes the factors we are interested in predicting or comparing. The response variable is measured with different values of independent variables (representing those factors that are assumed to be the causes of the outcome) and analyzed to determine whether the independent variables have any effect. For example, in an agricultural experiment, the crop yield could be the response variable, whereas the type of soil, temperature, and rainfall could be the independent variables. We would also like to identify known or expected sources of variability in the experimental units, because one of the main aims of the designed experiment is to reduce the effect of these sources of variability on the answers to questions of interest. Hence, we must make a list of the factors that may affect the value of the response variable. We must also decide how many observations should be included and what values should be chosen for each independent variable in each individual test run. The independent variables are sometimes referred to as the *attributable variables* or *risk factors* that cause of response.

Definition 8.2.1. *The variables that an experimenter is able to completely control in the DOE are called* **independent variables** *or* **treatment variables**. *These are also called* **input variables, explanatory variables**, *or* **factors**.

Basically, *factors* are independent variables whose effects on the response variable are the main objective of the study. The control variables are selected by the analyst for comparison. A factor is a general category or type of treatment. Factors can be either quantitative or qualitative, based on whether the variable is measured on a numerical scale or not. For example, a rice field is divided into six parts, and each part is treated with a different fertilizer to see which fertilizer produces the most rice. Here, the response variable is the amount of rice that is produced. The objective of the study is to compare the effects of different fertilizers on the rice produced. Thus, the type of fertilizer is the factor.

Definition 8.2.2. *Independent variables that are unknown or known but nonmanipulable are called* **nuisance variables**.

The factors that we could change, but deliberately kept fixed, are called the constants in the experiment. A factor can have different levels referred to as the *treatment* or *factor levels*. Different treatments constitute different levels of a factor. Levels are the values at which the factors are set in an experiment. The level of a variable or treatment means its amount or magnitude. For example, if the experimental units of medication were given as 2.5, 5, and 10 mg, these amounts would be three levels of the treatment. *Level* is also used for categorical variables, such as drugs I, II, and III, where the three are different kinds of drugs, not different amounts of the same drug. Suppose that four different groups of students are subjected to four different teaching methods. The students are the experimental units, the teaching methods are the treatments, and the four types of teaching methods constitute four levels of the factor "type of teaching." Note that this is a single-factor experiment, the factor being the method of teaching.

Definition 8.2.3. **Noise** *is the effect of all the uncontrolled factors in an experiment.*

In some experiments, all the noise factors are known; however, in most cases, only some of them are known. When an analyst controls the specification of the treatments and the method of allocating the experimental units to each of the treatments, the experiment is called designed. For example, n rats are randomly assigned to one of the five dose levels of an experimental drug under investigation. The analyst can also decide on the number n_i of the rats for each dose level such that $\sum_{i=1}^{5} n_i = n$.

Sometimes, conducting a designed experiment may not be practical or ethical. For example, if an analyst wants to know the relationship between fat content in a diet and the cholesterol level, it would be unethical and costly, as well as time-consuming, to subject human volunteers to different fat-content diets. However, it is possible to observe the cholesterol levels of people who consume different diets. Care must be taken to record various other factors, such as exercise habits, age, and gender, before reporting any association between cholesterol levels and the fat content of diets. The experiment is called observational if the analyst is just an observer of the treatments on a sample of experimental units. Note that the experimental units are objects to which treatments are applied.

The crucial difference between an experiment and an observational study for comparing the effects of treatments is that, in an experiment, the researcher decides which experimental units receive which treatments, whereas in an observational study, the researcher simply compares experimental units that happen to be there that have received each of the treatments. Observational studies are often useful for identifying possible causes of treatment effects, and they are often cheaper. Their main disadvantage is that they are less conclusive. Only properly designed and executed experiments can lead to reliable conclusions. Hence, in general, designed experiments are preferred over *observational* experiments. In designing the experiment, there are almost always going to be constraints such as budget, time, and availability of *experimental units*.

The following example illustrates an *observational experiment*, where the analyst has control over the random sampling from the treatment populations as well as the size of each sample, but has no control over the assignment of the experimental units to the treatments.

EXAMPLE 8.2.1

In order to compare the risk-taking tendency of the people who invest in mutual funds, samples are taken of individuals from three income groups—low-income class, middle-income class, and high-income class. A score is given based on the percentage of their investment allocation to different types of mutual funds, such as large-cap, mid-cap, small-cap, hybrid, and specialty. The mean score for each income group is calculated. Identify each of the following elements: *response, factors and factor type(s), treatments, and experimental units.*

Solution

The response is the variable of interest, which is the score given to each individual investor. The only factor investigated is the income class. This is a qualitative variable. The three income classes represent the levels of this factor. The treatment is the percentage of investments in the different types of mutual funds, such as large-cap, mid-cap, small-cap, hybrid, and specialty. The experimental unit is an individual investor.

There are *single-factor experiments* and *multifactor experiments*. The previous example was a case of a single-factor experiment. Single-factor experiments have only one independent variable. Another example of a single-factor experiment is when we are interested in the effect of size of the screen of a computer monitor on the reading speed. In this case, the size of the screen is the single factor. If there are only two sizes, say 15-in. and 17-in. monitors, that we wish to compare, tests such as the two-sample t-test could be used to compare the average reading speed. If there are more than two sizes of monitors, then the one-way ANOVA methods described in Chapter 9 could be used for analysis of the resulting data.

Even though the single-factor experiments are simple and elegant, they are costly and not very effective when there is more than one independent variable. Efficient use of resources is achieved through multifactor experiments in comparison to conducting many single-factor experiments. A *multifactor experiment* involves two or more independent variables and a dependent variable. Also, a greater range of questions could be answered using multifactor experiments. While multifactor experiments are efficient, care should be taken to identify and interpret the main effects and interaction effects, as well as nonlinear effects. For details on these concepts, we refer to books totally devoted on experimental designs. The resulting data are analyzed using ANOVA as described in Chapter 9. The following is an example of a multifactor experiment.

EXAMPLE 8.2.2

In order to study the conditions under which a particular type of commercially raised fish reaches maximum weight, an experiment is conducted at four water temperatures (60°F, 70°F, 80°F, 90°F) and four water salinity levels (1%, 5%, 10%, 15%). Fish are raised in tanks with specific salinity levels and temperature levels. There are 32 tanks, with one of the four temperatures and one of the four salinity levels assigned randomly to each tank. The weights are recorded at the beginning of the experiment and after 2 months. Identify each of the following elements: response, factors, and factor type(s). Write all the treatments from the factor-level combinations.

Solution

The response is the variable of interest, which is the weight gain of a fish. This experiment has two factors: water temperatures at four levels and water salinity at four levels. There are $4 \times 4 = 16$ *possible treatments:*

(60°F, 1%)	(60°F, 5%)	(60°F, 10%)	(60°F, 15%)
(70°F, 1%)	(70°F, 5%)	(70°F, 10%)	(70°F, 15%)
(80°F, 1%)	(80°F, 5%)	(80°F, 10%)	(80°F, 15%)
(90°F, 1%)	(90°F, 5%)	(90°F, 10%)	(90°F, 15%).

It should be noted that there may be other factors, such as the density of the fish population, the initial size of the fish, and the type of feeding, that may affect the weight gain of the fish.

Definition 8.2.4. *The* **experimental error** *explains the variation in the responses among experimental units that are assigned the same treatment and observed under identical experimental conditions.*

Experimental error can occur for many reasons, among them (1) the difference in the devices that record the measurements, (2) the natural dissimilarities in the experimental units prior to their receiving the treatment, (3) the variation in setting the treatment conditions, and (4) the effect on the response variable of all extraneous factors other than the treatment factors.

In order to construct confidence intervals on the treatment population means and to test hypotheses, it is necessary to obtain an estimate of the variance of the experimental design. In a single-factor experiment with k levels, the estimate of the variance of experimental design could be taken as the pooled variance of responses from experimental units receiving the identical treatments. A large variance of experimental errors will compromise the accuracy of inferences made from the experiments. Also, large amounts of experimental error make it difficult to determine whether the treatment has produced an effect or not, so one of the design goals is to reduce the experimental error. Bad execution of a design can lead to the whole experiment becoming a waste of time and resources. It is necessary to implement techniques to reduce experimental error in order to obtain more accurate inferences. One approach to reducing experimental error is to take extra care in conducting the experiment. The effect of experimental error can be reduced by using more homogeneous experimental materials (if available), and using the fundamental principles of replication, randomization, and blocking (see Section 8.2.2).

The *one-way analysis of variance* (in a single-factor experiment at several levels) enables one to compare several groups of observations, all of which are independent, with the possibility of a different mean for each group. A test of significance is whether all the means are equal. *Two-way analysis of variance* is a method of studying the effects of two factors on the response variable.

There are other terms that are important in different applications. For example, in the medical field, the terms *blinding*, *double-blind*, and *placebo* are used. In a medical experiment, the comparison of treatments may be distorted if the patient, the person administering the treatment, and those evaluating it know which treatment is being allocated to which patient. It is therefore necessary to ensure that the patient, and/or the person administering the treatment, and/or the trial evaluators do not know (are blind to) which treatment is allocated to whom. If only the patient is unaware of the treatment, it is called *blinding*, and if both the patient and the person administering the treatment are blind to which treatment is being allocated, it is called *double-blinding*. In order to study the effect of a particular drug, experimenters divide the study population into two groups and treat one group with the drug and the other group with a so-called placebo, which could be just sugar pills. In order to clarify the objective of a design, it is necessary for an experimental designer to consult a wide range of people, especially those affected by the problem to be solved.

8.2.2 Fundamental Principles: Replication, Randomization, and Blocking

A good design of an experiment makes efficient use of resources to gather the data needed to meet the goals of the study. There are three fundamental principles that need to be considered in a good experimental design. They are *replication*, *randomization*, and *blocking*. Replication and blocking increase precision in the experiment, whereas randomization reduces the bias.

Definition 8.2.5. Replication *means that the same treatment is applied (i) several times to the same experimental units, or (ii) one time to several similar experimental units, called replicate units.*

Replications are necessary for the estimation of the error variance in an experiment against which the differences among treatments are assessed. If an experiment is intended to test whether a number of treatments differ in their effects, these treatments must be applied to replicate the units of the experiment. In order to show that two treatments have different mean effects, we need to measure several samples given the same treatment. For example, observing that one plant of a particular genotype is more resistant to a disease than another plant of a different genotype does not convey anything about the difference between the mean disease resistances of the two genotypes. This difference could have been caused by the environment where the experiment was conducted or the inoculation procedure affecting the two plants differently. Hence, to make any inference about the mean difference between the genotypes, we must test several plants of each type. Thus, increasing the number of replications increases the reliability of inferences drawn from the observed data. It is necessary to increase the number of replications with varied experimental conditions to decrease the variance of the treatment effect estimates and also to provide more power for detecting differences in treatment effects. We should not confuse multiple observations of the same experimental unit with replication. Replication involves applying the treatment to a number of experimental units.

Definition 8.2.6. A **block** *is a portion of the experimental unit that is more likely to be homogeneous within itself than with other units.*

Blocking refers to the distribution of the experimental units into blocks in such a way that the units within each block are more or less homogeneous. The experimenter uses information about the possible variability among units to group them in such a way that most of the unwanted experimental errors can be removed through the block effect.

For blocking to be effective, the units should be arranged so that within-block variation is much smaller than between-block variation. As an example, suppose a researcher wishes to compare the yields of rice for four different kinds of fertilizers. In order to minimize the effect of environmental and soil conditions, the field may be divided into smaller blocks, and each block is further parceled into four plots. Each variety of fertilizer is applied in each block with one in each parcel. This method ensures that the external conditions from plot to plot within a block will be relatively uniform. Then we can use the ANOVA methods to pool from block to block to obtain the within-block information about the treatment differences while avoiding between-block differences. The relevant analysis is given in Section 9.5. Time could also be a block factor, because concentration or expertise could alter as one carries out a task, such as determining disease levels or scoring microscope slides.

Definition 8.2.7. Randomization *is the process of assigning experimental units to treatment conditions in an entirely chance manner.*

The main objective of randomization is to negate the effects of all uncontrolled extraneous variables. Usually, randomization is associated with design functions such as random sampling or selection, random assignment, and random order. Random assignment of experimental units to groups tends to spread out the differences between subjects in unsymmetrical or random ways so that there is no tendency to give an edge to any group. In any well-conducted experiment, randomization eliminates bias from the experiment, enables us to use statistical tests of significance, and creates valid estimates of experimental error. For instance, suppose we are measuring the time of flowering of plants in a glasshouse or in a growth cabinet. If the pots are arranged so that all the plants of one variety are next to each other, and we observe that one variety flowers earlier than the rest, does this imply that this variety is inherently earlier-flowering, or does it suggest that the light and temperature conditions in that part of the cabinet or glasshouse cause plants to flower early? It is not possible to tell from an experiment designed in this manner. Randomizing the treatments in time or space is an insurance policy to take account of variation that we may or may not know to exist under the conditions of our experiment. For instance, the levels of light in growth cabinets vary considerably, so randomizing the layout of the plants of different types is essential to make sure that no one type is consistently exposed to light and temperature levels that are particularly high or low. Another way of selecting experimental units is simply to use intact groups, such as all students in a particular statistics classroom. Results obtained this way may be highly biased and hence not desirable. In general, the process of randomization ensures independent observations, and it should be noted that random assignment does not completely eliminate the problem of correlated data values.

Now we study some steps that can be used for randomization. Suppose there are N homogeneous experimental units and k treatments. In order to randomly assign r_i experimental units to the ith treatment with $\sum_{i=1}^{k} r_i = N$, we could use the following steps.

Procedure for random assignment

1. Number the experimental units from 1 to N.
2. Use a random number table or statistical software to get a list of numbers that are random permutations of the numbers 1 to N.
3. Give treatment 1 to the experimental units having the first r_1 numbers in the list. Treatment 2 will be given to the next r_2 numbers in the list, and so on; give treatment k to the last r_k units in the list.

The following example illustrates the random assignment procedure.

EXAMPLE 8.2.3

In order to study the number of hours of relief provided by five different brands (A, B, C, D, E) of pain reliever, doses are administered to 25 subjects numbered 1 through 25, with each brand administered to five subjects. Develop a design using the random assignment procedure.

Solution

Using SoftwareSoftware, we obtained the following random permutations of the numbers from 1 to 25.

1 8 7 12 10 25 23 4 6 3
9 21 5 24 18 16 22 14 17 15
20 13 2 11 19

Using the randomized procedure, we obtain the design given in Table 8.1.

TABLE 8.1 The Number of Hours of Pain Relief by Brand.

Subject	1	8	7	12	10	25	23	4	6	3	9	21	
Brand	A	A	A	A	A	B	B	B	B	B	C	C	
Subject	5	24	18	16	22	14	17	15	20	13	2	11	19
Brand	C	C	C	D	D	D	D	D	E	E	E	E	E

That is, subject number 8 will get brand A pain reliever, subject 23 will get brand B pain reliever, and so forth. We can rewrite Table 8.1 *as shown in* Table 8.2.

TABLE 8.2 Rearrangement of the Data of Table 8.1.

Brand	Subject				
A	1	8	7	12	10
B	25	23	4	6	3
C	9	21	5	24	18
D	16	22	14	17	15
E	20	13	2	11	19

It should be noted that once we create the design, the actual data will contain the number of hours of relief for each individual.

It is important to note that randomization may not be possible in some cases. Observational studies may be necessary whenever the researcher cannot use controlled randomized experiments. For example, if we want to study the effect of smoking on lung cancer, randomization will mean that we should be able to select a group of people and tell a randomly selected subgroup to smoke and the other subgroup not to smoke. This is not only practically impossible; it is also unethical to deliberately expose people to a potentially hazardous substance.

8.2.3 Some Specific Designs

In this subsection, we will introduce three specific designs: *completely randomized design*, *randomized complete block design*, and *Latin square design*. The structure of the experiment in a *completely randomized design* is presumed to be such that the treatments are assigned to the experimental units completely at random. Example 8.2.1 is one such design. In order to create a completely randomized design, follow the procedure given in Section 8.2.2.

The *randomized complete block design* is a design in which the subjects are matched according to a variable that the experimenter wants to control. The subjects are put into groups (blocks) of the same size as the number of treatments. The elements of each block are then randomly assigned to different treatment groups so as to reduce the influence of unknown variables. For example, a researcher is carrying out a study of three different drugs for the treatment of high cholesterol. Suppose she has 45 patients and divides them into three treatment groups of 15 patients each. Using a randomized block design, the patients are rated and put in blocks of three, based on their cholesterol level. The three patients with the highest cholesterol are put in the first block, those with the next highest levels are put in the second block, and so on. The three members of each block are then randomly assigned, one to each of the three treatment groups. If there is very little extraneous, systematic variation, complete randomization allows for differences between the mean effects of the treatments to be estimated with a higher precision than other designs. However, it does not allow for the possibility that there could be some unknown extraneous factors, so if in doubt, use a randomized complete block design.

Suppose we have k treatments and N experimental units. Further, assume that the experimental units can be grouped into b groups containing k experimental units, so that $N = bk$. We could use the following steps for a randomized complete block design.

Procedure for randomization in a randomized complete block design

1. Group the experimental units into b groups (blocks) containing k homogeneous experimental units.
2. In group 1, number the experimental units from 1 to k and obtain a random permutation of numbers 1 to k using a random number generator.
3. In group 1, the experimental unit corresponding to the first number in the permutation receives treatment 1, the experimental unit corresponding to the second number in the permutation receives treatment 2, and so on.
4. Repeat steps 2 and 3 for each of the remaining blocks.

We illustrate the step-by-step procedure given above in the following example.

EXAMPLE 8.2.4

In order to study the number of hours to relief provided by five different brands (A, B, C, D, E) of pain relievers for pain resulting from different causes [headache (H), muscle pain (M), pain due to cuts and bruises (CB)], doses are administered to five subjects each having similar types of pain. Create a randomized complete block design. Choose, as blocks, the different types of pain (H, M, or CB).

Solution

Using SoftwareSoftware with $k = 5$ we have generated the random permutations shown in Table 8.3 *for each of the $b = 3$ blocks of five numbers and assigned the treatments according to the procedure just explained. As the table indicates, among the persons with headache, subject number 3 is treated with brand A painkiller, and so forth.*

TABLE 8.3 Hours to Relief Provided by Different Pain Relievers for Different Causes of Pain.

H	M	CB
3(A)	5(A)	1(A)
1(B)	4(B)	2(B)
2(C)	3(C)	4(C)
5(D) 4(E)	1(D) 2(E)	3(D)5(E)

In the previous example, we had only one replication of each treatment per block. This idea can be generalized to have r replications of each treatment per block. Then the generalized randomized complete block design with k treatments, b blocks, and r replications, with $N = kbr$ which has kr homogeneous experimental units, can be randomized as follows.

Procedure for a randomized complete block design with r replications

1. Group the experimental units into b groups (called blocks), each containing rk homogeneous experimental units.
2. In group 1, number the experimental units from 1 to rk and generate a list of numbers that are random permutations of the numbers 1 to rk.
3. In group 1, assign treatment 1 to the experimental units having numbers given by the first r numbers in the list. Assign treatment 2 to the experiments having the next r numbers in the list, and so on until treatment k receives r experimental units.
4. Repeat steps 2 and 3 for the remaining blocks of experimental units.

The following example illustrates this procedure.

EXAMPLE 8.2.5

With the following modifications, consider Example 8.2.2. Three groups of subjects are considered, with each group having 15 subjects. Group I consists of subjects with only headache (H), group II of subjects only with muscle pain (M), and group III of subjects with only pain due to cuts and bruises (CB). Of the 15 subjects with headache (group I), three are treated with brand A painkiller, three with brand B, and so forth. Subjects with other types of pain are treated similarly. Here, the number of replications is three for each type of drug and for each type of pain. Create a randomized complete block design with three replications.

Solution

Using software, for the group with headache (H), we generate a random permutation of numbers 1–15. The first three are given painkiller A, the next three B, and so forth. The process is repeated for other types of painkillers. The design is given in Table 8.4, *where "2(A)" means that patient 2 is given brand A painkiller.*

TABLE 8.4 Randomized Complete Block Design of Example 8.2.5.

H	M	CB	H	M	CB
2(A)	8(A)	3(A)	15(C)	9(C)	11(C)
14(A)	13(A)	8(A)	7(D)	4(D)	2(D)
10(A)	5(A)	14(A)	5(D)	11(D)	13(D)
8(B)	2(B)	6(B)	6(D)	15(D)	5(D)
12(B)	1(B)	15(B)	3(E)	7(E)	1(E)
11(B)	10(B)	12(B)	9(E)	12(E)	4(E)
4(C)	3(C)	10(C)	13(E)	6(E)	9(E)
1(C)	14(C)	7(C)			

By increasing the number of replications, we can increase the accuracy of estimators of treatment means and the power of the tests of hypotheses regarding differences between treatment means. However, because of constraints such as cost, time needed to handle a large number of experimental units, and even the availability of experimental units, it is not realistic to have a large number of replications. It is then necessary to determine the minimum number of replications needed to meet reasonable specifications on the accuracy of estimators or on the power of tests of hypotheses. We give a simple procedure for determining the number of replications needed.

Let r be the number of replications that we need to determine. Let σ be the experimental standard deviation, and E be the desired accuracy of the estimator. Then the sample size required to be $(1 - \alpha)$ 100% confident that the estimator is within E units of the true treatment mean, μ, is

$$r = \frac{(z_{\alpha/2})^2\hat{\sigma}^2}{E^2}.$$

The values of $\hat{\sigma}$ could be obtained from past experiments, from a pilot study, or by using a rough estimator

$$\hat{\sigma} = (\text{largest observation} - \text{smallest observation})/4.$$

Following is an example for determining the appropriate number of replications.

EXAMPLE 8.2.6

A researcher wants to know the effect of class sizes on the mean score in a standardized test. She wants to estimate the treatment means μ_1, μ_2, μ_3, and μ_4 such that she will be 95% confident that the estimates are within 10 points of the true mean score. What is the necessary number of replications to achieve this goal? It is known from the previous experiments that scores have ranged from 46 to 98.

Solution

A rough estimator of σ is

$$\hat{\sigma} = \frac{Range}{4} = \frac{98 - 46}{4} = 13.$$

From the normal table, $z_{0.025} = 1.96$. The value of $E = 10$. Thus, the number of replications necessary is

$$r = \frac{(z_{\alpha/2})^2\hat{\sigma}^2}{E^2} = \frac{(1.96)^2(13)^2}{(10)^2} = 6.4923 \cong 7.$$

Thus, the researcher should use seven replications of each of the treatments to obtain the desired precision.

We have used the randomized complete block design when we wanted to control a single source of extraneous variation, and there is only one factor of interest. When we need to compare k treatment means and there are two possible sources of extraneous variation, a Latin square design is the appropriate DOE.

Definition 8.2.8. *A $k \times k$* **Latin square design** *contains k rows and k columns. The k treatments are randomly assigned to the rows and columns so that each treatment appears in every row and column of the design.*

It was the famous mathematician Leonhard Euler who introduced Latin squares in 1783 as a new kind of magic squares. Even though the idea is fairly elementary, a systematic use of Latin squares in the DOE was advanced by Ronald A. Fisher only around 1921. Fisher realized that in a two-dimensional plot of land, the systematic error due to variation in soil and other factors could be minimized by a suitable Latin square partition of the plot.

The following example illustrates a case in which the experimental problems are affected by two sources of extraneous variation, the type of car and the type of driver used.

EXAMPLE 8.2.7

A gasoline company is interested in comparing the effect of four gasoline additives (A, B, C, D) on the gas mileage achieved per gallon. Four cars (I, II, III, IV) and four drivers (1, 2, 3, 4) will be used in the experiment. Create a Latin square design.

Solution

We can filter out the variability due to the type of cars used by ensuring that in each row only one of the additive types appears. Also, to filter the driver effect, use each additive only once for each driver. One such randomization results in the Latin square design given in Table 8.5.

TABLE 8.5 Latin Square Design for Gas Mileage Due to Different Gas Additives.

Cars	Drivers			
	1	2	3	4
I	D	B	A	C
II	C	A	D	B
III	B	D	C	A
IV	A	C	B	D

To construct a basic Latin square, one can use the following method, which we will present only for the 4 × 4 *Latin square of Example 8.2.7.*

Procedure for constructing a 4 × 4 Latin square

1. Begin with the first row as A, B, C, D.
2. Generate each succeeding row by taking the first letter of the preceding row and placing it last, which has the effect of moving the other letters one position to the left.
3. Randomly assign one block factor to the rows and the other to the columns.
4. Randomly assign levels of the row factor, column factor, and treatment to row positions, column positions, and letters, respectively.

In step 2 of the foregoing procedure, instead of using the cyclic placement of rows, we can perform a cyclic placement for the columns. Accordingly, change the procedures in steps 3 and 4.

The following example illustrates a 4 × 4 Latin square design.

EXAMPLE 8.2.8

Using the previous procedure, construct a Latin square for the case of Example 8.2.7.

Solution

Following the procedure just given, the Latin square in Example 8.2.7, the basic Latin square is represented by Table 8.6.

TABLE 8.6 **The Basic Latin Square for Example 8.2.7.**

Cars	Drivers			
	1	2	3	4
I	A	B	C	D
II	B	C	D	A
III	C	D	A	B
IV	D	A	B	C

Now, one random assignment of cars, I, II, III, IV, is to the rows 4, 3, 2, 1 (this is a random order of numbers 1, 2, 3, 4) of Table 8.6. *This is given in* Table 8.7.

TABLE 8.7 **Random Assignment of Cars in a Latin Square.**

Cars	Drivers			
	1	2	3	4
I	D	A	B	C
II	C	D	A	B
III	B	C	D	A
IV	A	B	C	D

Now, one random assignment of the drivers 1, 2, 3, 4 is to the columns 1, 2, 4, 3 (this is a random order of numbers 1, 2, 3, 4) of Table 8.7, *resulting in the Latin square shown in* Table 8.8.

TABLE 8.8 **Random Assignment of Drivers in a Latin Square.**

Cars	Drivers			
	1	2	3	4
I	D	A	C	B
II	C	D	B	A
III	B	C	A	D
IV	A	B	D	C

Now, along with this Latin square, we can represent the corresponding observations (numbers in parentheses are the gas mileage in miles per gallon) as shown in Table 8.9.

TABLE 8.9 The Latin Square for Cars vs. Drivers.

Cars	Drivers			
	1	2	3	4
I	D(18)	A(22)	C(25)	B(19)
II	C(22)	D(24)	B(26)	A(24)
III	B(21)	C(20)	A(22)	D(23)
IV	A(17)	B(24)	D(23)	C(21)

Note that if we use notation 1 for additive A, 2 for additive B, 3 for additive C, and 4 for additive D, the Latin square in the previous example can be rewritten as shown in Table 8.10, below.

TABLE 8.10 Rewritten Results of Table 8.9.

Cars	Drivers			
	1	2	3	4
I	4	1	3	2
II	3	4	2	1
III	2	3	1	4
IV	1	2	4	3

This representation will be convenient if we need to write down a model. In order to test the treatment effects, one could use the ANOVA methods discussed in Chapter 9 see (Table 8.10).

For Latin square experiments involving k treatments, it is necessary to include k observations for each treatment resulting in a total of k^2 observations. Table 8.11 shows two examples of Latin squares for $n = 3$, and $n = 5$.

We have used the Latin square design to eliminate two extraneous sources of variability. In order to eliminate three extraneous sources of variability, we can use a design called the *Greco-Latin square*. Greco-Latin squares are also called *orthogonal Latin squares*. This design consists of k Latin and k Greek letters. In this design, we take a Latin square and superimpose upon it a second square with treatments denoted by Greek letters. In this superimposed square, each Latin letter coincides with exactly one of each Greek letter. In our gasoline example, if we introduce the effect of, say, four different days, represented by Greek letters, then Table 8.12 shows the 4×4 Greco-Latin square.

We will not go into more detail about this design, or on many other similar designs.

When developing an experimental design, it is important for the researcher to learn more about the terminology as well as the intricacies of the field in which the experiment will be performed. It is also important to observe that there are many other practical constraints affecting the DOE. For example, experiments are done by organizations and individuals that have limited resources of money and time. Appropriating these resources within the constraints is an integral part of planning an experiment. Also, many problems are approached sequentially in several stages. Planning for each stage is built on what has been learned before. Dealing with these types of issues is beyond the scope of this book.

TABLE 8.11 Latin Square for n=3 and n=5.

A	B	C
C	A	B
B	C	A

3 × 3

A	B	C	D	E
B	A	E	C	D
C	D	A	E	B
D	E	B	A	C
E	C	D	B	A

5 × 5

TABLE 8.12 A Greco-Latin Square for Example 8.2.7.

$A\alpha$	$B\beta$	$C\gamma$	$D\delta$
$B\delta$	$A\gamma$	$D\beta$	$C\alpha$
$C\beta$	$D\alpha$	$A\delta$	$B\gamma$
$D\gamma$	$C\delta$	$B\alpha$	$A\beta$

EXERCISES 8.2

8.2.1 In order to study the conditions under which hash-brown potatoes will absorb the least amount of fat, an experiment is conducted with four frying durations (2, 3, 4, 5 min) and using four different types of fats (animal fat I, animal fat II, vegetable fat I, vegetable fat II). The amount of fat absorbed is recorded. Identify each of the following elements: response, factors, and factor type(s). Write all the treatments from the factor-level combinations.

8.2.2 A team of scientists is interested in the effects of vitamin A, vitamin C, and vitamin D on the number of offspring born for a specific species of mice. An experiment is set up using the same species of mice. The mice are randomly assigned to three groups. Each mouse in the study gets the same amount of food and daily exercise and is kept at the same temperature. One group of mice gets extra vitamin A, another group gets extra vitamin C, and the remaining group gets extra vitamin D. The supplements are added to their food. The number of offspring is counted and recorded for each group.

(a) What is the response variable?

(b) What is the factor?

8.2.3 Thirty rose bushes are numbered 1–30. Three different fertilizers are to be applied to 10 bushes each. Develop a design using the random assignment procedure.

8.2.4 Three different fertilizers are to be applied to five bushes, each for three varieties of flower plants: gardenia (G), rose (R), and jasmine (J). Create a randomized complete block design. Choose as blocks the different types of plants (G, R, or J).

8.2.5 With the following modifications, consider Exercise 8.2.4. Three groups of flower plants are considered, with each group having nine plants. Group I consists of gardenia (G), group II consists of rose (R), and group III consists of jasmine (J). Of the nine gardenias (group I), three are treated with brand A fertilizer, three with brand B, and three with brand C fertilizer. Other plant types are treated similarly. Here the number of replications is three for each type of fertilizer and for each type of plants. Create a randomized complete block design with three replications.

8.2.6 What are the reasons for using randomization in Exercises 8.2.3–8.2.5?

8.2.7 Suppose a food processing company wants to package sliced pineapples in cans. They have four different processing plants, say, A, B, C, and D. Suppose they have 56 truckloads (numbered 1–56) of pineapples collected from different parts of the country. In order to get some uniformity in taste, it is better to randomly assign the trucks to the four plants. Develop a design using the random assignment procedure.

8.2.8 In Exercise 8.2.1, suppose there are four pans and 25 packets of hash-brown potatoes. Randomly select six of the 25 packets to be fried with each of the fat types.

(a) Create a randomized complete block design.

(b) Create a Latin square design.

8.2.9 A chemist is interested in the effects of five different catalysts (A, B, C, D, E) on the reaction time of a chemical process. There are five batches of new material (1, 2, 3, 4, 5). She decides to study the effect of each catalyst on each material for 5 days (1, 2, 3, 4, 5). Construct a Latin square design for this experiment.

8.2.10 Suppose a dating service wants to schedule dates for four women, Anna, Carol, Judy, and Nancy, with Ed, John, Marcus, and Richard on Thursday, Friday, Saturday, and Sunday in such a way that each man dates each woman in the 4 days. Create a Latin square design displaying a schedule that the dating service could follow.

8.2.11 In order to test the relative effectiveness of four different fertilizer mixtures on an orange crop, a Florida farmer applies the fertilizer and measures the yield per unit area when it harvests. The four experiments cannot be carried out on the same plot of land. Devise a Latin square arrangement of dividing a single plot into a 4×4 grid of subplots for administering the fertilizers (labeled randomly A, B, C, D).

8.2.12 A researcher wants to know the effect of four different types of fertilizers on the mean number of tomatoes produced. He wants to estimate the treatment means μ_1, μ_2, μ_3, and μ_4 such that he will be 90% confident that the estimates are within five tomatoes of the true mean number of tomatoes. What is the necessary number of replications to achieve this goal? It is known from previous experiments that the numbers of tomatoes per plant have ranged from 20 to 60.

8.3 FACTORIAL DESIGN

In this section, we introduce a treatment design where the treatments are constructed from several factors rather than just being k levels of a single factor. The treatments are combinations of levels of the factors. A *factorial experiment* can be defined as an experiment in which the response variable is observed at all factor-level combinations of the independent variables. A *factorial design* is used to evaluate two or more factors simultaneously. In general, there are three ways to obtain experimental data, one-factor-at-a-time, full factorial, and fractional factorial. The most efficient design is the fractional factorials. A simple approach for examining the effect of multiple factors is the one-at-a-time approach. The advantages of factorial designs over one-factor-at-a-time experiments is that they allow interactions to be identified. An interaction occurs when the effect of one factor varies with the level of another factor or with some combination of levels of other factors when there are multiple factors.

The *one-way analysis of variance*, discussed in the next chapter, enables us to compare several groups of observations, all of which are independent with the possibility of a different mean for each group. A test of significance is whether all the means are equal. *Two-way analysis of variance* is a way of studying the effects of two factors separately, such as their main effects, and together, with their interaction effect.

8.3.1 One-Factor-at-a-Time Design

In one-factor-at-a-time design, one conducts the experiment with one factor at a time. Here we hold all factors constant except one and take measurements on the response variable for several levels of this one factor, then choose another factor to vary, keeping all others constant, and so forth. We are familiar with this type of experiment from undergraduate chemistry or physics labs. One of the drawbacks of this method is that all factors are evaluated while the other factors are at a single setting. For example, in the case of Example 8.2.2, we would set a fixed temperature and study the effect of water salinity on the fish weight gains and then set a fixed water salinity and vary the temperature. All this is time-consuming.

Definition 8.3.1. *Two factors I and II are said to* **interact** *if the difference in mean responses for different levels of one factor is not constant across all the levels of the second factor.*

EXAMPLE 8.3.1

Consider the following hypothetical data, in which two types of diet (fat, carbohydrates) in two levels (high, medium) were administered for a week in a sample of individuals. At the end of the week, each subject was put on a treadmill and recorded the time of exhaustion, in seconds. The objective is to determine the factor-level combination that will identify maximum time of exhaustion. Table 8.13 gives average time of exhaustion for each combination of diets.

TABLE 8.13 Average Time of Exhaustion for Each Combination of Diets.

Average Time to Exhaustion	Fat	Carbohydrate
88	High	Medium
98	Medium	Medium
77	Medium	High
74	High	High

Discuss this as a one-factor-at-a-time experiment to predict the average time of exhaustion.

Solution

We can see that the average time of exhaustion decreases when the fat content is increased from medium to high while holding carbohydrate at medium. The average time of exhaustion also decreases when carbohydrate content is increased from medium to high while holding fat at medium. Thus, it is tempting to predict that increasing both fat and carbohydrate consumption will result in a lower average time of exhaustion. The problem with this reasoning is that the prediction is based on the assumption that the effect of one factor is the same for both levels of the other factor. Changing the fat content from medium to high, keeping carbohydrate at medium, and the carbohydrate content from medium to high, keeping fat at medium, reduced the average time of exhaustion by approximately 10 seconds. The question then is, can we predict that increasing both fat and carbohydrate content to high will lower the average time of exhaustion to approximately 67 seconds? To answer this question, we need to administer high levels of both diets to a sample and observe the average time of exhaustion. If it is 67 seconds, then our observation is correct. However, what if the observation is 74 seconds? The average time of exhaustion has been lowered, but not as much. If this happens, we say that the two factors interact. When factors interact, the effect of one factor on the response is not the same for different levels of the other factor. Hence, the information obtained from the one-factor-at-a-time approach would lead to an invalid prediction.

The factor-level combination for a one-factor-at-a-time approach of Example 8.3.1 can be seen from Fig. 8.1.

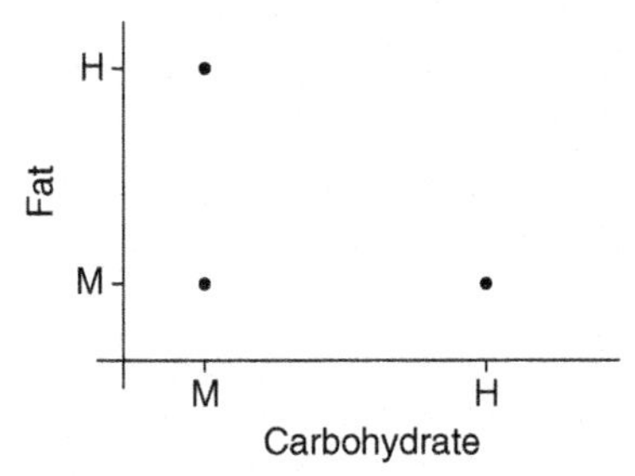

FIGURE 8.1 One-factor-at-a-time approach. The plot depicts Fat versus Carbohydrate with points at combinations of M and H categories.

If there is no interaction, we get Fig. 8.2, which shows average time to exhaustion with three given points and a possible point of around 67 seconds.

If there is interaction, the lines in Fig. 8.2 might cross each other, in which case a one-factor-at-a-time approach may not be the appropriate design. In that case, the following alternative designs will give more accurate data.

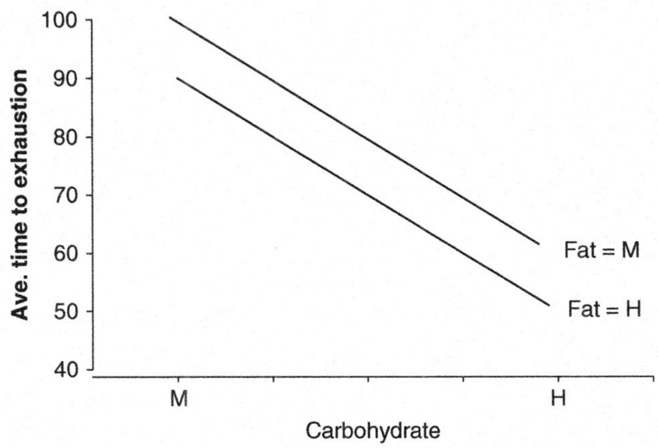

FIGURE 8.2 No interaction. These line graphs depict average time to exhaustion versus carbohydrate with two lines for fat levels M and H.

8.3.2 Full Factorial Design

One way to get around the problem of interaction in one-factor-at-a-time design is to evaluate all possible combinations of factors in a single experiment. This is called a *full factorial experiment.* The main benefit of a full factorial design is that every possible data point is collected. The choice of optimum conditions becomes easier. For example, in an experiment such as the one in Example 8.2.2, one could conduct a full factorial design. The simplest form of factorial experiment involves two factors only and is called a *two-way layout.* A full factorial experiment with n factors and two levels for each factor is called a 2^n *factorial experiment.* A full factorial experiment is practical if only a few factors (say, fewer than five) are being investigated. Beyond that, this design becomes time-consuming and expensive.

8.3.3 Fractional Factorial Design

In a fractional factorial experiment, only a fraction of the possible treatments are actually used in the experiment. A full factorial design is the ideal design, through which we could obtain information on all main effects and interactions. But because of the prohibitive size of the experiments, such designs are not practical to run. For instance, consider Example 8.2.2. Now, if we were to add say, two different densities, three sizes of fish, and three types of food, the number of factors becomes five, and total number of distinct treatments will be $4 \times 4 \times 2 \times 3 \times 3 = 288$. This method becomes very time-consuming and expensive. The number of relatively significant effects in a factorial design is relatively small. In these types of situations, fractional factorial experiments are used in which trials are conducted on only a well-balanced subset of the possible combinations of levels of factors. This allows the experimenter to obtain information about all main effects and interactions while keeping the size of the experiment manageable. The experiment is carried out in a single systematic effort. However, care should be taken in the selection of treatments in the experiment so as to be able to answer as many relevant questions as possible. Fractional factorial design is useful when the number of factors is large. Because we are reducing the number of factors, a fractional factorial design will not be able to evaluate the influence of some of the factors independently. Of course, the question is how to choose the factors and levels we should use in a fractional factorial design. The question of how fractional factorial designs are constructed is beyond the scope of this book.

EXERCISES 8.3

8.3.1 Suppose a large retail chain decides to introduce clothing in two types of materials' (ordinary, fine) qualities. Each store will have two different proportions (40%, 60%) displayed. At the end of the month, profits from each store for these two types of clothing are recorded. Table 8.14 represents the average profits for each of the quality–proportion combinations.

TABLE 8.14 Average Profits for Each of the Quality-proportion Combinations.

Average Profit	Quality	Proportion
10,000	Fine	40%
\$25,000	Ordinary	40%
\$9500	Ordinary	60%
\$8000	Fine	60%

Discuss this as a one-factor-at-a-time experiment to predict the average amount of profit.

8.3.2 In Problem 8.3.1, draw graphs for the data to represent quality–proportion combinations (a) for the one-factor-at-a-time approach, and (b) for the case where there is no interaction.

8.3.3 Discuss how a fractional factorial design can be performed for the problem in Exercise 8.3.1.

8.3.4 Suppose a researcher wants to conduct a series of experiments to study the effect of fertilizer and temperature on plant growth. She uses four different brands of fertilizers in three different settings for the rose plants of the same age and of similar growth.

(a) How many factor-level combinations are possible in this experiment?

(b) Each experiment makes use of one fertilizer–temperature combination (one-actor-at-a-time design). How should she implement randomization in this experiment?

8.4 OPTIMAL DESIGN

In 1959, J. Kiefer presented a paper to the Royal Statistical Society about his work on the theory of optimal design. He was trying to answer the major question, "How do we find the best design?" This work initiated a whole new field of optimal design. Optimal designs are a class of experimental designs that are optimal with respect to certain statistical criterion. For instance, in estimation problems, these designs allow parameters to be estimated without bias and minimal variance. The methods of optimal experimental design provide the technical tools for building experimental designs to attain well-defined objectives with efficiency and with minimum cost. The cost can be the monetary cost, time, number of experimental runs, and so on. There are many methods of achieving optimal designs such as sequential (simplex) or simultaneous experimental designs. In sequential design, experiments are performed in succession in the direction of improvement until the optimum is reached. Simultaneous experimental designs such as response surface designs are used to build empirical models. A survey by Atkinson in 1988 contains many references on optimal design.

In this section, we will focus only on one simple example to illustrate the ideas of optimal design in terms of choosing appropriate sample size. It is not possible to have a single design that is best for securing information concerning all types of population parameters. Indeed, it is beyond the scope of this section to present a general theory of optimal design.

8.4.1 Choice of Optimal Sample Size

The sample size estimation is an essential part of experimental design; otherwise, sample size may be very high or very low. If sample size is too low, the experiment will lack the accuracy to provide dependable answers to the questions we are investigating. If sample size is too large, time and resources will be wasted, often for insignificant gain. We now illustrate a simple case of optimal sample size determination.

Let $X_{11}, \ldots, X_{1n_1}$ be a random sample from population 1 with mean μ_1 and variance σ_1^2 and $X_{21}, \ldots, X_{2n_2}$ be random samples from population 2 with mean μ_2 and variance σ_2^2. Assume that the two samples are independent. Then we know that $\overline{X}_1 - \overline{X}_2$ is an unbiased estimator of $\mu_1 - \mu_2$ with standard error

$$\sigma^2_{(\overline{X}_1-\overline{X}_2)} = Var(\overline{X}_1 - \overline{X}_2)$$
$$= \frac{\sigma_1^2}{n_1} + \frac{\sigma_2^2}{n_2}.$$

Suppose that there is a restriction that the total observations should be n, that is, $n_1\ n_2 = n$. Such a restriction may be due to cost factors or a shortage of available subjects. An important design question is how to choose the sample sizes n_1 and n_2 so as to maximize the information in the data relevant to the parameter $\mu_1 - \mu_2$. We know that the samples contain maximum information when the standard error is minimum. Hence, the problem reduces to minimization of $Var(\overline{X}_1 - \overline{X}_2)$. Let $a = \frac{n_1}{n}$ be the fraction on n observations that are assigned to sample 1. Then $n_1 = na$ and $n_2 = n(1 - a)$, and we have

$$Var(\overline{X}_1 - \overline{X}_2) = \frac{\sigma_1^2}{n_1} + \frac{\sigma_2^2}{n_2}$$
$$= \frac{\sigma_1^2}{na} + \frac{\sigma_2^2}{n(1-a)}.$$

The problem is now reduced to finding an a that minimizes the function $g(a) = \frac{\sigma_1^2}{na} + \frac{\sigma_2^2}{n(1-a)}$. This problem can be solved using calculus. By taking the derivative with respect to a, $\frac{d}{da}g(a)$ and equating it to zero, we have

$$-\frac{\sigma_1^2}{na^2} + \frac{\sigma_2^2}{n(1-a)^2} = 0.$$

Multiplying throughout by $na^2(1-a)^2$, we have

$$-\sigma_1^2(1-a)^2 + \sigma_2^2 a^2 = 0,$$

which results in the quadratic equation

$$(\sigma_2^2 - \sigma_1^2)a^2 + 2\sigma_1^2 a - \sigma_1^2 = 0.$$

Using the quadratic formula, we obtain the two roots, that is,

$$a_1 = \frac{\sigma_1}{\sigma_1 + \sigma_2}$$

and

$$a_2 = \frac{\sigma_1}{\sigma_1 - \sigma_2}.$$

However, a_2 cannot be the solution because, if $\sigma_1 > \sigma_2$, then $a_2 > 1$, otherwise $a_2 < 0$; both are not admissible because a is a fraction. Hence,

$$a = \frac{\sigma_1}{\sigma_1 + \sigma_2} \quad \text{and} \quad 1 - a = \frac{\sigma_2}{\sigma_1 + \sigma_2}.$$

Using the second derivative test, we can verify that this indeed is a minimum for $\text{var}(\overline{X}_1 - \overline{X}_2)$. From this analysis we can see that the sample sizes that maximize the information in the data are relevant to the parameter $\mu_1 - \mu_2$ subject to the constraint $n_1\ n_2 = n$, are

$$n_1 = \frac{\sigma_1}{\sigma_1 + \sigma_2}n \quad \text{and} \quad n_2 = \frac{\sigma_2}{\sigma_1 + \sigma_2}n.$$

As a special case, we can see that when $\sigma_1^2 = \sigma_2^2$, the optimal design is to take $n_1 = n_2$.

EXERCISES 8.4

8.4.1 A total of 100 sample points were taken from two populations with variances $\sigma_1^2 = 4$ and $\sigma_2^2 = 9$. Find n_1 and n_2 that will result in the maximum amount of information about $(\mu_1 - \mu_2)$.

8.4.2 Suppose in Exercise 8.4.1 we want to take $n = n_1 = n_2$. How large should n be to obtain the same information as that implied by the solution of Exercise 8.4.1?

8.5 THE TAGUCHI METHODS

Taguchi methods were developed by Deming prize winner Dr. Genichi Taguchi to improve the implementation of total quality control in Japan. These *continuous quality improvement* (CQI) methods are claimed to have provided as much as 80% of Japanese quality gains. They are based on the DOEs to provide near-optimal quality characteristics for a specific objective. A special feature of Taguchi methods is that they integrate the methods of statistical DOEs into a powerful engineering process. The Taguchi methods are in general simpler to implement.

Taguchi methods are often applied on the Japanese manufacturing floor by technicians to improve their processes and their product. The goal is not just to optimize an arbitrary objective function, but also to reduce the sensitivity of engineering designs to uncontrollable factors or noise. The objective function used is the signal-to-noise ratio, which is then maximized. This moves design targets toward the middle of the design space so that external variation affects the behavior of the design as little as possible. This permits large reductions in both part and assembly tolerances, which are major drivers of manufacturing cost. Linking quality characteristics to cost through the Taguchi loss function (Taguchi and Yokoyama, 1994) was a major advance in quality engineering, as well as in the ability to design experiments to minimize the cost. Taguchi methods are also called robust design. In 1982, the American Supplier Institute introduced Dr. Taguchi and his methods to the US market.

Using a well-planned experimental design, such as a fractional factorial design, it is possible to efficiently obtain information about the model and the underlying process. Clearly, the purpose of these methods is to control and ensure the quality of the end product. In the conventional approach, this is achieved by further testing a few end products that are randomly chosen or using control charts and making decisions based on certain preset criteria, such as acceptable or unacceptable. Thus, "quality" of the product is thought of as inside or outside of specifications. Instead, Taguchi suggested that we should specify a target value, and the quality should be thought of as the variation from the target.

As an example, suppose we make n observations of the output $x_1, \ldots, x_n$ of a process at times $1, 2, \ldots, n$, as shown in Fig. 8.3.

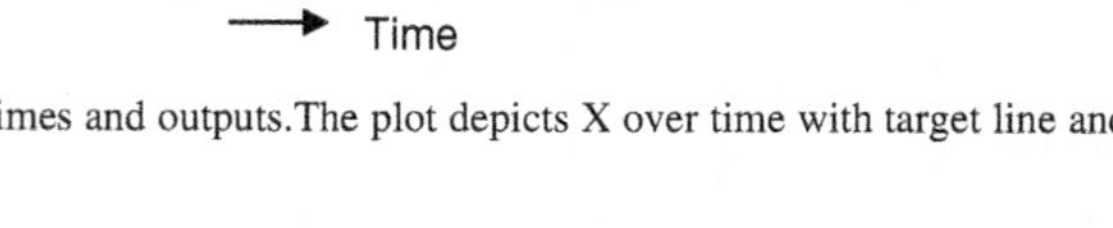

FIGURE 8.3 Control plot of processing times and outputs.The plot depicts X over time with target line and upper and lower limits labeled T_U and T_L.

The control chart consists of a plot of observed output values (x_i's) on the y-axis and the times of observation, $1, 2, \ldots, n$ on the x-axis, as shown in the figure. The letter T represents the target value. If the output value is between T_L and T_U, the process is deemed to be operating satisfactorily; otherwise, the process is said to be out of control and the output value is considered unsatisfactory.

Some other examples are (1) defining specification limits for acceptance, such as stating that the diameter of bolts must be between 8.8 and 10.2 mm with mean 10 mm, and (2) that the waiting time in a line should be less than 30 minutes for at least 90% of customers.

In all these situations, the specifications partition is that the state of the process is acceptable or unacceptable, that is, it classifies the state as a dichotomy. This is often called the "goal post mentality."

The basic idea of the Taguchi approach is a shift in mindset from demarcating the quality as acceptable or unacceptable to a more flexible and realistic classification. The traditional approach to quality control does not consider the size of departure from the target value. To accommodate the size of such departure as a significant factor in quality control, let us introduce the concept of a loss function (see Chapter 11). If an output value x differs from the target value T, let $L(T, x)$ denote the loss incurred, say in dollars. Other possible losses could also be reputation or customer satisfaction.

For the control chart example, we can assign the loss function

$$L(T, x) = \begin{cases} 0, \text{ if } T_U < x < T_L \\ L, \text{ if } x > T_L \text{ or } x < T_U \end{cases}$$

where L is a constant and x is the measured value. This is schematically shown in Fig. 8.4.

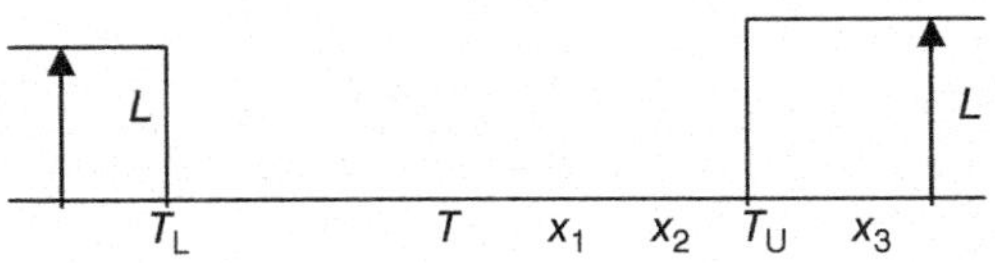

FIGURE 8.4 Loss function. The plot represents 0, L loss function.

From Fig. 8.4, it is seen that we view outputs x_1 and x_2 as having equal quality, whereas x_2 and x_3 are considered to have vastly differing quality (x_2 is acceptable and x_3 is not acceptable). A more reasonable conclusion would be that x_1 has excellent quality, whereas x_2 and x_3 are similar, both being of poor quality.

In Taguchi's approach, the loss function takes into account the size of departure from the target value. For example, a popular choice for the loss function is

$$L(T, X) = k(X - T)^2,$$

where

L = loss incurred,
k = constant,
X = actual value of the measured output, and
T = target value.

If loss is in terms of money, we can interpret the constant k as the constant that converts deviation into money. We can schematically represent the behavior of the loss function as shown in Fig. 8.5, below.

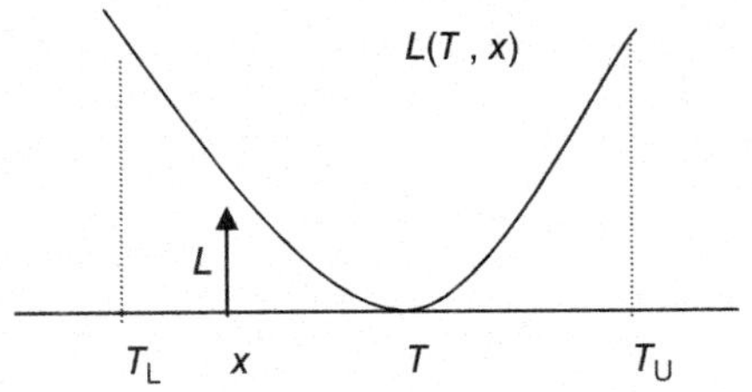

FIGURE 8.5 Quadratic loss function. The plot shows a quadratic loss function with minimum cost at the target, T.

This form of loss function is called the *quadratic loss function*. The choice of k depends on the particular problem. For example, the scaling factor k can be used to convert loss into monetary units to accommodate comparisons of systems with different capital loss. Or, in product manufacturing, let D denote the allowed deviation from the target, and let A denote the loss due to a defective product. Then a choice of k can be $k = (A / D)^2$. As shown earlier, the average loss is $E(L)$ and is given by

$$E(L) = k[(E(X) - T)^2 + \sigma^2] = k[(bias)^2 + variance],$$

where σ^2 is the variance of X (measured quality, which is assumed to be random).

In the Taguchi approach, the variation from the target can be broken into components containing bias and product variation. Thus, if our aim is to minimize the expected loss, $E(L)$, we should not only require $E(X) = \mu$ to be close to T but also should reduce the variance. It turns out that often these requirements are contradictory. The objective is to choose the design parameters (the factors that influence the quality) optimally to obtain the best quality product. In practice, the parameters μ and σ^2 are not known and are being estimated by $\overline{X}$ and S^2, respectively. This results in the Taguchi loss function, that is,

$$\overline{L} = k[(\overline{X} - T)^2 + S^2].$$

This loss function penalizes small deviations from T only slightly, while assessing a larger penalty for responses far from the target. The expected loss is similar to a mean squared error loss, which we have seen in regression analysis in the form of the least squares.

Why is controlling both bias and variance important? Suppose you want your community swimming pool temperature at 80°F, which is the T here. Suppose the temperature varies between 60°F and 100°F. Clearly the average (bias) is zero; however, it will be pretty uncomfortable to swim at 60°F or 100°F. Here, the bias takes the ideal value of zero, but the variance is large. In another scenario, the variance may be small, but the average temperature may be farther away from the target value of 80°F (e.g., the temperature is constant at 60°F). Hence, we want the pool temperature to be near to the target value of 80°F, with as small variance as possible (say, within 1°F–2°F).

Taguchi coined the term *design parameters* as the generic description for factors that may influence the quality and whose levels we want to optimize. Taguchi's philosophy is to "design quality in" rather than to weed out the defective items after manufacturing. In order to obtain an optimal set of design parameters that affect the quality of the end product, the Taguchi method utilizes appropriately designed experiments. More specifically, orthogonal arrays are used for fractional factorial designs. Orthogonal arrays provide a set of well-balanced experiments. Taguchi provides tables for these designs so that even a nonspecialist can use them. For two-level designs (high, low), we have a table for an L_4 orthogonal array up to three factors; a table for an L_8 orthogonal array up to seven factors; and so forth. Similar tables are available for three-level designs. We will not describe these design issues in this section. We refer the reader to specialized books on the subject matter for further details.

We will give an example to show how Taguchi loss function converts deviation from a target into dollar.

EXAMPLE 8.5.1

Suppose the following data, with a target value of 100.

97	101	102	99	100

Suppose that a unit which is off by 5 of the target value causes a loss of \$300. Find k and determine the Taguchi loss function.

Solution

Since a deviation of 5 units from the target 100 (that is a deviation, D=5) causes loss of \$300, we have,

$$k = \frac{A}{D^2} = \frac{300}{5^2} = 12.$$

Thus, the general form of Taguchi loss function is

$$L(T, X) = 12(X - T)^2.$$

For the given data, we have sample mean, $\overline{x} = 99.8$, *and variance,* $s^2 = 3.7$. *Thus, for the given data, the Taguchi loss function is,*

$$\begin{aligned}\overline{L} &= k[(\overline{X} - T)^2 + S^2]\\ &=12[(99.8 - 100)^2 + 3.7]\\ &=44.88.\end{aligned}$$

Thus, we have an average loss of \$44.88.

We can summarize the Taguchi approach to quality design as follows:

1. Taguchi's methods for experimental design are ready made and simple to use in the design of efficient experiments, even by nonexperts.
2. Taguchi's approach to total quality management is holistic and tries to design quality into a product rather than inspecting defects in the final product.
3. Taguchi's techniques can readily be applied to other fields such as management problems.

It should be noted that around 1980, Six Sigma, a new way of improving the quality process by identifying and eliminating defects and minimizing variability, became popular. Originally developed by Motorola, the Six Sigma methods have been adapted by many large manufacturing companies around the world to reduce operating costs, increase sales and revenue, and incorporate innovation in products. The Six Sigma method integrates statistical analysis with process management principles to achieve near-perfect performance. This method will not be pursued further in this book.

EXERCISES 8.5

8.5.1 Suppose the following data represents thickness between and within silicon wafers (in microns), with a target value of 14.5 microns.

13.688	13.788	14.173	14.557
13.925	14.545	13.797	14.778

Compute the Taguchi loss function.

8.5.2 One of the commonly used performance measures in the Taguchi method is

$$\log\left(\frac{(mean)^2}{s^2}\right),$$

where s^2 is the sample variance. In general, the higher the performance measure, the better the design. This measure is called robustness statistics. For the problem of Exercise 8.5.1, suppose that we run the experiment by controlling various factors affecting the thickness. Table 8.15 shows the data obtained in four different runs.

TABLE 8.15 Data of Thickness Between and Within Silicon Wafers.

Run 1:	**14.158**	**14.754**	**14.412**	**14.065**	**13.802**	**14.424**	**14.898**	**14.187**
Run 2:	13.676	14.177	14.201	14.557	13.827	14.514	13.897	14.278
Run 3:	13.868	13.898	14.773	13.597	13.628	14.655	14.597	14.978
Run 4:	13.668	13.788	14.173	14.557	13.925	14.545	13.797	14.778

(a) Using the robustness statistics given earlier, which of the processes gives us an improved performance?
(b) Another commonly used performance in the Taguchi method is

$$-\log(s^2).$$

Using this robustness statistic, which of the processes used gives us an improved performance? Compare this with the results of part (a).

8.6 CHAPTER SUMMARY

In this Chapter, we have learned some basic aspects of experimental design. Some fundamental definitions and tools for developing experimental designs such as *randomization*, *replication*, and *blocking* were introduced in Section 8.2. Basic concepts of factorial design were given in Section 8.3. In Section 8.4, we study an example of optimal design. The Taguchi method was introduced in Section 8.5. In the next chapter, we introduce the analysis component. We have discussed only a very small collection of experimental designs in this chapter. There exists a wide variety of experimental designs to deal with a large number of treatments and to address specific needs of research experiments in diverse fields. It is an exciting and growing area for the interested student to study and explore.

We list some of the key definitions introduced in this chapter:

- Response variable (output variable)
- Independent variables (treatment variables or input variables or factors)
- Nuisance variables
- Noise
- Observational
- Experimental units
- Single-factor experiments
- Multifactor experiments
- Experimental error
- Blinding, double-blinding, and placebo
- Replication
- Block
- Randomization
- Completely randomized design
- Randomized complete block design
- $k \times k$ Latin square design
- Greco-Latin square
- Design parameters

In this chapter, we have also learned the following important concepts and procedures.

- Procedure for random assignment
- Procedure for randomization in a randomized complete block design
- Procedure for a randomized complete block design with r replications
- Procedure for constructing a 4×4 Latin square
- One-factor-at-a-time design
- Full factorial design
- Fractional factorial design
- Choice of optimal sample size
- The Taguchi methods

8.7 COMPUTER EXAMPLES

In this section, we present R and SAS commands only.

8.7.1 Examples Using R

EXAMPLE 8.7.1 Permutation

Obtain a random perturbation of the numbers 1 to n. Where $n = 10$.

R Code:

```
sample(c(1:10));
```

Output:

```
This output will be a random sample without replacement, your output will look similar.
```

```
6 7 9 2 10 1 5 3 8 4
```

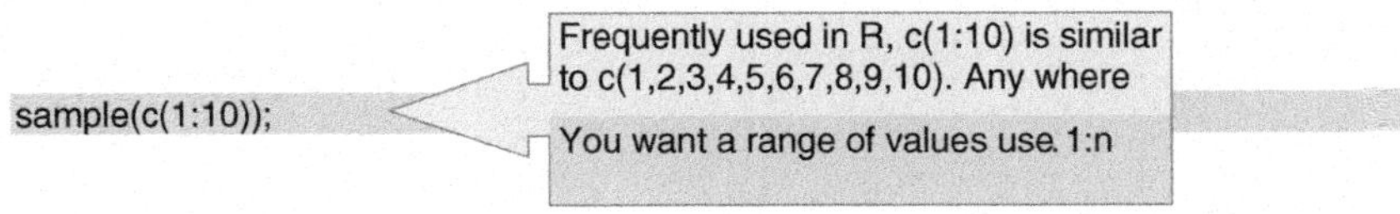

EXAMPLE 8.7.2 Randomized Block Design

In order to study the number of hours to relief provided by five different brands (A, B, C, D, E) of pain relievers for pain resulting from different causes [headache (H), muscle pain (M), pain due to cuts and bruises (CB)], doses are administered to five subjects each having similar types of pain. Create a randomized complete block design. Choose the different types of pain (H, M, CB) as the blocks.

R Code:

```
h=sample(c(1:5));
m=sample(c(1:5));
cb=sample(c(1:5));
table=cbind(h,m,cb);
table=as.data.frame(table);
colnames(table)=c("H,""M,""CB");
rownames(table)=c("A,""B,""C,""D,""E");
print(table);
```

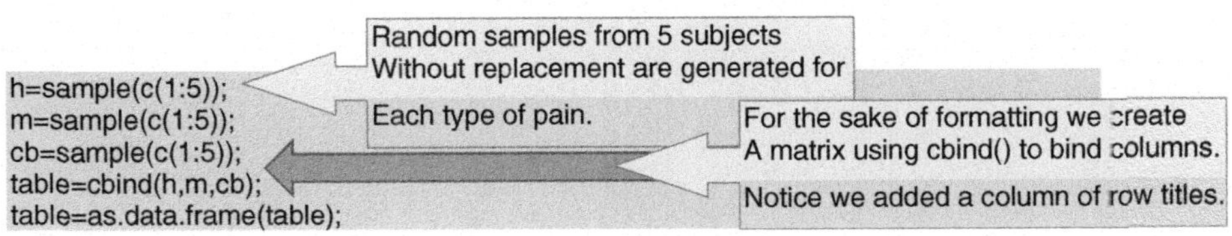

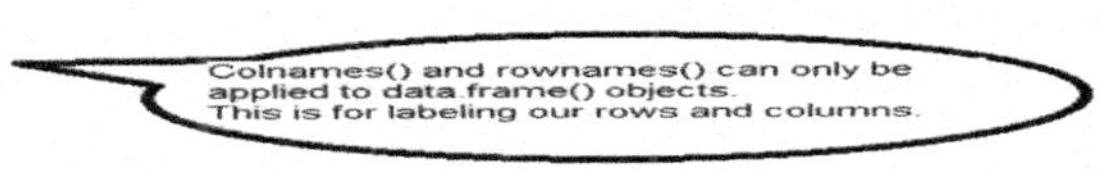

Output:

```
This output will be a random sample without replacement, your output will look similar.
H M CB
A 4 2 1
B 2 1 4
C 3 4 5
D 5 5 3
E 1 3 2
```

EXAMPLE 8.7.3: Latin Squares

A gasoline company is interested in comparing the effect of four gasoline additives (A, B, C, D) on the gas mileage achieved per gallon. Four cars (1, 2, 3, 4) and four drivers (I,II,III,IV) will be used in the experiment. Create a Latin square design.

R Code:

```
gasadd=c("A","B","C","D");
table=c();
for(i in 1:4) {
table=cbind(table,c(gasadd[i:4],gasadd[0:(i-1)]));
} table=as.data.frame(table);
colnames(table)=c(1:4);
rownames(table)=c("I","II","III","IV");
print(table[,sample(c(1:4))]);
```

Output: This output will be a random sample without replacement; your output will look similar.

```
4 1 3 2
I D A C B
II A B D C
III B C A D
IV C D B A
```

8.7.2 SAS Examples

EXAMPLE 8.7.4

For the data of Example 8.2.4, conduct a randomized complete block design using SAS.

Solution

We represent blocks that are reasons for pain by $H = 1$, $M = 2$, and $CB = 3$. Similarly five brands which are treatments by $A = 1$, $B = 2$, $C = 3$, $D = 4$, and $E = 5$. Then we can use the following code to generate a randomized complete block design.

```
options nodate nonumber;
data a;
    do block = 1 to 3;
        do subject = 1 to5;
            x = ranuni(0);
            output;
        end;
    end;
proc sort; by block x;
data c; set a;
trt = 1 mod(N – 1,5); /* mod = remainder of N/5 */
proc sort; by block subject;
proc print;
    var block subject trt;
run;
```

We obtain the following output.

Completely randomized 2 × 3 design, 4 subjects per cell

Obs	block	subject	Trt
1	1	1	5
2	1	2	4
3	1	3	3
4	1	4	2
5	1	5	1
6	2	1	2
7	2	2	5
8	2	3	3
9	2	4	4
10	2	5	1
11	3	1	4

12	3	2	5
13	3	3	1
14	3	4	2
15	3	5	3

Note that the numbers in the column corresponding to a block identify the type of pain, the numbers in the subject column correspond to the subjects, and the numbers in the column corresponding to trt identify the brands. Using the corresponding letters, we can rewrite the foregoing table in the familiar form shown in Table 8.14.

The PLAN procedure constructs experimental designs. The PLAN procedure does not have a DATA=option in the PROC statement; in this procedure, both the input and output data sets are specified in the OUTPUT statement. We will use this to construct a Latin square design. Table 8.16.

TABLE 8.16 Structure of a Latin Square Design.

H	M	CB
1(E)	1(B)	1(D)
2(D)	2(E)	2(E)
3(C)	3(C)	3(A)
4(B)	4(D)	4(B)
5(A)	4(A)	5(C)

EXAMPLE 8.7.5

A gasoline company is interested in comparing the effect of four gasoline additives (A, B, C, D) on the gas mileage achieved per gallon. Four cars (1, 2, 3, 4) and four drivers (I, II, III, IV) will be used in the experiment. Create a Latin square design.

Solution

We can use the following program, where we represent the additives by 1 = *A*, 2 = *B*, 3 = *C*, and 4 = *D*.

```
Options nodate nonumber;
title 'Latin Square design for 4 additives';
proc plan seed=37432;
      factors rows=4 ordered cols=4 ordered/NOPRINT;
      treatments tmts=4 cyclic;
      output out=g
        rows cvals=('car 1' 'car 2' 'car 3' 'car 4')
            random
        cols cvals=('Driver 1' 'Driver 2' 'Driver 3'
            'Driver 4') random
        tmts nvals=(1 2 3 4) random;
run;
proc tabulate;
      class rows cols;
      var tmts;
      table rows, cols*(tmts*f=1.);
      keylabel sum=' ';
run;
```

PROJECTS FOR CHAPTER 8

8A. Sample Size and Power

Suppose that the experimenter is interested in comparing the true means of two independent populations. If two similar treatments are to be compared, the assumption of equality of variances is not unreasonable. Hence, assume that the common variance of the two populations is σ^2, and the experimenter has a prior estimate of the variance. We learned in Section 8.4 that in this case, the optimal design will be to take sample sizes n_1 and n_2 to be equal. Let $n = n_1 = n_2$ be the size of the random sample that the experimenter should take from each population.

Now, suppose that the experimenter has decided to use the one-sided large sample test, H_0: $\mu_1 = \mu_2$ versus H_a: $\mu_1 > \mu_2$ with a fixed $\alpha = P(\text{Type I error})$. He wants to choose n to be so large that if $\mu_1 = \mu_2\ k\sigma$, he will get a fixed power $(1 - \beta)$ of deciding $\mu_1 > \mu_2$. Recall that power of a test is the probability of (correctly) rejecting H_0 when H_0 is false. Find the approximate value of n. Note that, for a given α, this will be an optimal sample size with a desired value of the power.

In particular, what should be the sample size in the hypothesis testing problem, H_0: $\mu_1 - \mu_2 = 0$ versus H_a: $\mu_1 - \mu_2 = 3$, if $\alpha = \beta = 0.05$? Assume that $\sigma = 7$.

8B. Effect of Temperature on Spoilage of Milk

Suppose you have observed that milk in your refrigerator spoils very fast. You may be wondering whether it has anything to do with the temperature settings. Design an experiment to study the effect of temperature on spoiled milk, with at least three meaningful settings of the temperature. (i) Write a possible hypothesis for your experiment. (ii) What are the independent and dependent variables? (iii) Which variables are being controlled in this experiment? (iv) Discuss how you used the three basic principles of replication, blocking, and randomization. (v) What conclusions can you draw? Think through any possible flaws in the design that may affect the integrity of your findings.

Chapter 9

Analysis of Variance

Chapter outline

OBJECTIVE

To analyze the means of several populations by identifying the sources of variability of the data.

John Wilder Tukey
(Source: http://en.wikipedia.org/wiki/john_tukey.)

John W. Tukey (1915–2000), a chemist-turned-topologist-turned statistician, was one of the most influential statisticians of the past 50 years. He is credited with inventing the word *software*. He worked as a Professor at Princeton University and a senior researcher at AT&T's Bell Laboratories. He made significant contributions to the fields of exploratory data analysis and robust estimation. His works on the spectrum analysis of time series and other aspects of digital signal processing have been widely used in engineering and science. He coined the word *bit*, which refers to a unit of information processed by a computer. In collaboration with Cooley, in 1965, Tukey introduced the fast Fourier transform (FFT) algorithm that greatly simplified computation for Fourier series and integrals. Tukey authored or coauthored many books in statistics and wrote more than 500 technical papers. Among Tukey's most far-reaching contributions was his development of techniques for "robust analysis," an approach to statistics that guards against wrong answers in situations where a randomly chosen sample of data happens to poorly represent the rest of the data set. Tukey also made significant contributions to the analysis of variance.

Mathematical Statistics with Applications in R. DOI: https://doi.org/10.1016/B978-0-443-27548-7.00009-8

9.1 INTRODUCTION

Suppose that we are interested in the effect of four different types of chemical fertilizers on the yield of rice, measured in pounds per acre. If there is no difference between the different types of fertilizers, then we would expect all the mean yields to be approximately equal. Otherwise, we would expect the mean yields to differ. The different types of fertilizers are called treatments, and their effects are the treatment effects. The yield is called the response. Typically, we have a model with a response variable that is possibly affected by one or more treatments. The study of these types of models falls under the purview of the design of experiments, which we discussed in Chapter 8. In this chapter we concentrate on the analysis aspect of the data obtained from the designed experiments. If the data came from one or two populations, we could use the techniques learned in Chapters 5 and 6. Here, we introduce some tests that are used to analyze the data from more than two populations. These tests are used to deal with treatment effects, including tests that take into account other factors that may affect the response. The hypothesis that the population means are equal is considered equivalent to the hypothesis that there is no difference in treatment effects. The analytical method we will use in such problems is called the analysis of variance (ANOVA). Initial development of this method could be credited to Sir Ronald A. Fisher, who introduced this method for the analysis of agricultural field experiments. The "green revolution" in agriculture would have been impossible without the development of the theory of experimental design and the methods of ANOVA.

ANOVA is one of the most flexible and practical techniques for comparing several means. It is important to observe that ANOVA is not about analyzing the population variance. In fact, we are analyzing treatment means by identifying sources of variability in the data. In its simplest form, ANOVA can be considered as an extension of the test of hypothesis for the equality of two means that we learned in Chapter 6. Actually, the so-called one-way ANOVA is a generalization of the two-means procedure to a test of equality of the means of more than two independent, normally distributed populations.

Recall that the methods of testing H_0: $\mu_1 - \mu_2 = 0$, such as the t-test, were discussed earlier. In this chapter, we are concerned with studying situations involving the comparison of more than two population or treatment means. For example, we may be interested in the question, "Do the rates of heart attack and stroke differ for three different groups of people with high cholesterol levels (borderline high such as 150–199 mg/dL, high such as 200–239 mg/dL, very high such as greater than 240 mg/dL) and a control group given different dosage levels of a particular cholesterol-lowering drug (say, a particular statin drug)?" Let us consider four populations with means μ_1, μ_2, μ_3, and μ_4, and say that we wish to test the hypotheses $\mu_1 = \mu_2 = \mu_3 = \mu_4$. That is, the true mean rate is the same for all four groups. The question here is: Why do we need a new method to test for differences among the four procedure population means? Why not use z- or t-tests for all possible pairs and test for differences in each pair? If any one of these tests leads to the rejection of the hypothesis of equal means, then we might conclude that at least two of the four population means differ. The problem with this approach is that our final decision is based on the results of $\binom{4}{2} = 6$ different tests, and any one of them can be wrong. For each of the six tests, let $\alpha = 0.10$ be the probability of being wrong (type I error). Then the probability that at least one of the six tests leads to the conclusion that there is a difference leads to an error of $1 - (0.9)^6 = 0.46856$, which clearly is much larger than 0.10, thus resulting in a large increase in the type I error rate. Hence, if an ordinary t-test is used to make several treatment comparisons from the same data, the actual α-value applying to the tests taken as a group will be larger than the specified value of α, and one is likely to declare significance when there is none.

ANOVA procedures were developed to eliminate the increase in error rates resulting from multiple t-tests. With ANOVA, we are able to set one alpha level and test whether any of the group means differ from the others. Given a sample from each of the populations, our interest is to answer the question: Are the observed discrepancies among the different sample means merely due to chance fluctuations, or are they due to inherent differences among the populations? ANOVA separates the effect of purely random variations from those caused by existing differences among population means: The phrase "analysis of variance" springs from the idea of analyzing variability in the data to see how much can be attributed to differences in μ and how much is due to variability in the individual populations. The ANOVA method incorporates information on variability from all of the samples simultaneously. At the heart of ANOVA is the fact that variances can be partitioned, with each partition attributable to a specific source. The method inspects various sums of squares (which are measures of variation in a sample) calculated from the data. ANOVA looks at two types of sums of squares: sums of squares within groups and sums of squares between groups. That is, it looks at each of the distributions and compares the between-group differences (variation in group means) with the within-group differences (variation in individuals' scores within groups).

9.2 ANALYSIS OF VARIANCE METHOD FOR TWO TREATMENTS (OPTIONAL)

In this section, we present the simplest form of the analysis of variance procedure, the case of studying the means of two populations I and II. For comparing only two means, the ANOVA will result in the same conclusions as the t-test for independent random samples. The basic purpose of this section is to introduce the concept of ANOVA in simpler terms. Let us consider two random samples of size n_1 and n_2, respectively. That is, $y_{11}, y_{12}, \ldots, y_{1n_1}$ from population I and $y_{21}, y_{22}, \ldots, y_{2n_2}$ from population II. Let

$$\bar{y}_1 = \frac{y_{11} + y_{12} + \cdots + y_{1n_1}}{n_1} \text{(sample mean from population I)},$$

and

$$\bar{y}_2 = \frac{y_{21} + y_{22} + \cdots + y_{2n_2}}{n_2} \text{(sample mean from population II)}.$$

These samples are assumed to be independent and come from normal populations with respective means μ_1, μ_2, and variances $\sigma_1^2 = \sigma_2^2$. We wish to test the hypothesis

$$H_0: \mu_1 = \mu_2 \text{ versus } H_a: \mu_1 \neq \mu_2.$$

The total variation of the two combined response measurements about $\bar{y}$ (the sample mean of all $n = n_1 + n_2$ observations) is (SS is used for sum of squares) defined by

$$Total\ SS = \sum_{i=1}^{2} \sum_{j=1}^{n_i} (y_{ij} - \bar{y})^2. \tag{9.1}$$

That is,

$$\bar{y} = \frac{y_{11} + y_{12} + \cdots + y_{1n_1} + y_{21} + y_{22} + \cdots + y_{2n_2}}{n} = \frac{1}{n} \sum_{ij} y_{ij}.$$

The total sums of squares measure the total spread of scores around the grand mean, $\bar{y}$. We can rewrite Eq. (9.1) as

$$\begin{aligned} Total\ SS &= \sum_{i=1}^{2} \sum_{j=1}^{n_i} (y_{ij} - \bar{y})^2 \\ &= \sum_{j=1}^{n_1} (y_{1j} - \bar{y})^2 + \sum_{j=1}^{n_2} (y_{2j} - \bar{y})^2 \\ &= \sum_{j=1}^{n_1} (y_{1j} - \bar{y}_1 + \bar{y}_1 - \bar{y})^2 + \sum_{j=1}^{n_2} (y_{2j} - \bar{y}_2 + \bar{y}_2 - \bar{y})^2 \\ &= \sum_{j=1}^{n_1} (y_{1j} - \bar{y}_1)^2 + n_1(\bar{y}_1 - \bar{y})^2 + 2(\bar{y}_1 - \bar{y}) \sum_{j=1}^{n_1} (y_{1j} - \bar{y}_1) \\ &+ \sum_{j=1}^{n_2} (y_{2j} - \bar{y}_2)^2 + n_2(\bar{y}_2 - \bar{y})^2 + 2(\bar{y}_2 - \bar{y}) \sum_{j=1}^{n_2} (y_{2j} - \bar{y}_2). \end{aligned}$$

Note that

$$\sum_{j=1}^{n_1} \left(y_{1j} - \bar{y}_1\right) = 0 = \sum_{j=1}^{n_2} \left(y_{2j} - \bar{y}_2\right).$$

Thus we obtain

$$\begin{aligned} Total\ SS &= \sum_{j=1}^{n_1} (y_{1j} - \bar{y}_1)^2 + \sum_{j=1}^{n_2} (y_{2j} - \bar{y}_2)^2 \\ &+ n_1(\bar{y}_1 - \bar{y})^2 + n_2(\bar{y}_2 - \bar{y})^2 \\ &= \sum_{i=1}^{2} \sum_{j=1}^{n_i} (y_{ij} - \bar{y}_i)^2 + \sum_{i=1}^{2} n_i(\bar{y}_i - \bar{y})^2. \end{aligned} \tag{9.2}$$

Define SST, the sum of squares for a treatment by,

$$SST = \sum_{i=1}^{2} n_i(\overline{y}_i - \overline{y})^2.$$

The SST measures the total spread of the group means $\overline{y}_i$ with respect to the grand mean, $\overline{y}$. Also, SSE represents the *sum of squares of errors* given by,

$$\begin{aligned} SSE &= \textstyle\sum_{i=1}^{2}\sum_{j=1}^{n_i}(y_{ij} - \overline{y}_i)^2 \\ &= \textstyle\sum_{j=1}^{n_1}(y_{1j} - \overline{y}_1)^2 + \sum_{j=1}^{n_2}(y_{2j} - \overline{y}_2)^2 \\ &= (n_1 - 1)s_1^2 + (n_2 - 1)s_2^2, \end{aligned}$$

where s_1^2 and s_2^2 are the unbiased sample variances of the two random samples. Note that this connects the sum of squares to the concept of variance we have been using in previous chapters. We can now rewrite Eq. (9.2) as

$$Total\ SS = SSE + SST.$$

It should be clear that the SSE measures the within-sample variation of the y-values (effects), whereas SST measures the variation among the two-sample means. The logic by which the analysis of variance tests is as follows: If the null hypothesis is true, then SST, as compared to SSE, should be about the same or less. The larger SST, the greater will be the weight of evidence to indicate a difference in the means μ_1 and μ_2. The question then is, how large?

To answer this question, let us suppose we have two populations that are normal. That is, let Y_{ij} be $N(\mu_i, \sigma^2)$ distributed with values y_{ij}. Then, the pooled unbiased estimate of σ^2 is given by

$$s_p^2 = \frac{(n_1 - 1)s_1^2 + (n_2 - 1)s_2^2}{n_1 + n_2 - 2} = \frac{SSE}{n_1 + n_2 - 2}.$$

Hence,

$$\sigma^2 = E\left(S_p^2\right) = E\left(\frac{SSE}{n_1 + n_2 - 2}\right).$$

Also, we can write

$$\frac{SSE}{\sigma^2} = \sum_{j=1}^{n_1}\frac{(Y_{1j} - \overline{Y}_1)^2}{\sigma^2} + \sum_{j=1}^{n_2}\frac{(Y_{2j} - \overline{Y}_2)^2}{\sigma^2},$$

which has a χ^2-distribution with $(n_1 + n_2 - 2)$ degrees of freedom.
Under the hypothesis that $\mu_1 = \mu_2$, $E(SST) = \sigma^2$. Furthermore,

$$Z = \frac{\overline{Y}_1 - \overline{Y}_2}{\sqrt{\sigma^2\left(\frac{1}{n_1} + \frac{1}{n_2}\right)}} \sim N(0,1).$$

This implies that

Theorem 9.2.1. *If χ_1^2 has υ_1 degrees of freedom, χ_2^2 has υ_2 degrees of freedom, and χ_1^2 and χ_2^2 are independent, then $F = \frac{\chi_1^2/\upsilon_1}{\chi_2^2/\upsilon_2}$ has an F-distribution with υ_1 numerator degrees of freedom and υ_2 denominator degrees of freedom.*

$$Z^2 = \left(\frac{1}{n_1} + \frac{1}{n_2}\right)\left[\frac{\overline{Y}_1 - \overline{Y}_2}{\sigma^2}\right] = \frac{SST}{\sigma^2},$$

has a χ^2–distribution with 1 degree of freedom. It can be shown that SST and SSE are independent. From Chapter 4, we restate the following result.
Using the foregoing result, we have

$$\frac{SST/(1)\sigma^2}{SSE/(n_1 + n_2 - 2)\sigma^2} = \frac{SST/1}{SSE/(n_1 + n_2 - 2)},$$

which has an F-distribution with $v_1 = 1$ numerator degrees of freedom and $v_2 = (n_1 + n_2 - 2)$ denominator degrees of freedom.

Now, we introduce the mean square error (MSE), defined by

$$MSE = \frac{SSE}{(n_1 + n_2 - 2)}$$
$$= \frac{(n_1 - 1)s_1^2 + (n_2 - 1)s_2^2}{(n_1 + n_2 - 2)},$$

and the mean square treatment (MST) given by

$$MST = \frac{SST}{1}$$
$$= \left[n_1(\bar{y}_1 - \bar{y})^2 + n_2(\bar{y}_2 - \bar{y})^2\right].$$

Under the null hypothesis, H_0: $\mu_1 = \mu_2$, both MST and MSE estimate σ^2 without bias. When H_0 is false and $\mu_1 \neq \mu_2$, MST estimates something larger than σ^2 and will be larger than MSE. That is, if H_0 is false, then $E(MST) > E(MSE)$ and the greater the differences among the values of μ, the larger $E(MST)$ will be relative to $E(MSE)$.

Hence, to test H_0: $\mu_1 = \mu_2$ versus H_a: $\mu_1 \neq \mu_2$, we use the F-test given by

$$F = \frac{MST}{MSE},$$

as the test statistic. Thus, for given α, the rejection region is $\{F > F_\alpha\}$. It is important to observe that, compared to the small sample t-test, here we work with variability. Now we summarize the ANOVA procedure for the two-sample case.

Analysis of variance procedure for two treatments

For equal sample sizes $n = n_1 = n_2$, assume $\sigma_1^2 = \sigma_2^2$.

We test

$$H_0: \mu_1 = \mu_2 \text{ versus } H_a: \mu_1 \neq \mu_2.$$

1. Calculate: $\bar{y}_1$, $\bar{y}_2$, $\Sigma_{ij} y_{ij}^2$, $\Sigma_{ij} y_{ij}$, and find

$$SST = \sum_{i=1}^{2} n_i(\bar{y}_i - \bar{y})^2.$$

Also calculate

$$Total\ SS = \sum_i \sum_j y_{ij}^2 - \frac{\left(\Sigma_i \Sigma_j y_{ij}\right)^2}{n_1 + n_2}.$$

Then

$$SSE = Total\ SS - SST.$$

2. Compute

$$MST = \frac{SST}{1},$$

and

$$MSE = \frac{SSE}{n_1 + n_2 - 2}.$$

3. Compute the test statistic,

$$F = \frac{MST}{MSE}.$$

4. For a given α, find the rejection region as

$$RR: F > F_\alpha,$$

based on 1 numerator and $(n_1 + n_2 - 2)$ denominator degrees of freedom.

5. Conclusion: If the test statistic F falls in the rejection region, conclude that the sample evidence supports the alternative hypothesis that the means are indeed different for the two treatments.

Assumptions: Populations are normal with equal but unknown variances.

EXAMPLE 9.2.1

The following data represents a random sample of end-of-year bonuses for lower-level managerial personnel employed by a large firm. Bonuses are expressed in percentage of yearly salary.

Female	6.2	9.2	8.0	7.7	8.4	9.1	7.4	6.7
Male	8.9	10.0	9.4	8.8	12.0	9.9	11.7	9.8

The objective is to determine whether the male and female bonuses are the same. We can answer this question by connecting the following.

(a) Use the ANOVA approach to test the appropriate hypotheses. Use α=0.05.
(b) What assumptions are necessary for the test in part (a)?
(c) Test the appropriate hypothesis by using the two-sample t-test for comparing population means. Compare the value of the t-statistic to the value of the F-statistic calculated in part (a).

Solution

(a) We need to test

$$H_0: \mu_1 = \mu_2 \text{ vs. } H_a: \mu_1 \neq \mu_2.$$

From the random samples, we obtain the following needed estimates, $n_1 = n_2 = 8$:

$$\bar{y}_1 = 7.8375,\ \bar{y}_2 = 10.0625,\ \sum_{ij} y_{ij}^2 = 1319.34,\ \sum_{ij} y_{ij} = 143.20,\ \bar{y} = 8.95$$

and

$$SST = \sum_{i=1}^{2} n_i(\bar{y}_i - \bar{y})^2 = 19.8025.$$

Therefore,

$$\begin{aligned} Total\ SS &= \Sigma_i \Sigma_j y_{ij}^2 - \frac{\left(\Sigma_i \Sigma_j y_{ij}\right)^2}{(n_1 + n_2)} \\ &= 1391.34 - \frac{(143.2)^2}{16} = 109.70. \end{aligned}$$

Then

$$\begin{aligned} SSE &= Total\ SS - SST \\ &= 109.7 - 19.8025 = 89.8975, \\ MST &= \frac{SST}{1} = 19.8025 \end{aligned}$$

and

$$\begin{aligned} MSE &= \frac{SSE}{2n_1 - 2} = \frac{89.8975}{14} \\ &= 6.42125. \end{aligned}$$

Hence, the test statistic

$$F = \frac{MST}{MSE} = \frac{19.8025}{6.42125}$$
$$= 3.0839.$$

For $\alpha = 0.05$, $F_{0.05,1,14} = 4.60$. Hence the rejection region is $\{F > 4.60\}$. Because 3.0839 is not greater than 4.60, H_0 is not rejected. There is not enough evidence to indicate that the average percentage bonuses are different for men and women at $\alpha = 0.05$.

(b) To solve the problem, we assumed that the samples are random and independent with $n_1 = n_2 = 8$, drawn from two normal populations with means μ_1 and μ_2 and common variance σ^2.

(c) The value of MSE is the same as $s^2 = s_p^2 = 6.42125$. Also, $\bar{y}_1 = 7.8375$ and $\bar{y}_2 = 10.0625$. Then, the t-statistic is

$$t = \frac{\bar{y}_1 - \bar{y}_2}{\sqrt{s^2\left(\frac{1}{n_1} + \frac{1}{n_2}\right)}} = \frac{7.8375 - 10.0625}{\sqrt{6.42125\left(\frac{1}{8} + \frac{1}{8}\right)}} = -1.756.$$

Now, $t_{0.025,14} = 2.145$ and the rejection region is $\{t < -2.145\}$.

Because -1.756 *is not less than* -2.145, H_0 *is not rejected, which implies that there is no significant difference between the bonuses for the males and the females.*

Note also that $t^2 = F$, *that is,* $(-1.756)^2 = 3.083$ *implying that in the two-sample case, the t-test and F-test lead to the same result.*

It is not surprising that in the previous example, the conclusions reached using ANOVA and two-sample *t*-tests are the same. In fact, it can be shown that for two sets of independent and normally distributed random variables, the two procedures are entirely equivalent for a two-sided hypothesis. However, a *t*-test can also be applied to a one-sided hypothesis, whereas ANOVA cannot. The purpose of this section is only to illustrate the computations involved in the ANOVA procedures as opposed to simple *t*-tests. The ANOVA procedure is effectively used for three or more populations, which is described in the next section.

EXERCISES 9.2

9.2.1 The following information was obtained from two independent samples selected from two normally distributed populations with unknown but equal standard deviations. Do the data present sufficient evidence to indicate that there is a difference in the mean for the two populations?

Sample 1	1	2	3	3	1	2	1	3	1
Sample 2	2	5	2	4	3	1	2	3	3

(a) Use the ANOVA approach to test the appropriate hypotheses. Use $\alpha = 0.05$.

(b) Test the appropriate hypothesis by using the two-sample t-test for comparing population means. Compare the value of the t-statistic to the value of the F-statistic calculated in part (a).

9.2.2 The following information was obtained from two independent samples selected from two normally distributed populations with unknown but equal standard deviations. Do the data present sufficient evidence to indicate that there is a difference in the mean for the two populations?

Sample 1:	15	13	11	14	10	12	7	12	11	14	15	
Sample 2:	18	16	13	21	16	19	15	18	19	20	21	14

(a) Use the ANOVA approach to test the appropriate hypotheses. Use $\alpha = 0.01$.

(b) Test the appropriate hypothesis by using the two-sample t-test for comparing population means. Compare the value of the t-statistic to the value of the F-statistic calculated in part (a).

9.2.3 A company claims that its medicine, brand A, provides faster relief from pain than another company's medicine, brand B. A random sample from each brand gave the following times (in minutes) for relief. Do the data present sufficient evidence to indicate that there is a difference in the mean time to relief for the two populations?

Brand A:	47	51	45	53	41	55	50	46	45	51	53	50	48
Brand B:	44	48	42	45	44	42	49	46	45	48	39	49	

(a) Use the ANOVA approach to test the appropriate hypotheses. Use $\alpha = 0.01$.
(b) What assumptions are necessary for the conclusion in part (a)?
(c) Test the appropriate hypothesis by using the two-sample t-test for comparing population means. Compare the value of the t-statistic to the value of the F-statistic calculated in part (a).

9.2.4 Table 9.1 gives mean SAT scores for math by state from 1989 and 1999 for 20 randomly selected states.

TABLE 9.1 SAT Scores.

State	1989	1999
Arizona	523	525
Connecticut	498	509
Alabama	539	555
Indiana	487	498
Kansas	561	576
Oregon	509	525
Nebraska	560	571
New York	496	502
Virginia	507	499
Washington	515	526
Illinois	539	585
North Carolina	469	493
Georgia	475	482
Nevada	512	517
Ohio	520	568
New Hampshire	510	518

Source: The World Almanac and Book of Facts 2000.

Using the ANOVA procedure, test that the mean SAT score for math in 1999 is greater than that in 1989 at $\alpha = 0.05$. Assume that the variances are equal and the samples come from a normal distribution.

9.2.5 Let $X_1, \ldots, X_{n_1}$ and $Y_1, \ldots, Y_{n_2}$ be two sets of independent, normally distributed random variables with means μ_1 and μ_2, and the common variance σ^2. Show that the two-sample t-test and the ANOVA are equivalent for testing H_0: $\mu_1 = \mu_2$ versus H_a: $\mu_1 > \mu_2$.

9.3 ANALYSIS OF VARIANCE FOR COMPLETELY RANDOMIZED DESIGN

In this section, we study the hypothesis testing problem of comparing population means for more than two independent populations, where the data are about several independent groups (different treatments being applied, or different populations being sampled). We have seen in Chapter 8 that the random selection of independent samples from k populations is known as a completely randomized experimental design or one-way classification.

Let $\mu_1, \ldots, \mu_k$ be the means of k normal populations with unknown but equal variance σ^2. The question is whether the means of these groups are different or are all equal. The idea is to consider the overall variability in the data. We partition the variability into two parts: (1) between-groups variability and (2) within-groups variability. If between groups is much larger than that within groups, this will indicate that differences between the groups are real, not merely due to the random nature of sampling. Let independent samples be drawn of sizes n_i, $i = 1, 2, \ldots, k$, and let $N = n_1 + \cdots + n_k$. Let y_{ij} be the measured response on the jth experimental unit in the ith sample. That is, Y_{ij} is the jth observation from population i, $i = 1, 2, \ldots, k$, and $j = 1, 2, \ldots, n_i$. Let $\bar{y}$ be the overall mean of all observations. The problem can be formulated as a hypothesis testing problem, where we need to test

$$H_0\text{: } \mu_1 = \mu_2 = \ldots = \mu_k \text{ versus } H_a\text{: Not all the } \mu'_i\, s \text{ are equal.}$$

The method of ANOVA tests the null hypothesis H_0 by comparing two unbiased estimates of the variance, σ^2, an estimate based on variations from sample to sample, and the other one based on variations within the samples. We will be rejecting H_0 if the first estimate is significantly larger than the second, so that the samples cannot be assumed to come from the same population. That is, the variances influence the decision.

We can write the total sum of squares of deviations of the response measurements about their overall mean for the k samples into two parts, from the treatment (SST) and from the error (SSE). This partition gives the fundamental relationship in ANOVA, where total variation is divided into two portions: between-sample variation and within-sample variation. That is,

$$Total\ SS = SST + SSE.$$

The following derivations will make computation of these quantities simpler. The total SS can be written as

$$Total\ SS = \sum_{i=1}^{k}\sum_{j=1}^{n_i}(y_{ij} - \bar{y})^2 = \sum_{i=1}^{k}\sum_{j=1}^{n_i} y_{ij}^2 - 2\bar{y}\sum_{i=1}^{k}\sum_{j=1}^{n_i} y_{ij} + N\bar{y}^2.$$

Note that

$$\bar{y} = \frac{\sum_{i=1}^{k}\sum_{j=1}^{n_i} y_{ij}}{N},$$

and then we have

$$Total\ SS = \sum_{i=1}^{k}\sum_{j=1}^{n_i} y_{ij}^2 - CM,$$

where CM is the correction factor for the correction for the means and is given by

$$CM = \frac{\left(\sum_{i=1}^{k}\sum_{j=1}^{n_i} y_{ij}\right)^2}{N} = N\bar{y}^2.$$

Let

$$T_i = \sum_{j=1}^{n_i} y_{ij}, \text{ be the sum of all the observations in the } i\text{th sample}$$

and

$$\overline{T_i} = \frac{\sum_{j=1}^{n_i} y_{ij}}{n_i}, \text{ the mean of the observations in the } i\text{th sample.}$$

We can rewrite $\bar{y}$ as

$$\bar{y} = \frac{\sum_{i=1}^{k}\sum_{j=1}^{n_i} y_{ij}}{N} = \frac{\sum_{i=1}^{k} n_i \overline{T_i}}{N}.$$

Now, we introduce SST, the sum of squares for treatment (sometimes known as the between-group sum of squares, SSB) by

$$SST = \sum_{i=1}^{k} n_i(\overline{T}_i - \overline{y})^2.$$

We note that $(\overline{T}_i)$ is the mean response due to its ith treatment, and $\overline{y}$ is the overall mean. A large value of $(\overline{T}_i - \overline{y})$ is likely to be caused by the ith treatment effect being much different from the rest. Hence, SST can be used to measure the differences in the treatment effects.

Thus the sum of squares of errors (SSE) is given by

$$SSE = Total\ SS - SST.$$

We must state that the *SSE* is the sum of squares within groups (thus sometimes *SSE* is referred to as the *within-group sum of squares, SSW*), and this can be seen from rewriting the expression as

$$SSE = \sum_{i=1}^{k} \sum_{j=1}^{n_i} \left(y_{ij} - \overline{T}_i\right)^2.$$

The decomposition of the total sum of squares can be easily seen in Fig. 9.1.

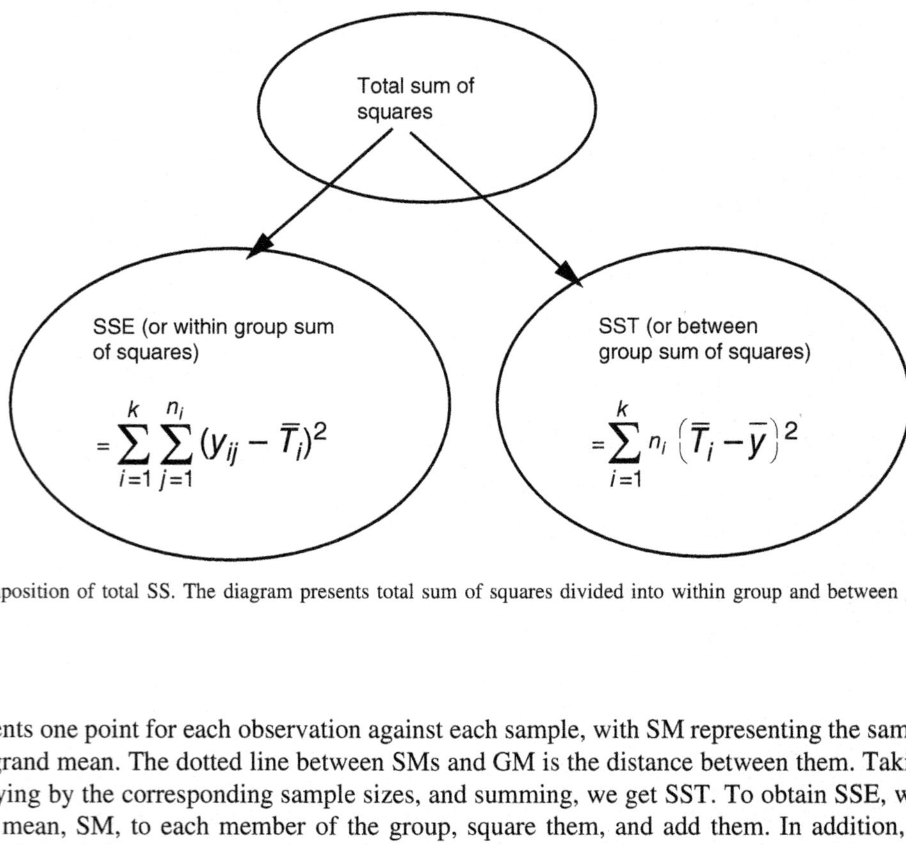

FIGURE 9.1 Decomposition of total SS. The diagram presents total sum of squares divided into within group and between group components with formulas.

Fig. 9.2 represents one point for each observation against each sample, with SM representing the sample means and GM representing the grand mean. The dotted line between SMs and GM is the distance between them. Taking these distances, squaring, multiplying by the corresponding sample sizes, and summing, we get SST. To obtain SSE, we take the distance from each group mean, SM, to each member of the group, square them, and add them. In addition, to give an idea of within-group variations, it is customary to draw side-by-side box plots.

As mentioned earlier, *SST* estimates the variation among the $\mu'_i\,s$, and hence if all the $\mu'_i\,s$ were equal, the $\overline{T}_i\,s$ would be similar, and the SST would be small. It can be verified that the unbiased estimator of σ^2 based on $(n_1 + n_2 + \cdots + n_k - k)$ degrees of freedom is

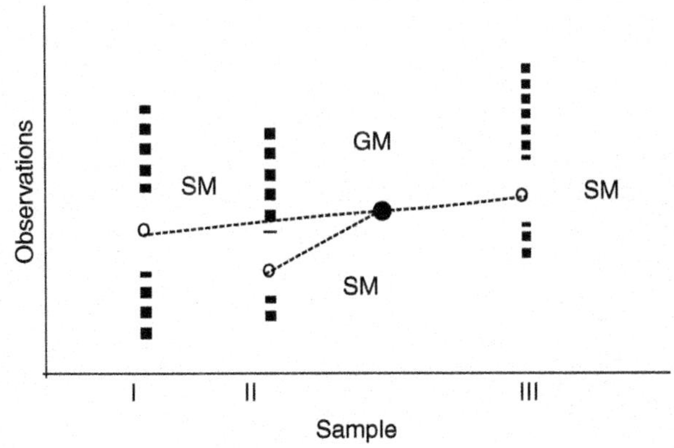

FIGURE 9.2 ANOVA decomposition. This figure presents observations across samples with group means and grand mean connected by lines.

$$S^2 = MSE = \frac{SSE}{(n_1 + n_2 + \cdots + n_k - k)}$$
$$= \frac{SSE}{N - k}.$$

Note that the quantity *MSE* is a measure of variability within the groups. If there were only one group with n observations, then the *MSE* is nothing but the sample variance, s^2. The fact that ANOVA deals simultaneously with all the k groups can be seen by rewriting MSE in the following form:

$$MSE = \frac{(n_1 - 1)s_1^2 + (n_2 - 1)s_2^2 + \cdots + (n_k - 1)s_k^2}{(n_1 - 1) + (n_2 - 1) + \cdots + (n_k - 1)}.$$

The *mean square for treatments* with $(k - 1)$ degrees of freedom is

$$MST = \frac{SST}{k - 1}.$$

The *MST* is a measure of the variability between the sample means of the groups. We now summarize the ANOVA hypothesis testing method for two or more populations.

One-way analysis of variance for $k \geq 2$ populations

We test

$$H_0: \mu_1 = \mu_2 = \ldots = \mu_k \quad \text{versus}$$

H_a: At least two of the μ'_i s are different.

When H_0 is true, we have

$$E(MST) = E(MSE).$$

The greater the differences among the μ's, the larger the $E(MST)$ will be relative to $E(MSE)$.

Test statistic:

$$F = \frac{MST}{MSE}.$$

Rejection region is

$$RR: F > F_\alpha$$

with $\upsilon_1 = (k - 1)$ numerator degrees of freedom and $\upsilon_2 = \sum_{i=1}^{k} n_i - k = N - k$ denominator degrees of freedom, where $N = \sum_{i=1}^{k} n_i$.

Assumptions: The observations $Y'_{ij}s$ are assumed to be independent and normally distributed with mean μ_i, $i=1, 2, \ldots, k$, and variance σ^2.

Now we give a five-step computational procedure that we could follow for the ANOVA for the completely randomized design.

One-way analysis of variance computational procedure for $k \geq 2$ populations

We test

$$H_0: \mu_1 = \mu_2 = \ldots = \mu_k \quad \text{versus}$$

H_a: At least two of the $\mu'_i s$ are different.

1. Compute

$$T_i = \sum_{j=1}^{n_i} y_{ij}, \; T = \sum_{i=1}^{k}\sum_{j=1}^{n_i} y_{ij}, \text{ and } \sum_{i=1}^{k}\sum_{j=1}^{n_i} y_{ij}^2,$$

$$CM = \frac{\left(\sum_{i=1}^{k}\sum_{j=1}^{n_i} y_{ij}\right)^2}{N} = \frac{T^2}{N}, \text{ where } N = \sum_{i=1}^{k} n_i,$$

$$\overline{T}_i = \frac{T_i}{n_i},$$

and

$$Total\ SS = \sum_{i=1}^{k}\sum_{j=1}^{n_i} y_{ij}^2 - CM.$$

2. Compute the sum of squares between samples (treatments) and the sum of squares within samples,

$$SST = \sum_{i=1}^{k} \frac{T_i^2}{n_i} - CM,$$

$$SSE = Total\ SS - SST.$$

Let

$$MST = \frac{SST}{k-1},$$

and

$$MSE = \frac{SSE}{n-k}.$$

3. Compute the test statistic:

$$F = \frac{MST}{MSE}.$$

4. For a given α, find the rejection region as

$$RR: F > F_\alpha$$

with $\upsilon_1 = (k - 1)$ numerator degrees of freedom and $\upsilon_2 = (\sum_{i=1}^{k} n_i) - k = N - k$ denominator degrees of freedom, where $N = \sum_{i=1}^{k} n_i$.

5. **Conclusion:** If the test statistic F falls in the rejection region, conclude that the sample evidence supports the alternative hypothesis that at least one pair of the means is indeed different for the k treatments and is not all equal.

Assumptions: The samples are randomly selected from the k populations in an independent manner. The populations are assumed to be normally distributed with equal variances σ^2 and means $\mu_1, \ldots, \mu_k$.

Even though the completely randomized design is extremely easy to construct and the calculations described above are relatively easy, the homogeneity of the treatments is crucial. Any extraneous sources of variability will make it more difficult to detect differences among treatment means due to inflation of the error term.

9.3.1 The *p*-Value Approach

Note that if we are using statistical software packages, the p-value approach can be used for the testing. Just compare the p-value and α to arrive at a conclusion. Refer to the computer examples in Section 9.7.

The following example illustrates the ANOVA procedure.

EXAMPLE 9.3.1

We are given three random samples as shown in Table 9.2 that represent test scores from three classes of statistics taught by three different instructors and are independently sampled from each class. Assume that the three different populations are normal with equal variances.

TABLE 9.2 Class Scores for Different Instructors.

Sample 1	Sample 2	Sample 3
64	56	81
84	74	92
75	69	84
77		
80		

At the $\alpha = 0.05$ level of significance, test for equality of the population means.

Solution

We test

$$H_0: \mu_1 = \mu_2 = \mu_3 \text{ versus } H_a: \text{At least two of the } \mu\text{'s are different.}$$

Here, $k = 3$, $n_1 = 5$, $n_2 = 3$, *and* $N = n_1 + n_2 + n_3 = 11$.

Also,

T_i	380	199	257
n_i	5	3	3
$\overline{T}_i$	76	66.33	85.67

Clearly, the sample means are different. The question we are going to answer is: Is this difference due to just chance, or is it due to a real difference caused by different teaching styles? For this, we now compute the following:

$$CM = \frac{\left(\Sigma_i \Sigma_j y_{ij}\right)^2}{N} = \frac{(836)^2}{11} = 63{,}536,$$

$$Total\ SS = \Sigma_i \Sigma_j y_{ij}^2 - CM$$

$$= 64{,}560 - 63{,}536 = 1024,$$

$$SST = \Sigma_i \frac{T_i^2}{n_i} - CM$$

$$= \frac{(380)^2}{5} + \frac{(199)^2}{3} + \frac{(257)^2}{3} - CM$$

$$= 64{,}096.66 - 63{,}536 = 560.66,$$

and

$$SSE = Total\ SS - SST$$

$$= 1024 - 560.66 = 463.34.$$

Hence,

$$MST = \frac{SST}{k-1} = \frac{560.66}{2} = 280.33,$$

and

$$MSE = \frac{SSE}{N-k} = \frac{463.34}{8} = 57.9175.$$

The test statistic is

$$F = \frac{MST}{MSE} = \frac{280.33}{57.9175} = 4.84.$$

From the F-table, $F_{0.05,2,8}=4.46$
Therefore the rejection region is given by

$$RR: F > 4.46.$$

Decision: Because the observed value of $F = 4.84$ *falls in the rejection region, we do reject* H_0 *and conclude that there is sufficient evidence to indicate a difference in the true means.*

If we want the p-value, we can see from the F-table that $0.025 < p\text{-value} < 0.05$, *indicating the rejection of the null hypothesis with* $\alpha = 0.05$. *Using statistical software packages, we can get the exact p-value.*

■

TABLE 9.3 ANOVA Table.

Source of Variation	Degree of Freedom	Sum of Squares	Mean Squares	F-Statistic	p-value
Treatments	$k-1$	$SST = \sum_{i=1}^{k} \frac{T_i^2}{n_i} - CM$	$MST = \frac{SST}{k-1}$	$\frac{MST}{MSE}$	$P(Fk-1,n-k \geq \text{observed } F)$
Error	$N-k$	$SSE = Total\ SS - SST$	$MSE = \frac{SSE}{N-k}$		
Total	$N-1$	$Total\ SS = \sum_{i=1}^{k}\sum_{j=1}^{n_i}\left(y_{ij} - \bar{y}\right)^2$			

Sometimes, one may also add a column for the *p*-value, $P(F_{k-1,n-k} \geq observed\ F)$, in the ANOVA table.
For the previous example, we can summarize the computations by the ANOVA table shown in Table 9.3.

9.3.2 Testing the Assumptions for One-Way ANOVA

The randomness assumption could be tested using the Wald–Wolfowitz test (see Project 12B). The assumption of independence of the samples is hard to test without knowing how the data are collected, and should be implemented during the collection of data at the design stage. Normality can be tested (this should be performed separately for each sample, not for the total data set) using probability plots or other tests such as the chi-square goodness-of-fit-test. ANOVA is fairly robust against violation of this assumption if the sample sizes are equal. Also, if the sample sizes are fairly large, the central limit theorem helps. The presence of outliers is likely to increase the sample variance, thus decreasing the value of the *F*-statistic for ANOVA, which will result in a lower power of the test. Box plots or probability plots could be used to identify the outliers. If the normality test fails, transforming the data (see Section 15.4.2), or a nonparametric test such as the Kruskal–Wallis test described in Section 12.5.1, may be more appropriate. If the sample sizes of each sample are equal, ANOVA is mostly robust for violation of homogeneity of the variances. A rule of thumb used for robustness for this condition is that the ratio of sample variance of the largest sample variance s^2 to the smallest sample variance s^2 should be no more than 3:1. Another popular rule of thumb used in one-way ANOVA to verify the requirement of equality of variances is that the largest sample standard deviation not be larger than two times the smallest sample standard deviation. Graphically, representing side-by-side box plots of the samples can also reveal lack of homogeneity of variances if some box plots are much longer than others (see Fig. 9.3E). For a significance test on the homogeneity of variances (Levene's test), refer to Section 15.4.3. If these tests reveal that the variances are different, then the populations are different, in spite of what ANOVA concludes about differences of the means. But this itself is significant, because it shows that the treatments had an effect (Table 9.3).

The calculations obtained in analyzing the total sum of squares into its components are usually summarized by the ANOVA table given in Table 9.4.

TABLE 9.4 ANOVA Table for Class Scores.

Source of Variation	Degree of Freedom	Sum of Squares	Mean Square	*F*-Statistic	*p*-Value
Treatments	2	560.66	280.33	4.84	0.042
Error	8	463.34	57.917		
Total	10	1024			

EXAMPLE 9.3.2

In order to study the effect of automobile size on the noise pollution, the following data are randomly chosen from the air pollution data (source: Lewin, A.Y., Shakun, M.F., 1976. Policy Sciences: Methodology and Cases. Pergamon Press, , p. 313). The automobiles are categorized as small, medium, large, and noise level reading (decibels) is given in Table 9.5.

TABLE 9.5 Automobile Size and Noise Pollution.

	Size of Automobile		
	Small	Medium	Large
Noise level (decibels)	820	840	785
	820	825	775
	825	815	770
	835	855	760
	825	840	770

At the $\alpha = 0.05$ level of significance, test for equality of population mean noise levels for different sizes of the automobiles. Comment on the assumptions.

Solution

Let μ_1, μ_2, μ_3 be population mean noise levels for small, medium, and large automobiles, respectively. First, we test the assumptions. Using R, run tests for each of the samples; we can justify the assumption of randomness of the sample values. A normality test for each column gives the graphs shown in Figs. 9.3A through 9.3C, *through which we can reasonably assume normality. Because the sample sizes are equal, we will use the one-way ANOVA method to analyze these data.*

Fig. 9.3D *indicates that the relative positions of the sample means are different, and* Fig. 9.3E *gives an indication of within-group variations; perhaps the group 2 (medium-size) variance is larger. Now, we will do the analytic testing.*

We test

$$H_0: \mu_1 = \mu_2 = \mu_3 \text{ versus } H_a: \text{At least two of the } \mu\text{'s are different.}$$

Here, $k=3$, $n_1=5$, $n_2=5$, $n_3=5$ and $N=n_1 + n_2 + n_3=15$.

Also,

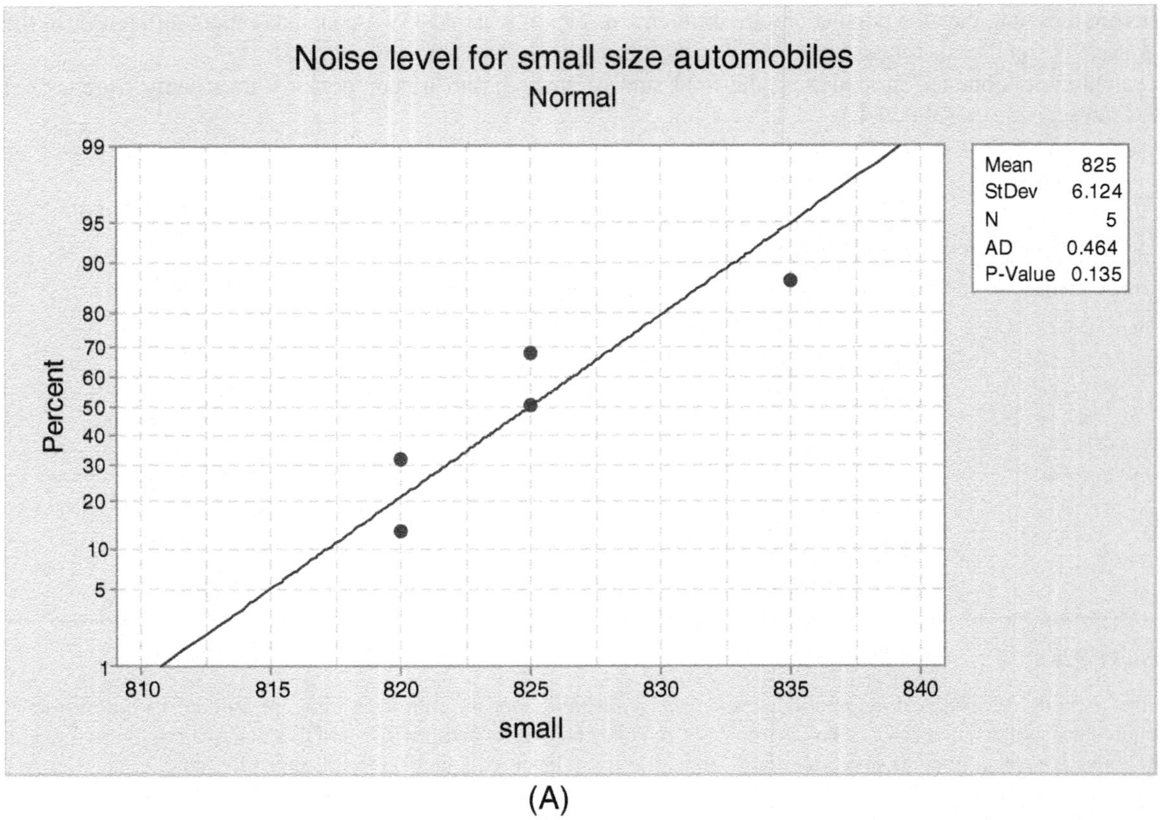

FIGURE 9.3A Normal plot for noise level of small automobiles.

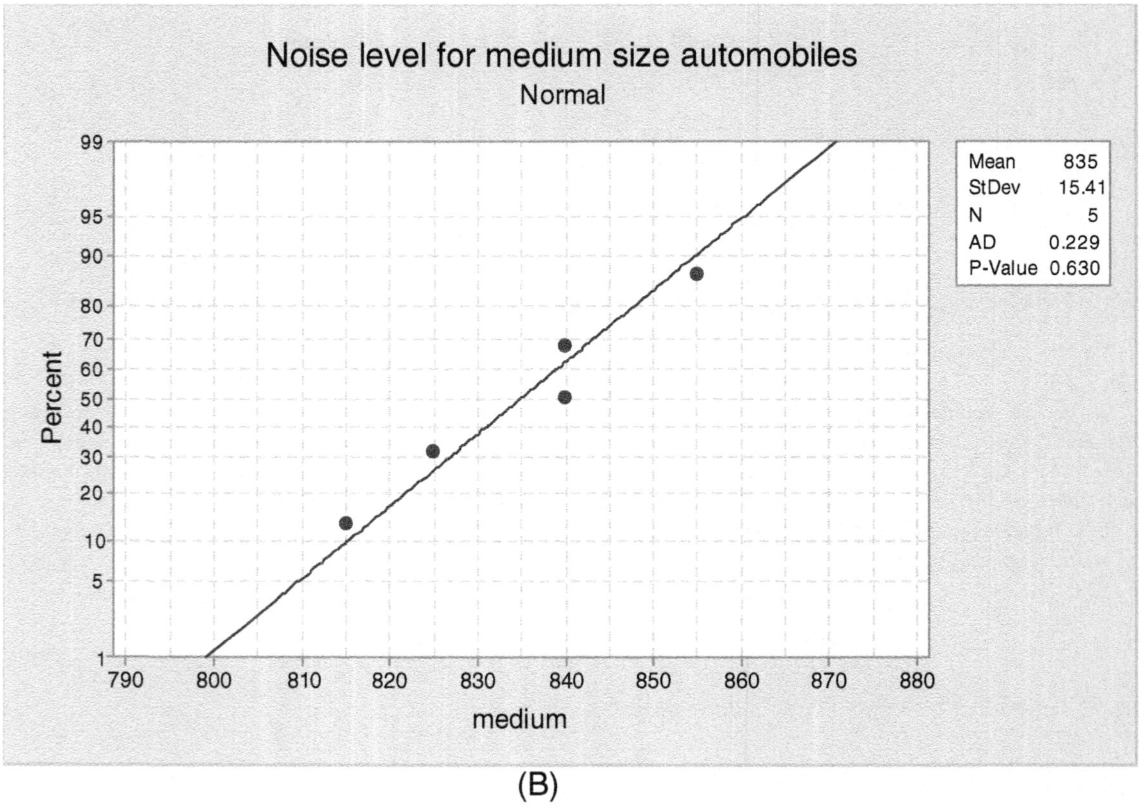

FIGURE 9.3B Normal plot for noise level of medium-sized automobiles.

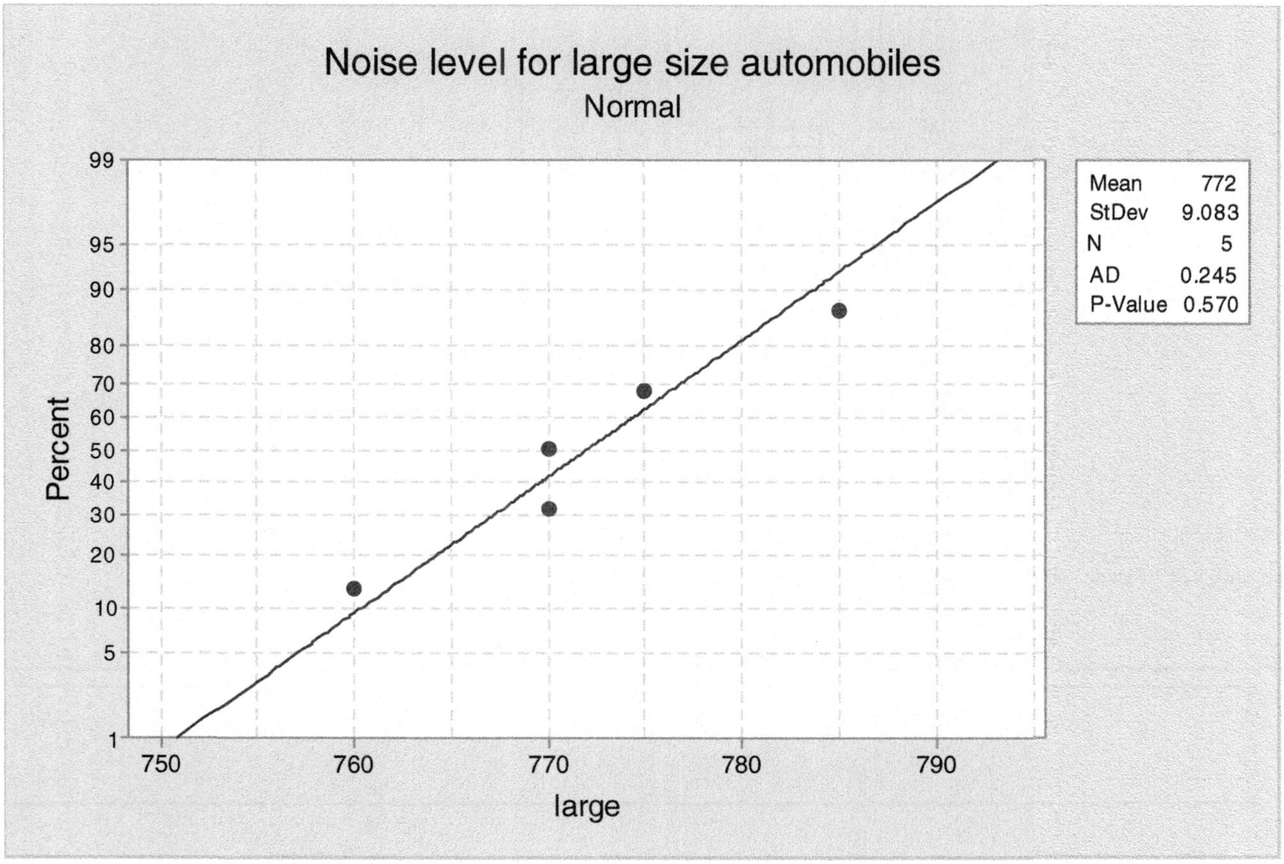

(C)

FIGURE 9.3C Normal plot for noise level of large automobiles.

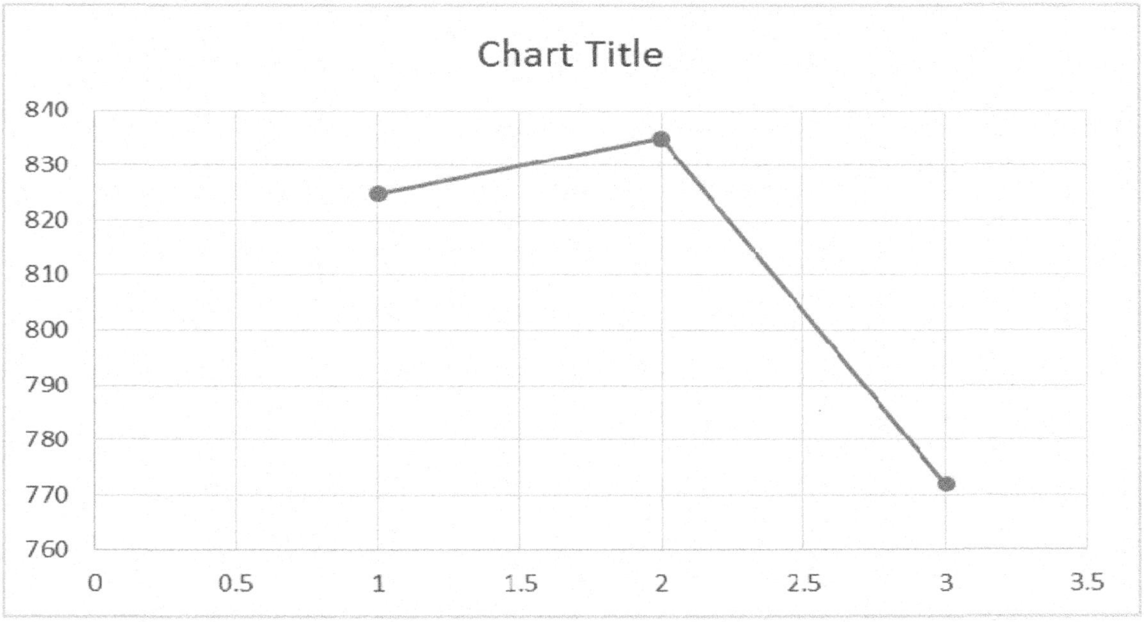

(D)

FIGURE 9.3D Mean decibel levels for three sizes of automobiles. The box plot shows decibels for three auto sizes with varying medians and spreads.

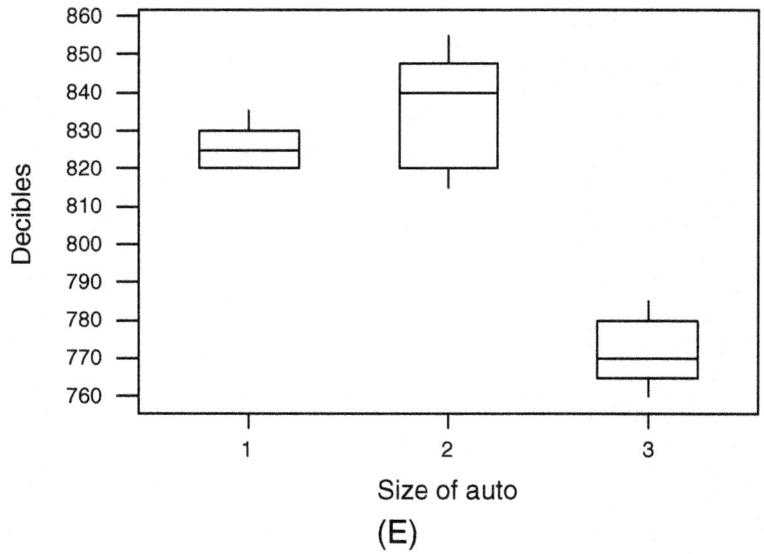

FIGURE 9.3E Side-by-side box plots for decibel levels for three sizes of automobiles.

T_i	4125	4175	3860
n_i	5	5	5
$\bar{T}_i$	825	835	772

In the following calculations, for convenience we will approximate all values to the nearest integer.

$$CM = \frac{\left(\Sigma_i \Sigma_j y_{ij}\right)^2}{N} = \frac{(12{,}160)^2}{15} = 9{,}857{,}707,$$

$$\text{Total } SS = \Sigma_i \Sigma_j y_{ij}^2 - CM$$

$$= 12{,}893,$$

$$SST = \Sigma_i \frac{T_i^2}{n_i} - CM$$

$$= 11{,}463,$$

and

$$SSE = \text{Total } SS - SST$$
$$= 1430.$$

Hence,

$$MST = \frac{SST}{k-1} = \frac{11{,}463}{2} = 5732,$$

and

$$MSE = \frac{SSE}{N-k} = \frac{1430}{12} = 119.$$

The test statistic is

$$F = \frac{MST}{MSE} = \frac{5732}{119} = 48.10.$$

From the table, we get $F_{0.05,2,12} = 3.89$. *Because the test statistic falls in the rejection region, we reject at* $\alpha = 0.05$ *the null hypothesis that the mean noise levels are the same. We conclude that size of the automobile does affect the mean noise level.*

It should be noted that the alternative hypothesis H_a in this section covers a wide range of situations, from the case where all but one of the population means are equal to the case where they are all different. Hence, with such an alternative, if the samples lead us to reject the null hypothesis, we are left with a lot of unsettled questions about the means of the k populations. These are called post hoc testing. This problem of multiple comparisons is the topic of Section 9.5.

9.3.3 Model for One-Way ANOVA

We conclude this section by presenting the classical model for one-way ANOVA. Because the variables Y_{ij} values are random samples from normal populations with $E(Y_{ij}) = \mu_i$ and with common variance $Var(Y_{ij}) = \sigma^2$, for $i = 1, \ldots, k$ and $j = 1, \ldots, n_i$, we can write a model as

$$Y_{ij} = \mu_i + \varepsilon_{ij},\ j = 1, \ldots, n_i$$

where the error terms ε_{ij} are independent normally distributed random variables with $E(\varepsilon_{ij}) = 0$ and $Var(\varepsilon_{ij}) = \sigma^2$. Let $\alpha_i = \mu_i - \mu$ be the difference of μ_i (ith population mean) from the grand mean μ. Then α_i can be considered as the ith treatment effect. Note that the α_i values are nonrandom. Because $\mu=\Sigma_i (n_i \mu_i/N)$, it follows that $\Sigma_{i=1}^{k} \alpha_i = 0$. This will result in the following classical model for one-way layout:

$$Y_{ij} = \mu + \alpha_i + \varepsilon_{ij},\ i = 1, \ldots, k,\ j = 1, \ldots, n_i.$$

With this representation, the test H_0: $\mu_1 = \mu_2 = \mu_k$ reduces to testing the null hypothesis that there is no treatment effect, H_0: $\alpha_i = 0$, for $i = 1, \ldots, k$.

EXERCISES 9.3

9.3.1 In an effort to investigate the premium charged by insurance companies for auto insurance, an agency randomly selects a few drivers who are insured by one of three different companies. These individuals have similar cars, driving records, and levels of coverage. Table 9.6 gives the premiums paid per 6 months by these drivers with the three companies.

TABLE 9.6 Insurance Charges for Companies I, II, and III.

Company I	Company II	Company III
396	348	378
438	360	330
336	522	294
318	474	432

(a) Construct an analysis-of-variance table and interpret the results.
(b) Using the 5% significance level, test the null hypothesis that the mean auto insurance premium paid per 6 months by all drivers insured for each of these companies is the same. Assume that the conditions of completely randomized design are met.

9.3.2 Three classes in elementary statistics are taught by three different persons: a regular faculty member, a graduate teaching assistant, and an adjunct from outside the university. At the end of the semester, each student is given a standardized test. Five students are randomly picked from each of these classes, and their scores are as shown in Table 9.7.
(a) Construct an analysis-of-variance table and interpret your results.
(b) Test at the 0.05 level whether there is a difference between the mean scores for the three persons teaching. Assume that the conditions of completely randomized design are met.

TABLE 9.7 Students' scores.

Faculty	Teaching Assistant	Adjunct
93	88	86
61	90	56
87	76	73
75	82	90
92	58	47

9.3.3 Let $n_1=n_2= n_k=n'$. Show that

$$\sum_{i=1}^{k}\sum_{j=1}^{n'}(y_{ij}-\bar{y})^2=\sum_{i=1}^{k}\sum_{j=1}^{n'}(y_{ij}-\bar{T}_i)^2+n\sum_{i=1}^{k}(\bar{T}_i-\bar{y})^2.$$

9.3.4 For the sum of squares for treatment

$$SST=\sum_{i=1}^{k}n_i(\bar{T}_i-\bar{y})^2,$$

show that

$$E(SST)=(k-1)\sigma^2+\sum_{i=1}^{k}n_i(\mu_i-\mu)^2$$

where $\mu=\frac{1}{N}\sum_{i=1}^{k}n_i\mu_i$.

[This exercise shows that the expected value of SST increases as the differences among the $\mu'_i\,s$ increase.]

9.3.5 **(a)** Show that

$$SSE=\sum_{i=1}^{k}(n_i-1)S_i^2=\sum_{i=1}^{k}\sum_{j=1}^{n_i}(Y_{ij}-\bar{T}_i)^2,$$

where $S_i^2=\frac{1}{n-1}\Sigma_{j=1}^{n_i}(Y_{ij}-\bar{T}_i)^2$ provides an independent, unbiased estimator for σ2 in each of the k samples.

(b) Show that SSE/σ^2 has a chi-square distribution with $N-k$ degrees of freedom, where $N=\Sigma_{i=1}^{k}n_i$.

9.3.6 Let each observation in a set of k independent random samples be normally distributed with means $\mu_1, \ldots, \mu_k$ and common variance σ^2. If $H_0=\mu_1=\mu_2=\ldots=\mu_k$ is true, show that has an F-distribution with $k-1$ numerator and $n-k$ denominator degrees of freedom.

$$F=\frac{SST/(k-1)}{SSE/(n-k)}=\frac{MST}{MSE},$$

9.3.7 The management of a grocery store observes various employees for work productivity. Table 9.8 gives the number of customers served by each of its four checkout lanes per hour.

(a) Construct an analysis-of-variance table and interpret the results. Indicate any assumptions that were necessary.

TABLE 9.8 Number of Customers Each Line Served.

Lane 1	Lane 2	Lane 3	Lane 4
16	11	8	21
18	14	12	16
22	10	17	17
21	10	10	23
15	14	13	17
	10	15	

(b) Test whether there is a difference between the mean numbers of customers served by the four employees at the 0.05 level. Assume that the conditions of a completely randomized design are met.

9.3.8 Table 9.9 represents immunoglobulin levels (with each observation being the IgA immunoglobulin level measured in international units) of children under 10 years of age of a particular group. The children are grouped as follows: A: ages 1 to less than 3, B: ages 3 to less than 6, C: ages 6 to less than 8, and D: ages 8 to less than 9. Test whether there is a difference between the means for each of the age groups. Use $\alpha = 0.05$. Interpret your results, and state any assumptions that were necessary to solve the problem.

TABLE 9.9 Data of IgA Immunoglobulin Levels for Each Group.

A	35	8	12	19	56	64	75	25		
B	31	79	60	45	39	44	45	62	20	66
C	74	56	77	35	95	81	28			
D	80	42	48	69	95	40	86	79	51	

9.3.9 Table 9.10 gives rental and homeowner vacancy rates by US region (source: US Census Bureau) for 5 years. Test at the 0.01 level whether the true rental and homeowner vacancy rates by area are the same for all 5 years. Interpret your results, and state any assumptions that were necessary to perform the analysis.

TABLE 9.10 Homeowner Vacancy Rates by U.S. Region.

Rental Units	1995	1996	1997	1998	1999
Northeast	7.2	7.4	6.7	6.7	6.3
Midwest	7.2	7.9	8.0	7.9	8.6
South	8.3	8.6	9.1	9.6	10.3
West	7.5	7.2	6.6	6.7	6.2

9.3.10 Table 9.11 gives lower limits of income (approximated to the nearest \$1000 and calculated as of March of the following year) of the top 5% of US households by race from 1994 to 1998 (source: US Census Bureau). Test at the 0.05 level whether the true lower limits of income for the top 5% of US households for each race are the same for all 5 years.

TABLE 9.11 Lower Limits of Income by Race.

Race	Year				
	1994	**1995**	**1996**	**1997**	**1998**
All races	110	113	120	127	132
White	113	117	123	130	136
Black	81	80	85	87	94
Hispanic	82	80	86	93	98

9.3.11 Table 9.12 gives mean serum cholesterol levels (given in milligrams per deciliter) by race and age for the adult population in the United States between 1978 and 1980 (source: "Report of the National Cholesterol Education Program Expert Panel on Detection, Evaluation, and Treatment of High Blood Cholesterol in Adults," Arch. Intern. Med. 148, January 1988).

TABLE 9.12 Cholesterol Levels by Age and Race.

Race	Age					
	20–24	**25–34**	**35–44**	**45–54**	**55–64**	**65–74**
All races	180	199	217	227	229	221
White	180	199	217	227	230	222
Black	171	199	218	229	223	217

Test at the 0.01 level whether the true mean cholesterol levels for the adult population in the United States between 1978 and 1980 are the same.

9.4 TWO-WAY ANALYSIS OF VARIANCE, RANDOMIZED COMPLETE BLOCK DESIGN

A *randomized block design*, or the *two-way ANOVA*, consists of b blocks of k experimental units each. In many cases, we may be required to measure response at combinations of levels of two or more factors considered simultaneously. For example, we might be interested in gas mileage per gallon among four different makers of cars for both in-city and highway driving, or to examine weight loss comparing five different diet programs among Whites, African Americans, Hispanics, and Asians according to their gender. In studies involving various factors, the effect of each factor on the response variable may be analyzed using one-way classification. However, such an analysis will not be efficient with respect to time, effort, and cost. Also, such a procedure would give no knowledge about the likely interactions that may exist among different factors. In such cases, the two-way ANOVA is an appropriate statistical method to use.

In a randomized block design, the treatments are randomly assigned to the units in each block, with each treatment appearing exactly once in every block (i.e., there is no interaction between factors). Thus the total number of observations obtained in a randomized block design is $n = bk$. The purpose of subdividing experiments into blocks is to eliminate as much variability as possible, that is, to reduce the experimental error or the variability due to extraneous causes. Refer to Section 9.2.3 for a procedure to obtain completely randomized block design. The goal of such an experiment is to test the equality of levels for the treatment effect. Sometimes, it may also be of interest to test for a difference among blocks. We proceed to give a formal statistical model for the completely randomized block design.

For $i = 1, 2, \ldots, k$ and $j = 1, 2, \ldots, b$, let $Y_{ij} = \mu + \alpha_i + \beta_j + \varepsilon_{ij}$, where Y_{ij} is the observation on treatment i in block j, μ is the overall mean, α_i is the nonrandom effect of treatment i, β_j is the nonrandom effect of block j, and ε_{ij} are the random error terms such that ε_{ij} are independent normally distributed random variables with $E(\varepsilon_{ij}) = 0$ and $Var(\varepsilon_{ij}) = \sigma^2$. In this case, $\Sigma\alpha_i=0$, and $\Sigma\beta_j=0$.

The ANOVA for a randomized block design proceeds similarly to that for a completely randomized design, the main difference being that the total sum of squares of deviations of the response measurements from their means may be partitioned into three parts: the sum of squares of blocks (SSB), treatments (SST), and error (SSE).

Let $B_j = \Sigma_{i=1}^{k} y_{ij}$ and $\overline{B}_j$ denote, respectively, the total sum and mean of all observations in block j. Represent the total for all observations receiving treatment i by $T_i = \Sigma_{j=1}^{b} y_{ij}$, and mean and $\overline{T}_i$, respectively. Let

$$\begin{aligned}\bar{y} &= \text{average of } n = bk \text{ observations}\\ &= \frac{1}{n}\Sigma_{j=1}^{b}\Sigma_{i=1}^{k}y_{ij}\end{aligned}$$

and

$$\begin{aligned}CM &= \frac{1}{n}(\text{total of all observations})^2\\ &= \frac{1}{n}\left(\Sigma_{j=1}^{b}\Sigma_{i=1}^{k}y_{ij}\right)^2.\end{aligned}$$

For convenience, we can represent the two-way classification as in Table 9.13.

TABLE 9.13 Summary of Two-way Classification.

	Blocks							
	1	**2**	**...**	***j***	**...**	***b***	**Total T_i**	**Mean $\overline{T}_i$**
Treatment 1	y_{11}	y_{12}	...	y_{1j}	...	y_{1b}	T_1	$\overline{T}_1$
Treatment 2	y_{21}	y_{22}	...	y_{2j}	...	y_{2b}	T_2	$\overline{T}_2$
.	.		.					
.	.		.					
.	.		.					
Treatment i	y_{i1}	y_{i2}	...	y_{ij}	...	y_{ib}	T_i	$\overline{T}_i$
.					.			
.					.			
.					.			
Treatment k	y_{k1}	y_{k2}	...	y_{kj}	...	y_{kb}	T_k	$\overline{T}_k$
Total B_j	B_1	B_2	...	B_j	...	B_b		
Mean $\overline{B}_j$	$\overline{B}_1$	$\overline{B}_2$	...	$\overline{B}_j$	...	$\overline{B}_b$		$\bar{y}$

Note that from the table we can obtain $\sum_{j=1}^{b}\sum_{i=1}^{k} y_{ij} = \sum_{j=1}^{b} B_j$. Hence, $CM = (1/n)\left(\sum_{j=1}^{b} B_j\right)^2$.

Then, for a randomized block design with b blocks and k treatments, we need to compute the following sums of squares. They are

$$\begin{aligned} Total\ SS &= SSB + SST + SSE \\ &= \sum_{j=1}^{b}\sum_{i=1}^{k}(y_{ij} - \overline{y})^2 = \sum_{j=1}^{b}\sum_{i=1}^{k} y_{ij}^2 - CM, \\ SSB &= k\sum_{j=1}^{b}(\overline{B}_j - \overline{y})^2 = \frac{\sum_{j=1}^{b} B_j^2}{k} - CM, \end{aligned}$$

and

$$SSB = b\sum_{i=1}^{k}(\overline{T}_i - \overline{y})^2 = \frac{\sum_{i=1}^{k} T_i^2}{b} - CM,$$

$$SSE = Total\ SS - SSB - SST.$$

We define

$$MSB = \frac{SSB}{b-1},$$
$$MST = \frac{SST}{k-1},$$

and

$$MSE = \frac{SSE}{n-b-k+1}.$$

The ANOVA for the randomized block design is presented in Table 9.14. The column corresponding to d.f. represents the degrees of freedom associated with each sum of squares. MS denotes the mean square.

To test the null hypothesis that there is no difference in treatment means, that is, to test

TABLE 9.14 Summary of Randomized Block Design.

Source	d.f.	SS	MS
Blocks	$b-1$	SSB	$\frac{SSB}{b-1}$
Treatmen-ts	$k-1$	SST	$\frac{SST}{k-1}$
Error	$(b-1)(k-1) = n-b-k+1$	SSE	$\frac{SST}{n-b-k+1}$
Total	$n-1$	$Total\ SS$	

$$H_0: \alpha_i = 0,\ i = 1, \ldots, k \text{ vs. } H_a: \text{Not all } \alpha'_i s \text{ are zero.}$$

We use the F-statistic where $S_i^2 = \frac{1}{n-1}\sum_{j=1}^{n_i}(Y_{ij} - \overline{T}_i)^2$ provides an independent, unbiased estimator for σ^2 in each of the k samples.

$$F = \frac{MST}{MSE},$$

where $S_i^2 = \frac{1}{n-1}\Sigma_{j=1}^{n_i}(Y_{ij} - \overline{T}_i)^2$ provides an independent, unbiased estimator for σ^2 in each of the k samples.

Although blocking lowers the experimental error, it also furnishes a chance to see whether evidence exists to indicate a difference in the mean response for blocks. In this case, we will be testing the hypothesis

$$H_0: \beta_j = 0, j = 1, \ldots, b \text{ versus } H_a: \text{Not all } \beta'_j s \text{ are zero.}$$

Under the assumption that there is no difference in the mean response for blocks, MSB provides an unbiased estimator for σ^2 based on $(b - 1)$ degrees of freedom. If there is a real difference that exists among block means, *MSB* will be larger in comparison with *MSE* and

$$F = \frac{MSB}{MSE},$$

will be used as a test statistic. The rejection region will be if $F > F_\alpha$ based on $(b - 1)$ numerator and $(n - b - k + 1)$ denominator degrees of freedom.

We now summarize the foregoing methodology in a step-by-step computational procedure. For a reasonable data size, we could use scientific calculators for handling the ANOVA calculations. For larger data sets, the use of statistical software packages is recommended.

Computational procedure for randomized block design

1. Calculate the following quantities:

(i) Sum the observations for each row to form row totals:

$$T_1, T_2, \ldots, T_k, \text{ where } T_i = \sum_{j=1}^{b} y_{ij}.$$

(ii) Sum the observations for each column to form column totals:

$B_1, B_2, \ldots, B_b$, where $B_j = \Sigma_{i=1}^{k} y_{ij}$.

(iii) Find the sum of all observations:

$$\sum_{j=1}^{b}\sum_{i=1}^{k} y_{ij} = \sum_{j=1}^{b} B_j.$$

2 Calculate the following quantities:

(i) Square the sum of the totals for each column and divide it by $n = bk$ to obtain

$$CM = \frac{1}{n}\left(\sum_{j=1}^{b} B_j\right)^2.$$

(ii) Find the sum of squares of the totals of each column and divide it by k to obtain

$$\frac{1}{k}\sum_{j=1}^{b} B_j^2$$

and

$$SSB = \frac{\Sigma_{j=1}^{b} B_j^2}{k} - CM \quad \text{and} \quad MSB = \frac{SSB}{b-1}.$$

(iii) Find the sum of squares of the totals of each row and divide it by b to obtain

$$\frac{\Sigma_{i=1}^{k} T_i^2}{b},$$

and

$$SST = \frac{\Sigma_{i=1}^{k} T_j^2}{b} - CM, \text{ and } MSB = \frac{SST}{k-1}.$$

(iv) Find the sum of squares of individual observations:

$$\sum_{j=1}^{b}\sum_{i=1}^{k} y_{ij}^2.$$

Also compute

$$Total\ SS = \sum_{j=1}^{b}\sum_{i=1}^{k} y_{ij}^2 - CM.$$

(cont'd)

(v) Using (ii), (iii), and (iv), find

$$SSE = Total\ SS - SSB - SST \quad \text{and} \quad MSE = \frac{SSE}{n-b-k+1}.$$

3 To test the null hypothesis that there is no difference in treatment means:

(i) Compute the F-statistic,

$$F = \frac{MST}{MSE}.$$

(ii) From the F-table, find the value of $F_{\alpha,\upsilon_1,\upsilon_2}$, where $\upsilon_1 = (k-1)$ is the numerator and $\upsilon_2 = (n-b-k+1)$ the denominator degrees of freedom.

(iii) Decision: Reject H_0 if $F > F_{\alpha,\upsilon_1,\upsilon_2}$ and conclude that there is evidence to conclude that there is a difference in treatment means at level α.

4 To test the null hypothesis that there is no difference in the mean response for blocks,

(i) Compute the F-statistic,

$$F = \frac{MSB}{MSE}.$$

(ii) From the F-table, find the value of $F_{\alpha,\upsilon_1,\upsilon_2}$, where $\upsilon_1 = (b-1)$ is the numerator and $\upsilon_2 = (n-b-k+1)$ the denominator degrees of freedom.

(iii) Decision: Reject H_0 if $F > F_{\alpha,\upsilon_1,\upsilon_2}$ and conclude that there is evidence to conclude there is a difference in the mean response for blocks at level α.

Assumptions: The samples are randomly selected in an independent manner from $n = bk$ populations. The populations are assumed to be normally distributed with equal variances σ^2. Also, there are no interactions between the variables (two factors).

We have already discussed the assumptions and how to verify those assumptions in one-way analysis. The only new assumption in the randomized blocked design is about the interactions. One of the ways to verify the assumption of no interaction is to plot the observed values against the sample number. If there is no interaction, the line segments (one for each block) will be parallel or nearly parallel; see Fig. 9.2. If the lines are not approximately parallel, then there is likely to be interaction between blocks and treatments. In the presence of interactions, the analysis of this section needs to be modified. For details on those procedures, refer to more specialized books on ANOVA methods.

We illustrate the randomized block design procedure with the following example.

EXAMPLE 9.4.1

A furniture company wants to know whether there are differences in stain resistance among the four chemicals used to treat three different fabrics. Table 9.15 shows the yields of resistance to stain (a low value indicates good stain resistance).

At the $\alpha = 0.05$ level of significance, is there evidence to conclude that there is a difference in mean resistance among the four chemicals? Is there any difference in the mean resistance among the materials? Give bounds for the p-values in each case.

Solution

Here $T_1 = 16$, $T_2 = 28$, $T_3 = 14$, *and* $T_4 = 24$. *Also,* $B_1 = 21$, $B_2 = 32$, *and* $B_3 = 29$. *In addition,* $b = 3$, $k = 4$, *and* $n = bk = 12$. *Now*

$$CM = \frac{1}{n}\left(\sum_{j=1}^{b} B_j\right)^2 = \frac{1}{12}(82)^2 = 560.3333.$$

We can compute the following quantities:

$$SSB = \frac{\sum_{j=1}^{b} B_j^2}{k} - CM = \frac{2306}{4} - 560.3333 = 16.1667,$$

$$MSB = \frac{SSB}{b-1} = \frac{16.1667}{2} = 8.0834,$$

$$SST = \frac{\sum_{i=1}^{k} T_i^2}{b} - CM = \frac{1812}{3} - 560.3333 = 43.6667,$$

and

$$MST = \frac{SST}{k-1} = \frac{43.6667}{3} = 14.5556.$$

We have

$$\sum_{j=1}^{b}\sum_{i=1}^{k} y_{ij}^2 = 632.$$

Thus,

$$Total\ SS = \sum_{j=1}^{b}\sum_{i=1}^{k} y_{ij}^2 - CM = 632 - 560.3333 = 71.666,$$

$$SSE = Total\ SS - SSB - SST = 71.6667 - 16.1667 - 43.6667 = 11.8333,$$

TABLE 9.15 Stain Resistance Data of Four Chemicals and Three Different Fabrics.

Chemical	Material			
	I	II	III	Total
C_1	3	7	6	16
C_2	9	11	8	28
C_3	2	5	7	14
C_4	7	9	8	24
Total	21	32	29	82

(cont'd)

and

$$MSE = \frac{SSE}{n-b-k+1} = \frac{11.8333}{6} = 1.9722.$$

The F-statistic is

$$F = \frac{MST}{MSE} = \frac{14.5556}{1.9722} = 7.3804.$$

From the F-table, $F_{0.05,3,6} = 4.76$. *Because the observed value* $F = 7.3804 > 4.76$, *we reject the null hypothesis and conclude that there is a difference in mean resistance among the four chemicals. Because the F-value falls between* $\alpha = 0.025$ *and* $\alpha = 0.01$, *the p-value falls between 0.01 and 0.025. To test for the difference in the mean resistance among the materials,*

$$F = \frac{MSB}{MSE} = \frac{8.0834}{1.9722} = 4.0987.$$

From the F-table, $F_{0.05,2,6} = 5.14$. *Because the observed value of* $F = 4.098 < 5.14$, *we conclude that there is no difference in the mean resistance among the materials. Because the F-value falls between* $\alpha = 0.10$ *and 0.05, the p-value falls between 0.05 and 0.9.*

EXERCISES 9.4

9.4.1. Show that

$$\sum_{j=1}^{b}\sum_{i=1}^{k}(y_{ij}-\bar{y})^2 = \sum_{i=1}^{k}\sum_{j=1}^{b}(y_{ij}-\bar{T}_i-\bar{B}_j-\bar{y})^2 + b\sum_{i=1}^{k}(\bar{T}_i-\bar{y})^2 + k\sum_{j=1}^{b}(\bar{B}_j-\bar{y})^2.$$

[Hint: Use the identity $y_{ij}-\bar{y} = \left(y_{ij}-\bar{T}_i-\bar{B}_j-\bar{y}\right) + (\bar{T}_i-\bar{y}) + \left(\bar{B}_j-\bar{y}\right)$.]

9.4.2. Show the following:

(a) $E(MSE) = \sigma^2$,

(b) $E(MSB) = \frac{k}{b-1}\sum_{j=1}^{b} B_j^2 + \sigma^2$,

(c) $E(MST) = \frac{b}{k-1}\sum_{i=1}^{k} \tau_i^2 + \sigma^2$.

9.4.3. The least-square estimators of the parameters μ, τ_i's, and β_j's are obtained by minimizing the sum of squares

$$W = \sum_{i=1}^{k} \sum_{j=1}^{b} (y_{ij} - \mu - \tau_i - \beta_j)^2,$$

with respect to μ, τ_i's, and β_j's; subject to the restrictions: $\Sigma_{i=1}^{k} \tau_i = \Sigma_{j=1}^{b} \beta_j = 0$. Show that the resulting estimators are

$$\hat{\mu} = \bar{y},$$
$$\hat{\tau}_i = \overline{T}_i - \bar{y},\ i = 1,2,\ldots,k,$$

and

$$\hat{\beta}_j = \overline{B}_j - \bar{y},\ j = 1,\ \ldots,\ b.$$

9.4.4. In order to test the wear on four hyperalloys, a test piece of each alloy was extracted from each of the three positions of a test machine. The reduction of weight in milligrams due to wear was determined on each piece, and the data are given in Table 9.16.

TABLE 9.16 Loss of Weights Due to Wear Testing of Four Materials (in mg).

	Position		
Type of Alloy	1	2	3
1	241	270	274
2	195	241	218
3	235	273	230
4	234	236	227

At α = 0.05, test the following hypotheses, regarding the positions as blocks:
(a) There is no difference in average wear for each material.
(b) There is no difference in average wear for each position.
(c) Interpret your final result and state any assumptions that were necessary to solve the problem.

9.4.5. For the data of Exercise 9.3.10, test at the 0.05 level that the true income lower limits of the top 5% of US households for each race are the same for all 5 years. Also, test at the 0.05 level that the true income lower limits of the top 5% of US households for each year between 1994 and 1998 are the same.

9.4.6. For the data of Exercise 9.3.11, test at the 0.01 level that the true mean cholesterol levels for all races in the United States during 1978–80 are the same. Also, test at the 0.01 level that the true mean cholesterol levels for all ages in the United States during 1978–1980 are the same.

9.4.7. In order to see the effect of hours of sleep on tests of different skill categories (vocabulary, reasoning, and arithmetic), tests consisting of 20 questions each in each category were given to 16 students, four each based on the hours of sleep they had on the previous night. Each right answer is given one point. Table 9.17 gives the cumulative scores of each of the four students in each category.
Test at the 0.05 level whether the true mean performance for different hours of sleep is the same. Also, test at the 0.05 level whether the true mean performance for each category of the test is the same.

9.4.8. We compare recognition accuracy (%) obtained from three different AI models (Factor A), Convolution Neural Network (CNN), Support Vector Machine (SVM), and Human, and four different emotion types (Factor B), Happy, Sad, Angry, Neutral. The data is given in Table 9.18.

TABLE 9.17 Scores Due to Sleeping for Different Categories.

Hours of Sleep	Category		
	Vocabulary	Reasoning	Arithmetic
0	44	33	35
4	54	38	18
6	48	42	43
8	55	52	50

TABLE 9.18 Percent Accuracy of Emotions for Different Models.

Model\Emotion	Happy	Sad	Angry	Neutral
CNN	88	80	75	85
SVM	82	76	70	80
Human	92	86	81	89

Test at the 0.05 level whether the true mean performance for different models is the same. Also, test at the 0.05 level whether the true mean performance for each emotional category of the test is the same.

9.5 MULTIPLE COMPARISONS

The ANOVA procedures that we have used so far showed whether differences among several means are significant. However, if the equality of means is rejected, the F-test did not pinpoint for us which of the given means or group of means differ significantly from another given mean or group of means. With ANOVA, when the null hypothesis of equality of means is rejected, the problem is to see whether there is some way to follow up (post hoc) this initial test H_0: $\mu_1 = \mu_2 = \ldots = \mu_k$ by looking at subhypotheses, such as H_0: $\mu_1 = \mu_2$.

This involves multiple tests. However, the solution is not to use a simple t-test repeatedly for every possible combination taken two at a time. That, apart from introducing many tests, will considerably increase the significance level, the probability of type I error. For example, to test four samples, we will need $\binom{4}{2} = 6$ tests. If each one of the comparisons is tested with the same value of $\alpha = P$ (type I error), and if all the null hypotheses involving six comparisons are true, then the probability of rejecting at least one of them is

$$P(\text{at least one type I error}) = 1 - (1 - \alpha)^6.$$

In particular, if $\alpha = 0.01$, then P(at least one type I error) = 0.077181, which is significantly higher than the original specified error value of 0.01.

One way to investigate the problem is to use a multiple comparison procedure. A good deal of work has been done on problems of multiple comparisons. There are a variety of methods available in the literature, such as the Bonferroni procedure, Tukey's method, and Scheffe's method. We now describe one of the more popular procedures called Tukey's method for completely randomized, one-factor design.

In this multiple comparison problem, we would like to test H_0: $\mu_i = \mu_j$ versus H_a: $\mu_i \neq \mu_j$, for all $i \neq j$. Tukey's method will be used to test all possible differences of means to decide whether at least one of the differences $\mu_i - \mu_j$ is considerably different from zero. In this comparison problem, Tukey's method makes use of confidence intervals for $\mu_i - \mu_j$. If each confidence interval has a confidence level $1 - \alpha$, then the probability that all confidence intervals include their respective parameters is less than $1 - \alpha$. We now describe this method where each of the k sample means is based on the common number of observations, n.

Let $N = kn$ be the total number of observations, and let

$$S^2 = \frac{1}{N-k} \sum_{i=1}^{k} \sum_{j=1}^{n_i=n} (Y_{ij} - \overline{T}_i)^2.$$

Let $\overline{T}_{\max} = \max(\overline{T}_1, \ldots, \overline{T}_k)$ and $\overline{T}_{\min} = \min(\overline{T}_1, \ldots, \overline{T}_k)$. Define the random variable

$$Q = \frac{\overline{T}_{\max} - \overline{T}_{\min}}{S\sqrt{n}}.$$

The distribution of Q under the null hypothesis H_0: $\mu_1 = \mu_k$ is called the Studentized range distribution, which depends on the number of samples k and the degrees of freedom $\upsilon = N - k = (n-1)k$. We denote the upper α critical value by $q_{\alpha,k,\upsilon}$. The Studentized range distribution table gives values for selected values of k, υ, and $\alpha = 0.01, 0.05$, and 0.9. The following theorem, due to Tukey, defines the test procedure.

Theorem 9.5.1. *Let $\overline{T}_i$, $i = 1, 2, \ldots, k$ be the k sample means in a completely randomized design. Let μ_i, $i = 1, 2, \ldots, k$ be the true means and let $n_i = n$ be the common sample size. Then the probability that all $\binom{k}{2}$ differences $\mu_i - \mu_j$ will simultaneously satisfy the inequalities.*

$$(\overline{T}_i - \overline{T}_j) - q_{\alpha,k,\upsilon}\frac{s}{\sqrt{n}} \le \mu_i - \mu_j \le (\overline{T}_i - \overline{T}_j) + q_{\alpha,k,\upsilon}\frac{s}{\sqrt{n}},$$

is $(1 - \alpha)$, where $q_{\alpha,k,\upsilon}$ is the upper α critical value of the Studentized range distribution. If, for a given i and j, zero is not contained in the preceding inequality, H_0: $\mu_i = \mu_j$ can be rejected in favor of H_a: $\mu_i \neq \mu_j$, at the significance level of α.

Now we give a step-by-step approach to implementing Tukey's method discussed above.

Procedure to find $(1-\alpha)100\%$ confidence intervals for difference of means with common sample size n: Tukey's method

1 There are $\binom{k}{2}$ comparisons of μ_i versus μ_j.

2 Compute the following quantities:

$$\overline{T}_i = \frac{\sum_{j=1}^{n_i} Y_{ij}}{n_i}, \quad i = 1,2, \ldots, k,$$

and

$$s^2 = \frac{1}{N-k} \sum_{i=1}^{k} \sum_{j=1}^{n_i=n} (Y_{ij} - \overline{T}_i)^2, \text{ where } N = kn.$$

3 From the Studentized range distribution table, find the upper α critical value, $q_{\alpha,k,\upsilon}$, where $\upsilon = N - k = (n-1)k$.

4 For each of $\binom{k}{2}$ pair (i, j), $i \neq j$, compute the Tukey's interval

$$\left(\left(\overline{T}_i - \overline{T}_j\right) - q_{\alpha,k,\upsilon}\frac{s}{\sqrt{n}}, \left(\overline{T}_i - \overline{T}_j\right) + q_{\alpha,k,\upsilon}\frac{s}{\sqrt{n}}\right).$$

5 Let NR denote insufficient evidence for rejecting H_0. Create the following table for each of $\binom{k}{2}$ pairwise difference $\mu_i - \mu_j$, $i \neq j$, and do not reject if the Tukey interval contains the number 0. Otherwise reject.

Table 9.19 is used to summarize the final calculations of the Tukey method.

In practice, there are now numerous statistical packages available for Tukey's purpose. The following example is solved using software.

TABLE 9.19 Summary of Tukey Method.

$\mu_i - \mu_j$		$\bar{T}_i - \bar{T}_j$	Tukey Interval	Observation	Conclusion
$\mu_1 - \mu_2$		$\bar{T}_1 - \bar{T}_2$	...	Doesn't contain 0	Reject
$\mu_1 - \mu_3$		$\bar{T}_1 - \bar{T}_3$	...	Contains 0	Do not reject
.		.	.	.	.
.		.	.	.	.
.		.	.	.	.

EXAMPLE 9.5.1

Table 9.20 shows the 1-year percentage total return of the top five stock funds for five different categories (source: *Money*, July 2000). Which categories have similar top returns, and which are different? Use 95% Tukey's confidence intervals.

TABLE 9.20 1-year Percentage Total Return of Five Stock Funds.

Large-cap	Mid-cap	Small-cap	Hybrid	Specialty
19.1	299.8	153.8	68.3	181.6
102.9	139.0	139.8	67.1	159.3
93.1	131.2	138.3	42.5	138.3
83.0	19.5	121.4	40.0	132.6
83.3	129.2	135.9	41.0	135.7

Solution

For simplicity of computation, we will use SPSS. The following is the output.

One-way

ANOVA

RETURN

	Sum of Squares	df	Mean Square	*F*	Sig.
Between Groups	41243.698	4	1039.925	7.397	.001
Within Groups	27877.580	20	1393.879		
Total	69121.278	24			

Note that since the p-value is 0.001, we are rejecting the null hypothesis that all means are equal. To find out which of the means might be different, we use the multiple comparison output.

Post Hoc Tests
Multiple Comparisons
Dependent Variable: RETURN
Tukey HSD

(I) FUND	(J) FUND	Mean Difference (I-J)	Std. Error	Sig.	95% Confidence Interval Lower Bound	Upper Bound
1.00	2.00	−67.4600	23.61253	.066	−138.1175	3.1975
	3.00	−43.3600	23.61253	.382	−114.0175	27.2975
	4.00	42.7000	23.61253	.396	−27.9575	113.3575
	5.00	−55.0200	23.61253	.177	−125.6775	15.6375
2.00	1.00	67.4600	23.61253	.066	−3.1975	138.1175
	3.00	24.1000	23.61253	.843	−46.5575	94.7575
	4.00	19.1600*	23.61253	.001	39.5025	180.8175
	5.00	12.4400	23.61253	.984	−58.2175	83.0975
3.00	1.00	43.3600	23.61253	.382	−27.2975	114.0175
	2.00	−24.1000	23.61253	.843	−94.7575	46.5575
	4.00	86.0600*	23.61253	.012	15.4025	156.7175
	5.00	−11.6600	23.61253	.987	−82.3175	58.9975
4.00	1.00	−42.7000	23.61253	.396	−113.3575	27.9575
	2.00	−19.1600*	23.61253	.001	−180.8175	−39.5025
	3.00	−86.0600*	23.61253	.012	−156.7175	−15.4025
	5.00	−97.7200*	23.61253	.004	−168.3775	−27.0625
5.00	1.00	55.0200	23.61253	.177	−15.6375	125.6775
	2.00	−12.4400	23.61253	.984	−83.0975	58.2175
	3.00	11.6600	23.61253	.987	−58.9975	82.3175
	4.00	97.7200*	23.61253	.004	27.0625	168.3775

*The mean difference is significant at the 0.05 level.

Homogeneous Subsets
RETURN
Tukey HSD[a]

FUND	N	Subset for alpha = .05 1	2
4.00	5	51.7800	
1.00	5	94.4800	94.4800
3.00	5		137.8400
5.00	5		149.5000
2.00	5		161.9400
Sig.		.396	.066

Means for groups in homogeneous subsets are displayed.
[a]Uses Harmonic Mean Sample Size = 5.000.

The Tukey intervals for pairwise differences ($\mu_i - \mu_j$) are in the foregoing computer printout. For example, the Tukey interval for ($\mu_1 - \mu_2$) is (−138.1, 3.2) and for ($\mu_2 - \mu_4$) is (39.5, 180.8). Also, sample mean and standard deviation are given in the output. For example, 94.48 is the sample mean of the five data points of large-cap funds, and 11.97 is the sample standard deviation of the five data points of large-cap funds.

If the Tukey interval for a particular difference ($\mu_j - \mu_i$) contains the number zero, we do not reject the H_0: $\mu_i = \mu_j$. Otherwise, we reject the H_0: $\mu_i = \mu_j$. For example, the interval for ($\mu_4 - \mu_2$) is (39.5–180.8) and does not contain zero. Hence, we reject H_0: $\mu_4 = \mu_2$.

The complete table corresponding to step 5 is produced in Table 9.21, *where N.R. represents "not reject."*

TABLE 9.21 Tukey Table for Stock Fund Returns.

$\mu_i - \mu_j$	$\bar{T}_i - \bar{T}_j$	Tukey Interval	Reject or N.R.	Conclusion
$\mu_1 - \mu_2$	161.94–94.48	(−138.1, 3.2)	N.R.	$\mu_1 = \mu_2$
$\mu_1 - \mu_3$	137.84–94.48	(−114.0, 27.3)	N.R.	$\mu_1 = \mu_3$
$\mu_2 - \mu_3$	137.84–161.94	(−46.6, 94.8)	N.R.	$\mu_3 = \mu_2$
$\mu_1 - \mu_4$	51.78–94.48	(−27.9, 113.3)	N.R.	$\mu_4 = \mu_1$
$\mu_2 - \mu_4$	51.78–161.94	(39.5, 180.8)	R	$\mu_4 \neq \mu_1$
$\mu_3 - \mu_4$	51.78–137.84	(15.4, 156.7)	R	$\mu_4 \neq \mu_3$
$\mu_1 - \mu_5$	149.50–94.98	(−125.6, 15.6)	N.R.	$\mu_5 = \mu_1$
$\mu_2 - \mu_5$	149.50–161.94	(−58.2, 83.1)	N.R.	$\mu_5 = \mu_2$
$\mu_3 - \mu_5$	149.50–137.84	(−82.3, 59.0)	N.R.	$\mu_5 = \mu_3$
$\mu_4 - \mu_5$	149.50–51.78	(−168.3, −27.1)	R	$\mu_5 \neq \mu_4$

Based on the 95% Tukey intervals, the average top return of hybrid funds is different from those for mid-cap, small-cap, and specialty funds. All other returns are similar.

In Tukey's method, the confidence coefficient for the set of all pairwise comparisons $\{\mu_i - \mu_j\}$ is exactly equal to $1 - \alpha$ when all sample sizes are equal. For unequal sample sizes, the confidence coefficient is greater than $1 - \alpha$. In this sense, Tukey's procedure is conservative when the sample sizes are not equal. In the case of unequal sample sizes, one has to estimate the standard deviation for each pairwise comparison. Tukey's procedure for unequal sample sizes is sometimes referred to as the Tukey–Kramer method.

EXERCISES 9.5

9.5.1 A large insurance company wants to determine whether there is a difference in the average time to process claim forms among its four different processing facilities. The data in Table 9.22 represent weekly average number of days to process a form over a period of 4 weeks.

TABLE 9.22 Average Number of Days to Process a Form Among Four Different Facilities.

Facility 1	Facility 2	Facility 3	Facility 4
1.50	2.25	1.30	2.0
0.9	1.85	2.75	1.5
1.12	1.45	2.15	2.85
1.95	2.15	1.55	1.15

(a) Test whether there is a difference in the average processing times at the 0.05 level.
(b) Test whether there is a difference using Tukey's method to find which facilities are different.
(c) Interpret your results and state any assumptions you have made in solving the problem.

9.5.2 Table 9.23 gives the rental vacancy rates by US region (source: US Census Bureau) for 5 years.
(a) Test at the 0.01 level whether the true rental vacancy rates by region are the same for all 5 years.

TABLE 9.23 Vacancy Rates in the U.S. by Region.

Rental Units	1995	1996	1997	1998	1999
Northeast	7.2	7.4	6.7	6.7	6.3
Midwest	7.2	7.9	8.0	7.9	8.6
South	8.3	8.6	9.1	9.6	10.3
West	7.5	7.2	6.6	6.7	6.2

(b) If there is a difference, use Tukey's method to find which regions are different.

9.5.3 Table 9.24 gives lower limits of income (approximated to nearest $1000 and calculated as of March of the following year) by race for the top 5% of US households from 1994 to 1998. (Source: US Census Bureau.)

TABLE 9.24 Lower limits of income by race for the top 5% of households (1994–1998).

Race	1994	1995	1996	1997	1998
All races	110	113	120	127	132
White	113	117	123	130	136
Black	81	80	85	87	94
Hispanic	82	80	86	93	98

(a) Test at the 0.05 level whether the true lower limits of income for the top 5% of US households for each race are the same for all 5 years.

(b) If there is a difference, use Tukey's method to find which is different.

(c) Interpret your results and state any assumptions you have made in solving the problem.

9.5.4 The data in Table 9.25 represent the mean serum cholesterol levels (given in milligrams per deciliter) by race and age in the United States from 1978 to 1980 (source: "Report of the National Cholesterol Education Program Expert Panel on Detection, Evaluation, and Treatment of High Blood Cholesterol in Adults," Arch. Intern. Med. 148, Jan. 1988).

TABLE 9.25 Mean cholesterol levels by race and age.

Race	Age					
	20–24	25–34	35–44	45–54	55–64	65–74
All races	180	199	217	227	229	221
White	180	199	217	227	230	222
Black	171	199	218	229	223	217

(a) Test at the 0.01 level whether the true mean cholesterol levels for all races in the United States during 1978–1980 are the same.

(b) If there is a difference, use Tukey's method to find which of the races are different with respect to the mean cholesterol levels.

9.6 CHAPTER SUMMARY

In this chapter, we have introduced the basic idea of analyzing various experimental designs. In Section 9.3, we explained the one-way ANOVA for the hypothesis testing problem for more than two means (different treatments being applied, or different populations being sampled). The two-way ANOVA, having *b* blocks and *k* treatments consisting of *b* blocks of *k* experimental units each, is discussed in Section 9.5. We also describe one popular procedure called Tukey's method for completely randomized, one-factor design for multiple comparisons. We saw in Chapter 8 that there are other possible designs, such as the Latin square design or Taguchi methods. We refer to specialized books on experimental design (Hicks and Turner) for more details on how to conduct ANOVA on such designs. In the final section, we give some computational examples.

We now list some of the key definitions introduced in this chapter:

Completely randomized experimental design
Randomized block design
Studentized range distribution
Tukey–Kramer method

In this chapter, we also learned the following important concepts and procedures:

ANOVA procedure for two treatments
One-way ANOVA for $k \geq 2$ populations
One-way ANOVA procedure for $k \geq 2$ populations
Procedure to find $(1 - \alpha)$ 100% confidence intervals for difference of means with common sample size n; Tukey's method
Computational procedure for randomized block design

9.7 COMPUTER EXAMPLES

SPSS, SAS, and other statistical programming packages are especially useful when we perform an ANOVA. As we have experienced in earlier sections, an ANOVA computation is very tedious to complete by hand.

9.7.1 Examples Using R

EXAMPLE 9.7.1 One-way ANOVA

The three following random samples in the table are independently obtained from three different normal populations with equal variances. At the $\alpha = 0.05$ level of significance, test for equality of means.

Sample	64	84	75	77	80	56	74	69	81	92	84
Group	1	1	1	1	1	2	2	2	3	3	3

This example assumes you've stored the data into two variables, sample and group. Please modify your code appropriately.
R Code:

```
model=lm(sample ~as.factor(group ));
anova(model);
```

Notice we must use as.factor() to get the proper degrees of freedom

Output:

```
Analysis of Variance Table
Response: sample
                    Df   Sum      Sq Mean    Sq  F  value    Pr(>F)
as.factor(group)     2   560.67  280.333   4.8403         0.04192  *
Residuals                8 463.33 57.917
--
Signif. codes:       0 '***' 0.001 '**' 0.01 '*' 0.05 '.' 0.1 ' ' 1
```

Since p-value is less than 0.05, we reject H_0 of equal means.

EXAMPLE 9.7.2 Two-way ANOVA

A furniture company wants to know whether there are differences in stain resistance among the four chemicals used to treat three different fabrics. The following table shows the yields on resistance to stain (a low value indicates good stain resistance). At the $\alpha = 0.05$ level of significance, is there evidence to conclude that there is a difference in mean resistance among the four chemicals? Is there any difference in the mean resistance among the materials?

Chemical	1	2	3	4	1	2	3	4	1	2	3	4
Resistance	3	9	2	7	7	11	5	9	6	8	7	8
Material	1	1	1	1	2	2	2	2	3	3	3	3

This example assumes you've stored data into three variables: chemical, resistance, and material. Please modify your code appropriately.

R Code:

```
model=lm(resistance~as.factor(chemical)+as.factor(material));
anova(model);
```

Output:

```
Analysis of Variance Table

Response: resistance
                     Df Sum Sq Mean Sq F value  Pr(>F)
as.factor(chemical)   3 43.667 14.5556  7.3803 0.01943 *
as.factor(material)   2 16.167  8.0833  4.0986 0.07548 .
Residuals             6 11.833  1.9722
---
Signif. codes:  0 '***' 0.001 '**' 0.01 '*' 0.05 '.' 0.1 ' ' 1
```

P-values suggest that the chemical is a significant factor but the material is not.

EXAMPLE 9.7.3 Tukey's Method

The following table shows the 1-year percentage total return of the top five stock funds for five different categories (source: *Money*, July 2000). Which categories have similar top returns, and which are different? Use 95% Tukey's confidence intervals. This example assumes you've stored your data into "stocks" and "groups" variables that pair.

Large-cap	Mid-cap	Small-cap	Hybrid	Specialty
19.1	299.8	153.8	68.3	181.6
102.9	139.0	139.8	67.1	159.3
93.1	131.2	138.3	42.5	138.3
83.0	19.5	121.4	40.0	132.6
83.3	129.2	135.9	41.0	135.7

This assumes your "stocks" variable is typed in column by column, top to bottom.

R Code:

```
groups=c(rep("Large-cap",5),rep("Mid-cap",5),rep("Small-
cap",5),rep("Hybrid",5),rep("Specialty",5));
model=aov(stocks~as.factor(groups));
TukeyHSD(model);
```

Output:

```
  Tukey multiple comparisons of means
    95% family-wise confidence level

Fit: aov(formula = stocks ~ as.factor(groups))

$`as.factor(groups)`
                      diff        lwr       upr     p adj
Large-cap-Hybrid      42.70   -27.957536  113.35754 0.3963504
Mid-cap-Hybrid       19.16    39.502464  180.81754 0.0012546*
Small-cap-Hybrid      86.06    15.402464  156.71754 0.0124242*
Specialty-Hybrid      97.72    27.062464  168.37754 0.0041271*
Mid-cap-Large-cap     67.46    -3.197536  138.11754 0.0657451
Small-cap-Large-cap   43.36   -27.297536  114.01754 0.3816028
Specialty-Large-cap   55.02   -15.637536  125.67754 0.1765264
Small-cap-Mid-cap    -24.10   -94.757536  46.55754  0.8429013
Specialty-Mid-cap    -12.44   -83.097536  58.21754  0.9835150
Specialty-Small-cap   11.66   -58.997536  82.31754 0.9870429
```

This shows that mean returns for hybrid to mid-cap, small-cap, and specialty are different.

9.7.2 SPSS Examples

EXAMPLE 9.7.4

Conduct a one-way ANOVA for the data of Example 9.7.1. Use $\alpha = 0.05$ level of significance, and test for equality of means.

Solution

In SPSS, we need to enter the data in a special way. First name column ***C1*** *as* **Sample**, *and column* **C2** *as* **Values**. *In the Sample column, enter the numbers to identify from which group the data comes. In this case, enter 1 in the first five rows, 2 in the next three rows, and 3 in the last three rows. In the Values column, enter sample 1 data in the first five rows, sample 2 data in the next five rows, and sample 3 data in the last three rows. Then:*

Analyze > Compare Means > One-way ANOVA... > Bring ***Values*** to ***Dependent List***: and **Sample** to ***Factor***: **>** ***OK***

EXAMPLE 9.7.5

Give the SPSS steps for using Tukey's method for the data of Example 9.5.1.

Solution

First name column C1 as ***Fund*** *and column C2 as* ***Return***. *In the Fund column, enter the numbers to identify from which group the data comes. In this case, the first four numbers are 1 (identifying 1 as the data belonging to large-cap), the next four numbers are 2, and so on, until the last four numbers are 5. In the Return column, enter large-cap return data in the first four rows, mid-cap data in the next four rows, and so on; the last four are from specialty. Then:*

Analyze > Compare Means > One-way ANOVA... > Bring ***Return*** to ***Dependent List***: and ***Fund*** to ***Factor***: **>** Click ***Post Hoc... >*** click ***Tukey*** **>** click ***Continue > OK***

We will get the output as in Example 9.5.1.

Interpretation of output is given in Example 9.5.1. When the treatment effects are significant, as in this example where the p-value is 0.001, the means must then be further examined to determine the nature of the effects. There are procedures called post hoc tests to assist the researcher in this task. For example, looking at the output column Sig., we could observe that there are significant differences in the mean returns between funds 2 and 4 and funds 4 and 5.

9.7.3 SAS Examples

EXAMPLE 9.7.6

Using SAS, conduct a one-way ANOVA for the data of Example 9.7.1. Use $\alpha = 0.05$ level of significance, and test for equality of means.

Solution

We could use the following code.

```
Options nodate nonumber;
options ls=80 ps=50;
DATA Scores;
INPUT Sample Value @@;
DATALINES;
```

1	64	1	84	1	75	1	77	1	80
2	56	2	74	2	69				
3	81	3	92	3	84				

```
;
PROC ANOVA DATA=Scores;
TITLE 'ANOVA for Scores';
CLASS Sample;
MODEL Value=Sample;
MEANS Sample;
RUN;
```

We could have used PROC GLM instead of PROC ANOVA to perform the ANOVA procedure. Usually, PROC ANOVA is used when the sizes of the samples are equal; otherwise PROC GLM is more desirable. The next example will show how to do the multiple comparison using Tukey's procedure.

EXAMPLE 9.7.7

Give the SAS commands for using Tukey's method for the data of Example 9.5.1.

Solution

We could use the following code.

```
Options nodate nonumber;
options ls=80 ps=50;
DATA Mfundrtn;
INPUT Fund Return @@;
DATALINES;
```

1 19.1	2 299.8	3 153.8	4 68.3	5 181.6
1 102.9	2 139.0	3 139.8	4 67.1	5 159.3
1 93.1	2 131.2	3 138.3	4 42.5	5 138.3
1 83.3	2 129.2	3 135.9	4 41.0	5 135.7
1 83.0	2 19.5	3 121.4	4 40.0	5 132.6

```
PROC GLM DATA=Mfundrtn;
TITLE 'ANOVA for Mutual fund returns';
CLASS Fund;
MODEL Return=Fund;
MEANS Fund / tukey;
RUN;
```

ANOVA for Mutual fund returns
The GLM Procedure
Class Level Information

Class		Levels	Values
Fund		5	1 2 3 4 5
	Number of observations		25

ANOVA for Mutual fund returns
The GLM Procedure
Dependent Variable: Return

Source	DF	Sum of Squares	Mean Square	F Value	Pr > F
Model	4	41243.69840	1039.92460	7.40	0.0008

Error	20	27877.58000	1393.87900
Corrected Total	24	69121.27840	

R-Square	Coeff Var	Root MSE	Return Mean
0.596686	31.34524	37.33469	119.1080

Source	DF	Type I SS	Mean Square	F Value	Pr > F
Fund	4	41243.69840	1039.92460	7.40	0.0008
Source	DF	Type III SS	Mean Square	F Value	Pr > F
Fund	4	41243.69840	1039.92460	7.40	0.0008

ANOVA for Mutual fund returns

The GLM Procedure

Tukey's Studentized Range (HSD) Test for Return

NOTE: This test controls the Type I experiment wise error rate, but it generally has a higher Type II error rate than REGWQ.

Alpha	0.05
Error Degrees of Freedom	20
Error Mean Square	1393.879
Critical Value of Studentized Range	4.23186
Minimum Significant Difference	70.658

Means with the same letter are not significantly different.

Tukey Grouping	Mean	N	Fund
A	161.94	5	2
A			
A	149.50	5	5
A			
A	137.84	5	3
A			
A	94.48	5	1
B			
B	51.78	5	4

The GLM Procedure

Tukey's Studentized Range (HSD) Test for Value

NOTE: This test controls the Type I experiment wise error rate, but it generally has a higher Type II error rate than REGWQ.

Alpha	0.05
Error Degrees of Freedom	20
Error Mean Square	1393.879
Critical Value of Studentized Range	4.23186
Minimum Significant Difference	70.658

Means with the same letter are not significantly different.

Tukey Grouping		Mean	N	Sample
	A	161.94	5	2
A				
	A	149.50	5	5
A				
	A	137.84	5	3
A				
B	A	94.48	5	1
B				
	B	51.78	5	4

Looking at the p-value of 0.008, which is less than α = 0.05, we conclude that there is a difference in mutual fund returns.

In the previous example, we used the post hoc test Tukey. We could have used other options such as DUNCAN, SNK, LSD, and SCHEFFE. The test is performed at the default value of α = 0.05. If we want to specify, say, α = 0.01, or 0.1, we could have done so by using the command MEANS Fund / Tukey ALPHA=0.01.

If we need all the confidence intervals in the Tukey method, in the code just given, we have to modify "MEANS Fund / Tukey;" to "MEANS Fund / LSD TUKEY CLDIFF;"
which will result in the following output.

```
ANOVA for Mutual fund returns
The GLM Procedure
Class Level Information
```

Class	levels	Values
Fund	5	1 2 3 4 5

```
Number of observations 25
ANOVA for Mutual fund returns
The GLM Procedure
Dependent Variable: Return
```

Source	DF	Sum of Squares	Mean Square	F Value	Pr > F
Model	4	41243.69840	1039.92460	7.40	0.0008
Error	20	27877.58000	1393.87900		

Corrected Total 24	69121.27840

R-Square	Coeff Var	Root MSE	Return Mean
0.596686	31.34524	37.33469	119.1080

Source	DF	Type I SS	Mean Square	F Value	Pr > F
Fund	4	41243.69840	1039.92460	7.40	0.0008
Source	DF	Type III SS	Mean Square	F Value	Pr > F
Fund	4	41243.69840	1039.92460	7.40	0.0008

```
ANOVA for Mutual fund returns
The GLM Procedure
```

t-tests (LSD) for Return
NOTE: This test controls the Type I comparison wise error rate, not the experiment wise error rate.

Alpha	0.05
Error Degrees of Freedom	20
Error Mean Square	1393.879
Critical Value of t	2.08596
Least Significant Difference	49.255

Comparisons significant at the 0.05 level are indicated by ***.

Fund Comparison	Between Means	Difference 95% Confidence Limits		
2 - 5	12.44	−36.81	61.69	
2 - 3	24.10	−25.15	73.35	
2 - 1	67.46	18.21	116.71	***
2 - 4	19.16	60.91	159.41	***
5 - 2	−12.44	−61.69	36.81	
5 - 3	11.66	−37.59	60.91	
5 - 1	55.02	5.77	104.27	***
5 - 4	97.72	48.47	146.97	***
3 - 2	−24.10	−73.35	25.15	
3 - 5	−11.66	−60.91	37.59	
3 - 1	43.36	−5.89	92.61	
3 - 4	86.06	36.81	135.31	***
1 - 2	−67.46	−116.71	−18.21	***
1 - 5	−55.02	−104.27	−5.77	***
1 - 3	−43.36	−92.61	5.89	
1 - 4	42.70	−6.55	91.95	
4 - 2	−19.16	−159.41	−60.91	***
4 - 5	−97.72	−146.97	−48.47	***
4 - 3	−86.06	−135.31	−36.81	***
4 - 1	−42.70	−91.95	6.55	

ANOVA for Mutual fund returns
The GLM Procedure
Tukey's Studentized Range (HSD) Test for Return
NOTE: This test controls the Type I experiment wise error rate.

Alpha	0.05
Error Degrees of Freedom	20
Error Mean Square	1393.879
Critical Value of Studentized Range	4.23186
Least Significant Difference	70.658

Comparisons significant at the 0.05 level are indicated by ***.

Fund Comparison	Difference Between Means	Simultaneous 95% Confidence Limits		
2 - 5	12.44	−58.22	83.10	
2 - 3	24.10	−46.56	94.76	
2 - 1	67.46	-3.20	138.12	
2 - 4	19.16	39.50	180.82	***
5 - 2	−12.44	−83.10	58.22	
5 - 3	11.66	−59.00	82.32	
5 - 1	55.02	-15.64	125.68	
5 - 4	97.72	27.06	168.38	***
3 - 2	−24.10	−94.76	46.56	

3 - 5	−11.66	−82.32	59.00	
3 - 1	43.36	−27.30	114.02	
3 - 4	86.06	15.40	156.72	***
1 - 2	−67.46	−138.12	3.20	
1 - 5	−55.02	−125.68	15.64	
1 - 3	−43.36	−114.02	27.30	
1 - 4	42.70	−27.96	113.36	
4 - 2	−19.16	−180.82	−39.50	***
4 - 5	−97.72	−168.38	−27.06	***
4 - 3	−86.06	−156.72	−15.40	***
4 - 1	−42.70	−113.36	27.96	

EXERCISES 9.7

9.7.1. For the data of Exercise 9.5.4, perform a one-way ANOVA using any of the software (R, Minitab, SPSS, or SAS).
9.7.2. For the data of Exercise 9.5.2, perform Tukey's test using any of the software (R, Minitab, SPSS, or SAS).
9.7.3. For the data of Exercise 9.5.4, perform Tukey's test using any of the software (R, Minitab, SPSS, or SAS).

PROJECTS FOR CHAPTER 9

9A. Transformations

The basic model for the ANOVA requires that the independent observations come from normal populations with equal variances. These requirements are rarely met in practice, and the extent to which they are violated affects the validity of the subsequent inference. Therefore it is important for the investigator to decide whether the assumptions are at least approximately satisfied and, if not, what can be done to rectify the situation. Hence it is necessary to (a) examine the data for marked departures from the model and, if necessary, (b) apply an appropriate transformation tc the data to bring it more in line with the basic assumptions.

A simple way to check for the equality of the population variances is to calculate the sample variances and plot against mean as in Fig. 9.3. If the graph suggests a relation between sample mean and variance, then the relation very likely exists between population mean and variance, and hence the population from which the samples are taken may very well be nonnormal and/or data is heterogenous. A simple visual check of heterogeneity can be done using the following type of scatter plots: mean versus variance across replicates.

If a study of sample means and variances reveals a marked departure from the model, the observations may be transformed into a new set to which the methods of ANOVA are better suited. Three commonly used transformations are the following:

(a) The logarithmic transformation: If the graph of sample means against sample variance suggests a relation of the form

$$s^2 = C(\overline{X}^2),$$

That is, $\sigma^2 = k\mu^2$, replace each observation X with its logarithm to the base 10,

$$Y = log_{10}X;$$

or, if some X-values are zero, by $Y = \log_{10}(X + 1)$.

(b) The square root transformation: If the relation is of the form

$$s^2 = C\overline{X}$$

That is, $\sigma^2 = k\mu$, replace X by its square root,

$$Y = \sqrt{X}$$

or, if the values of X are very close to zero, by the square root of (X + ½). This relation is found in data from Poisson populations, where the variance is equal to the mean.

(c) The angular transformation: If the observations are counts of a binomial nature, and $\hat{p}$ is the observed proportion, replace $\hat{p}$ by

$$\theta = \arcsin\sqrt{\hat{p}},$$

which is the principal angle (in degrees or radians) whose sine is the square root of $\hat{p}$.

(i) To check for the equality of the population variances, calculate the sample variances for each of the data sets given in the exercises of Section 9.3 and plot against the corresponding mean.

(ii) If there is an assumption violation, perform one of the transformations described earlier and do the ANOVA procedure for the transformed data.

9B. ANOVA With Missing Observations

In the two-way ANOVA, we assumed that each block cell has one treatment value. However, it is possible that some observations in some block cells may be missing for various reasons, such as that the investigator failed to record the observations, the subject discontinued participation in the experiment, or the subject moved to a different place or died prior to completion of the experiment. In those cases, this project gives a method of inserting estimates of the missing values.

Let $y_{..}$ denote the total of all kb observations. If the observation corresponding to the ith row and the jth column, which is denoted by $y_{ij\cdot}$, is missing, then all the sums of squares are calculated as before, except that the y_{ij} term is replaced by

$$\hat{y}_{ij} = \frac{bB'_j + kT'_i - y'_{..}}{(k-1)(b-1)},$$

where T'_i denotes the total of $b - 1$ observations in the ith row, B'_j denotes the total of $k - 1$ observations in the jth column, and $y'_{..}$ denotes the sum of all $kb - 1$ observations. Using calculus, one can show that $\hat{y}_{ij}$ minimizes the error sum of squares. One should not include these estimates when computing relevant degrees of freedom. With these changes, proceed to perform the analysis as in Section 9.4. For more details on the method, refer to Sahai and Ageel (2000, p. 145).

Perform the test of Example 9.4.1, now with a missing value for material III and chemical C_4. Does the conclusion change?

9C. ANOVA in Linear Models

In order to determine whether the multiple regression model introduced in Section 7.6 is adequate for predicting values of the dependent variable y, one can use the ANOVA F-test. The model is

$$Y = \beta_0 + \beta_1 x_1 + \beta_2 x_2 + \ldots + \beta_k x_k + \varepsilon,$$

where $\varepsilon = (\varepsilon_1, \varepsilon_2, \ldots, \varepsilon_n) \sim N(0, \sigma^2)$ and ε_i and ε_j are uncorrected if $i \neq j$. Define the multiple coefficient of determination, R^2, by

$$R^2 = 1 - \frac{\Sigma(y_i - \hat{y}_i)^2}{\Sigma(y_i - \bar{y})^2}.$$

The ANOVA F-Test

$$H_0: \beta_1 = \beta_2 = \ldots = \beta_k = 0 \text{ versus}$$
$$H_a: \text{At least one of the parameters, } \beta_1, \beta_2, \ldots, \beta_k, \text{ differs from 0.}$$

Test statistic:

$$F = \frac{\text{Mean square for model}}{\text{Mean square for error}}$$
$$= \frac{SS\,(\text{model})\,/\,k}{SSE\,/\,[n-(k+1)]}$$
$$= \frac{R^2\,/\,k}{(1-R^2)\,/\,[n-(k+1)]},$$

where,

n = number of observations
k = number of parameters in the model excluding β_0.

From the F-table, determine the value of F_α with k numerator d.f. and $n-(k+1)$ denominator d.f. Then the rejection region is $\{F > F_\alpha\}$.

If we reject the null hypothesis, then the model can be taken as useful in predicting values of y.

For the data of Example 7.6.2, test the overall utility of the fitted model

$$y = 66.12 - 0.3794X_1 + 21.4365X_2$$

using the F-test described earlier.

Chapter 10

Bayesian Estimation and Inference

Chapter outline

OBJECTIVE

To study Bayesian analysis methods and procedures that are becoming very popular in building statistical models for real-world problems.

The Reverend Thomas Bayes
(Source: http://en.wikipedia.org/wiki/Thomas_Bayes)

The Reverend Thomas Bayes (1702–61) was a Nonconformist minister. In the 1720s, Bayes started working on the theory of probability. Even though he did not publish any of his works on mathematics during his lifetime, Bayes was elected a Fellow of the Royal Society in 1742. His famous work titled "Essay toward solving a problem in the doctrine of

Mathematical Statistics with Applications in R. DOI: https://doi.org/10.1016/B978-0-443-27548-7.00010-4

chances" was published in the *Philosophical Transactions of the Royal Society of London* in 1764, after his death. The paper was sent to the Royal Society by Richard Price, a friend of Bayes. Another mathematical publication on asymptotic series also appeared after his death.

10.1 INTRODUCTION

Bayesian procedures are becoming increasingly popular in building statistical models for real-world problems. In recent years, Bayesian statistical methods have been increasingly used in scientific fields ranging from archeology to computing. Bayesian inference is a method of analysis that combines information collected from experimental data with the knowledge one has prior to performing the experiment. Bayesian and classical (frequentist) methods take basically different outlooks toward statistical inference. In this approach to statistics, the uncertainties are expressed in terms of probabilities. In the Bayesian approach, we combine any new information that is available with the prior information we have to form the basis for the statistical procedure. The classical approach to statistical inference that we have studied so far is based on the random sample alone. That is, if a probability distribution depends on a set of parameters θ, the classical approach makes inferences about θ solely on the basis of a sample $X_1,\ldots,X_n$. This approach to inference is based on the concept of sampling distribution. To correctly interpret traditional inferential procedures, it is necessary to fully understand the notion of a sampling distribution. In this approach, we analyze only one set of sample values. However, we have to imagine what could happen if we drew a large number of random samples from the population. For example, consider a normal sample with known variance. We have seen that a 95% confidence interval for the population mean μ is given by the random interval $(\overline{X} - 1.96\sigma/\sqrt{n}, \overline{X} + 1.96\sigma/\sqrt{n})$. This means that when samples are repeatedly taken from the population, at least 95% of the random intervals contain the true mean μ. The classical inferential approach does not use any of the prior information we might have as a result of, say, our familiarity with the problem, or information from earlier studies. Scientists and engineers are faced with the problem that there is typically only a single data set, and they need to determine the value of the parameter at the time the data are taken. The basic question then is, "What is the best estimate of a parameter one can make from the data using one's prior information?" Statistical approaches that use prior knowledge, possibly subjective, in addition to the sample evidence to estimate the population parameters are known as Bayesian methods.

Bayesian statistics provides a natural method for updating uncertainty in the light of evidence. Data are still assumed to come from a distribution belonging to a known parametric family. However, the Bayesian outlook toward inference is founded on the subjective interpretation of probability. Subjective probability is a way of stating our belief in the validity of a random event. The following example will illustrate the idea. Suppose we are interested in the proportion of all undergraduate students at a particular university who take on out-of-campus jobs for at least 20 hours a week. Suppose we randomly select, say, 50 students from this university and obtain the proportion of students who have out-of-campus jobs for at least 20 hours a week. Let us assume that the sample proportion is 30/50 = 0.6. In a frequentist approach, all of the inferential procedures, such as point estimation, interval estimation, or hypothesis testing, are based on the sampling distribution.

That is, even though we are analyzing only one data set, it is necessary to know the mean, standard deviation, and shape of this sampling distribution of the proportion for the correct interpretation, in classical inferential procedures. In the subjective interpretation of probability, the proportion of undergraduates who work on an out-of-campus job for at least 20 hours a week is assumed to be unknown and random. A probability distribution, called the prior, represents our knowledge or belief about the location of this proportion before any data collected is used. For instance, the college placement office already may have an opinion on this proportion based on its earlier experience. The classical approach ignores this prior knowledge, whereas the Bayesian approach incorporates this knowledge with the current observed data to update the value of this proportion. That is, after the data are collected, our opinion about the proportion may change. Using Bayes' rule, we will compute the posterior probability distribution for the proportion, based on our prior belief and evidence from the data. All of our inferences about the proportion are made by computing appropriate statistics of the posterior distribution.

The Bayesian approach seeks to optimally merge information from two sources: (1) knowledge that is known from theory or opinion formed at the beginning of the research in the form of a prior and (2) information contained in the data in the form of likelihood functions. Basically, the prior distribution represents our initial belief, whereas the information in the data is expressed by the likelihood function. Combining prior distribution and likelihood function, we can obtain the posterior distribution. This expresses our revised uncertainty in light of the data. The main difference between the Bayesian approach and the classical approach is that in the Bayesian setting, the parameter is viewed as a random variable, whereas the classical approach considers the parameter to be fixed but unknown. The parameter is random in the sense that we can assign to it a subjective probability distribution that describes our confidence about the actual value of the parameter.

Some of the reasons for Bayesian approaches are as follows: (1) Most Bayesian inferential conclusions are made conditional on the observed data. Unlike the traditional approach, one need not be concerned with data sets other than the one that is observed. There is no need to discuss sampling distributions using the Bayesian approach. Also, (2) from a Bayesian viewpoint, it is legitimate to talk about the probability that the proportion falls in a specific interval, say (0.2, 0.6), or the probability that a hypothesis is true. Too often, traditional inferential conclusions are misstated; for example, if a confidence interval computed from a sample for a parameter is (0.2, 0.6), it is common for the student to incorrectly state that the population parameter falls in the interval (0.2, 0.6) with probability at least 0.90. The Bayesian viewpoint provides a convenient model for implementing the scientific method. The prior probability distribution can be used to state initial beliefs about the population of interest, relevant sample data are collected, and the posterior probability distribution reflects one's new updated beliefs about the population parameter in light of the new data that were collected. All inferences about the parameter are made by computing appropriate summaries of the posterior probability distribution. Because of formidable theoretical and computational challenges, the Bayesian approach has found relatively limited use. Recent advances in Bayesian analysis, combined with the growing power of computers, are making Bayesian methods practical and increasingly popular. The Markov chain Monte Carlo (MCMC) method described in Section 13.5 is one of the computationally intensive methods that is often useful in Bayesian estimation.

10.2 BAYESIAN POINT ESTIMATION

The cornerstone of Bayesian methodology is the Bayes theorem. It helps us to update our beliefs in the form of probability statements about the parameters after the sample has been taken. The conditional distribution of the parameters after observing the data is called the *posterior distribution*, which integrates the prior and the sample information. Suppose we have two discrete random variables, X and Y. Then the joint probability function (pmf) can be written as $p(x, y) = p(x|y)p_Y(y)$, and the marginal probability density function of X is $p_X(x) = \Sigma_y\, p(x, y) = \Sigma_y\, p(x|y)p_Y(y)$. Then Bayes' rule for the conditional $p(y|x)$ is

$$p(y|x) = \frac{p(x, y)}{p_X(x)} = \frac{p(x|y)p_Y(y)}{p_X(x)} = \frac{p(x|y)p_Y(y)}{\Sigma_y p(x|y)p_Y(y)}.$$

The denominator in this expression is a fixed normalizing factor that ensures that the $\Sigma_y p(y|x) = 1$. If Y is continuous, the Bayes theorem can be stated as where the integral is over the range of values of y. These two equations are the Bayes formulas for random variables.

$$p(y|x) = \frac{p(x|y)p_Y(y)}{\int p(x|y)p_Y(y)dy},$$

In Bayesian terminology, $p_Y(y)$ represents the probability statement of our prior belief, $p(x|y)$ is the probability of the data x given our prior beliefs, which is called the likelihood, and the updated probability $p(y|x)$ is the posterior. Because $p_X(x)$ (which is the likelihood accumulated over all possible prior values) is independent of y, we can express the posterior distribution as proportional ($\propto$) to [(likelihood) × (prior distribution)], that is,

$$p(y|x) \propto p(x|y)p(y).$$

We use the notation $f(x|\theta)$ to represent a probability distribution whose population parameter is considered to be a random variable. One of the problems is of finding a point estimate of the parameter θ (possibly a vector) for the population with distribution $f(x|\theta)$, given θ. Because Θ is assumed to be a random variable, we can talk of the distribution of Θ. Assume that $\pi(\theta)$ is the prior distribution of θ, which reflected the experimenter's prior belief about θ. We will not distinguish between the scalars and vectors, which will be clear based on the specific situation. Suppose that we have a random sample $X = (X_1, \ldots, X_n)$ of size n from $f(x|\theta)$. Then the posterior distribution of Θ can be written as where $L(X_1,\ldots, X_n|\theta)$ is the likelihood function. Letting C represent all terms that do not involve θ (in this case, $C = 1/f(X_1, \ldots, X_n)$), we have

$$f\left(\theta \middle| X_1, \ldots, X_n\right) = \frac{f(\theta, X_1, \ldots, X_n)}{f(X_1, \ldots, X_n)} = \frac{L(X_1, \ldots, X_n|\theta)\pi(\theta)}{f(X_1, \ldots, X_n)},$$

$$f(\theta|X_1, \ldots, X_n) = CL(X_1, \ldots, X_n|\theta)\pi(\theta).$$

For specific sample values $X_1 = x_1, X_2 = x_2, \ldots, X_n = x_n$, the foregoing equation can be written in a compact form as

$$f(\theta|x) \propto f(x|\theta)\pi(\theta), \text{ where } \quad x = (x_1, x_2, \ldots, x_n).$$

This can be expressed as

(posterior distribution) ∝ (prior distribution) × (likelihood).

The full result, including the normalization, can be written as

(posterior distribution) = [(prior distribution) × (likelihood)]/[Σ(prior × likelihood)],

where the denominator is a fixed normalizing factor obtained by the likelihood accumulated over all possible prior values. We can now give a formal definition.

Definition 10.2.1. *The distribution of θ, given data $x_1, \ldots, x_n$, is called the* **posterior distribution**, *which is given by*

$$\pi(\theta|x) = \frac{f(x|\theta)\pi(\theta)}{g(x)}, \tag{10.1}$$

where $g(x)$ is the marginal distribution of X. The **Bayes estimate** *of the parameter θ is the posterior mean.*

The marginal distribution $g(x)$ can be calculated using the formulawhere $\pi(\theta)$ is the prior distribution of θ. Here, the marginal distribution $g(x)$ is also called the predictive distribution of X, because it represents our current predictions of the values of X, taking into account both the uncertainty about the value of θ and the residual uncertainty about the random variable X when θ is known.

$$g(x) = \begin{cases} \sum_{\theta} f(x|\theta)\pi(\theta), & \text{in discrete case} \\ \int_{-\infty}^{\infty} f(x|\theta)\pi(\theta)d\theta, & \text{in continuous case} \end{cases}$$

In a Bayesian setting, all the information about θ from the observed data and from prior knowledge is contained in the posterior distribution, $\pi(\theta|x)$. In almost all practical cases, because we are combining our prior information with the information contained in the data, the posterior distribution provides a more refined estimation of θ than the prior. All inferences from Bayesian methods are based on the posterior probability distribution of the parameter θ. Using the explanation given later, we will take the Bayes estimate of a parameter as the posterior mean.

Furthermore, consider a Bayesian statistical inference problem where the parameter is a population proportion. In the Bernoulli trials, the population contains two types called "successes" and "failures." The proportion of successes in the population is denoted by θ. We take a random sample of size n from the population and observe s successes and f failures. The goal is to learn about the unknown proportion θ on the basis of these data.

In this situation, a model is represented by the population proportion θ. We do not know its value. In Chapter 5, we have seen that we could use the maximum likelihood estimator (MLE) for estimating θ, which did not use any prior knowledge we may have about θ. Note that the maximum likelihood estimate is broadly equivalent to finding the mode of the likelihood. In a Bayesian setting, we represent our beliefs about the location of θ in terms of a prior probability distribution. We introduce proportion inference by using a discrete prior distribution for θ. We can construct a prior by specifying a list of possible values for the proportion θ, and then assigning probabilities to these values that reflect our knowledge about θ. Then the posterior probabilities can be computed using the Bayes theorem. The following example illustrates this concept.

EXAMPLE 10.2.1

It is believed that cross-fertilized plants produce taller offspring than self-fertilized plants. In order to obtain an estimate on the proportion θ of cross-fertilized plants that are taller, an experimenter observes a random sample of 15 pairs of plants that are exactly the same age. Each pair is grown in the same conditions with some cross-fertilized and the others self-fertilized. Based on previous experience, the experimenter believes that the following are possible values of θ and that the prior probability for each value of θ (prior weight) is $\pi(\theta)$.

θ: 0.80 0.82 0.84 0.86 0.88 0.90
$\pi(\theta)$: 0.13 0.15 0.22 0.25 0.15 0.10

From the experiment, it is observed that in 13 of 15 pairs, cross-fertilized is taller. Create a table with columns of the prior $\pi(\theta)$, likelihood of $L(X_1, X_2, \ldots, X_n|\theta)$ for different values of θ and for the given sample, prior times likelihood, and posterior probability of θ. Based on the posterior probabilities, what value of θ has the highest support? Also, find $E(\theta)$ based on the posterior probabilities.

Solution

The likelihood of obtaining 13 of 15 taller plants to the different prior values of π are given using the binomial pdf $\binom{15}{13}\theta^{13}(1-\theta)^2$. For example, if the prior value of θ is 0.80, then the likelihood of θ given the sample is

$$f(x|\theta) = \binom{15}{13}(0.8)^{13}(0.2)^2 = 0.2309.$$

TABLE 10.1 Bayesian calculation for Cross-fertilized and Offspring Plants.

Prior Values of θ	Prior Probability $\pi(\theta)$	Likelihood of θ Given Sample	Prior Times Likelihood	Posterior Probability of θ
0.80	0.13	0.2309	3.0017 × 10–2	0.11029
0.82	0.15	0.2578	0.03867	0.14208
0.84	0.22	0.2787	6.1314 × 10–2	0.22528
0.86	0.25	0.2897	7.2425 × 10–2	0.2661
0.88	0.15	0.2870	0.4305	0.15817
0.90	0.10	0.2669	0.02669	0.098064
		Total 0.27217		0.9998 ≈ 1.0

From Table 10.1, *we obtain Σ(prior × likelihood) = 0.27217. Hence, the normalized value corresponding to θ = 0.80 is the posterior probability $f(\theta|x)$, which is equal to (0.030017/0.27217) = 0.11029. Now, we can obtain the table of posterior distribution of a proportion π using the discrete prior given in* Table 10.1. *When we substitute in Bayes' rule, the factor $\binom{15}{13}$ would be canceled. Hence, in the calculation of the likelihood function, we could have just used $\theta^{13}(1-\theta)^2$ instead of the full expression $\binom{15}{13}\theta^{13}(1-\theta)^2$. Thus, the Bayesian estimate of θ is*

$$\begin{aligned} E(\theta) &= (0.8)(0.11029) + (0.82)(0.14028) + (0.84)(0.22528) \\ &\quad + (0.86)(0.2661) + (0.88)(0.15817) + (0.9)(0.098065) \\ &= 0.84879 \approx 0.85. \end{aligned}$$

It may be noted that the MLE of θ is 13/15 = 0.867.

In Example 10.2.1, the priors are called *informative priors*, because it favored certain values of θ; for example, for the value $\theta = 0.86$, the prior value of $\pi(\theta)$ is 0.25, which is higher than all the rest of the values. If there were no information or no strong prior opinions, then we could select a *noninformative prior*, which would have assigned equal prior probability of 1/6 to each of the possible values of θ. A noninformative prior (also called a *flat or uniform prior*) provides little or no information. In practice, a noninformative prior is used when we do not have any prior information, and still want to use the Bayes method. Thus, a uniform prior amounts to random choice from possible values of the parameter. Based on the situation, noninformative priors may be quite disperse, may avoid only impossible values of the parameter, and oftentimes give results similar to those obtained by classical frequentist methods.

EXAMPLE 10.2.2

Repeat the Example 10.2.1 using a noninformative prior, $\pi(\theta) = 1/6$, for each given value of θ.

Solution

Here $\pi(\theta) = 1/6$ for each value of θ. See Table 10.2.

TABLE 10.2 Bayesian calculation for Noninformative Priors.

Prior Values of θ	Prior Probability $\pi(\theta)$	Likelihood of θ Given Sample	Prior Times Likelihood	Posterior Probability of θ
0.80	1/6	0.2309	3.8483×10^{-2}	0.14333
0.82	1/6	0.2578	4.2967×10^{-2}	0.16003
0.84	1/6	0.2787	0.04645	0.173
0.86	1/6	0.2897	4.8283×10^{-2}	0.17982
0.88	1/6	0.2870	4.7833×10^{-2}	0.17815
0.90	1/6	0.2669	4.4483×10^{-2}	0.16567
		Total 0.2685		1.0

The Bayesian estimate for the noninformative prior is

$$\begin{aligned} E(\theta) &= (0.8)(0.14333) + (0.82)(0.16003) + (0.84)(0.173) \\ &+ (0.86)(0.17982) + (0.88)(0.17815) \\ &+ (0.9)(0.16567) = 0.85173. \end{aligned}$$

It should be noted that because the choice of priors in Example 10.2.1 is only mildly informative, we do not see much difference in the values of Bayesian estimates. In general, it is difficult to construct an acceptable prior, because most often it has to be based on subjective experiences. Therefore, it is relatively easy to use a "noninformative" prior. For example, if we have no information on the values of proportion θ, then one type of standard "noninformative" prior is to take the proportion θ as one of the equally spaced values $0, 0.1, 0.2, \ldots, 0.9, 1$. We can assign for each value of θ the same probability, $\pi(\theta) = 1/11$. This prior is convenient and may work reasonably well when we do not have many data. It is fairly easy to construct a prior when there exists considerable prior information about the proportion of interest.

The posterior distribution gives us information regarding the likelihood of values of θ given sample data. Then the question is how to use this information to estimate θ. Instead of having explicit probabilities, the prior may be given through an assumed probability distribution. We illustrate the calculations involved to find the posterior distribution in the following example.

EXAMPLE 10.2.3

Let X be a binomial random variable with parameters n and p. Assume that the prior distribution of p is uniform on [0,1]. Find the posterior distribution, $f(p|x)$.

Solution

Because X is binomial, the likelihood function is given by

$$f(x|p) = \binom{n}{x} p^x (1-p)^{n-x}.$$

Because p is uniform on [0,1], $\pi(p) = 1$, $0 \le p \le 1$.
Then the posterior distribution is given by

$$f(p|x) \propto f(x|p)\pi(p) = \binom{n}{x} p^x (1-p)^{n-x},\ x = 0,1,\ldots,n,$$

which is the same as the likelihood.

Note that in the previous example, the form of the pmf in both $f(x|p)$ and $f(p|x)$ are same; however, in $f(p|x)$, p is considered random and in $f(x|p)$, p is not random. This particular form of $f(p|x)$ is also called *beta-binomial distribution for p* with parameters $\alpha = x + 1$ and $\beta = n - x + 1$. This example illustrates that if the prior is noninformative (uniform), then the posterior is essentially the likelihood function. In the case where the prior and posterior are of the same functional form, we call it a conjugate prior. Bayesian inference becomes simpler when the prior density has the same functional form as the likelihood (which is the case for the *conjugate prior*) or when data are an independent sample from an exponential family (such as normal, Poisson, or binomial). Bayesian priors act just like adding pseudo observations to the data.

The following example demonstrates the method of finding posterior distribution for a continuous random variable.

EXAMPLE 10.2.4

Suppose that X is a normal random variable with mean μ and variance σ^2, where σ^2 is known and μ is unknown. Suppose that μ behaves as a random variable whose probability distribution (prior) is $\pi(\mu)$ and is also normally distributed with mean μ_p and variance σ_p^2, both assumed to be known or estimated. Find the posterior distribution $f(\mu|x)$.

Solution

Using the Bayes theorem, we have

$$\begin{aligned} f(\mu|x) &= \frac{f(x\mid\mu)\pi(\mu)}{\int f(x\mid\mu)\pi(\mu)d\mu} \\ &= \frac{\frac{1}{\sqrt{2\pi}\sigma}e^{-(x-\mu)^2|2\sigma^2}\frac{1}{\sqrt{2\pi}\sigma_p}e^{-(\mu-\mu_p)^2|2\sigma_p^2}}{\int\frac{1}{\sqrt{2\pi}\sigma}e^{-(x-\mu)^2|2\sigma^2}\frac{1}{\sqrt{2\pi}\sigma_p}e^{-(\mu-\mu_p)^2|2\sigma_p^2}d\mu} \\ &= \frac{1}{2\pi\sigma\sigma_p}e^{-\left[\frac{(x-\mu)^2}{2\sigma^2}+\frac{(\mu-\mu_p)^2}{2\sigma_p^2}\right]}. \end{aligned} \tag{10.2}$$

Consider the exponential term in (10.2), namely, $\frac{(x-\mu)^2}{2\sigma^2} + \frac{(\mu-\mu_p)^2}{2\sigma_p^2}$.

$$\begin{aligned} \frac{(x-\mu)^2}{2\sigma^2} + \frac{(\mu-\mu_p)^2}{2\sigma_p^2} &= \frac{1}{2}\left[\frac{(x-\mu)^2}{\sigma^2} + \frac{(\mu-\mu_p)^2}{\sigma_p^2}\right] \\ &= \frac{1}{2}\left[\left(\frac{1}{\sigma^2}+\frac{1}{\sigma_p^2}\right)\mu^2 - 2\left(\frac{\mu_p}{\sigma_p^2}+\frac{x}{\sigma^2}\right)\mu + \left(\frac{x^2}{\sigma^2}+\frac{\mu_p^2}{\sigma_p^2}\right)\right] \\ &= \frac{1}{2}\left[\frac{\sigma_p^2+\sigma^2}{\sigma^2\sigma_p^2}\mu^2 - 2\left(\frac{\mu_p}{\sigma_p^2}+\frac{x}{\sigma^2}\right)\mu + \left(\frac{x^2}{\sigma^2}+\frac{\mu_p^2}{\sigma_p^2}\right)\right] \\ &= \frac{1}{2}\frac{\sigma_p^2+\sigma^2}{\sigma^2\sigma_p^2}\left[\mu^2 - 2\frac{\sigma^2\sigma_p^2}{\sigma_p^2+\sigma^2}\left(\frac{\mu_p}{\sigma_p^2}+\frac{x}{\sigma^2}\right)\mu \right. \\ &\quad \left. + \frac{\sigma^2\sigma_p^2}{\sigma_p^2+\sigma^2}\left(\frac{x^2}{\sigma^2}+\frac{\mu_p^2}{\sigma_p^2}\right)\right] \end{aligned}$$

$$= \frac{1}{2}\frac{\sigma_p^2+\sigma^2}{\sigma^2\sigma_p^2}\left[\mu^2 - 2\left(\frac{\sigma^2}{\sigma_p^2+\sigma^2}\mu_p + \frac{\sigma_p^2}{\sigma_p^2+\sigma^2}x\right)\mu + \left(\frac{\sigma^2}{\sigma_p^2+\sigma^2}\mu_p + \frac{\sigma_p^2}{\sigma_p^2+\sigma^2}x\right)^2\right]$$

$$+ \frac{1}{2}\frac{\sigma_p^2+\sigma^2}{\sigma^2\sigma_p^2}\left[\frac{x^2}{\sigma^2} + \frac{\mu_p^2}{\sigma_p^2} - \left(\frac{\sigma_p^2}{\sigma_p^2+\sigma^2}x + \frac{\sigma^2}{\sigma_p^2+\sigma^2}\mu_p\right)^2\right]$$

$$= \frac{1}{2}\frac{\sigma_p^2+\sigma^2}{\sigma^2\sigma_p^2}\left[\mu - \left(\frac{\sigma^2}{\sigma_p^2+\sigma^2}\mu_p + \frac{\sigma_p^2}{\sigma_p^2+\sigma^2}x\right)\right]^2 + \tilde{K},$$

where

$$\tilde{K} = \frac{1}{2}\frac{\sigma_p^2+\sigma^2}{\sigma^2\sigma_p^2}\left[\frac{x^2}{\sigma^2} + \frac{\mu_p^2}{\sigma_p^2} - \left(\frac{\sigma^2}{\sigma^2+\sigma_p^2}\mu_p + \frac{\sigma_p^2}{\sigma^2+\sigma_p^2}x\right)^2\right].$$

From the foregoing derivation, we obtain

$$f(\mu|x) = Ke^{-\frac{1}{2}\frac{\sigma_p^2+\sigma^2}{\sigma^2\sigma_p^2}\left[\mu - \left(\frac{\sigma^2}{\sigma_p^2+\sigma^2}\mu_p + \frac{\sigma_p^2}{\sigma_p^2+\sigma^2}x\right)\right]^2},$$

where K does not contain μ.

This implies that the posterior density $f(\mu|x)$ is the pdf of normal random variable with mean

$$\left(\frac{\sigma^2}{\sigma_p^2+\sigma^2}\mu_p + \frac{\sigma_p^2}{\sigma_p^2+\sigma^2}x\right)$$

and variance

$$\frac{\sigma^2\sigma_p^2}{\sigma_p^2+\sigma^2}.$$

If we let $\tau_p = \frac{1}{\sigma_p^2}$ and $\tau = \frac{1}{\sigma^2}$, then the posterior density can be rewritten as the pdf of normal random variable with mean $\frac{1}{\tau_p+\tau}(\tau_p\mu_p + \tau x)$ and variance $\frac{1}{\tau_p+\tau}$.

As an example, suppose that $\mu_p = 100$, $\sigma_p = 15$, and $\sigma = 10$, $x = 115$. Then $f(\mu|x)$ is the pdf of a normal random variable with

$$Mean = \frac{100}{100+225}(100) + \frac{225}{100+225}(115) = 110.4$$

and

$$Variance = \frac{(100)(225)}{100+225} = 69.2.$$

10.2.1 Criteria for Finding the Bayesian Estimate

In the Bayesian approach to parameter estimation, we use both the prior and observations. This leads to an estimation strategy based on the posterior distribution. How do we know that the estimate thus obtained is "good?" To assess the quality of likely estimators, we use a loss function $L(\theta, a)$ that measures the loss incurred by using a as an estimate of θ. Here θ is the parameter being estimated (in real-world problems it is not known), and a is the estimate of θ. Then the "optimal" or "best" estimate $a = \hat{\theta}$ is chosen so as to minimize the expected loss $E[L(\theta, \hat{\theta})]$, where the expectation is taken over θ with respect to the posterior distribution $f(\theta|x)$. Here we mention two types of commonly used loss functions: quadratic and absolute error loss functions, and the resulting estimates.

(1) A *quadratic* (or *squared error*) *loss function* is of the form $L(\theta,a) = (a - \theta)^2$. In this case,

$$E[L(\theta, a)] = \int L(\theta, a) f(\theta|x_1, \ldots, x_n) d\theta$$
$$= \int (a-\theta)^2 f(\theta|x_1, \ldots, x_n) d\theta.$$

Differentiating with respect to a and equating to zero, we obtain

$$2\int (a-\theta) f(\theta|x_1, \ldots, x_n) d\theta = 0.$$

This implies

$$a = \int \theta f(\theta|x_1, \ldots, x_n) d\theta.$$

This is the *posterior mean* (expected value) of θ, $E(\theta|x_1, \ldots, x_n)$. Hence the quadratic loss function is minimized by taking the estimate of θ, that is, $\hat{\theta}$, to be the posterior mean. In previous examples in this section, we used this value as the estimate $\hat{\theta}$. Note that what the quadratic loss function displays is that if the estimate $\hat{\theta}$ and the true parameter θ are close to each other, the loss we expect is very small. Likewise, if the difference is larger, the expected loss in estimating θ with $\hat{\theta}$ is going to be large.

(2) An *absolute error loss function is* of the form $L(\theta, a) = |a - \theta|$. In this case,

$$E[L(\theta, a)] = \int L(\theta, a) f(\theta|x_1, \ldots, x_n) d\theta$$
$$= \int_{\theta=-\infty}^{a} (a-\theta) f(\theta|x_1, \ldots, x_n) d\theta$$
$$+ \int_{\theta=a}^{\infty} (\theta-a) f(\theta|x_1, \ldots, x_n) d\theta$$

Differentiating with respect to a and equating to zero, we obtain

$$\int_{\theta=-\infty}^{a} f(\theta|x_1, \ldots, x_n) d\theta - \int_{\theta=a}^{\infty} f(\theta|x_1, \ldots, x_n) d\theta = 0.$$

The minimum loss is attained when the values of both integrals are equal to 1/2. This can be achieved by taking $\hat{\theta}$ to be the *posterior median.*

There are other loss functions, such as the *all-or-nothing (or 0–1) loss function* given by

$$L(a, \theta) = 1 - \delta_{a\theta} = \begin{cases} 0, & \textit{if } \theta = a \\ 1, & \textit{otherwise,} \end{cases}$$

where δ is the Kronecker Delta function. This loss function is used mostly when values of θ are assumed to be discrete. In this case, it can be shown that the expected loss is minimized when $\hat{\theta}$ is the maximum of the posterior (MAP) distribution, or the mode.

The following can be considered as a general Bayesian procedure for point parameter estimation.

Bayesian parameter estimation procedure

1. Consider the unknown parameter θ as a random variable.
2. Use a probability distribution (prior) to describe the uncertainty about the unknown parameter.
3. Update the parameter distribution using the Bayes theorem:

$$P(\theta|Data) \propto P(\theta)P(Data|\theta),$$

that is,

$$(\text{posterior of } \theta) \propto (\text{prior of } \theta)(\text{likelihood}).$$

4. The Bayes estimator of θ is set to be the expected value of the posterior distribution $P(\theta|Data)$ under a quadratic loss function.
5. The Bayes estimator of θ is set to be the posterior median under the absolute error loss function.

From the procedure of Bayesian estimation, it is clear that a bad choice of prior may result in a bad estimate. Generally, if the priors are based on a previous and trustworthy sample, Bayesian estimation methods are desirable. A schematic figure of the steps involved in the Bayesian estimate is given in Fig. 10.1.

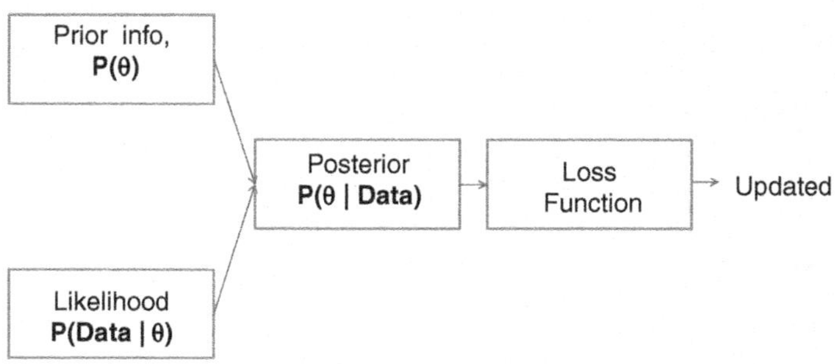

FIGURE 10.1 Bayesian estimation procedure. A flowchart illustrating Bayesian updating with prior info, likelihood, posterior, and loss function.

In this chapter, we use only the quadratic loss function unless it is explicitly stated otherwise. We also mention that this loss function is very popular because of its analytic tractability. We now derive Bayesian point estimates for some specific distributions.

Whereas uniform priors are useful in the noninformative situations, the beta family of distributions is one of the commonly taken informative priors. Distributions in the beta family take values in the interval (0, 1). Recall that if $X \sim beta(\alpha, \beta)$, then the pdf of X is given by

$$f(x) = \begin{cases} \frac{\Gamma(\alpha+\beta)}{\Gamma(\alpha)\Gamma(\beta)} x^{\alpha-1}(1-x)^{\beta-1}, & 0 \le x < 1 \\ 0, & \text{otherwise}, \ \alpha > 0, \ \beta > 0. \end{cases}$$

The beta pdf can be written as

$$f(x) = Cx^{\alpha-1}(1-x)^{\beta-1} \propto x^{\alpha-1}(1-x)^{\beta-1},$$

where $C = \frac{\Gamma(\alpha+\beta)}{\Gamma(\alpha)\Gamma(\beta)}$. We also know thatWhen using beta prior, we will take the number of successes as $\alpha - 1$ and the number of failures as $\alpha - 1$ and the number of failures as $\beta - 1$.

$$E(X) = \frac{\alpha}{\alpha+\beta}, \text{ and } \quad Var(X) = \frac{\alpha\beta}{(\alpha+\beta)^2(\alpha+\beta+1)}.$$

EXAMPLE 10.2.5

Let $X_1, \ldots, X_n$ be a sample from a geometric distribution with parameter p, $0 \le p \le 1$. Assume that the prior distribution of p is beta with $\alpha = 4$, and $\beta = 4$.

(a) Find the posterior distribution of p.
(b) Find the Bayes estimate under a quadratic loss function.

Solution

(a) *Because p is Beta(4,4), the prior density is*

$$\frac{\Gamma(8)}{\Gamma(4)\Gamma(4)} p^3(1-p)^3 = 140p^3(1-p)^3.$$

Because the r.v.'s X'_i s have a geometric distribution with parameter p, the likelihood is given by

$$L(X_1, \ldots, X_n|\theta) = \prod_{i=1}^{n} p(1-p)^{x_i - 1} = p^n(1-p)^{\sum_{i=1}^{n} x_i - n}.$$

The product of the likelihood function and the prior is given by

$$p^n(1-p)^{\sum_{i=1}^{n} x_i - n}[140p^3(1-p)^3] = 140p^{n+3}(1-p)^{\sum_{i=1}^{n} x_i - n + 3}.$$

Because (posterior of p) ∝ (prior of p) (likelihood), rewriting the normalizing constant in the denominator of Eq. (10.1) as C, and letting $C_1 = 140C$, the posterior distribution (because and $\alpha - 1 = n + 3$, and *$\beta - 1 = \sum_{i=1}^{n} x_i - n + 3$) is $Beta(n + 4, \sum_{i=1}^{n} x_i - n + 4)$.*

(b) *Recall that for a Beta(α, β) random variable, the mean is $[\alpha/(\alpha+\beta)]$. Because the Bayes estimate is the posterior mean, the mean of $Beta(n + 4, \sum_{i=1}^{n} x_i - n + 4)$ is*

$$\frac{n+4}{[\sum_{i=1}^{n} x_i - n + 4] + (n+4)} = \frac{n+4}{\sum_{i=1}^{n} x_i + 8}.$$

Note that for large n, the Bayes estimate is approximately $n/\sum_{i=1}^{n} x_i$, which is the MLE of p.

In general, for a Bernoulli random variable with unknown probability of success p in [0,1], the usual conjugate prior is the beta distribution, where the parameters of the beta distribution are chosen to reflect any prior information that we have.

We will follow the idea of the previous example in a binomial experiment of tossing a coin.

EXAMPLE 10.2.6

Suppose we are flipping a biased coin, where the probability of heads p could be any value between 0 and 1. Given a sequence of toss samples $x_1, \ldots, x_n$, we want to estimate $P(H) = p$. We may have two sources of information: our prior belief, which we will express as a beta distribution, and the data, which could come from counts of heads x in $n = 20$ independent flips of the coin, say $x = 13$. Suppose that in six prior tosses, we observed three heads and three tails, which lead us to believe that the value of p is near 0.5. Obtain the posterior distribution of p.

Solution

Here our prior belief or assumption can be written in terms of beta distribution aswhere $\alpha = 4$ and $\beta = 4$. That is (noting $\Gamma(n) = (n-1)!$) Hence, $\pi(p) \propto p^3(1-p)^3$. Because the mean of a beta distribution is $\alpha/(\alpha+\beta)$ and the variance is $\alpha\beta/((\alpha+\beta)^2(\alpha+\beta+1))$, for the prior,

$$\pi(p) = \frac{\Gamma(\alpha+\beta)}{\Gamma(\alpha)\Gamma(\beta)} p^{\alpha-1}(1-p)^{\beta-1}$$

$$\pi(p) = \frac{7!}{(3!)(3!)} p^3 (1-p)^3.$$

$$Mean(p) = \frac{4}{4+4} = 0.5,$$

and

$$Var(p) = \frac{(4)(4)}{(4+4)^2(4+4+1)} = 0.028.$$

Let X denote the number of heads in 20 flips of this coin. Then X has a binomial distribution, and the pmf is given by

$$f(x|p) = \binom{20}{x} p^x (1-p)^{20-x},\ x = 0,1,\ldots,20.$$

This we can write as

$$f(x|p) \propto p^x (1-p)^{20-x}.$$

In the 20 flips we have observed 13 heads. Then fix x = 13, and we are interested in the likelihood, which is the relative value of the function at different values of p:

$$f(13|p, 20) \propto p^{13}(1-p)^7.$$

The posterior probability of p, given x = 13, is

$$\begin{aligned} \pi(p|x=13) &\propto f(x|p)\pi(p) \\ &= \left(p^{13}(1-p)^{20-13}\right)p^3(1-p)^3 \\ &= p^{16}(1-p)^{10}. \end{aligned}$$

Thus, the posterior is a beta distribution with $\alpha = 17$ and $\beta = 11$. Consequently, we can now obtain the mean and variance of p as

$$Mean(p) = \frac{17}{17+11} = 0.607$$

and

$$Var(p) = \frac{(17)(11)}{(17+11)^2(17+11+1)} = 0.008.$$

Note that the prior was beta distribution with mean 0.5 and variance 0.028. Fig. 10.2 *gives the prior and posterior densities.*

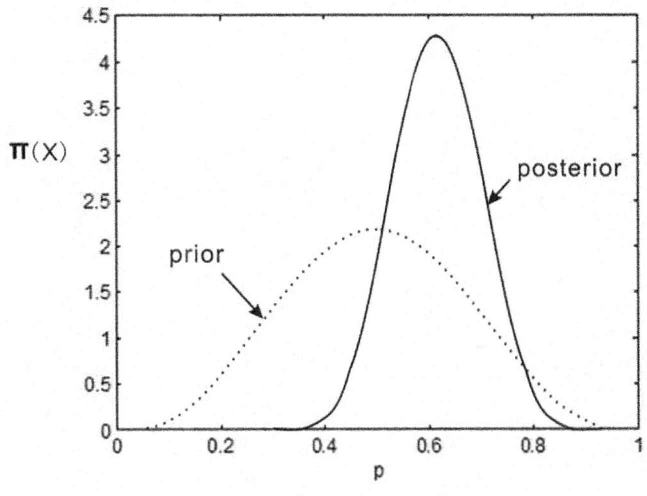

FIGURE 10.2 Prior and posterior distributions for the proportions.

Note that if we had ignored the prior and just took the point estimation, then the MLE of p is MLE (p) = $\hat{p} = \frac{13}{20} = 0.65$. *Compare this with the Bayesian estimate of p = 0.607. Because Beta(1,1) is the Uniform* [0, 1], *the method of the previous example can be used for noninformative priors. The method could also be used in many applications. For example, suppose p represents the proportion of infected individuals in a population, and x is the number of infected individuals in a sample of size n. Then with a noninformative prior, we can show that the posterior of p is Beta(x + 1, n − x + 1). This type of setting can be used for estimating the true proportion of infected individuals in the population.*

EXAMPLE 10.2.7

Suppose for the past million days we have been predicting whether the sun will rise the next morning or not. Each evening we say that the sun will rise the next morning ($\hat{R}$), and we were right (R) all these days. Suppose on the 10^6 evenings we predicted that the sun will rise on the next day. What is the probability that the sun will rise the next day?

Solution

The problem can be cast in the following table form.

1	2	…	10^6	$10^6 + 1$
$\hat{R}$	$\hat{R}$	…	$\hat{R}$	$\hat{R}$
R	R	…	R	?

$P(R|\hat{R}) = 1$ *if we use the frequency method of estimation (e.g., the MLE). Let us now consider the Bayes method. Suppose the prior is uniform on [0,1]. That is,*

$$\pi(p) = \begin{cases} 1, & \text{if } 0 \le p \le 1 \\ 0, & \text{otherwise.} \end{cases}$$

Suppose we predict n times and we succeed x times. Then

$$f(x|p) = \binom{n}{x} p^x (1-p)^{n-x}.$$

The joint pdf is given by

$$f(x, p) = f(x|p)\pi(p)$$
$$= \binom{n}{x} p^x (1-p)^{n-x},\ x = 0,1,\ldots,n;\ 0 \le p \le 1.$$

By the Bayes theorem, the posterior pdf $\pi(p|x)$ *is*

$$\pi(p|x) = \frac{f(x \mid p)\pi(p)}{\int_0^1 (x \mid p)\pi(p)dp}$$
$$= K(n, x)p^x(1-p)^{n-x},\ 0 \le p \le 1,\ 0 \le x \le n,$$

which is a beta probability distribution. Recall that the beta density is given by

$$f(y) = \frac{1}{B(\alpha, \beta)} y^{\alpha-1}(1-y)^{\beta-1}$$

and $E(Y) = \frac{\alpha}{\alpha+\beta}$. *Thus,*

$$E[\pi(p|x)] = \frac{x+1}{(x+1)+(n-x)+1} = \frac{x+1}{n+2}.$$

In our example, $x = 10^6$, $n = 10^6$, *which implies that the posterior mean is given by*

$$\hat{p}_B = \frac{10^6+1}{10^6+2} \approx 1.$$

EXAMPLE 10.2.8

Let $X_1, X_2, \ldots, X_n$ *be* $N(\mu,\sigma^2)$ *random variables with prior* $\pi(\mu)$ *having* $N(\mu_0, \sigma_0^2)$ *distribution with known* σ^2.

(a) Obtain the posterior distribution of μ.
(b) Suppose it is known from past experience that the weight loss for a particular combination of diet and exercise program (if followed for a month) is normally distributed with mean 10 lb and standard deviation of 2 lb. A random sample of five persons who went through this program for a month produced the following weight loss in pounds:

What is the point estimate of the mean, μ? Assume $\sigma^2 = 4$.

14 8 11 7 11

Solution

(a) *Because* $\pi(\mu) \sim N(\mu_0, \sigma_0^2)$, $\pi(\mu) \propto \exp[(\mu-\mu_0)^2/\sigma_0^2]$ *and we omit the terms that do not depend on* μ. *We have from the data* $x = (x_1, \ldots, x_n)$, *the likelihood function, where* μ *is determined by the posterior distribution. The product of the likelihood function and the prior give the posterior, which is obtained (after some algebra) as follows:*

$$L(x_1, \ldots, x_n|\mu) = f(x|\mu) \propto \prod_{i=1}^n \exp\left\{-\frac{(x_i-\mu)^2}{2\sigma^2}\right\}$$
$$= \exp\left\{-\sum_{i=1}^n \left[(x_i-\mu)^2/2\sigma^2\right]\right\},$$

where

$$f(\mu|x) \propto \pi(\mu) \propto \exp\left[-(\mu-\mu_1)^2/2\sigma_1^2\right]$$

$$\mu_1 = \frac{\frac{n}{\sigma^2}x + \frac{1}{\sigma_0^2}\mu_0}{\frac{n}{\sigma^2} + \frac{1}{\sigma_0^2}}$$

and

$$\sigma_1^2 = \frac{1}{\frac{n}{\sigma^2} + \frac{1}{\sigma_0^2}}.$$

Thus, the posterior distribution of μ is $N(\mu_1, \sigma_1^2)$.

(b) *Note that the sample mean $\bar{x} = 10.2$ lb, and sample standard deviation $s = 2.77$ lb. Now from part (a), the posterior distribution of μ is normal with mean*

$$\mu_1 = \frac{\frac{n}{\sigma^2}\bar{x} + \frac{1}{\sigma_0^2}\mu_0}{\frac{n}{\sigma^2} + \frac{1}{\sigma_0^2}} = \frac{\frac{5}{2^2}(10.2) + \frac{1}{2^2}(10)}{\frac{5}{2^2} + \frac{1}{2^2}} = 10.167$$

and variance

$$\sigma_1^2 = \frac{1}{\frac{n}{\sigma^2} + \frac{1}{\sigma_0^2}} = \frac{1}{\frac{5}{2^2} + \frac{1}{2^2}} = 0.66667.$$

Thus, the point estimate of μ is the posterior mean, 10.167. Fig. 10.3 represents the prior and posterior densities of μ.

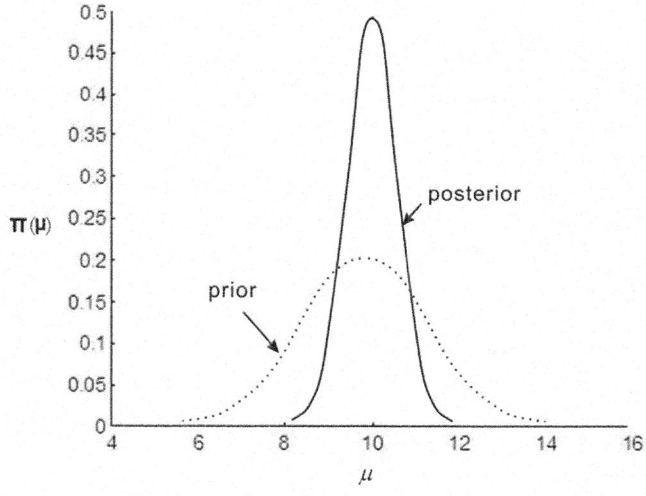

FIGURE 10.3 Prior and posterior densities of μ.

Sometimes, the inverse of variance in the normal distribution is called the precision of the normal distribution and denoted by $\tau = 1/\sigma^2$. Also note that in part (a) of the previous example, if the prior variance $\sigma_0^2 \to \propto$, then the prior flattens out; $\pi(\mu) \propto c$, is a constant. This basically amounts to saying that prior information on μ decreases, that is, all μ are equally probable. This corresponds to a noninformative prior. Also, in this case as $\sigma_0^2 \to \infty$, $\sigma_1^2 \to \frac{\sigma^2}{n}$ and $\mu_1 \to \bar{x}$. Hence, in the limit (i.e., for noninformative priors), the posterior $f(\mu|x)$ will have an $N(\bar{x}, \sigma^2/n)$ distribution, which is exactly the same inference as in classical statistics.

In Bayesian inference problems, one of the questions is, which will have relatively more influence, prior or likelihood? As we observe a large amount of data, it can be shown that the posterior distribution is almost exclusively determined by the data. That is, asymptotically, observed data will have a larger influence compared to the choice of prior, and thus the prior will be irrelevant. Hence, we can make the following general observations. If the prior is noninformative and we have a large data set, then we can expect that the likelihood will have greater influence. Whereas, if we have a small data set and an informative prior, then the prior will have a larger influence on the updated posterior distribution. Bayesian estimators are more complicated to compute than calculating the maximum likelihood estimates in simple cases. However, in complex settings, Bayesian statistics are often relatively easier to compute.

One of the problems in using Bayesian analysis is choosing an appropriate prior. There are no specific rules available for this purpose. For instance, the following priors are commonly used in literature. If data are in [0,1], we could use a uniform or beta distribution. If the data are in $[0, \infty)$, normal (with nonnegative and relatively large μ), gamma or log-normal distributions are used. If the data are in $(-\infty, \infty)$, normal or t-distributions are commonly used. In Section 10.6, we will learn the empirical Bayes method for choosing priors based on the data itself.

EXERCISES 10.2

10.2.1 Suppose in a casino, two kinds of dice are used, one kind of which 98% are fair, and 2% are loaded such that five comes up 60% of the time and the rest of the numbers are equally probable. We pick a die at random and roll it three times. We get three consecutive fives. What is the probability that the die is loaded?

10.2.2 It is believed that cross-fertilized plants produce taller offspring than self-fertilized plants. In order to obtain an estimate on the proportion θ of cross-fertilized plants that are taller, an experimenter observes a random sample of 15 pairs of plants exactly the same age, with each pair grown in the same conditions, with one cross-fertilized and the other self-fertilized. Based on previous experience, the experimenter believes that the following are possible values of π and prior probabilities for each value (prior weight), $\pi(\theta)$:

θ:	0.80	0.82	0.84	0.86	0.88	0.90
$\pi(\theta)$:	0.03	0.40	0.22	0.15	0.15	0.05

From the experiment, it is observed that in 13 of 15 pairs, the cross-fertilized is taller.

(a) Create a table with columns for prior, likelihood of θ given sample, prior times likelihood, and posterior probability of θ. Based on the posterior probabilities, what value of θ has the highest support? Also, find $E(\theta)$ based on the posterior probabilities.

(b) Redo part (a) with a completely noninformative prior; that is, take the prior for the proportion θ as one of the equally spaced values 0, 0.1, 0.2, ..., 0.9, 1. Also, assign for each value of θ the same probability, $\pi(\theta) = 1/11$.

(c) Calculate the MLE of θ and compare it with the Bayesian estimate.

10.2.3 Suppose you have three coins in your wallet; coin 1 is biased 3:1 in favor of tails, coin 2 is a fair coin, and coin 3 is biased in favor of heads. Suppose you randomly select one coin and flip it once and you observe a heads. What is the probability that you have chosen coin 3?

10.2.4 Consider the problem of estimating p in a binomial distribution. Let X be the number of successes in a sample of size n.

(a) Let the prior distribution of p be given by $Beta(3,1)$, that is
Find the posterior distribution of p.

$$\pi(p) = \begin{cases} 3p^2, & 0 < p < 1 \\ 0, & \text{otherwise.} \end{cases}$$

$$\left[\text{Hint: } f(x|p) = \begin{cases} \binom{n}{x} p^x (1-p)^{n-x}, & x = 0,1,2,\ldots,n \\ 0, & \text{otherwise.} \end{cases} \right]$$

(b) Let the prior distribution of p be given by $Beta(a,b)$ (i.e., $\pi(p) \propto p^{a-1}(1-p)^{b-1}$). Find the posterior distribution of p.

10.2.5 A biased coin is tossed n times. Let x_i be 1 if the ith toss is heads and 0 if it is tails. Assume a noninformative prior, $p(\theta) = 1, 0 \le \theta \le 1$. Let t be the number of heads obtained. Show that the posterior distribution of θ is *Beta* $(t + 1, n - t + 1)$.

10.2.6 Let $X_1, X_2, \ldots, X_n$ be exponential random variables with parameter λ. Let the prior $\pi(\lambda)$ be exponentially distributed with parameter μ, which is a fixed and known constant.
(a) Show that the posterior distribution of λ is *Gamma* $(n + 1, \mu + \sum_{i=1}^{n} x_i)$.
(b) Obtain the Bayes estimate of λ.

10.2.7 Let $X_1, X_2, \ldots, X_n$ be Poisson random variables with parameter λ. Assume that λ has a *Gamma* (α, β) prior.
(a) Compute the posterior distribution of λ.
(b) Obtain the Bayes estimate of λ.
(c) Compare the MLE of λ with the Bayes estimate of λ.
(d) Which of the two estimates is better? Why?

10.2.8 Let $X_1, X_2, \ldots, X_n$ be Poisson random variables with parameter λ. Assume that λ has an exponential distribution with
$\theta = 1$ prior.
(a) Compute the posterior distribution of λ and show that it is *Gamma* $((\sum_{i=1}^{n} x_i + 1), (n + 1))$.
(b) Find the Bayes estimate of λ.

10.2.9 It is known that a certain disease has affected 10% of the population. In a random sample of 50 patients typical of the disease group who are exposed to a new treatment, we observe that 12 patients were hospitalized in a year. Let μ be the rate of population that need hospitalization. Assume that

$$\mu \sim Gamma(0.1,2) \text{ and } f(x|\mu) \sim Poi(50\mu).$$

Given that 0.24 is an observation from $f(x|\mu)$, find the Bayesian estimator of μ (i.e., obtain $E(u|x)$).

10.2.10 Let $X_1, \ldots, X_n$ be an $N(\mu, 2)$ random sample with prior $\pi(\mu)$ having $N(0, \sigma^2)$ distribution with known σ^2. Obtain the posterior distribution of μ.

10.2.11 Let $X_1, \ldots, X_n$ be an $N(\mu, 1)$ random sample with prior $\pi(\mu)$ having the pdf $[1/\pi\ (1 + \mu^2)]$. Show that the posterior

$$\pi(\mu|x) \propto \exp\left\{-\frac{n(\mu - \bar{x})^2}{2}\right\} \times \frac{1}{1 + \mu^2}.$$

10.2.12 Suppose we are interested in the true mortality risk θ, in a hospital, A, for an open-heart surgery. On average in the country around 2.5% of people die, but mortality rates in different hospitals vary from around 1% to around 5%. Hospital A has no deaths in their first 12 open-heart surgeries. What should be the estimator of θ under a quadratic loss function?

10.3 BAYESIAN CONFIDENCE INTERVAL OR CREDIBLE INTERVALS

In this section, we want to study the question, "Can we construct an interval where we are confident that the interval contains the unknown true value of θ?" We have seen how, in many situations, it may be preferable to use an interval estimate instead of a point estimate for a population parameter θ. Such intervals in classical statistics were called confidence intervals. We can extend the concept of interval estimation to a Bayesian setting. The Bayesian analog of a confidence interval is called a credible interval and is defined as follows.

Definition 10.3.1. *A* $100(1 - \alpha)\%$ **credible interval** *for θ is an interval (a, b) such that here α is given a small positive number between 0 and 1, and $x_1, \ldots, x_n$ are the sample values.*

$$p(a \le \theta \le b|x_1, \ldots, x_n) \ge (1 - \alpha)$$

Note that we read this definition backwards, that is, we are at least $(1 - \alpha)100\%$ confident that the true value of θ is between a and b, given the sampled information.

Because the conditional distribution of θ given $X_1, \ldots, X_n$ is actually a probability distribution, it makes sense to talk about the probability that θ is in the interval (a, b). Once we have observed data, the credible interval is fixed while θ is random. This is in contrast to the classical confidence interval, where the interval is random but θ is a fixed parameter. In the classical case, we would say, "In the long run, $100(1 - \alpha)\%$ of all such intervals will contain the true parameter θ." In the Bayesian approach, we would say, "The probability is at least $(1 - \alpha)$ that θ lies within the specified interval (a, b)."

As in the classical case, it would be desirable to minimize the length of the credible interval. This entails choosing only those points with the highest values in the posterior density of $f(\theta|x_1, \ldots, x_n)$, as shown in Fig. 10.4. This will be better, especially if the density is not symmetric.

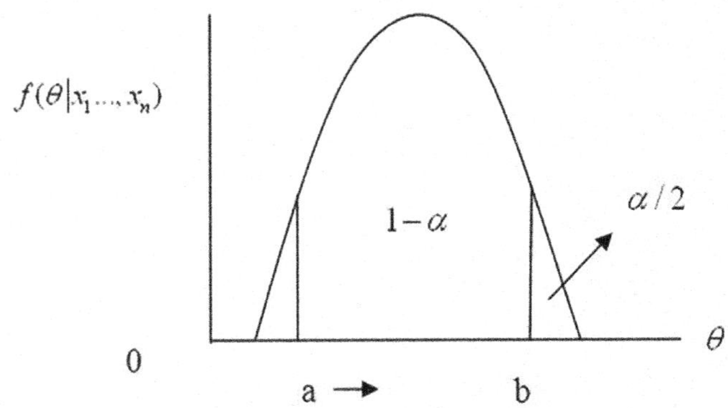

FIGURE 10.4 Credible interval for θ.

Definition 10.3.1 can be rephrased as follows using the posterior distribution of θ.

Definition 10.3.2. *A* $100(1 - \alpha)\%$ **credible interval** *for* θ *is an interval* (a, b) *such that*

1. $\int_a^b f(\theta/x_1,\ldots,x_n)d\theta \geq 1 - q$, *if* θ *is continuous, and the posterior pdf of* θ *is* $f(\theta|x_1, \ldots, x_n)$.
2. $\Sigma^b f(\theta/x_1,\ldots,x_n \geq 1 - \alpha$, if θ is discrete.

We will now give some examples for computing credible intervals.

EXAMPLE 10.3.1

Suppose $X_1, \ldots, X_n$ is a random sample from $N(\mu, \sigma^2)$ with $\sigma^2 = 4$. Suppose the prior pdf of μ is $N(0, 1)$, that is, $\pi(\mu) \sim N(0, 1)$. Find a 95% credible interval for μ.

Solution

We have seen from Example 10.2.8 that the posterior distribution of μ *given* $x_1, \ldots, x_n$, *is normally distributed with*

$$\text{Mean} = \frac{1}{1 + \frac{4}{n}}\bar{X}$$

and

$$\text{Variance} = \frac{1}{1 + \frac{n}{4}}.$$

Fig. 10.3 *represents the posterior distribution of* μ.

■

To find the 95% credible interval for μ, we have to find two numbers, a and b, such that (Fig. 10.5)

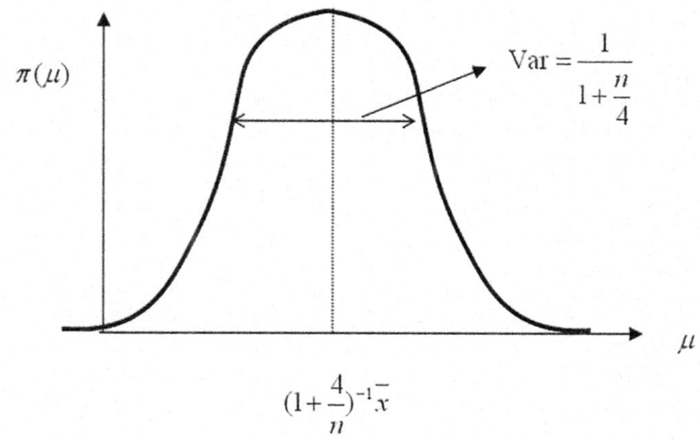

FIGURE 10.5 Posterior distribution of μ.

$$p(a \le \mu \le b) = 0.95,$$

where

$$\mu \sim N\left(\mu = \frac{\overline{x}}{1+\frac{4}{n}},\ \sigma^2 = \frac{1}{1+\frac{n}{4}}\right).$$

We choose a to be $-b$ (b is positive). Using z-scores, we get (X is continuous),

$$p\left(-z_{\alpha|2} < \frac{\mu - \frac{1}{1+\frac{4}{n}}\overline{x}}{\sqrt{\frac{1}{1+\frac{n}{4}}}} < z_{\alpha|2}\right) = 1 - \alpha$$

which can be rearranged as

$$p\left(\frac{1}{1+\frac{4}{n}}\overline{x} - \frac{1}{\sqrt{1+\frac{n}{4}}}z_{\alpha|2} < \mu < \frac{1}{1+\frac{4}{n}}\overline{x} + \frac{1}{\sqrt{1+\frac{n}{4}}}z_{\alpha|2}\right) = 1 - \alpha.$$

Thus, a 95% credible interval for μ is

$$\left(\frac{1}{1+\frac{4}{n}}\overline{x} - \frac{1}{\sqrt{1+\frac{n}{4}}}z_{\alpha/2},\ \frac{1}{1+\frac{4}{n}}\overline{x} + \frac{1}{\sqrt{1+\frac{n}{4}}}z_{\alpha/2}\right).$$

For convenience, we summarize this procedure in the following steps.

Bayesian credible interval procedure

1. Consider θ as a random variable with prior pdf (or pmf) $\pi(\theta)$.
2. Update the prior distribution $\pi(\theta)$ using the Bayes theorem. That is, find the posterior distribution of θ by the formula

$$\pi(\theta|data) = \begin{cases} \frac{f(data \mid \theta)\pi(\theta)}{\int f(data \mid \theta)\pi(\theta)d\theta}, & \text{if continuous} \\ \frac{f(data \mid \theta)\pi(\theta)}{\Sigma f(data \mid \theta)\pi(\theta)}, & \text{if discrete.} \end{cases}$$

3. Find two numbers, a and b, such that

$$\int_a^b \pi(\theta|data)d\theta \ge 1 - \alpha, \text{ if continuous}$$

$$\Sigma_{\theta=a}^{b} \pi(\theta|data) \ge 1 - \alpha, \text{ if discrete.}$$

Note: The numbers a and b are found such that

$$\int_{-\infty}^{a} \pi(\theta|data)d\theta = \alpha/2, \text{ if continuous}$$

$$\Sigma_{\theta \le a} \pi(\theta|data) = \alpha/2, \text{ if discrete.}$$

and

$$\int_b^{\infty} \pi(\theta|data)d\theta = \alpha/2, \text{ if continuous}$$

$$\Sigma_{\theta \ge b} \pi(\theta|data) = \alpha/2, \text{ if discrete.}$$

4. The $(1 - \alpha)100\%$ credible interval for θ is the interval (a, b).

In the discrete case, an easy way of finding a credible interval of smallest length is to arrange the values of θ from most likely to least likely (i.e., in the order of the magnitude of the posterior probabilities), and then put values of θ into the interval until the cumulative posterior probability of the set exceeds $(1 - \alpha)100\%$. Such an interval is called the highest posterior density (HPD) interval. It can be shown that the HPD interval always exists, and it is unique, so long as for all intervals of probability $(1 - \alpha)$, the posterior density is never uniform in any interval of values of θ.

EXAMPLE 10.3.2

For the data of Example 10.2.1, find a 90% credible interval for θ.

Solution

Arranging the values of θ from most likely to least likely, we have Table 10.3. *Looking at the "cumulative probability" column, we see that the probability that θ is in the set {0.86, 0.84, 0.88, 0.82, 0.80} is 0.90192. So this set is a 90% probability (or credible) interval for θ.*

EXERCISES 10.3

10.3.1.

(a) Suppose $X_1, \ldots, X_n$ is a random sample from $N(\mu, \sigma^2)$ with $\sigma^2 = 9$. Suppose the prior pdf of μ is $N(0,1)$; that is, $\pi(\mu) \sim N(0,1)$. Find a 95% credible interval for μ.

(b) The following is a set of random data from a normal distribution with variance 9:

0.92	1.05	5.53	3.64	− 4.47	− 2.60	0.71	− 3.66	1.38	3.87
7.42	1.76	0.01	2.69	1.54	3.97	1.34	− 1.63	− 1.24	− 4.78

TABLE 10.3 Normal distribution Sample Data Calculations.

Prior Values of θ	Posterior Probability of θ	Cumulative Probability
0.86	0.2661	0.2661
0.84	0.22528	0.49138
0.88	0.15817	0.64955
0.82	0.14208	0.79163
0.80	0.11029	0.90192
0.90	9.8064×10^{-2}	0.99984

Using the results of part (a), compute a 95% credible interval for μ, interpret its meaning, and state any assumptions you have made.

10.3.2. Suppose that a person believes that his last year's weight was normally distributed with mean of 165 lb and standard deviation of 5 lb. That is, the prior pdf of μ is $N(165, 25)$, or $\pi(\mu) \sim N(165, 25)$. He expects his current weight X is normally distributed with mean μ and standard deviation 7 lb. The following are 10 random measurements (in pounds) from this year:

176	165	180	172	175
179	166	177	184	183

Find a 95% credible interval for μ.

10.3.3. It is known that a certain disease affects 10% of a population. In a random sample of 50 patients in the disease group who are exposed to a new treatment, we observe that 12 patients were hospitalized in a year. Let μ be the population rate that needs hospitalization in a year. Assume μ has a *Gamma* (0.1, 2) prior. Let $\mu \sim$ *Gamma* (0.1,2) and $f(x|\mu) \sim$ *Poi* (50μ). Given that $x = 0.24$ is an observation of X, find a 95% credible interval for μ. Obtain a Bayesian credible interval for μ. (If X is the number of patients admitted in a year, assume $X \sim$ *Poi* (50μ), the Poisson approximation of the binomial.) How can we improve on this estimate?

10.3.4. For an upcoming congressional election, suppose we want to estimate the amount of support for a particular candidate in a district. By previous experience and voter registration data, we can assume that the prior distribution of the proportion of support, p, is a beta distribution with $\mu = 10$, and $\beta = 8$ (i.e., $\pi(p) \sim$ *Beta* (10, 8)). We conducted a survey of 1000 randomly selected voters, of whom 600 support the candidate. Obtain a 95% credible interval for p. What will happen to the credible interval if we reduce the confidence interval? What will happen to the 95% credible interval if we increase the sample size?

10.3.5. It is recommended that the daily intake of sodium be 2400 mg per day. From a previous study on a particular ethnic group, the prior distribution of sodium intake is believed to be normal with mean 2700 mg and standard deviation 250 mg. If a recent survey for this group resulted in a mean of 3000 mg and standard deviation of 300 mg, obtain a 95% credible interval for the mean intake of sodium for this ethnic group.

10.3.6. Suppose we have a coin (not necessarily balanced) with p being the probability of heads. Assume a uniform prior for p. Suppose in 20 tosses of this coin, we obtained 12 heads. Obtain a 90% credible interval for p.

10.3.7. Suppose that in a particular telephone exchange, the number of calls received per minute has a Poisson distribution with parameter λ. Assume an exponential prior for λ with parameter 2. Suppose this exchange received 25 calls in 5 minutes. Obtain a 95% credible interval for λ.

10.3.8. Consider flipping a coin 20 times with the probability of heads being θ. Suppose we observed four heads. Assuming a uniform prior for θ over [0, 1] (i.e., $p(\theta) = 1, 0 \le \theta \le 1$), obtain a 95% credible interval for θ.

10.4 BAYESIAN HYPOTHESIS TESTING

The Bayesian approach to hypothesis testing for simple hypotheses is pretty straightforward. Deciding between two hypotheses for a given set of data x reduces to computing their posterior probabilities. If an explicit loss function is available, the Bayes rule is chosen to minimize the expected value of the loss function with respect to the posterior distribution. In the absence of a loss function, the probabilities of type I and type II errors are of little interest to the Bayesian.

In classical hypothesis testing, we test a null hypothesis (denoted by H_0) against an alternative hypothesis (denoted by H_1 or H_a). The test procedure is based on controlling the two types of errors—type I and type II. The classical test procedures limit the type I error to α and minimize the type II error. If the type II error is unacceptably high, it is reduced by increasing the sample size.

In the Bayesian approach, the problem of deciding between the null and alternative is rather straightforward. Consider the problem of hypothesis testing with.

$$H_0\colon \theta \in \Theta_0 \text{ versus } H_1\colon \theta \in \Theta_1, \tag{10.3}$$

where θ_0, θ_1 are subsets of the real line. Let $X_1,\ldots,X_n$ be the sample from a population with pdf $f\theta(x)$.
In the Bayesian hypothesis testing approach, we compute the following posterior probabilities:

$$\alpha_0 = P(\theta \in \Theta_0|x_1, \ldots, x_n) \tag{10.4}$$

and

$$\alpha_1 = P(\theta \in \Theta_1|x_1, \ldots, x_n). \tag{10.5}$$

If $\alpha_0 > \alpha_1$, we accept the null hypothesis, and if $\alpha_0 < \alpha_1$, we reject the null hypothesis. That is, if the posterior probability under null hypothesis is larger than the posterior probability under the alternative, we accept the null hypothesis, and vice versa. We now outline the Bayes hypothesis testing procedure for testing hypotheses (10.3).

Let $\pi(\theta)$ be the prior. Also,

$$\pi_0 = P(\theta \in \Theta_0) = P(\theta \in \Theta_0)$$

and

$$\pi_1 = P(\theta \in \Theta_1) = P(H_1).$$

Definition 10.4.1. *The ratio π_0/π_1 is called the* **prior odds ratio**. *The ratio α_0/α_1 (see* Eqs. 10.4 and 10.5*) is called the* **posterior odds ratio.**

The posterior odds ratio is the ratio of the posterior probabilities, given the data, of the null and alternate hypotheses. The posterior odds ratio will be used in decision-making for testing the hypotheses. We now compute α_0 and α_1 using the Bayes theorem. That is,

$$\alpha_0 = p(\theta \in \Theta_0|x_1,\ldots,x_n) = \begin{cases} \int_{\Theta_0} f(\theta|x_1,\ldots,x_n)d\theta, \text{ if continuous} \\ \Sigma_{\theta\in\Theta_0} f(\theta|x_1,\ldots,x_n), \text{ if discrete.} \end{cases}$$

Similarly,

$$\alpha_1 = p(\theta \in \Theta_1|x_1,\ldots,x_n) = \begin{cases} \int_{\Theta_1} f(\theta|x_1,\ldots,x_n)d\theta, \text{ if continuous} \\ \Sigma_{\theta\in\Theta_1} f(\theta|x_1,\ldots,x_n), \text{ if discrete.} \end{cases}$$

We reject H_0 if the odds ratio $(\alpha_0/\alpha_1) < 1$ and accept H_0 if $(\alpha_0/\alpha_1) > 1$.
This method of hypothesis testing is called Jeffreys' hypothesis testing criterion. It basically says that if the posterior odds ratio is greater than 1, we accept the null hypothesis; otherwise, we reject the null in favor of the alternative hypothesis.

Because we cannot determine the probability of a single value in the continuous variable case, it should be noted that for a simple null hypothesis of the form θ equals some specified value cannot be dealt with easily in the Bayesian framework. Hence, unlike the classical framework, here we mostly deal with the composite hypotheses for both null and alternative.

EXAMPLE 10.4.1

A student taking a standardized test is classified as gifted if they score at least 100 out of a possible score of 150. Otherwise, the student is classified as not gifted. Suppose the prior distribution of the scores of all students is normal with mean 100 and standard deviation 15. It is believed that scores will vary each time the student takes the test and that these scores can be modeled as a normal distribution with mean μ and variance 100. Suppose the student takes the test and scores 115. Test the hypothesis that the student can be classified as a gifted student.

Solution

The hypothesis testing problem can be phrased as

$$H_0: \theta < 100 \text{ versus } H_a: \theta \geq 100.$$

Referring to Example 10.2.8, we know that the posterior distribution $f(\theta|x)$ is a normal with mean 110.4 and variance 69.2. Because the prior is an N(100,225), we have $\pi_0 = P(\theta < 100) = 1/2$ and $\pi_1 = P(\theta \geq 100) = 1/2$.

We can now compute

$$\begin{aligned}\alpha_0 &= p(\theta < 100|x = 115)\\ &= p\left(\frac{\theta - 110.4}{\sqrt{69.2}} < \frac{100 - 110.4}{\sqrt{69.2}}\right)\\ &= p\left(z \leq -\frac{10.4}{\sqrt{69.2}}\right) = 0.106\end{aligned}$$

and

$$\begin{aligned}\alpha_1 &= p(\theta \geq 100|x = 115)\\ &= 1 - p(\theta < 100|x = 115)\\ &= 1 - 0.106 = 0.894.\end{aligned}$$

Thus, $\alpha_0/\alpha_1 = (0.106/0.894) = 0.119 < 1$, and we reject H_0.

Bayesian hypothesis testing procedure

To test H_0: $\theta \in \Theta_0$ versus H_1: $\theta \in \Theta_1$, where Θ_0 and Θ_1 are given sets:

1. Consider θ as a random variable with prior distribution $\pi(\theta)$.
2. Compute the posterior distribution $f(\theta|x_1,\ldots,x_n)$ of θ given $x_1,\ldots,x_n$, using Bayes' theorem.
3. Compute α_0 and α_1 using the following formulas:

$$\alpha_0 = p(\theta \in \Theta_0|x_1,\ldots,x_n) = \begin{cases}\int_{\Theta_0} f(\theta|x_1,\ldots,x_n)d\theta, & \text{if continuous}\\ \sum_{\theta\in\Theta_0} f(\theta|x_1,\ldots,x_n), & \text{if discrete}\end{cases}$$

and

$$\alpha_1 = p(\theta \in \Theta_1|x_1,\ldots,x_n) = \begin{cases}\int_{\Theta_1} f(\theta|x_1,\ldots,x_n)d\theta & \text{if continuous}\\ \sum_{\theta\in\Theta_1} f(\theta|x_1,\ldots,x_n), & \text{if discrete.}\end{cases}$$

4. Reject H_0 if the posterior odds ratio, $\frac{\alpha_0}{\alpha_1} < 1$. Otherwise accept.

In the foregoing procedure, we assume that $P(\theta \in \Theta_0)$ and $P(\theta \in \Theta_1)$ are both greater than zero.

EXAMPLE 10.4.2

Suppose a random variable,is transmitted over a communication channel. The signal received is noise corrupted. Assume that the received signal is given by $Y = X + \varepsilon$, where $\varepsilon \sim N(0, \sigma^2)$. The goal is to decide between $X = 1$ and $X = -1$ by observing the random variable Y only. Find the Bayesian hypothesis test for this problem.

$$X = \begin{cases} 1, & w.p.\ p \\ -1, & w.p.\ 1-p. \end{cases}$$

Solution

The hypothesis testing problem can be phrased as

$$H_0: X = 1 \text{ versus } H_a: X = -1.$$

Under the null hypothesis, $Y = 1 + \varepsilon$. *Hence,* $Y \sim N(1, \sigma^2)$. *Also, under* H_a, $Y = -1 + \varepsilon$, so, $Y \sim N(-1, \sigma^2)$. *Therefore,*

$$f_Y(y|H_0) = \frac{1}{\sqrt{2\pi}\sigma} e^{-\frac{(y-1)^2}{2\sigma^2}},$$

and

$$f_Y(y|H_1) = \frac{1}{\sqrt{2\pi}\sigma} e^{-\frac{(y+1)}{2\sigma^2}}.$$

Thus, accept H_a *if and only if*

$$\frac{1}{\sqrt{2\pi}\sigma} e^{-\frac{(y+1)}{2\sigma^2}} P(H_a) \geq \frac{1}{\sqrt{2\pi}\sigma} e^{-\frac{(y-1)^2}{2\sigma^2}} P(H_0).$$

We have $P(H_0) = p$, and $P(H_1) = 1 - p$. After a simple algebra, we can see that, we accept H_a if and only if

$$\exp\left(\frac{-2y}{\sigma^2}\right) \geq \frac{p}{1-p}.$$

Equivalently, we choose H_1 *if and only if*

$$y \leq -\frac{\sigma^2}{2} \ln\left(\frac{p}{1-p}\right).$$

This type of test is called the maximum a posteriori (MAP) test.

EXERCISES 10.4

10.4.1 The following is random data from a normal distribution with variance 9.

(a) Test the hypothesis, $H_0: \mu \leq 0$ versus $H_a: \mu > 0$. Assume that the prior is $N(0, 4)$, so that $\mu \leq 0$ and $\mu > 0$ are equally probable.

(b) Compare your decision with classical hypothesis testing, with $\alpha = 0.05$.

0.92	1.05	5.53	3.64	−4.47	−2.60	0.71	−3.66	1.38	3.87
7.42	1.76	0.01	2.69	1.54	3.97	1.34	−1.63	−1.24	−4.78

10.4.2

(a) For the data of Exercise 10.3.2, using the Bayesian method, test the hypothesis $H_0: \mu \leq 170$ versus $H_a: \mu > 170$.

(b) Compare your decision with classical hypothesis testing, with $\alpha = 0.05$.

10.4.3 It is known that a certain disease affects 10% of the population. Of a random sample of 50 patients in the disease group who are exposed to a new treatment, we observe that 12 patients were hospitalized in a year. Let μ be the population rate that needs hospitalization in a year. Assume μ has a $Gamma(0.1,2)$ prior. Let $\mu \sim Gamma(0.1,2)$ and $f(x|\mu) \sim Poi(50\mu)$. Given that $x = 0.24$ is an observation of X, test the hypothesis H_0: $p \leq 0.10$ versus H_a: $p > 0.10$. (If X is the number of patients admitted in a year, assume $X \sim Poi(50\mu)$, the Poisson approximation of the binomial.)

10.4.4 For an upcoming congressional election, suppose we want to estimate the amount of support for a particular candidate in a district. By previous experience and voter registration data, we can assume that the prior distribution, the proportion of support, p, is a beta distribution with $\alpha = 10$, and $\beta = 8$ (i.e., $\pi(p) \sim Beta(10, 8)$). We conducted a survey of 1000 randomly selected voters, of whom 600 support the candidate. Test the hypothesis H_0: $p \geq 0.60$ versus H_a: $p < 0.60$.

10.4.5 For the data of Exercise 10.3.5, test the hypothesis H_0: $\mu \leq 2400$ mg versus H_a: $\mu > 2400$ mg for this ethnic group.

10.4.6 Suppose we have a coin (not necessarily balanced) with p being the probability of heads. Assume a uniform prior for p. Suppose in 20 tosses of this coin, we obtained 12 heads. Test the hypothesis H_0: $p \geq 0.50$ versus H_a: $p > 0.50$.

10.4.7 Suppose we have two image classifiers, Model A is the baseline, and Model B is a new model with improved architecture. For Model A, we tested $n_A = 100$ test images of which it classified $s_A = 78$ images correctly. Model B had $n_B = 80$ test images of which it classified $s_B = 66$ images correctly. Assume independent uniform priors for the two proportions, that is, p_A, and $p_B \sim Beta(1,1)$.

(a) What is the posterior probability that Model B is more accurate than Model A, that is, $P(p_B > p_A|\text{data})$?

(b) Test the null hypothesis H_0: $p_A = p_B$ versus H_a: $p_A \neq p_B$.

10.5 BAYESIAN DECISION THEORY

Bayesian methods in general are more concerned with problems of decision-making than with problems of inference. Decision theory, as the name implies, is concerned with the problem of making decisions. Statistical decision theory is concerned with optimal decision-making under uncertainty or when statistical knowledge is available only on some of the uncertainties involved in the decision problem. Uncertainty could be about the true value related to the decision, or, uncertainty could be about the actual state of the nature. Abraham Wald (1902–50) laid the foundation for statistical decision theory. Original works on the decision theory emerged out of game theory considerations. Many books and articles have been written on various aspects of decision theory. The Bayesian approach to the decision theory was introduced by Leonard Jimmie Savage in 1954. In this section, we introduce the general idea of decision theory. We basically deal with analytical procedures for the decision-making process. This will involve selecting an optimum decision from a choice of courses of action among two or more alternatives. The Bayesian decision theory quantifies the trade-offs between different decisions using costs and probabilities that accompany such decisions.

Consider, as an example, a company deciding whether to market a new brand of toothpaste with a whitening agent. Clearly many factors will affect the decision (e.g., the proportion of people who are likely to switch to the new brand, and the likelihood of other competing companies introducing similar toothpastes). These factors are generally unknown, but estimates can be obtained from statistical investigations.

The classical statistical approach relies exclusively on the data obtained from these statistical investigations, ignoring other relevant information such as the company's past experiences in marketing similar products. Statistical decision theory tries to combine other relevant information with the sample information to arrive at the optimal decision. Therefore, a Bayesian setting seems to be more appropriate for decision theory.

One piece of relevant information that decision theory considers is the possible consequences of the decisions. Often these consequences can be quantified. That is, the loss or gain of each decision can be expressed as a number (called the loss or utility). A loss or utility to a decision maker is the effect of the interaction of two factors: (1) the decision or action selected by the decision maker and (2) the event or state of the world that actually occurs. Classical statistics does not explicitly use a loss function or a utility (payoff) function.

A second source of information that decision theory utilizes is prior information. Prior information could be based on past experiences of similar situations, or on expert opinion. We can follow the procedure explained next as a guideline for decision-making.

General decision theory procedure

1. Identify the objectives of the decision-making process.
2. Identify the set of actions and set of possible events (states of nature).
3. Assign probabilities to the occurrence of each possible state of nature (prior). If more observations are available, calculate the posterior probabilities to the occurrence of each possible state of nature.
4. For each possible event, assign a numerical value to the anticipated payoff (or loss) of each course of action.
5. Compute the expected value of the payoffs (utility or loss function). This could be done by either using the prior probabilities if there are no observations, or using the posterior probabilities.
6. Select the optimum decision among the available alternative courses of action that maximizes the expected value of the payoffs.

There are many other decision criteria available in literature. In this section, we only consider the expected utility or loss function approach. We now consider an example to illustrate the idea of statistical decision-making.

EXAMPLE 10.5.1

Suppose you own a small stall at a flea market that is open only on weekends. If the weather is good, you make a profit of \$200, and if it is bad, you close your stall and you make no (zero) profit. However, you have the option of buying, from an insurance company, weather insurance that costs \$75. The company pays you \$210 if the weather is bad. Suppose you believe that the probability of good weather on a particular weekend is p. Compute the expected gain if you insure and if you do not. What is the best course of action? Arrive at a decision.

Solution

From the information in the problem, we can obtain the utility gain or profit table shown in Table 10.4, *based on our decision to insure or not insure. Suppose that we model the state of weather as good or bad by means of a random variable defined as follows.*

TABLE 10.4 Flea Market Data Calculations.

	Weather	
Parameter space → Decision space ↓D	Good (θ_1)	Bad (θ_2)
Insurance (I)(d1)	\$125 (200–75)	\$135 (210–75)
No insurance (NI)(d2)	\$200	\$0

$$\theta = \begin{cases} 1, & \text{if the weather is good} \\ 0, & \text{if the weather is bad.} \end{cases}$$

Suppose for our example we believe that during a particular weekend $P(\theta = 1) = p$, and $P(\theta = 0) = 1 - p$. This can be considered as prior information. The different values of θ are called states of nature. We assign (perhaps subjectively) a probability structure for the states of nature defined by a prior distribution $\pi(\theta)$. Now we can compute the expected gain when we insure and when we do not.

Using the values in the table,

$$\begin{aligned}\text{Expected gain given we insure} &= (125)p + (135)(1 - p) \\ &= 135 - 10p\end{aligned}$$

$$\begin{aligned}\text{Expected gain when we do not insure} &= (200)p + (0)(1 - p) \\ &= 200p\end{aligned}$$

Hence, insurance is preferable if

$$135 - 10p > 200p$$

or

$$p < \frac{135}{210} = 0.643.$$

That is, we should take the insurance if we believe the probability of good weather is less than 0.643.

In general, the states of the nature are represented by $\theta_1, \ldots, \theta_n$ and the possible decisions (actions) are represented by $d_1, \ldots, d_m$. Let $U(d_j, \theta_i)$ represent the net gain when the true states of nature is θ_i and the decision d_j is made. Then we can construct the general utility table shown in Table 10.5.

TABLE 10.5 Net Gains and Decision Calculations.

		States of Nature					
		θ_1	θ_2	...	θ_i	...	θ_n
	d_1	$U(d_1, \theta_1)$	$U(d_1, \theta_2)$		$U(d_1, \theta_i)$		$U(d_1, \theta_n)$
	d_2						
Decision	·						
States							
	d_j				$U(d_j, \theta_i)$		
	·						
	d_m	$U(d_1, \theta_1)$					$U(d_m, \theta_n)$

In Bayesian decision theory, we assume a probability distribution on the states of nature called the prior distribution. Using this probability distribution, we can find the decision that maximizes the expected utility. That is, let the states of nature be initially modeled by a random variable θ with probability function $\pi(\theta)$ such that $P(\theta = \theta_i) = \pi(\theta_i)$, $i = 1, \ldots, n$. Let U denote the utility. Then the expected utility for decision d_j is given by

$$E(U|d_j) = \sum_{i=1}^{n} U(d_j, \theta_i)\pi(\theta_i).$$

The optimal decision, called the Bayes decision, denoted by d^*, is that which maximizes the expected utility. That is, d^* satisfies the following equation:

$$\max_{d_j} \sum_{i=1}^{n} U(d_j, \theta_i)\pi(\theta_i) = \sum_{i=1}^{n} U(d^*, \theta_i)\pi(\theta_i).$$

This procedure is called the *Bayes decision procedure* with respect to the assumed or given prior $\pi(\theta_i)$, $i = 1,2, \ldots, n$.

Procedure to find optimal decision

1. For each decision di, compute $\sum_{i=1}^{n} U(d_j, \theta_i)\pi(\theta_i)$
2. Find a decision d^* from the decision space that maximizes the sum in step 1. This is the Bayes decision.

In determining the Bayes decision, we have assumed a prior distribution $\pi(\theta)$ for the states of nature $\{\theta_i\}$. Naturally the question arises: Can there be information or observations that will help us to determine $\pi(\theta)$?

Definition 10.5.1. Observations that can aid us in determining the relative likelihoods of the possible states of nature are called **observables.**

We remark that observables enable us to refine and update our initial prior $\pi(\theta)$. The updated prior is the conditional distribution $\pi(\theta|\text{observables})$, which clearly depends on the observables as well as the initial prior $\pi(\theta)$. The updated prior is also called the posterior.

For example, to determine the nature of weather we may hear the weather forecast (80% chance of rain), in which case we may assume $P(G)=0.2$, and $P(B)=0.8$. However, the weather forecast is not perfect. Let $\hat{G}$ and $\hat{B}$ denote the meteorologist's prediction. We may like to know $P(G|\hat{G})$ and $P(G|\hat{B})$. That is, what is the probability of the weather being good when the meteorologist predicts the weather will be good, and what is the probability that the weather is good when the meteorologist predicts the weather will be bad?

It may be noted that there is no direct cause–effect relation in $G|\hat{G}$. That is, the prediction of the weather forecast does not influence the weather. If a probability distribution depends on a set of parameters θ, the classical approach estimates θ on the basis of an observed sample $X_1, \ldots, X_n$. The samples $X_1, \ldots, X_n$ are the observables. Thus, observables are used to estimate the parameters; that is, we want the distribution of θ given $X_1, \ldots, X_n$ or $p(\theta|X_1, \ldots, X_n)$. In our weather situation, the observable is the weather forecast, whereas the parameter is one of the weather conditions, good or bad. In $P(\hat{G}|G)$, we are asking, "Given that the weather is good, what is the probability that the weather forecast is correct?" We can imagine that meteorological conditions such as the barometric pressure determine the weather (i.e., $G=f(m_1,\ldots,m_k)$, m_i = meteorological factor), and in this sense we can consider that G is a parameter. We thus want $P(G|\hat{G})$.

To compute the posterior $P(G|\hat{G})$, we use the Bayes theorem (which needs a prior distribution, $P(G)$). That is,

$$P(G|\hat{G}) = \frac{P(\hat{G}|G)P(G)}{P(\hat{G}|G)P(G) + P(\hat{G}|B)P(B)}.$$

Similarly, we can compute $P(B|\hat{B})$.

Coming back to our weather situation, if $P(G)$ is known and $P(\hat{G}|G)$, $P(\hat{B}|B)$ are known, we could obtain the required posterior distributions $P(G|\hat{G})$and $P(B|\hat{B})$. We can now use this distribution to calculate the expected utilities and choose the decision that maximizes the expected utility.

We now consider an example.

EXAMPLE 10.5.2

Let us initially assume $P(\theta = 1) = P(\theta = 0) = \frac{1}{2}$. That is,

$$P(\text{good weather}) = P(\text{bad weather}) = \frac{1}{2}.$$

Suppose we have the following record on the meteorologist's predictions. The meteorologist predicts good weather ($\hat{G}$), given the weather is good, two-thirds of the time, that is, $P(\hat{G}|G) = 2/3$, and predicts bad weather, given the weather is bad, three-fourths of the time, that is, $P(\hat{B}|B) = 3/4$. Thus, given that the meteorologist predicts good weather, what is the probability that the weather will turn out to be good, and given the meteorologist predicts bad weather, what is the probability that the weather will turn out to be bad?

Solution

To compute the true probabilities, we use the Bayes theorem.

We are given $P(\hat{G}|G) = \frac{2}{3}$ and $P(\hat{B}|B) = \frac{3}{4}$, *which imply* $P(\hat{B}|G) = \frac{1}{3}$ *and* $P(\hat{G}|B) = \frac{1}{4}$. *Using the Bayes theorem, we obtain the likelihood of G as*

$$P(G|\hat{G}) = \frac{P(\hat{G} \mid G)P(G)}{P(\hat{G} \mid G)P(G) + P(\hat{G} \mid B)P(B)}$$
$$= \frac{\left(\frac{2}{3}\right)\left(\frac{1}{2}\right)}{\left(\frac{2}{3}\right)\left(\frac{1}{2}\right) + \left(\frac{1}{4}\right)\left(\frac{1}{2}\right)} = \frac{8}{11}$$

and the likelihood of B is

$$P(B|\hat{B}) = \frac{P(\hat{B} \mid B)P(B)}{P(\hat{B} \mid B)P(B) + P(\hat{B} \mid G)P(G)}$$
$$= \frac{\left(\frac{3}{4}\right)\left(\frac{1}{2}\right)}{\left(\frac{3}{4}\right)\left(\frac{1}{2}\right) + \left(\frac{1}{3}\right)\left(\frac{1}{2}\right)} = \frac{9}{13}.$$

Thus, we have the following updated prior depending upon the meteorologist's prediction. The updated prior when the meteorologist predicts good weather is

$$\pi(G) = P(G|\hat{G}) = \frac{8}{11}; \pi(B) = 1 - \pi(G) = \frac{3}{11}.$$

Thus, the updated $\pi(G)$ is actually $\pi_{\hat{G}}(G)$. Similarly, the updated prior when the meteorologist predicts bad weather (i. e., $\pi_{\hat{B}}(G)$) is

$$\pi(G) = P(G|\hat{B}) = \frac{4}{13}; \pi(B) = P(B|\hat{B}) = \frac{9}{13}.$$

That is, if the meteorologist predicts good weather, he will be right about 72.7% of the time, and if he predicts bad weather, he will be right about 69.2% of the time.

EXAMPLE 10.5.3

Consider Example 10.5.2, with the additional information that the meteorologist has predicted that the weather will be good on a given weekend. Referring to the utility table (Table 10.5) given in Example 10.5.1, we ask, what should be our decision—to insure or not to insure—in light of this prediction?

Solution

From Example 10.5.2, we know that the updated prior, given that the meteorologist predicts good weather, is

$$\pi(G) = P(G|\hat{G}) = \frac{8}{11} \text{ and } \pi(B) = P(B|\hat{G}) = \frac{3}{11}.$$

Using the foregoing prior and the utility table in Example 10.5.2, we can compute the following expected gains:

$$\text{Expected gain if we insure} = (125)\pi(G) + (135)\pi(B)$$
$$= (125)\tfrac{8}{11} + (135)\tfrac{3}{11} = 127.73$$

and

$$\text{Expected gain if we do not insure} = (200)\frac{8}{11} = 145.45.$$

Therefore, our decision, given that the meteorologist predicts good weather, is not to insure.

EXERCISES 10.5

10.5.1 Suppose that we will receive \$25 if we get two consecutive heads (H) on two flips of a balanced coin. If only one head appears, we will get \$10. On the other hand, if there is no heads, we will lose \$15. If monetary return is the only concern, should we play this game? Why?

10.5.2 In the previous problem, suppose we suspect the coin is not balanced. We feel that $P(H)$ is only 0.4. In our last 10 observations, we counted three heads and seven tails. Should we play the game? Defend your answer.

10.5.3 The owner of a small structural engineering firm in Tampa wants to open a new branch office in Orlando. The single most influential factor is the projected state of the economy for the next 4 years. If the economy keeps expanding or at least does not take a turn for the worse, the owner expects an annual profit of \$300,000 by opening the new office. If the economy experiences a downward trend, then the owner forecasts an annual loss of \$200,000. If he just continues to operate his business in Tampa, he expects a \$50,000 annual profit. Suppose a government forecast indicates that there is a 70% chance of economic expansion or status quo in the next 4 years, and there is a 30% chance that the economy will show a decline. What is the optimal decision in this problem? Did you make any assumptions in obtaining this optimal decision?

10.5.4 In Exercise 10.5.3, suppose the owner decides to look at the accuracy of past forecasts by the government. Suppose his study indicates that a forecast of economic expansion came true only 2/3 of the time, whereas an economic downturn came true 4/5 of the time. Based on this new evidence, what is the optimal option for the owner?

10.5.5 Consider the weather Example 10.5.1, discussed earlier. The meteorologist's prediction record over the past 15 days is as follows:

Weather person's prediction	G	B	B	G	G	G	B	G	G	B	B	G	B	G	G
How the weather turned out to be	B	B	B	G	G	B	B	G	B	G	B	G	G	G	G

(a) Assuming a uniform distribution for the states of nature, obtain an updated prior (posterior) based on the meteorologist's record.

(b) Obtain the Bayes decision.

10.5.6 A coin (not necessarily fair) will be tossed once, and you have to predict the outcome. If you predict the outcome correctly you win \$1000. Otherwise, you lose \$5.

(a) What are the states of nature? What is the decision space? Write the utility table.

(b) Suppose that you believe that the probability of heads is two-third. What is your price for the states of nature? Find the expected gains.

(c) Suppose that you are allowed to toss the coin twice and you find that the first toss results in heads and the second in tails. What are the observables?

(d) Assume the situation in (c). The coin is going to be tossed again and you have to predict the outcome. What is your updated prior?

(e) What are your expected gains, and what is your decision for the situation in (d)?

10.5.7 We are given the following utility table:

States of Nature			
	θ_1	θ_2	θ_3
d_1	0	10	4
d_2	−2	5	1

Determine the Bayes decision assuming a uniform prior for the states of nature.

10.5.8 Suppose that we have an observable X that can take only two values, X_1 and X_2, for the situation in Exercise 10.5.7. The distribution of X depends on the states of nature and is as follows:

	θ_1	θ_2	θ_3
X_1	0.1	0.5	0.6
X_2	0.9	0.5	0.4

That is, $P(X = x_1|\theta_1) = 0.1$ or $P(X = x_2|\theta_3) = 0.4$, and so forth.

Suppose you observe X_1; what is the updated prior? What is the Bayes decision?

10.5.9 A large lot has $p\%$ defectives and you have to predict p. If you predict p correctly, you gain \$$g$, and if the prediction is wrong, you lose \$l. It is known that the possible values of p are $p_1, p_2, \ldots, p_k$.
(a) Set up a utility table.
(b) Suppose you assume a uniform prior for p. That is, $\pi(p_i) = \frac{1}{k}$, $i = 1,2, \ldots, k$. Find an expression for the Bayes decision.
(c) Suppose you have an observable X such that $P(X = x_1|p_i) = a_i$, $i = 1, 2 \ldots, k$ and $P(X = x_1|p_i) = 1 - a_i$, $i = 1, 2, \ldots, k$. Find the updated prior for p. What is the Bayes decision in this case?

10.6 EMPIRICAL BAYES ESTIMATES

Empirical Bayes methods are techniques for statistical inference in which the prior distribution is estimated from the data, instead of assuming a specific fixed prior distribution. Thus, the essential empirical Bayes task is to learn an appropriate prior distribution from ongoing statistical experience, rather than knowing it by assumption. The empirical Bayes model often provides superior estimates of parameters in comparison to ordinary Bayes model.

The roots of empirical Bayes can be traced back to a work by von Mises in the 1940s. However, the first major work was developed by Herbert Robbins who introduced the concept of empirical Bayes to estimate the parameter that behaves as a random variable in the pdf of a given set of data, $f(x|\theta)$. The Robbins framework is considered nonparametric empirical Bayes, in which the prior distribution is completely unspecified. In this section, we will mostly study parametric empirical Bayes. In this approach, we specify a parametric family of distributions. Major works on (parametric) empirical Bayes were done by Efron and Morris in 1970s.

What we have learned in the previous sections of this chapter we refer to as ordinary or standard Bayesian analysis of a given set of data, $X_1, \ldots, X_n$, that has been drawn from a population or follows the probability density function pdf, $f(x|\theta)$. Usually, once we are given the random sample of size n, we perform a goodness-of-fit test to identify the pdf that probabilistically characterizes the behavior of the given data. Sometimes, it is not possible to define the pdf of difficult data; then we proceed to analyze the subject data using nonparametric methods that we present in Chapter 12.

In ordinary Bayes estimation, we assume that we have identified the pdf, $f(x|\theta)$, that characterizes the behavior of the given data. In such a situation, Bayes theory assumes that the parameter θ in $f(x|\theta)$ behaves as a random variable rather than a fixed point estimate. Thus, we need the pdf of the parameter θ, since we assumed it behaves as a random variable. We refer to pdf of θ, say, $\pi(\theta)$ as the prior pdf of θ. In summary, to perform standard Bayesian estimation, we need the following:

1. Identify through the goodness-of-fit methods the pdf of the given data, $f(x|\theta)$.
2. Assume a prior pdf that defines the random parameter θ, $\pi(\theta)$.
3. Assume a loss function, $L(\theta, \hat{\theta})$, to be used in the analysis.

Thus, using the above information, we develop the posterior pdf and its expected value in the ordinary (standard) Bayesian estimate of the parameter θ, that was assumed to behave as a random variable. The major problem that we have in performing ordinary Bayesian analysis is that we must assume or guess the prior pdf. In analyzing real-world data from health sciences, business, and engineering, among others, we cannot assume a prior pdf. That is, if we obtain ordinary Bayesian results from an assumed prior pdf, and we check our result using a different prior pdf, the results will be different. Thus, ordinary Bayesian estimation is very sensitive to the choice of the assumed *prior* pdf. To address the issue of not assuming the *prior* pdf, we will study some basic aspects of **empirical base estimates**. There are several methods that have been introduced to empirically estimate the prior pdf of θ, $\pi(\theta)$ so that we do not have to assume or guess it as in the standard Bayes method. You recall that the Bayes theorem can be stated as follows:

$$p(\theta|y) = \frac{\pi(\theta)p(y|\theta)}{p(y)},$$

where $p(\theta|y)$ is posterior pdf, $\pi(\theta)$ and $p(y|\theta)$ are the prior and sampling pdf, respectively. For θ has a discrete distribution, the marginal pdf of Y is given by

$$p(y) = \textstyle\sum_\theta \pi(\theta)p(y|\theta),$$

where the sum is over all possible value of θ. In the continuous case of θ we have

$$p(y) = \int \pi(\theta)p(y|\theta)d\theta.$$

In standard Bayes, in general we assume a prior pdf $\pi(\theta)$ depends on another parameter η, that is $\pi(\theta|\eta)$, where η is called a hyperparameter (η could be a vector). Because the prior of θ depends on another parameter η, the posterior density is,

$$p(\theta|y) \propto p(\eta)\pi(\theta|\eta)p(y|\theta).$$

In general, it is difficult if not impossible to directly calculate $\pi(\theta/\eta)$. In the empirical Bayes, we consider

$$p(\theta|y) \approx p(\theta|y, \hat{\eta}(y)), \text{ with } \hat{\eta}(y) = \arg\max_{\eta} \pi(\theta|\eta).$$

Note that we use argmax; because $\pi(\theta|\eta)$ is function of both θ and η, we are maximizing only with respect to η. This way, we could reduce the complexity choosing the hyper parameter (prior) by replacing it, in most of the cases, with the MLE of $\hat{\eta}$ of η based on observed data y. This is why in empirical Bayes, the choice of the prior itself is based on the observed data. The main difference between ordinary Bayes and empirical Bayes method is that, in standard Bayes, the hyperparameter η is assumed to be known, that is, a hyperprior pdf has been placed as η. In contrast, in the empirical Bayes approach, the hyperparameter remains unknown. Thus, η needs to be estimated from the given data. This approach is not really Bayesian because we use the same data to identify the prior pdf and the estimation is done using non-Bayesian methods like MLE or method of moments. Empirical Bayes methods are often considered as a bridge between classical and Bayesian inference.

It is more common to use Bootstrap methodology to estimate prior in the empirical Bayes approach, in the present approach, we shall introduce two resampling methods of the given data to estimate (identify) the prior pdf. The resampling methods that we shall use are the Jackknife and Bootstrap. These resampling methods are discussed in Chapter 13; however, we will give here a brief discussion.

10.6.1 Jackknife Resampling

Quenouille, M. H. in 1949 introduced this resampling method, and in 1956, John Tukey refined the method and named it Jackknife, after the Swiss Jackknife, that has multiple useful tools. Given a random sample of size n, $X = (X_1, \ldots, X_n)$, the Jackknife samples are computed by omitting one observation x_i at a time, that is,

$$x_i = (x_1, \ldots, x_{i-1}, x_{i+1}, \ldots, x_n).$$

The dimension of the Jackknife sample x_i is $m = n - 1$, that is, n different Jackknife sample, $\{x_{(i)}\}_{i=1,\ldots,n}$. The following diagram illustrates the process of Jackknife resampling, and for each new sample, we obtain the MLE of θ, that is, $\hat{\theta}_1^*, \hat{\theta}_2^*, \ldots, \hat{\theta}_m^*$, $m = n - 1$.

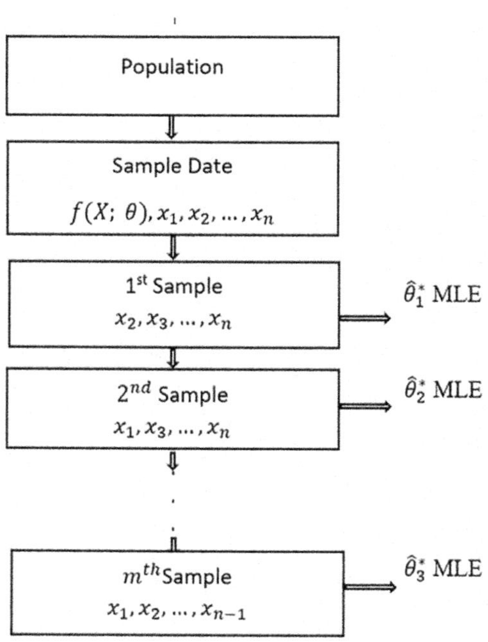

Our objective is to use this sequence of Jackknife resampling MLE of θ, $\hat{\theta}_1^*, \hat{\theta}_2^*, \ldots, \hat{\theta}_m^*$, $m = n - 1$ to obtain, if possible, the pdf of these estimates and use it as our prior pdf, $\pi(\theta)$ and proceed to obtain the Bayesian estimate of θ, without guessing it.

10.6.2 Bootstrap Resampling

Bradley Efron in 1979 introduced the Bootstrap resampling method for estimating the sampling distribution of an estimator. Given a set of data n, using the subject method, we generate k samples with replacement from the given data with $k < n$. The pdf of the k samples will follow the original pdf of the n independent and identically distributed observations. Consider the observation $x_1, x_2, \ldots, x_n$. By bootstrapping, we obtain different subsets of our original sample, that is, subsample of size k. There are several uses of this method, but in our present study of empirical Bayes, we shall use bootstrapping resampling to obtain an estimate of the prior pdf. For a given set of data $x_1, x_2, \ldots, x_n$, we will proceed if possible to identify the pdf, $f(x|\theta)$, that follows the observations or the population that is drawn from and through Bootstrap resampling to obtain a sequence of estimate, $\hat{\theta}_1^*, \hat{\theta}_2^*, \ldots, \hat{\theta}_k^*$, and, through goodness-of-fit methods, we will proceed to obtain an estimate of the prior pdf, $\pi(\theta)$, if possible. The following diagram illustrates the process we follow to resample using the Bootstrap method:

A FLOWCHART ILLUSTRATING THE SAMPLING PROCESS FROM A POPULATION TO MULTIPLE SAMPLES WITH MAXIMUM LIKELIHOOD ESTIMATION.

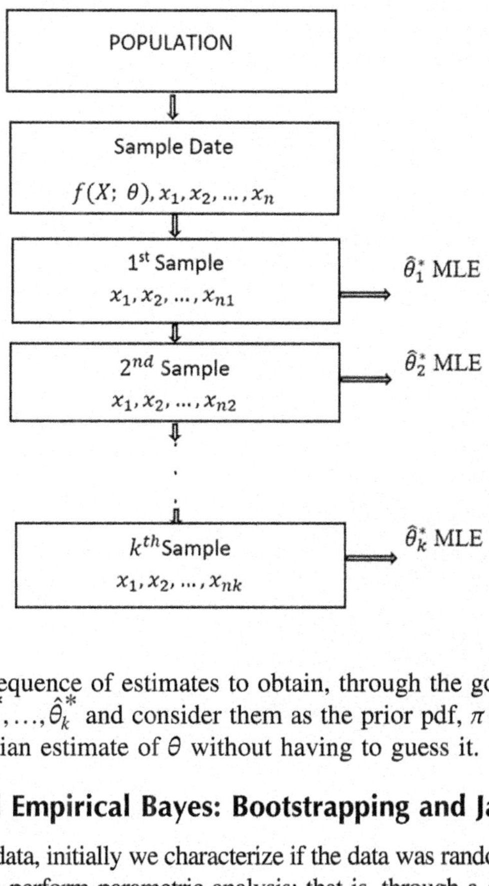

Thus, our objective is to use this sequence of estimates to obtain, through the goodness-of-fit method, if possible, the pdf that drives these estimates $\hat{\theta}_1^*, \hat{\theta}_2^*, \ldots, \hat{\theta}_k^*$ and consider them as the prior pdf, $\pi(\theta)$ of the parameter θ; that is, $\pi(\hat{\theta}^*)$. We then proceed to obtain the Bayesian estimate of θ without having to guess it.

10.6.3 Parametric, Bayes, and Empirical Bayes: Bootstrapping and Jackknife

In statistics, when we are given a set of data, initially we characterize if the data was randomly collected and test and, if necessary, remove any outliers. Our next step is to perform parametric analysis; that is, through a goodness-of-fit test we try to identify, if possible, the pdf that probabilistically characterizes the behavior of the given data. If we cannot identify a well-defined pdf we must rely on nonparametric methods. The parametric analysis is the underlying pdf, say $f(x|\theta)$. A better estimate than the parametric estimate, usually the MLE, is the Bayesian estimate. In the Bayesian estimate, we ask for more information such as the prior pdf, therefore we expect to get more about the estimate of the true parameter θ. In Bayesian analysis, we proceed with standard and empirical Bayes method to study θ. In the following Example 10.6.1, we shall use the data given in the Table 10.6 below that represent a certain phenomenon of interest to illustrate the parametric, standard, and empirical Bayesian estimate of the true parameter θ.

TABLE 10.6 The Data.

0.46	0.36	0.05	1.55	0.31	0.59	0.59	0.05	0.87	0.12	0.10	0.17	1.01	0.56	0.57
0.19	0.04	0.21	0.04	0.25	0.69	0.69	0.88	0.27	0.10	0.47	0.06	0.05	0.28	0.33
0.10	0.42	0.46	0.51	0.99	0.79	0.79	0.35	1.11	0.57	0.18	0.43	0.67	0.50	0.07
0.22	0.27	0.33	1.27	0.55	0.01	0.01	0.77	0.56	0.48	0.02	0.85	0.63	1.54	0.57
0.07	0.18	0	0.34	0.31	1.19	0.19	0.71	0.07	0.34	0.64	0.47	2.06	0.05	1.36

EXAMPLE 10.6.1

(a) Parametric Analysis

We would like to find, if possible, the pdf, say $f(x|\theta)$, that follow the given data in Table 10.1. We shall use the three commonly used good-of-fit tests in search for the pdf of the given data. These tests are

1. Kolmogorov–Smirnov
2. Anderson–Darling
3. Cramer–Von Mises Criterion

More information about the definition and structure of these goodness-of-fit tests will be found in Chapter 11.

To obtain a visual idea of what type of pdf we are looking for, we structure the histogram of the given data as shown below to obtain some idea of what type of pdf we are looking for (Fig. 10.6).

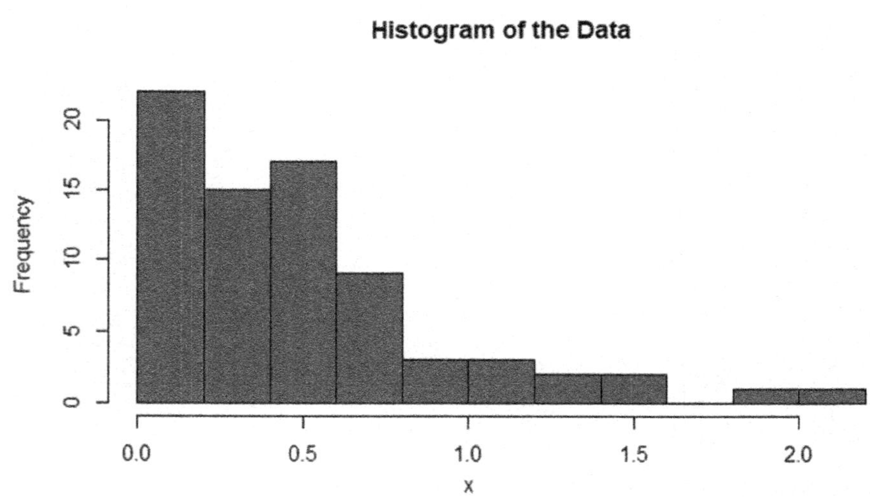

FIGURE 10.6 Histogram of the data.

The histogram suggests some sort of exponential decay of a pdf. Thus, we believe that exponential pdf is a good candidate and we begin to perform goodness-of-fit, using the three tests we mentioned. The table below shows the goodness-of-fit result for the exponential pdf:

Below is the SAS code for producing the goodness-of-fit result that are given in Table 10.7.

```
data random;
input var @@;
cards;
```

TABLE 10.7 Goodness-of-Fit Result.

Goodness-of-Fit Tests for Exponential Distribution				
Test	**Statistic**		***P*-Value**	
Kolmogorov–Smirnov	D	0.09398415	Pr > D	>0.250
Cramer–Von Mises	W-Sq	0.14110857	Pr > W-Sq	0.168
Anderson–Darling	A-Sq	0.71224681	Pr > A-Sc	>0.250

```
0.46 0.36 0.05 1.55......
;
run;
proc univariate data = random;
var var; histogram/exp; run;
```

Thus, all three tests identify the one parameter exponential pdf that fits the given data, that is, we used the MLE of θ in performing the above tests, which is given by

$$f(x|\theta) = \begin{cases} \frac{1}{\theta}\exp\left(-\frac{x}{\theta}\right), & x \geq 0,\ \theta > 0 \\ 0, & \text{elsewhere.} \end{cases}$$

$$\hat{\theta}_{MLE} = \overline{X} = \frac{1}{75}\sum_{i=1}^{75} x_i = 0.492.$$

The graph of the exponential pdf, $f(x|\theta = 0.492)$ is given below in. Fig. 10.7

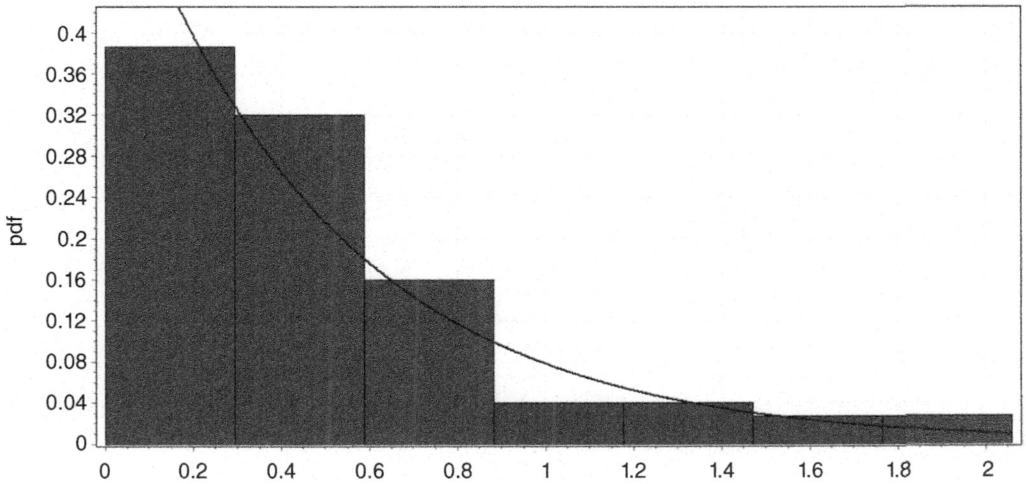

FIGURE 10.7 PDF of $f(x;0.492)$. A histogram with an overlaid curve showing probability density function distribution values.

In addition to the MLE of the true parameter, we can obtain $100(1-\alpha)\%$ confidence limits for the parameter θ, which will be used to compare with standard and empirical Bayes estimates. The confidence limit is based on the χ^2 distribution. That is,

$$P\left[\frac{2n}{\overline{X}\,(\chi^2_{\frac{\alpha}{2},2n})} \le \theta \le \frac{2n}{\overline{X}\,(\chi^2_{1-\frac{\alpha}{2},2n})}\right] \ge 100(1-\alpha)\%.$$

For 90% and 95% confidence limits, we have

$$P[0.41 \le \theta \le 0.58] \ge 0.90$$

and

$$P[0.40 \le \theta \le 0.60] \ge 0.95.$$

Thus, the confidence range of θ for 90% and 95% confidence limit is 0.17 and 0.2, respectively.

(b) Standard Bayes Estimate
Now that we have identified the pdf of the given data to be exponential distribution, we shall assume that the parameter θ behave as a random variable and we denote the pdf by $f(x|\theta)$. Here we will assume or guess that the pdf of θ, that is, the prior pdf $\pi(\theta)$, is given by the inverted gamma, that is, where α and β are hyper parameters.

$$\pi(\theta|\alpha,\beta) = \begin{cases} \frac{\beta^\alpha}{\Gamma(\alpha)}\theta^{\alpha-1}\exp\left(-\frac{\beta}{\theta}\right), & \theta > 0,\ \alpha,\ \beta > 0 \\ 0, & \text{elsewhere.} \end{cases}$$

Now, we have identified the pdf of the data, we can assume the prior pdf and, assuming a mean squared error loss function, we can obtain a Bayesian estimate of the true parameter θ.
The squared error loss function is given by

$$L(\theta,\hat{\theta}) = c(\theta)(\theta-\hat{\theta})^2.$$

and assume the $c(\theta) = 1$. We choose the estimate of θ, $\hat{\theta}$ so as to minimize the expected loss, $E[L(\theta,\hat{\theta})]$.
The posterior pdf is given by

$$P(\theta|X) = \frac{\pi(\theta)L(\theta|X)}{m(x)}. \tag{10.6}$$

The likelihood function $L(\theta|X)$ and prior pdf, $\pi(\theta)$ for a single observation in given by

$$L(\theta|X) = \frac{1}{\theta}e^{\frac{-X}{\theta}}\frac{\beta^\alpha}{\Gamma\alpha}\theta^{-(\alpha+1)}e^{\frac{-\beta}{\theta}}$$

$$= \frac{\beta^\alpha}{\Gamma\alpha}\frac{1}{\theta^{2+\alpha}}e^{\frac{-(\beta+x)}{\theta}}.$$

The marginal pdf, $m(x)$ is given by

$$m(x) = \int_0^\infty \frac{\beta^\alpha}{\Gamma\alpha}\frac{1}{\theta^{2+\alpha}}e^{\frac{-(\beta+x)}{\theta}}d\theta.$$

To compute the above integral, we make a transformation from θ to Y and assume $\theta = \frac{1}{y}$, then where $|J| = |\frac{d\theta}{dy}|$ is the absolute value of the Jacobian of transformation. Simplifying $m(x)$, we have

$$m(x) = \frac{\beta^\alpha}{\Gamma\alpha}\int_0^\infty y^{\alpha+2}\, e^{-y(\beta+x)}\, |J|\, dy,$$

$$m(x) = \frac{\beta^\alpha}{\Gamma\alpha}\int_0^\infty y^{\alpha+2}\, e^{-y(\beta+x)} y^{-2} dy$$

$$\frac{\beta^\alpha}{\Gamma\alpha}\int_0^\infty y^{(\alpha+1)-1}\, e^{-y(\beta+x)} dy$$

$$= \frac{\beta^\alpha}{\Gamma\alpha}\frac{\Gamma(\alpha+1)}{\Gamma(\beta+x)^{\alpha+1}}.$$

Thus, the posterior pdf, Eq. (10.6.1), $P(\theta|X)$, is given by

$$P(\theta|X) = \frac{(\beta+x)^{\alpha+1}}{\Gamma(\alpha+1)}\theta^{-(\alpha+2)}\, e^{\frac{-(\beta+x)}{\theta}}.$$

Note that for a single observation x, $P(\theta|X)$ is the same as an inverted gamma pdf with hyperparameter $\alpha + 1$ and $\beta + x$. Thus, in general, $\underset{\sim}{x} = (x_1, x_2, \ldots, x_n)$ the posterior pdf is given by

$$P(\theta|X = \underset{\sim}{x}) = \frac{(\beta + \sum_{i=1}^n x_i)^{\alpha+n}}{\Gamma(\alpha+n)}\theta^{-(\alpha+n)}\, e^{\frac{-\beta+\sum_{i=1}^n x_i}{\theta}}.$$

Using the R-output given below, we have estimated a and β for the given data, $\hat{\alpha} = 0.57$ and $\hat{\beta} = 0.06$.

```
> library(fitdistrplus)
> ob = fitdist(1/ro, "gamma")
> ob
Fitting of the distribution 'gamma' by maximum likelihood parameters:
                  Estimate                std. Error
Shape           0.57263897             0.07816841
rate            0.05705247             0.01167617
```

Thus, for $n = 75$, $\hat{\alpha} = 0.57$, and $\hat{\beta} = 0.06$, the posterior pdf, $P(\theta|X)$ is given by

$$P(\theta|X) = \frac{(36.9)^{75.57}}{\Gamma(75.57)}\theta^{-(75.57)} e^{\frac{-(36.9)}{\theta}},\ 0 < \theta.$$

Recall that the inverted gamma pdf is of the form

$$\pi(\theta; \alpha, \beta) = \frac{\beta^\alpha}{\Gamma(\alpha)}\theta^{\alpha-1}\exp^{\left(\frac{-\beta}{\theta}\right)},\ \theta > 0,\ \alpha, \beta > 0$$

that has mean, $E[\theta] = \frac{\beta}{\alpha-1}$ and $\mathrm{Var}(\theta) = \frac{\beta^2}{(\alpha-1)^2(\alpha-2)}$, $\alpha > 0$.

Thus, for a squared loss function, the expected value of the posterior pdf is the standard Bayesian estimate of θ. That is,

$$E(P(\theta|x)) = \frac{\hat{\beta}}{\hat{\alpha} - 1} = \frac{36.9}{75.57 - 1} \approx 0.495.$$

Thus,

$$\hat{\theta}_{OB} \approx 0.495.$$

Now, to calculate Bayesian credible interval (upper and lower confidence limits), say a and b, we need to find the

$$P[a \leq \theta \leq b|\ x_1, x_2, \ldots, x_n] \geq (1 - \alpha)100\%.$$

That is, we need to integrate

$$\int_0^a \frac{(36.9)^{75.57}}{\Gamma(75.57)} \theta^{-(75.57)} e^{\frac{-(36.9)}{\theta}} d\theta = \frac{\alpha}{2},$$

and

$$\int_0^b \frac{(36.9)^{75.57}}{\Gamma(75.57)} \theta^{-(75.57)} e^{\frac{-(36.9)}{\theta}} d\theta = \frac{\alpha}{2}.$$

The following R-code gives us a 95% and 90% confidence limits on true θ:

```
> library(pscl)
> c (qigamma (0.025, alpha=75.57, beta=36.9), qigamma(0.975, alpha=75.57, beta=36.9))
[1] 0.3945061      0.6201909
> c (qigamma (0.05, alpha=75.57, beta=36.9), qigamma(0.95, alpha=75.57, beta=36.9))
[1] 0.4081207      0.5964911
```

That is, the Bayesian 95% and 90% confidence limits (credible interval) are
$P[0.4 \leq \theta \leq 0.62] \geq .95$
and
$P[0.41 \leq \theta \leq 0.6] \geq .90$.
Thus, we have a 95% and 90% confidence range of 0.22 and 0.19, respectively. While these ranges are slightly wider than the non-Bayesian range, we need not assume the sampling distribution.
(c) Empirical Bayes: Bootstrap

Here we will use Bootstrap resampling to obtain MLE of each of the sample of the true parameter θ that behaves as a random variable. We follow the resampling procedure of Bootstrap that we discussed and obtain 50 samples from the original data $n = 75$ that was given and, through goodness-of-fit, we find they follow the exponential pdf, $P(x|\theta)$. That is, for each of the 50 samples of size 75, we obtained 50 estimates of θ; $\hat{\theta}_1^*, \hat{\theta}_2^*, \ldots, \hat{\theta}_{50}^*$ as shown in the table below (Table 10.8).

TABLE 10.8 Bootstrap Estimates of θ.

0.52	0.46	0.48	0.44	0.53	0.52	0.54	0.54	0.47	0.49
0.45	0.52	0.63	0.49	0.54	0.44	0.46	0.49	0.49	0.54
0.45	0.52	0.53	0.52	0.47	0.51	0.50	0.45	0.47	0.54
0.42	0.53	0.43	0.54	0.51	0.48	0.57	0.49	0.57	0.47
0.55	0.42	0.44	0.44	0.50	0.52	0.52	0.44	0.42	0.48

The R-codes for obtaining the Bootstrap of θ are

```
> set.seed(100)
>N ←length (ro)
> nboots ← 50
> boot.result ← numeric (nboots)
> for (i in 1:nboots)
+{
+      boot.samp ←sample(ro, N, replace= TRUE)
+      boot.result [i] ← mean (boot.samp)
+}
```

To obtain, if possible, the pdf of θ, we started with a histogram to obtain a visual indication of a possible pdf. Given below in Fig. 10.8 is the diagram of the 50 MLE of θ.

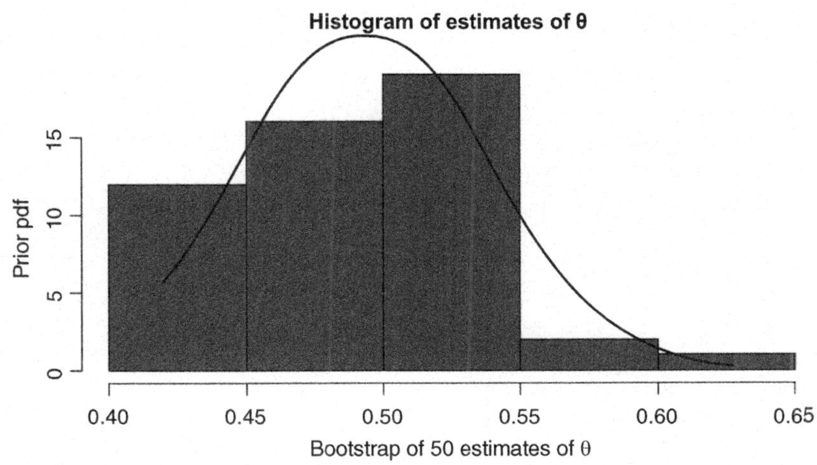

FIGURE 10.8 Histogram of the estimate of $\theta, \hat{\theta}^*_{50}$. A histogram titled Histogram of estimates of theta showing a bootstrap distribution of estimates of theta.

The histogram indicates a gamma pdf for $\hat{\theta}^*_{50}$. We performed a goodness-of-fit test to confirm that indeed the 50 Bootstrap estimates of θ, $\hat{\theta}^*_1, \hat{\theta}^*_2, \ldots, \hat{\theta}^*_{50}$ follow the gamma pdf. That is,

$$\pi(\hat{\theta}; \alpha, \beta) = \pi(\hat{\theta}^*|\alpha, \beta) = \begin{cases} \frac{1}{\Gamma(\alpha)\beta^\alpha}\theta^{\alpha-1}\exp\left(-\frac{\theta}{\beta}\right), & \hat{\theta} > 0, \alpha, \beta > 0 \\ 0, & \text{elsewhere.} \end{cases}$$

The goodness-of-fit result using the three tests are given below (Table 10.9).

TABLE 10.9 Goodness-of-Fit Result.

Goodness-of-Fit Tests for Exponential Distribution				
Test	**Statistic**		***P*-Value**	
Kolmogorov–Smirnov	D	0.10861215	Pr > D	0.146
Cramer–Von Mises	W-Sq	0.07346459	Pr > W-Sq	>0.250
Anderson–Darling	A-Sq	0.49233552	Pr > A-Sq	0.220

All three tests confirm the gamma prior pdf for $\hat{\theta}^*$ the estimate of θ using Bootstrap resampling. Through the goodness-of-fit testing, we obtained the MLE of the hyperparameter α and β of the identified prior pdf, that is, $\hat{\alpha} = 125.23$ and $\hat{\beta} = \frac{1}{253.5} = 0.0039$. Thus,

$$\pi(\hat{\theta}^*; \hat{\alpha}, \hat{\beta}) = \frac{1}{\Gamma(125.23)(0.0039)^{125.23}} \hat{\theta}^{*124.23} \exp\left(\frac{-\hat{\theta}^*}{0.0039}\right), \ \hat{\theta}^* > 0.$$

The posterior pdf, $\pi(\hat{\theta}^*; \hat{\alpha}, \hat{\beta} \mid \underline{X})$ is given by

$$\pi(\hat{\theta}^*; \hat{\alpha}, \hat{\beta} \mid \underline{X}) = \frac{(\hat{\beta}^{-1} + n\overline{X})^{\hat{\alpha}+n}}{\Gamma(\hat{\alpha}+n)} \hat{\theta}^{*\hat{\alpha}+n-1} \exp^{-\hat{\theta}^*(\hat{\beta}^{-1} + n\overline{X})},$$

$$= \frac{(355)^{175.23}}{\Gamma(175.23)} \hat{\theta}^{*174.23} \exp^{-\hat{\theta}^*(355)}, \ \hat{\theta}^* > 0.$$

We know, under squared error loss function, the Bayes estimate of θ is the posterior mean. That is, for $\alpha = n + \hat{\alpha}$ and $\beta = \hat{\beta}^{-1} + n\overline{X} = 355$, we have

$$E[\pi(\hat{\theta}^*; \hat{\alpha}, \hat{\beta} | \underline{X})] = \frac{n + \hat{\alpha}}{\hat{\beta}^{-1} + n\overline{X}} = \frac{50 + 125.23}{254.5 + 101.5} = 0.494.$$

We used the SAS code given below to obtain the necessary calculations.

```
data
input bootvar @@;
cards;
   0.52 0.46 0.48 0.44 0.53 0.52......
;
run;
proc univariate data= boot;
var bootvar;
histogram/ gamma odstitle = " fitting gamma distribution on 50 bootstap estimates" VAXISLABEL=
"prior"; inset n mean (5.3) std='Std Dev' (5.3)
skewness (5.3) kurtosis (5.3)
/ pos = ne header = 'Summary Statistics'; run;
```

Thus, the Bayesian estimate of θ under Bootstrap resampling to determine the prior pdf of θ is

$$\hat{\theta}^*_{empboot} = 0.494.$$

The analytical form of $100(1-\alpha)\%$ credible interval for the true parameter θ, under the Bootstrap Bayesian estimate for the true θ is given by

$$\int_0^a \frac{(355)^{175.23}}{\Gamma(175.23)} \theta^{(174.23)} e^{-\theta(355)} d\theta = \frac{\alpha}{2},$$

and

$$\int_b^\infty \frac{(355)^{175.23}}{\Gamma(175.23)} \theta^{(174.23)} e^{-\theta(355)} d\theta = \frac{\alpha}{2}.$$

Using the following R-code, we obtain (a, b) for 90% and 95% credible interval using the Bootstrap Bayes estimate of θ, $\hat{\theta}^*_{empboot}$:

```
> bootconf← rgamma(50, shape = 175.23, scale = .0028)
> library(EnvStats)
> eqgamma ( bootconf, p = 0.5, method = "mle," ci = TRUE,    ci.type = "two-sided,"
 +                    conf.level = 0.95, normal.approx.transform = "Kulkarni.powar,"
digits = 0)
```

Thus, the 90% credible interval for the true θ is [0.476, 0.495] and

$$P[0.476 \le \theta \le 0.495] \ge 90\%$$

with confidence range of 0.19. Similarly, the 95% credible interval for the true θ is [0.481, 0.502]; thus,

$$P[0.481 \le \theta \le 0.502] \ge 95\%$$

with a confidence range of 0.021. This is a much smaller confidence range compared to parts (a) and (b).

(d) Empirical Bayes Estimate: Jackknife

Here we shall use the Jackknife resampling method on the given data in Table 11.6.1. Recall that we have identified that the one parameter exponential pdf with the MLE of $\hat{\theta} = 0.492$, that is $f(x;0.492)$, the underlying pdf of the data.

Now, we will follow the Jackknife resampling procedure and obtain 50 samples and for each sample we will calculate the MLE of the true θ, that is, $\hat{\theta}_1^*, \hat{\theta}_2^*, \ldots, \hat{\theta}_{50}^*$. These Jackknife MLE estimates of θ are given below by Table 6.1.3 (Table 10.10).

TABLE 10.10 Jackknife Estimates of θ.

0.57	0.10	0.46	0.77	0.05	0.04	0.69	0.69	0.27	0.46
1.27	0.67	0.12	0.7	0.06	0.46	0.55	0.25	0.33	0.00
0.31	0.27	1.54	0.06	0.10	0.56	0.00	0.21	0.63	0.06
0.47	0.05	0.01	0.07	0.42	1.85	0.18	0.47	0.77	1.11
0.69	0.21	0.36	0.02	0.04	1.01	0.36	0.35	0.87	0.07

The R-code for obtaining the Jackknife estimates of θ is

```
> set.seed(10)
> jack ←numeric(length (ro)-1)
> pseudo ←numeric(length (ro))
> for (i in 1:length(ro))
+{  for (j in 1:length(ro))
+  {     if(j<i) jack[j] ←ro[j] else if(j>i) jack[j-1] ←ro[j]}
+     pseudo[i] ←length(ro)*mean(ro) − (length(ro)-1)*mean(jack)}
> samj ← sample(pseudo, 50, replace = T)
```

Now, we are interested in finding, if possible, the pdf of the 50 Jackknife estimates through goodness-of-fit testing. Using the three commonly used goodness-of-fit tests, the results are given in Table 10.11, below:

TABLE 10.11 Goodness-of-Fit Result.

Goodness-of-Fit Tests for Exponential Distribution				
Test	**Statistic**		***P*-Value**	
Kolmogorov–Smirnov	D	0.09045389	Pr > D	>0.500
Cramer–Von Mises	W-Sq	0.08260139	Pr > W-Sq	>0.250
Anderson–Darling	A-Sq	0.51924948	Pr > A-Sq	>0.250

All three tests at the level of significance, $\alpha = 0.05$, identified the one parameter exponential pdf fit the Jackknife estimates with the MLE of the hyperparameter being 0.46. That is, the estimated prior pdf of θ, $\pi(\theta; 0.46)$ is given by

$$\pi(\hat{\theta}^*; 0.46) = \frac{1}{0.46}\exp\left(-\frac{\hat{\theta}^*}{0.46}\right),\ \hat{\theta}^* \ge 0.$$

The histogram along with the graph of the estimated prior, $\pi(\theta; 0.46)$, is given below by Fig. 10.9.

The above figure shows the pdf of 50 Jackknife estimates which has been identified as the exponential pdf. Thus, we will use the exponential pdf as our prior to obtain the empirical Bayes estimate of the true θ.

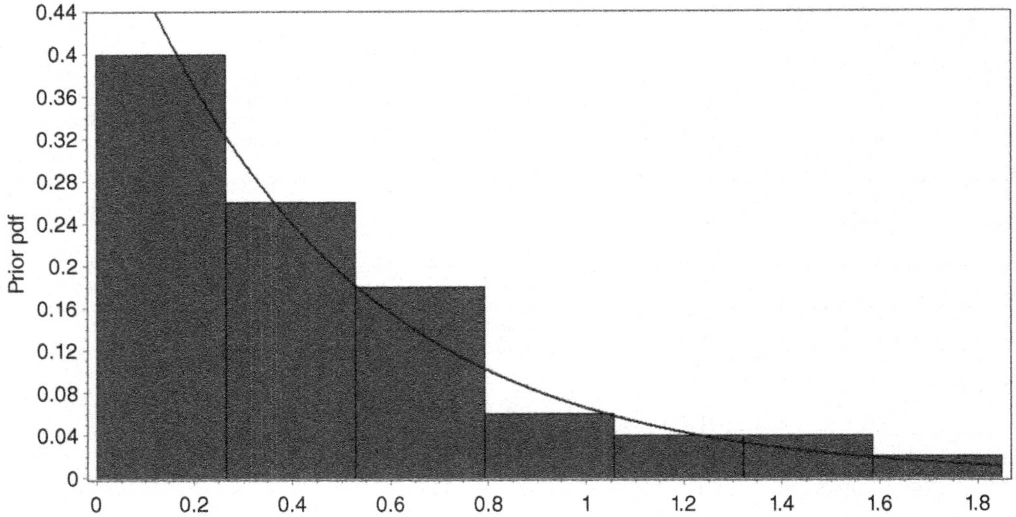

FIGURE 10.9 Histogram and prior pdf of $\hat{\theta}_1^*, \hat{\theta}_2^*, \ldots, \hat{\theta}_{50}^*$.

The SAS codes for producing the above results are given below:

```
data
input jackvar @@;
cards;
        0.57    0.10    0.46    0.77    0.05    0.04 ......
;
run;
proc univariate data= jack; var jackvar;
histogram/ exp odstitle= "fitting an Exponential distribution on 50 Jackknife estimates" VAXISLABEL= "Prior
PDF"; inset n mean (5.3) std='Std Dev' (5.3)
skewness (5.3) kurtosis (5.3)
/ pos = ne header = "Summary Statistics"; run;
```

The posterior pdf $\hat{\theta}^*$ with hyper parameter β, and its MLE $\hat{\beta}$, $\pi(\hat{\theta}^*; \hat{\beta} \mid x_1, x_2, \ldots, x_n)$, is given by

$$\pi(\hat{\theta}^*; \hat{\beta} \mid \underline{X}) = \frac{(\hat{\beta}^{-1} + n\overline{X})^{n+1}}{\Gamma(n+1)} \hat{\theta}^{*(n+1)-1} \exp^{-}\hat{\theta}^*(\hat{\beta}^{-1} + n\overline{X}),\ 0 \le \theta.$$

For $n = 50$ and $\hat{\beta} = 0.46$, we have

$$\pi(\hat{\theta}^*; 0.49 \mid 2.03) = \frac{(103.67)^{51}}{\Gamma(51)} \hat{\theta}^{*50} \exp^{-}\hat{\theta}^*(103.67),\ 0 \le \theta.$$

Thus, the posterior pdf of $\hat{\theta}^*$ is the gamma pdf with shape parameter equal to 51 and scale parameter equal to $\frac{1}{103.67}$. We know the Bayes estimate under the square error loss function is the posterior mean.

Thus, the mean or expected value of the gamma pdf. That is,

$$E\left[\pi\left(\hat{\theta}^{*}; 0.46|\underline{X}\right)\right] = \frac{n+1}{\hat{\beta}^{-1} + n\overline{X}} = \frac{50+1}{2.17+101.5} = 0.492,$$

which is fairly close to the MLE of the parameter θ, the standard Bayes and empirical Bootstrap estimate.

The analytical form of $100(1-\alpha)\%$ credible interval for the true parameter θ, under the Jackknife empirical Bayes estimate is given by

$$\int_{0}^{a} \frac{(103.67)^{51}}{\Gamma(51)} \theta^{(50)} e^{-\theta(103.67)} d\theta = \frac{\alpha}{2},$$

and

$$\int_{b}^{0} \frac{(103.67)^{51}}{\Gamma(51)} \theta^{(50)} e^{-\theta(103.67)} d\theta = \frac{\alpha}{2},$$

where (a, b) is the credible interval and α is the level of significance.

The following R-code give the 90% and 95% credible interval.

```
> jackconf← rgamma(50, shape = 51, scale = .0096)

> library(EnvStats)

> eqgamma (jackconf, p = 0.5, method = "mle," ci = TRUE,  ci.type = "two-sided,"
+                conf.level = 0.95, normal.approx.transform = "Kulkarn-.powar," digits = 0)
```

Thus, the 90% credible interval under Jackknife Bayes estimate $\hat{\theta}$ is [0.47, 0.5]. That is,

$$P[0.47 \le \theta \le 0.5] \ge 90\%.$$

The confidence range is 0.5–0.47=0.03. Similarly, the 95% credible interval is (0.457, 0.495), that is,with a confidence range of 0.038.

$$P[0.459 \le \theta \le 0.495] \ge 95\%.$$

Note that all the estimates of the true parameter θ, MLE, standard Bayes, empirical Bayes Bootstrapping and Jackknife are all very close.

In the literature, there are many different empirical Bayes models and various applications are available. The purpose of this section is mainly to introduce the concept of empirical Bayes.

EXERCISES 10.6

10.6.1 You are given the following observation, $n = 60$, in Table 10.12 below, characterizing the behavior of certain phenomenon, A, that we are interested to analyze and learn as much as possible about. Assume that the given data was randomly obtained.

- **(a)** Through goodness-of-fit testing identify, if possible, the pdf that characterizes probabilistically the behavior of phenomenon A, say $f(x;\theta)$.
- **(b)** In part (a), obtain the MLE of the parameter, or the parameters that drive the pdf, $f(x;\theta)$.
- **(c)** Determine the 90% and 95% confidence limit of the two parameters in $f(x;\theta_1, \theta_2)$.
- **(d)** Interpret the meanings and usefulness of part (a), (b), and (c), with respect to the phenomenon A.

10.6.2 For the data given in Exercise 10.6.1, assume that the parameter θ, in the pdf, $f(x;\theta)$ that you have identified, assume that it behave as a random variable with prior pdf, $\pi(\theta)$, that follows the exponential pdf. Assuming a mean squared error loss function, proceed to answer the following questions:

- **(a)** Obtain the standard Bayes estimate of the true parameter θ.

TABLE 10.12 The Data.

0.69	0.54	0.08	2.33	0.47	0.88	0.07	1.31	0.19	0.15
0.39	0.25	1.52	0.84	0.85	0.29	0.05	0.32	0.06	0.37
1.03	1.32	0.41	0.14	0.70	0.29	0.09	0.07	0.42	0.50
0.16	0.63	0.70	0.76	1.49	1.19	0.53	1.67	0.86	0.27
0.71	0.65	1.01	0.75	0.11	0.33	0.41	0.50	1.91	0.83
0.02	1.15	0.85	0.72	0.03	1.04	2.78	0.94	2.32	0.86

(b) Obtain 90% and 95% credible interval of the true parameter θ.
(c) Interpret the meaning of your results in comparison to those found in Exercise 10.6.1.

10.6.3
(a) Obtain an empirical Bayes estimate of the true θ, using the data in exercise 10.6.1, by estimating the prior $\pi(\theta)$ of the parameter θ, in $f(x \mid \theta)$; using the bootstrapping resampling method with $n = 50$.
(b) Obtain 90% and 95% credible interval for the true parameter θ.
(c) Interpret your results.
(d) Compare your findings with the MLE of θ, the standard Bayesian estimate of θ, and the empirical Bayes by bootstrapping of θ.

10.6.4 Repeat exercise 10.6.3, instead of using bootstrapping use Jackknife resampling and compare the four estimates of θ, that is, parametric, standard Bayes, and empirical Bayes using bootstrapping and the Jackknife estimate of the prior.

10.6.5 We were told by a laboratory scientist that they conducted an experiment and measured the behavior of a certain characteristic of 60 mice and the data they collected is given in the table below (Table 10.13).
The laboratory scientist also told us that the data follow a 2-parameter gamma pdf, $f(x;\alpha, \beta)$.
(a) Goodness-of-fit testing confirms the fact that the data follow a gamma pdf. Through the process, you have identified the MLE of the shape parameter α and location parameter β.
(b) If in part (a) the data follows the gamma pdf, $f(x; \alpha, \beta)$, find the 95% confidence limit on the true parameter α.
(c) Plot the pdf and its cumulative probability distribution of part (a).

TABLE 10.13 Laboratory Mice Data.

0.56	2.06	1.12	1.34	0.50	1.49	0.67	1.09	1.10	0.81
2.54	0.50	0.65	2.60	1.36	0.19	1.91	1.28	1.92	0.35
0.33	1.24	0.18	0.36	3.53	0.87	0.87	0.80	3.68	1.34
1.90	0.11	2.90	0.77	0.87	1.04	6.37	1.54	1.60	1.09
2.05	0.41	2.86	0.34	0.75	0.66	3.47	0.13	1.73	1.21
0.93	1.36	0.10	0.18	4.88	0.95	0.26	1.84	0.85	2.15

10.6.6 **(a)** If the given data follows the gamma pdf, assume the shape parameter behaves as a random variable with exponential pdf. Using mean squared error, obtain the standard Bayesian estimate of α.

(b) Obtain a 95% credible interval for the true parameter α and its confidence range. Interpret the meaning of your results.

(c) Compare and discuss the results of the parametric analysis in Exercise 10.6.5 with the standard Bayesian results.

10.6.7 **(a)** We want to estimate the prior pdf, $\pi(\alpha)$. Rather than assume or guess it, as in Exercise 10.6.6, use Bootstrap resampling from the given data, that is, estimate the prior, $\pi(\hat{\alpha})$ and proceed to obtain an empirical Bayes estimate of the true α.

(b) Obtain a 95% credible interval of the true parameter α and its confidence range.

(c) Compare the results of the parametric analysis, Exercise 10.6.5, standard Bayesian, Exercise 10.6.6 and empirical Bayes estimates using bootstrapping.

10.6.8 Repeat Exercise 10.6.7 (a), (b), and (c) using Jackknife resampling to obtain the empirical Bayesian estimate of the shape parameter α. Compare and discuss your current results with the results of Exercises 10.6.5, 10.6.6, and 10.6.7.

10.7 CHAPTER SUMMARY

In this chapter, we introduced the basic philosophy, definitions, and methods of performing statistical analysis in a Bayesian setting. The treatment of unknown parameters as if they are random variables provides a feedback mechanism to update our original beliefs about the parameter(s). The posterior distribution of the parameter(s) represents our revised belief and is calculated by combining data and prior knowledge. We also saw a brief explanation of Bayesian decision theory. It should be noted that there are various other aspects of Bayesian analysis, such as Bayesian regression, in which priors are used about the regression coefficients as well as about the error variance. It is beyond the scope of one chapter to deal with all aspects of Bayesian analysis. There are many publications on Bayesian statistics. We have also briefly studied some elements of decision theory, which has a natural base in the Bayesian approach. Empirical Bayes method calculations are illustrated through an example.

We now list some of the key definitions introduced in this chapter:

- Posterior distribution
- Quadratic loss function
- Absolute error loss function
- $100(1-\alpha)$ % credible interval
- Prior odds ratio
- Posterior odds ratio
- Observable

In this chapter, we have also learned the following important concepts and procedures:

- Bayesian parameter estimation procedure
- Bayesian credible interval procedure
- General decision theory procedure
- Procedure to find optimal decision
- Empirical Bayes

10.8 COMPUTER EXAMPLES

A very popular software (and it is free) for the Bayesian computation is WinBUGS, which can be obtained from http://www.mrc-bsu.cam.ac.uk/bugs/. Computing posterior probability for proportions using the steps, we learned in Section 10.2 that it can be performed using Minitab. Refer to the book, Bayesian Computation Using Minitab, by Jim Albert (Wadsworth, 1996). For R help, we suggest the book, Bayesian Computation with R (Second Edition), by Jim Albert, Springer, 2009. The methods explained in this book can also be used in Chapter 13.

EXAMPLES with R

In order to do the R-codes in this section, download the R-package "LearnBayes."

Example 10.8.1: For the data of example 10.2.1, write an R-code to obtain posterior.

Solution: We use $p = \theta$.

```
p=seq(0.8, 0.9, by = 0.02)
prior=c(0.13, 0.15, 0.22, 0.25, 0.15, 0.10)
prior=prior/sum(prior)
plot(p, prior, type="h", ylab="Prior Probability")
data=c(13, 2)
post=pdisc(p, prior, data)
post=pdisc(p, prior, data)
round(cbind(p, prior, post), 2)
```

Output:

Figure: discrete prior distribution for a proportion *p*

p prior post
[1,] 0.80 0.13 0.11
[2,] 0.82 0.15 0.14
[3,] 0.84 0.22 0.23
[4,] 0.86 0.25 0.27
[5,] 0.88 0.15 0.16
[6,] 0.90 0.10 0.10

EXAMPLE 10.8.2

(Posterior calculation). Consider Example 10.2.4 with $\mu_p = 100$, $\sigma_p = 15$, and $x = 115$. Write an R-code to find the posterior.

Solution

R-code:

```
library (LearnBayes)
mup=100
sigmp=15
sigma=10
x=115
post=rnorm(1000,((sigma^2*mup/(sigmp^2+sigma^2))+(sigma^2*x/(sigmp^2+sigma^2))),(sigma^2*sigmp^2/(sigmp^2+-
sigma^2)))
post
hist(post)
```

Output

Along with many posterior sample values, we will get the following histogram for the posterior:

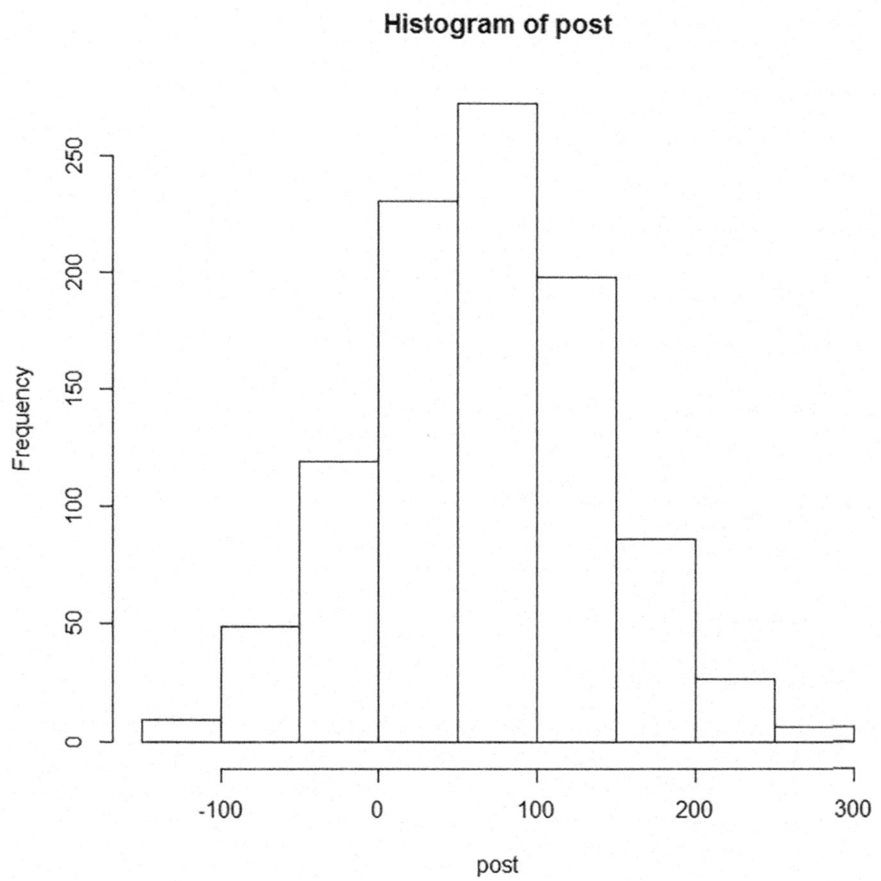

EXAMPLE 10.8.3

(Credible interval) Obtain a 95% credible interval for the posterior obtained in Example 10.8.2.

Solution

Once we have posterior stored in post, then the following will give us the credible interval:

R-code

```
quantile(post, c(0.025,0.5,0.975))
```

Output
2.5% 50% 97.5%
-76.84277 66.83870 207.86700

Example 10.8.4

(Bayesian Hypothesis testing)

The following is random data from a normal distribution with variance 9:

0.92	1.05	5.53	3.64	−4.47	−2.60	0.71	−3.66	1.38	3.87
7.42	1.76	0.01	2.69	1.54	3.97	1.34	−1.63	−1.24	−4.78

Test the hypothesis, H_0: $\mu \le 0$ versus H_a: $\mu > 0$. Assume that the prior is N(0, 4), so that $\mu \le 0$ and $\mu > 0$ are equally probable.

Solution

R-code

```
y=c(.92, 7.42, 1.05, 1.76, 5.53,.01, 3.64, 2.69, -4.47, 1.54,
+ -2.60, 3.97,.71, 1.34, -3.66, -1.63, 1.38, -1.24, 3.87, -4.78)
pop.s=3
norpar=c(0,4) # vector of mean and standard deviation of the normal prior distribution
m0=0 # value of the normal mean to be tested
mnormt.onesided(m0,normpar,data)
```

Output

$BF (*Bayes factor in support of the null hypothesis)*
[1] 0
$prior.odds (*prior odds of the null hypothesis)*
[1] 0.7621303
$post.odds (*posterior odds of the null hypothesis)*
[1] 0
$postH (posterior probability of null hypothesis)
[1] 0

Post. Odds <1
reject the null hypothesis

PROJECTS FOR CHAPTER 10

10A. Predicting Future Observations

Suppose we want to predict the value of future observations based on the prior and observed data. In addition to the posterior distribution $f(\theta|x)$, in Bayesian statistics we are interested in the marginal density of the observations (note that because both θ and x are random, it makes sense to speak about their joint, marginal, and conditional densities). Using the Bayes theorem, we have seen that g(x) is at $x = (x_1, \ldots, x_n)$ (for the continuous case) to be where $f(x|\theta)\pi(\theta)$ is the joint density of x and θ. This can also be written as

$$g(x) = \int f(x|\theta)\pi(\theta)d\theta$$

$$g(x) = E[f(x|\theta)],$$

the expected density of observations with respect to the prior distribution $\pi(\theta)$. With the help of $g(x)$, we can predict observations.

We are more interested in the density of future observations y, given present data x. However, because we have already updated the value of θ using the posterior density, this should be reflected in our prediction:

$$
\begin{aligned}
f(y|x) &= \int f(y, \theta|x)\,d\theta \\
&= \int f(y|\theta, x)\cdot\pi(\theta|x)\,d\theta \\
&= \int f(y|\theta)\pi(\theta|x)\,d\theta,
\end{aligned}
$$

if y and x are conditionally independent given θ. Conditional independence is achieved, for example, when $x = (x_1,\ldots, x_n)'$ and $y = (x_{n+1},\ldots, x_{n+m})'$ both are samples from $f(x|\theta)$.

We see that the density of future observations is the expected density of observations with respect to the posterior distribution. Consider two different priors for θ.

Uniform [0,2], (2) $N(1, 1/16)$. Assume $f(x|\theta) \sim N(\theta, 1)$. Find the predictive distributions given the sample $X_1, X_2, \ldots, X_n$.

Chapter 11

Categorical Data Analysis and Goodness of Fit Tests and Applications

Chapter outline

OBJECTIVE

In this chapter, we will study various methods of categorical data analysis, including goodness of fit tests, to determine if a given set of data follows a particular probability distribution.

Karl Pearson
(Source: http://www-history.mcs.st-and.ac.uk/~history/PictDisplay/Pearson.html).

Karl Pearson (1857–1936) is considered the founder of the 20th-century science of statistics. Pearson has contributed to several different fields, such as anthropology, biometry, genetics, scientific methods, and statistical theory. He applied statistics to biological problems of heredity and evolution. In 1911 he founded the world's first university statistics department at University College London.

Mathematical Statistics with Applications in R. DOI: https://doi.org/10.1016/B978-0-443-27548-7.00011-6

He is the author of *The Grammar of Science*, the three volumes of *The Life, Letters and Labours of Francis Galton*, and *The Ethic of Freethought: A Selection of Essays and Lectures*. Pearson was the founder of the statistical journal *Biometrika*. In 1900, he published a paper on the chi-square goodness of fit test that we will study in this chapter. This is one of Pearson's most significant contributions to statistics. In 1893, Pearson coined the term "standard deviation."

11.1 INTRODUCTION

Techniques presented in the previous chapters are mostly designed for quantitative or numerical data that include both discrete and continuous data. In general, there are two types of data, namely, quantitative and categorical. This chapter provides an introduction to categorical data analysis. Categorical (or qualitative) data is the outcome of an experiment or a process that can be categorized into a finite number of mutually exclusive groups or categories. The categorical variables are measured on a scale that is nominal or ordinal. These data are represented through contingency tables. Examples of categorical variables are political philosophy of a person (such as liberal, conservative, or moderate), sex of an individual, make and model of a new auto, education level of an individual, or customer satisfaction surveys with categories such as poor, fair, good, great, excellent, etc. Categories may either be unordered (*nominal*) or ordered (*ordinal*). Telephone numbers, zip codes, blood types, occupation, gender, race/ethnicity, etc., have no particular order. Age group, degree of agreement with a statement on a questionnaire (strongly agree, agree, neutral, disagree, strongly disagree, etc.), grade in an exam (such as A, B, C, etc.), or patient condition (poor, fair, good, excellent) has a natural ordering of categories. Binary variables such as success and failure are those for which nominal-ordinal distinction is unimportant. Categorical variables can be analyzed with the chi-square goodness of fit test. Counts and percentages are the basic statistics available for categorical variables. Hence, goodness of fit tests consist of determining whether the frequency counts in the categories of the variable agree with a specific distribution. For regression analysis with a contingency table, we will use logistic regression.

Categorical data can be summarized using a frequency table. We can use a bar graph, Pareto chart, or pie chart for graphically representing the categorical data. There are many other effective graphical representations available in practice; however, they are beyond the level of this book. For a detailed account on categorical analysis, we refer to other books, such as Agresti's, on the topic.

11.2 CONTINGENCY TABLES AND PROBABILITY CALCULATIONS

In categorical data, the observed frequencies are organized in rows and columns like a spreadsheet. The table of observed cell frequencies is called a *contingency table*. Basics of two-way contingency tables are introduced in this section. Categorical data are often summarized by reporting the proportion or percentage of each category. Contingency tables are used in recording counts or percentages for categorical data. We might be interested in whether the new medicine's effectiveness depends on sex. Contingency tables are very useful for figuring out whether two events are dependent or independent. In this section, we will study two-way contingency tables with N rows and M columns. We will now give examples where $N = 2$ and $M = 2$, as well as where M and N are both greater than 2. In many applications, 2×2 contingency tables are very common, where binary variables (yes-no or success-failure) play an important role. In the following example, the effectiveness is almost the same; hence, the treatment can be considered gender-neutral (Table 11.1).

EXAMPLE 11.2.1

In a medical trial to study effectiveness of a new medication for a specific illness, 180 patients were included in the study, among whom 80 were females and 100 were males. Out of these people, 55 females and 68 males responded positively to the medication.

(a) Create a contingency table.
(b) What is the probability that the medication gives a positive (success) result for males?
(c) What is the overall probability that the medication gives a positive result?
(d) Is a positive response to the new medication independent of gender?

Solution

a) *The contingency table is given by*

TABLE 11.1 Effect of Medication Based on Sex.

	Male	*Female*	*Total*
Positive	*68*	*55*	*123*
Negative	*32*	*25*	*57*
Total	*100*	*80*	*180*

b) *The probability that the medication gives a positive result for males is*

$$P(\textit{positive if male}) = \frac{68}{100} = 0.68.$$

c) *The overall probability that the medication gives a positive result is*

$$P(\textit{overall positive}) = \frac{123}{180} = 0.6833.$$

d) *Recall that two events A and B are independent if and only if* $P(A \cap B) = P(A)P(B)$. *Let the event A represent female and event B represent positive response to medication.*
Then, $P(A) = \frac{80}{180} = 0.44$, *and* $P(B) = \frac{123}{180} = 0.68$. *Also,*

$$P(A \cap B) = \frac{55}{180} = 0.305$$
$$\neq P(A)P(B) = 0.300.$$

Hence, positive response to the new medication and gender may not be independent events. However, to make a statistical conclusion, we need to perform the chi-square test described later in the chapter.

In general, a 2 × 2 contingency table can be written as in Table 11.2.

TABLE 11.2 Notation for Joint Outcomes.

		Y		
		1	2	Total
X	1	n_{11}	n_{12}	n_1
	2	n_{21}	n_{22}	n_2
		n^1	n^2	n

where n_{11}, n_{12}, n_{21}, and n_{22} are random variables that have a multinomial distribution with sample size $n = (n_{11} + n_{12} + n_{21} + n_{22})$, and we can create a corresponding probability table for the joint distribution as in Table 11.3.

TABLE 11.3 Notation for Joint Probabilities.

		Y	
		1	2
X	1	π_{11}	π_{12}
	2	π_{21}	π_{22}

Thus, π_{11}, π_{12}, π_{21}, and π_{22} define the probability structure of the contingency table, where $\pi_{ij}'s$ can be estimated using observed data, $p_{ij} = n_{ij}/n$. From Table 11.2, we can estimate the following probabilities:

$$\hat{P}(Y = 1) = \frac{n_{11} + n_{21}}{n} = \frac{n^1}{n}$$

$$\hat{P}(X = 1) = \frac{n_{11} + n_{12}}{n} = \frac{n_1}{n}$$

$$\hat{P}(Y = 1|X = 1) = \frac{n_{11}}{n_{11} + n_{12}} = \frac{n_{11}}{n_1}$$

$$\hat{P}(X = 1|Y = 1) = \frac{n_{11}}{n_{11} + n_{21}} = \frac{n_{11}}{n^1}.$$

The marginal probability distributions of X and Y are the sums of cell probabilities across the columns and rows, respectively. In disease diagnostic tests, usually Y is taken as the outcome of the test (positive [1], negative [2]), and X is the actual condition (has disease [1], no disease [2]). With this interpretation in Table 11.2, we can define sensitivity and specificity.

Definition 11.2.1. *Sensitivity* (also called *true positive rate*) is defined as the probability that a patient gets a positive test result when they have the disease, that is, the proportion of actual positives that are correctly identified:

$$Sensitivity = P(Y = 1|X = 1).$$

Specificity (or *true negative rate*) is the probability that the patient gets a negative test result when they don't have the disease, that is, specificity measures the proportion of actual negatives that are correctly identified. Thus,

$$Specificity = P(Y = 2|X = 2).$$

In diagnostic tests, it is better to rewrite the contingency table in the following manner.

In many of the categorical data analyses, the *odds ratio* plays an important role, appearing as a parameter in the models, as a measure of association, or as a relative measure of effect. Thus, an odds ratio is a relative measure of the effect of a treatment. In Table 11.4, a 2 × 2 table, within row i, the odds of success instead of failure are $L_i = \pi_i/(1 - \pi_i)$. The ratio of odds L_1 and L_2 is the odds ratio given by

$$\theta = \frac{L_1}{L_2} = \frac{\pi_1/(1 - \pi_1)}{\pi_2/(1 - \pi_2)}.$$

TABLE 11.4 Notation for Probabilities.

		Test result (Y)	
		Positive (1)	Negative (2)
True state (X)	Positive (1)	π_1 (sensitivity)	$1 - \pi_1$ (false negative)
	Negative (2)	π_2 (false positive)	$1 - \pi_2$ (specificity)

For joint distributions with cell probabilities (π_{ij}) in Table 11.3, the odds in row i are $O_i = \frac{\pi_{i1}}{\pi_{i2}}$, where $i = 1,2$. Then the *odds ratio* is defined as

$$\theta = \frac{\pi_{11}/\pi_{12}}{\pi_{21}/\pi_{22}} = \frac{\pi_{11}\pi_{22}}{\pi_{12}\pi_{21}}.$$

Thus we have the following interpretations based on the values of the odds ratio. $\theta = 1$ will correspond to the independence of X and Y. When $1 < \theta < \infty$, subjects in row 1 are more likely to have a success than are the subjects in row 2, and for values of $0 < \theta < 1$, subjects in row 2 are more likely to have success than are subjects in row 2 (control—placebo or no treatment—is better than intervention). The odds ratio is always nonnegative. When values of θ are further away from 1, it represents a stronger association in that direction.

EXAMPLE 11.2.2

For the data of Example 11.2.1, we rewrite the table in terms of probabilities and obtain the odds ratio.

Solution

The contingency table in terms of probabilities is given by (after approximating to two decimal places)

	Male	Female	Totals
Positive	*0.38*	*0.31*	*0.69*
Negative	*0.17*	*0.14*	*0.31*
Total	*0.55*	*0.45*	*1.00*

Now, the odds ratio is given by

$$\theta = \frac{(0.38)(0.14)}{(0.31)(0.17)} = 1.009.$$

That is, for males, the test is slightly more likely to give a correct result than for females.

Contingency tables can have more than two categories, as can be seen in Example 11.2.2.

EXAMPLE 11.2.3

Fruit trees are subject to a bacteria-caused disease commonly called fire blight (because the resulting dead branches look like they have been burned). One can imagine several different treatments for this disease: treatment *A*, no action (a control group); treatment *B*, careful removal of clearly affected branches; and treatment *C*, frequent spraying of the foliage with an antibiotic in addition to careful removal of clearly affected branches. One can also imagine several different outcomes from the disease:

outcome 1, tree dies in the same year as the disease was noticed; outcome 2, tree dies 2–4 years after the disease was noticed; and outcome 3: tree survives beyond 4 years. A group of N trees is sorted into one of the treatments (i.e., every tree falls into exactly one of the following treatment categories [A | B | C]), and over the next few years the outcome is recorded (i.e., every tree falls into exactly one of the following outcome categories [1 | 2 | 3]). If we count the number of trees in a particular treatment/outcome pair (e.g., the number of trees that received treatment B and lived beyond 4 years: $B3 = 6$), we can display the results in a contingency table:

		Treatment		
Outcome	A	B	C	Total
1	8	5	3	16
2	4	3	3	10
3	3	6	7	16
Total	15	14	13	42

a) What is the probability that a randomly selected tree was given treatment B?
b) What is the probability that a randomly selected tree received treatment B given it had outcome 2?
c) What is the probability that a randomly selected tree received treatment B or will have outcome 2?

Solution

(a) *From the table,* $P(B) = \frac{14}{42} = 0.33$.
(b) *From the table,* $P(B|2) = \frac{3}{10} = 0.3$.
(c) *Thus we have*

$$P(B \cup 2) = P(B) + P(2) - P(B \cap 2)$$
$$= \frac{14}{42} + \frac{10}{42} - \frac{3}{42} = 0.5.$$

The general representation of a two-way $r \times c$ contingency table, with cells representing counts of outcomes with n_{ij} representing observed cell frequency at cell (i, j), can be represented by a table of the following form (Table 11.5).

TABLE 11.5 Two-Way $r \times c$ Contingency Table.

$X\backslash Y\rightarrow$ $\downarrow$	**1**	**2**	**...**	**c**	**Total**
1	n_{11}	n_{12}	...	n_{1c}	n_1
2	n_{21}	n_{22}	...	n_{2c}	n_2
. . .	. . .	. . .	. . . ...	. . .	
r	n_{r1}	n_{r2}	...	n_{rc}	n_r
Total	n^1	n^2	...	n^c	n

Here, $n = \sum_{i=1}^{r} n_i = \sum_{j=1}^{c} n^j$ is the total number of observations, $n_i = \sum_{j=1}^{r} n_{ij}$ is the marginal frequency of row i, and $n^j = \sum_{i=1}^{c} n_{ij}$ is the marginal frequency of column j, $i = 1, \ldots, r$, $j = 1, \ldots, c$. From this table, we can calculate various probabilities. For example, the joint distribution of X and Y can be expressed using the multinomial distribution given by

$$P(N_{11} = n_{11}, \ldots, N_{rc} = n_{rc}) = \frac{n!}{\prod_{i=1}^{r} \prod_{j=1}^{c} n_{ij}!} \prod_{i=1}^{r} \prod_{j=1}^{c} p_{ij}^{n_{ij}}$$

where the probability of having an outcome with $X = i$ and $Y = j$ is denoted as

$$p_{ij} = P(X = i, Y = j) = \frac{n_{ij}}{n}, \; i = 1,2, \ldots, r, \; j = 1,2, \ldots, c.$$

By restricting to a particular column or row, we can also obtain the conditional probabilities.

EXERCISES 11.2

11.2.1. In a random sample of 120 females and 110 males, 80 females and 70 males own iPhones; the rest own other brands. (a) Create a contingency table. (b) What is the probability that a chosen female doesn't own an iPhone? (c) What is the probability that a randomly chosen person from this group owns an iPhone? (d) Are gender and iPhone ownership independent?

11.2.2. In order to study the association between mortality and treatment, a sample of 150 mice was divided into two groups; 110 were given a standard dose of pathogenic bacteria followed by an antiserum, and a control group of 40, after receiving pathogenic bacteria, was not given the antiserum. After a month the numbers of alive and dead mice in each group are given in Table 11.6.

Compare the probabilities of survival between the two groups.

TABLE 11.6 Contingency: Treatment and Mortality Rate.

	Outcome		
	Alive	**Dead**	**Total**
Antiserum	80	30	110
Control	15	25	40
Total	95	55	150

11.2.3. Among teen drivers, two major causes of accidents are texting while driving and drunk driving. In a sample of 1000 accidents, Table 11.7 lists the numbers of fatal and nonfatal accidents based on the two causes of accidents.

TABLE 11.7 Contingency Table: Accidents Due to Texting While Driving and Drunk Driving.

Type of Accident↓	Reason		
	Texting While Driving	**Drunk Driving**	**Total**
Fatal accident	210	42	252
Nonfatal accident	140	608	748
Total	350	650	1000

(a) What is the probability that a randomly chosen teen involved in an accident was texting while driving, given that they were involved in a nonfatal accident?
(b) What is the probability that a randomly chosen teen was driving while drunk, given that they were involved in a nonfatal accident?
(c) What is the probability that a randomly chosen teen is involved in an accident?

11.2.4. In two simple random samples of 100 males and 100 females, the color of their eyes was recorded. Here, you are now sampling from two different populations that may have different response probabilities. The actual data of the experiment is summarized in Table 11.8.

TABLE 11.8 Contingency Table: Sex by Eye Color.

	Eye Color			
Sex	**Blue**	**Green**	**Brown**	**Total**
Female	40	25	35	100
Male	45	20	35	100
Total	85	45	70	200

(a) What is the probability that a chosen female doesn't have Brown eye color?
(b) What is the probability that a randomly chosen person from this group has brown eyes?
(c) What is the probability that a randomly chosen person from this group has brown eyes, given that he is a male?
(d) Create a relative frequency table and interpret its contents.

11.3 ESTIMATION IN CATEGORICAL DATA

Estimation in categorical data generally involves the proportion of "successes" in a given population. This may consist of estimating a single population proportion, comparing two population proportions, or investigating the potential relationship between two or more categorical variables. Thus, if X is a binary response from a trial with two possible outcomes (success/failure), and if we are interested in obtaining a confidence interval for the true value of p, the proportion of successes in a given population. In Sections 5.5.2 and 5.5.7, we studied single-population and two-population cases, respectively. We will summarize the results here.

11.3.1 Large Sample Confidence Intervals for *P*

For a random sample of size n from a given population, the point estimate of the population parameter p is given by

$$\hat{p} = \frac{\text{the number of ``successes''}}{n} = \frac{X}{n}.$$

The statistic $\hat{p}$ is the key entity in the binomial probability estimation, with true mean p and variance $(p(1-p))/n$, respectively. For large sample size n (if both $np \geq 5$ and $n(1-p) \geq 5$), we use the normal probability density function (PDF) to obtain an approximate $100(1-\alpha)\%$ confidence interval for p, which is given by

$$\left(\hat{p} - z_{\alpha/2}\sqrt{\frac{\hat{p}(1-\hat{p})}{n}}, \hat{p} + z_{\alpha/2}\sqrt{\frac{\hat{p}(1-\hat{p})}{n}}\right).$$

That is,

$$P\left[\hat{p} - z_{\alpha/2}\sqrt{\frac{\hat{p}(1-\hat{p})}{n}} \leq p \leq \hat{p} + z_{\alpha/2}\sqrt{\frac{\hat{p}(1-\hat{p})}{n}}\right] \approx (1-\alpha)$$

and based on the random sample of size n, we are about $100(1-\alpha)\%$ certain that the true value of p is in the interval $\left(\hat{p} - z_{\alpha/2}\sqrt{\frac{\hat{p}(1-\hat{p})}{n}},\ \hat{p} + z_{\alpha/2}\sqrt{\frac{\hat{p}(1-\hat{p})}{n}}\right)$.

We now restate a procedure from Section 5.7 for a large sample confidence interval for the difference of the true proportions, $p_1 - p_2$, in two binomial distributed populations.

Large Sample Confidence Interval for p_1–p_2

The $(1-\alpha)100\%$ large sample confidence interval for p_1–p_2 is given by

$$(\hat{p}_1 - \hat{p}_2) \pm z_{\alpha/2}\sqrt{\left(\frac{\hat{p}_1(1-\hat{p}_1)}{n_1} + \frac{\hat{p}_2(1-\hat{p}_2)}{n_2}\right)},$$

where $\hat{p}_1$ and $\hat{p}_2$ are the point estimators of p_1 and p_2. This approximation is applicable if $\hat{p}_i n_i \geq 5$, $i = 1,2$ and $(1-\hat{p}_i)n_i \geq 5$, $i = 1,2$. The two samples are independent.

The Wald confidence interval can be obtained using the following R commands:

```
1-sample proportion test (Wald)
library(epitools)
binom.approx(9,20)
```

We will get following output:

X	n	Proportion	Lower	Upper conf. level	
9	20	0.45	0.2319678	0.6680322	0.95

From this, we can see that about 95% confidence interval for the true proportion p is (0.2319678, 0.6680322).

11.4 HYPOTHESIS TESTING IN CATEGORICAL DATA ANALYSIS

Hypothesis testing on population proportion is the same as in Sections 6.4 and 6.5 for one and two proportions, respectively, and we will refer to those sections. Now we will explain the chi-square tests. There are two kinds of chi-square tests: one-way and two-way analyses. For example, if we are interested in comparing the effectiveness of two or more types of drugs in treating a particular disease, we will have a one-way analysis. Note that in order to use ANOVA or a t-test, we need at least one of the variables to be continuous. Thus we resort to Pearson's chi-square test. In addition, if we are interested to find out whether these drugs affect differently for men and women, then we need a two-way analysis, and in two-way analyses, we will use data from contingency tables. The purpose of both is to determine if the observed frequencies are significantly different from the frequencies that we would expect by chance or from a hypothesized distribution. In both one-way and two-way analyses, chi-square tests are commonly used.

11.4.1 The Chi-Square Tests for Count Data: One-Way Analysis

A chi-square test is used to analyze categorical data, and it is intended to test how likely it is that an observed probability distribution is due to chance, that is, to test whether a frequency distribution observed for categories fits an expected probability distribution. In this section, we will study several commonly used tests for count data, where observations are obtained by counting and assume nonnegative integer values, $\{0, 1, 2, \ldots\}$ (these tests can be considered as one-way tests). These are basically large sample tests based on a chi-square approximation. Suppose that we have outcomes of a multinomial experiment that consists of k mutually exclusive and exhaustive events, $A_1, \ldots, A_k$. Let $P(A_i) = p_i$, $i = 1, 2, \ldots, k$. Then $\sum_{i=1}^{k} p_i = 1$. Let the experiment be repeated n times, and let X_i (i=1, 2,…,k) represent the number of times the event A_i occurs. Then $(X_1, \ldots, X_k)$ have a multinomial distribution with parameters n, $p_1, \ldots, p_k$. Recall that if n is the total number of trials, that is, $x_i \in \{0,1,\ldots,n\}$ with $\sum_{i=1}^{k} x_i = n$, then the probability mass function of the multinomial distribution is given by

$$\frac{n!}{x_1!\ldots x_k!}p_1^{x_1}\ldots p_k^{x_k},$$

with $E(X_i) = np_i$.

Now, let

$$Q^2 = \sum_{i=1}^{k} \frac{(X_i - np_i)^2}{np_i}.$$

It can be shown that for large n, the random variable Q^2 is approximately χ^2-distributed with $(k - 1)$ degrees of freedom. It is required that $np_i \geq 5$ $(i = 1, 2, \ldots, k)$ for the approximation to be valid, although the approximation generally works well if we only have a few values of i (no more than 20% of the total cells), $np_i \geq 1$, and the rest (about 80%) satisfy the condition that $np_i \geq 5$. This statistic was proposed by Karl Pearson in his 1900 paper.

It should be noted that the chi-square tests that we are studying in this section are approximate tests valid for large samples. Often X_i is called the observed frequency and is denoted by O_i (this is the observed value in class i), and np_i is called the expected frequency and is denoted by E_i (this is the theoretical distribution frequency under the null hypothesis). Thus, with these notations, we can calculate

$$Q^2 = \sum_{i=1}^{k} \frac{(O_i - E_i)^2}{E_i} = \sum \frac{(\text{Observed} - \text{Expected})^2}{\text{Expected}}.$$

The example given below illustrates how we apply this goodness of fit test.

EXAMPLE 11.4.1

A plant geneticist grows 200 progeny from a cross that is hypothesized to result in a 3:1 phenotypic ratio of red-flowered to white-flowered plants. Suppose the cross process produces 170 red-flowered to 30 white-flowered plants. (a) Calculate Q^2 for this experiment. (b) Does the given data support the 3:1 ratio at $\alpha = 0.05$?

Solution

There are two categories of data totaling $n = 200$. Hence, $k = 2$. Let $i = 1$ represent red-flowered and $i = 2$ represent white-flowered plants. Then $O_1 = 170$, and $O_2 = 30$.

Here, we want to test the hypothesis to answer the questions posed.

H_0: The flower color population ratio is 3:1,

Vs.

H_a: The sampled flower color population has a ratio that is not 3 red:1 white.

a) *We are given that the probability of red flowers is $p_1 = 3/4$ and the probability of white flowers is $p_2 = 1/4$ and the condition that $np_1 \geq 5$ and $np_2 \geq 5$, are satisfied. Thus we can proceed to calculate Q^2 for the information that is given.*

Thus

$E_1 = np_1 = (200)(3/4) = 150$, *and* $E_2 = np_2 = (200)(1/4) = 50$

and

$$Q^2 = \sum_{i=1}^{2} \frac{(O_i - E_i)^2}{E_i}$$

$$= \frac{(170 - 150)^2}{150} + \frac{(30 - 50)^2}{50} = 10.667.$$

Since $k = 2$, from the χ^2 - table with 1 degree of freedom and $\alpha = 0.05$, the rejection region is $\{Q^2 > \chi^2_{1,0.05} = 3.841\}$. Since 10.667 is greater than 3.841, we reject the null hypothesis and conclude that the color ratio is not 3:1. The data supports the alternative hypothesis that the ratio is not 3 red:1 white.

The type of calculation in Example 11.4.1 gives a measure of how close our observed frequencies compared to the expected frequencies. The smaller the values of Q^2, the better the fit of the data. The test is also called a **"goodness of fit"** test statistic, because this measures how well the observed distribution of the data fits with the distribution that is expected if data are consistent with the assumed distribution. Note that this is equivalent to testing the parameters of a multinomial distribution. Let an experiment have k mutually exclusive and exhaustive outcomes $A_1, A_2, \ldots, A_k$. We would like to test the null hypothesis that all the $p_i = p(A_i)$, $i=1, 2, \ldots, k$ are equal to known numbers p_{i0}, $i = 1, \ldots, k$.

The procedure that we use to test the subject hypothesis is summarized below.

Testing the Parameters of a Multinomial Distribution (Summary)

To test

H_0: $p_1 = p_{10}, \ldots, p_k = p_{k0}$

vs.

H_a: At least one of the probabilities is different from the hypothesized values.

The test is always a one-sided upper tail test.

Let O_i be the observed frequency, $E_i = np_{i0}$ be the expected frequency (frequency under the null hypothesis), and k be the number of classes. The test statistic is

$$Q^2 = \sum_{i=1}^{k} \frac{(O_i - E_i)^2}{E_i}.$$

The test statistic Q^2 has an approximate chi-square probability distribution with $k - 1$ degrees of freedom.

The rejection region is given by

$$Q^2 \geq \chi^2_{k-1,\alpha}.$$

Assumption: $E_i \geq 5$ for all k and no more than 20% cells have $5 > E_i \geq 1$.

Note that the chi-square test will tell us if there is a significant difference between the observed data and the hypothesis distribution. However, it cannot test the strength of dependence or direction of the difference. This test is known as the χ^2– *goodness of fit test.* It implies that if the observed data are very close to the expected data, we have a very good fit and we do not reject the null hypothesis. That is, for small Q^2 values, we don't have enough evidence to reject H_0 and hence we will not reject H_0.

The following examples illustrate how we apply the chi-square goodness of fit test.

EXAMPLE 11.4.2

A TV station broadcasts a series of programs on the ill effects of smoking marijuana. After the series, the station wants to know whether people have changed their opinion about legalizing marijuana. Historical data from before the series shows the proportions of different categories of opinions are what is shown in the first table below, and then the second table below shows the sample proportion from the 500 randomly selected people:

Before the Series Was Shown

For legalization	Decriminalization	Existing law (fine or imprisonment)	No opinion
7%	18%	65%	10%

After the Series Was Shown

For legalization	Decriminalization	Existing law (fine or imprisonment)	No opinion
39%	9%	36%	16%

Here, $k = 4$, and we wish to test the following hypothesis

$$H_0: p_1 = 0.07; \quad p_2 = 0.18; \quad p_3 = 0.65; \quad p_4 = 0.1$$

Vs.

Ha: At least one of the probabilities is different from the hypothesized value.

The test is always an upper tail test. We will test this hypothesis using $\alpha = 0.01$.

Solution

We have the expected frequencies,

$$E_1 = (500)(0.07) = 35; \quad E_2 = 90; \quad E_3 = 325; \quad E_4 = 50.$$

The observed frequencies are

$$O_1 = (500)(0.39) = 195; \ O_2 = 45; \ O_3 = 180; \ O_4 = 80.$$

The value of the test statistic is given by

$$Q^2 = \sum_{i=1}^{4} \frac{(O_i - E_i)^2}{E_i}$$

$$= \left[\frac{(195-35)^2}{35} + \frac{(45-90)^2}{90} + \frac{(180-325)^2}{325} + \frac{(80-50)^2}{50}\right]$$

$$= 836.62.$$

From the chi-square table, $\chi^2_{0.01,3} = 11.3449$. *Because the test statistic* $Q^2=836.62 > 11.3449$, *we reject* H_0 *at* $\alpha=0.01$. *Hence, the data suggest that people have changed their opinion after watching the series on the ill effects of smoking marijuana. That is, the TV's station broadcast did change the opinion of the audience.*

EXAMPLE 11.4.3

A die is rolled 60 times and the face values are recorded. The results of this experiment are:

Up face	1	2	3	4	5	6
Frequency	8	11	5	12	15	9

Is the die balanced fair? Test this question using $\alpha = 0.05$.

Solution

If the die is fair, we must have

$$p_1 = p_2 = \ldots = p_6 = \frac{1}{6}$$

where $p_i = P$(face value on the die is i), $i = 1, 2, \ldots, 6$. *This experimental outcome follows the discrete uniform probability distribution. Hence,*

$$H_0: p_1 = p_2 = \ldots = p_6 = \frac{1}{6}$$

Vs.

H_a: At least one of the probabilities is different from the hypothesized value of 1/6.

Note that $E_1 = n_1 p_1 = (60)(1/6) = 10, \ldots, E_6 = 10$, *and the condition of using this test is satisfied.*

We summarize the calculations in the following table:

Face value	1	2	3	4	5	6
Frequency, O_i	8	11	5	12	15	9
Expected value, E_i	10	10	10	10	10	10

The test statistic value is given by

$$Q^2 = \sum_{i=1}^{6} \frac{(O_i - E_i)^2}{E_i} = 6.$$

From the chi-square table with 5 d.f., $\chi^2_{0.05,5} = 11.070$.

Thus, $\chi^2_{0.05,5} = 11.070 = 11.07 > Q^2 = 6$ *and since the value of the test statistic does not fall in the rejection region, we do not reject* H_0. *Therefore we do not have enough evidence to conclude that the die is not fair.*

The tests that we will study here are approximate tests, but they are very useful in performing statistical analysis. Let the random variables $(X_1, \ldots, X_k)$ have a multinomial distribution with parameters $n, p_1, \ldots, p_k$. Let n be known. We will now present some important tests based on the chi-square statistic.

11.4.2 Two-Way Contingency Table: Test for Independence

Another important use of the chi-square statistic is testing for dependencies or associations between the rows and the columns in a contingency table. That is, if we have two categorical variables, is there convincing evidence of association between the variables in the population? Here, we have seen that n randomly selected items are classified according to two different criteria, or two factors (row factor and column factor) where the row factor has r levels and the column factor has c levels. The data obtained are displayed in a contingency table as shown in Table 11.9, where n_{ij} represents the number of data values in row i and column j. Our interest here is to test the independence of the two-way classifications of observed events. For example, we might classify a sample of students by male or female and by their grade on a statistics course in order to test the hypothesis that the grades are independent of gender. More generally, the problem is to investigate a *dependency* (or *contingency*) between two classification criteria.

In this study, the given data of a problem are presented in a tabular form, as illustrated in the table below.

TABLE 11.9 Two-Way Contingency Table.

	Levels of c				
	1	2	...	c	Row Total
1	n_{11}	n_{12}		n_{1c}	$n_{1.}$
2	n_{21}	n_{22}		n_{2c}	$n_{2.}$
.					
.					
r	n_{r1}	n_{r2}		n_{rc}	$n_{r.}$
Column totals	$n_{.1}$	$n_{.2}$		$n_{.c}$	N

where $N = \sum_{j=1}^{c} n_{.j} = \sum_{i=1}^{r} n_{i.} = \sum_{i=1}^{r}\sum_{j=1}^{c} n_{ij}$ is the grand total.

Here, we wish to test the hypothesis that the two factors (rows and columns) are independent. We summarize the procedure below to test whether the factors represented by the rows are independent of those represented by the columns.

Testing for the Independence of Two Factors

To test

H_0: The factors are independent

vs.

H_a: The factors are dependent

the test statistic is

$$Q^2 = \sum_{i=1}^{r}\sum_{j=1}^{c} \frac{(O_{ij} - E_{ij})^2}{E_{ij}},$$

where

$$O_{ij} = n_{ij}$$

and

$$E_{ij} = \frac{n_{i.}n_{.j}}{N}$$

Then under the null hypothesis the test statistic Q^2 has an approximate chi-square probability distribution with $(r - 1)(c - 1)$ degrees of freedom.

Hence, the rejection region is $Q^2 > \chi^2_{\alpha,(r-1)(c-1)}$

Assumption: $E_{ij} \geq 5$.

EXAMPLE 11.4.4

Table 11.10 gives a classification according to religious affiliation and marital status for 500 randomly selected individuals.

TABLE 11.10 Marital Status and Religious Affiliation.

		Religious Affiliation					
		A	B	C	D	None	**Total**
Marital status	Single	39	19	12	28	18	116
	With spouse	172	61	44	70	37	384
	Total	211	80	56	98	55	**500**

Using a level of significance, $\alpha = 0.01$, test the null hypothesis that marital status and religious affiliation are independent.

Solution

We need to test the hypotheses

H_0: Marital status and religious affiliation are independent

vs.

H_a: Marital status and religious affiliation are dependent.

Here, $c = 5$, and $r = 2$. For $\alpha = 0.01$, and for $(c - 1)(r - 1)=4$ degrees of freedom, we have

$$\chi^2_{0.01,4} = 13.2767.$$

Hence, the rejection region is $Q^2 > 13.2767$.
We have $E_{ij} = \frac{n_i n_j}{N}$. Thus

$$E_{11} = \frac{(116)(211)}{500} = 48.952;\ E_{12} = \frac{(116)(80)}{500} = 18.5;$$
$$E_{13} = \frac{(116)(56)}{500} = 12.992;\ E_{14} = \frac{(116)(98)}{500} = 22.736;$$
$$E_{15} = \frac{(116)(55)}{500} = 12.76;\ E_{21} = \frac{(384)(211)}{500} = 162.05;$$
$$E_{22} = \frac{(384)(80)}{500} = 61.44;\ E_{23} = \frac{(384)(56)}{500} = 43.008;$$

and

$$E_{24} = \frac{(384)(98)}{500} = 75.264;\ E_{25} = \frac{(384)(55)}{500} = 42.24.$$

The value of the test statistic is

$$Q^2 = \Sigma_{i=1}^{r}\Sigma_{j=1}^{c}\frac{(O_{ij} - E_{ij})^2}{E_{ij}}$$
$$= [\frac{(39 - 48.952)^2}{48.952} + \frac{(19 - 18.5)^2}{18.5} + \frac{(12 - 12.992)^2}{12.992} + \frac{(28 - 22.736)^2}{22.736}$$
$$+ \frac{(18 - 12.76)^2}{12.76} + \frac{(172 - 162.05)^2}{162.05} + \frac{(61 - 61.44)^2}{61.44} + \frac{(44 - 43.08)^2}{43.08}$$
$$+ \frac{(70 - 75.264)^2}{75.264} + \frac{(37 - 42.24)^2}{42.24}]$$
$$= 7.1351.$$

Because the observed value of Q^2 does not fall in the rejection region, we do not reject the null hypothesis at $\alpha = 0.01$. Therefore, based on the given data, the marital status and religious affiliation are independent. Note that the assumption of $E_{ij} \geq 5$ is satisfied.

It should be noted that the chi-square test becomes inaccurate when used to analyze 2 × 2 contingency tables and when the large sample conditions, $E_i \geq 5$ for all cells and no more than 20% cells with $E_i \geq 1$, are not met. Fisher's exact test is used in these cases, and we refer the reader to the book by Agresti, among other sources.

EXERCISES 11.4

11.4.1. If we toss a coin a few times, we expect half heads and half tails. Suppose we tossed a coin 200 times and obtained 104 heads. Can we assume the coin is fair? Use $\alpha = 0.05$.

11.4.2. The following table presents the opinions on collective bargaining of a random sample of 200 employees of a school system belonging to a teachers' union.

Opinion on Collective Bargaining by the Teachers' Union

	For	Against	Undecided	**Total**
Staff	30	15	15	60
Faculty	50	10	40	100
Administration	10	25	5	40
Column totals	90	50	60	**200**

Test the hypotheses

H_0: Opinion on collective bargaining is independent of employee classification

vs.

H_a: Opinion on collective bargaining is dependent on employee classification using $\alpha = 0.05$.

11.4.3. A random sample of 300 undergraduate students from a university was taken. The students in the sample were classified according to their gender and chosen major. The results are presented in Table 11.11.

TABLE 11.11 Gender and Major Contingency Table.

	College				
Gender	**Arts and Sciences**	**Engineering**	**Business**	**Other**	**Total**
Male	75	40	24	66	205
Female	45	12	15	23	95
Total	120	52	39	89	300

Test the hypothesis that undergraduate students' choice of major in this university is independent of their gender. Use $\alpha = 0.01$.

11.4.4. A presidential candidate advertises on TV by comparing his positions on some important issues with those of his opponent. After a series of advertisements, a pollster wants to know whether people have changed their opinion about the candidate. Historical data from before the advertisement show the proportions of different opinion categories, as shown in the first table below, while the second table below shows data from a survey of 950 randomly chosen people.

Before the Advertisement Was Shown

Support the candidate	Oppose the candidate	Need to know more about the candidate	Undecided
40%	20%	5%	35%

After the Advertisement Was Shown

Support the candidate	Oppose the candidate	Need to know more about the candidate	Undecided
45%	25%	2%	28%

Let p_i, $i = 1, 2, 3, 4$, represent the respective true proportions.
Test

H_0: $p_1 = 0.35$; $p_2 = 0.20$; $p_3 = 0.15$; $p_4 = 0.3$
vs.
H_a: At least one of the probabilities is different from the hypothesized value.

Test this hypothesis using $\alpha = 0.05$.

11.4.5. A survey of footwear preferences among a random sample of 100 undergraduate students (50 females and 50 males) from a large university resulted in the following data, as presented in Table 11.12.

TABLE 11.12 Gender and Footwear Table.

	Boots	Leather Shoes	Sneakers	Sandals	Other
Female	12	9	12	10	7
Male	10	12	17	7	4

(a) Let p_i, $i = 1, 2, 3, 4, 5$, represent the respective true proportions of students with a particular footwear preference, and let

H_0: $p_1 = 0.20$; $p_2 = 0.20$; $p_3 = 0.30$; $p_4 = 0.20$; $p_5 = 0.10$
Vs.
H_a: At least one of the probabilities is different from the hypothesized value.

Test this hypothesis using $\alpha = 0.05$.

(b) Test the hypothesis that the choice of footwear by undergraduate students in this university is independent of their gender using $\alpha = 0.05$.

11.4.6. A casino game involves rolling three dice. The winnings are directly proportional to the total number of sixes rolled. Suppose a gambler plays the game 150 times, with the following observed counts:

Number of sixes	0	1	2	3
Number of rolls	72	51	21	6

Assuming that the roll of one die does not affect the roll of others, test to determine if the dice are fair, at $\alpha = 0.05$.

11.4.7. Criminologists are interested in determining whether there is any relationship between homicides and the seasons of the year. In the paper "Is Crime Seasonal?" (https://bjs.gov/content/pub/pdf/ics.pdf), the following data for 1361 homicides are presented by seasons.

Winter	Spring	Summer	Fall
328	334	372	327

Does this data support the theory that homicide rates are not the same across seasons?

11.4.8. In order to find out the relationship between packaging preferences (in terms of size) and economic status, a manufacturing company of pain medication conducted a survey. Table 11.13 presents the results of this survey.

Is there a significant relationship between packaging preferences and economic status? Use $\alpha = 0.05$.

TABLE 11.13 Economic Status and Size of Purchase.

	Lower	Middle	Upper
Small	23	24	19
Medium	22	26	20
Large	16	28	18
Jumbo	15	21	30

11.5 GOODNESS OF FIT TESTS TO IDENTIFY THE PROBABILITY DISTRIBUTION

In studying various real-world phenomena, we begin with a random sample of data $X_1, \ldots ,X_n$ that represents values of some sort of subject of interest. These measurements could represent the amount of carbon dioxide, CO_2, in the atmosphere on a daily basis, the sizes of cancerous breast tumors, the monthly average rainfall in the state of Florida, the average monthly unemployment rate in the United States, the hourly wind forces of a hurricane, etc. In order for us to probabilistically understand the behavior of these phenomena, we will need to identify the probability distribution that characterizes the probabilistic behavior of the given data, that is, the PDF of the random sample the data was drawn from. For example, at a certain time point, we say that this data follows or comes from the normal or exponential probability distribution. One of the important questions then is whether the observed data is representative or follows a particular probability distribution. The goodness of fit tests are used to determine whether a sample fits a particular distribution. In fact, there is nothing we can do parametrically or statistically (unless through goodness of fit testing) to identify the PDFs that probabilistically characterize the behavior of the given data—the phenomenon of interest.

To accomplish this objective of identifying the underlying probability distribution, we will discuss four statistical tests (methods) that we can use to determine how well the data fits a particular well-defined probability distribution. These four tests are *Pearson's chi-square test*, *Kolmogorov-Smirnov test*, *Anderson-Darling test*, and *Shapiro-Wilk test*. Even though we will give theoretical steps on how to calculate the quantities for most of the tests, it should be noted that, in practice, most of the goodness of fit tests will be done using statistical software. For large data sets, it is tedious to do goodness of fit analysis by hand. There are other methods we can follow, if we are not able to identify the appropriate PDF, such as nonparametric or probability distribution-free analysis, which will be discussed in Chapter 12.

11.5.1 Pearson's Chi-Square Test

When we are interested in studying the behavior of a given unknown phenomenon, we begin by obtaining, through experimentation or other means, a set of data, the random sample. The initial step of studying this phenomenon is to try to identify the probability distribution that characterizes the behavior of the given data. The method that we use is called the goodness of fit test. That is, if we assume that a given set of data follows the normal or Gaussian probability distribution, the data must be a good fit to this distribution with a high degree of assurance. Historically, the first statistical method to test the fit of a particular distribution to a given set of data was Pearson's chi-square Goodness of fit test.

In hypothesis testing problems, we often assume that the form of the population distribution is known. For example, in a chi-square test for variance, we assume that the population is normal. The goodness of fit tests examine the validity of such an assumption if we have a large enough sample. We now describe the goodness of fit test procedure for such an application. This test uses a measure of goodness of fit which is the mean of the differences between the observed and expected outcome frequencies (counts of observations), each squared and divided by the expected frequencies. That is,
the test statistic is given by

$$Q^2 = \sum_{i=1}^{k} \frac{(O_i - E_i)^2}{E_i}.$$

Here, O_i is the ith observed outcome frequency (in class i); E_i is the ith expected (theoretical) frequency, $(E(O_i))$; and $i = 1,2,\ldots,k$ is the number of classes. The expected frequency E_i, is calculated by

$$E_i = [F_0(y_u) - F_0(y_l)]n,$$

where F_0 is the cumulative probability distribution that is being tested (assumed) to determine if the given data follows (fits) this probability distribution; Y_u and Y_l are the upper and lower limits of class i, respectively; and n is the sample size. Thus we proceed to set up the hypothesis:

H_0: *The given data follows a specific probability distribution* (F)

Vs.

H_a: *The data does not follow the specified probability distribution.*

We proceed to calculate the value of the Q^2 statistic, and if it is greater than the value we obtain from the $\chi^2_{\alpha,k-1}$ tables for a given level of significance α and $k-1$ degrees of freedom, we reject the hypothesis. Note that for $(k-1)$ degrees of freedom, we need to know that the F distribution is completely defined. If there are any unknown parameters that need to be estimated, we need to reduce that many degrees of freedom. That is, the data does not follow or fit the specified probability distribution. Thus, if the calculated value of the chi-square test statistic is less than the $\chi^2_{\alpha,k-1}$ value that we obtain from the tables, indeed the specified data fits the specified probability distribution at a level of significance α. That is, the rejection region is given by

$$P(Q^2 \geq \chi^2_{\alpha,k-1}) = \alpha.$$

The basic assumptions for applying this test are as follows:

i. The observed frequencies in the k classes should be independent.
ii. $\sum_{i=1}^{k} E_i = \sum_{i=1}^{k} O_i = n$
iii. The total frequency, n, should be approximately greater than 50.
iv. Each expected frequency, E_i, in each class should be at least 5.

In testing the above hypothesis, we usually assume a value of the level of significance α, like $\alpha = .01, 0.05, 0.1$, etc., and proceed to make the decision of accepting or rejecting the null hypothesis based on the assumed α. However, by using statistical packages such as R, it gives you a p-value, in contrast to fixed α value, that is calculated based on the test statistic and denotes the threshold value of the significance level in the sense that the null hypothesis will be accepted at all significance α level less than the calculated p-value. For example, if p-value = 0.05, the null hypothesis will not be rejected for all values of assumed $\alpha <$ p-value of 0.05 and will be rejected for higher levels. Recall that the p-value is the probability of observing a sample statistic as extreme as the test statistic. Since, here, the test statistic has chi-square distribution, use the package to calculate the p-value. Note that recently, using the p-value has created some useful criticisms of its applicability and we will not discuss these issues here. Below is a summary of the step-by-step procedure for applying the subject test.

Goodness of Fit Test Procedures for Identifying the Probability Distributions

Let $X_1,..X_n$ be a sample from a population with cumulative distribution function (CDF) $F(x)$. We wish to test H_o: $F(x)=F_o(x)$, where $F_o(x)$ is completely specified (assumed) PDF.

1. Divide the range of values of the random variables X_1 into k nonoverlapping intervals $I_1, I_2, \ldots, I_k$. Let O_j be the number of sample values that fall in the interval I_j ($j = 1, 2, \ldots, k$).
2. Assuming the probability distribution of X to be $F_o(x)$, find $P(X \in I_j)$. Let $P(X \in I_j)=\pi_j$. Let $E_j = n\pi_j$ be the expected frequency.
3. Compute the test statistic Q^2 given by

$$Q^2 = \sum_{i=1}^{k} \frac{(O_i - E_i)^2}{E_i}.$$

The test statistic Q^2 has an approximate chi-square distribution with $(k - 1)$ degrees of freedom.

4. Reject the H_0 if $Q^2 \geq \chi^2_{\alpha,(k-1)}$.

Assumptions: $E_j \geq 5$, $j=1, 2, \ldots, k$.

It should be noted that when the hypothesis distribution does not involve any extra parameters, the df is $(k - 1)$. If the hypothesis distribution involves extra parameters (e.g., in the exponential distribution example given below, because the exponential distribution has one rate parameter involved which needs to be estimated from the data), the degrees of freedom for the chi-square test needs to be adjusted to subtract one degrees of freedom used for estimating each of the unknown parameters. Also note that if the observed data O_i, is very close to the expected value E_i, the difference $O_i - E_i$ is going to be very small which implies the Q^2 statistic will be small and thus, a good fit of the given data to the assumed PDF. It should be noted that when data is numerical, we don't have natural categories. We need to create categories (similar to the way we create intervals for histogram) such that for each category the condition $E_j = n\pi_j \geq 5$ is satisfied.

EXAMPLE 11.5.1

We are given a random sample of $n = 30$ observations of a given experiment of a certain phenomenon of interest, that is,

1.79	2.62	11.92	9.77	12.13	15.04	16.14	20.74	22.73	23.29	24.97	26.12
211.06	29.60	32.47	36.32	42.18	45.06	45.64	48.34	48.87	64.99	66.28	68.00
68.60	75.34	99.32	162.48	164.38	235.95						

We believe that this data may follow the exponential PDF. Test our belief at $\alpha = 0.05$.

Solution

We need to test

H_0: *The given data follows an expontential probability distribution*

Vs.

H_a: *The data does not follow the specified probability distribution.*

We will now present the steps on how to solve this problem analytically and then illustrate how this can be implemented in R.

Recall that the PDF of the exponential distribution with the rate parameter λ is $f(x) = \lambda \exp(-\lambda x)$, for $x \geq 0$, and $\lambda > 0$. The MLE of λ is given by, $\hat{\lambda} = \frac{1}{\bar{x}}$. Since λ is unknown, we can calculate bin probabilities using $\hat{\lambda}$ in place of λ.

For the exponential random variable X with the rate parameter λ, we know that

$$P(a \leq X \leq b) = \int_a^b \lambda e^{-\lambda x} dx = e^{-\lambda a} - e^{-\lambda b}.$$

Based on this, we can now calculate the probability of the exponential random variable falling in each individual interval (bin).

Note that the minimum value for this data is 1.79, and the maximum is 235.16. The sample mean is $\bar{x} = 57.738$. Thus $\hat{\lambda} = \frac{1}{57.738} \approx 0.017$. Considering the observed range of data and the size of each bin to ensure the large sample approximation is valid for the grouped data, we divide the data into four unequal-width bins (intervals) as [0, 25], [25, 50], [50, 80], and [80, ∞] (since the exponential is continuous, open or closed interval will not change the probabilities). Then we can calculate each cell/bin probability as follows (you could also use R to calculate the CDF at a certain value, a, by using *pexp(a, rate=λ)*) and then calculate the difference between the CDFs at the lower and upper bounds of each interval to calculate these cell probabilities:

$P(0 \leq X \leq 25) = 0.351$, $P(25 \leq X \leq 50) = 0.228$, $P(50 \leq X \leq 80) = 0.170$ and $P(X \geq 80) = 0.250$.

Thus, the expected cell frequencies under the assumed exponential distribution are calculated as $E_1 = 0.351 \times 30 = 10.54$, $E_2 = 0.228 \times 30 = 6.84$, $E_3 = 0.170 \times 30 = 5.11$, and $E_4 = 0.250 \times 30 = 7.51$.

Note that the condition, $E_i \geq 5$, *for each i*, is satisfied, and hence the test is appropriate as the approximate chi-square distribution is satisfied. Now the observed and expected frequencies, as well as $\frac{(O_i - E_i)^2}{E_i}$, $i = 1, \ldots, 4$, needed for calculating the chi-square test statistic are given in the following table.

Data interval (bin)	**Observed frequency (O_i)**	**Expected frequency (E_i)**	$\frac{(O_i - E_i)^2}{E_i}$
0–25	11	10.54	0.0198
25–50	9	6.84	0.6837
50–80	5	5.11	0.0025
≥ 80	5	7.51	0.8364

Thus the test statistic Q^2 is given by

$$Q^2 = \sum_{i=1}^{k=4} \frac{(O_i - E_i)^2}{E_i} = 1.5424.$$

From χ^2–table with $k - 2 = 2$ degrees of freedom (one additional degree of freedom is lost because we had to estimate λ), and with $\alpha = 0.05$, rejection region is $\{Q^2 \geq 5.991\}$. Since 1.5424 is less than 5.991, we fail to reject H_0. Thus we can conclude that the observed data fit well with the exponential distribution.

Below we provide the R code and output for implementing the above-described goodness of fit test.

R-code & output:

```
> x=c(1.79,2.62,11.92,9.77,12.13,15.04,16.14,20.74,22.73,23.29,24.97,
+ 26.12,211.06,29.60,32.47,36.32,42.18,45.06,45.64,48.34,48.87,64.99,
+ 66.28,68.00,68.60,75.34,99.32,162.48,164.38,235.95)
> # estimate the rate parameter
> lambda <- 1/mean(x)
> lambda
[1] 0.01731962
> # define the bin boundaries
> bounds <- c(25,50,80, Inf)
> # the cummulative bin frequencies
> Ocum <-
c(sum(x<=bounds[1]), sum(x<=bounds[2]), sum(x<=bounds[3]), sum(x<=bounds[4]))
> # the observed bin frequencies
> O <- Ocum-c(0, Ocum[-4])
> # CDF
> cp<-pexp(bounds, rate=lambda)
> # bin probabilities
> bps<-cp-c(0, cp[-4])
> # chi-square test
> res<-chisq.test(x=O, p=bps)
> res
# p-value
> pv<-1-pchisq(as.numeric(res[1]),2)
> pv
[1] 0.4624616
```

Thus, for a p-value of 0.4624616, we do not reject the null hypothesis and we conclude that the given data are consistent with the exponential distribution.

EXAMPLE 11.5.2

The grades of students in a class of 200 are given in the following table. Test the hypothesis that the grades are normally distributed with a mean of 75 and a standard deviation of 8. Use $\alpha = 0.05$.

Range	0–59	60–69	70–79	80–89	90–100
Number of students	12	36	90	44	18

Solution

To test the hypothesis,

H_0: *Student grades follow a* $N(\mu = 75, \sigma^2 = 64)$ *distribution.*
Vs.
H_a: *Student grades do not follow the* $N(\mu = 75, \sigma^2 = 64)$ *distribution.*

We have O_1=12, O_2=36, O_3=90, O_4=44, O_5=18.
We now compute π_i $(i = 1, 2, \ldots, 5)$, *using the continuity correction factor,*

$$\pi_1 = P\{X \le 59.5|H_0\} = P\{z \le \frac{59.5 - 75}{8}\} = 0.0262,$$

$$\pi_2 = 0.2189,\ \pi_3 = 0.4722,\ \pi_4 = 0.2476,\ \pi_5 = 0.0351,$$

and

$$E_1 = 5.24,\ E_2 = 43.78,\ E_3 = 94.44,\ E_4 = 49.52,\ E_5 = 7.02.$$

The test statistic results in

$$Q^2 = \sum_{i=1}^{n} \frac{(O_i - e_i)^2}{e_i}$$

$$= \frac{(12 - 5.74)^2}{5.74} + \frac{(36 - 43.78)^2}{43.78} + \frac{(90 - 94.44)^2}{94.44} + \frac{(44 - 49.52)^2}{49.52} + \frac{(18 - 7.02)^2}{7.02}$$

$$= 26.22.$$

Q^2 has a chi-square distribution with (5–1)=4 degrees of freedom. The critical value is $\chi^2_{0.05,4} = 7.11$. *Hence, the rejection region is* $Q^2 > 11.11$. *Because the observed value of* Q^2=26.22 > 11.11, *we reject* H_0 *at* α=0.05. *Thus we conclude that the given data does not follow (or draw) from the normal PDF.*

11.5.2 The Kolmogorov-Smirnov Test (One Population)

Let X_i, $i = 1,2,\ldots,n$ be a random sample of n observations and we shall assume it is drawn (it follows) from a probability distribution whose cumulative distribution is specified to be $F_0(x)$. Our objective now is to determine if the actual (correct) cumulative probability is $F(x)$ based on the assumed $F_0(x)$. That is, we wish to test the following hypothesis:

H_0: *The true probability distribution that follows the given data,* $F(x)$ *is actually the assumed distribution* $F_0(x)$, for all x
Vs.
H_a: *The actual cumulative distrubution,* $F(x)$, *is not* $F_0(x)$, *for at least one x*

based on a level of significance α.
The Kolmogorov-Smirnov goodness of fit test to test the above hypothesis is based on the following test statistic:

$$D = \max_{-\infty < x < \infty} \{|F_n(x) - F_0(x)|\},$$

where $F_n(x)$ is the sample (empirical) distribution function given by

$$F_n(x) = \frac{\text{number of } X'\text{s in the sample} \le x}{n}.$$

Note that for an ordered data $X_{(1)}, \ldots, X_{(n)}$, $F_n(x)$ can be given as

$$F_n(x) = \begin{cases} 0, & x < X_{(1)}, \\ \frac{i}{n}, & X_{(i)} \le x < X_{(i+1)}. \\ 1, & x > X_{(n)}. \end{cases}$$

If $F_0(x)$ and $F_n(x)$ are plotted against x-axis, D is the value of the largest vertical distance between $F_0(x)$ and $F_n(x)$. In order to compute D, we can use the following. If the n observations are distinct, then define

$$K_i = \max\left\{\left|\frac{i}{n} - F_0(X_{(i)})\right|, \left|\frac{(i-1)}{n} - F_0(X_{(i)})\right|\right\},$$

and

$$D = \max_{i=1,\ldots,n} K_i.$$

If there are tied observations, let l be the number of distinct observations and let $Y_{(1)} < \ldots < Y_{(l)}$ be ordered distinct observations. Then, let

$$K_i' = \max\{|F_n(Y_{(i)}) - F_0(Y_{(i)})|, |F_n(Y_{(i-1)}) - F_0(Y_{(i)})|\},$$

and

$$D = \max_{i=1,\ldots,l} K_i'.$$

Procedure to Calculate D

To calculate the value of the test statistic D, we follow the following three steps:

1. We calculate the assumed cumulative distribution, $F_0(x)$, based on the given data of observations and the specified population distribution.
2. We proceed to obtain the cumulative distribution of the sample, $F_n(x)$ is the empirical distribution function defined as a step function,

$$F_n(x) = \frac{\#X_i \le x}{n},$$

the number of observations $X_i \le x$ divided by n.

3. We find the absolute difference:

$$|F_0(x) - F_n(x)|$$

Thus we have a value of the test statistic D, and if

$$D \le D_\alpha,$$

we will not reject the hypothesis, H_0, at level of significance α.

That is, we accept the hypothesis, where D_α is the critical value from the Kolmogorov-Smirnov tables that is based on a given α *and* n. The following example illustrates how we apply this test (Fig. 11.1).

EXAMPLE 11.5.3

From a large statistics class, we have taken a random sample of 55 students, n=55, and recorded their ages. The resulted data is:

27	25	24	24	22	20	21	22	21	25	24
26	25	24	23	22	20	21	19	21	25	24
26	25	22	23	22	22	21	19	21	23	21
26	24	22	23	22	22	20	19	21	23	21
26	24	22	23	21	19	20	18	20	20	18

We believe that this data follows the normal PDF and wish to use the Kolmogorov-Smirnov goodness of fit test, given above, to test our belief. That is, test

H_0: The ages of the students follow the normal probability distribution

Vs.

H_a: The ages of students do not follow the normal probability distribution

Solution

It usually helps to obtain a possible visual indication of the PDF by structuring a histogram of the given data. That is,

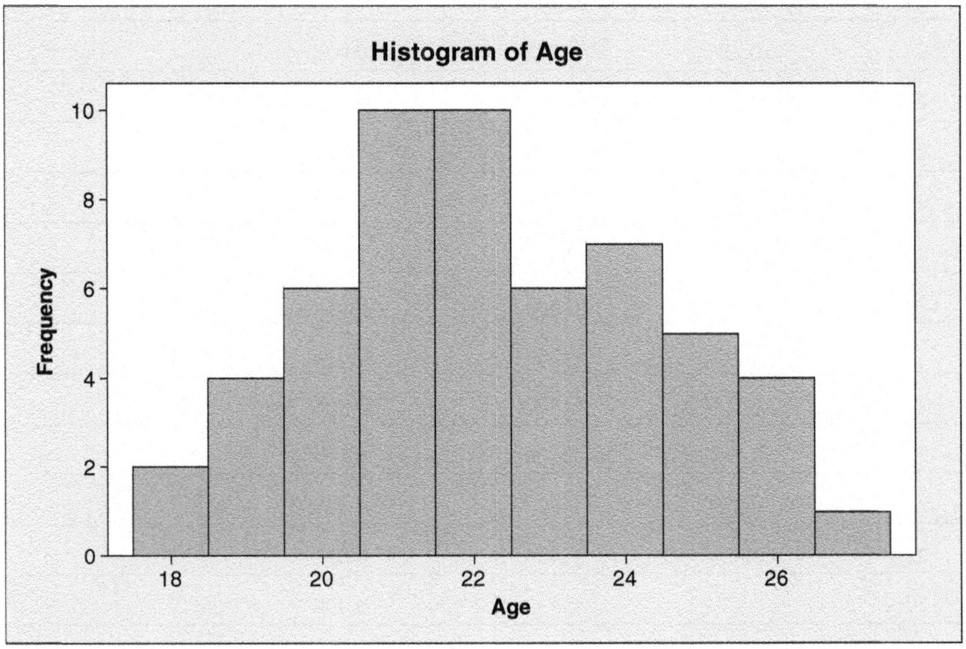

FIGURE 11.1 Histogram of the ages.

Visually it seems that the normal PDF is a good possibility. We shall now test it statistically.

The sample mean is $\bar{x} = 22$ and the sample standard deviation is $s = 2.08$. The three-step procedure of the subject test to obtain the value of the test statistic D can be easily calculated using the following table and letting D as max of the column, $|F_0(x) - F_n(x)|$.

Row	Age	$F_0(x)$	$F_n(x)$	$\lvert F_0(x) - F_n(x) \rvert$	D	Critical Value
1	18	0.028	0.018	0.010	0.127	0.183
2	18	0.028	0.036	0.009		
3	19	0.071	0.055	0.017		
4	19	0.071	0.073	0.001		
5	19	0.071	0.091	0.019		
6	19	0.071	0.109	0.038		

7	20	0.155	0.127	0.028
8	20	0.155	0.145	0.010
9	20	0.155	0.164	0.009
10	20	0.155	0.182	0.027
11	20	0.155	0.200	0.045
12	20	0.155	0.218	0.063
13	21	0.286	0.236	0.050
14	21	0.286	0.255	0.032
15	21	0.286	0.273	0.013
16	21	0.286	0.291	0.005
17	21	0.286	0.309	0.023
18	21	0.286	0.327	0.041
19	21	0.286	0.345	0.059
20	21	0.286	0.364	0.078
21	21	0.286	0.382	0.096
22	21	0.286	0.400	0.114
23	22	0.454	0.418	0.036
24	22	0.454	0.436	0.018
25	22	0.454	0.455	0.000
26	22	0.454	0.473	0.018
27	22	0.454	0.491	0.037
28	22	0.454	0.509	0.055
29	22	0.454	0.527	0.073
30	22	0.454	0.545	0.091
31	22	0.454	0.564	0.109
32	22	0.454	0.582	0.127
33	23	0.631	0.600	0.031
34	23	0.631	0.618	0.013
35	23	0.631	0.636	0.005
36	23	0.631	0.655	0.023
37	23	0.631	0.673	0.041

38	23	0.631	0.691	0.059
39	24	0.784	0.709	0.075
40	24	0.784	0.727	0.057
41	24	0.784	0.745	0.039
42	24	0.784	0.764	0.020
43	24	0.784	0.782	0.002
44	24	0.784	0.800	0.016
45	24	0.784	0.818	0.034
46	25	0.892	0.836	0.055
47	25	0.892	0.855	0.037
48	25	0.892	0.873	0.019
49	25	0.892	0.891	0.001
50	25	0.892	0.909	0.017
51	26	0.954	0.927	0.027
52	26	0.954	0.945	0.009
53	26	0.954	0.964	0.010
54	26	0.954	0.982	0.028
55	27	0.984	1.000	0.016

Since the D-statistic = 0.127 < $D_{\alpha=0.05}$ = 0.183 (from the K-S table), we fail to reject the null hypothesis at the level of significance α = 0.05. Thus the ages of the students in the class indeed follow the normal PDF.

Also, we can easily calculate the Kolmogorov-Smirnov test statistics and the p-value using R code and the output is given below:

```
x=c(27,25,24,24,22,20,21,22,21,25,24,
+ 26,25,24,23,22,20,21,19,21,25,24,
+ 26,25,22,23,22,22,21,19,21,23,21,
+ 26,24,22,23,22,22,20,19,21,23,21,
+ 26,24,22,23,21,19,20,18,20,20,18)
```

`ks.test(x, pnorm, mean(x), sd(x))` obtain the P-P Plot as shown by Figure 11.2.

Output

One-sample Kolmogorov-Smirnov test

```
data: x
D=0.1274, p-value=0.3336
alternative hypothesis: two-sided
```

Since p-value is large, we cannot reject the null hypothesis.

11.5.3 The Anderson-Darling Test

The Anderson-Darling goodness of fit test is also used to determine if a given set of data is drawn from a population that follows a specific probability distribution. This is a modification of the Kolmogorov-Smirnov test and gives more weight to the tails than that test. However, critical values for the Anderson-Darling test make use of a particular distribution resulting in the need to calculate critical values for each distribution. As a result, we will not give critical values for this test; instead, we will use the software. There are Anderson-Darling tables available for many popular distributions, such as normal, lognormal, exponential, Weibull, etc. Let X_i, $i = 1,2,\ldots,n$ be a random sample of observations and Y_i, $i = 1,2,\ldots,n$ be the corresponding ordered value according to size. The hypothesis that we wish to test is

H_0: *The given data follow a specific probability distribution*

Vs.

H_a: *The given data does not follow the specified probability distribution*

The Anderson-Darling test statistic for testing the above hypothesis is given by $A^2 = -n - S$ where $s = \sum_{i=1}^{n} \frac{(2i-1)}{n}[\ln F(Y_i) + \ln(1 - F(Y_{n+1-i})]$, n is the random sample size, Y_i is the ordered data, and F is the specified probability distribution that we are testing. For a given level of significance α, the hypothesis is rejected if the value of the test statistic An is greater than the critical value A_α, that is, if

$$A > A_\alpha.$$

Thus we reject the null hypothesis in favor of the alternative hypothesis; the specified probability distribution does not fit the distribution of the drawn data from the population. The A_α is obtained from the Anderson-Darling tables for a given α. The following example illustrates how we apply the subject test.

EXAMPLE 11.5.4

Use ages of the 55 students given in Example 11.5.3 to illustrate the applicability of the Anderson-Darling goodness of fit test.

Solution

The data is given in Example 11.3.3, and we proceed to test our belief that the students' ages follow the normal PDF.

```
install.packages('nortest')
library(nortest)
ad.test(x,"pnorm")
Output
Anderson—Darling normality test
data: x
A=0.6456, p-value=0.08743
```

Thus the Anderson-Darling statistic is A = 0.6456 with a p-value of 0.08743. Thus, at 5% level of significance, we fail to reject the null hypothesis. The data fits the normal probability distribution with mean of 22 and standard deviation of 2.

11.5.4 Shapiro-Wilk Normality Test

The Shapiro-Wilk goodness of fit test is used to determine if a random sample X_i, $i = 1,2,\ldots,n$ is drawn from a normal (Gaussian) probability distribution with true mean and variance, μ and σ^2, respectively. That is, $X \sim N(\mu, \sigma^2)$. Thus we wish to test the following hypothesis:

H_0: *The random sample was drawn from a normal population,* $N(\mu, \sigma^2)$

Vs.

H_a: *The random sample does not follow* $N(\mu, \sigma^2)$

To test this hypothesis, we use the Shapiro-Wilk test statistic, which is given by

$$W = \frac{\left(\sum_{i=1}^{n} a_i x_{(i)}\right)^2}{\sum_{i=1}^{n} (x_i - \bar{x})^2}$$

where $x_{(i)}$ are the ordered sample values and a_i are constants generated by the expression

$$(a_1, a_2, \ldots, a_n) = \frac{m^T V^{-1}}{(m^T V^{-1} m)^{1/2}}$$

where $m = (m_1, m_2, \ldots, m_n)^T$ are the expected values of the ordered statistics that are independent and identically distributed random variables that follow the standard normal, $N(0,1)$, and V is the covariance matrix of the order statistics.

EXAMPLE 11.5.5

Proceed to use the Shapiro-Wilk normality test for the data in **Example 11.5.3**, which we previously analyzed using the Anderson-Darling goodness of fit test, to determine whether the ages of the students follow the normal PDF.

Use $\alpha = 0.05$.

Solution

The R code for the subject test is

Shapiro.test(x)

Output

Shapiro-Wilk normality test

Data: x

W=0.9683, p-value=0.1551

Thus, since the p-value is larger than 0.05, we fail to reject the null hypothesis, and the ages of the students indeed follow the normal PDF. This result is the same as the one obtained using the Anderson-Darling test.

11.5.5 The P-P Plots and Q-Q Plots

We commonly use a visual interpretation of graphs (plots) to determine if a given random sample of data follows or is drawn from a well-known probability distribution. These graphs are the probability, *P-P*, plots and the quantile, Q-Q, plots.

The *P-P plot* is a graphical tool used to determine how well a given data set fits a specific probability distribution that we are testing. This plot compares the empirical CDFs of the given data with that of the assumed true cumulative probability distribution functions. If the plot of these two distributions is approximately linear, it indicates that the assumed true PDF gives a reasonably good fit to the given data that we seek to find its true PDF.

11.5.5.1 Steps to Construct the P-P Plot

Let $F(x)$ be the cumulative PDF of the random variable, X with a random sample $x_{(1)}, x_{(2)}, \ldots, x_{(n)}$ of ordered data values with associated probabilities $\hat{C}_{(i)} = \frac{i}{n+1}$, the scattered P-P plot is the plot of $\hat{C}_{(i)}$ versus $C_{(i)} = F[X = x_{(i)}]$, of the possibly true cumulative PDF that we are testing. The step-by-step procedure that we follow to structure the P-P plot is given below.

Steps for P-P Plot

Step 1. Given a random sample $x_1, x_2, \ldots, x_n$ sort the data in ascending order:

$$x_{(1)}, x_{(2)}, \ldots, x_{(n)}$$

Step 2. Associate with each of the ordered data values $x_{(1)}$ a cumulative probability:

$$\hat{C}_{(i)} = \frac{i}{n+1}$$

Step 3. Determine the hypothetical probabilities associated with the probability distribution we are testing:

$$C_{(i)} = F[X = x_{(i)}],$$

$$F(x) = P[X \le x],$$

where $F(x)$ is the cumulative PDF.

Step 4. Construct the scatterplot of $\hat{C}_{(i)}$ **versus** $C_{(i)} = F[X = x_{(i)}]$.

(cont'd)

Step 5. Interpret the plot: if the overall pattern follows approximately a straight line, then the data follows the assumed probability distribution; if the overall pattern has curvature or shelves, then the data has skewed behavior, and therefore it does not follow the assumed PDF.

The following example illustrates how we obtain and interpret the subject plot.

EXAMPLE 11.5.6

Using the data in Example 11.5.3 obtain the P-P plot as

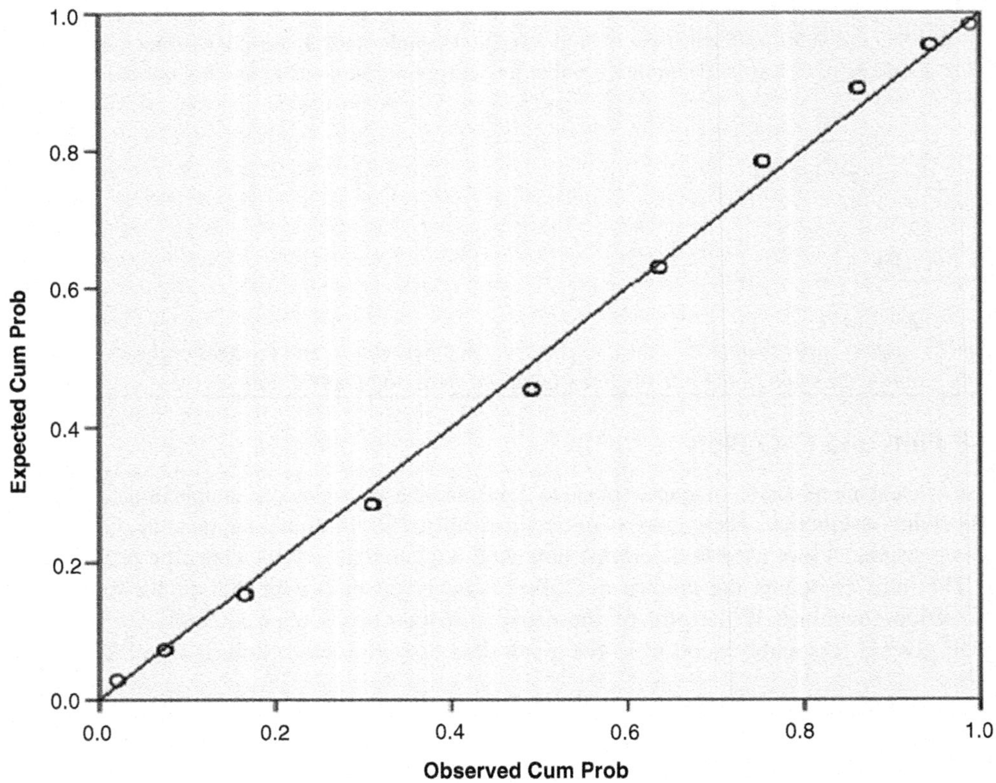

FIGURE 11.2 P-P plot of the ages.

Thus the data falls on a straight line, and we can conclude that the information on the ages of the students follows the normal PDF, which is consistent with our previous test. Again, the P-P plot is a visual decision and we cannot associate with it a degree of confidence.

The Q-Q plot is another graphical method that is commonly used to obtain a graphical (visual) indication of the true PDF that the given data comes from. This method is a graph of the quantiles of the empirical distribution of the given data versus the quantiles of the assumed true PDF that we are testing. If the resulting graph of these two distributions follows a linear pattern, it indicates that the assumed PDF fits the given data reasonably well. A step-by-step procedure of obtaining the Q-Q plot is given below.

Steps to Obtain Q-Q Plots

Let $F(x)$ be the assumed cumulative PDF of the random variable X, with a random sample $x_{(1)}, x_{(2)}, \ldots, x_{(n)}$ of ordered data values with associated probabilities $\hat{C}_{(i)} = \frac{i}{n+1}$, the Q-Q plot is the $x_{(i)} = F^{-1}(\hat{C}_{(i)})$, the inverse function of $F(x)$.

Step 1. Given a random sample $x_1, x_2, \ldots, x_n$ sort the data in ascending order:

$$x_{(1)}, x_{(2)}, \ldots, x_{(n)}$$

Step 2. Associate with each of the order data value $x_{(1)}$ a cumulative probability:

$$\hat{C}_{(i)} = \frac{i}{n+1}$$

Step 3. Determine the estimated value of the random variable associated with the assumed probability distribution:

$$x_{(i)} = F^{-1}(\hat{C}_{(i)})$$

where $F(x)$ is the CDF.

Step 4. Construct the scatterplot of $x_{(i)}$ versus $\hat{x}_{(i)} = F^{-1}[\hat{C}_{(i)}]$.

Step 5. If the overall pattern follows approximately a straight line, then the data follows the assumed probability distribution. If the overall pattern has curvature or shelves, then the data has skewed behavior, and it does not follow the assumed probability distribution.

The following example illustrates how we structure the Q-Q plot.

EXAMPLE 11.5.7

We shall use the data given in Example 11.5.3, the ages of 55 students, to construct the Q-Q plot to verify normality.

Solution

The results are given below as shown by Figure 11.3.

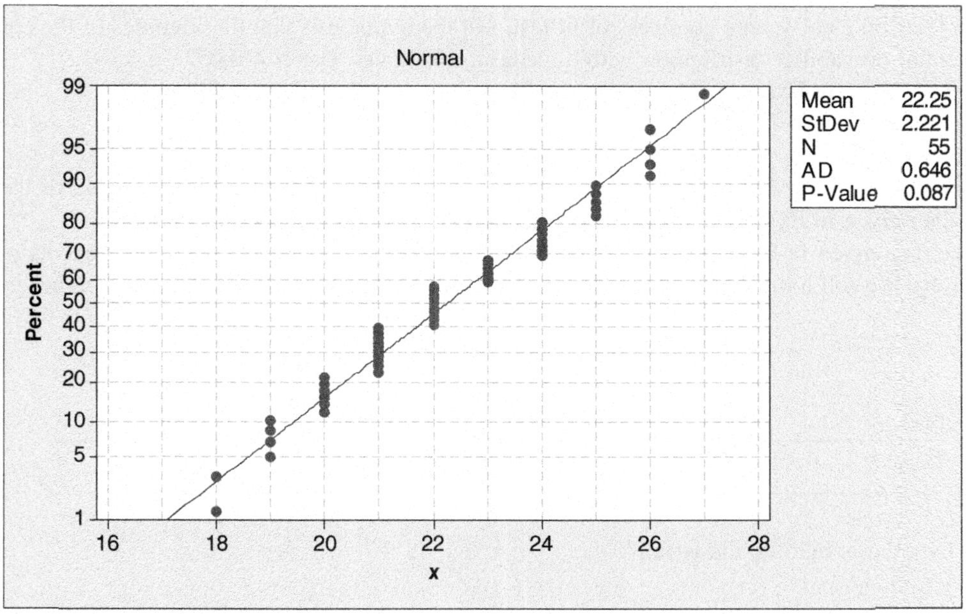

FIGURE 11.3 Q-Q plot for the ages.

Note that the plot follows approximately a straight line, which suggests that the data follow the normal PDF, which we have also proven using two other goodness of fit tests.

EXERCISES 11.5

11.5.1. The speeds of vehicles (in mph) passing through a section of Highway 75 are recorded for a random sample of 150 vehicles and are presented below. Test the hypothesis using the Anderson-Darling test that the speeds are normally distributed with a mean of 70 mph and a standard deviation of 4. Use $a = 0.01$.

Range	40–55	56–65	66–75	76–85	> 85
Number	12	14	78	40	6

11.5.2. The temperature in degrees Fahrenheit is recorded for a randomly selected set of 50 days in the City of Tampa, Florida, in 2018. The data collected is presented below.

City of Tampa					
Temperature	46–55	56–65	66–75	76–85	86–95
Number of days	4	6	13	23	4

Using one of the tests introduced in this section, test the hypothesis that the data follow the normal PDF with a mean of 77^0F and a standard deviation of 6. Use $\alpha = 0.05$.

11.5.3. A sample of 30 electronic circuit components is randomly selected from a production process. The lifetime, in hours, of each component is precisely measured by testing it until it fails. The time in hours that it took each component to fail is given below.

268.276	420.559	6.590	78.389	14.123	85.507	216.594	39.892	9.468	83.088
519.682	315.754	139.046	4.522	81.480	209.099	170.128	711.794	115.778	108.640
226.053	443.029	35.662	115.668	5.032	111.357	331.462	184.734	79.502	611.019

Using the Pearson's chi-square goodness of fit test, test the hypothesis that the lifetimes of the components follow an exponential probability distribution with a mean of 200 hours. Use $\alpha = 0.05$.

11.5.4. For the data given in Example 11.5.3, test the goodness of fit that the data follows:

a) the gamma PDF

b) the Weibull PDF

11.5.5. Using the data given in Example 11.5.1, construct the P-P plot and interpret the meaning of the graph.

11.5.6. For the data given in Example 11.5.2, construct the P-P plot and interpret its meaning.

11.5.7. Using the data given in Example 11.5.1, construct the graph of the Q-Q plot and interpret its meaning.

11.5.8. Suppose that we roll a die 36 times and observe the following table with the number of times each face ends up on top.

Face	1	2	3	4	5	6
Count	4	8	6	11	3	4

Test the null hypothesis that the die is fair.

11.6 CHAPTER SUMMARY

In this chapter, we learned different aspects of categorical data analysis, including estimation and hypothesis testing problems. We also looked at goodness of fit methods and how we use them to attempt to identify the PDF that characterizes the probabilistic behavior of a given set of data. These are the methods: chi-square, Kolmogorov-Smirnov, Anderson-Darling, and Shapiro-Wilk tests.

Below are some of the key definitions introduced in this chapter:

- Categorical data analysis
- Estimation in categorical data
- Hypothesis testing in categorical data
- Test of independence
- Chi-square tests for count data
- Goodness of fit test
- Test for independence
- Contingency table
- P-P plot
- Q-Q plot
- Shapiro-Wilk normality test

In this chapter, we have also learned the following important concepts and procedures:

- Pearson's chi-square test
- Kolmogorov-Smirnov test
- Anderson-Darling test
- Shapiro-Wilk Test
- P-P plot construction
- Q-Q plot construction

11.7 COMPUTER EXAMPLES

11.7.1 R Commands

Since most of the R codes are already given in the chapter, we will only give the R codes for selecting a random sample from a large data set.

In R, the sample() function can be used to take a random sample of size n. Suppose we want to take a random sample of size 40 from a data set named *mydata* without replacement.

R code:

```
Mysample <- mydata[sample(1:nrow(mydata), 40, replace=FALSE),]
```

When multiple distributions fit well with a data set based on the goodness of fit tests, then we may select the best-fit distribution based on maximizing the log-likelihood value. The fitdistr() function in the MASS package in R can be used to calculate maximum likelihood fitting of a univariate distribution. Then the distribution with the largest log likelihood can be chosen as the best fit. Download package "MASS." Then do the following:

```
library(MASS)
fitdistr(mydata, 't')$loglik
> fitdistr(mydata, 'normal')$loglik
> fitdistr(mydata, 'logistic')$loglik
> fitdistr(mydata, 'weibull')$loglik
> fitdistr(mydata, 'gamma')$loglik
> fitdistr(mydata, 'lognormal')$loglik
> fitdistr(mydata, 'exponential')$loglik
```

Some other distributions, such as beta, may need specification of additional parameters. We suggest you look at R help.

PROJECTS FOR CHAPTER 11

11A. Fitting a Distribution to Data

A common problem in statistical modeling is fitting a probability distribution to a set of observations (data set) for a given variable. Using a graphic (e.g., a histogram), we may have some rough idea. If we do goodness of fit tests, with say two different distributions, it can happen that both hypotheses may not be rejected. So which one should we choose? This is mainly important in forecasting. Do a short paper on fitting a distribution to data and apply your results to each of the data in Section 11.4 to check if the chosen distributions are best possible. Some references are:

(1) *Fitting distributions with r* (http://cran.r-project.org/doc/contrib/Ricci-distributions-en.pdf)
(2) Fitting distributions to data and why you are probably doing it wrong, by David Vose (http://www.vosesoftware.com/whitepapers/Fitting%20distributions%20to%20data.pdf).

11B. Simpson's Paradox

Simpson's paradox refers to a phenomena whereby the association between a pair of variables (X, Y) reverses sign upon conditioning of a third variable, Z, regardless of the values taken by Z. Confounding factors play very important role in categorical data, resulting in Simpson's paradox, if we are not careful. As an example, consider the data below from two hospitals on an emergency surgical procedure:

	Lived	Died	Survival Rate
Hospital 1	120	180	40%
Hospital 2	60	140	30%

From this contingency table, it looks like Hospital 1 is significantly better than Hospital 2.

(a) Use hypothesis testing procedure to test if the survival rates are different at the two hospitals. Use $\alpha = 0.05$.
Now, we are given the information that Hospital 1 is situated in a rich area and as a result, patients arrive there relatively in good condition. Hospital 2 is in a poor neighborhood, thus resulting in patients arriving in a much worse condition. Now let us see what happens when we break down the above data by patient condition when they reached the hospital.

	Good Condition			Bad Condition		
	Lived	Died	Survival Rate	Lived	Died	Survival Rate
Hospital 1	120	150	44.44%	0	30	0%
Hospital 2	30	30	50%	30	110	21.42%

Now, Hospital 2 is better in both good and bad conditions! This is an example of Simpson's paradox. Here, the confounding factor is the patient's condition.

(b) Find two more real examples for Simpson's paradox.

11C. Test whether Bitcoin (or any other crypto of your interest) shows a *seasonal pattern* in its monthly closing prices

Here, "seasonal" means certain months of the year tend to have systematically different behavior (positive return or negative return) in Bitcoin price (or returns) than other months. We will now describe a possible sequence of steps you can follow for this project.

Step1: Get data. Download monthly return data for Bitcoin from the internet (such as Bitcoin Historical Data on Investing.com).

Step 2: Create categories by choosing "Return>0" versus "Return<0." For each month observation label which category each month return (1 to 12 months) falls into.

Step 3: Build a contingency table.

Step 4: State hypotheses as H_0: The category is independent of month (thus no seasonal effect) versus H_a: The category depends on the month (i.e., there is seasonality).

Step 5: Follow the chi-square test described in this chapter to test the hypothesis and make appropriate decision. Then, interpret the result.

Chapter 12

Nonparametric Statistics

Chapter outline

OBJECTIVE

In this chapter, we shall introduce several classical nonparametric or distribution-free tests. These tests do not require distributional assumptions about the population, such as normality.

Jacob Wolfowitz
(Source: http://www-groups.dcs.st-and.ac.uk/~history/mathematicians/wolfowitz.html).

Jacob Wolfowitz was born on 19 March 1910 in Warsaw, Russian Empire (now Poland), and died on 16 July 1981 in Tampa, Florida, USA. Wolfowitz's earliest interest was nonparametric inference, and the first joint paper he wrote with Abraham Wald introduced methods of calculating confidence intervals that are not necessarily of fixed width. It is in this paper by Wolfowitz in 1942 that the term *nonparametric* appears for the first time. Later, he worked on the area of sequential analysis and published work

Mathematical Statistics with Applications in R. DOI: https://doi.org/10.1016/B978-0-443-27548-7.00012-8

on sequential estimators of a Bernoulli parameter and results on the efficiency of certain sequential estimators. He also studied asymptotic statistical theory and worked on many aspects of the maximum likelihood method. Information theory pioneered by Shannon was another area to which Wolfowitz made important contributions, culminating in a classic book titled *Coding Theorems of Information Theory* (3rd [ed.] 1978). After working at different places such as the Statistical Research Group at Columbia University, the University of North Carolina, and the University of Illinois at Urbana, in 1978 he joined the faculty of the University of South Florida in Tampa. Wolfowitz was elected to the National Academy of Sciences and the American Academy of Arts and Sciences. He was also elected a Fellow of the Econometric Society, the International Statistics Institute, and the Institute of Mathematical Statistics. In 1979 he was Shannon Lecturer of the Institute of Electrical and Electronic Engineers.

12.1 INTRODUCTION

Most of the tests that we have learned up to this point are based on the assumption that the sample(s) came from a normal population, or at least that the population probability distribution(s) is specified except for a set of free parameters. Such tests are called parametric tests. In general, a parametric test is known to be generally more powerful than other procedures when the underlying assumptions are met. Usually, the assumption of normality or any other distributional assumption about the population is hard to verify, especially when the sample sizes are small, or the data are measured on an ordinal scale, such as the letter grades of a student, in which case we do not have a precise measurement. For example, incidence rates of rare diseases, data from gene-expression microarrays, and the number of car accidents in a given time interval are not normally distributed. Nonparametric tests are tests that do not make such distributional assumptions, particularly the usual assumption of normality. In situations where a distributional model for a set of data is unavailable, nonparametric tests are ideal. Even if the data are distributed normally, nonparametric methods are frequently almost as powerful as parametric methods. These tests involve only order relationships among observations and are based on the ranks of the variables and analyzing the ranks, instead of the original values. Nonparametric methods include tests that do not involve population parameters at all, such as testing whether the population is normal. Distribution-free tests generally do make some weak assumptions, such as equality of population variances and/or the distribution, and are of the continuous type.

Sometimes we may be required to make inferences about models that are difficult to parameterize, or we may have data in a form that makes, say, the normal theory tests unsuitable. For example, incomes of families generally follow a skewed distribution. If we do a sample survey of a large number of families in a feeder area, the income distribution may look as in Fig. 12.1.

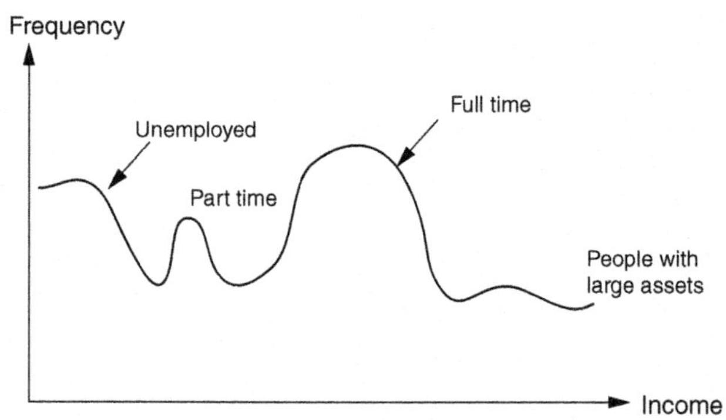

FIGURE 12.1 Income distribution of families.

This distribution is clearly difficult to parameterize, that is, to identify a classical probability distribution that will characterize the data's behavior. Moreover, the mean income of this sample may be misleading. A better measure of the central tendency is the median income. At least we know that 50% of families are below the median and 50% above. Appropriate techniques of inference in these situations are based on distribution-free methods. Most of the nonparametric methods use only the order of magnitude of observations, known as order statistics, in a random sample, rather than the observed values of the random variables.

In general, nonparametric methods are appropriate for estimation or hypothesis-testing problems when the population distributions can only be specified in general terms. The conditions may be specified as being continuous, symmetric, or identical, differing only in median or mean.

The distributions need not belong to specific families, such as normal or gamma, and so forth. Because most of the nonparametric procedures depend on a minimum number of assumptions, the chance of their being improperly used is relatively small. Most of the nonparametric procedures involve ranking data values and developing testing methods based on the ranks. Because of this, nonparametric procedures may be used when the data are measured on a weak scale, such as only count data or rank data. We may ask: Why not use nonparametric methods all the time? The answer lies in the fact that when the assumptions of the parametric tests can be verified as true, parametric tests are generally more powerful than nonparametric tests. Because only ranks are used in nonparametric methods, and even though the ranks preserve information about the order of the data, because the actual values are not used, some information is lost. Because of this, nonparametric procedures cannot be as powerful as their parametric counterparts when parametric tests can be used. For brevity and clarity, this chapter is presented without much theoretical explanation to focus on the methods. Theoretical developments can be found in many specialized books on the subject.

In this chapter, we study some of the commonly used classical nonparametric methods that are based on ordering, ranking, and permutations. The modern approaches are based on resampling methods such as bootstrap and will be discussed in Chapter 13.

12.2 NONPARAMETRIC CONFIDENCE INTERVAL

We have seen that for a large sample, using the Central Limit Theorem, we can obtain a confidence interval for a parameter within a well-defined probability distribution. However, for small samples, we need to make distributional assumptions that are often difficult to verify. For this reason, in practice, it is often advisable to construct confidence intervals or interval estimates of population quantities that are not parameters of a particular family of distributions. In a nonparametric setting, we need procedures where the sample statistics used have distributions that do not depend on the population distribution. The median is commonly used as a parameter in nonparametric settings. One of the intuitive reasons for using the sample median as a central tendency is that the sample median depends only on the middle order statistic(s). If we change a small fraction of observations—even if those changed values are huge outliers—as long as fewer than half of the observations are altered, the median must remain at or near the original middle. In contrast, the sample mean averages every observation, so a single arbitrarily large outlier can drive the mean arbitrarily far. Thus, we could say that the sample median is a robust estimator of the population median. We assume that the population distribution is continuous.

Let M denote the median of a distribution, and X (assumed to be continuous), be any observation from that distribution. Then

$$P(X \leq M) = P(X \geq M) = \frac{1}{2}.$$

This implies that, for a given random sample $X_1, \ldots, X_n$ from a population with median M, the distribution of the number of observations falling below M will follow a binomial distribution with parameters n and $p = 1/2$, irrespective of the population distribution. That is, let N^- be the number of observations less than M. Then the distribution of N^- is binomial with parameters n and $p = 1/2$ for a sample of size n. Hence, we can construct a confidence interval for the median using binomial distribution.

For a given probability value α, we can determine a and b such that

$$\begin{aligned} P(N^- \leq a) &= \sum_{i=0}^{a} \binom{n}{i}\left(\frac{1}{2}\right)^i\left(\frac{1}{2}\right)^{n-i} \\ &= \sum_{i=0}^{a} \binom{n}{i}\left(\frac{1}{2}\right)^n = \frac{\alpha}{2} \end{aligned}$$

and

$$\begin{aligned} P(N^- \geq b) &= \sum_{i=b}^{n} \binom{n}{i}\left(\frac{1}{2}\right)^i\left(\frac{1}{2}\right)^{n-i} \\ &= \sum_{i=b}^{n} \binom{n}{i}\left(\frac{1}{2}\right)^n = \frac{\alpha}{2}. \end{aligned}$$

If exact probabilities cannot be achieved, choose a and b such that the probabilities are as close as possible to the value of $\alpha/2$. Furthermore, let $X_{(1)}, X_{(2)}, \ldots, X_{(a)}, \ldots, X_{(b)}, \ldots, X_{(n)}$ be the order statistics of $X_1, \ldots, X_n$ as in Fig. 12.2.

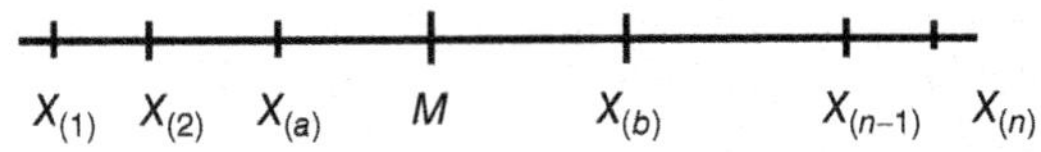

FIGURE 12.2 Ordered sample.

Then the population median will be above the order statistic, $X_{(b)}$, $\left(\frac{\alpha}{2}\right)100\%$ of the time and below the order statistic, $X_{(a)}$, $\left(\frac{\alpha}{2}\right)100\%$ of the time. Hence, a $(1-\alpha)100\%$ confidence interval for the median of a population distribution will be

$$X_{(a)} < M < X_{(b)}.$$

We can write this result as $P\left(X_{(a)} < M < X_{(b)}\right) \geq 1-\alpha$.

By dividing the upper and lower tail probabilities equally, we find that $b = n + 1 - \alpha$. Therefore, the confidence interval becomes

$$X_{(a)} < M < X_{(n+1-a)}.$$

In practice, a will be chosen so as to come as close to attaining $\frac{\alpha}{2}$ as possible.

We can summarize the nonparametric procedure for finding the confidence interval for the population median as follows.

Procedure for finding $(1-\alpha)100\%$ confidence interval for the median m

For a sample of size n:

1. Arrange the data in ascending order.
2. From the binomial table with n and $p = \frac{1}{2}$, find the value of a such that

$$p(X \leq a) = \frac{\alpha}{2} \text{ or nearest to } \frac{\alpha}{2}.$$

3. Set $b = n + 1 - a$.
4. Then the confidence interval is such that the lower limit is the ath value and the upper limit is the bth value of the observations in step 1.

Assumptions: Population distribution is continuous; the sample is a simple random sample.

We illustrate this four-step procedure with an example.

EXAMPLE 12.2.1

In a large company, the following data represents a random sample of the ages of 20 employees.

24	31	28	43	28	56	48	39	52	32
38	49	51	49	62	33	41	58	63	56.

Construct a 95% confidence interval for the population median M of the ages of the employees of this company.

Solution

For a 95% confidence interval, $\alpha = 0.05$. Hence, $\alpha/2 = 0.025$. The ordered data are

24	*28*	*28*	*31*	*32*	*33*	*38*	*39*	*41*	*43*
48	*49*	*49*	*51*	*52*	*56*	*56*	*58*	*62*	*63.*

Looking at the binomial table with $n = 20$ and $p = 1/2$, we see that $P(X \leq 5) = 0.0207$. Hence, $a = 5$ comes closest to achieving $\alpha/2 = 0.025$. Hence, in the ordered data, we should use the fifth observation, 32, for the lower confidence limit and the 16th *observation* ($n + 1 - a = 21 - 5 = 16$), 56, *for the upper confidence limit. Therefore, an approximate 95% confidence interval for M is*

$$32 < M < 56.$$

Thus, we are at least 95% certain that the true median of the employee ages of this company will be greater than 32 and less than 56, that is,

$$P(32 < M < 56) \geq 0.95.$$

The data of Example 12.2.1 passes the normality test, and we can calculate the 95% parametric confidence interval as (38.40, 49.70). Comparing this to the nonparametric confidence interval, the length of the parametric confidence interval, in general, is smaller whenever parametric assumptions can be made.

EXAMPLE 12.2.2

A drug is suspected of causing an elevated heart rate in a certain group of high-risk patients. Twenty patients from this group were given the drug. The changes in heart rates were found to be as follows,

−1	8	5	10	2	12	7	9	1	3
4	6	4	20	11	2	−1	10	2	8.

Construct a 98% confidence interval for the mean change in heart rate. Can we assume that the population has a normal distribution? Interpret your answer.

Solution

First testing for normality, we get the normal probability plot as shown in Fig. 12.3.

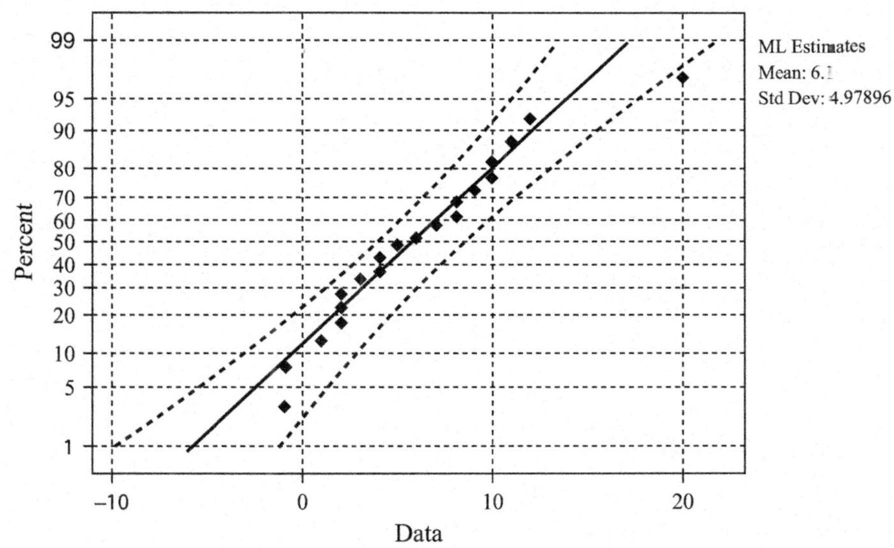

FIGURE 12.3

This shows that the normality assumption may not be satisfied, and thus the nonparametric method is more suitable (this conclusion is based strictly on the normal probability plot, which is a visual interpretation). Using a box plot, we can also test for outliers. The ordered data are

$$\begin{matrix} -1 & -1 & 1 & 2 & 2 & 2 & 3 & 4 & 4 & 5 \\ 6 & 7 & 8 & 8 & 9 & 10 & 10 & 11 & 12 & 20. \end{matrix}$$

Looking at the binomial table with $n = 20$ *and* $p = \frac{1}{2}$, *we see that* $P(X \le 4) = 0.006$. *Hence,* $a = 4$ *comes closest to achieving* $\alpha/2 = 0.01$. *Hence, in the ordered data, we should use the fourth observation, 2, for the lower confidence limit and the 17th observation* ($n + 1 - a = 21 - 4 = 17$), *10, for the upper confidence limit. Therefore, an approximate 98% confidence interval for* M *is*

$$2 < M < 10.$$

That is, we are at least 98% certain that the true median of the mean change in heart rate will be greater than 2 and less than 10.

If we perform the usual t-test, we will get the 98% confidence interval as (3.20, 9.0). However, such an interval is not valid, because the normality assumptions are not satisfied and will lead to misinterpretation of the facts.

EXERCISES 12.2

12.2.1 For the following random sample values, construct a 95% confidence interval for the population median M:

7.2 5.7 4.9 6.2 8.5 2.7 5.9 6.0 8.2.

12.2.2 The following data represent a random sample of end-of-year bonuses for the lower-level managerial personnel employed by a large firm. Bonuses are expressed in percentage of yearly salary.

6.2	9.2	8.0	7.7	8.4	9.1	7.4	6.7	8.6	6.9
8.9	10.0	9.4	8.8	12.0	9.9	11.7	9.8	3.2	4.6.

Construct a 98% confidence interval for the median bonus expressed in percentage of yearly salary of this firm. Also, draw a probability plot and test for normality. Can this be considered a random sample?

12.2.3 Air pollution in large US cities is monitored to see if it conforms to requirements set by the Environmental Protection Agency. The following data, expressed as an air pollution index, gives the air quality of a city for 10 randomly selected days.

57.3 58.1 58.7 66.7 58.6 61.9 59.0 64.4 62.6 64.9.

(a) Draw a probability plot and test for normality.

(b) Construct a 95% confidence interval for the actual median air pollution index for this city and interpret its meaning.

12.2.4 A random sample from a population yields the following 25 values:

90	87	121	96	106	107	89	107	83	92
117	93	98	120	97	109	78	87	99	79
104	85	91	107	89.					

Obtain a 99% confidence interval for the population median.

12.2.5 In an experiment on the uptake of solutes by liver cells, a researcher found that six determinations of the radiation, measured in counts per minute after 20 minutes of immersion, were:

2728 2585 2769 2662 2876 2777.

Construct a 99% confidence interval for the population median and interpret its meaning.

12.2.6 The nominal resistance of a wire is 0.20 ohm. A testing of the wire randomly chosen from a large collection of such wires yields the following resistance data.

0.199 0.211 0.198 0.201 0.197 0.200 0.198 0.208.

Obtain a 95% confidence interval for the population median.

12.2.7 In order to measure the effectiveness of a new procedure for pruning grapes, 15 workers are assigned to prune an acre of grapes. The effectiveness is measured in worker hours per acre for each person.

5.2 5.0 4.8 4.5 3.9 6.1 4.2 4.4 5.5 5.8
4.2 5.3 4.9 4.7 4.9.

Obtain a 99% confidence interval for the median time required to prune an acre of grapes for this procedure and interpret its meaning.

12.2.8 The following data give the exercise capacity (in minutes) for 10 randomly chosen patients being treated for chronic heart failure.

15 27 11 19 12 21 11 17 13 22.

Obtain a 95% confidence interval for the median exercise capacity for patients being treated for chronic heart failure.

12.2.9 The data given below refer to the in-state tuition costs (in dollars) of 15 randomly selected colleges from a list of the 100 best values in public colleges (source: *Kiplinger's Magazine*, October 2000).

3788 4065 2196 7360 5212 4137 4060 3956
3975 7395 4058 3683 3999 3156 4354.

Obtain a 95% confidence interval for the median in-state tuition costs and interpret its meaning.

12.2.10 Sepsis is an extreme immune system response to an infection that has spread throughout the blood and tissues. Sepsis can reduce blood flow to kidneys resulting in acute renal failure (also called acute kidney injury). Relative risk of mortality associated with developing acute renal failure due to sepsis in sixteen studies is given below (*Crit Care*. 2002: 6(6): 509–513).

0.75	2.03	2.29	2.11	0.80	1.50	0.79	1.01
1.23	1.48	2.45	1.02	1.03	1.30	1.54	1.27

Obtain a 95% confidence interval for the median relative risk of mortality.

12.3 NONPARAMETRIC HYPOTHESIS TESTS FOR ONE SAMPLE

In this section, we study two popular tests for testing hypotheses about population location, or median using the *sign test* and the *Wilcoxon signed rank test*. The comparison of medians rather than means is a technicality that is not important unless the data are skewed substantially. In such cases, medians are somewhat more accurate than means for comparing the locations of probability distributions. Further discussions on nonparametric tests can be found in many references, such as those by W. J. Conover and by E. L. Lehmann. Before using nonparametric tests, it is desirable to test for normality of the data using normal probability plots, and for the existence of outliers using box plots, and run tests for test of randomness of the data. When we make any particular choice of method, we test for the assumptions made. These assumption checks are relatively easier using statistical software packages. Many of the examples in this chapter are given more for illustration of the nonparametric methods than for assumption violations of parametric tests or for comprehensive assumption testing techniques. Also, when we use statistical software packages, generally, the p-value of the test will be given in the output. In order to make a decision on a particular hypothesis, we just need to compare the p-value with the chosen value of α. We are going to explain a more traditional approach instead of using the p-value approach in the discussion; however, the computer example section will illustrate the p-value approach.

12.3.1 The Sign Test

In this section, we describe a test that is the nonparametric alternative to the one-sample t-test and to the paired-sample t-test. Let M be the median of a certain population. Then we know that

$$P(X \le M) = 0.5 = P(X > M).$$

We consider the problem of testing the null hypothesis

$$H_0: M = m_0 \quad \text{versus} \quad H_a: M > m_0.$$

Assume that the underlying population distribution is continuous. Let X_i be the ith observation, and let N^+ be the number of observations that are greater than m_0. N^+ will be our test statistic. We will reject H_0 if, n^+ the observed value of N^+, is too large. This test is called the sign test. A test at a significance level α will reject H_0 if $n^+ \ge k$, where k is chosen such that

$$P(N^+ \ge k \text{ when } M = m_0) = \alpha.$$

Similarly, if the alternative is of the form H_a: $M \ne m_0$, the critical region is of the form $N^+ \le k$ or $N^+ \ge k_1$, where $P(N^+ \le k) + P(N^+ \ge k_1) = \alpha$.

In order to determine such a k and k_1, we need to determine the distribution of N^+. The test works on the principle that if the sample were to come from a population with a continuous distribution, then each of the observations falls above the median or below the median with probability $\frac{1}{2}$. Hence, the number of sample values falling below the median follows a binomial distribution with parameters n and $p = \frac{1}{2}$, n being the sample size. If a sample value equals the hypothesized median m_0, that observation will be discarded and the sample size will be adjusted accordingly (we remark that such values should be very few). Thus, when H_0 is true, N^+ will have a binomial distribution with parameters n and $p = \frac{1}{2}$. For this reason, some authors call this test the *binomial test*. The following procedure summarizes the test statistic and the corresponding critical regions.

Sign test

H_0: $M = m_0$

Alternative Hypothesis	Critical Region
H_a: $M > m_0$	$N^+ \ge k$, where $\sum_{i=k}^{n} \binom{n}{i} \left(\frac{1}{2}\right)^n = \alpha$
H_a: $M < m_0$	$N^+ \le k$, where $\sum_{i=0}^{k} \binom{n}{i} \left(\frac{1}{2}\right)^n = \alpha$
And H_a: $M \ne m_0$	$N^+ \le k_1$, where $\sum_{i=k_1}^{n} \binom{n}{i} \left(\frac{1}{2}\right)^n = \frac{\alpha}{2}$ or $N^+ \le k$, where $\sum_{i=0}^{k} \binom{n}{i} \left(\frac{1}{2}\right)^n = \frac{\alpha}{2}$.

If α or $\alpha/2$ cannot be achieved exactly, choose k (or k and k_1) so that the probability comes as close to α (or $\alpha/2$) as possible.

We now summarize the procedure of the sign test in the case of an upper tail alternative. The other two cases are similar.

Hypothesis-testing procedure using the sign test

We test

$$H_0: M = m_0 \text{ versus } H_a: M > m_0.$$

1. Replace each value of the observation that is greater than m_0 by a plus sign and each sample value less than m_0 by a minus sign. If the sample value is equal to m_0, discard the observation and adjust the sample size n accordingly.
2. Let n^+ be the number of +'s in the sample. For n and $p = \frac{1}{2}$, from the binomial table, find

$$\gamma = P(N^+ \geq n^+).$$

3. **Decision:** If γ is less than α, H_0 must be rejected. Based on the sample, we will conclude that the median of the population is greater than m_0 at the significance level α. Otherwise, do not reject H_0.

Assumptions: The population distr bution is continuous. The number of ties is small (less than ˆ0% of the sample).

Note that the approach described in the foregoing procedure is nothing but the p-value method for hypothesis testing regarding a median using the sign test. Recall that the p-value is the probability of observing a test statistic as extreme or more extreme than what was really observed, under the assumption that the null hypothesis is true. In the sign test, we had assumed that the median is $M = m_0$, so 50% of the data should be less than m_0 and 50% of the data greater than m_0. Thus, we expect half of the data to result in plus signs and half to result in minus signs. Hence, we can think of the data as following a binomial distribution with $p = 1/2$ under the null hypothesis. The p-value is computed from its definition given by the formula

$$p\text{−}value = P(N^+ \geq n^+) = \sum_{i=k}^{n} \binom{n}{i} \left(\frac{1}{2}\right)^n = \gamma.$$

The p-value method is to reject the null hypothesis if the computed p-value is greater than α. These binomial probabilities can be obtained from the binomial tables, or statistical software packages. The following example illustrates how we apply the three-step procedure.

EXAMPLE 12.3.1

For the given data from an experiment

1.51 1.35 1.69 1.48 1.29 1.27 1.54 1.39 1.45.

test the hypothesis that H_0: $M = 1.4$ versus H_a: $M > 1.4$ at $\alpha = 0.05$.

Solution

We test

$$H_0: M = 1.4 \text{ versus } H_a: M > 1.4.$$

Replacing each value greater than 1.4 with a plus sign and each value less than 1.4 with a minus sign, we have

$$+ - + + - - + - +.$$

Thus, $n^+ = 5$. *From the binomial table with* $n = 9$ *and* $p = \frac{1}{2}$, *we have*

$$P(N^+ \geq 5) = 0.50.$$

Hence, the p-value is 0.5. Because $\alpha = 0.05 < 0.50$, *the null hypothesis is not rejected. We conclude that the median does not exceed* 1.4.

When the sample size n is large, we can apply the normal approximation to the binomial distribution. That is, the test statistic N^+ is approximately normally distributed. Thus, under H_0, N^+ will have an approximate normal distribution with mean $np = n/2$ and variance of $np(1-p) = n/4$. By the z-transform, we have

$$Z = \frac{N^+ - n/2}{\sqrt{n/4}} = \frac{2N^+ - n}{\sqrt{n}} \sim N(0,1).$$

We could utilize this test if n is large, that is, if $np \geq 5$ and $n(1-p) \geq 5$. Hence, under H_0, because $p = 1/2$, if $n \geq 10$, we could use the large sample test. The following table summarizes the method for the large sample sign test.

A sign test for a large random sample

When the sample size is large ($n \geq 10$), we can use the normal approximation to a binomial. This leads to the large sample sign test:

$$H_0: M = m_0$$

versus

Alternative Hypothesis	Rejection Region
$H_a: M > m_0$	$z \geq z_\alpha$
$H_a: M < m_0$	$z \leq -z_\alpha$
$H_a: M \neq m_0$	$\lvert z \rvert \geq z_{\alpha/2}$.

The test statistic is

$$Z = \frac{2N^+ - n}{\sqrt{n}}.$$

Decision: Reject H_0 if the test statistic falls in the rejection region, and conclude that H_a is true with at least $(1-\alpha)100\%$ confidence. Otherwise, do not reject H_0 because there is not enough evidence to conclude that H_a is true for a given α, and more data is needed.

Assumptions: (i) Population distribution is continuous. **(ii)** Sample size greater than or equal to 10 (after the removal of ties). **(iii)** The number of ties is small (less than 10% of the sample size).

We illustrate this procedure with the following example.

EXAMPLE 12.3.2

In order to measure the effectiveness of a new procedure for pruning grapes, 15 workers are assigned to prune an acre of grapes. The effectiveness is measured in worker hours/acre for each person. The results are given below:

5.2 5.0 4.8 3.9 6.1 4.2 4.4 5.5 5.8 4.5
4.2 5.3 4.9 4.7 4.9.

Test the null hypothesis that the median time to prune an acre of grapes with this method is 4.5 hours against the alternative that it is larger. Use $\alpha = 0.05$.

Solution

We test

$$H_0: M = 4.5 \text{ versus } H_a: M > 4.5.$$

Replacing each value greater than 4.5 with a plus sign and each value less than 4.5 with a minus sign, we have

$$+ ++-+--++-++++.$$

Because there is one observation that is equal to 4.5, we must discard it and take n = 14.
Thus $N^+ = 10$, using the large sample approximation, the test statistic is

$$Z = \frac{2N^+ - n}{\sqrt{n}} = \frac{20 - 14}{\sqrt{14}} = 1.6.$$

For $\alpha = 0.05$, from the standard normal table, the value of $z_{0.05} = 1.645$. Hence, the rejection region is $z > 1.645$. Because the observed value of the test statistic does not fall in the rejection region, we do not reject the null hypothesis at $\alpha = 0.05$ and conclude that the median time to prune an acre of grapes is 4.5 hours.

12.3.2 Wilcoxon Signed Rank Test

In the sign test, we have considered only whether each observation is greater than m_0 or less than m_0 without giving any importance to the magnitude of the difference from m_0. Neglecting information on the magnitude of the observations is rather inefficient and may reduce the statistical power of the test. An improved version of the sign test is the Wilcoxon signed rank test, in which one replaces the observations by their ranks of the ordered magnitudes of differences, $|x_i - m_0|$. The smallest observation is ranked as 1, the next smallest will be 2, and so on. However, the Wilcoxon signed rank test requires an additional assumption that the continuous population distribution is symmetric with respect to its center. Thus, if the data are ordinal, the Wilcoxon test cannot be used.

Hypothesis-testing procedure using Wilcoxon signed rank test

We wish to test

$$H_0: M = m_0 \text{ versus } H_a: M \neq m_0.$$

1. Compute the absolute differences $z_i = |x_i - m_0|$ for each observation. Replace each value of the observation that is greater than m_0 by a plus sign and each sample value that is less than m_0 by a minus sign. If the sample value is equal to m_0, discard the observation and adjust the sample size n accordingly.
2. Assign each z_i a value equal to its rank. If two values of z_i are equal, assign each z_i a rank equal to the average of the ranks each should receive if there were not a tie.
3. Let W^+ be the sum of the ranks associated with plus signs and W^- be the sum of ranks with negative signs.
4. **Decision:** If m_0 is the true median, then the observations should be evenly distributed about m_0. For a given α critical region, reject H_0 if

$$W^+ \leq c_1, \text{ where } P(W^+ \leq c_1) = \frac{\alpha}{2},$$

or

$$W^+ \geq c_2, \text{ where } P(W^+ \geq c_2) = \frac{\alpha}{2}.$$

Assumptions: The population distribution is continuous and symmetrical. The number of ties is small, less than 10% of the sample size.

The exact distribution of W^+ is considerably complicated, and we will not derive it. However, for certain values of n, the distribution is given in the *Wilcoxon signed rank test table*.

For the Wilcoxon signed rank test, the rejection region based on the alternative hypothesis is given next.

For

$$H_a: M > m_0, \text{ rejection region is } W^+ \geq c, \text{ where } P(W^+ \geq c) = \alpha.$$

and for

$$H_a: M < m_0, \text{ rejection region is } W^+ \leq c, \text{ where } P(W^+ \leq c) = \alpha.$$

We illustrate the Wilcoxon signed rank test with the following examples.

EXAMPLE 12.3.3

For the given data that resulted from an experiment

1.51 1.35 1.69 1.48 1.29 1.27 1.54 1.39 1.45

test the hypothesis that H_0: $M = 1.4$ versus H_a: $M \neq 1.4$. Use $\alpha = 0.05$.

Solution

We wish to test H_0: $M = 1.4$ versus H_a: $M \neq 1.4$.
Here, $\alpha = 0.05$, *and* $m_0 = 1.4$. *The results of steps 1 to 3 are given in* Table 12.1.

TABLE 12.1 Results of Experimental Data From Steps 1 to 3.

x_i	$z_i = \|x_i - 1.4\|$	Sign	Rank
1.51	0.11	+	5.5
1.35	0.05	–	3
1.69	0.29	+	9
1.48	0.08	+	4
1.29	0.11	–	5.5
1.27	0.13	–	7
1.54	0.14	+	8
1.39	0.01	–	1.5
1.45	0.01	+	1.5

$$H_0: M = 1.4 \text{ versus } H_a: M \neq 1.4.$$

Thus we have $W^+ = 29$ *and* $n = 9$. *From the Wilcoxon signed rank test table in the appendix, we should reject* H_0 *if* $W^+ \leq 6$ *or* $W^+ \geq 38$ *with actual level of* $\alpha = 0.054$. *Because* $W^+ = 29$ *does not fall in the rejection region, we do not reject the null hypothesis that* $M = 1.4$.

EXAMPLE 12.3.4

Air pollution in large US cities is monitored to see whether it conforms with requirements set by the Environmental Protection Agency. The following data, expressed as an air pollution index, gives the air quality of a city for 10 randomly selected days.

57.3 58.1 58.7 66.7 58.6 61.9 59.0 64.4 62.6 64.9.

Test the hypothesis that H_0: $M = 65$ versus H_a: $M < 65$. Use $\alpha = 0.05$.

Solution

We will test

$$H_0: M = 65 \text{ versus } H_a: M < 65.$$

Here, $\alpha = 0.05$, and $m_0 = 65$.
The results of steps 1 to 3 are given in Table 12.2.
Thus, $W^+ = 3$, *and* $n = 10$. *Using the Wilcoxon signed rank test table, we should reject* H_0 *if* $W^+ \leq 10$ *with a level of significance* $\alpha = 0.042$. *Because the observed value of* W^+ *falls in the rejection region, we reject* H_0 *and conclude that the sample evidence suggests that we conclude that the median air pollution index is less than 65.*

The Wilcoxon signed rank test is a nonparametric alternative to the one-sample t-test. The question then is, how do we decide which one to choose? Choose the one-sample t-test if it is reasonable to assume that the population follows a

TABLE 12.2 Results of Steps 1 to 3 Using Air Pollution Index Data.

x_i	$z_i = \lvert x_i - 65 \rvert$	Sign	Rank
57.3	7.7	−	10
58.1	6.9	−	9
58.7	6.3	−	8
66.7	1.7	+	3
58.8	6.2	−	7
61.9	4.1	−	5
59.0	6.0	−	6
64.4	0.6	−	2
62.6	2.4	−	4
64.9	0.1	−	1

normal distribution. Otherwise, choose the Wilcoxon nonparametric test. However, the Wilcoxon test will have less power. For example, a normal probability plot of the data of Example 12.3.4 is given in Fig. 12.4. Looking at this figure, we can see that the normality assumption is suspected (Also, the p-value is $0.081 > 0.05$). It may make more sense to use the nonparametric method.

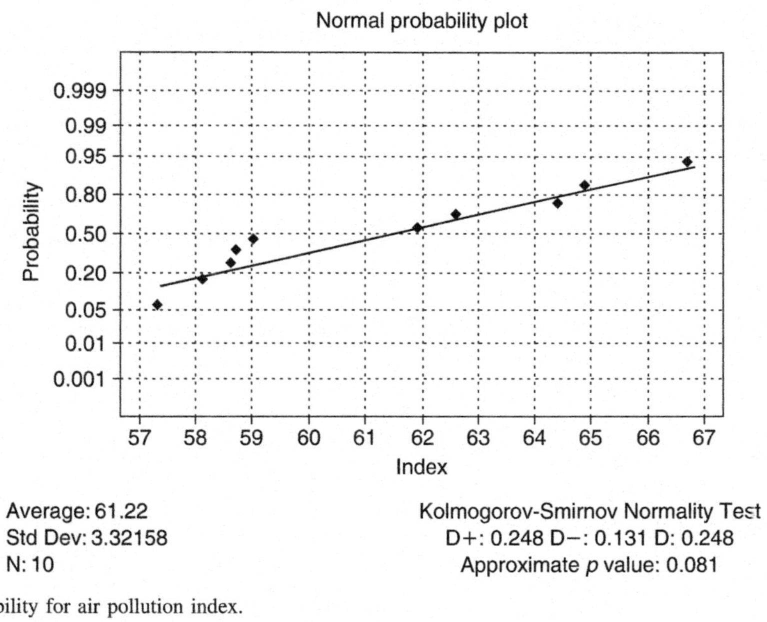

FIGURE 12.4 Normal probability for air pollution index.

When sample size n is sufficiently large, under the assumption of H_0 being true, the distribution of W^+ is approximately normal with mean

$$E(W^+) = \frac{1}{4}n(n+1)$$

and variance

$$Var(W^+) = \frac{n(n+1)(2n+1)}{24}.$$

Hence, the test statistic is given by

$$Z = \frac{W^+ - \frac{1}{4}n(n+1)}{\sqrt{n(n+1)(2n+1)/24}}$$

which is approximately the standard normal distribution. This approximation can be used when $n > 20$. We summarize the test procedure below:

Summary of the Wilcoxon signed rank test for large samples ($n > 20$)

We test

$$H_0: M = m_0$$

versus

$$H_a: \begin{cases} M > m_0, & \text{upper tailed test} \\ M < m_0, & \text{lower tailed test} \\ M \neq m_0, & \text{two–tailed test.} \end{cases}$$

The test statistic:

$$Z = \frac{W^+ - \frac{1}{4}n(n+1)}{\sqrt{n(n+1)(2n+1)/24}}.$$

Rejection region:

$$\begin{cases} z > z_\alpha, & \text{upper tail RR} \\ z < -z_\alpha, & \text{lower tail RR} \\ |z| > z_{\alpha/2}, & \text{two tail RR.} \end{cases}$$

Decision: Reject H_0 if the test statistic falls in the RR, and conclude that H_a is true with $(1 - \alpha)100\%$ confidence. Otherwise, do not reject H_0, because there is not enough evidence to conclude that H_a is true for a given α, and more data is needed.

Assumptions: **(i)** The population distribution is continuous and symmetric about 0. **(ii)** Sample size is greater than or equal to 20. **(iii)** The number of ties is small, <10% of the sample size.

We illustrate the Wilcoxon signed rank test with the following example.

Example 12.3.5

The following data give the monthly rents (in dollars) paid by a random sample of 25 households selected from a large city.

425 960 1450 655 1025 750 670 975 660 880
1250 780 870 930 550 575 425 900 525 1800
545 840 765 950 1080.

Using the large sample Wilcoxon signed rank test, test the hypotheses that the median rent in this city is \$750 against the alternative that it is higher with $\alpha = 0.05$.

Solution

We test

$$H_0: M = 750 \text{ versus } H_a: M > 750.$$

Here $\alpha = 0.05$, *and* $m_0 = 750$. *The results of steps 1 to 3 are given in* Table 12.3 *(where the asterisk indicates* $z_i = 0$*). Here, for* $n = 24$, $W^+ = 172.5$, *and the test statistic is*

$$Z = \frac{W^+ - \frac{1}{4}n(n+1)}{\sqrt{n(n+1)(2n+1)/24}} = \frac{172.5 - \left(\frac{1}{4}\right)(24)(25)}{\sqrt{\frac{(24)(25)(49)}{24}}} = 0.64286.$$

TABLE 12.3 Results of Steps 1 to 3 Using Monthly Rent (in Dollars).

x_i	$z_i = \|x_i - 750\|$	Sign	Rank
425	325	−	19.5
960	210	+	15
1450	700	+	23
655	95	−	6
1025	302	+	18
750	0	*	ignore
670	80	−	3
975	225	+	16.5
660	90	−	4.5
880	130	+	8
1250	500	+	22
780	30	+	2
870	120	+	7
930	180	+	11
550	200	−	12.5
575	175	−	10
425	325	−	19.5
900	150	+	9
525	225	−	16.5
1800	1050	+	24
545	205	−	14
840	90	+	4.5
765	15	+	1
950	200	+	12.5
1080	330	+	21

For $\alpha = 0.05$, the rejection region is $z > 1.645$. Because the observed value of the test statistic does not fall in the rejection region, we do not reject the null hypothesis. There is not enough evidence to conclude that the median rent in this city is more than \$750.

The rank tests are useful for situations when you suspect that the data do not follow the normal population. It is important to note that ignoring the tied observations reduces the effective sample size, which in turn reduces the power of the test (see Example 6.1.4 for the effect of n on the value of β). This loss is not significant if there are only a few ties. However, if the ties are 10% or more, hypothesis testing using rank tests becomes considerably conservative. That is, they yield error probabilities that are significantly high.

12.3.3 Dependent Samples: Paired Comparison Tests

The sign test and the Wilcoxon signed rank test can also be used for paired comparisons. The experimental procedure typically consists of taking "before" and "after" type or otherwise matched as in the paired t-test case readings for each unit. Suppose there are n pairs of before and after observations, and we are interested in testing the equality of the two medians. One way to test such observations is to consider the difference between the two observations for a unit to be a single observation on that unit. Thus, we can treat the sample as being n observations on a population of differences. For this new sample of differences, the testing problem becomes

$$H_0: M = 0 \text{ versus } H_a: M > 0 \text{ (or } M < 0, \text{ or } M \neq 0).$$

Hence, the basic procedure could be summarized to first find the difference between the two units for each of the observations and then follow the testing procedures explained earlier for the sign test or the Wilcoxon signed rank test. Both small sample and large sample cases can be handled as before. In the following example, we illustrate this concept for a large sample sign test.

EXAMPLE 12.3.6

A dietary program claims that 3 months of its diet will reduce weight. In order to test this claim, a random sample of eight individuals who went through this program for 3 months is taken. The following table gives weight in pounds.

Before	180	199	175	226	189	205	169	211
After	172	191	172	230	178	199	171	201

Using a 5% significance level, is there evidence to conclude that the program really reduces the population median weight?

Solution

Let M denote the median of the population of difference of weights. We will use the difference as "after"–"before." Then we will test

$$H_0: M = 0 \text{ versus } H_a: M < 0.$$

We will use the large sample sign test. Replacing each value of the difference that is greater than zero by a + sign, and less than zero by a − sign, we have

Difference	−8	−8	−3	4	−11	−6	2	−10
Sign	−	−	−	+	−	−	+	−

For $n = 8$ and $N^+ = 2$, the test statistic is given by

$$Z = \frac{2N^+ - n}{\sqrt{n}} = \frac{4 - 8}{\sqrt{8}} = -1.414.$$

For $\alpha = 0.05$, $z_{0.05} = 1.645$, and the rejection region is $z < -1.645$. Because the observed value of the test statistic does not fall in the rejection region, we do not reject the null hypothesis. Thus, there is not enough evidence to conclude that the new program reduces weight. Note that even though $n = 8$ is small, here we are using the large sample test only for demonstration purposes.

EXERCISES 12.3

12.3.1 It was reported that the median interest rate on 30-year fixed mortgages in a certain large city is 7.75% on a particular day, with zero points. A random sample of nine lenders produced the following data of interest rates in percentage.

7.625 7.375 8.00 7.50 7.875 8.00 7.625 7.75 7.25

Test the hypothesis that the median interest rate in this city is different from 7.75%, using (a) the sign test, and (b) the Wilcoxon signed rank test. Use $\alpha = 0.01$. Compare the two results.

12.3.2 It is believed that a typical family spends 35% of its income on food and groceries. A sample of eight randomly selected families yielded the following data.

30	29	39	49	36	33	37	35

Test the hypothesis that the median percentage of family income spent for food and groceries is 35 against the alternative that it is less than 35. Use $\alpha = 0.05$.

12.3.3 The SAT scores (out of a maximum possible score of 1600) for a random sample of 10 students who took this test recently are:

1355	765	890	1089	986	1128	1157	1065	1224	567

Test the hypothesis that the median SAT score is 1000 against the alternative that it is greater using $\alpha = 0.05$. Use both the sign test and the Wilcoxon signed rank test. Explain if the conclusions are different.

12.3.4 The regulatory board of health in a particular state specifies that the fluoride levels in water must not exceed 1.5 parts per million (ppm). The 20 measurements given here represent the randomly selected daily early morning readings on fluoride levels in water at a certain city.

0.88	0.82	0.97	0.95	0.84	0.90	0.87	0.78	0.75	0.83
0.71	0.92	1.11	0.81	0.97	0.85	0.97	0.91	0.78	0.87

Test the hypothesis that the median fluoride level for this city is 0.90 against the alternative that the median is different from 0.9 at $\alpha = 0.01$, using (a) the large sample sign test, and (b) the Wilcoxon signed rank test. Interpret the results.

12.3.5 The following data give the weights (in pounds) for a random sample of 20 NFL players.

285	178	311	276	192	232	259	189	298	211
269	285	296	193	288	254	246	234	274	229

Test the hypothesis that the median weight of NFL players is 250 pounds against the alternative that it is greater at $\alpha = 0.05$, using (a) the large sample sign test and (b) the Wilcoxon signed rank test.

12.3.6 The following data give the amount of money (in dollars) spent on textbooks by 18 students for the last academic year at a large university.

510	425	190	298	157	260	320	615	455
490	188	115	230	610	220	155	315	110

Test the hypothesis that the median amount spent on books at this university is \$325 against the alternative that it is different using the large-sample sign test. Use $\alpha = 0.05$.

12.3.7 It is desired to study the effect of a special diet on systolic blood pressure. The following sample data are obtained for eight adults over 40 years of age before and after 6 months of this diet.

Before	185	222	235	198	224	197	228	234
After	188	217	229	190	226	185	225	231

At 95% confidence level, is there evidence to conclude that the new diet reduces the systolic blood pressure in individuals over 40 years old? Test (a) using the sign test, and (b) using the Wilcoxon signed rank test. Interpret the results.

12.3.8 In an effort to study the effect on absenteeism of having a day-care facility at the workplace for women with newborn babies (less than 1 year old), a large company compared the number of absent days for a year for seven women with newborn children before and after instituting a day-care facility.

Before	20	18	35	22	17	24	15
After	16	9	22	28	19	13	10

At a 99% confidence level, is there evidence to conclude that having a day-care facility at the workplace reduces absenteeism for women with newborn children?

12.3.9 For a popular computer tablet, the user ratings (1 through 5 stars, with 5 stars being the highest rating) of 10 randomly selected are given as follows

5,5,1,4,3,5,4,4,5,4

At the 0.05 level, is there evidence that the median rating is at least 4?

12.3.10 For the data given in Exercise 12.2.10, does the combined evidence from all 16 studies suggest that developing acute renal failure as a complication of sepsis impacts mortality? Use $\alpha = 0.05$. Do both the sign test and the Wilcoxon signed rank test.

12.3.11 Suppose we want to test whether a new machine-learning model improves Bitcoin daily returns compared to the old model. We have the following seven-day data (Table 12.4).

TABLE 12.4 Daily Bitcoin Returns in %.

Day	Old Model Return (%)	New Model Return (%)
1	2.1	2.5
2	1.8	1.6
3	3.2	3.6
4	2.9	2.7
5	3.0	3.8
6	2.5	2.9
7	1.9	2.3

Use $\alpha = 0.05$. Do both the sign test and the Wilcoxon signed rank test. Also do the large sample approximation.

12.4 NONPARAMETRIC HYPOTHESIS TESTS FOR TWO INDEPENDENT SAMPLES

In this section, we learn how to test the equality of the medians of two independent samples from two populations. This is especially useful when one studies the treatment effects, such as the effect of a certain drug to treat a given medical condition when we have two groups—an experimental group and a control group—or the effect of a particular type of teaching method. Even though this test can be used for more than two samples, here, we will restrict it to two samples. We will describe the *median test*, which corresponds to the sign test, and the *Wilcoxon rank sum test.*

12.4.1 Median Test

Let m_1 and m_2 be the medians of two populations 1 and 2, respectively, both with continuous distributions. Assume that we have a random sample of size n_1 from population 1 and a random sample of size n_2 from population 2. The median test can be summarized as follows.

Hypothesis-testing procedure using median test

We test

H_0: $m_1 = m_2$ versus

$$H_a: \begin{array}{ll} m_1 > m_2, & \text{upper tailed test} \\ m_1 < m_2, & \text{lower tailed test} \\ m_1 \neq m_2, & \text{two–tailed test.} \end{array}$$

1. Combine the two samples into a single sample of size $n = n_1 + n_2$, keeping track of each observation's original population. Arrange the $n_1 + n_2$ observations in increasing order and find the median of this combined sample. If the median is one of the sample values, discard those observations and adjust the sample size accordingly.
2. Define N_{1b} to be the number of observations of a sample from population 1.
3. **Decision:** If H_0 is true, then we would expect N_{1b} to be equal to some number around $n_1/2$. For H_a: $m_1 > m_2$, rejection region is $N_{1b} \le c$, where $P(N_{1b} \le c) = \alpha$, for H_a: $m_1 < m_2$, rejection region is $N_{1b} \ge c$, where $P(N_{1b} \ge c) = \alpha$, and for H_a: $m_1 = m_2$, rejection region is $N_{1b} \ge c_1$, or $N_{1b} \le c_2$, where

$$P(N_{1b} \ge c_1) = \frac{\alpha}{2} \text{ and } P(N_{1b} \le c_2) = \frac{\alpha}{2}.$$

Assumptions: **(i)** Population distribution is continuous. **(ii)** Samples are independent.

Note that since some observations can be equal to the overall median, and those values will be discarded, N_{1b} need not be equal to n_1. Let $n_1 + n_2 = 2k$. Under H_0, N_{1b} has a hypergeometric distribution given by

$$P(N_{1b} = n_{1b}) = \frac{\binom{n_1}{n_{1b}}\binom{n_2}{k - n_{1b}}}{\binom{n_1 + n_2}{k}}, \quad n_{1b} = 0,1,2,\ldots,n_1.$$

with the assumption that $\binom{i}{j} = 0$, if $j > i$. Note that the hypergeometric distribution is a discrete distribution that describes the number of "successes" in a sequence of n draws from a finite population without replacement. Thus, we can find the values of c, c_1, and c_2, required earlier. This calculation can be tedious. To overcome this, we can use the following large sample approximation valid for $n_1 > 5$ and $n_2 > 5$. First, classify each observation as above or below the sample median as shown in Table 12.5.

TABLE 12.5 Classification of the Data.

	Below	Above	Totals
Sample 1	N_{1b}	N_{1a}	n_1
Sample 2	N_{2b}	N_{2a}	n_2
Total	N_b	N_a	$n_1 + n_2 = n$

It can be verified that the expected value and variance of N_{1a} (similarly for N_{1b}) are given by

$$E(N_{1a}) = \frac{N_a n_1}{n}, \text{ and } Var(N_{1a}) = \frac{N_a n_1 n_2 N_b}{n^2(n-1)}.$$

Thus, for a large sample, we can write

$$z = \frac{N_{1a} - E(N_{1a})}{\sqrt{Var(N_{1a})}} \sim N(0,1).$$

Hence, we can follow the usual large sample rejection region procedure, which is summarized next.

Summary of large sample median sum test ($n_1 > 5$ and $n_2 > 5$)

We test

$$H_0: m_1 = m_2 \text{ versus } H_a: \begin{cases} m_1 > m_2, \text{ upper tailed test} \\ m_1 < m_2, \text{ lower tailed test} \\ m_1 \neq m_2, \text{ two-tailed test.} \end{cases}$$

The test statistic:

$$z = \frac{N_{1a} - E(N_{1a})}{\sqrt{Var(N_{1a})}},$$

where

$$E(N_{1a}) = \frac{N_a n_1}{n}$$

and

$$Var(N_{1a}) = \frac{N_a n_1 n_2 N_b}{n^2(n-1)}.$$

Rejection region:

$$\begin{cases} z > z_\alpha, \text{ upper tail RR} \\ z < -z_\alpha, \text{ lower tail RR} \\ |z| > z_{\alpha/2}, \text{ two tail RR.} \end{cases}$$

Decision: Reject H_0 if the test statistic falls in the *RR*, and conclude that H_a is true with $(1 - \alpha)100\%$ confidence. Otherwise, do not reject H_0, because there is not enough evidence to conclude that H_a is true for a given α, and more data is needed.

Assumptions: (i) Population distributions are continuous. **(ii)** $n_1 > 5$ and $n_2 > 5$.

We illustrate this procedure with the following example.

EXAMPLE 12.4.1

Given below are the mileages (in thousands of miles) of two samples of automobile tires of two different brands, say I and II, before they wear out.

Tire I:	34	32	37	35	42	43	47	58	59	62	69	71	78	84
Tire II:	39	48	54	65	70	76	87	90	111	118	126	127.		

Use the median test to see whether the tire II gives more median mileage than tire I. Use $\alpha = 0.05$.

Solution

We will test

$$H_0: m_1 = m_2 \text{ versus } H_0: m_1 < m_2.$$

Because the sample size assumption is satisfied, we will use the large sample normal approximation. The results of steps 1 and 2, using notation A for above the median and B for below the median, are given in Table 12.6.

The median is 63.5. Thus, we obtain Table 12.7.

Also

$$EN_{1a} = \frac{N_a n_1}{n} = \frac{(13)(14)}{26} = 7,$$

TABLE 12.6 Mileage of Automobiles are Either Above (A) or Below (B) the Median.

Sample Values	Population	Above/Below the Median
32	I	B
34	I	B
35	I	B
37	I	B
39	II	B
42	I	B
43	I	B
47	I	B
48	II	B
54	II	B
58	I	B
59	I	B
62	I	B
65	II	A
69	I	A
70	II	A
71	I	A
76	II	A
78	I	A
84	I	A
87	II	A
90	II	A
111	II	A
118	II	A
126	II	A
127	II	A

TABLE 12.7 Number of Data Below and Above the Median.

	Below	Above	Totals
Sample 1	$N_{1b} = 10$	$N_{1a} = 4$	$n_1 = 14$
Sample 2	$N_{2b} = 3$	$N_{2a} = 9$	$n_2 = 12$
Total	$N_b = 13$	$N_a = 13$	$n_1 + n_2 = n = 26$

and

$$Var(N_{1a}) = \frac{N_a n_1 n_2 N_b}{n^2(n-1)} = \frac{(13)(13)(14)(12)}{16{,}900} = 1.68.$$

Hence, the test statistic is

$$z = \frac{N_{1a} - E(N_{1a})}{\sqrt{Var(N_{1a})}} = \frac{4-7}{\sqrt{1.68}} = -2.31.$$

For $\alpha = 0.05$, $z_{0.05} = 1.645$. Hence, the rejection region is $\{z < -1.645\}$. Because the observed value of z does fall in the rejection region, we reject H_0 and conclude that there is enough evidence to conclude that there is a difference in the median mileage for the two types of tires.

12.4.2 The Wilcoxon Rank Sum Test

The Wilcoxon rank sum test is used for comparing the medians of two independent populations, as in the two-sample t-test in the parametric case. For accurate results, it is necessary to assume that the variances of the populations are equal. This test is quite similar to the Wilcoxon signed rank test. Whereas the one-sample Wilcoxon signed rank test requires an additional assumption that the population distribution is symmetric, such an assumption is not necessary for the two-sample Wilcoxon rank sum test. This test can be applied for skewed distributions. The test is almost as powerful as the parametric version when the population distributions are close to normal. Many statistical software packages do not give the Wilcoxon rank sum test; instead, the Mann–Whitney test is given. It should be noted that the Wilcoxon rank sum test is equivalent to the Mann–Whitney U-test. We will not separately describe the Mann–Whitney test; however, in practice, just perform the Mann–Whitney test if the software has only that test.

Assume that we have n_1 observations randomly sampled from population I and n_2 observations randomly sampled from population II with $n_1 \le n_2$. The Wilcoxon rank sum test procedure can be summarized as follows.

Hypothesis-testing procedure using the Wilcoxon rank sum test

We test

$$H_0: m_1 = m_2 \text{ versus } H_1: m_1 \neq m_2.$$

1. Combine the two samples into a single sample of size $n_1 + n_2$, keeping track of each observation's original population. Arrange the $n_1 + n_2$ observations in ascending order and assign ranks.
2. Sum the ranks of observations from population II and call it R.
3. Let the test statistic be $W = R - \frac{1}{2}n_2(n_2+1)$.
4. **Decision:** If H_0 is false, one would expect that the value of W would be very small or very large. For a size α critical region, reject H_0 if

$$W \le c_1, \text{ where } P(W \le c_1) = \frac{\alpha}{2},$$

or

$$W \ge c_2, \text{ where } P(W \ge c_2) = \frac{\alpha}{2}.$$

Note: The exact distribution of W is given in the Wilcoxon rank sum test table in the appendix for small values of n_1 and n_2.

In the Wilcoxon rank sum test, based on the alternative hypothesis, we have the following rejection regions. For

$$H_a: m_1 > m_2, \text{ rejection region is } W \ge c, \text{ where } P(W \ge c) = \alpha.$$

and for

$$H_a: m_1 < m_2, \text{ rejection region is } W \le c, \text{ where } P(W \le c) = \alpha.$$

We will illustrate the foregoing procedure with the following example.

EXAMPLE 12.4.2

Comparison of the prices (in dollars) of two brands of similar automobile tires resulted in the data in Table 12.8.

TABLE 12.8 Price Comparisons of Two Brands of Tires.

Tire I:	85	99	100	110	105	87		
Tire II:	67	69	70	93	105	90	110	115

Use the Wilcoxon rank sum test with $\alpha = 0.05$ to test the null hypothesis that the two population medians are the same against the alternative hypothesis that the population medians are different.

Solution

Here, we need to test

$$H_0: m_1 = m_2 \text{ versus } H_a: m_1 \neq m_2.$$

The sample sizes are $n_1 = 6$, *and* $n_2 = 8$. *Combining step 1 and step 2, we have the results shown in* Table 12.9.

TABLE 12.9 Price Comparison Analysis.

Value	67	69	70	85	87	90	93	99	100	105	105	110	110	115
Population	II	II	II	I	I	II	II	I	I	I	II	I	II	II
Rank	1	2	3	4	5	6	7	8	9	10.5	10.5	12.5	12.5	14

The sum of ranks of observations from population II is $R = 56$. *Hence, the test statistic is*

$$\begin{aligned} W &= R - \frac{1}{2}n_2(n_2 + 1) \\ &= 56 - \frac{1}{2}(8)(9) = 20. \end{aligned}$$

For $\alpha = 0.05$, *the rejection region is* $W \leq 9$ *or* $W > 38$, *with the actual* α *being* 0.0592. *Because the observed value of the test statistic does not fall in the rejection region,* H_0 *is not rejected. Thus, we do not have enough evidence to conclude that the median prices are different for these two brands of automobile tires.*

When the sample sizes are large and when H_0 is true, the distribution of the Wilcoxon rank sum test can be approximated by the normal distribution. It can be shown that under H_0, when both n_1 and n_2 are greater than 10, the distribution of W is approximately normal with

$$E(W) = \frac{n_1 n_2}{2} \text{ and } Var(W) = \frac{n_1 n_2 (n_1 + n_2 + 1)}{12}.$$

For a large random sample, we can summarize the test procedure as follows.

Summary of large sample median sum test ($n_1 > 10$ and $n_2 > 10$)

We test

$$H_0: m_1 = m_2 \text{ versus } H_a: \begin{cases} m_1 > m_2, \text{ upper tailed test} \\ m_1 < m_2, \text{ lower tailed test} \\ m_1 \neq m_2, \text{ two–tailed test.} \end{cases}$$

The test statistic:

$$z = \frac{W - n_1n_2/2}{\sqrt{n_1n_2(n_1+n_2+1)/12}}.$$

Rejection region:

$$\begin{cases} z > z_\alpha, \text{ upper tail RR} \\ z < -z_\alpha, \text{ lower tail RR} \\ |z| > z_{\alpha/2}, \text{ two tail RR.} \end{cases}$$

Assumption: The samples are independent and $n_1 > 10$ and $n_2 > 10$.

Decision: Reject H_0 if the test statistic falls in the *RR*, and conclude that H_a is true with $(1 - \alpha)100\%$ confidence. Otherwise, do not reject H_0, because there is not enough evidence to conclude that H_a is true for a given α, and more data is needed.

We will use the foregoing procedure to solve the following problem.

Example 12.4.3

In an effort to determine the immunoglobulin D (IgD) levels of a certain ethnic group, a large number of blood samples representing both sexes for 12-year-olds were taken. The following sample data give the IgD levels (in mg/100 mL).

Male:	9.3	0.0	12.2	8.1	5.7	6.8	3.6	9.4	8.5	7.3	9.7	
Female:	7.1	0.0	5.9	7.6	2.8	5.8	7.2	7.4	3.5	3.3	7.5	7.0

Use the large sample Wilcoxon rank sum test with the significance level $\alpha = 0.01$ to test the hypothesis that there is no difference between the sexes in the median level of IgD.

Solution

We need to test

$$H_0: m_1 = m_2 \text{ versus } H_1: m_1 \neq m_2.$$

Here, n_1=11, and n_2=12, and the results of step 1 and step 2 are given in Table 12.10, *where we use M or F to identify the population from which the data are coming.*

The sum of the ranks for females is R = 114.5, and

$$W = R - \frac{1}{2}n_2(n_2+1)$$
$$= 114.5 - \frac{1}{2}(12)(13) = 36.5.$$

TABLE 12.10 Blood Samples of IgD.

Value	0	0	2.8	3.3	3.5	3.6	5.7	5.8	5.9	6.8	7	7.1
M or F	M	F	F	F	F	M	M	F	F	M	F	F
Rank	1.5	1.5	3	4	5	6	7	8	9	10	11	12
Value	7.2	7.3	7.4	7.5	7.6	8.1	8.5	9.3	9.4	9.7	12.2	
M or F	F	M	F	F	F	M	M	M	M	M	M	
Rank	13	14	15	16	17	18	19	20	21	22	23	

Therefore, the test statistic results in

$$Z = \frac{W - n_1n_2/2}{\sqrt{n_1n_2(n_1+n_2+1)/12}}$$
$$= \frac{36.5-(11)(12)/2}{\sqrt{(11)(12)(24)/12}} = -1.815 \approx -1.82.$$

For $\alpha = 0.01$, we have $z_{\alpha/2} = z_{0.005} = 2.575$. Hence, the rejection region is $z < -2.575$ or $z > 2.575$. Because the test statistic does not fall in the rejection region, we do not reject H_0 at $\alpha = 0.01$ and conclude that there is not enough evidence to conclude that there is any difference between the sexes in the median level of IgD.

With a slight modification of the ranking system in the Wilcoxon rank sum test, we could test for the equality of variances when the normality assumption of the F-test fails.

EXERCISES 12.4

12.4.1 The following data give the winning proportions of the top six football teams from each of the two conferences of the NFL.

American Conference	0.818	0.727	0.909	0.818	0.727	0.545
National Conference	0.636	0.545	0.636	0.636	0.818	0.455

Use the Wilcoxon rank sum test at the significance level of 0.05 to test the null hypothesis that the two samples contain populations with identical medians against the alternative hypothesis that the medians are not equal. State any assumptions you have made to solve the problem.

12.4.2 Comparison of two protective methods against corrosion yielded the following maximum depths of pits (in thousandths of an inch) in pieces of similar metals subjected to the respective treatments:

Method I:	68	75	69	75	70	69	72
Method II:	61	65	57	63	58		

Use the Wilcoxon rank sum test at the significance level of 0.01 to test the null hypothesis that the two samples have identical medians against the alternative hypothesis that the medians are not equal.

12.4.3 Show that when H_0 is true, the mean and variance of the Wilcoxon rank sum test with sample sizes n_1 and n_2 are

$$E(W) = \frac{n_1n_2}{2} \text{ and } Var(W) = \frac{n_1n_2(n_1+n_2+1)}{12}.$$

12.4.4 In order to make inferences about the temporal muscles of the cat, a certain dose of tubocurarine is injected into a random sample of nine cats. The following data give the tetanus frequency (in hertz) in the temporal (T) muscles before and after injection of tubocurarine.

T before	24	33	27	23	31	28	31	24	19
T after	27	38	34	32	37	28	35	28	41

Use the Wilcoxon rank sum test at the significance level of 0.05 to test the null hypothesis that the median tetanus frequency (in hertz) in the temporal (T) muscles is larger after injection of tubocurarine. State any assumptions you made to solve the problem.

12.4.5 In a study of the net conversion of progesterone in rat liver, the following samples were attained for the net conversion in rats 3 to 4 weeks old:

Male:	16.9	16.0	13.5	13.1	14.2	11.6	12.8	17.3	13.8	9.8	16.0	15.9	16.7	15.1
Female:	13.8	11.2	7.5	10.4	15.8	14.5	9.5	9.8	5.1	5.5	6.5	7.2		

Use the large sample Wilcoxon rank sum test at the significance level of 0.05 to test the hypothesis that the median net conversion of progesterone in male rats is larger than that in female rats. What would be your conclusion if you were to use the median test?

12.4.6 Two groups of randomly selected 1-acre plots were treated with two different brands of fertilizer. The following data give the yields of corn (in bushels) from each of these plots.

Fertilizer I:	89	93	105	94	92	96	93	101
Fertilizer II:	85	88	94	87	86	91		

Use the data to determine whether there is a difference in yields for two brands of fertilizers. Use $\alpha = 0.01$. State any assumptions you made to solve the problem.

12.4.7 The following information is obtained from two independent samples.

Sample 1:	15	8	12	4	10	8	13	7	12	6	14	11
Sample 2:	18	13	15	19	17	13	17	16				

Test at 1% significance level that the median for sample 1 is less than the median for sample 2, and interpret the meaning of your result.

12.4.8 In order to determine if a new hybrid seedling produces a bushier flowering plant, data is collected on shrub girth (in inches) for both current variety and hybrid plants resulting in the following values.

Current variety	27.7	25.1	35.4	36.5	22.0	30.5	
Hybrid	35.8	30.0	34.6	37.5	31.9	32.6	39.7

Test at 1% significance level that the median for sample 1 is different from the median for sample 2, and interpret the meaning of your result.

12.4.9 You have two independently trained classifiers, and you evaluate them on 5 independent holdout datasets each (or on 5 independent tasks). You obtain the following accuracy results.

Model A accuracies	0.75	0.78	0.79	0.80	0.82
Model A accuracies	0.77	0.80	0.81	0.84	0.85

(a) Test whether Model B tends to produce larger accuracies than Model A, without assuming that accuracy values are normally distributed, using the Wilcoxon rank sum test. Use $\alpha = 0.05$.

(b) Test whether Model B tends to produce larger accuracies than Model A, using large sample approximations.

12.4.10 AI systems are increasingly used to classify emotions (e.g., happiness, sadness, anger) using facial recognition. However, their accuracy compared to humans can vary. Ten images were shown to humans and AI models for emotion classification. The score (out of 100) represents the percent accuracy in recognizing the emotion correctly.

Image	1	2	3	4	5	6	7	8	9	10
Human Accuracy	92	85	90	80	88	91	84	87	89	86
AI Accuracy	88	75	83	77	84	86	70	81	82	79

Determine whether there is a significant difference between human evaluators and the AI emotion detection system in recognizing emotions correctly from facial expressions. Use $\alpha = 0.05$.

12.5 NONPARAMETRIC HYPOTHESIS TESTS FOR $k \geq 2$ SAMPLES

In this section, we learn how to compare the medians of more than two independent samples and determine whether the medians of the groups differ. These tests are nonparametric alternatives to the ANOVA methods discussed in Chapter 9. We study the *Kruskal–Wallis test* and the *Friedman test.* Both of these methods test the equality of the treatment medians.

12.5.1 The Kruskal–Wallis Test

The Kruskal–Wallis test is a generalization of the Wilcoxon rank sum test for two independent samples to several independent samples. This test is a nonparametric alternative to one-way ANOVA. The Kruskal–Wallis test is almost as powerful as the one-way ANOVA when the data are from a normal distribution, and more powerful in case of nonnormality or in the presence of outliers. We now describe this test.

Suppose that we have k populations, with θ_i being the median of the population i and k independent random samples from these populations. Let the samples from the ith population be n_i. We wish to test the equality of the medians of different groups—that is, to test the hypothesis

$$H_0\text{: } \theta_1 = \theta_2 = \cdots = \theta_k = 0 \text{ versus } H_a\text{: Not all } \theta\text{'s equal } 0.$$

We shall show that the hypothesis $\theta_1 = \cdots = \theta_k$ is equivalent to the hypothesis H_0: $\theta_1 = \theta_2 = \cdots = \theta_k = 0$. Let $\theta_1 = \cdots = \theta_k = t$ (same number). Then the observations $y_{ij} - t$ $(i = 1, 2, \ldots, k)$ will be from a population with a median of zero. Because the Kruskal–Wallis test procedure depends only on the ranks of y_{ij} values in the combined sample and the ranks of $(y_{ij}-t)$ values are identical to those of y_{ij} values, the two hypotheses are equivalent.

We summarize that the Kruskal–Wallis procedure to solve this type of problem is given in the following steps.

Kruskal–Wallis test procedure

1. Combine and rank all $N = \sum_{i=1}^{n} n_i$ observations y_{ij} in ascending order. Also, keep track of the groups from which the observations came. Assign average ranks in case of ties. Let

$$r_{ij} = \text{rank}(y_{ij}).$$

2. Calculate the group sum,

$$r_i = \sum_{i=1}^{n_i} r_{ij}, \; i = 1,2,\ldots,k.$$

and the group averages

$$\bar{r}_i = \frac{r_i}{n_i}, \; i = 1,2,\ldots,k.$$

3. Let

$$r = \sum_{i=1}^{k} r_i = \frac{N(N+1)}{2}$$

(this can be used as a check on the accuracy of your calculation of$r'_i s$) and let

$$\bar{r} = \frac{r}{N} = \frac{N+1}{2}.$$

4. Calculate the Kruskal–Wallis test statistic

(cont'd)

$$H = \frac{12}{N(N+1)} \sum_{i=1}^{k} n_i(\bar{r}_i - \bar{r})^2$$

or the convenient computational form of H,

$$H = \frac{12}{N(N+1)} \sum_{i=1}^{k} \frac{r_i^2}{n_i} - 3(N+1).$$

Note that to compute the convenient form of H, there is no need to calculate $\bar{r}_i$ and $\bar{r}$.

5. Reject H_0 if

$$H \geq c,$$

where the constant c is chosen to achieve a specified value for α.

The exact distribution of H is complicated. It depends on the sample sizes, $n_1, n_2, \ldots, n_k$, and so it is not practical to tabulate its values beyond a small number of cases. When k or N is large, the exact distribution of H under the null hypothesis can be approximated by the chi-square distribution with $(k - 1)$ degrees of freedom. To this effect, we state the Kruskal–Wallis theorem without proof.

Theorem 12.5.1. *When H_0: $\theta_1 = \theta_2 = \ldots = \theta_k$ is true, then as N becomes large, the statistic*

$$H = \frac{12}{N(N+1)} \sum_{i=1}^{k} n_i(\bar{r}_i - \bar{r})^2$$

has an asymptotic distribution that is chi-square with (k – 1) degrees of freedom.

Thus, for approximate large samples, the Kruskal–Wallis test for a given α is to reject H_0 if

$$H > \chi^2_\alpha(k-1).$$

The chi-square approximation is acceptable when the group sample sizes $n_i > 5$ with $k \geq 3$. However, for convenience, we will use the chi-square approximation for all values of n_i. For this test, we follow the procedure described earlier, except that for finding the rejection region, we use the chi-square table.

The following example illustrates how we use the foregoing procedure to test the appropriate hypothesis for three populations.

EXAMPLE 12.5.1

In an effort to investigate the premium charged by insurance companies for auto insurance, an agency randomly selects a few drivers who are insured from three different companies. Assume that these persons have similar autos, driving records, and level of coverage. Table 12.11 gives the premiums paid for every 6 months by these drivers with these three companies. Using the 5% level of significance, test the null hypothesis that the median auto insurance premium paid per 6 months by all drivers insured with each of these companies is the same.

Solution

Here, we need to test where M_i is the true median of the auto insurance premium paid to company i, i = 1, 2, 3.

$$H_0: M_1 = M_2 = M_3 = 0 \textit{ versus } H_a: \text{Not all } M'_i\text{s equal 0},$$

Here $n_1 = 4$, $n_2 = 3$, and $n_3 = 5$. Hence, there are $N = \sum_{i=1}^{3} n_i = 12$ observations. Let Y denote the observations in ascending order. Table 12.12 *gives the combined data in ascending order while keeping track of the groups and their ranks.*

Thus, the group rank sums are

$$r_1 = 24,\ r_2 = 23, \textit{ and } r_3 = 31.$$

TABLE 12.11 Auto Insurance Premium

Company I	Company II	Company III
396	348	378
438	360	330
336	522	294
318		474
		432

TABLE 12.12 Combined Insurance Data in Ascending Order.

Premium	294	318	330	336	348	360	378	396	432	438	474	522
Group	3	1	3	1	2	2	3	1	3	1	3	2
Rank	1	2	3	4	5	6	7	8	9	10	11	12

As a check for accuracy of these calculations, note that

$$r_1 + r_2 + r_3 = 78 = \frac{N(N+1)}{2} = \frac{(12)(13)}{2}.$$

The test statistic is given by

$$\begin{aligned} H &= \frac{12}{N(N+1)} \sum_{i=1}^{k} \frac{r_i^2}{n_i} - 3(N+1) \\ &= \frac{12}{(12)(13)} \left(\frac{(24)^2}{4} + \frac{(23)^2}{3} + \frac{(31)^2}{5} \right) - 3(13) \\ &= 0.42564. \end{aligned}$$

From the chi-square table, $\chi^2_{0.05}(2) = 5.991$, and hence the rejection region is $H \geq 5.991$. Because the observed value of H does not fall in the rejection region, we do not reject H_0 and conclude that there is no evidence to show that the median auto insurance premiums paid per 6 months by all drivers insured in each of these companies are different.

12.5.2 The Friedman Test

The Friedman test, named after the Nobel laureate economist Milton Friedman, tests whether several treatment effects (measured as locations) are equal for data in a two-way layout. We will assume that there are k different treatment levels and l blocks. In each block, assign one experimental unit to each treatment level. We want to test whether the true medians for different treatment levels are the same in each block—that is, to test

$$H_0\text{: True medians at different levels are all equal}$$

versus

$$H_a\text{: Not all the medians are equal.}$$

Rather than combine the entire sample as in the Kruskal–Wallis statistic, here we order the y-values within each block and then assign each its rank. In order to eliminate the differences due to blocks, we take the sum of ranks for each treatment level. The following gives a summary of the procedure.

The Friedman test procedure

1. Rank observations from k treatments separately within each block. Assign average ranks in case of ties. Let R_{ij} = rank(Y_{ij}), the rank of the observation for treatment level i in block j.
2. Calculate the rank sums

$$R_i = \sum_{j=1}^{l} R_{ij},\ i = 1, 2,\ldots,k.$$

3. Calculate the Friedman statistic

$$S = \frac{12}{lk(k+1)} \sum_{i=1}^{k} \left(R_i - \frac{l(k+1)}{2} \right)^2$$

or a convenient computational form,

$$S = \frac{12}{lk(k+1)} \sum_{i=1}^{k} R_i^2 - 3l(k+1).$$

4. Reject H_0 if $S \ge c$, where the constant c is chosen to achieve a specified value for α.

The exact distribution of S is complicated. Here, for $k = 3, 4, 5$, and for various values of l, the Friedman distribution has been calculated, and its values are given in the table in the Appendix A7. We will illustrate this four-step procedure with an example.

EXAMPLE 12.5.2

Three classes in elementary statistics are taught by three different persons: a regular faculty member, a graduate teaching assistant, and an adjunct from outside the university. At the end of the semester, each student is given a standardized test. Five students are randomly picked from each of these classes, and their scores are given in Table 12.13. Test whether there is a difference between the scores for the three persons teaching with $\alpha = 0.05$.

TABLE 12.13 Grades for Different Teacher Types.

Faculty	Teaching Assistant	Adjunct
93	88	86
61	90	56
87	76	73
75	82	90
92	58	47

Solution

Here we need to test

H_0: The medians for the three persons' scores are all equal

H_a: The medians are not equal

We are given $\alpha = 0.05$, $k = 3$, and $l = 5$. To compute the value of the statistic S, we first assign ranks for each student as shown in Table 12.14. H_a: *Note that they are not all equal.*

Thus we have

$$R_1 = 12,\ R_2 = 11,\ \text{and}\ R_3 = 7,$$

TABLE 12.14 Standardized Results of Students.

Faculty	Teaching Assistant	Adjunct
3	2	1
2	3	1
3	2	1
1	2	3
3	2	1

and the test statistic is given by

$$S = \frac{12}{lk(k+1)} \sum_{i=1}^{k} R_i^2 - 3l(k+1)$$
$$= \frac{12}{(5)(3)(4)}((12)^2 + (11)^2 + (7)^2) - (3)(5)(4) = 2.8.$$

From the Friedman table, the rejection region is $S \geq 5.20$ *at an exact significance level of* 0.092. *Because the computed value of the test statistic does not fall in the rejection region, we do not reject* H_0 *and conclude that there is no difference in scores based on who teaches the course.*

When the number of blocks, l, becomes large, the Friedman test statistic has an approximate chi-square distribution under the null hypothesis. That is:

Theorem 12.5.2. *When* H_0: $\theta_1 = \theta_2 = \cdots = \theta_3$ *is true then, as* l *becomes large,*

$$S = \frac{12}{lk(k+1)} \sum_{i=1}^{k} \left(R_i - \frac{l(k+1)}{2}\right)^2$$

has an asymptotic distribution that is chi-squared with $(k-1)$ *degrees of freedom.*

Thus, for an approximate large random sample, the Friedman test for given α is to reject H_0 if $S > \chi^2_\alpha(k-1)$.
When the values of k and l exceed the values given in the Friedman table, we could use the chi-square approximation, which gives acceptable results. We proceed to illustrate the Friedman test with the following example.

EXAMPLE 12.5.3

From the previous example, we now randomly select 10 student grades from each class, resulting in the data shown in Table 12.15.

Test whether there is a difference between the scores for the three persons' teaching. Use $\alpha = 0.05$.

Solution

Here we need to test

H_0: The true median scores for the three instructors are all equal

versus

H_a: They are not all equal.

We are given $\alpha = 0.05$, $k = 3$, *and* $l = 10$. *We use the chi-square approximation to solve the problem. To compute the value of the statistic S, we first assign ranks for each student as shown in* Table 12.16. *The Friedman test statistic is*

TABLE 12.15 Students' Grades for Each Class.

Faculty	Teaching Assistant	Adjunct
93	88	86
61	90	56
87	76	73
75	82	90
92	58	47
45	74	88
99	23	77
86	61	18
82	60	66
74	77	55

TABLE 12.16 Assigned Ranks for Each Student.

	Faculty	Teaching Assistant	Adjunct
	3	2	1
	2	3	1
	3	2	1
	1	2	3
	3	2	1
	1	2	3
	3	1	2
	3	2	1
	3	1	2
	2	3	1
Total	24	20	16

$$S = \frac{12}{lk(k+1)} \sum_{i=1}^{k} R_i^2 - 3l(k+1)$$
$$= \frac{12}{(10)(3)(4)}((24)^2 + (20)^2 + (16)^2) - (3)(10)(4) = 3.2.$$

From the chi-square table, $\chi^2_{0.05}(2) = 5.992$. Hence, the rejection region is $S \geq 5.992$. The computed value of the test statistic does not fall in the rejection region, and we do not reject H_0. We conclude that there is no difference in scores based on who teaches the course.

Friedman's test is an alternative to the repeated measures ANOVA, when assumptions such as that of normality or equality of variance are not satisfied. Because this test, like many other nonparametric tests, does not make a distribution assumption, it is not as powerful as the ANOVA.

EXERCISES 12.5

12.5.1 Table 12.17 shows a random sample of observations on children under 10 years of age, each observation being the IgA immunoglobulin level measured in international units from a large number of blood samples, and the population is studied in blocks in terms of age groups (the upper value is not included) as I: (1 to 3), II: (3 to 6), III: (6 to 8), and IV: (8 to 10). Test for the hypothesis of equality of true medians for IgA level in each block (age level), (**a**) with the 5% level and (**b**) with the 1% level of significance. Compare the results obtained.

TABLE 12.17 Blood Measurements of IgA.

I	6	37	19	14	51	68	27	75
II	32	65	76	42	45	41	38	63
III	73	75	59	90	37	32	63	80
IV	81	42	48	60	98	100	79	45

12.5.2 In an effort to study the effect of four different preventive maintenance programs on downtimes (in minutes) for a certain period of time in a production line, a factory runs four parallel production lines, and each line has five different types of machines. The different maintenance programs are randomly assigned to each of the four production lines so as to treat the various machines as blocks. Results are shown in Table 12.18.

TABLE 12.18 Downtimes (in Minutes) of Production Lines.

Machine	Method 1	Method 2	Method 3	Method 4
I	181	124	126	181
II	185	122	125	160
III	67	65	68	69
IV	121	66	120	68
V	62	60	62	65

Test the hypothesis at $\alpha = 0.05$, H_0: True medians of the four maintenance programs are equal versus H_a: Not all are equal. [Hint: In the Friedman test, $k = 4$, and $l = 6$.] State any assumptions you have made to solve this problem.

12.5.3 Show that, when $k = 2$, the Kruskal–Wallis statistic,

$$H = \frac{12}{N(N+1)} \sum_{i=1}^{k} \frac{r_i^2}{n_i} - 3(N+1)$$

becomes equivalent to the Wilcoxon rank sum test.

12.5.4 A consumer testing agency is interested in determining whether there is a difference in the mileage for three brands of gasoline. To test this, four different vehicles are driven with each of these gasolines. Results are shown in Table 12.19.

TABLE 12.19 Mileage for Three Brands of Gasoline Used in Four Different Vehicles.

	Gasoline		
Vehicle	A	B	C
I	19	25	22
II	26	33	39
III	20	28	25
IV	18	30	21

Test whether there is a difference between the three gasoline medians at the 0.05 level.

12.5.5 In order to study the effect of fertilizers, five groups of 1-acre plots were randomly selected. One group was not treated with any fertilizers, and the remaining four groups were treated with four different brands of fertilizers. Table 12.20 gives the yields of corn (in bushels) from each of these plots.

TABLE 12.20 Data of the Effect of Four Different Fertilizers.

None:	58	27	36	41	48	36	50	50	39
Fertilizer I:	69	67	57	63	49	65	78	69	
Fertilizer II:	95	92	92	89	100	88	79	97	75
Fertilizer III:	102	111	92	103	102	94	100	112	96
Fertilizer IV:	127	115	112	122	114	107	116	112	108

Use the data to determine whether there is a difference in yields for different fertilizers. Use $\alpha = 0.01$.

12.5.6 In order to compare grocery prices of four different grocery stores on a particular day in November 1999, 11 randomly selected items with the same brands are given in Table 12.21.

Use the data to determine whether there is a difference in prices at these four grocery store chains. Use $\alpha = 0.01$. State any assumptions you have made to solve this problem.

12.5.7 Use the Kruskal–Wallis method to test whether the median daily returns differ across three cryptocurrencies Bitcoin (BTC), Ethereum (ETH), and XRP. The Table 12.22 gives daily return in % for six days.

TABLE 12.21 Grocery Prices of Four Different Stores.

Product	Store A	Store B	Store C	Store D
Bread (20 oz)	\$1.39	\$1.39	\$1.39	\$1.39
Red apple (1 lb)	1.29	1.29	0.99	0.68
Large eggs (1 dozen)	0.69	0.88	0.89	0.89
Orange juice (64 oz)	3.29	2.99	2.79	2.69
Cereal (15 oz)	3.59	3.19	3.19	3.58
Canned corn (15.25 oz)	0.50	0.53	0.50	0.49
Crystals sugar (5 lb)	1.99	2.09	1.99	1.89
2% milk (1 gal)	3.19	3.19	3.09	3.09
Frozen pizza (21.5 oz)	3.00	4.59	3.50	3.50
Puppy Chow (4.4 lb)	4.59	3.69	3.69	3.99
Diapers (56-pack)	12.99	12.99	12.99	11.88

TABLE 12.22 Crypto Return (%).

BTC	1.2	0.8	1.5	1.0	0.5	1.3
ETH	0.4	0.6	0.2	0.5	0.3	0.4
XRP	-0.2	-0.1	0.0	0.0	0.1	0.2

12.6 CHAPTER SUMMARY

In this chapter, we first learned about nonparametric approaches to interval estimation and nonparametric hypothesis tests for one sample, such as the sign test, the Wilcoxon signed rank test, and dependent sample paired comparison tests. Then nonparametric hypothesis tests for two independent samples, such as the median test and Wilcoxon rank sum test, were considered. Later, the Kruskal–Wallis test and the Friedman test were explained for more than two samples.

It is natural to ask, "Why do we substitute a set of nonnormal numbers, such as ranks, for the original data?" Few data are truly normal. Rank tests are sometimes called "approximate" tests. They are most useful in instances when we suspect that the data are not normal, and we either cannot transform the data to make them more normal, or do not like to do so. One of the simple ways to check for appropriateness of use of nonparametric tests is to simply construct a stem-and-leaf display or a histogram for the sample data and see whether they look symmetric and approximately bell-shaped. If this is not so, we may often be better off using a nonparametric approach.

Since the 1940s, many nonparametric procedures have been introduced, and the number of procedures continues to grow. The nonparametric tests presented in this chapter represent only a small portion of available nonparametric tests. There are many references available in the bibliography for further reading on the subject.

In summary, nonparametric statistics complement parametric approaches, without assuming specific parametric forms for underlying distributions, by providing tools that are flexible, robust, and widely applicable across diverse fields, making them essential in the statistician's and data scientist's toolkit. In this chapter, we have also learned the following important concepts and procedures.

- Procedure for finding $(1 - \alpha)$ 100% confidence interval for the median M

- Hypothesis-testing procedure by sign test
- A large sample sign test
- Hypothesis-testing procedure by Wilcoxon signed rank test
- Summary of large sample Wilcoxon signed rank test ($n > 20$)
- Summary of large sample median sum test ($n_1 > 5$ and $n_2 > 5$)
- Hypothesis-testing procedure by Wilcoxon rank sum test
- Summary of large sample Wilcoxon rank sum test ($n_1 > 10$ and $n_2 > 10$)
- Kruskal–Wallis test procedure
- Friedman test procedure

12.7 COMPUTER EXAMPLES

In this section, we illustrate some nonparametric procedures using statistical software packages.

12.7.1 Examples Using R

EXAMPLE 12.7.1 (Sign test)

Using the following data test H_0: $M = 1.4$ vs. H_a: $M > 1.4$, using the sign test.

Sample (x): 1.51 1.35 1.69 1.48 1.29 1.27 1.54 1.39 1.45

R Code:

```
y=length(which(x>1.4));
n=length(x);
binom.test(y,n,alternative="greater");
```

Our p value suggests that we fail to reject the null hypothesis for any reasonable level of significance and that the medians are equal.

Output:

```
            Exact binomial test
data:   y and n
number of successes = 5, number of trials = 9, p-value = 0.5
alternative hypothesis: true probability of success is greater than 0.5
95 percent confidence interval:
   0.2513676 1.0000000
sample estimates:
probability of success
            0.5555556
```

Example 12.7.2 (Wilcoxon test)

Using the data from the previous example test H_0: $M = 1.4$ vs. H_a: $M \neq 1.4$, using one-sample Wilcoxon test.

R Code:

```
wilcox.test(x,mu=1.4);
```

Output:

```
          Wilcoxon signed rank test
```

```
data:   x
```

We fail to reject the null hypothesis suggesting that the true mean is equal to 1.4 for any reasonable level of significance.

```
V = 30, p-value = 0.4258
alternative hypothesis: true location is not equal to 1.4
```

EXAMPLE 12.7.3 (Two-sample sign test)

Using the following data, test H_0: $M = 0$ vs. H_a: $M < 0$, using the two-sample sign test, where M is the median difference. Use $\alpha = 0.05$.

Sample (x): 180 199 175 226 189 205 169 211

Sample (y): 172 191 172 230 178 199 171 201

R Code:

```
z=x-y;
y=length(which(z<0));
n=length(z);
binom.test(y,n,alternative="less");
```

Output:

```
              Exact binomial test
data:   y and n
```

Fail to reject the null hypothesis since the p value is larger than our alpha. This suggests the median difference is zero.

```
number of successes = 2, number of trials = 8, p-value = 0.1445
alternative hypothesis: true probability of success is less than 0.5
95 percent confidence interval:
 0.0000000 0.5996894
sample estimates:
probability of success
              0.25
```

EXAMPLE 12.7.4 (Wilcoxon two-sample test)

Use the Wilcoxon rank sum test with $\alpha = 0.05$ to test the null hypothesis that the two population medians are the same against the alternative hypothesis that the population medians are different.

Sample (x): 85 99 100 110 105 87

Sample (y): 67 69 70 93 105 90 110 115

R Code:

wilcox.test(x,y);

Output

```
         Wilcoxon rank sum test with continuity
data:   x and y
```

Fail to reject the null hypothesis

```
W = 28, p-value = 0.6507
alternative hypothesis: true location shift is not equal to 0
```

EXAMPLE 12.7.5 (Kruskal–Wallis test)

In an effort to investigate the premiums charged by insurance companies for auto insurance, an agency randomly selects a few drivers who are insured by three different companies. Assume that these people have similar cars, driving records, and levels of coverage. The following data is the premiums paid per 6 months by these drivers with these three companies. Using α = 0.05, test the null hypothesis that the median auto insurance premium paid per 6 months by all drivers insured in each of these companies is the same.

Company	C1	C1	C1	C1	C2	C2	C2	C3	C3	C3	C3	C3
Value	396	438	336	318	348	360	522	378	330	294	474	432

R Code:

kruskal.test(value,company);

Output:

A large p value suggests we fail to reject the null hypothesis

```
         Kruskal-Wallis rank sum test
data:   value and company
Kruskal-Wallis chi-squared = 0.4256, df = 2, p-value = 0.8083
```

Example 12.7.6 (Friedman test)

Using the following data, conduct a Friedman test.

C1	93	61	87	75	92	45	99	86	82	74
C2	88	90	76	82	58	74	23	61	60	77
C3	86	56	73	90	47	88	77	18	66	55

R Code:

```
blocks=c(c(1:10),c(1:10),c(1:10));
```

A data set called blocks contains matching block data ranged 1 to 10

```
friedman.test(values,groups,blocks);
Output:
          Friedman rank sum test
data:  values, groups, and blocks
Friedman chi-squared=3.2, df=2, p-value=0.2019
```

12.7.2 SPSS Examples

EXAMPLE 12.7.7

(Wilcoxon rank sum test): For the data of Example 12.4.2, use the Wilcoxon rank sum test at the significance level of 0.05 to test the null hypothesis that the two population medians are the same against the alternative hypothesis that the population medians are different. Use an SPSS procedure.

Solution

Because the SPSS pull-down menu does not have the Wilcoxon rank sum test, we will use the Mann–Whitney U-test. The Mann–Whitney U-test is equivalent to the Wilcoxon rank sum test, although we calculate it in a slightly different way. For the same data set, any p-values generated from one test will be identical to those generated from the other. The following gives the steps to follow. Enter tire brands as I to identify brand 1 and 2 to identify brand ***II****, in C1. Enter the corresponding prices in C2. Name C1 as Brand and C2 as Price. Then click*

Analyze > Nonparametric Tests > 2 Independent Samples... > move ***Brand*** to ***Grouping Variable:*** and ***Price*** to ***Test Variable list:*** > click ***Define Groups...*** > enter ***1*** in ***Group 1:***, and **2** in ***Group 2:*** > click **continue** > choose ***Mann–Whitney U > OK***

We obtained the following output:

Mann–Whitney Test

Ranks

	BRAND	**N**	**Mean Rank**	**Sum of Ranks**
PRICE	1.00	6	8.17	49.00
	2.00	8	7.00	56.00
	Total	14		

Test Statistics

	PRICE
Mann–Whitney U	20.000
Wilcoxon W	56.000
Z	–0.518

Asymp. Sig. (2-tailed)	0.605
Exact Sig. [2*(1-tailed Sig.)]	0.662

(a) *Not corrected for ties.*
(b) *Grouping Variable: BRAND*

In the first table just shown, ranks show the mean ranking of tire brand I and tire brand II. The Mann–Whitney test is used to assess whether the distribution of ranks is statistically significant. Under the null hypothesis, the distribution of ranks *should be the same for both groups. Looking at the second table, the calculated value of the Mann–Whitney U is 20. The value U represents the amount by which the ranks for tire brand I and tire brand II deviate from what we would expect under the null hypothesis. For a 0.05 significance level, we can reject the null hypothesis if the 2-tailed significance (see Asymp. sig in the second table) is less than 0.05. In this case, because Asymp. Sig. (2-tailed) = 0.605, we do reject the null hypothesis.*

EXAMPLE 12.7.8

(Kruskal–Wallis test): For the data of Example 12.5.1, conduct the Kruskal–Wallis test using SPSS.

Solution

Enter insurance companies as 1 to identify company I, 2 to identify company II, and 3 to identify company III, in C1. Enter the corresponding premiums in C2. Name C1 as Company, and C2 as Premium. Then:

Analyze* > *Nonparametric Tests* > *K Independent Samples... > move ***Premium*** to ***Test Variable List*:** and ***Company*** to ***Grouping variable*:** > click ***Define Rage...*** > enter ***1*** in ***Minimum***, and **3** in ***Maximum*** > click **continue** > click ***Kruskal–Wallis H*** > ***OK***

If we need to do a Friedman test, say for the data of Example 12.7.5, enter each row of data in **C1, C2**, and **C3**, respectively. Then use the following sequence to obtain the appropriate output.

Analyze* > *Nonparametric Tests* > *K Related Samples... > move each of the three columns to ***Test Variables:*** > check in ***Test Type Friedman* > *OK***

12.7.3 SAS Examples

To perform the nonparametric tests, use the SAS statement PROC NPAR1WAY. In the procedure, if we include the EXACT statement, the program will compute the exact p-value computations for the Wilcoxon rank sum test.

EXAMPLE 12.7.9

(Wilcoxon rank sum test): Comparison of the prices (in dollars) of two brands of similar tires gave the following data.

Tire I:	85	99	100	110	105	87		
Tire II:	67	69	70	93	105	90	110	115

Use the Wilcoxon rank sum test at the significance level of 0.05 to test the null hypothesis that the two population medians are the same against the alternative hypothesis that the population medians are different. Use the SAS procedure.

Solution

We can use the following procedure.

```
options nodate nonumber;
DATA tprice;
INPUT Brand Price @@;
CARDS;
1 85 1 99 1 100 1 110 1 105 1 87
2 67 2 69 2  70 2  93 2 105 2 90 2 110 2 115
;
/* Nonparametric statistics/Wilcoxon Rank-
Sum */
PROC NPAR1WAY DATA=tprice WILCOXON;
CLASS Brand;
VAR Price;
EXACT WILCOXON;
run;
```

EXAMPLE 12.7.10

(Kruskal–Wallis test): For the data of Example 12.7.4, perform the Kruskal–Wallis test using SAS.

Solution

We can use the following code.

```
options nodate nonumber;
DATA insprice;
INPUT Company Price @@;
CARDS;
1 396 1 438 1 336 1 318
2 348 2 360 2 522
3 378 3 330 3 294 3 474 3 432
;
proc npar1way data=insprice;
class company;
var Price;
run;
```

12.8 PROJECTS FOR CHAPTER 12

12A. Comparison of Wilcoxon Tests With Normal Approximation

(i) For the Wilcoxon signed rank test, compare the results from the Wilcoxon signed rank test table with the normal approximation using several sets of data of various sample sizes. Also, if the sample size is very small, compare the results from the Wilcoxon signed rank test with a small sample t-test.

(ii) For the Wilcoxon rank sum test, compare the results from the Wilcoxon rank sum test table with the normal approximation using several sets of data (from pairs of samples) of various sample sizes. Also, if the sample sizes are very small, compare the results from the Wilcoxon rank sum test with the small sample t-test for two samples.

12B. Randomness Test (Wald–Wolfowitz Test)

When we have no control over the way in which the data are selected, it is useful to have a technique for testing whether the sample may be looked on as random. The condition of randomness is essential for all of the analysis explained in this book: that is, whether a sequence of random variables $X_1, \ldots, X_n$ is independent based on a set of observations $x_1, \ldots, x_n$ of these random variables. Here, we will give a method based on the number of runs displayed in the sample events. This is a nonparametric procedure. The run test is used to test the randomness of a sample at $100(1-\alpha)\%$ confidence level.

Given a sequence of two symbols, say H and T, a run is defined as a succession of identical symbols contained between different symbols or none at all. The total number of runs in a sequence of n trials serves as an indication whether the arrangement is random or not. If a sequence contains n_1 symbols of one kind and n_2 symbols of another kind and both n_1 and n_2 are greater than 10 (this is a rule of thumb; for more accuracy we can also take both n_1 and n_2 as greater than 20), then the sampling distribution of the total number of runs, R, has an asymptotic normal distribution with mean

$$\mu_R = \frac{2n_1n_2}{n_1+n_2} + 1$$

and variance

$$\sigma_R^2 = \frac{2n_1n_2(2n_1n_2 - n_1 - n_2)}{(n_1+n_2)^2(n_1+n_2-1)}.$$

For example, if we have the following symbols

$$\underline{HHH}\ \underline{T}\ \underline{HH}\ \underline{TTTT}\ \underline{HH}\ \underline{TTT}$$

there are six runs indicated by the underlines, and $n_1 = 7$ and $n_2 = 8$. If the sample contains numerical data, the run test is used by counting runs above and below the median. Denoting the observations above the median by the letter A and observations below the median by the letter B, we can determine the run as before. For example, if we have data values

2 5 11 13 7 22 6 8 15 9

then the median is 8.5. Hence, we get the following arrangement of values above and below the median:

$$\underline{BB}\ \underline{AA}\ \underline{B}\ \underline{A}\ \underline{BB}\ \underline{AA}.$$

Hence, there are six runs with $n_1 = 5$ and $n_2 = 5$.

Now we can formulate the test of randomness as a hypothesis-testing problem as described in the following procedure.

Procedure for test of randomness using the run test

To test

H_0: Arrangement of sample values is random

versus

H_a: Data is not random.

1. Compute the median of the sample.
2. Going through the sample values, replace any observation with A if the value is above the median, or B if the value is below the median. Discard any ties.
3. Compute n_1, n_2, and R. Also, compute the mean and variance of R.

$$\mu_R = \frac{2n_1n_2}{n_1+n_2} + 1,$$

and

$$\sigma_R^2 = \frac{2n_1n_2(2n_1n_2 - n_1 - n_2)}{(n_1+n_2)^2(n_1+n_2-1)}.$$

4. Compute the test statistic:

$$Z = \frac{R-\mu_R}{\sigma_R}.$$

5. Rejection region: $|Z| > Z_{\alpha/2}$.
6. **Decision:** If the test statistic falls in the rejection region, reject H_0 and conclude that the sample is not random with $(1-\alpha)$ 100% confidence.

Assumption: $n_1 \geq 10$ and $n_2 \geq 10$.

Note 1: Sometimes the same procedure is used with the median replaced by the mean of the sample. That is, if the observation is above the sample, use A, and if it is below the sample, use B. We use this procedure for large samples. For small sample sizes, to determine the upper and lower critical values, a special table is needed. Some statistical software packages have the ability to use the run test for randomness.

Example 12.B.1

The following table gives radon concentration in pCi/L obtained from 40 houses in a certain area.

2.9	0.6	13.5	17.1	2.8	3.8	16.0	2.1	6.4	17.2
7.9	0.5	13.7	11.5	2.9	3.6	6.1	8.8	2.2	9.4
15.9	8.8	9.8	11.5	12.3	3.7	8.9	13.0	7.9	11.7
6.2	6.9	12.8	13.7	2.7	3.5	8.3	15.9	5.1	6.0

Test using software whether the data are random at a 95% confidence level.

Solution

Running the data with software, we get the following output.

```
radon
K=8.3400
The observed number of runs=17
The expected number of runs=20.9500
19 Observations above K 21 below
The test is significant at 0.2046
Cannot reject at alpha=0.05
```

Thus, the data set is a random sample at a 95% confidence level.

Note 2: If the large samples assumption is not satisfied (that is, $n_1 < 10$ and $n_2 < 10$, for more accuracy use 20 instead of 10), then use the total number of runs, R, itself as the test statistic, and we can find lower and upper critical values for a given α from Frieda S. Swed and C. Eisenhart Tables for testing randomness of grouping in a sequence of alternatives, Annals of Mathematical Statistics, 14, 83–86, 1943. We will not be giving this table in this book.

12.9 EXERCISE

Pick a couple of data sets from this book or your own and test for randomness using **(i)** hand calculations, and **(ii)** a statistical software package.

Chapter 13

Empirical Methods

Chapter outline

OBJECTIVE

In this chapter we introduce several empirical methods that are being increasingly used in statistical computations as an alternative or as an improvement to classical statistical methods.

Stanislaw Ulam
(Source: http://scienceworld.wolfram.com/biography/Ulam.html).

Stanislaw Ulam (1909–86) was a Polish-American mathematician who was born in Lwów, Poland, and came to the United States in 1936. He worked at Princeton University. He was involved with the Manhattan Project to build the first atomic bomb. Ulam solved the problem of how to initiate fusion in the hydrogen bomb. Ulam was interested in astronomy, physics, and mathematics from an early age. He obtained his Ph.D. from the Polytechnic Institute in Lwów in 1933, where he studied under a famous mathematician named Banach. Ulam's writings included *A Collection of Mathematical Problems* (1960), *Sets, Numbers and Universes* (1974), and *Adventures of a Mathematician* (1976). His major contribution to statistics is through the introduction of the Monte Carlo methods along with Metropolis in 1949.

Mathematical Statistics with Applications in R. DOI: https://doi.org/10.1016/B978-0-443-27548-7.00013-X

These methods are widely used in solving mathematical problems using statistical sampling. Monte Carlo methods became widely popular with the ever-increasing power of computers and the development of specialized mathematical and statistical software.

13.1 INTRODUCTION

In statistics, major efforts are made to develop and study accurate statistical models that are able to describe natural phenomena. The dilemma is whether to use the standard model that may allow closed-form solutions or to describe the phenomena more accurately that rely on observed data rather than purely theoretical assumptions about population distributions, which would often preclude the computation of explicit answers. Obtaining methods that result in useful qualitative and quantitative understanding of realistic complex systems is difficult, and obtaining exact analytical tools is not practical either. Because of this problem, practitioners have relied on simulation-based methods. Computer simulation methods are becoming tools of choice for problems in statistics. Most of the empirical methods discussed in this chapter had been in existence in statistical literature as possible numerical methods for some time. Because of the difficulty of computing by hand, these methods did not gain much popularity. These numerical techniques became popular and practical with the advent of high-quality pseudorandom number generators and high-speed computers. Modern statistics is increasingly equipped with theoretical concepts complemented with effective computational tools to handle the challenges that arise in science and technology. The methods presented in this chapter could be effectively used for Bayesian computation and for problems arising in such diverse areas as environmental modeling, epidemiology, finance, genetics, image analysis, and statistical physics.

It is important to note that the literature on these simulation methods is growing, and it is impossible to present the whole picture in a single chapter. The purpose of this chapter is only to introduce some basic and popular computational methods. There are many specialized books for further study.

13.2 THE JACKKNIFE METHOD

Resampling methods are data-driven techniques that involve repeatedly drawing subsamples from observed data (with or without replacement) to estimate sampling distributions, compute confidence intervals, or perform hypothesis tests. Instead of relying on theoretical probability distributions (like normal, t, χ^2), resampling lets the data approximate its own distribution. Two common resampling methods are *jackknife* and the *bootstrap*.

It was Tukey who in 1958 gave the name "jackknife" (sometimes also known as the Quenouille–Tukey jackknife) to a general statistical method, invented by Maurice Quenouille in 1956, for testing hypotheses and finding confidence intervals where traditional methods are not applicable or not well suited. In general usage, a jackknife is a large clasp knife that has a multitude of small pullout tools. Because this method could be used for small tasks without resorting to other tools, it was named the jackknife. The jackknife method can also be used with multivariate data. However, here we will only present the method for univariate data. The jackknife procedure is very useful when outliers are present in the data, or the dispersion of the distribution is wide. In the jackknife method, we systematically recompute the statistic, leaving out one observation at a time from the observed sample. This is used to estimate the variability of statistic from the variability of that statistic between subsamples. This avoids the parametric assumptions that we used in obtaining the sampling distribution of the statistic to calculate standard error. Thus, this can be considered as a nonparametric estimate of the parameter. Initially, the jackknife method was introduced for bias reduction (thus improving a given estimator) and is a useful method for variance estimation. In this section, we study only how to compute a jackknife estimate and a confidence interval. We do not discuss how it reduces bias or any other theoretical properties. Jackknife methods predate the bootstrap method discussed in the next section.

Let $X_1, ..., X_n$ be a random sample from a population with finite variance. Then, the sample mean is

$$\overline{X} = \frac{1}{n}\sum_{i=1}^{n} X_i.$$

If one of the observations, say, the kth observation, is taken out (or missing), then

$$\overline{X}_{-k} = \frac{1}{n-1}\left(\sum_{i=1}^{n} X_i - X_k\right) = \frac{1}{n-1}\sum_{k\neq i=1}^{n} X_i.$$

Now, if we know the overall sample mean $\overline{X}$ and we calculated $\overline{X}_{-k}$, then we can obtain the deleted observation X_k by using the formula

$$X_k = n\overline{X} - (n - 1)\overline{X}_{-k}.$$

In general, suppose that the population parameter θ is estimated by a function of the sample values $\widehat{\theta}$ $(X_1, \ldots, X_n)$, represented by $\widehat{\theta}$, and let $\widehat{\theta}_{-k}$ be the corresponding estimate by removing the kth observation. Note that here θ is any parameter; it need not be the population mean. Then the set of "pseudo-values" $\widehat{\theta}_k^*$, $k = 1, 2, \ldots, n$ is obtained by

$$\hat{\theta}_k^* = n\hat{\theta} - (n - 1)\hat{\theta}_{-k}^*.$$

The average of these pseudo-values

$$\hat{\theta}^* = \frac{1}{n}\sum_{k=1}^{n}\hat{\theta}_{-k}^*,$$

is the *jackknife estimate* of the parameter θ.

If $\widehat{\theta}$ is an estimator of θ based on the entire sample $X_1, \ldots, X_n$, then the estimate of the bias of $\widehat{\theta}$ is given by

$$\widehat{bias} = (n - 1)(\widehat{\theta}^* - \widehat{\theta}).$$

The factor $(n - 1)$ is to scale up the spread of subsample estimates around the true estimate. The variance of the jackknife estimates is given by. In here, we first take the variance of jackknife estimates and then we scale it up by the same factor $(n - 1)$. We will not discuss about the need for the factor $(n - 1)$. Interested reader should consult advanced books on resampling methods.

$$\widehat{Var} = (n - 1)\frac{1}{n}\sum_{i=1}^{n}\left(\hat{\theta}_{-k}^* - \hat{\theta}^*\right)^2.$$

Let s^{*2} be the sample variance of these pseudo-values. Then, the variance of $\hat{\theta}^*$ is estimated by s^{*2}/n, and a $(1 - \alpha)$ 100% *jackknife confidence interval* for θ is given by

$$\hat{\theta}^* \pm t_{\alpha/2}\frac{s^*}{\sqrt{n}}$$

where $t_{\alpha/2}$ is evaluated with $(n - 1)$ degrees of freedom.

A procedure for jackknife point and interval estimation

1. Generate a random sample $X_1, \ldots, X_n$ from a population.
2. First remove X_1 from the sample (so the new sample will be $X_2, \ldots, X_n$) and compute the estimator $\hat{\theta}_{-1}$ (such as the sample mean); then remove X_2 (the resulting sample will be $X_1, X_3, \ldots, X_n$) and compute the estimator $\hat{\theta}_{-2}$, and so on until the last sample is $X_1, \ldots, X_{n-1}$, with the estimator being $\hat{\theta}_{-n}$.
3. The jackknife point estimate of θ is

$$\hat{\theta}^* = \frac{1}{n}\sum_{k=1}^{n}\hat{\theta}_{-k}^*.$$

4. Calculate the sample variance of the values $\hat{\theta}_{-i}$, $i = 1, \ldots, n$, and denote the variance by s^{*2}.
5. A $(1 - \alpha)100\%$ jackknife confidence interval for θ is given by

$$\hat{\theta}^* \pm t_{\alpha/2}\frac{s^*}{\sqrt{r}}.$$

EXAMPLE 13.2.1

A random sample of $n = 6$ from a given population resulted in the following data:

$$7.2 \quad 5.7 \quad 4.9 \quad 6.2 \quad 8.5 \quad 2.8$$

(a) Find a jackknife point estimate of the population mean μ.
(b) Construct a 95% jackknife confidence interval for the population mean μ.

Solution
(a) *Here* $n = 6$. Table 13.1 *represents the original sample and the six jackknife samples.*

TABLE 13.1 Original and Six Jackknife Samples.

Original	Sample 1	Sample 2	Sample 3	Sample 4	Sample 5	Sample 6
7.2	5.7	7.2	7.2	7.2	7.2	7.2
5.7	4.9	4.9	5.7	4.9	4.9	4.9
4.9	6.2	6.2	6.2	5.7	6.2	6.2
6.2	8.5	8.5	8.5	8.5	5.7	8.5
8.5	2.8	2.8	2.8	2.8	2.8	5.7
2.8						

Using software descriptive statistics, we obtained the summary of the analysis given in Table 13.2. *Now taking the mean and standard deviation of the means of the six jackknife samples, we get* $\hat{\mu}^* = 5.883$, *and the standard deviation* $s^* = 0.392$. *Thus, the jackknife point estimate of* μ *is* $\hat{\mu}^* = 5.883$, *that is the same as the mean of the original sample. However, we can see that the standard deviation resulting from the jackknife is reduced to only 0.392, compared to 1.959 for the original sample.*

TABLE 13.2 Descriptive Statistics of Table 13.1.

Variable	N	Mean	Median	TrMean	StDev	SE Mean
Original	6	5.883	5.950	5.883	1.959	0.800
Sample 1	5	5.620	5.700	5.620	2.068	0.925
Sample 2	5	5.920	6.200	5.920	2.188	0.978
Sample 3	5	6.080	6.200	6.080	2.123	0.949
Sample 4	5	5.820	5.700	5.820	2.183	0.976
Sample 5	5	5.360	5.700	5.360	1.656	0.741
Sample 6	5	6.500	6.200	6.500	1.395	0.624

(b) *A 95% jackknife confidence interval for* μ *is*

$$\hat{\mu}^* \pm t_{\alpha/2}\frac{s^*}{\sqrt{n}} = 5.883 \pm 2.571\frac{0.392}{\sqrt{6}}$$

resulting in (5.471, 6.2944). *Compare this with Example 5.5.7, where we got the confidence interval as* (3.827, 7.939). *Thus, through the jackknife method, we get a much tighter confidence interval for* μ.

The jackknife method of resampling is also known as the "leave-one-out" method because it uses all observations but one in each subsample. Here, every observation is left out exactly once. Note that in the jackknife method, sampling is done without replacement. This procedure can also be used for other statistical procedures, such as hypothesis testing and regression. We use the jackknife resampling method to estimate the prior pdf of true parameters from the pdf of the given data, $f(x|\theta)$, in order to obtain empirical Bayes estimates of θ, in Chapter 10. In addition, in Bayesian, we often summarize posterior distribution by point summaries such as posterior mean or posterior median. We can use the jackknife to (i) estimate the bias of a posterior summary, (ii) estimate the standard error of that posterior summary, and (iii) assess influence of single observation by inspecting leave-one-out posterior summaries. We will now give an example to demonstrate these concepts.

EXAMPLE 13.2.2

Suppose we analyze the monthly returns (%) of a small portfolio for 6 months:

2.5,	3.0,	2.8,	3.5,	2.9,	3.2.

(a) Find the jackknife estimate mean return.
(b) Find its standard error.

Solution

(a) *Here $n = 6$.* Table 13.3 *represents the original sample and the six jackknife samples.*

TABLE 13.3 Original and Six Jackknife Samples.

Original	Sample 1	Sample 2	Sample 3	Sample 4	Sample 5	Sample 6
2.5	3.0	2.5	2.5	2.5	2.5	2.5
3.0	2.8	2.8	3.0	3.0	3.0	3.0
2.8	3.5	3.5	3.5	2.8	2.8	2.8
3.5	2.9	2.9	2.9	2.9	3.5	3.5
2.9	3.2	3.2	3.2	3.2	3.2	2.9
3.2						
$\bar{x} = 2.98$	$\bar{x}_{-1} = 3.08$	$\bar{x}_{-2} = 2.98$	$\bar{x}_{-3} = 3.02$	$\bar{x}_{-4} = 2.88$	$\bar{x}_{-5} = 3.00$	$\bar{x}_{-6} = 2.94$

Now taking the mean of the means of the six jackknife samples, we get $\hat{\mu}^* = 2.98$.

(b) *We can also obtain the jackknife variance estimate as 0.0197. Thus, the jackknife estimate of the standard error is approximately 0.14.*

EXAMPLE 13.2.3

Assume that the data in Example 13.2.2, $X_i \sim N(\mu, \sigma^2 = 0.25)$ and the prior of $\mu \sim N(\mu_p, \sigma_p^2)$, where $\mu_p = 0$ and $\sigma_p^2 = 10$. Obtain the jackknife posterior mean.

Solution

Using Example 10.2.4, we can derive the posterior mean as

$$\hat{\mu} = \frac{n\sigma^{-2}\,\bar{x} + \sigma_p^{-2}\mu_p}{n\sigma^{-2} + \sigma_p^{-2}}.$$

From Example 13.2.2, $\bar{x} = 2.98$.

Also, $\sigma^{-2} = \frac{1}{0.25} = 4$, $n\sigma^{-2} = 6 \times 4 = 24$, and $\sigma_p^{-2} = \frac{1}{10} = 0.1$.

Thus, the posterior mean for the original sample is $\hat{\mu} = 2.97$.

Now, using jackknife (leave-one-out) sample means, we will get the posterior mean formula for each subsample with $n' = 5$ as

$$\hat{\mu}_i = \frac{n'\sigma^{-2}\,\bar{x}_{-i} + \sigma_p^{-2}\mu_p}{n'\sigma^{-2} + \sigma_p^{-2}} = \frac{20\bar{x}_{-i}}{20.1}.$$

Using this, we get,

$$\hat{\mu}_{-1} = \frac{20 \times 3.08}{20.1} \approx 3.06,\ \hat{\mu}_{-2} \approx 2.97,\ \hat{\mu}_{-3} \approx 3.00,\ \hat{\mu}_{-4} \approx 2.87,\ \hat{\mu}_{-5} \approx 2.99,\ \hat{\mu}_{-6} \approx 2.93.$$

Thus, the jackknife estimate of posterior mean is,

$$\hat{\mu}^* = \frac{1}{6}\sum_{i=1}^{6} \hat{\mu}_{(-i)} \approx 2.97.$$

Similar to Example 13.2.2, we can also obtain a standard error estimate.

EXERCISES 13.2

13.2.1 The following data represent the total ozone levels measured in Dobson units at randomly selected locations on Earth on a particular day.

269	246	388	354	266	303
295	259	274	249	271	254

(a) Find a jackknife point estimate of the population mean μ ozone level.
(b) Construct a 95% jackknife confidence interval for the population mean μ.
(c) Compare the confidence interval obtained in part (b) with that in Example 6.3.3.

13.2.2 A drug is suspected of causing an elevated heart rate in a certain group of high-risk patients. Twenty patients from those groups were given the drug. The changes in heart rates were found to be as follows.

−1	8	5	10	2	12	7	9	1	3
4	6	4	12	11	2	−1	10	2	8

Construct a 98% jackknife confidence interval for the mean change in heart rate. Interpret your answer.

13.2.3 Air pollution in large US cities is monitored to see whether it conforms to requirements set by the Environmental Protection Agency. The following data, expressed as an air pollution index, gives the air quality of a city for 10 randomly selected days.

57.3 58.1 58.7 66.7 58.6 61.9 59.0 64.4 62.6 64.9

Construct a 95% jackknife confidence interval for the actual average air pollution index for this city and interpret.

13.2.4 The mileage (in thousands) for a random sample of 10 rental cars from a large rental company's fleet is listed.

7 13 5 5 11 15 7 9 13 8

Find a 95% jackknife confidence interval for the population mean mileage of the rental cars of this company.

13.2.5 The following data represent cholesterol levels (in mg/dL) of 10 randomly selected patients from a large hospital on a particular day.

360 352 294 160 146 142 318 200 142 116

Determine a 95% jackknife confidence interval for σ^2. Compare this with the confidence interval obtained in Example 6.4.2.

13.2.6 Air pollution in large US cities is monitored to see whether it conforms to requirements set by the Environmental Protection Agency. The following data, expressed as an air pollution index, gives the air quality of a city for five randomly selected days.

56.23 57.12 57.7 63.92 59.40

Construct a 99% jackknife confidence interval for the actual variance of the air pollution index for this city and interpret.

13.2.7 It is known that some brands of peanut butter contain impurities within an acceptable level. A test conducted on 12 randomly selected jars of a certain brand of peanut butter resulted in the following percentages of impurities:

1.9 2.7 2.1 2.8 2.3 3.6 1.4 1.8 2.1 3.2 2.0 1.9

(a) Construct a 95% jackknife confidence interval for the average percentage of impurities in this brand of peanut butter.
(b) Give an approximate 95% jackknife confidence interval for the population variance.
(c) Interpret your results.

13.2.8 The following is a random sample taken from the data that represents the time intervals in days between earthquakes that either registered magnitudes greater than 7.5 on the Richter scale or produced more than 1000 fatalities during the time period December 1902 to March 1977.

263 1901 121 832 150 99

(a) Construct a 95% jackknife confidence interval for the average number of days between earthquakes of this type.
(b) Give an approximate 95% jackknife confidence interval for the population variance of number of days between earthquakes of this type.

13.3 AN INTRODUCTION TO BOOTSTRAP METHODS

In this section, we describe some aspects of a relatively recent statistical technique known as the bootstrap method that can be used when the statistical distribution is unknown, or the assumptions of normality are not satisfied and especially when the samples are small. Thus, bootstrap is a nonparametric technique; however, it can also be used for inference about parameters in parametric models. This is a general method for estimating sampling distributions. The concept of the bootstrap was introduced by Bradley Efron in 1979 and further developed by Efron and Tibishirani in 1993. We often try to determine the exact (sampling) distribution in an inferential procedure, such as the sampling distribution of the sample mean, the median, or the variance, to be used in computing confidence intervals and for testing hypotheses. However, as we have seen, this is often the most difficult part of the work, because the sampling distribution depends on the population distribution, which is often unknown. This is the reason why asymptotic methods are quite frequently used for hypothesis testing and interval estimation. The bootstrap procedure provides us with a simple method for obtaining an approximate sampling distribution of the statistic, conditional on the observed data. However, it should be noted that the distribution thus obtained is only approximate. It is not as "good" as the exact distribution, because we have only a sample from the population. However, often, bootstrap sampling distribution is easier to compute. Bootstrap methods are computer-intensive methods that use simulation to calculate standard errors, confidence intervals, and significance tests. The methods are applied by researchers in business, econometrics, life sciences, medical sciences, social sciences, and other areas where statistics is being utilized. The bootstrap method uses computer-generated pseudorandom numbers. So, the same situations might give similar but possibly different results. Also, it is computationally more involved to obtain results than by using the asymptotic distribution. The advantage is that the results are conditional on observed data, not based on large sample approximations. How does bootstrap help in reality? For instance, suppose we have 10 years of monthly return data on a particular stock. If we were to use these data to predict the future return, say through linear regression, we would be assuming that the future is going to behave similarly to what happened in the past. We know from experience that such an assumption may not give us a good prediction, and the underlying parametric assumptions may not hold. By creating bootstrap samples from these available data, what we are creating is not what happened, but rather what could have happened in the past from what did happen. For example, to see how resampling affects a sample mean, a particular mutual fund had the following total return (in percentage) for the past five 5 years:

Year	1	2	3	4	5
Total return	40.7	10.8	29.2	9.9	0.7

In this case, the average return for the past 5 years is 18.26%. A two-times resampling (what could have happened) resulted in the following outcomes.

Year	1	2	3	4	5
Total return	29.2	40.7	9.9	10.8	10.8

Here, the average is 20.28%. The next resampling gave the following:

Year	1	2	3	4	5
Total return	0.7	0.7	40.7	0.7	9.9

The resulting average return is 10.54%. A realistic future prediction method should depend on these possible fluctuations that could have happened in different scenarios.

Most of the inferential procedures we learned are based on a single sample drawn from the population. Bootstrap methods, in contrast, generate repeated subsamples from the single original sample itself and make inferences without assuming any particular functional form for the population distribution. Because this has the effect of sampling with replacement, we can create as many subsamples as we wish. These subsamples will have the same sample size and values as the original sample, except that many values in each of the subsamples will be repeated because of sampling with replacement. It should be noted that the effectiveness of a bootstrap procedure depends on the original sample being representative of the population. If the original sample is not representative, the conclusions drawn from the bootstrap methods will be completely inappropriate.

Using the jackknife method, the size of resamples is confined to $(n - 1)$, and the number of total possible samples is only n, the original sample size. The resampling strategy based on bootstrap has no such limitations in terms of the number and magnitude of replications possible. The only limitation comes from the computing resources, and these new sets of samples can be treated as a virtual population.

EXAMPLE 13.3.1

Suppose that the population distribution is an $N(1, \sigma^2)$. Estimate σ^2.

Solution

Because we know the functional form of the distribution, we could use the estimation procedures discussed in Chapter 5. There is no need for the bootstrap method. These steps are as follows.

Step 1. *If we have a random sample from $N(1, \sigma^2)$ of size n, use it. Otherwise, generate a random sample $X_1, \ldots, X_n$ from $N(\mu, \sigma^2)$. This could be done using the method described in Project 4A of Chapter 4.*

Step 2. *Estimate σ^2 by using the method of maximum likelihood, yielding*

$$\hat{\sigma}^2_{ml} = \frac{1}{n}\sum_{i=1}^{n}(X_i - \overline{X})^2.$$

Note that the maximum likelihood procedure requires the knowledge of the functional form of the distribution; see the derivation in Chapter 5. Suppose the form of the population distribution is not known, but we do have a random sample $X_1, \ldots, X_n$ from a distribution. Now we will describe how we can estimate σ^2 using the bootstrap method.

Let $X_1, \ldots, X_n$ be a random sample from a probability distribution F with $\mu = E(X_i)$ and $\sigma^2 = Var(X_i)$. Then the *standard error* of $\overline{X}$ is defined as σ^2/n. In general, the population distribution F is unknown. A simple estimate of F is the *empirical (or sample) cumulative distribution function* defined by

$$\hat{F}(x) = \frac{\#\{X_i \le x\}}{n} = \text{Proportion of } X_i's \le x.$$

This $\hat{F}$ is a step function with the size of the jump being $1/n$ at each ordered X_i.

Summary of the bootstrap method of estimating the standard error of $\overline{X}$

Step 1. Use the sample $X, \ldots, X_n$ and find $\hat{F}$, the empirical cumulative distribution function of F.

Step 2. Generate a sample $\{X^*_{11}, X^*_{12}, \ldots, X^*_{1n}\}$ from $\hat{F}$. From this sample, compute $\overline{X}^*_1$.

Step 3. Repeat step 2, $(N - 1)$ times to obtain samples $\{X^*_{i1}, X^*_{i2}, \ldots, X^*_{in}\}$, $i = 1, 2, \ldots, N$ and find $\overline{X}^*_2, \overline{X}^*_3, \ldots, \overline{X}^*_N$. Now calculate $\overline{X}^* = \frac{1}{N}\sum_{i=1}^{N}\overline{X}^*_i$. This is the bootstrap mean.

Step 4. Then the bootstrap estimate of $Var(\overline{X})$, denoted by $\hat{\sigma}^2_{bs}$, is given by

$$\hat{\sigma}^2_{bs} = \frac{1}{N-1}\sum_{i=1}^{N}(\overline{X}^*_i - \overline{X}^*)^2.$$

Observe that once we have the subsample means $\overline{X}^*_1, \ldots, \overline{X}^*_N$, the formulas for calculating the bootstrap mean and bootstrap variance are the same as those for calculating the mean and variance of a given sample.

Note that when $\hat{F}$ is taken to be the empirical cumulative distribution function, generating a sample from $\hat{F}$ is equivalent to generating a sample from $\{X_1, \ldots, X_n\}$ with replacement. As a result, we obtain the following algorithm.

Bootstrap algorithm for estimating the standard error of $\overline{X}$

1. Draw N random samples *with replacement* from the original sample $X_1, \ldots, X_n$, with each observation having the same probability of being drawn $(1/n)$. Let these bootstrap samples be denoted by $\{\{X^*_{i1}, X^*_{i2}, \ldots, X^*_{in}\}, i = 1,2, \ldots, N\}$.
2. Calculate the sample means of each of these bootstrap samples and the overall sample mean by

$$\overline{X}^*_i = \frac{1}{n}\sum_{j=1}^{n}X^*_{ij} \quad \text{and} \quad \overline{X}^* = \frac{1}{N}\sum_{i=1}^{N}\overline{X}^*_{ij}.$$

3. Compute

$$\hat{\sigma}^2_{bs} = \frac{1}{N-1}\sum_{i=1}^{N}(\overline{X}^*_i - \overline{X}^*)^2.$$

4. Then the bootstrap estimate of $Var(\overline{X})$ is $\hat{\sigma}^2_{bs}$, or equivalently, the standard error of $\overline{X}$ is $\sqrt{\hat{\sigma}^2_{bs}}$.

It is not necessary that the size of the bootstrap sample also must be n or the samples have to be obtained with replacement. However, it is suggested that the best results are obtained when the repeated samples are the same size n as the original sample and when the samples are obtained with replacement. The number of bootstrap samples N could be in the hundreds or more, depending only on the capacity of the software that we are using to generate these samples.

EXAMPLE 13.3.2

The following data represent the total ozone levels measured in Dobson units at randomly selected locations on Earth on a particular day.

269	246	388	354	266	303
295	259	274	249	271	254

Generate $N = 6$ bootstrap samples of size 12 each and find the bootstrap mean and standard deviation (standard error).

Solution

We have created 200 bootstrap samples of size 12. We obtain the following summary results.

$$\overline{X}^* = 285.74$$

and

$$\hat{\sigma}_{bs}^2 = 153.02 \text{ and } \hat{\sigma}_{bs} = 12.37.$$

Note that the mean of the original sample is 285.7, but the standard deviation is 43.9 (see Example 5.5.9). Even though the means of the original sample and the bootstrap means are very close, their standard deviations are substantially different.

In real applications, one of the difficulties is to estimate the standard errors of more complicated statistics. We can now generalize the bootstrap method for those situations. Let $\hat{\theta} = \hat{\theta}(X_1, \ldots, X_n)$ be a sample statistic that is an estimate of the parameter θ of an unknown distribution F using some procedure. We wish to estimate the standard error of $\hat{\theta}$ using the bootstrap procedure, which is summarized next.

General bootstrap procedure to estimate the standard error of $\hat{\theta}$

1. Draw N samples *with replacement* from the original sample, $(X_1, \ldots, X_n)$. Denote these bootstrap samples by $\{X_{i1}^*, X_{i2}^*, \ldots, X_{in}^*\}$, $i = 1, 2, \ldots, N$.

2. Compute $\hat{\theta}_1, \hat{\theta}_2, \ldots, \hat{\theta}_N$, where

$$\hat{\theta}_i^* = \hat{\theta}_i(X_{i1}, X_{i2}, \ldots, X_{in}).$$

The procedure for computing $\hat{\theta}_i^*$ is the same procedure as that used to compute $\hat{\theta}$ original sample $X_1, \ldots, X_n$. Also, compute

$$\hat{\theta}^* = \frac{1}{N}\sum_{i=1}^{N} \hat{\theta}_i^*.$$

3. The bootstrap estimator of standard error (BSE) of $\hat{\theta}$ is given by

$$\left[\widehat{BSE(\hat{\theta})}\right] = \sqrt{\frac{\sum_{i=1}^{N}\left(\hat{\theta}_i^* - \hat{\theta}^*\right)^2}{N-1}}.$$

It is clear that these algorithms are considerably computer intensive, and it is necessary to have suitable software to implement them. The accuracy of the bootstrap approximation depends on the accuracy of $\hat{F}$ as an estimate of F and how large a bootstrap sample is used to estimate the standard error of $\hat{\theta}$. We will leave the computation to Project 13A. We now give a theoretical example.

EXAMPLE 13.3.3

Let $X_1, \ldots, X_n$ be a sample from a Poisson distribution with parameter λ. Let

$$\theta = P\{X \le 1\} = e^{-\lambda}(1 + \lambda).$$

Obtain a bootstrap estimate of θ.

Solution

It can be shown that the maximum likelihood estimator of θ is

$$\hat{\theta}_{ml} = e^{-\overline{X}(1+\overline{X})}.$$

In order to estimate the bias of θ, take N bootstrap samples from $\{X_1, \ldots, X_n\}$. Let

$$\hat{\theta}_i = e^{-\overline{X}_i}(1 + \overline{X}_i) - \frac{\{\#X_i' s \le 1\}}{n}.$$

Then the bootstrap estimate of the bias of θ is

$$\hat{\theta}_{bias} = \frac{\hat{\theta}_1 + \cdots + \hat{\theta}_N}{N}.$$

One might now use

$$e^{-\overline{X}_i}(1 + \overline{X}_i) - \hat{\theta}_{bias}$$

as an estimator of θ.

13.3.1 Bootstrap Confidence Intervals

We could use the repeated sampling method to construct bootstrap confidence intervals. We now give a procedure to obtain this.

Procedure to find bootstrap confidence interval for the mean

1. Draw N samples (N will be in the hundreds, and if the software allows, in the thousands) from the original sample with replacement.
2. For each of the samples, find the sample mean.
3. Arrange these sample means in order of magnitude.
4. To obtain, say, a 95% confidence interval, we will find the middle 95% of the sample means. For this, find the means at the 2.5% and 97.5% quartile. The 2.5th percentile will be at the position $(0.025)(N + 1)$, and the 97.5th percentile will be at the position $(0.975)(N + 1)$. If any of these numbers are not integers, round to the nearest integer. The values of these positions are the lower and upper limits of the 95% bootstrap interval for the true mean.

It should be noted that every time we do this procedure, we may get a slightly different bootstrap interval. We now give an example.

EXAMPLE 13.3.4

For the data given in Example 13.3.2, obtain a 95% bootstrap confidence interval for μ.

Solution

We took $N = 200$ samples of size 12. Thus $0.025 \times 201 = 5.025 \approx 5$ and $0.975 \times 201 = 195.975 \approx 196$. *Thus, taking the 5th and 196th values of sorted (in ascending order) sample means, we get the 95% bootstrap confidence interval for μ is* (263.8, 311.5).

1. *Comparing the classical confidence interval, we obtained in Example 6.3.3, which is (257.81, 313.59), the bootstrap confidence interval of Example 13.3.4 has smaller length and thus less variability. In addition, we saw in Example 6.3.3 that the normality assumption necessary for the confidence interval there was suspected. In the bootstrap method, we did not have any distributional assumptions.*
2. *Because the bootstrap methods are more in tune with nonparametric methods, sometimes it makes sense to obtain confidence intervals about the median rather than the mean. With a slight modification of the procedure that we have described for the bootstrap confidence interval for the mean, we can obtain the bootstrap confidence interval for the median.*

Procedure to find bootstrap confidence interval for the median

1. Draw N samples (N will be in the hundreds, and if the software allows, in the thousands) from the original sample with replacement.
2. For each of the samples, find the sample median.
3. Arrange these sample medians in order of magnitude.
4. To obtain, say, a 95% confidence interval, we will find the middle 95% of the sample medians. For this, find the medians at the 2.5% and 97.5% quartile. The 2.5th percentile will be at the position $(0.025)(N + 1)$, and the 97.5th percentile will be at the position $(0.975)(N + 1)$. If any of these numbers are not integers, round to the nearest integer. The values of these positions are the lower and upper limits of the 95% bootstrap interval for the median.

In practice, how many bootstrap samples should be taken? The answer depends on two things: how much the result matters and what type of computing power is available. In general, it is better to start with 1000 subsamples. With the computational power available now, even taking 10,000 replications is not much of a problem. There are many works in the literature on bootstrap hypothesis testing and regression. These are beyond the scope of this chapter. In Chapter 10 we use bootstrap resampling to obtain empirical Bayes estimates.

EXERCISES 13.3

13.3.1 For the data of Exercise 13.2.2, generate $N = 8$ bootstrap samples of size 20 each and find the bootstrap mean and standard deviation (standard error).

13.3.2 For the data of Exercise 13.2.5, generate $N = 12$ bootstrap samples of size 10 each and find the bootstrap mean and standard deviation (standard error).

13.3.3 For the data of Exercise 13.3.3, obtain a 95% bootstrap confidence interval for μ.

13.3.4 For the data of Exercise 13.2.6, (a) obtain a 95% bootstrap confidence interval for μ, and (b) obtain a 95% bootstrap confidence interval for the population median.

13.3.5 For the data of Exercise 13.2.8, (a) obtain a 95% bootstrap confidence interval for μ, and (b) obtain a 95% bootstrap confidence interval for the population median.

13.4 THE EXPECTATION–MAXIMIZATION ALGORITHM

In this section, we introduce an algorithm, called the *expectation–maximization* (EM) algorithm, that is widely used to compute maximum likelihood estimates when some elements of the data set are either missing, unobservable, or incomplete. In real-life problems, observing the complete data is the exception rather than the rule. For example, in lifetime studies, when n items are placed on a given test, we may have the failure times of only $n_1 < n$ items, while for the rest of $(n - n_1)$ items, we only know the censored failure time, that they survived a particular failure time T (fixed beforehand). For example, we may want to know whether the lifetime of a certain brand of fluorescent light bulbs is at least 24 months. For this purpose, let us say we randomly test 100 light bulbs of this brand. In this case, our data will contain all the months within which the bulbs burned out, and some that survived for 24 months. After 24 months, we do not know when these bulbs will eventually burn out; all we know is that these bulbs lasted for 24 months. Such data is an example of censored data. We can consider the censored failure times of $(n - n_1)$ items as the unobservable data values.

Another common problem is the missing data. For example, suppose we were to take a survey on some socioeconomic problems from a random sample of families from a city in 2009 and then a follow-up study on the same families in 2014. This may result in many missing values in the follow-up study, because it is possible that we may not be able to locate some of the families. Missing values can also occur if some of the respondents refuse to answer certain questions. We have seen in Section 5.3 that sometimes it is not possible to obtain closed-form solutions for MLE. In the completely observed case, there are other algorithms, such as Newton–Raphson, that can be used to numerically obtain appropriate estimates. With missing values, those algorithms cannot be used. The name *EM algorithm* was coined by Dempster, Laird, and Rubin in 1977. This is a general iterative algorithm to obtain the MLE when the data set is incomplete. The EM algorithm is a formalization of an intuitive idea of obtaining an approximate estimate of the parameters with missing data: (i) replace missing values with estimated values as true values, (ii) estimate the parameters, (iii) and repeat.

Let $X_1, \ldots, X_{n_1}$ be the n_1 observed data values and let $y_1, \ldots, y_{n-n_1}$ be the $(n - n_1)$ unobserved data values. Assume that X_i' s are iid random variables with pdf $f(x|\theta)$. Also, assume X_i' s and Y_i' s are independent, that is, data is missing at random.

We denote the random vector by **X** and the corresponding data vector by **x**.

The joint pdf of $X_1, \ldots, X_{n_1}$ is represented by $f(\mathbf{x}|\theta)$, where θ is the parameter vector with values in $\Theta \subset R^p$, a p-dimensional Euclidean space. Let $g(\mathbf{x}, \mathbf{y}|\theta)$ denote the pdf of the complete data set **x** and **y**, that is, the vector (**x**, **y**) represents the conceptualized complete data set. Let $h(\mathbf{y}|\theta, \mathbf{x})$ be the conditional pdf of the unobserved data **y** given θ and the observed data **x**. The likelihood function for the observed data **x** is, by definition,

$$L(\theta; \mathbf{x}) = f(\mathbf{x}|\theta).$$

The likelihood function for the combined data (**x**, **y**) is again, by definition, given by

$$L_c(\theta; \mathbf{x}, \mathbf{y}) = g(\mathbf{x}, \mathbf{y}|\theta).$$

The problem is to find the maximum likelihood estimator that maximizes the likelihood function $L(\theta, \mathbf{x})$, at the same time using $L_c(\theta; \mathbf{x}, \mathbf{y})$.

From the foregoing definitions, we know that

$$g(x, y|\theta) = f(x|\theta)h(y|\theta, x).$$

Thus, we have the conditional pdf of the missing (or unobserved) data **y**, given **x**:

$$h(\mathbf{y}|\theta, \mathbf{x}) = \frac{g(\mathbf{x}, \mathbf{y}|\theta)}{f(\mathbf{x}|\theta)}$$

or equivalently

$$f(\mathbf{x}|\theta) = \frac{g(\mathbf{x}, \mathbf{y}|\theta)}{h(\mathbf{y}|\theta, \mathbf{x})}. \tag{13.1}$$

Let $\theta_0 \in \Theta$ be a given θ-value. Because $h(\mathbf{y}|\theta_0, \mathbf{x})$ is a pdf, we have

$$\int h(\mathbf{y}|\theta_0, \mathbf{x})d\mathbf{y} = 1.$$

Thus ln of observed likelihood,

$$\begin{aligned}\ln L(\theta; \mathbf{x}) &= ln\ L(\theta; \mathbf{x}) \int h(\mathbf{y}|\theta_0, \mathbf{x})d\mathbf{y} \\ &= \int \ln L(\theta; \mathbf{x})h(\mathbf{y}|\theta_0, \mathbf{x})d\mathbf{y} \ \text{(as } \ln L(\theta; \mathbf{x}) \text{is independent of } \mathbf{y}).\end{aligned}$$

Because $L(\theta, \mathbf{x}) = f(\mathbf{x}|\theta)$, we have

$$\begin{aligned}\ln L(\theta; \mathbf{x}) &= \int \ln f(\mathbf{x}|\theta)h(\mathbf{y}|\theta_0, \mathbf{x})d\mathbf{y} \\ &= [\ln g(\mathbf{x}, \mathbf{y}|\theta) - \ln h(\mathbf{y}|\theta, \mathbf{x})]h(\mathbf{y}|\theta_0, \mathbf{x})d\mathbf{y} \ \text{(from(1))} \\ &= \int \ln g(\mathbf{x}, \mathbf{y}|\theta)h(\mathbf{y}|\theta_0, \mathbf{x})d\mathbf{y} - \int \ln h(\mathbf{y}|\theta, \mathbf{x})h(\mathbf{y}|\theta_0, \mathbf{x})d\mathbf{y} \\ &= E_{\theta_0}[\ln g(\mathbf{x}, \mathbf{y}|\theta)] - E_{\theta_0}[\ln h(\mathbf{y}|\theta, \mathbf{x})],\end{aligned} \tag{13.2}$$

where the expectation is taken with respect to the conditional distribution of $\mathbf{y}$ given θ_0 and $\mathbf{x}$. Let us now consider maximizing this with respect to θ. This maximization is the maximization step (M-step) in the EM algorithm. It is important to note that EM-based estimates are approximate estimates.

Let θ_0 be an initial estimate of θ. The choice of this initial value θ_0 could be done randomly or heuristically based on any prior knowledge about the optimal value of the parameter. For instance, suppose we have to estimate mean and variance of a normal distribution. One good starting point could be to take the sample mean $\bar{x}$ and sample variance s^2 based on a subset of the data containing no missing values.

Let

$$\begin{aligned}Q(\theta|\theta_0, \mathbf{x}) &= E_{\theta_0}[\ln L_c(\theta; \mathbf{x}, \mathbf{y})] \\ &= E_{\theta_0}[\ln g(\mathbf{x}, \mathbf{y}|\theta)]\end{aligned}$$

Here, θ_0 is used only to compute the expectation; we should not substitute for θ in the complete data log-likelihood. Let $\hat{\theta}_{(1)}$ be the maximizer that maximizes $Q(\theta|\theta_0, \mathbf{x})$ with respect to θ. That is, $Q\left(\hat{\theta}_{(1)}|\theta_0, \mathbf{x}\right) \geq Q(\theta|\theta_0, \mathbf{x})$ for all $\theta_0 \in \Theta$. Then $\hat{\theta}_{(1)}$ is the first-step estimator of θ. Continuing the procedure, we obtain a sequence of approximate estimators $\hat{\theta}_{(m)}$, which under appropriate conditions converges to the maximum likelihood estimate with likelihood $L_c(\theta; \mathbf{x}, \mathbf{y})$.

Steps for using the expectation–maximization algorithm

1. $\hat{\theta}_{(n)}$ is the estimate of the parameter θ on the nth step.
2. Expectation step (E-step). Compute

$$Q\left(\theta\middle|\hat{\theta}_{(n)}, \mathbf{x}\right) = E_{\hat{\theta}_{(n)}}\left[\ln g\left(\mathbf{x}; \mathbf{y}\middle|\theta\right)\right]$$

where the expectation is with respect to the conditional pdf of $\mathbf{y}$ given $\hat{\theta}_{(n)}$ and $\mathbf{x}$ (i.e., with respect to $h\left(\mathbf{y}\middle|\hat{\theta}_{(n)}, \mathbf{x}\right)$).

3. Maximization step (M-step). Find $\hat{\theta}_{(n+1)} \in \Theta$ such that

$$\hat{\theta}_{(n+1)} = \max_{\theta} Q\left(\theta\middle|\hat{\theta}_{(n)}, \mathbf{x}\right).$$

4. Repeat until specified convergence criteria are met.

Thus, in the EM algorithm, each iteration involves two steps: the expectation step (E-step), followed by the maximization step (M-step). In the E-step, we find the conditional expectation of the unobserved or missing data given the observed data and the current estimated parameters. Thus, in E-step, missing data is estimated given the observed data and the present estimate of the model parameters. That is, the E-step constitutes the calculation of

$$Q\left(\theta|\hat{\theta}_{(n)}, \mathbf{x}\right) = E_{\hat{\theta}_{(n)}}[\ln g(\mathbf{x}, \mathbf{y}|\theta)]$$
$$= \int \ln g(\mathbf{x}, \mathbf{y}|\theta) h\left(\mathbf{y}|\theta_{(n)}, \mathbf{x}\right) d\mathbf{y}$$

(which is the sum, if discrete), where the integration is over the range of values that $\mathbf{y}$ can assume. The M-step constitutes the maximization of $Q(\theta|\hat{\theta}_{(n)}, \mathbf{x})$ with respect to θ. Thus in M-step, the likelihood function is maximized under the assumption that the missing or hidden data are known. The estimate of the missing data from E-step is used in place of the actual missing data. This procedure improves the log-likelihood at every iteration; that is, the log-likelihood is nondecreasing for every iteration. Thus, for the sequence $(\hat{\theta}_{(n)})$ obtained through the EM algorithm, we have $L(\hat{\theta}_{(n+1)}; \mathrm{x}) \geq L(\hat{\theta}_{(n)}; \mathrm{x})$ with equality holding if and only if $Q(\hat{\theta}_{(n+1)}|\hat{\theta}_{(n)}, \mathbf{x}) = Q(\hat{\theta}_{(n)}|\hat{\theta}_{(n)}, \mathbf{x})$. When we have filled the completed data set, the parameter θ can be estimated by maximizing the log-likelihood estimating procedure (M-step). It can be shown that under some conditions (such as that $\ln f(\mathbf{x}|\theta)$ is bounded, or that $Q(\theta|\theta_0, \boldsymbol{x})$ is continuous in both θ and θ_0), $\hat{\theta}_{(n)}$ converges in probability as $n \to \infty$ to the maximum likelihood estimate based on the complete likelihood $L_c(\theta; \boldsymbol{x}, \boldsymbol{y})$.

For computational convergence purposes, the E-step and M-step are alternated repeatedly until the difference $L(\hat{\theta}_{(n+1)}, \mathbf{x}) - L(\hat{\theta}_{(n)}, \mathbf{x})$ is less than δ, a small but specified quantity. Another possible convergence criterion is to stop the iteration when the distance between $\hat{\theta}_{(n+1)}$ and $\hat{\theta}_n$ becomes arbitrarily small. In practice, it may be necessary to run the EM algorithm a number of times with different (random) starting points to ensure that the global maximum is obtained.

In general, the E-step and M-step could be complex. Even though the EM algorithm is applicable to any model, it is particularly effective if the data come from an exponential family. It turns out that, in this case, the log-likelihood is linear in the sufficient statistic for θ. For the E-step, simply compute the expectation of the complete data sufficient statistic given the observed data. By substituting the conditional expectations of the sufficient statistics computed in the E-step for the sufficient statistics that occur in the expression obtained for the complete data maximum likelihood estimators of θ, we can obtain the next iterate in the M-step. Thus, when the complete data set is from an exponential family, both the E-step and the M-step are simplified.

Let $\mathbf{z} = (\mathbf{x}, \mathbf{y})$ be the complete observation vector. A particular case in which $g(\mathbf{x}, \mathbf{y}|\theta) = g(\mathbf{z}, \theta)$ is from an exponential family:

$$g(\mathbf{z}, \theta) = a(\mathbf{x}) \exp\{\mathbf{k}'(\theta)\mathbf{t}(\mathbf{x})\}/c(\theta)$$

where $\mathbf{t}(\mathbf{x})$ is a vector of sufficient statistics with complete data, $\mathbf{k}'(\theta)$ is a vector function of the parameter vector θ, and $a(\mathbf{x})$ and $c(\theta)$ are scalar functions. Recall that the members of the exponential family include many popular distributions, such as the normal, multivariate normal, Poisson, and multinomial distributions. In this case, the E-step can be written as

$$Q(\theta|\theta_{(n)}, \mathbf{x}) = E_{\theta_{(n)}}[\ln a(\mathbf{x})|\mathbf{x}] + \mathbf{k}'(\theta)\mathbf{t}_{(n)} - \ln c(\theta)$$

where $\mathbf{t}_{(n)} = E_{\theta(n)}[\mathbf{t}(\mathbf{Z})|\mathbf{x}]$ is an estimator of the sufficient statistic. The M-step maximizes the Q-function with respect to θ. Because $E_{\theta(n)}[\ln a(\mathbf{x})|\mathbf{x}]$ does not depend on θ, we can rewrite the steps as follows:

E-step: Compute $\mathrm{t}(n) = E_{\theta(n)}[\mathbf{t}(\mathbf{Z})|\mathrm{x}]$.

M-step: Find $\hat{\theta}_{(n+1)} \in \Theta$, such that,

$$\hat{\theta}_{(n+1)} = \max_{\theta}\left[\mathbf{k}'(\theta)\mathbf{t}_{(n)} - \ln c(\theta)\right].$$

The following example gives an EM algorithm for a special case of censored survival times. In the following example, the survival function is defined as the probability that an individual survives beyond time y, that is, $S(y) = P(Y > y)$.

EXAMPLE 13.4.1

Let $\mathbf{x} = (x_1, \ldots, x_{n_1})$ be the observed data and the censored observations at T are $\mathbf{y} = (y_1, \ldots, y_{n_2})$ (i.e., the survival time is at least T). Let the mean survival time be θ and the probability density be given by

$$f(x|\theta) = \theta^{-1}\exp(-x/\theta),\ x > 0.$$

(a) Obtain the MLE, $\hat{\theta}_{ML}$.
(b) Obtain an EM algorithm.
(c) Consider the following censored data, which represent the number of years 20 patients survived after a major surgery, where a + symbol represents that we know only that this patient survived for 4 years, with no further information. That is,

4+	12	12	1	4+	3	3	5	2	0
5	1	4+	0	3	13	13	1	0	4

Using the algorithm developed in part (b), run for 50 iterations with initial value of θ_0 being the observed sample mean, $\bar{x}$, and with $\theta_0 = 0$. Comment on the results.

Solution
The joint pdf of the uncensored observation, **x**, *is*

$$f(\mathbf{x}|\theta) = \frac{1}{\theta^n}\exp\left(-\sum_{i=1}^{n_1} x_i/\theta\right)$$

For the right censored (at T) observations y_i, $i = 1,\ldots,n_2$, the pdf can be calculated as follows:

$$K\int_T^\infty \frac{1}{\theta}e^{-y/\theta}dy = 1$$

implies that $K = e^{T/\theta}$. Thus, the pdf of y_i is given by

$$h(y|\theta, x) = \frac{e^{T/\theta}}{\theta}e^{-y/\theta} = \frac{1}{\theta}e^{\frac{1}{\theta}(T-y)},\ y \geq T.$$

(a) *The likelihood, $L_c(\theta, x, y)$, can also be written in the form*

$$L_c(\theta, x, y) = \frac{1}{\theta^{n_1}}e^{-\sum_{i=1}^{n_1}(x_i/\theta)}[1 - F(T)]^{n_2}$$
$$= \frac{1}{\theta^{n_1}}e^{-\sum_{i=1}^{n_1}(x_i/\theta)}e^{-\frac{n_2 T}{\theta}}.$$

Thus,

$$\ln L_c(\theta, x, y) = -n_1\ln\theta - \frac{\sum_{i=1}^{n_1}x_i}{\theta} - \frac{n_2 T}{\theta}.$$

Differentiating with respect to θ, and equating to zero, we have

$$\frac{\partial}{\partial\theta}\ln L_c(\theta, x, y) = -\frac{n_1}{\theta} + \frac{\sum_{i=1}^{n_1}x_i}{\theta^2} + \frac{n_2 T}{\theta^2} = 0.$$

This implies, that

$$n_1\theta = \sum_{i=1}^{n_1} x_i + n_2 T$$

or

$$\hat{\theta} = \frac{1}{n_1}\sum_{i=1}^{n_1} x_i + \frac{n_2}{n_1}T = \overline{x} + \frac{n_2}{n_1}T.$$

Hence, the MLE of θ *is,*

$$\hat{\theta}_{ML} = \overline{X} + \frac{n_2}{n_1}T.$$

(b) *Because* $g(\mathbf{X}, \mathbf{Y}|\theta)$ *denotes the pdf of the complete data, and we assumed the pdf of all the data (censored or not) follows exponential distribution, we have*

$$g\left(\mathbf{x}, \mathbf{y}\middle|\theta\right) = \frac{1}{\theta^{n_1}}e^{-\sum_{i=1}^{n_1}(x_i/\theta)}\frac{1}{\theta^{n_2}}e^{-\sum_{i=1}^{n_2} y_i/\theta},$$

we get

$$\ln g\left(\mathbf{x}, \mathbf{y}\middle|\theta\right) = -n_1 \ln\theta - \sum_{i=1}^{n_1}\frac{x_i}{\theta} - n_2\ln\theta - \sum_{i=1}^{n_2}\frac{y_i}{\theta}.$$

For the E-step of the EM algorithm, we first compute

$$\begin{aligned} E_{\theta_0}Y &= e^{T/\theta_0}\int_T^{\infty} y\frac{1}{\theta_0}e^{-y/\theta_0}dy \\ &= T + \theta_0 \text{ (using the integration by parts).} \end{aligned}$$

Thus, we get

$$\begin{aligned} Q\left(\theta\middle|\theta_0, \mathbf{x}\right) &= E_{\theta_0}\left[g\left(\mathbf{x}, \mathbf{y}\middle|\theta\right)\right] \\ &= E_{\theta_0}\left[-n_1\ln\theta - \sum_{i=1}^{n_1}\frac{x_i}{\theta} - n_2\ln\theta - \sum_{i=1}^{n_2}\frac{y_i}{\theta}\right] \\ &= -n_1\ln\theta - \sum_{i=1}^{n_1}\frac{x_i}{\theta} - n_2\ln\theta - \frac{1}{\theta}\sum_{i=1}^{n_2}E_{\theta_0}\left(y_i\right) \\ &= -n_1\ln\theta - \sum_{i=1}^{n_1}\frac{x_i}{\theta} - n_2\ln\theta - \frac{1}{\theta}n_2(T+\theta_0) \\ &= -n_1\ln\theta - \sum_{i=1}^{n_1}\frac{x_i}{\theta} - n_2\ln\theta - \frac{n_2T + n_2\theta_0}{\theta}. \end{aligned}$$

For the M-step, we differentiate $Q(\theta|\theta_0, x)$ with respect to θ and set it equal to zero,

$$\frac{\partial}{\partial\theta}Q\left(\theta\middle|\theta_0, x\right) = \frac{\partial}{\partial\theta}\left[-n_1 \ln\theta - \sum_{i=1}^{n_1}\frac{x_i}{\theta} - n_2\ln\theta - \frac{n_2T + n_2\theta_0}{\theta}\right]$$

$$= -\frac{n_1}{\theta} + \frac{\sum_{n_1}^{i=1}x_i}{\theta^2} - \frac{n_2}{\theta} + \frac{n_2T + n_2\theta_0}{\theta^2} = 0$$

$$[n_1 + n_2]\theta = \sum_{i=1}^{n_1} x_i + n_2T + n_2\theta_0$$

$$\hat{\theta}_1 = \frac{1}{[n_1 + n_2]}\sum_{i=1}^{n_1} x_i + \frac{n_2T}{[n_1 + n_2]} + \frac{n_2}{[n_1 + n_2]}\theta_0$$

$$= \frac{n_1}{[n_1 + n_2]}\bar{x} + \frac{n_2T}{[n_1 + n_2]} + \frac{n_2}{[n_1 + n_2]}\theta_0.$$

Thus, for the general n, the algorithm is

$$\hat{\theta}_{(n+1)} = \frac{n_1}{[n_1 + n_2]}\bar{x} + \frac{n_2T}{[n_1 + n_2]} + \frac{n_2}{[n_1 + n_2]}\hat{\theta}_{(n)}.$$

Now putting $\theta_{(k+1)} = \theta_{(k)} = \theta^*$ in the previous equation and solving for θ^*, we have the EM sequence $\{\theta_{(k)}\}$ has the MLE $\hat{\theta}_{ML}$ as its unique limit point, as $k \to \infty$. That is, $\theta^* = \hat{\theta}_{ML}$.

(c) *We used the following MATLAB code to run the algorithm with starting value θ_0 as the sample mean, that is, 4.5. Here $T = 4$. We run it for 50 iterations.*

```
A(1) = 4.5
for n=2: 50
A(n) = 4.41*(17./20)+3*4/20+(3./20)*A(n-1)
end
```

Following is the output.

4.5000	5.0235	5.1020	5.1138	5.1156	5.1158	5.1159
5.1159	5.1159	5.1159	5.1159	5.1159	5.1159	5.1159
5.1159	5.1159	5.1159	5.1159	5.1159	5.1159	5.1159
5.1159	5.1159	5.1159	5.1159	5.1159	5.1159	5.1159
5.1159	5.1159	5.1159	5.1159	5.1159	5.1159	5.1159
5.1159	5.1159	5.1159	5.1159	5.1159	5.1159	5.1159
5.1159	5.1159	5.1159	5.1159	5.1159	5.1159	5.1159
		5.1159				

Thus $\hat{\theta}$ = 5.1159.

To run with $\theta_0 = 0$, in the previous code, just change $A(1) = 0$. We get the following output.

0.0000	4.3485	5.0008	5.0986	5.1133	5.1155
5.1158	5.1159	5.1159	5.1159	5.1159	5.1159
5.1159	5.1159	5.1159	5.1159	5.1159	5.1159
5.1159	5.1159	5.1159	5.1159	5.1159	5.1159
5.1159	5.1159	5.1159	5.1159	5.1159	5.1159
5.1159	5.1159	5.1159	5.1159	5.1159	5.1159
5.1159	5.1159	5.1159	5.1159	5.1159	5.1159
5.1159	5.1159	5.1159	5.1159	5.1159	5.1159
		5.1159	5.1159		

With $\theta_0 = \bar{x} = 4.5$, it took six iteration steps to converge, whereas with $\theta_0 = 0$, it took seven steps to converge. Note that in both cases, $\hat{\theta} = 5.1159 = \hat{\theta}_{ML}$.

Example 13.4.1 is a simple case, where there is no need for iterative computation of $\hat{\theta}_{ML}$. However, this demonstrates how the EM algorithm would work. These types of problems are abundant in the medical field. For example, we may be interested in the survival times of n patients after a treatment. For practical reasons, we may be observing only for a fixed duration, such as 10 years. In Example 13.4.1, the vector **x** will represent the time of death for the n_1 individuals. For the remaining $n_2 = n - n_1$ individuals, the only data we have states that they survived for more than 4 years. Thus, the value of T is 4. There is a possibility that during these experimental times, we may lose contact with some individuals, perhaps because they moved to some other place or they simply refused to participate in this experiment. In those cases, we will know only that the individual survived until we lost contact. This generalization of Example 13.4.1 to where the survival time data are different for each observation is given in Exercise 13.4.5.

We now give a similar example with a normal sample.

EXAMPLE 13.4.2

Let $\mathbf{x} = (x_1, \ldots, x_{n_1})$ be observed data from a normal population with mean θ and variance 1. Let the censored observations at T be $y = (y_1, \ldots, y_{n_2})$ (i.e., the survival time is at least T) from the same population. Assume that the two sets of observations $\{x_i\}$ and $\{y_i\}$ are independent. Write down an EM algorithm to estimate θ.

Solution

For the uncensored observed sample $x_1, \ldots, x_{n_1}$, the likelihood function is

$$L(\theta|\mathbf{x}) = f_x(\mathbf{x}|\theta) = \frac{1}{(\sqrt{2\pi})^{n_1}} e^{-\frac{1}{2}\sum_{i=1}^{n_1}(x_i-\theta)^2}.$$

Furthermore, the complete likelihood for both the samples is

$$L(\theta|\mathbf{x}, \mathbf{y}) = \frac{1}{(\sqrt{2\pi})^{n_1}} e^{-\frac{1}{2}\sum_{i=1}^{n_1}(x_i-\theta)^2} \frac{1}{(\sqrt{2\pi})^{n_2}} e^{-\frac{1}{2}\sum_{i=1}^{n_2}(y_i-\theta)^2}. \tag{13.3}$$

From the definition of $Q(\theta|\theta_0, \mathbf{x})$, we obtain

$$Q(\theta|\theta_0, \mathbf{x}) = E_{\theta_0}\left[\ln L_c(\theta|\mathbf{x}, \mathbf{y})\right], \tag{13.4}$$

where the expectation is taken with respect to the conditional pdf

$$\begin{aligned} h(y|\theta_0, x, T) &= \frac{1}{\sqrt{2\pi}} e^{-(y-\theta_0)^2/2} \frac{1}{1 - F_Y(T, \theta_0)} \\ &= \frac{1}{\sqrt{2\pi}} e^{-(y-\theta_0)^2/2} \frac{1}{1 - \Phi(T - \theta_0)}, \end{aligned}$$

where

$$F_Y(T, \theta_0) = \int_{-\infty}^{T} \frac{1}{\sqrt{2\pi}} e^{-(y-\theta_0)^2/2} dy = \int_{-\infty}^{T-\theta_0} \frac{1}{\sqrt{2\pi}} e^{-u^2/2} du = \Phi(T - \theta_0).$$

Thus, from (13.4) and (13.5),

$$\begin{aligned} Q(\theta|\theta_0, \mathbf{x}) &= E_{\theta_0} \sum_{i=1}^{n_1} \ln\left[\frac{1}{\sqrt{2\pi}} e^{-\frac{(x_i-\theta)^2}{2}}\right] + E_{\theta_0} \ln\left[\frac{1}{\sqrt{2\pi}^{n_2}} e^{-\frac{(y_i-\theta)^2}{2}}\right] \\ &= -\frac{n_1}{2}\ln(2\pi) - \sum_{i=1}^{n_1} \frac{(x_i-\theta)^2}{2} \\ &+ n_2 \int_T^{\infty} \ln\left[\frac{1}{(\sqrt{2\pi})^{n_2}} e^{-\frac{(y_i-\theta)^2}{2}}\right] \times \frac{1}{\sqrt{2\pi}} e^{-(y-\theta_0)^2/2} \frac{1}{1-\Phi(T-\theta_0)} dy. \end{aligned}$$

Now taking the derivative with respect to θ, we have

$$\begin{aligned} \frac{\partial Q}{\partial \theta} &= \sum_{i=1}^{n_1} (x_i - \theta)^2 + \frac{n_2}{\sqrt{2\pi}} \int_T^{\infty} (y-\theta) \frac{e^{-(y-\theta_0)^2/2}}{1-\Phi(T-\theta_0)} dy \\ &= \sum_{i=1}^{n_1} x_i - n_1\theta + \frac{n_2}{[1-\Phi(T-\theta_0)]} \Phi(T-\theta_0) - n_2(\theta - \theta_0). \end{aligned}$$

Solving $\frac{\partial Q}{\partial \theta} = 0$, and letting $n = n_1 + n_2$, we obtain

$$\theta = \frac{\sum_{i=1}^{n_1} x_i}{n} + \frac{n_2}{n}\theta_0 + \frac{n_2 \Phi(T-\theta_0)}{1-\Phi(T-\theta_0)}. \tag{13.5}$$

From (13.5), we obtain the EM algorithm as

$$\hat{\theta}_{m+1} = \frac{\sum_{i=1}^{n_1} x_i}{n} + \frac{n_2}{n}\hat{\theta}_m + \frac{n_2 \Phi(T-\hat{\theta}_m)}{1-\Phi(T-\hat{\theta}_m)},$$

where Φ is the cumulative distribution function of a standard normal random variable.

We have seen that the incomplete data could occur as a result of missing data, or the complete data may contain variables that are not observable (hidden). The following is an example of the latter situation.

EXAMPLE 13.4.3

Suppose that in a set of n twin pairs of children, n_1 are male twin pairs, n_2 are female twin pairs, and $n_3 = n - (n_1 + n_2)$ are opposite-sex twin pairs. Let p be the probability that a twin pair is identical and q be the probability that a child is male. It is not known which pairs of same-sex twins are identical. Obtain an EM sequence for $\theta = (p, q)$.

Solution

We have $n = (n_1 + n_2 + n_3)$, and $\theta = (p, q)$ is the parameter vector. Let $x = (n_1, n_2, n_3)$ be the observed data. Because we don't know which pairs of the same sex are identical, postulate the complete data set as $z = (n_{11}, n_{12}, n_{21}, n_{22}, n_3)$, where n_{11} is the number of male identical pairs, n_{21} is the number of female identical pairs, and n_{12} and n_{22} are the nonidentical pairs for males and females, respectively. Here, the complete data, z, has a multinomial distribution with the likelihood given by

$$\begin{aligned} L(\mathbf{z}, \theta) &= f(\mathbf{z}|\theta) \\ &= \binom{n}{n_{11}, n_{12}, n_{21}, n_{22}, n_3} (pq)^{n_{11}} \left[(1-p)q^2\right]^{n_{12}} \left[p(1-q)\right]^{n_{21}} \\ &\times \left[(1-p)(1-q)^2\right]^{n_{22}} \left[2(1-p)(1-q)q\right]^{n_3} \end{aligned}$$

where the identical twins involve one choice of sex and the nonidentical twins involve two choices of sex. The log-likelihood for the complete data is

$$\ln f(x|\theta) = (n_{11} + n_{21})\ln p + (n_{12} + n_{22} + n_3)\ln(1-p)$$
$$+ (n_{11} + 2n_{12} + n_3)\ln q + (n_{21} + 2n_{22} + n_3)$$
$$\times \ln(1-q) + \text{constant}.$$

For the E-step, use Bayes rule to obtain the following:

$$n_{11}^{(k)} = E\left(n_{11}|x, \theta_{(k)}\right) = n_1 \frac{p_{(k)}q_{(k)}}{p_{(k)}q_{(k)} + \left(1-p_{(k)}\right)\left(q_{(k)}\right)^2}$$
$$n_{12}^{(k)} = E\left(n_{12}|x, \theta_{(k)}\right) = n_1 \frac{\left(1-p_{(k)}\right)\left(q_{(k)}\right)^2}{p_{(k)}q_{(k)} + \left(1-p_{(k)}\right)\left(q_{(k)}\right)^2}$$
$$n_{21}^{(k)} = E\left(n_{21}|x, \theta_{(k)}\right) = n_2 \frac{p_{(k)}\left(1-q_{(k)}\right)}{p_{(k)}\left(1-q_{(k)}\right) + \left(1-p_{(k)}\right)\left(1-q_{(k)}\right)^2}$$
$$n_{22}^{(k)} = E\left(n_{22}|x, \theta_{(k)}\right) = n_2 \frac{\left(1-p_{(k)}\right)\left(1-q_{(k)}\right)^2}{p_{(k)}\left(1-q_{(k)}\right) + \left(1-p_{(k)}\right)\left(1-q_{(k)}\right)^2}.$$

*Thus, the **Q**-function is given by*

$$Q(\theta, \theta_{(k)}) = \left(n_{11}^{(k)} + n_{21}^{(k)}\right)\ln p + \left(n_{12}^{(k)} + n_{22}^{(k)} + n_3\right)\ln(1-p)$$
$$+\left(n_{11}^{(k)} + 2n_{21}^{(k)} + n_3\right)\ln q + \left(n_{21}^{(k)} + 2n_{22}^{(k)} + n_3\right)$$
$$\times \ln(1-q) + \text{constant}.$$

It can be verified that the M-step gives the following:

$$p_{(k+1)} = \frac{n_{11}^{(k)} + n_{21}^{(k)}}{n}$$
$$q_{(k+1)} = \frac{n_{11}^{(k)} + 2n_{12}^{(k)} + n_3}{n + n_{12}^{(k)} + n_{22}^{(k)}}.$$

Substituting for the log-likelihoods by log-posteriors, the EM algorithm can also be used for computations related to Bayesian analysis to find the posterior mode of θ. In the context of incomplete data coming from mixtures of parametric families, the EM algorithm provides a very powerful numerical technique. In this book, we will not go into the mixture models. The steps necessary to compute the required quantities depend on the particular application, and thus in general how to code the EM algorithm is not clear. There are special cases available in some software packages such as SAS using PROC MI with EM option when the data comes from a multivariate normal distribution. It is desirable to search the literature on the particular software you are using to find out the availability of "EM codes" to suit the particular application in which you are interested. Also, another difficulty with implementation of EM algorithm is that in each E-step, we require computation of the conditional expectation. To overcome this difficulty, Wei and Tanner in 1990 proposed an algorithm called MCEM (Monte Carlo EM) based on the Monte Carlo approach explained in Section 13.5. This basically involves simulating m variables, $Y_1,\ldots,Y_m$, from the conditional distribution $h(y|\theta_{(n)}, \mathrm{x})$ and then maximizing the approximate complete data likelihood, that is,

$$\hat{Q}\left(\theta\middle|\hat{\theta}_{(n)}, \mathbf{x}\right) = \frac{1}{m}\sum_{i=1}^{m}\left[\ln g\left(\mathbf{x}, \mathbf{y}|\theta\right)\right].$$

We will not go into the details of these methods. The student may refer to Wei and Tanner's paper for further details.

EXERCISES 13.4

13.4.1 Suppose that Y is a noise-corrupted observation of a signal S. That is, $Y = S + N$, where S is independent of N. Assume that for a known σ, $N \sim N(0, \sigma^2)$ and $S \sim N(0, \theta^2)$, where θ is unknown. Given the observation $Y = y$:
(a) Obtain the MLE, $\hat{\theta}_{ML}$.
(b) Obtain an EM algorithm.

13.4.2 Let $X_1, \ldots, X_n$ be an observed random sample and $X_{(n_1+1)}, \ldots, X_n$ be the missing (at random) observations. Assume that X_i are iid from an $N(\mu, \sigma^2)$ distribution.
(a) Show that $(\Sigma_{i=1}^{n} X_i, \Sigma_{i=1}^{n} X_i^2)$ are sufficient statistics for $\theta = (\mu, \sigma^2)$.
(b) Obtain the EM sequence for $\theta = (\mu, \sigma^2)$.
(c) Consider a censored normal sample with $n = 10$, with the largest three being censored.

1.613 1.644 1.663 1.732 1.740 1.763 1.778

Using the results of part (a), obtain an EM estimate of $\theta = (\mu, \sigma^2)$ with an arbitrary starting point.

13.4.3 In Example 13.4.3, suppose that q is the probability that a child is a female. Obtain an EM sequence for $\theta = (p, q)$.

13.4.4 Let $\mathbf{x} = (x_1, \ldots, x_{n_1})$ and censored observations $(x_{n_1} + 1, \ldots, x_n)$ (i.e., ith experiment, if $i > n_1$, the survival time is at least y_i). Let the new complete censored data y_i be such that,

$$y_i = \begin{cases} x_i, & i \le n_1 \\ y_i, & i > n_1. \end{cases}$$

Let the mean survival time be θ and the probability density of y be

$$f(y|\theta) = \theta^{-1}\exp(-y/\theta),\ y > 0$$

and let the survival function be defined as the probability that an individual survives beyond time y, that is, $S(y) = P(Y > y)$. Thus,

$$S(y) = \exp(-y/\theta),\ y > 0.$$

(a) Obtain the MLE, $\hat{\theta}_{ML}$.
(b) Obtain an EM algorithm.

13.4.5 Let $\mathbf{x} = (x_1, \ldots, x_{n_1})$ be observed data and the censored observations be $\mathbf{y} = (y_1, \ldots, y_{n_2})$ (i.e., in the ith experiment, if $i > n_1$, the survival time is at least y_i). Let the mean survival time be 9 and the probability density be given by

$$f(x|\theta) = \frac{1}{\sqrt{2\pi}}\exp\left(-\frac{1}{2}(x - \theta)^2\right).$$

(a) Obtain the MLE, $\hat{\theta}_{ML}$.
(b) Obtain an EM algorithm.

13.5 INTRODUCTION TO MARKOV CHAIN MONTE CARLO

In this section, we give a brief introduction to Markov chain Monte Carlo (MCMC) methods. Among the computational simulation methods, MCMC is enormously useful for realistic statistical modeling. MCMC methods were initially developed and used in physics. These methods have had a profound influence on statistics over the past two decades, especially in Bayesian inference. MCMC is a computer-driven sampling method that allows us to characterize a distribution without the knowledge of the distribution's mathematical properties. MCMC methods are used to solve problems in many diverse areas such as archeology, biology, biophysics, computational chemistry, computer graphics, finance, nuclear medicine, transport theory, and zoology. These methods have enabled researchers to exploit a degree of complexity and realism in modeling and analysis of problems in these areas that were previously beyond reach. The name *Monte Carlo method* was coined by Stan Ulam and John von Neumann, who introduced this method to solve neutron shielding and other related problems at Los Alamos in the early 1940s. MCMC originated with the now classic paper of Metropolis et al. in 1953, where it was used to simulate the distribution of states for a system of idealized molecules.

The popular MCMC procedures make use of two standard algorithms: the Metropolis algorithm and the Gibbs sampler. In the Metropolis approach, all the parameters are varied at once. In the Gibbs method, each variable of the target pdf is changed one at a time. An improvement on Metropolis, called the Metropolis–Hastings algorithm, was introduced by Hastings in 1970. There are other efficient hybrid methods, such as the Hamiltonian Monte Carlo (HMC) method that alternates between Gibbs and Metropolis procedures. In our present study, we will explain only the first three methods, namely, the Metropolis algorithm, the Metropolis–Hastings algorithm, and the Gibbs sampler.

The objective of MCMC techniques is to generate random variables having certain distributions called target distributions with pdf $\pi(x)$. The simulation of standard distributions is readily available in many statistical software packages. In cases where the functional form of $\pi(x)$ is not known, MCMC techniques become very useful. The basic idea of MCMC methods is to find a Markov chain with a stationary distribution that is the same as the desired probability distribution $\pi(x)$; this is the target distribution. Run the Markov chain for a long time (say, K iterations), and observe in which state the chain is after these K iterations. The probability that the chain is in state x will be approximately the same as the probability that the discrete random variable equals x.

In Bayesian analysis, whether we are finding a posterior distribution or a Bayesian estimate (usually, the posterior mean), integration is involved. We know from calculus that obtaining closed-form solutions for integrations becomes almost impossible (too difficult) for all but some simple functions. A standard approach to numerical integration of a function $f(x)$ is to first divide the range of integration R into n segments $x_1, ..., x_n$, calculate the value of $f(x)$ at each of these points $f(x_1), ..., f(x_n)$, multiply the values by the length of each segment, and sum these rectangles to approximate the integral, which is the area under the curve. The error in this approximation is reduced by increasing the number of segments n.

In *Monte Carlo integration*, instead of taking $x_1, ..., x_n$ as fixed deterministic numbers, we proceed to draw a random sample from a uniform distribution over the range of integration R, then evaluate $f(x_i)$ for each x_i, and take the average. This assumes that the range R is bounded. If R is not bounded, then $f(x)$ can be integrated when it can be written as the product of another function $h(x)$ and a distribution function $\pi(x)$ from which we can draw values of x (i.e., $x_1, ..., x_n$ is drawn from the distribution $\pi(x)$). That is,

$$\int f(x)dx = \int h(x)\pi(x)dx$$

where integration is over the range R. Then, the integral can be approximated with averaging the $f(x_i)$ values, that is,

$$\int f(x)dx \approx \frac{1}{n}\sum_{i=1}^{n} h(x_i),$$

where we assume that x_i values are a random sample from $\pi(x)$ and in the range R. When $\pi(x)$ is a standard distribution, many statistical software packages can generate random samples from this distribution. In those cases, a general coding to evaluate this integral can be written as

```
sum ≐ 0
For i =1 to n
{Draw x_i from π(x)
sum ≐ sum + h(x_i)}
return sum/n
```

In the preceding coding, by multiplying $h(x_i)$ by the indicator function of R (i.e., $I_R(x_i) = 1$, if $x_i \in R$, and zero otherwise), we can avoid the assumption that x_i values are in the range R. For instance, let $X_1, ..., X_n$ be a random sample generated from a target pdf, $\pi(x)$. Then, the expectation of any function $f(X)$ can be estimated using the Monte Carlo method by

$$E_\pi f(X) = \int f(x)\pi(x)dx \approx \frac{1}{n}\sum_{i=1}^{n} f(x_i) = \overline{f}$$

where E_π denotes the expectation with respect to the pdf $\pi(x)$. By the law of large numbers, it follows that

$$\frac{1}{n}\sum_{i=1}^{n} f(X_i) \rightarrow E_\pi\left[f(X)\right] \text{ as } n \rightarrow \infty,$$

provided $X_1, \ldots, X_n$ are independent. We can verify that $\overline{f}$ is an unbiased estimate of $E_\pi f$. In addition, the sampling distribution of $\overline{f}$ is approximately normal, with variance σ^2/n, where σ^2 is estimated by

$$s^2 = \frac{1}{n}\sum_{i=1}^{n} (f(x_i) - \overline{f})^2.$$

For example, in a Bayesian setting, an estimate of the posterior mean can be obtained by taking $f(x) = x$, and the variance can be obtained by taking $f(x) = (x - \overline{x})^2$, if $\pi(x)$ is the posterior distribution (recall that in Chapter 10, we used the notation $\pi(\theta|x)$ for the posterior distribution). Using the sampling distribution of $\overline{f}$, we can also construct point and interval estimates for $E_\pi f$. In the following example, we will show an example of finding posterior mean analytically and give R-code for Monte Carlo integration.

EXAMPLE 13.5.1

Suppose we flip 20 times a biased coin with probability of heads p, and we got $x = 12$ heads. Assume that prior of $p \sim Beta(2,2)$.

(a) Obtain analytic posterior mean.
(b) Write and run R-code for Monte Carlo integration to approximate the posterior mean.

Solution

(a) *We have $n = 20$, with $x = 12$. Thus, $Y \sim Binomial(n, p)$. Using the results of Chapter 10, we get the posterior as $f(p|x) \sim Beta(x + 2, n - x + 2) = Beta(14,10)$.*
Thus, the analytic posterior mean is,
$E(p|x) = \frac{14}{14+10} = 0.5833.$

(b) *We run the following R-code:*

```
# Set seed for reproducibility
set.seed(134)
# Data: 20 flips, 12 heads
n <- 20
y <- 12
# Prior: Beta(2,2)
alpha_prior <- 2
beta_prior <- 2
# Posterior parameters
alpha_post <- alpha_prior + y
beta_post <- beta_prior + (n - y)
# Exact posterior mean
exact_mean <- alpha_post / (alpha_post + beta_post)
# Monte Carlo integration
M <- 10000
theta_samples <- rbeta(M, alpha_post, beta_post)
mc_mean <- mean(theta_samples)
# Results
cat("Exact posterior mean =", round(exact_mean, 4), "\n")
cat("Monte Carlo estimate =", round(mc_mean, 4), "\n")
```

We got Monte Carlo estimate = 0.5836. We can see the Monte Carlo estimate is pretty close to the analytic estimate. It should be noted that with a random seed, we may get slightly different estimate.

Observe that the heart of the Monte Carlo method is to obtain random samples from the target distribution $\pi(x)$. One of the problems encountered using this approach is that, while it is easy to generate samples from standard distributions using popular statistical software packages, it is very difficult (sometimes not feasible) to do so from any distribution that is not standard (see Project 4A for a method of generating random samples from a given distribution). For these reasons, the ordinary Monte Carlo method can be implemented in only a very few cases for Bayesian inference. That is where the MCMC method plays a crucial role. MCMC methods allow the data analyst to build and analyze more realistic statistical models that may be more complex than standard formulations.

Using the MCMC methods, we will construct a Markov chain $\{X_n\}$ with a limiting distribution as the target distribution, $\pi(x)$. Let us first introduce the concept of Markov chains. For a brief description of Markov chains, refer to Appendix A2. We call a sequence of random variables $\{X_n\}$ a *Markov chain* (MC) with state space S if

$$P\left(X_n = x_n \middle| X_{n-1} = x_{n-1}, \ldots, X_1 = x_1\right) = P\left(X_n = x_n \middle| X_{n-1} = x_{n-1}\right).$$

That is, the probability distribution of future states of an MC depends only on the present state and not on the past states. However, it is important to note that an MC $\{X_n\}$ is a dependent sequence of random variables; thus, the independence assumption inherent in a random sample cannot be used. The *transition probability function* of a discrete parameter MC is defined as

$$p_{m,n}(x, y) = P\left(X_n = y \middle| X_m = x\right),\ x,\ y \text{ in } S.$$

We simply denote this transition probability by $p(x, y)$. When the number of elements in the state space S is finite, we can form a matrix P with the (x, y)th element being $p(x, y)$. This matrix is called a one-step *transition probability matrix.* $\pi(x)$ is called an *invariant (limiting) distribution* if it satisfies the equation

$$\pi(x) = \sum_{y \in S} \pi(y) p(y, x).$$

We say that the chain satisfies the *reversibility or detailed balanced condition* if $\pi(x)p(x, y) = \pi(y)p(y, x)$ holds for some $\pi(.)$. It can be shown that such a $\pi(x)$ that satisfies the reversibility condition is invariant. Basically, if an MC is reversible and its limiting distribution exists, then the limiting distribution is the invariant distribution.

The results explained for discrete MCs can be extended to continuous time defined in a continuous state space. The stationary or the equilibrium distribution $\pi(x)$ of a continuous MC satisfies

$$\pi(x) = \int p(y, x) \pi(y) dy.$$

Assume that the samples are generated from an MC whose equilibrium distribution is the target distribution, $\pi(x)$. We know by the law of large numbers that

$$\frac{1}{n} \sum_{i=1}^{n} f(X_i) \to E_\pi\left[f(X)\right] \text{ as } n \to \infty$$

provided $X_1,\ldots,X_n$ are independent. It turns out that, if we generate an MC $X_1,\ldots,X_n$ from the target distribution $\pi(x)$, the result

$$\frac{1}{n} \sum_{i=1}^{n} f(X_i) \to E_\pi\left[f(X)\right] \text{ as } n \to \infty$$

still holds. In this sense, the chain $\{X_i\}$ resulting from an MCMC algorithm with stationary distribution π is similar to the use of a random sample from π. The analytical details are beyond the scope of this book. Instead, we focus on the question: How do we construct a Markov chain whose stationary distribution is our target distribution, $\pi(x)$? The answer is given by the Metropolis–Hastings algorithm and the two special cases: the Metropolis algorithm and the Gibbs sampler. An MCMC method for simulating a distribution π can be defined as any method that produces an ergodic (thus, forgets the initial starting point x_0). MC $\{X_i\}$ whose stationary distribution is π. We start with the Metropolis algorithm. Subsequently, we will explain both the Metropolis–Hastings algorithm and the Gibbs sampler. MCMC methods are increasingly being used for the simulation of complex probability models, for the computation of integrals, and optimization.

13.5.1 Metropolis Algorithm

One of the simplest algorithms in MCMC calculations is the Metropolis algorithm, introduced by the Greek-American mathematician Nicholas Constantine Metropolis and his colleagues in 1953. This work was mentioned in *Computing in Science and Engineering* as being among the top 10 algorithms having the "greatest influence on the development and practice of science and engineering in the 20th century." In this case, we make a trial perturbation from the current position in a parameter space by randomly selecting a trial step from a symmetric probability distribution called *candidate-generating density* or *proposal density* $q(x, y)$ (in the discrete case, it is a symmetric matrix called the *nominating matrix* $A = (a_{ij})$, with $a_{ij} = a_{ji}$, where $i, j \in S$, the state space of the MC). The $q(x, y)$ depends only on the current state x and the new proposed state y (i.e., $q(x, y) = q_x(y)$ is a function of the next proposed state y that is allowed to depend on the current state x). Thus, starting at x, $q(x, y)$ can be regarded as the conditional density of landing at y in one transition step. The trial step is either accepted or rejected on the basis of the probability of the new position relative to the previous one. The Metropolis algorithm is formulated as an instance of the rejection method used for generating steps in an MC. The idea of the rejection algorithm is that if we want to sample from a specific distribution, simply sample from any distribution that is convenient, but keep only the good samples.

We now give the Metropolis algorithm for a discrete distribution. We want to obtain a sample from a distribution $\{\pi_j\}$, where $\pi(j) = P(X_{k+1} = j)$, and we have a symmetric nominating matrix A; then we can write the Metropolis algorithm in five steps as follows.

Metropolis algorithm (discrete case)

For $k = 0$, start with an arbitrary point, $x_k = i$.

1. Generate j from the probability distribution $\{a_{ij}, j = 1, 2, \ldots\}$.
2. Set

$$r = \frac{\pi(j)}{\pi(i)}.$$

3. If $r \geq 1$, set $x_{k+1} = j$ (acceptance); otherwise generate u from Uniform (0, 1), if $u < r$, set $x_{k+1} = j$ (acceptance), else $x_{k+1} = x_k$ (rejection); (note that the value of x_{k+1} becomes the next state).
4. Set $k = k + 1$, and go to step 1.

Each of the accepted points is considered to be a sample value from the target distribution $\{\pi_j\}$.

The continuous case of the Metropolis algorithm is given next.

Metropolis algorithm (continuous case)

1. Start with an arbitrary point, x_0.
2. Select a new position $x^* = x_k + \Delta x$, where Δx is randomly chosen from a symmetric distribution.
3. Calculate the ratio

$$r = \frac{\pi(x^*)}{\pi(x_k)},$$

where $\pi(x)$ is the target distribution.

4. Accept the trial position, that is, set

$$x_{k+1} = x^*, \text{ if } r \geq 1.$$

Otherwise, generate u from Uniform (0, 1)
If $u < r$, set $x_{k+1} = x^*$
else set $x_{k+1} = x_k$.

5. Set $k = k + 1$, and go to step 2.

If the proposal step size is dx, we could use the proposal distribution as $U(-dx, dx)$; for example. if the step size is 1, then randomly choose $\Delta x \sim U(-1, 1)$. For further discussion on the selection of the proposal distribution, read Subsection 13.5.4. The Metropolis algorithm generates a set of states that is an MC because each state x_{k+1} depends only on the previous state x_k. Using MC techniques, it can be shown that the equilibrium distribution of the chain constructed by the Metropolis algorithm is indeed $\pi(x^*)$. Note that in the Metropolis algorithm, it is not necessary to have the pdf; instead, all that is necessary is to know the ratio $\pi(x^*)/\pi(x_k)$. Thus, none of the multiplicative constants in the pdf π plays a role in the algorithm.

This algorithm works well in most applications. The following is a simple example to show how the Metropolis algorithm works.

EXAMPLE 13.5.2

Using the Metropolis algorithm, generate a random sample from a Poisson distribution with mean λ. For the nominating matrix, use the symmetric matrix with elements

$$a_{00} = 1/2,\ a_{ij} = \begin{cases} 1/2, & j = i - 1 \\ 1/2, & j = i + 1 \\ 0, & \text{otherwise.} \end{cases}$$

Solution

The nominating probability matrix is an one-step transition matrix (see Appendix A2),

$$A = \begin{bmatrix} 1/2 & 1/2 & 0 & 0 & 0 & \cdots \\ 1/2 & 0 & 1/2 & 0 & 0 & \cdots \\ 0 & 1/2 & 0 & 1/2 & 0 & \cdots \\ 0 & 0 & 1/2 & 0 & 1/2 & \cdots \\ \vdots & \vdots & \vdots & \vdots & \vdots & \vdots \end{bmatrix}.$$

Now we apply the Metropolis algorithm for generating samples from Poisson (λ) in the following steps.

Step 1. *Start with $x_{n-1} = i$.*

Step 2. *Generate j from $A = \{a_{ij}\}$. How do we do it? We can do this using the following procedure:*

For $i \neq 0$,

Generate u_1 from $U(0, 1)$.

If $u_1 \geq \frac{1}{2}$, set $j = i + 1$ else set $j = i - 1$.

For $i = 0$,

If $u_1 < \frac{1}{2}$, set $j = 0$

else set $j = 1$.

Step 3. *Set*

$$r = \frac{\pi(j)}{\pi(i)} = \frac{e^{-\lambda}\lambda^j/j!}{e^{-\lambda}\lambda^i/i!} = \frac{i!\lambda^j}{j!\lambda^i} = \frac{i!\lambda^{j-i}}{j!}.$$

Set

$$r = \begin{cases} 1, & \text{if } i = 0, j = 0 \\ \frac{\lambda}{j}, & \text{if } j = i + 1 \\ \frac{i}{\lambda}, & \text{if } j = i - 1. \end{cases}$$

Step 4. *Acceptance/rejection:*
If $r \geq 1$, set $x_n = j$ (i.e., accept the new state j).
Otherwise, generate u_2 from $U(0, 1)$
if $u_2 < r$, set $x_n = j$ (i.e., accept the new state j)
else set $x_n = x_{n-1}$ (i.e., reject the new state j and keep the current state i).
Step 5. *Set $n = n + 1$, and go to step 2.*

In Example 13.5.2, let us say we want to generate a random sample from Poisson with $\lambda = 2$, and we are at state $i = 3$ in the iteration step $(n - 1)$. If our proposed new state is $j = 4$, then $r = 2/4 = 1/2$. Suppose we obtained the value of u_2 as 0.672772. Because this value is larger than 1/2, we reject the proposed new state 4 and stay at state 3 for the iteration step n (if you generate a new u_2, your decision might be different). Instead, suppose our proposed step was $j = 2$; then $r = i/\lambda = 3/2 > 1$, and we will immediately accept our new state as $j = 2$ (no need to generate a uniform random number; if you did, it would have been smaller than 3/2 anyway) for the iteration step n.

EXAMPLE 13.5.2

Let $\pi(x) = c \exp(-f(x))$ be the form of the target distribution function. Write a general Metropolis algorithm to generate a sample from π.

Solution

Let $q(x, y)$ be any symmetric distribution. Starting from an arbitrary $x_{(0)}$, we can write the Metropolis algorithm through the following steps.

Step 1. *Let $x_{(t)}$ be the current state.*
Step 2. *Generate y from the distribution $q(x, y)$.*
Because,

$$r = \frac{\pi(y)}{\pi(x_{(t)})} = \frac{c\exp(-f(y))}{c\exp(-f(x_{(t)}))} = \exp(-f(y) - f(x_{(t)})),$$

calculate the change in f, $\Delta f = f(y) - f(x_{(t)})$.

Step 3. *Generate a random number from the uniform distribution, $U(0, 1)$. If $u \leq \exp(-\Delta f)$, set $x_{(t+1)} = y$ (accept the proposed new state); otherwise set $x_{(t+1)} = x_{(t)}$ (reject the proposed new state).*
Step 4. *Continue (i.e., go to step 1).*

Note that in the previous example, the normalizing constant in $\pi(x)$ is not important, because it cancels in the ratio. In fact, this is true in all Metropolis and Metropolis–Hastings algorithms. In the special case, where $q(x, y) = q(|y - x|)$, the Metropolis algorithm is also called the *random-walk Metropolis*. Another special choice is $q(x, y) = q(y)$; this is called the *independence sampler*. In all of these cases, it is important to observe that whereas the target distribution is independent of the positions, the proposal functions depend on where we are. For example, let $\pi(x)$ be the standard normal density, and let the proposal density be of the form

$$q(x, y) \propto \exp\left(-\frac{(y - x)^2}{2(.25)^2}\right).$$

The Fig. 13.1 gives a representation of the target distribution and some representative proposals. For each point x of the target distribution, we generate a y from the corresponding proposal distribution. Then, according to the accept/reject rule that we specified earlier, we will make a decision whether to treat this new value y as being from the target distribution.

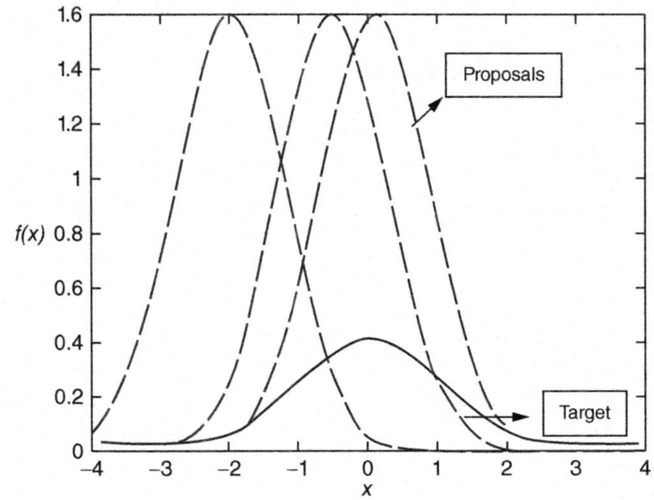

FIGURE 13.1 Target and proposal densities.

13.5.2 The Metropolis–Hastings Algorithm

The Metropolis–Hastings (M-H) algorithm is a generalization of the Metropolis algorithm, in which we need not assume symmetry of the nominating matrix A (or for proposal density $q(x, y)$). The acceptance probability is given by

$$\alpha(i, j) = \min\left\{\frac{\pi(j)a_{ji}}{\pi(i)a_{ij}}, 1\right\}.$$

This algorithm is the basic building block of MCMC methods. The Metropolis–Hastings algorithm is widely used in applied statistics and is very useful for sampling from complicated, high-dimensional probability distributions. Now we present the steps involved in the Metropolis–Hastings algorithm in the discrete case.

Metropolis–Hastings algorithm (discrete case)

For $k = 0$, start with an arbitrary point, $x_k = i$.

1. Generate j from the nominating distribution $\{a_{ij}, j = 1, 2, \ldots\}$.
2. Set

$$r = \frac{\pi(j)a_{ji}}{\pi(i)a_{ij}}.$$

3. If $r \geq 1$, set $x_{k+1} = j$.
   Otherwise generate u from $U(0, 1)$
   if $u < r$, set $x_{k+1} = j$
   else set $x_{k+1} = x_k$.
4. Set $k = k + 1$, and go to step 1.

In the preceding algorithm, if we calculate $\alpha(i, j) = \min\{r, 1\}$, basically, we accept the proposed new step j if $u < \alpha(i, j)$; otherwise, we stay at the current step i. The resulting MC from both Metropolis and Metropolis–Hastings algorithms would have the transition probability matrices defined by

$$p(i, j) = a_{ij}\alpha(i, j) \text{ for } i \neq j$$

and

$$p(i,i) = 1 - \sum_{j \neq i} a_{ij}\alpha(i,j).$$

In the continuous case, for any given $\pi(x)$, the Metropolis–Hastings algorithm takes the following form. To start the algorithm, we choose an arbitrary proposal distribution $q(x, y)$ so that it is easy to obtain a sample from this distribution. Define the acceptance/rejection function as

$$\alpha(x,y) = \min\left\{\frac{\pi(y)q(y,x)}{\pi(x)q(x,y)}, 1\right\}.$$

If both $\pi(x)$ and $\pi(y)$ are zero, set $\alpha(x, y) = 0$.

Metropolis–Hastings algorithm (continuous case)

Step 1. Start with an arbitrary point, x_0.
Step 2. Given a current state $x^{(t)}$, draw y from the proposal distribution $q(x, y)$.
Step 3. Draw u from $U[0, 1]$.
Step 4. If $u < \alpha(x^{(t)}, y)$, set $x^{(t+1)} = y$; otherwise set $x^{(t+1)} = x^{(t)}$.
Step 5. Set $t = t + 1$, and go to step 2.

Note that if the $q(x, y)$ is symmetric (i.e., $q(x, y) = q(y, x)$), then the Metropolis–Hastings algorithm reduces to the Metropolis algorithm. In practice, there are other forms of acceptance/rejection functions suggested. Observe that in the Metropolis–Hastings algorithm, as in the Metropolis algorithm, it is not necessary to have the pdf; instead, all that is necessary is to know the ratio $\pi(y)/\pi(x)$. Thus, none of the multiplicative constants in the pdf, $\pi(x)$, plays a role in the algorithm.

Because of the versatility of this method, there are many generalizations of the Metropolis–Hastings algorithm in the literature. It is also necessary to impose some conditions both on $\pi(x)$ and on the proposal distribution $q(x, y)$ for $\pi(x)$ to be the limiting distribution of the MC $\{X^{(t)}\}$ produced by the M-H algorithm. We do not want a large ratio of the proposed new values to be rejected. Discussion of these issues is beyond the scope of this book.

EXAMPLE 13.5.3

Using the Metropolis–Hastings algorithm, generate a sample from the following distribution. Let $\Omega = \{2, 3, \ldots, 11, 12\}$, which represents the sum of the up faces of two balanced dice, and let the distribution be given by

Sum i	2	3	4	5	6	7	8	9	10	11	12
$\pi(i)$	1/36	2/36	3/36	4/36	5/36	6/36	5/36	4/36	3/36	2/36	1/36

Using the nominating matrix

$$a_{22} = a_{(12)(12)} = 1/2,\ a_{ij} = \begin{cases} 1/2, & j = i-1 \\ 1/2, & j = i+1,\ i, j \in \Omega \\ 0, & \text{otherwise,} \end{cases}$$

write the M-H algorithm to generate samples from the distribution $\pi(x)$.

Solution

Suppose we start with state $i \in \Omega$, say at 5 (starting at any other state is ok).

Step 1. *Generate j from the nominating distribution $\{a_{ij}, j = 1, 2, \ldots\}$.* Thus, $j = i - 1$ or $i + 1$, *and in this case, j has to be 4 or 6. We can follow the same procedure as in Example 13.5.1 to choose between $i - 1$ and $i + 1$. Let us say, we got $j = i + 1$, here 6.*

Step 2. *Set* $r = \frac{\pi(j)a_{ji}}{\pi(j)a_{ij}}$. *In this case,* $r = \frac{\pi(6)}{\pi(5)} = \frac{5/36}{4/36} = \frac{5}{4}$. *(If we had chosen 4, then,* $r = \frac{\pi(4)}{\pi(5)} = \frac{3}{4}$. *)*

Step 3. *If* $r \geq 1$ *set* $x_n = j$. *Here* $r > 1$*; hence, we accept the new state,* $x_n = 6$. *Otherwise, generate u from* $U(0, 1)$ *if* $u < r$, *set* $x_n = j$ *else set* $x_n = x_{n-1}$.

Step 4. *Set* $n = n + 1$, *and go to step 1.*

EXAMPLE 13.5.4

Write a Metropolis–Hastings algorithm to generate samples from $N(0, 1)$ based on the proposal $U[-1, 1]$.

Solution

Note that in order for y to be generated based on $U[-1, 1]$, *we need* $y - x^{(t)} \sim U[-1, 1]$. *Thus,* $y \sim U\,[x^{(t)}-1, x^{(t)}+1]$. Fig. 13.2 *shows the target distribution as the standard normal and the representative proposals that are uniform at points* $x^{(t)} = -2$ *and* 2.

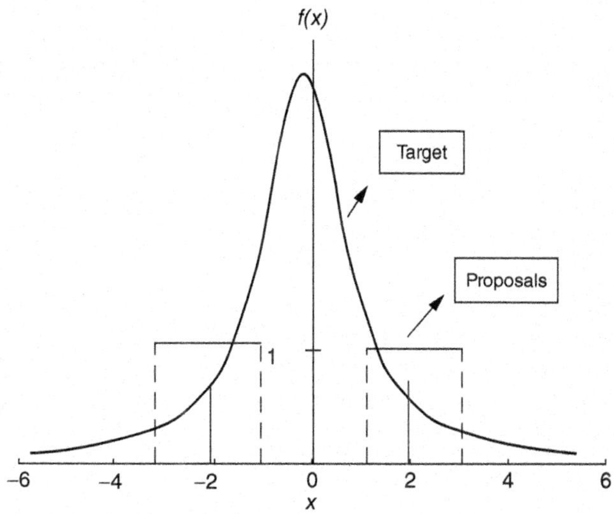

FIGURE 13.2 Normal target and uniform proposal distributions.

Now, the M-H algorithm can be obtained in the following way.
Set

$$\alpha\left(x^{(t)}, y\right) = \min\left\{\frac{\pi(y)q(y,x)}{\pi(x)q(x,y)}, 1\right\}$$

$$= \min\left\{\left(\exp\left\{x^{(t)2} - y^2/2\right\}\right)\frac{x+1}{y+1}, 1\right\}$$

Generate $u \sim U[0, 1]$
If $u < \alpha\,(x^{(t)}, y)$, *set* $x^{(t+1)} = y$; *otherwise set* $x^{(t+1)} = x^{(t)}$, *Continue.*

Observe that in order to generate normal random variables, it is not necessary to use M-H algorithms. Most of the statistical software packages will give us a random sample from the normal distribution. Example 6.5.2 (originally suggested by Hastings in 1970) is given for demonstration of the M-H algorithm. The algorithm is effective in general cases, for instance, to generate a sample from a gamma distribution. In *Gamma*(α, β), if α is an integer, we can use the method of Project 4A to generate a random sample. However, if α is not an integer, we could use *Gamma*([α], β) (here [α] denotes the integer part of α) as the proposal distribution and follow the steps of the M-H algorithm to generate a sample from *Gamma*(α, β) (see Exercise 13.5.3).

13.5.3 Gibbs Algorithm

The name ***Gibbs algorithm*** (or ***Gibbs sampler***) was coined by the brothers Stuart Geman and Donald Geman in 1984 and refers to Gibbs distributions in statistical physics. This is very useful in obtaining a sequence of observations from a specified multivariate probability distribution, when direct sampling is hard or the joint distribution is not known explicitly. The Gibbs sampler can be used in those situations when the conditional distribution of each variable is known and is relatively easier to sample from. In the Gibbs sampler, only one parameter is varied at a time, while all others are held fixed. The parameter then is randomly drawn from a conditional probability density function, the probability distribution of one parameter, given all other parameters, $\pi(x_i|x_{-i})$, where x_{-i} is the full set of parameters excluding only the single component x_i. Let $x = (x_1,\ldots,x_k)$ be $k(\geq 2)$-dimensional. Recall from Chapter 3 that these conditional densities can be obtained as follows:

$$\pi(x_i|x_{-i}) = \pi(x_i|x_1,\ldots,x_{i-1}, x_{i+1},\ldots,x_k)$$
$$= \frac{\pi(x_1,\ldots,x_{i-1}, x_i, x_{i+1},\ldots,x_k)}{\int \pi(x_1,\ldots,x_{i-1}, x_i, x_{i+1},\ldots,x_k)dx_i}.$$

The basic assumption under which the Gibbs algorithm works is that we could easily draw a random sample from these conditional pdfs. Thus, the Gibbs algorithm is a particular case of Metropolis–Hastings algorithms. For example, at the ith step, y_i is generated from the nominating density $q_i(x_i, y_i)$ where q_i depends on the current state x_i. The candidate y_i is accepted with probability

$$\alpha_i(x_i, y_i) = \min\left\{\frac{\pi_i(y_i)q_i(y_i, x_i)}{\pi_i(x_i)q_i(x_i, y_i)}, 1\right\}.$$

If y_i is accepted, we will set the ith component of x_n, $x_{n,i} = y_i$; otherwise set $x_{n,i} = x_{n\,i}$. The remaining components of $\mathbf{x}_n$ are not changed in step i. This is repeated for each i, at the end of which the entire vector $\boldsymbol{x}_n$ would have been updated. Thus, if we are in state $\mathbf{x}$ at time t, at time $t+1$ we either remain at $\mathbf{x}$ or go to $\mathbf{y}$ by modifying only one component of $\mathbf{x}$. It is important to use the most recent values of updated components to update the next component. That is, given $\mathbf{x}^{(t)} = \left(x_1^{(t)},\ldots,x_k^{(t)}\right)$ at time t, generate

$$x_1^{(t+1)}: \pi\left(x_1\middle|x_2^{(t)}, x_3^{(t)},\ldots,x_k^{(t)}\right)$$
$$x_2^{(t+1)}: \pi\left(x_2\middle|x_1^{(t+1)}, x_3^{(t)},\ldots,x_k^{(t)}\right)$$
$$x_3^{(t+1)}: \pi\left(x_3\middle|x_1^{(t+1)}, x_2^{(t+1)}, x_4^{(t)},\ldots,x_k^{(t)}\right)$$
$$\vdots$$
$$x_k^{(t+1)}: \pi\left(x_k\middle|x_1^{(t+1)}, x_2^{(t+1)},\ldots,x_{k-1}^{(t+1)}\right).$$

For instance, let $k = 2$. The Gibbs sampler updates in the following manner. Start at $\mathbf{x}^{(0)} = \left(x_1^{(0)}, x_2^{(0)}\right)$; first update $x_1^{(0)}$ to $x_1^{(1)}$, using this updated value $x_1^{(1)}$ and $x_2^{(0)}$, update $x_2^{(0)}$ to $x_2^{(1)}$, resulting in the updated vector $\boldsymbol{x}^{(\mathbf{1})}$. Repeat this procedure to obtain $\mathbf{x}^{(\mathbf{2})}$, $\mathbf{x}^{(\mathbf{3})}$, ... Fig. 13.3 depicts this updating procedure.

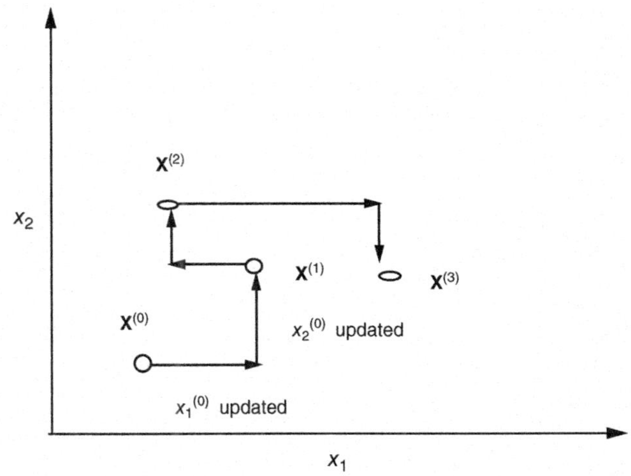

FIGURE 13.3 Gibbs updating procedure. Graph shows iterative updates of x_1 and x_2 coordinates showing stepwise movement between points.

The conditional densities $f_1,...,f_k$ are called the *full conditionals*. In the Gibbs sampler, only these conditional densities are needed for simulation. Thus, this procedure becomes very efficient when the vector **x** is large, because all of the simulations can be done as univariate.

The following example of bivariate density is popularly used in literature to illustrate the Gibbs sampler. It is the case where the joint density is complex, because one variable (x) is discrete, while the other variable (y) is continuous. However, the conditional densities are simple known distributions, binomial and beta distributions, respectively. It is then easier to simulate these distributions, thus demonstrating the power of the Gibbs sampler.

EXAMPLE 13.5.5

(a) Write a Gibbs sampler for generating samples from the following bivariate density:

$$f(x, y) = \binom{n}{x} y^{x+\alpha-1}(1-y)^{n-x+\beta-1}, \text{ for } x = 0,1,\ldots,n \text{ and } 0 \le y \le 1.$$

(b) Starting with $y_0 = 1/4$, $n = 15$, and $\alpha = 1$, $\beta = 2$, obtain the first three realizations of the Gibbs sequence.

Solution

(a) *From Exercise 3.3.14, we know that*

$$f(x|y) \propto \binom{n}{x} y^x (1-y)^{n-x}.$$

That is, the conditional distribution of x (treating y as a constant) is binomial with parameters n and y, $0 \le y \le 1$. *Also,*

$$f(x|y) \propto y^{x+\alpha-1}(1-y)^{n-x+\beta-1}.$$

Thus, the conditional distribution of y given x is a beta distribution with parameters $x + \alpha$ *and* $n - x + \beta$. *The Gibbs sampler for generating bivariate samples from f (x, y) is then given as follows: For i = 1, ..., n, repeat:*

1. *Generate* y_i *from* $f_{Y|X}(.|x^{(i-1)})$, *that is, from* **Beta** $(x_{i-1} + \alpha, n - x_{i-1} + \beta)$.
2. *Generate* x_i *from* $f_{X|Y}(.|y^{(i)})$, *that is, from* ***binomial (n, y_i)***.
3. *Return* (x_i, y_i).

(b) *We proceed with the following steps.*

(i) *For $y_0 = 1/4$, x_0 is obtained from generating a random variable from binomial with $n = 15$, $y_0 = 1/4$, that is, from $B(15, 1/4)$, resulting in a value of 4 (generated using Minitab; you may get a different value when you do it). Thus, $x_0 = 4$.*

(ii) *Generate y_1 randomly from*

$$\begin{aligned} Beta(x_0 + \alpha, n - x_0 + \beta) &= Beta(4 + 1, 15 - 4 + 2) \\ &= Beta(5, 13), \end{aligned}$$

resulting in $y_1 = 0.53$ (approximated to the second decimal place). Now $x_1 \sim B(15, 0.53)$, resulting in $x_1 = 6$.

(iii) *Generate y_2 randomly from*

$$Beta(x_1 + \alpha, n - x_1 + \beta) = Beta(7, 11)$$

resulting in $y_2 = 0.30$. Now $x_2 \sim B(15, 0.30)$, resulting in $x_2 = 3$.

Thus, a particular realization of the Gibbs sampler for the first three iterations is (4, 0.25), (6, 0.53), and (3, 0.30).

From Exercise 13.5.8, it can be observed that at the beginning, the values of the chain are highly dependent on the choice of the initial value y_0. In practice, it is necessary to run a sufficient number of iterations to remove the effect of the starting values. Even though the Gibbs sampler is a special case of the Metropolis–Hastings algorithm, it is important to observe that, unlike the M-H algorithm, every sample generated by the Gibbs algorithm is accepted. Also, we should have at least a two-dimensional problem for the Gibbs sampler to be used. Since Gibbs sampling (like other MCMC sampling) generates a Markov chain of samples, each sample is correlated with neighboring samples, and to obtain a random sample, one needs to perform *thinning* the resulting chain by only taking every k^{th} value (like taking every 50th value). There are some pros and cons in the practice of thinning; for that and some nice applications of the Gibbs method, we refer the reader to specialized books.

From the previous discussions, we can see that a general description of an MCMC method can be summarized in the following algorithm.

```
Initialize X0
  For i = 1, ..., N repeat
  x = X(i-1);
  Generate Y from a nominating density, q(x; y);
  Calculate the acceptance rate, α(x; y);
  Generate U from the uniform U(0; 1);
  If (U < α(x; y)) set X(i) = y,
  Else set X(i) = x;
  End;
```

If we choose a nominating density $q(x, y)$ and an acceptance rate $\alpha(x, y)$ such that the reversibility condition

$$\pi(x)\alpha(x, y)q(x, y) = \pi(y)\alpha(y, x)q(y, x),$$

is satisfied, then the foregoing procedure generates an MC with limiting distribution $\pi(x)$. In order to use Gibbs sampling for Bayesian analysis, we must have an explicit analytical posterior conditional distribution.

13.5.4 MCMC Issues

Two major issues in MCMC are convergence and burn-in. Because in all three MCMC algorithms we start the sequence from an arbitrary point, any particular sequence may take some time to pass through the transient stage, and the effect of the starting value is very small and can be ignored, that is, it attains convergence. In practice, we will have to run the algorithm for a few thousand iterations so that the effect of this initial state is negligible. The samples obtained during this burn-in period should be discarded for the subsequent analysis as they do not represent the target pdf. By monitoring the sequence itself, we can determine whether the sequence has reached the convergence. A simple way to decide how much burn-in is necessary is to create scatterplots of X_i versus X_j, $i \neq j$. When the wild variations stop, then it is safe to assume that the chain has reached stationarity.

Another major issue in the implementation of MCMC algorithms is the choice of proposal density. In the continuous case, popular choices among others are the multivariate normal density and the multivariate t with specified parameters. Even in these cases, there is the question of the appropriate size of the spread, or scale of the proposal density. The size of the acceptance ratio is another issue. If the ratio is too small, the samples will get stuck (because almost all proposed new states will be rejected), and if the ratio is too high, the samples will show tracking. A general rule of thumb is that the acceptance ratio should be within 30%–60%. If not, adjust the step size (for a small ratio, decrease the step size, and for a high ratio, increase the step size). There are many publications devoted to these issues.

For the Bayesian computation, MCMC allows us to sample from any posterior. Because of the availability of specialized software packages, such as Bayesian inference Using Gibbs Sampling (BUGS), it is practical to code up for a particular problem.

There are many books on MCMC methods; some of these are listed in the references at the end of this book. For a good discussion including some technical details, refer to http://mpdc.mae.cornell.edu/Courses/MAE714/Papers/wp9.pdf.

EXERCISES 13.5

13.5.1 For Example 13.5.1, let $\lambda = 3$. Starting with initial state $x_0 = 6$, compute relevant quantities by performing 10 iterations of the algorithm.

13.5.2 Using the Metropolis–Hastings algorithm, generate a random sample from a geometric distribution with mean θ. Use the nominating distribution $\{a_{ij}, j = 1, 2, \ldots\}$, such that,

$$a_{ij} = \begin{cases} \frac{1}{2} & j = i-1,\ i+1, \text{ and } i = 1, 2, 3, \ldots \\ \frac{1}{2} & j = 0,1 \text{ and } i = 0 \\ 0 & \text{otherwise.} \end{cases}$$

[Recall that if X is geometric with parameter θ, then $P(X = x) = (1 - \theta)^x\theta$, for $x = 0, 1, 2, \ldots$.]

13.5.3 Write down the Metropolis–Hastings algorithm to generate a sample from $Gamma(\alpha, \beta)$ using the proposal density as $Gamma([\alpha], [\alpha]/\alpha)$.

13.5.4 Write down the Metropolis–Hastings algorithm for simulating an MC with stationary distribution $\pi = (1/6, 2/3, 1/6)$, using the "proposal" transition matrix

$$Q = \begin{pmatrix} 1/2 & 1/2 & 0 \\ 1/2 & 0 & 1/2 \\ 0 & 1/2 & 1/2 \end{pmatrix}.$$

13.5.5 In tossing three fair coins, let the random variable X be defined as X = number of tails. Then the distribution of X is given by

x	0	1	2	3
$\pi(x)$	1/8	3/8	3/8	1/8

Write down the Metropolis or Metropolis–Hastings algorithm for simulating an MC with a stationary distribution $\pi(x)$. Use any nominating matrix.

13.5.6 Write a Metropolis algorithm to generate samples from a target distribution, $\pi(x) \propto \exp\left(-\frac{x^2}{2}\right)$, based on the proposal density

$$q_x(y) = \exp\left(-\frac{(y-x)^2}{2(0.4)^2}\right).$$

13.5.7 Write a general Metropolis or Metropolis–Hastings algorithm to generate a sample from a target distribution π, where π is an exponential random variable with parameter θ.

13.5.8 Write a general Metropolis or Metropolis–Hastings algorithm to generate a sample from a target distribution π, where $\pi(x)\ \alpha\ x^{34}(1-x)^{38}(2+x)^{125}$. Use the proposal density as $q(x, y) = 1$ on the interval [0, 1].

13.5.9 For the bivariate density given in Example 13.5.5, starting with three different values of y_0, say, 1 /3, 1 /2, and 2/3 and $n = 15$, $\alpha = 1$, and $\beta = 2$, obtain the first three realizations of the Gibbs sequence. Comment on the influence of the initial values.

13.5.10 Consider the problem of sampling bivariate random variables with joint density given by

$$f(x, y) = \begin{cases} ce^{-(x+y+4xy)}, & x \geq 0,\ y \geq 0 \\ 0, & \text{otherwise.} \end{cases}$$

(a) Find $f(x\,|y)$ and $f(y\,|x)$.

(b) Write a Gibbs procedure to generate samples from this distribution. Discuss why it is easier to use the Gibbs sampler for this case.

(c) Starting from an arbitrary point, obtain the first three sample points.

13.5.11 Suppose the target distribution is

$$(X, Y) \sim N\left(\begin{pmatrix} 0 \\ 0 \end{pmatrix}, \begin{pmatrix} 1 & \rho \\ \rho & 1 \end{pmatrix}\right).$$

Then write the Gibbs sampler to generate a sample from this distribution. In particular, say, we start with $(X, Y) = (12, 12)$ and $\rho = 0.7$. What is the Gibbs procedure to generate a sample from a binormal distribution? [The pdf of a bivariate normal distribution with

$$\mathbf{x} = \begin{pmatrix} x \\ y \end{pmatrix} \boldsymbol{\mu} = \begin{pmatrix} \mu_x \\ \mu_y \end{pmatrix} \boldsymbol{\Sigma} = \begin{pmatrix} \sigma_X^2 & \rho\sigma_X\sigma_Y \\ \rho\sigma_X\sigma_Y & \sigma_Y^2 \end{pmatrix} \text{ is given by}$$

$$f(\mathbf{x}) = \frac{1}{2\pi}(\det \boldsymbol{\Sigma})^{-1/2} \exp\left[-\frac{1}{2}(x-\boldsymbol{\mu})'\boldsymbol{\Sigma}^{-1}(x-\boldsymbol{\mu})\right],$$

where $'$ denotes the vector transpose.]

13.5.12 Suppose the target distribution is

$$(X, Y) \sim N\left(\begin{pmatrix} \mu \\ \mu \end{pmatrix}, \begin{pmatrix} 2 & 1 \\ 1 & 1 \end{pmatrix}\right).$$

Then write the Gibbs sampler to generate a sample from this distribution.

13.6 CHAPTER SUMMARY

In this chapter, we introduced some empirical methods that are becoming increasingly popular in modern statistical analysis. The methods presented must be viewed as introductory in nature and by no means most efficient or general. Because of ever-evolving applications and advancements in technology, most of the methods presented here also evolve.

Also, based on the situation, it is necessary to write computer codes to run the algorithms introduced in this chapter. Our hope is that students will explore these topics in more detail by referring to specialized books and publications.

In this chapter, we also learned the following important concepts and procedures:

- The jackknife method
- General bootstrap procedure to estimate the standard error of $\hat{\theta}$
- Bootstrap confidence intervals
- EM algorithm
- MCMC methods
- Metropolis algorithm
- Metropolis–Hastings algorithm
- Gibbs sampler

13.7 COMPUTER EXAMPLES

Most of the procedures described in this chapter could be implemented using Minitab, SAS, or SPSS. There are other specialized programs that will do a good job of implementing the methods discussed in this chapter. BUGS is a free software that has proven to be effective in MCMC computations, and the details are at the website: http://www.mrc-bsu.cam.ac.uk/bugs/. Most of the procedures discussed in this chapter can also be implemented in "R," which is also a free software that can be downloaded from http://www.rproject.org/. Few examples in R are given. However, we will not discuss SAS or SPSS examples. For SAS, there are %JACK and %BOOT macros available to do jackknife and bootstrap computations. A good site with example programs from the SAS institute is http://ftp.sas.com/techsup/download/stat/jackboot.html. Sometimes, PROC IML could also be used to bootstrap. In the case of multivariate normal data, PROC MI with the EM option will perform the EM algorithm in SAS. Refer to http://support.sas.com/documentation/cdl/en/statug/63033/HTML/default/viewer.htm#mcmc_toc.htm for the options available for the MCMC procedure. Example SAS codes could be obtained from a simple search of the web for almost all the procedures explained in this chapter.

13.7.1 Examples Using R

EXAMPLE 13.7.1 Bootstrap

Using the following data, perform a bootstrap point and interval estimate for the median. Generate 6 replications or bootstrap samples of size 12 each.

Sample (x): 269 246 388 354 266 303 295 259 274 249 271 254

R-code:

```
library('boot');        <- Load the boot library

mystatfun = function(data,index) {
  return(median(data[index]));      <- Create a function returning your parameter
                                       to be estimated. Notice this requires an index
}

mybs=boot(x,mystatfun,R=6);

print(mean(mybs$t));  <- mybs$t contains the values generated by your function from each bootstrap replication

print(sd(mybs$t));

boot.ci(mybs,type="basic");
```

```
Output:
269.6667
BOOTSTRAP CONFIDENCE INTERVAL CALCULATIONS
Based on 6 bootstrap replicates
CALL:
boot.ci(boot.out = mybs, type = "basic")
Intervals:
Level Basic
95% (241, 286 )
Calculations and Intervals on Original Scale
Warning: Basic Intervals used Extreme Quantiles
Some basic intervals may be unstable
```

EXAMPLE 13.7.2 Jackknife

Using the data from the previous example, perform a jackknife point estimate for the mean and standard deviation. Notice the jackknife computation is not simulated like the bootstrap and will have one answer.

R-code:

```
tmp=c();
for(i in 1:12) {
tmp=c(tmp,mean(x[-i]));
}
mean(tmp);
sd(tmp);
```

Output:

285.6667 ← Mean

3.988777 ← Standard Deviation

EXAMPLE 13.7.3 MCMC

MCMC is used to simulate random variables from distributions we cannot sample from. In this example, our **target** distribution is the chisq(4), and our **proposal** distribution is normal(i,1). Notice we can use the rchisq() function to do this and obtain a better result; however, we're going to compare the results of rchisq() to MCMC for learning purposes. Another important note, our proposal distribution's mean is the previous value in the MC.

The chain variable maybe treated as a generated random sample from our target distribution. Notice we can evaluate the target distribution, but perhaps we cannot sample or integrate the target.

Your means will be unique for both your rchisq() and chain, but they should be close. Observe the density curves over the histogram (Fig. 13.4).

R-code:

```
i=10;            #Step 1
chain=c();
for(c in 1:100000) {
 j=rnorm(1,i,1); #Step 2
 u=runif(1,0,1); #Step 3
 r=(dchisq(j,df=4)*dnorm(j,i,1))/(dchisq(i,df=4)*dnorm(i,j,1));
 a=min(c(r,1),na.rm=TRUE);
 if(u<a) {
  chain=c(chain,j);
  i=j;
 } else {
  chain=c(chain,i);
 }
}
mean(chain);
mean(rchisq(3000,df=4);
plot(density(chain),col="blue",type="l");
lines(density(rchisq(3000,df=4),col="red");
lines(seq(0,25,by=0.1),dchisq(seq(0,25,by=0.1),df=4),col="black");
```

Step 4

Step 5

Output:

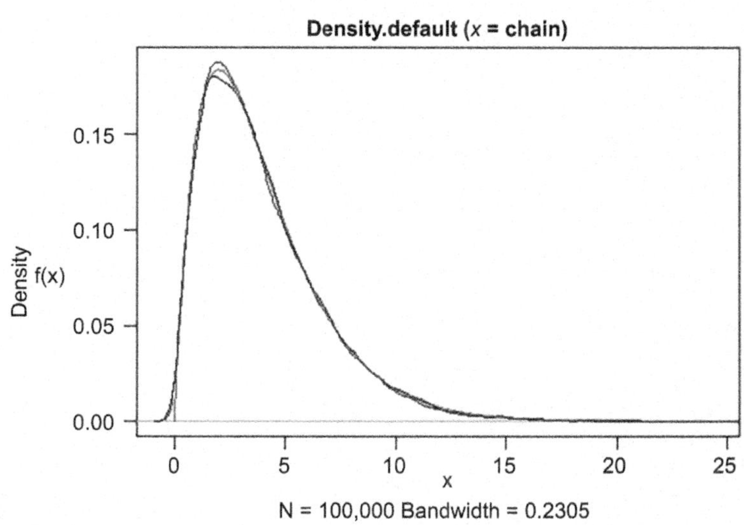

FIGURE 13.4 Simulated densities.

EXAMPLE 13.7.4 (EM algorithm)

For the data of Exercise 13.4.2 (c), give R-code.

Solution

We take arbitrary initial values for the parameters μ and σ.
#Change the values in (*) and (**) and put any arbitrary values as follows:

```
em.norm <- function(Y){
Yobs <- Y[!is.na(Y)]
Ymis <- Y[is.na(Y)]
n <- length(c(Yobs, Ymis))
r <- length(Yobs)
# initial values
mut <- 1 # (*)put arbitrary value for μ
sit <- 0.1 # (**)put arbitrary value for σ
# Define log-likelihood function
ll <- function(y, mu, sigma2, n){
-.5*n*log(2*pi*sigma2)-.5*sum((y-mu)^2)/sigma2
}
# Compute the log-likelihood for the initial values, and ignoring the missing data mechanism
lltm1 <- ll(Yobs, mut, sit, n)
repeat{
# E-step
EY <- sum(Yobs) + (n-r)*mut
EY2 <- sum(Yobs^2) + (n-r)*(mut^2 + sit)
# M-step
mut1 <- EY / n
sit1 <- EY2 / n - mut1^2
# Update parameter values
mut <- mut1
sit <- sit1
# compute log-likelihood using current estimates, and igoring the missing data mechanism
llt <- ll(Yobs, mut, sit, n)
# Print current parameter values and likelihood
cat(mut, sit, llt, "\n")
# Stop if converged
if ( abs(lltm1 - llt) < 0.001) break
lltm1 <- llt
}
# fill in missing values with new mu.
return(mut,sit)
}
```

EXAMPLE 13.7.5 (MCMC) Write MCMC algorithm for Example 13.5.4.

Solution

```
metrop3=function(n=1000,eps=0.5)
{
vec=vector("numeric", n)
x=0
oldll=dnorm(x,log=TRUE)
vec[1]=x
for (i in 2:n) {
can=x+runif(1,-eps,eps)
loglik=dnorm(can,log=TRUE)
loga=loglik-oldll
if (log(runif(1)) < loga) {
x=can
oldll=loglik
}
vec[i]=x
}
vec
}
```

In addition, if we want to plot the results, use the following code:

```
plot.mcmc<-function(mcmc.out)
{
op=par(mfrow=c(2,2))
plot(ts(mcmc.out),col=2)
hist(mcmc.out,30,col=3)
qqnorm(mcmc.out,col=4)
abline(0,1,col=2)
acf(mcmc.out,col=2,lag.max=100)
par(op)
}
metrop.out<-metrop3(10000,1)
plot.mcmc(metrop.out)
```

With the plot, we get the following output

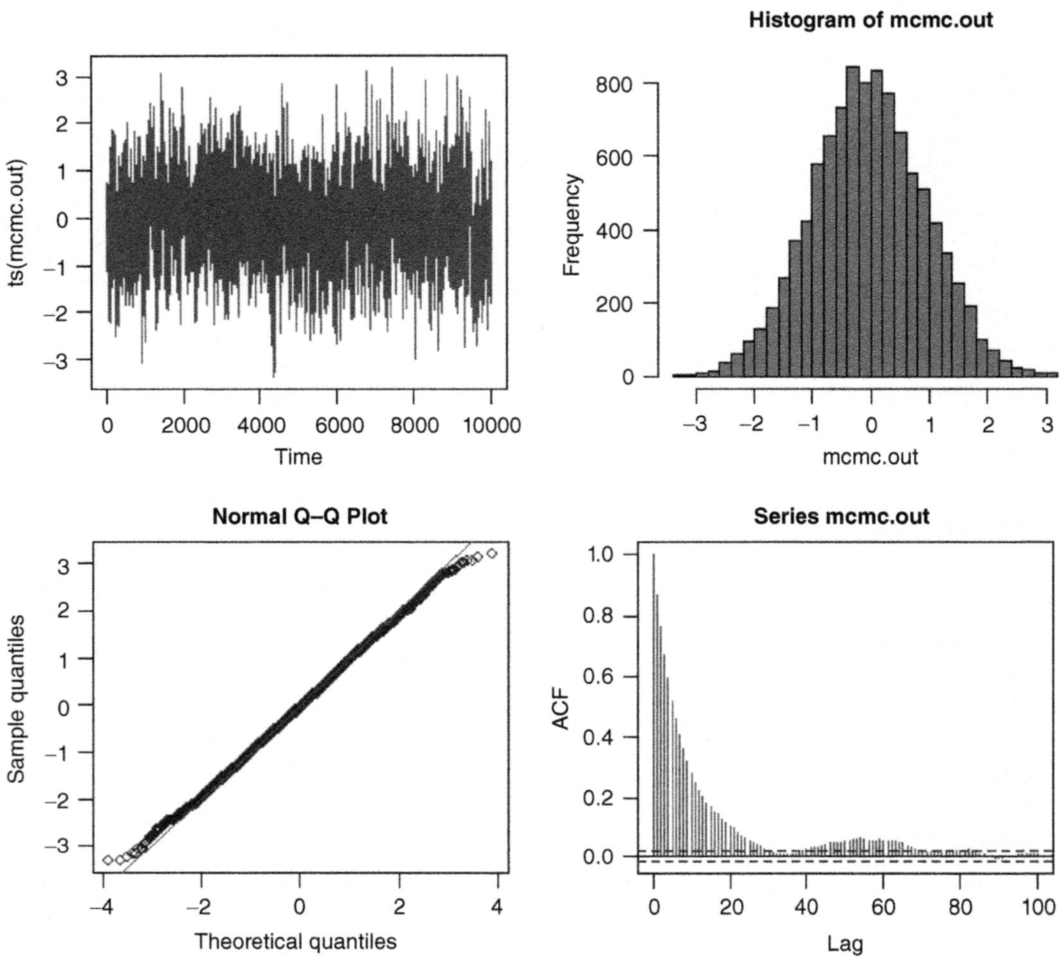

EXAMPLE 13.7.6 (Gibbs sampler): Write an R-code for Example 13.5.5 (b).

Solution

```
#R program for Gibbs sampling
>
> n=15
> y0=1/4
> p=y0
> x0=rbinom(1,n,p)
>
> a=1
> b=2
> A=x0+a
> B=n-x0+b
> X=matrix(x0,3);Y=matrix(y0,3)
>
> for(i in 2:3){#sample from f(y/x)
+ Y[i]=rbeta(1,A,B)
```

```
+ #sample from f(x/y)
+ X[i]=rbinom(1,n,Y[i])
+ }
> print(matrix(c(X,Y),3,2))
```

Output

```
[,1] [,2]
[1,] 5 0.2500000
[2,] 5 0.4011747
[3,] 4 0.2047587
```

It should be noted that each time we run the code, we may get a different output.

13.8 PROJECTS FOR CHAPTER 13

13A. Bootstrap Computation

Use any statistical computer programs to generate random numbers. By specifying a particular distribution, such as normal with mean 0 and variance 1 or other similar distributions, we can then generate numbers that follow this distribution. (This can be done either directly, if your software allows, or by the method described in Project 4A.)

(a) Use such a package to generate 200 numbers from an N (0, 1) distribution. Then calculate the sample mean and sample variance. (They will be slightly off from the actual mean and variance. From this, we can draw the conclusion that the estimates of data parameters which are computed using the data set are not necessarily the true parameters, but are often reasonable guesses.) Using these values, calculate an estimate of the standard error.

(b) Now, for the same data, pretend that we are not really sure what the distribution is. Then, we could consider letting the observed data specify what the distribution is. This is the essence of bootstrapping. In particular, sample with replacement from a distribution that we have observed (the empirical distribution of the data), in order to study the possible estimates that might have resulted from a similar sample (same data observations, but in possibly different quantities). Using the bootstrap algorithm described in Section 13.3, obtain a bootstrap estimate of the standard error and compare this with the estimate obtained in part (a).

Chapter 14

Extreme Value Theory With Applications

Chapter outline

OBJECTIVE

In this chapter, we model and quantify the behavior of the rare, extreme events (maximum or minimums) of a process or data set, rather than its mean or variance. This theory is useful in various fields in understanding and predicting rare events is crucial, such as finance, insurance, climate science, and engineering reliability.

Boris Vladimirovich Gnedenko
Source (https://mathshistory.st-andrews.ac.uk/biographies/gnedenko/).

Boris Vladimirovich Gnedenko (1912–95) was a Soviet mathematician and a student of Andrey Kolmogorov. Even though Ronald A. Fisher and Leonard H. C. Tipper in 1928 published the so-called Fisher–Tippett theorem about the distribution of maxima (or minima), Gnedenko in 1943 published a rigorous proof of Fisher–Tippett theorem, and since then often the result is called the Fisher–Tippett–Gnedenko theorem. His work established three possible types of extreme value distributions, namely, Gumble, Fréchet, and Weibull. In addition, Gnedenko is also known for his pioneering work on mathematical statistics, probability theory, and queueing theory. Many of his books and research publications remain influential in applied mathematics,

Mathematical Statistics with Applications in R. DOI: https://doi.org/10.1016/B978-0-443-27548-7.00014-1

statistics, and engineering sciences. His work on extreme value theory continues to underpin modern statistical modeling of rare, catastrophic events, making him one of the most influential figures in applied probability of the 20th century.

14.1 INTRODUCTION

In most real-world applications, we are not only interested in the *average* behavior of a system but also in its *extreme* outcomes. Thus, in real life situations we experience problems in which we are interested in the largest observation, **maximum extreme**, or the smallest observation, **minimum extreme**, such as: "What is the highest value, **maximum extreme**, or the lowest value, **minimum extreme**, the S&P 500 can reach in the next year?"; "How high can a given drug concentration in the bloodstream go before causing toxicity?"; "In Global Warming, what is the largest value of carbon dioxide in the atmosphere beyond which value it would be dangerous to our health?"; "How tall should we design a wall to prevent flooding in Honeymoon Island?"; among many other real-world problems that we encounter on a daily basis. Extreme value theory (EVT) is becoming increasingly important in AI and data science, specifically in areas where rare, extreme, or anomalous events matter. EVT provides theoretical tools to model tails distribution, letting AI systems quantify uncertainty in extreme events, detect anomalies, and improve robustness. For instance, in cybersecurity or fraud detection, EVT can model the distribution of maximum deviations and identify rare, extreme events. In autonomous driving, EVT can help avoid rare catastrophic crashes by modeling tail risks. These unusual and extreme conditions, such as large earthquakes, for example, tend to have much more substantial net impacts, even though they occur a much smaller proportion of the time.

To address such problems in understanding their extreme behavior, *maximum* or *minimum* extremes, we use probabilistic EVT. Such theory has been successfully applied to solve real-world problems like stock market crashes, rainfall floods, the behavior of wind gusts, hurricanes, earthquakes, reliability of systems, survival analysis in health sciences, evaluating the effectiveness of drugs, economic and financial systems, and natural disasters, among many other real-world problems. EVT is developed to focus on statistically analyzing the behavior of extremes, maximum, minimum, and deviation of the mean (median) of the probability distribution that probabilistically characterizes the particular phenomenon of interest. In this chapter, we shall explore EVT, the development of the **Generalized Extreme Value Probability Distribution** (**GEV**), and study some very important probability distributions that are byproducts of the GEV, namely, the **Gumbel**, **Fréchet**, **Weibull**, and **Generalized Pareto** Probability Distributions.

14.2 EXTREME VALUE THEORY

Let us consider a sequence of independent random variables with identical probability distribution, $X_1, X_2, \ldots, X_n$ with a joint cumulative probability distribution function, $F(x)$; the probability distribution of the maximum (extreme),

$$M_n = \max\{X_1, X_2, \ldots, X_n\}$$

is given by

$$P\{M_n \le z\} = P\{X_1 \le z, \ldots, X_n \le z\} = \prod_{i=1}^{n} P\{X_i \le z\} = F^n(z). \tag{14.1}$$

Because for any z such that $0 < F(z) < 1$, $F^n(z) \to 0$ as $n \to \infty$, the result in (14.1) by itself is of not much interest. The problem also is that we encounter in practice is that of finding the cumulative probability distribution, F. One approach is to use a standard statistical procedure (goodness-of-fit) to identify (estimate) F from the given data and then proceed to substitute this estimate of F into Eq. (14.1). However, very small discrepancies in estimating F can lead to substantial discrepancies in F^n. Also, in practice, the behavior of X_i is unknown. Thus, calculation of exact properties of M_n is impossible. These problems lead to an approach faced with an asymptotic argument that requires determining what are the possible limit probability distributions for

$$M_n \text{ as n} \to \infty,$$

that is, it leads to the question, what are the possible limit probability distributions in the extreme case?

Because M_n grows with n, we begin by studying the probability distribution of normalizing,

$$\frac{M_n - b_n}{a_n},$$

where $a_n > 0$ and $b_n \in \mathbb{R}$, are sequences of normalizing coefficients such that

$$F^n\left(\frac{M_n - b_n}{a_n}\right),$$

leads to a nongenerate probability distribution G, as $n \to \infty$. By nongenerate probability distribution, we mean a probability distribution where the random variable can take on a range of values, that is, it is not concentrated on a single value.

In this case, we seek normalizing sequences $\{a_n > 0\}$ and $\{b_n\}$, such that the limit as $n \to \infty$, and the cumulative probability distribution is,

$$P\left\{\frac{M_n - b_n}{a_n} \le z\right\} \to G(z),$$

where $G(z)$ does not depend on n, and it is a nongenerate probability distribution. We can summarize this in the following theorem:

Theorem 14.2.1. *If there exists a sequence of constants $\{a_n > 0\}$ and $\{b_n\}$ such that, as $n \to \infty$. $P\left\{\frac{M_n - b_n}{a_n}\right\} \to G(z)$, where G is a nongenerate cumulative probability distribution function and belongs to one of the following three families:*

$$G(z) = \exp\left\{-\exp\left[-\left(\frac{z-b}{a}\right)\right]\right\}, \ -\infty < z < \infty \tag{14.2}$$

$$G(z) = \begin{cases} 0, \ z < b \\ \exp\left\{-\left(\frac{z-b}{a}\right)\right\}^{-\xi}, \ z \ge b \end{cases} \tag{14.3}$$

$$G(z) = \begin{cases} \exp\left\{-\left[-\left(\frac{z-b}{a}\right)\right]\right\}^{\xi}, \ z \le \text{b}, \\ 1, \ z > b \end{cases} \tag{14.4}$$

where $a > 0$ is the scale parameter, b is the location parameter and $\xi > 0$ is the shape parameter.

These three types of limiting distributions are often called the Gumble, Fréchet, and Weibull types, respectively. This result is also called Fisher–Tippett–Gnedenk theorem and can be considered the Central Limit Theorem of EVT.

EXAMPLE 14.2.2

Let $X_1,\ldots,X_n$ be a random sample from *Exponential(1)* distribution with CDF $F(x) = 1 - e^{-x}$, $x \ge 0$. Let $M_n = \max\{X_1,\ldots,X_n\}$. Let $a_n = 1$, and $b_n = \ln n$. Find a nondegenerate limit of $(M_n - b_n)/a_n$.

Solution

Compute the normalized maximum,

$$\begin{aligned} P(M_n \le b_n + z) &= [F(b_n + z)]^n \\ &= (1 - e^{-(\ln n + z)})^n = (1 - n^{-1}e^{-z})^n. \end{aligned}$$

Now, use the fact that the limit $\left(1 - \frac{t}{n}\right)^n \to e^{-t}$ as $n \to \infty$ *for fixed* $t > 0$. *Put* $t = e^{-z}$. *Then,*

$$\lim_{n\to\infty}(1 - n^{-1}e^{-z})^n \to e^{-e^{-z}}.$$

Therefore,

$$P\left(\frac{M_n - \ln n}{1} \le z\right) \to G(z) = e^{-e^{-z}}, \ z \in \mathbb{R},$$

which is the Gumble distribution. So, for exponential samples,

$$M_n - \ln n \xrightarrow{d} \textit{Gumble}(0,1).$$

Utilizing this family of cumulative probability distributions, (14.2), (14.3), and (14.4), we will define the **Generalized Extreme Value** (GEV) probability distribution to unify the three families of distributions into a single parametric form.

EXERCISE 14.2

14.2.1 Let X be Pareto distributed with tail index $\alpha > 0$, with CDF for $x \geq 1$, $F(x) = 1 - x^{-\alpha}$. Let $M_n = \max\{X_1, \ldots, X_n\}$. Show that

$$\lim_{n\to\infty} P\left(\frac{M_n}{n^{1/\alpha}}\right) = e^{-z^{-\alpha}},\ z > 0,$$

which is the Fréchet CDF $G(z) = \exp(-z^{-\alpha})$

14.2.2 Let X be distributed as *Uniform(0,1)*, and let $M_n = \max\{X_1, \ldots, X_n\}$. Show that

$$\lim_{n\to\infty} P(n(1\_M_n) \leq z) = 1 - e^{-z},\ z > 0.$$

Thus, $n(1 - M_n) \xrightarrow{d} Exponential(1)$, which is a special case of Weibull type.

14.2.3 For each of the three distributions below, simulate many repetitions using R and show empirically that normalized maxima follow the claimed limit law.

a. If $X_i \sim Exp(1)$, use $b_n = \ln n$, $a_n = 1$ converges to Gumbel.

b. If X_i follows Parato CDF, $F(x) = 1 - \left(\frac{x_m}{x}\right)^{\alpha}$, $x \geq x_m > 0$, $\alpha > 0$. For simulation, use $\alpha = 2$, $x_m = 1$, $a_n = n^{1/2}$, and $b_n = 0$, converges to Fréchet.

c. If $X_i \sim Uniform(0,1)$, then $n(1 - M_n)$ converges to *Exponential*(1).

14.3 THE GENERALIZED EXTREME VALUE DISTRIBUTION

Combining the three classes of probability distributions, (14.2), (14.3), and (14.4), we define the **GEV** cumulative probability distribution by

$$G(z) = \exp\left\{-\left[1 + \xi\left(\frac{z-\mu}{\sigma}\right)\right]^{-\frac{1}{\xi}}\right\}, \tag{14.5}$$

where

$$1 + \xi\left(\frac{z-\mu}{\sigma}\right) > 0, \qquad \sigma > 0, \qquad -\infty < \xi < \infty.$$

Note that the scale parameter a is replaced with $\sigma > 0$, indicating the spread of the distribution, the location parameter b is replaced by μ indicating the location of the center of the distribution, and $\xi \in \mathbb{R}$ is the shape parameter that determines the type of tail (heavy, light, or bounded). The cumulative GEV probability distribution in Eq. (14.5) can also be written as

$$G(z) = \begin{cases} \exp\left\{-\left[1 + \xi\left(\frac{z-\mu}{\sigma}\right)\right]^{-1/\xi}\right\}, & 1 + \xi\left(\frac{z-\mu}{\sigma}\right) > 0,\ \xi \neq 0, \\ \exp\left\{-\exp\left[-\left(\frac{z-\mu}{\sigma}\right)\right]\right\}, & -\infty < z < \infty,\ \xi = 0. \end{cases} \tag{14.6}$$

To obtain the probability density function of the GEV, we take the first derivative with respect to z, that is,

$$g(z; \mu, \sigma, \xi) = \frac{1}{\sigma}\left[1 + \xi\left(\frac{z-\mu}{\sigma}\right)\right]^{-1/\xi-1} \exp\left\{-\left[1 + \xi\left(\frac{z-\mu}{\sigma}\right)\right]^{-1/\xi}\right\}, \tag{14.7}$$

where $\mu \in \mathbb{R}$ is the location parameter, σ is the scale parameter that measures the spread, ξ is the shape parameter that gives us information on the tails of the distribution and $\left[1+\xi\left(\frac{z-\mu}{\sigma}\right)\right] > 0$ and if $\xi < 0, -\infty < z \le \mu - \frac{\sigma}{\xi}$ and if $\xi > 0, \mu - \frac{\sigma}{\xi} \le z < \infty$. However, if $\xi = 0, -\infty < z < \infty$ and the pdf (14.7) reduces to

$$g(z;\mu,\sigma,\xi) = \frac{1}{\sigma}\exp\left[-\left(\frac{z-\mu}{\sigma}\right)\right]\exp\left\{-\exp\left[-\left(\frac{z-\mu}{\sigma}\right)\right]\right\}. \tag{14.8}$$

The pdf of Z given by (14.7) and (14.8) satisfies the necessary conditions of integrating to 1 and the results are greater or equal to zero under the given conditions of μ, σ, and ξ. Furthermore, the shape parameter ξ, if positive, indicates heavy tail, while negative ξ implies bounded tail, and $\xi \to 0$ implies exponential tail.

Given in Figs. 14.1 and 14.2 are the graphs of the probability density function of GEV and the cumulative distribution of GEV, for different values of the parameters, μ, σ, and ξ, respectively.

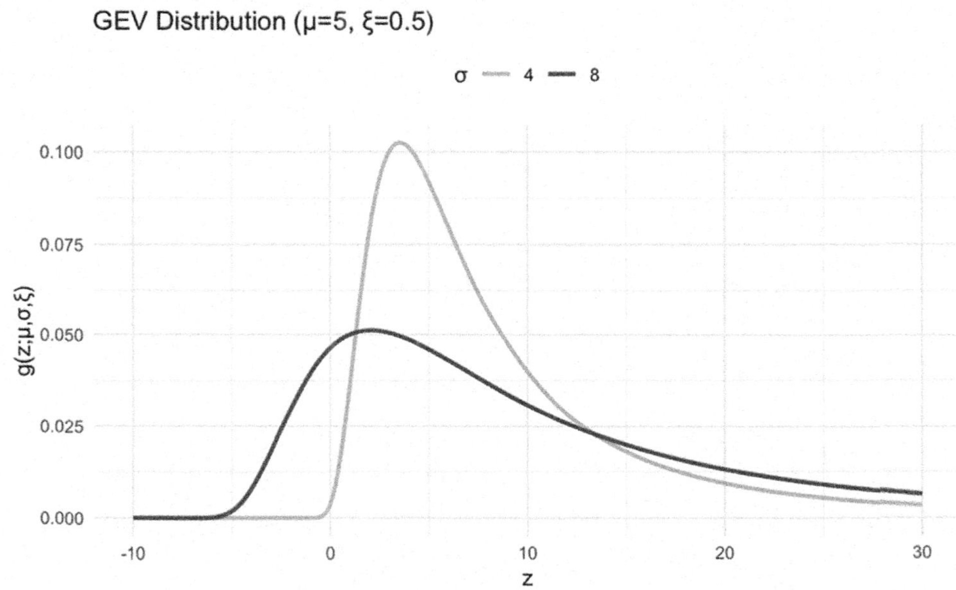

FIGURE 14.1 The pdf of GEV.

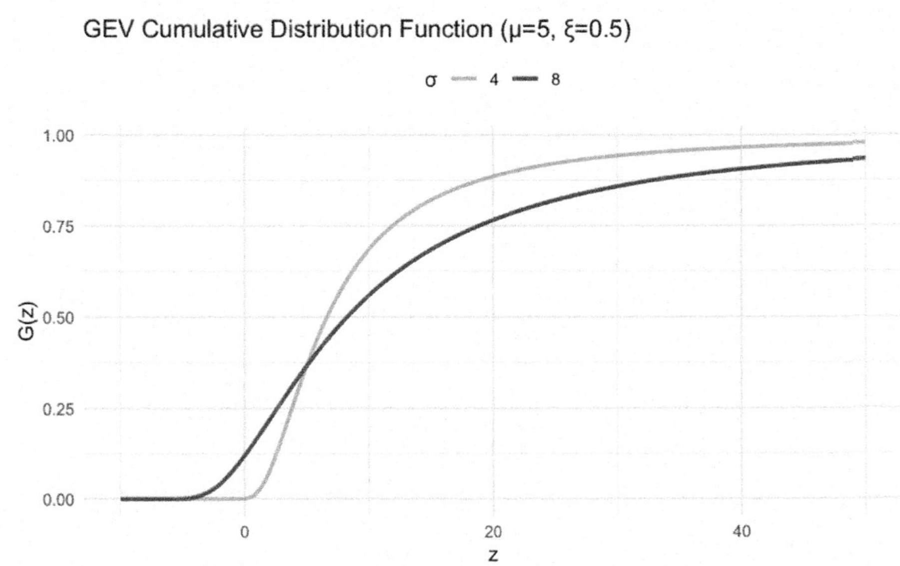

FIGURE 14.2 The cumulative distribution of GEV.

14.3.1 Estimating the Parameters of the GEV Distribution

Here we shall give a brief description on how we attain estimates of the parameters that define the GEV probability distribution. We will use the maximum likelihood method to achieve our estimates of the subject parameters. The probability density function of GEV distribution is given by Eq. (14.7). The likelihood function of the random sample $z_1, z_2, \ldots, z_m$ is given by

$$L_g(z;\ \mu, \sigma, \xi) = \frac{1}{\sigma_m}\prod_{i=1}^{m}\left[1 + \xi\left(\frac{z_i - \mu}{\sigma}\right)\right]^{-1/\xi - 1} \exp\left\{-\sum_{i=1}^{m}\left[1 + \xi\left(\frac{z_i - \mu}{\sigma}\right)\right]^{-1/\xi}\right\} \tag{14.9}$$

The natural log of the likelihood function of (14.9) for $\xi \neq 0$ is given by

$$\text{In } L_g(z;\ \mu, \sigma, \xi) = -\ m\ln\sigma - \left(1 + \frac{1}{\xi}\right)\sum_{i=1}^{m}\text{In}\left[1 + \xi\left(\frac{z_i - \mu}{\sigma}\right)\right] - \sum_{i=1}^{m}\left[1 + \xi\left(\frac{z_i - \mu}{\sigma}\right)\right]^{-1/\xi} \tag{14.10}$$

where $\left(1 + \xi\left(\frac{z_i - \mu}{\sigma}\right)\right) > 0$, for $i = 1,2,\ldots,m$. Note that if this condition is not satisfied, Eq. (14.10) results in minus infinity and the $\ln L[g(z;\mu, \sigma, \xi)]$ becomes zero.

When $\xi = 0$, the probability density function of GEV distribution is given by Eq. (14.8) and its likelihood function is given by

$$Lg(z;\mu, \sigma, \xi) = \frac{1}{\sigma^m}\exp\left[-\sum_{i=1}^{m}\left(\frac{z_i - \mu}{\sigma}\right)\right]\exp\left\{-\sum_{i=1}^{m}\exp\left[-\left(\frac{z_i - \mu}{\sigma}\right)\right]\right\}, \tag{14.11}$$

and the ln of (14.11) results in

$$\ln L[g(z;\mu, \sigma, \xi)] = -m\ln\sigma - \sum_{i=1}^{m}\left(\frac{z_i - \mu}{\sigma}\right) - \sum_{i=1}^{m}\exp\left[-\left(\frac{z_i - \mu}{\sigma}\right)\right]. \tag{14.12}$$

The maximum likelihood estimates are obtained for $\xi \neq 0$ by obtaining the partial derivative of the likelihood function (14.10) with respect to the parameters μ, σ, and ξ, and setting them equal to zero. That is,

$$\frac{\partial \ln L[g(z;\mu, \sigma, \xi)]}{\partial\mu} = \left(1 + \frac{1}{\xi}\right)\sum_{i=1}^{m}\left(1 + \xi\left(\frac{z_i - \mu}{\sigma}\right)\right)^{-1} +$$

$$-\sum_{i=1}^{m}\left(\left(1 + \xi\left(\frac{z_i - \mu}{\sigma}\right)\right)^{-\frac{1}{\xi}-1}\frac{\xi}{\sigma}\right) = 0,$$

$$\frac{\partial \ln L[g(z;\mu, \sigma, \xi)]}{\partial\sigma} = -\frac{m}{\sigma} + \left(1 + \frac{1}{\xi}\right)\sum_{i=1}^{m}\left(1 + \xi\left(\frac{z_i - \mu}{\sigma}\right)\right)^{-1}\xi\left(\frac{z_i - \mu}{\sigma^2}\right)$$

$$-\sum_{i=1}^{m}\left(1 + \xi\left(\frac{z_i - \mu}{\sigma}\right)\right)^{-\frac{1}{\xi}-1}\xi\left(\frac{z_i - \mu}{\sigma}\right) = 0,$$

and

$$\frac{\partial \ln L[g(z;\mu,\sigma,\xi)]}{\partial \xi} = \sum_{i=1}^{m}\left[-\frac{1}{\xi^2}\ln\left(1+\xi\left(\frac{z_i-\mu}{\sigma}\right)\right)-\frac{1}{\xi}\left(\frac{\frac{z_i-\mu}{\sigma}}{1+\xi\left(\frac{z_i-\mu}{\sigma}\right)}\right)\right.$$

$$\left.+\frac{1}{\xi^2}\left(\left(1+\xi\left(\frac{z_i-\mu}{\sigma}\right)\right)\right)^{-\frac{1}{\xi}}\ln\left(1+\xi\left(\frac{z_i-\mu}{\sigma}\right)\right)\right]=0.$$

Similarly, for $\xi = 0$, we obtain

$$\frac{\partial \ln L[g(z;\mu,\sigma,\xi)]}{\partial \mu} = \sum_{i=1}^{m}(z_i e^{-z_i} - e^{-z_i}) = 0$$

and

$$\frac{\partial \ln L[g(z;\mu,\sigma,\xi)]}{\partial \sigma} = -\frac{n}{\sigma} + \sum_{i=1}^{m} z_i e^{-z_i} = 0.$$

The above system of nonlinear equations cannot be solved mathematically, and thus we use numerical methods, such as the Newton–Raphson method, to obtain approximate estimates of the subject parameters.

We will now give an example of GEV applied to data with R-code.

Example 14.3.2

Suppose that we have 30 years of annual maximum daily rainfall (mm):
85, 90, 112, 105, 98, 123, 134, 150, 142, 160, 175, 168,
189, 200, 210, 195, 180, 178, 156, 170, 182, 190, 205, 215,
225, 210, 198, 205, 218, 230
Fit a GEV distribution using R.

Solution

We will use extRemes *or* evd *package. Following is the R-code used.*

```
rainfall <- c(85, 90, 112, 105, 98, 123, 134, 150, 142, 160, 175, 168,
189, 200, 210, 195, 180, 178, 156, 170, 182, 190, 205,
215, 225, 210, 198, 205, 218, 230)
# install.packages("evd")
library(evd)
# Fit GEV to data
fit <- fgev(rainfall)
summary(fit)
# Extract parameters
fit$estimate
# Plot fitted density and data histogram
hist(rainfall, breaks=10, freq=FALSE, col="lightblue",
main="GEV Fit to Annual Max Rainfall")
curve(dgev(x, loc=fit$estimate[1], scale=fit$estimate[2], shape=fit$estimate[3]),
add=TRUE, col="red", lwd=2)
return_100 <- qgev(0.99, loc=fit$estimate[1], scale=fit$estimate[2], shape=ft$estimate[3])
return_100
```

We will get following output:

```
Length Class Mode
estimate 3 -none- numeric
```

```
std.err 3 -none- numeric
fixed 0 -none- NULL
param 3 -none- numeric
deviance 1 -none- numeric
corr 0 -none- NULL
var.cov 9 -none- numeric
convergence 1 -none- character
counts 2 -none- numeric
message 0 -none- NULL
data 30 -none- numeric
tdata 30 -none- numeric
nsloc 0 -none- NULL
n 1 -none- numeric
prob 0 -none- NULL
loc 1 -none- numeric
call 2 -none- call
Extracted parameters
loc scale shape
163.3430010 46.3071768-0.6700617
By substituting these values in Eq. 14.8, we will obtain the desired pdf.
Plot of fitted density and data histogram is given in Figure
```

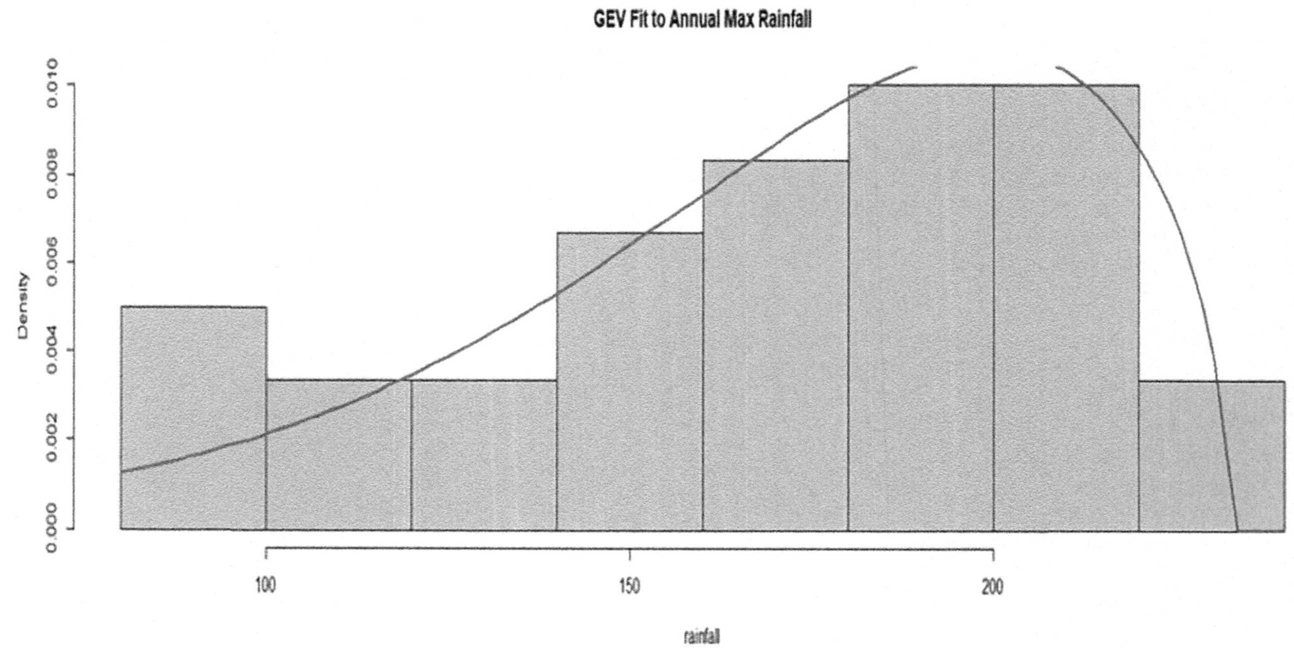

FIGURE 14.3 Density and histogram for rain fall data.

We can also get that 99% of data will fall below 229.2832. That is, a rainfall of about 229.28 mm has a 1% annual probability.

14.3.2 Properties of the GEV Distribution

The expected value of the GEV Distribution exists only if the shape parameter is $\xi < 1$ and $\xi = 0$, that is,

$$E(Z) = \begin{cases} \mu + \sigma \frac{\Gamma(1-\xi) - 1}{\xi}, & if\, \xi < 1 \\ \mu + \gamma\sigma, & if\, \xi = 0 \end{cases},$$

where $\Gamma(1-\xi)-1$ is the gamma function, and $\gamma \approx 0.5772$ is the approximate Euler–Mascheroni constant, and for $\xi \geq 1$, it does not exist.

The variance, σ^2, of the GEV distribution exists only if the shape parameter $\xi < \frac{1}{2}$ and it is given by

$$V(Z) = \begin{cases} \frac{\sigma^2}{\xi^2}\Gamma(1-2\xi) - (\Gamma(1-\xi))^2, & if\ \xi < \frac{1}{2} \\ \frac{\pi^2\sigma^2}{6} & , if\ \xi = 0. \end{cases}$$

It should be noted that higher moments exist only when $\xi < \frac{1}{k}$ for the k-th moment.

We conclude that the shape parameter ξ of the GEV distribution controls the behavior of the tails, that is,

$\xi = 0$, results in light-tailed behavior.

$\xi > 0$, results in heavy-tailed behavior.

$\xi < 0$, results in bounded upper tail behavior.

This flexibility in tail modeling is one of the main features of GEV distribution. The heavy-tailed ($\xi > 0$, Fréchet) models extreme floods, and financial crashes. Light-tailed extremes ($\xi = 0$, Gumbel) are often used to model in meteorology and hydrology, whereas bounded upper-tail behavior ($\xi < 0$, Weibull) is used to model physical or engineering lifetimes.

Some additional general properties of the GEV distribution are

i. The GEV distribution is defined only if

$$1 + \xi\left(\frac{z-\mu}{\sigma}\right) > 0.$$

ii. Regularity conditions of the maximum likelihood estimates (MLE) of the parameters of GEV do necessarily hold if

$$\frac{\mu-\sigma}{\xi} \leq z \text{ and } \xi > 0.$$

$$z \leq \frac{\mu-\sigma}{\xi} \text{ and } \xi < 0.$$

iii. Asymptotic normality and efficiency of the MLE can be obtained if the shape parameter is $\xi \geq -0.5$.

14.3.3 The Generalized Extreme Value Distribution, GEV: Application to Pancreatic Cancer

Pancreatic adenocarcinoma is one of the most fatal cancers and continues to be a major unsolved health problem at the start of the 21st century. It is the fourth leading cause of cancer deaths in the United States. From the Surveillance, Epidemiology and End Results, SEER, database we obtained the survival times in months of 1400 pancreatic cancer patients that were classified as stage 2. That is, the time that the patients were diagnosed with stage 2 pancreatic cancer until their death in months. The descriptive statistics of the patients are given below:

Patients Stage II	Mean	Median	Std. dev.	Skewness	Kurtosis
	21.44	14.00	23.50	2 14	5.10

A histogram of the survival times in months is given below by Fig. 14.3.

Our objective is to find the probability distribution that gives us the best fit of the survival times of patients with pancreatic cancer. Utilizing the Kolmogorov–Smirnov and Anderson–Darling tests, both confirmed that the GEV distribution fits the survival times the best. Thus, we proceeded to obtain the approximate MLE of the three parameters of GEV, that is,

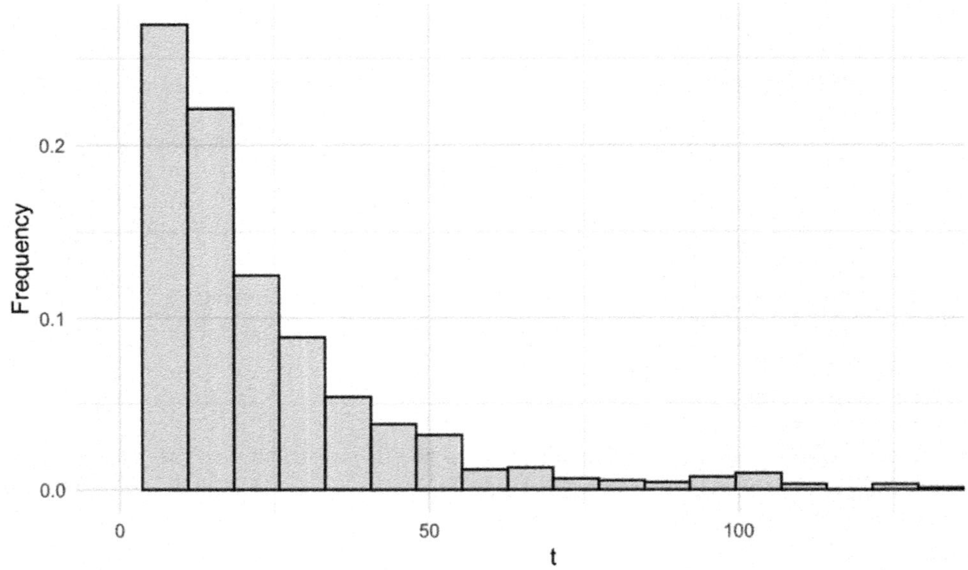

FIGURE 14.4 Histogram of the survival times of pancreatic cancer patients, stage II.

$$\text{Location: } \hat{\mu} = 10.18 \text{ months}$$

$$\text{Scale: } \hat{\sigma} = 10.83 \text{ months}$$

$$\text{Shape: } \hat{\xi} = 0.32.$$

The probability density of GEV is given by

$$g(t; 10.18, 10.83, 0.32) = \frac{1}{10.83} \exp\left\{-\left[1 + 0.32\left(\frac{t - 10.18}{10.83}\right)\right]^{3.125}\right\}$$

$$\left[1 + 0.32\left(\frac{t - 10.18}{10.83}\right)\right]^{2.15}.$$

A graph of $g(t; 10.18, 10.83, 0.32)$ is given below by Fig. 14.4.
The cumulative GEV distribution is given by

$$G(t) = \exp\left\{-\left[1 + 0.32\left(\frac{t - 10.18}{10.83}\right)^{-3.13}\right]\right\}.$$

A plot of $G(t)$ is given by Fig. 14.5, below:
We can use both the density and cumulative distributions of the GEV to obtain various probabilities of a patient with the subject disease that will survive up to time t or more than a specific time t. For example, that such a patient will survive 34 months, is approximately 14% from Fig. 14.6,

$$P(T \leq 34 \text{ months}) \approx 0.14, \text{ or } 14\%.$$

The probability that a patient with the subject disease will live less than 12 months is approximately 63%, that is,

$$G(t) = P[T \leq 12] \approx 0.63, \text{ or } 63\%.$$

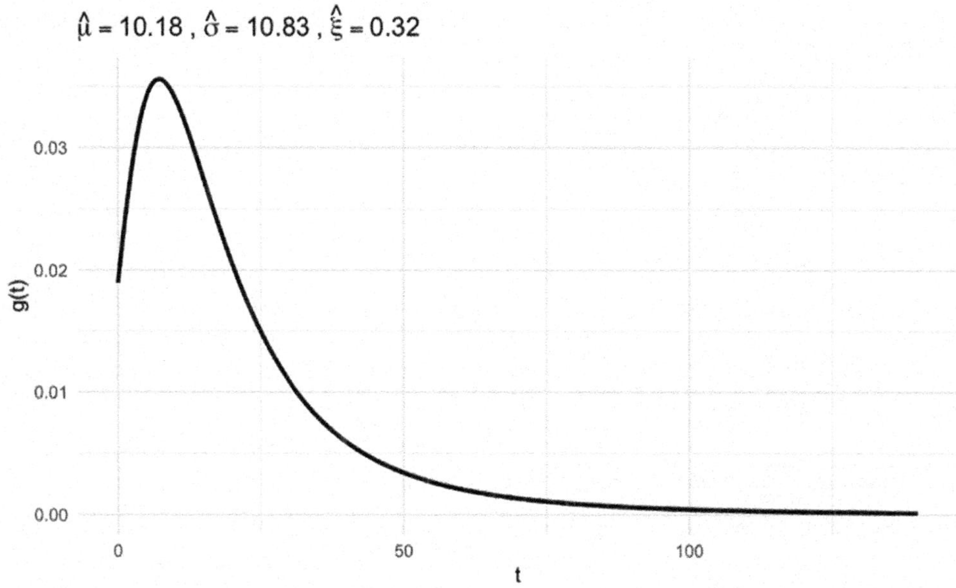

FIGURE 14.5 Cumulative distribution of GEV of survival time of pancreatic cancer patients, stage II.

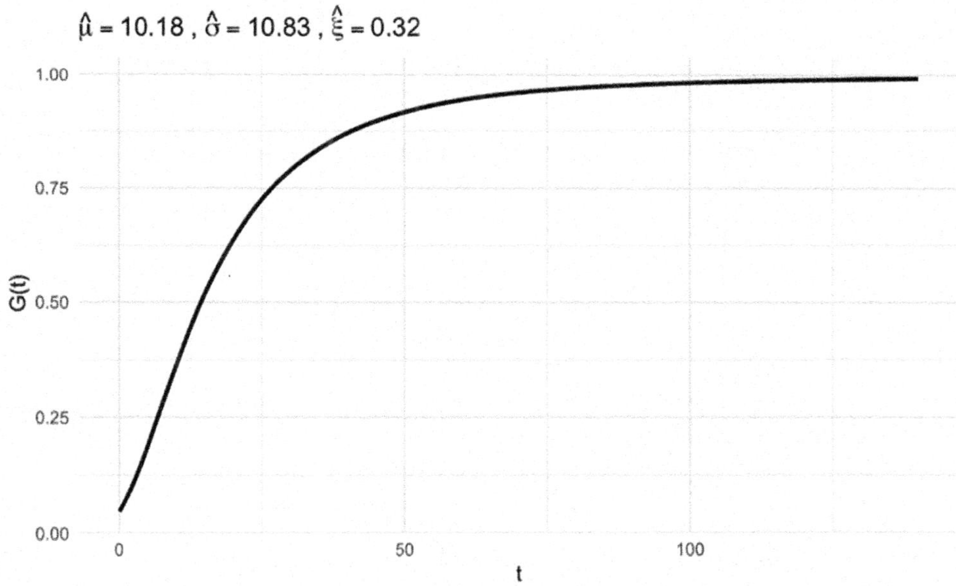

FIGURE 14.6 Probability density of GEV for survival time of pancreatic cancer patients, stage II.

The expected time of survival for Pancreatic cancer patients, stage II, is

$$E[T] = 10.18 + 10.83\frac{\Gamma(0.68) - 1}{0.32}$$

$$\approx 10.18 + 10.19$$

$$\approx 20.37 \text{ months.}$$

Thus, a patient with this type of cancer and classified in stage II will expect to survive **20.37** months. The variance of these survival times is

$$V[T] = \frac{(10.83)^2}{(0.32)^2}\Gamma(1 - 2(0.32)) - (\Gamma(1 - 0.32))^2$$

$$\approx 3{,}063.45.$$

Note that from these results, we see most patients' survival is clustered around low values (because pancreatic cancer is aggressive), and the standard deviation is approximately 55.4 months, which is much larger than the mean. That means, a few are long-term survivors, highlighting the heavy-tailed nature of extreme survival outcomes. This large variance is consistent with extreme survival cases, which GEV models are meant to capture.

14.4 THE GUMBEL PROBABILITY DISTRIBUTION

The Gumbel Probability Distribution was founded by **Emil Gumbel** (1891–1966). It is very useful in analyzing the *maximum* or *minimum* of a number of samples of a given phenomenon, such as the maximum level of water of a river in a particular year, predicting the chance that a level 7 (extreme) earthquake will occur, predicting the maximum or minimum financial returns of the S&P 500, among many other natural disasters and economic and financial issues. The Gumbel pdf is one of the families of distributions that constitute the GEV probability distribution as defined by Eq. (14.5), and by letting the shape parameter $\xi = 0$, we obtain the Gumbel cumulative pdf, that is,

$$G(z) = \exp\left\{-\exp\left[-\left(\frac{z-\mu}{\sigma}\right)\right]\right\}, \tag{14.13}$$

where $z \in \mathbb{R}$, μ is the location (center) parameter and $\sigma > 0$ is a scale (spread) parameter. The *Gumbel pdf* is also known as the *double exponential pdf* and a *Type I extreme value* distribution.

The Gumbel probability density function is obtained by taking the first derivative of (14.13) with respect to z and results in

$$g(z) = \frac{\exp\left[\pm\left(\frac{z-\mu}{\sigma}\right)\right]\exp\left\{-\exp\left[\pm\left(\frac{z-\mu}{\sigma}\right)\right]\right\}}{\sigma}, \tag{14.14}$$

where $\sigma > 0$, $\mu \in \mathbb{R}$, and when investigating the *maximum extreme*, we use the $(-)$ in (14.14) and analyzing the *minimum extreme* we use the $(+)$. There are standard tables available to obtain values of the Gumbel cumulative probability distribution of the maximum extreme. We also have the

$$G(z)_{\min} = 1 - G(z)_{\max}.$$

The minimum extreme can also be obtained from the tables. A graphical view of the Gumbel pdf and Gumbel cumulative pdf, for various values of the parameter μ and σ with $\xi = 0$, is given by Figs. 14.7 and 14.8 below:

14.4.1 Estimating the Parameters of the Gumbel Distribution

Estimation of the parameters μ and σ is essential in applying the subject pdf. We shall give a brief summary using the maximum likelihood method to obtain the estimators of μ and σ. The likelihood function of (14.14) based on a random sample $z_1, z_2, \ldots, z_m$ is given by

$$L[g(z;\mu,\sigma)] = \frac{1}{\sigma^m}\exp\left\{-\sum_{i=1}^{m}\left(\frac{z_i-\mu}{\sigma}\right)-\sum_{i=1}^{m}\exp\left(-\frac{z_i-\mu}{\sigma}\right)\right\}$$ and the $\ln L[g(z;\mu,\sigma)]$ is given by

$$\ln L[g(z;\mu,\sigma)] = mln\sigma - \left(1+\frac{1}{z}\right)\sum_{i=1}^{m}\ln\left(1+\frac{z(z_i-\mu)}{\sigma}\right),$$

parameter $\mu = z_m$ (maximum), $\sigma > 0$ and $1 + \frac{z(z_i - z_m)}{\sigma} > 0$.

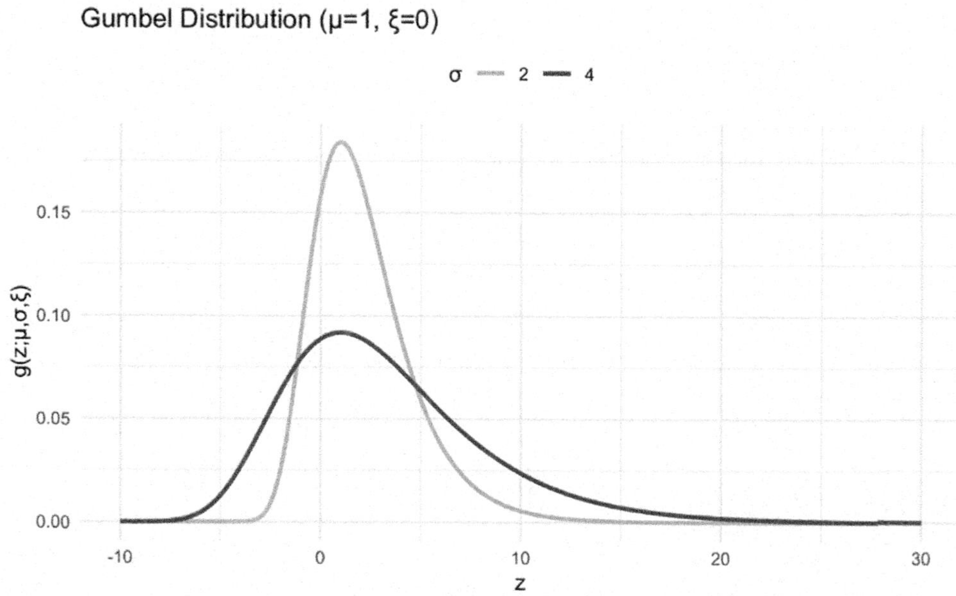

FIGURE 14.7 The Gumbel pdf.

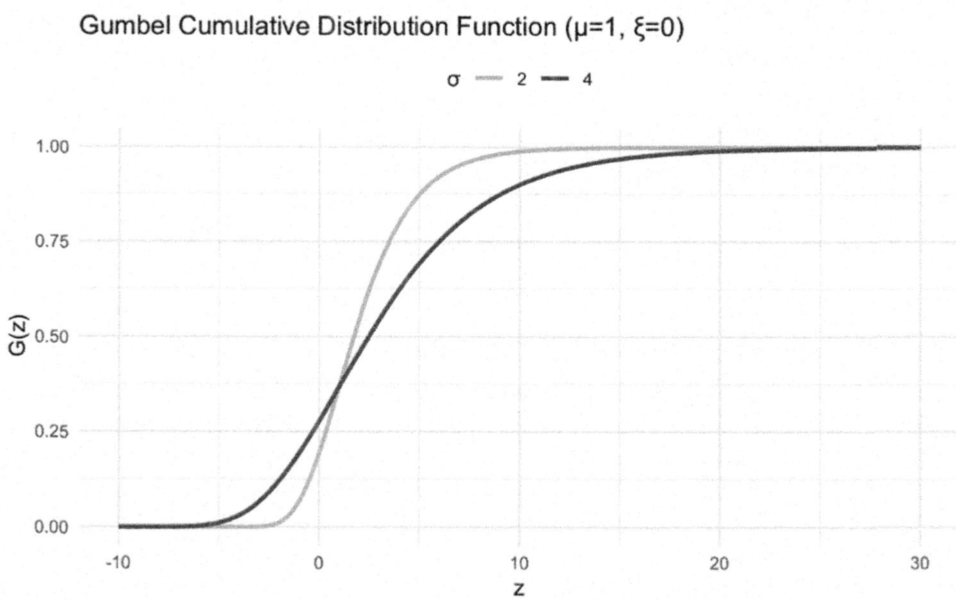

FIGURE 14.8 The Gumbel cumulative distribution.

We proceed to take the partial derivatives of $\ln L\,[g\,(z;\mu,\,\sigma)]$ with respect to μ and σ, and setting them equal to zero, we obtain the maximum likelihood equations

$$m = \sum_{i=1}^{m} \exp\left(-\frac{(z_i - \hat{\mu})}{\hat{\sigma}}\right),$$

and

$$\hat{\sigma} + \hat{\mu} + \frac{1}{m}\sum_{i=1}^{m} z_i \exp\left(-\frac{(z_i - \hat{\mu})}{\hat{\sigma}}\right) - \frac{\hat{\mu}}{m}\sum_{i=1}^{m} \exp\left(-\frac{(z_i - \hat{\mu})}{\hat{\sigma}}\right) = \overline{z},$$

where $\overline{z}$ is the sample average. Substituting the first equation into the second equation, we have an equation only in $\hat{\sigma}$, that is,

$$\hat{\sigma} - \overline{z} + \frac{\sum_{i=1}^{m} z_i \exp\left(-\frac{z_i}{\hat{\sigma}}\right)}{\sum_{i=1}^{m} \exp\left(-\frac{z_i}{\hat{\sigma}}\right)} = 0.$$

We can obtain the approximate estimates of σ, $\hat{\sigma}$ using iterative numerical procedures, and once we have the estimate $\hat{\sigma}$, the MLE of μ, $\hat{\mu}$ is then obtained from

$$\hat{\mu} = -\hat{\sigma} \ln\left\{\frac{1}{m}\right\} \sum_{i=1}^{m} \exp\left(-\frac{z_i}{\hat{\sigma}}\right).$$

Thus, having the estimates of μ and σ, we can perform goodness-of-fit tests to determine if a given sample of data follows the Gumbel pdf, and once we confirm that, we can probabilistically characterize the probabilistic behavior of the phenomenon of interest.

14.4.2 Properties of the Gumbel Distribution

Some properties of the Gumbel distribution are as follows:

i. $E(Z) = \mu + \sigma$, where γ is the Euler-Mascheroni constant and is approximately equal to 0.5772.

ii. $V(Z) = \frac{\pi^2}{6}\sigma^2 = 1.6449\sigma^2$.

iii. Median of $Z = \mu - \sigma \ln(\ln 2) = \mu - 0.3665\sigma$.

iv. Skewness and Kurtosis are constant regardless of the values of the parameter (μ, σ) and equal respectively, to 1.14 and 5.4.

14.4.3 The Gumbel Probability Distribution: Application to Colon Cancer

Colon cancer is one of the most common cancers, and with early detection and timely treatment, the outcome can improve significantly. Our objective is to obtain actual data for 200 patients, measure the malignant tumor size of colon cancer patients, and find the probability distribution that best fits this data. The subject data was taken from **SEER** database of the **National Institute of Health**. The malignant tumor size of the 200 colon cancer patients was measured in mm, and a histogram of the data is given below by Fig. 14.9.

Both of the goodness-of-fit tests, Kolmogorov–Smirnov and Anderson–Darling, confirm at the $\alpha = 0.05$, level of significance, that the Gumbel probability distribution is the best fit of the data. The approximate MLE of the parameters of the Gumbel pdf are

$$\text{Location: } \hat{\mu} = 43.71 \text{ mm}$$

$$\text{Scale: } \hat{\sigma} = 21.87 \text{ mm}$$

$$\text{Shape: } \hat{\xi} = 0.$$

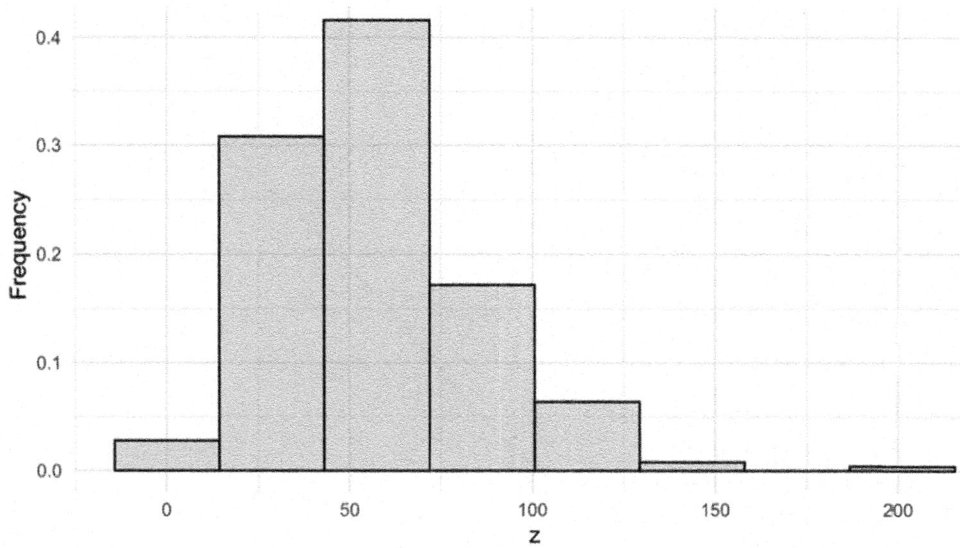

FIGURE 14.9 Histogram of malignant tumor size of colon cancer patients in mm.

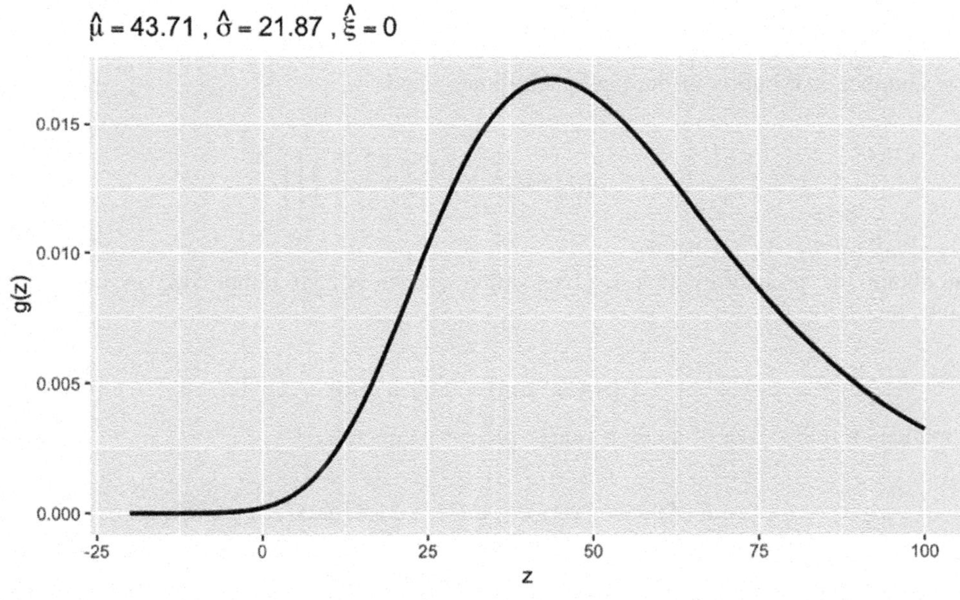

FIGURE 14.10 Probability density of the Gumbel distribution for colon cancer tumor size.

The probability density that characterizes the probabilistic behavior of the malignant tumor size of colon cancer patients is given by

$$g(z;43.71,21.87)=\frac{\exp\left[-\left(\frac{z-43.71}{21.87}\right)\right]\exp\left\{-\exp\left[-\left(\frac{z-43.71}{21.87}\right)\right]\right\}}{21.87}.$$

A graph of $g(z;\ 43.71, 21.87)$ is given below by Fig. 14.10, below:

Thus, the probability of the malignant tumor size in mm of a colon cancer patient between 20 mm and 30 mm is approximately 8.5%, that is,

$$P[20\,\text{mm} \le z \le 30\,\text{mm}] \approx 8.5\%.$$

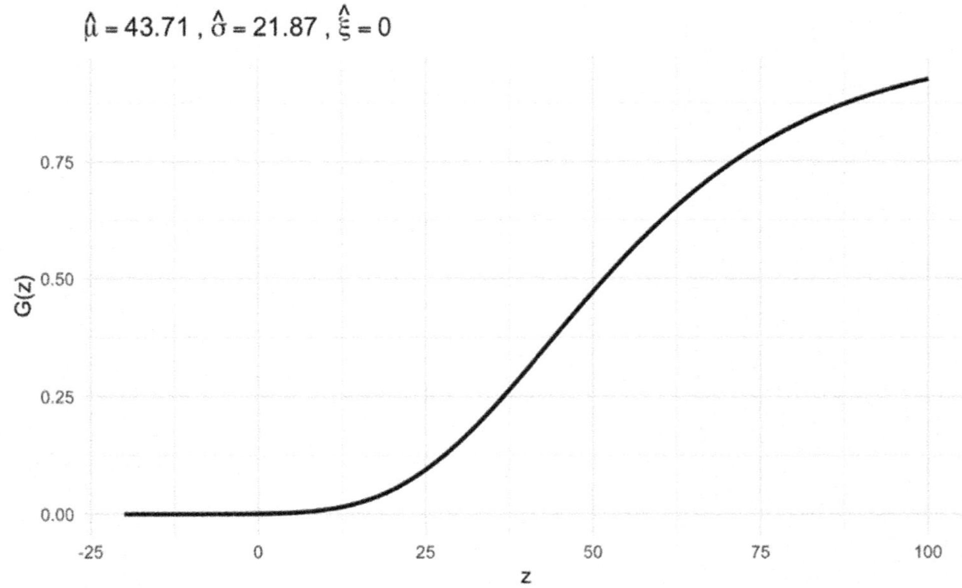

FIGURE 14.11 The cumulative Gumbel for colon cancer tumor size.

The cumulative Gumbel distribution of the malignant tumor size is

$$G(z) = \exp\left\{-\exp\left[-\left(\frac{z - 43.72}{21.87}\right)\right]\right\}.$$

A graph of z is given by Fig. 14.11, below:

Thus, we can obtain the probability that a given colon cancer patient tumor size in mm less than 60 mm is approximately 60%, that is,

$$P[z \leq 60\,\text{mm}] \approx 0.60 \text{ or } 60\%.$$

The expected malignant tumor size of a colon cancer patient is given by

$$\begin{aligned} E(Z) &= \hat{\mu} + \hat{\sigma}(\gamma) = 43.71 + 21.87(0.5772) \\ &= 56.33\text{ mm}. \end{aligned}$$

Thus, if a colon cancer patient was chosen at random, we expect that the malignant tumor size would be 56.33 mm. The variance of the subject data is

$$\begin{aligned} Var(Z) &= \frac{\pi^2}{6}\sigma^2 = 1.6449(21.87) \\ &= 35.97\text{ mm}. \end{aligned}$$

Thus, our model predicts that the typical colon cancer patient survives about 56 months, with most patients surviving 50–62 months. The variance shows there is moderate variability in survival.

14.5 THE FRÉCHET PROBABILITY DISTRIBUTION

The **Fréchet** probability distribution is a special case of the GEV distribution when the shape parameter $\xi > 0$. It is applicable to a heavy-tailed pdf. It has been identified by **Maurice Fréchet** (1878–1973), and it is also referred to as the **Type II** distribution. The Fréchet distribution has a wide range of applications in engineering, economic and financial systems, environmental problems, and hydrology, among others.

If the shape parameter $\xi > 0$ in the GEV distribution, it results in the Fréchet pdf, that is,

$$g(z; \mu, \sigma, \xi) = \frac{\xi}{\sigma}\left(\frac{\sigma}{z-\mu}\right)^{\xi+1} \exp\left\{-\left(\frac{\sigma}{z-\mu}\right)^{\xi}\right\}, \tag{14.15}$$

where $0 < \mu$, $0 < \sigma$ and $0 < \xi$.

The Fréchet cumulative distribution function is given by

$$\begin{aligned} P[Z \le z] &= G(z) \\ &= \exp\left\{-\left(\frac{\sigma}{z-\mu}\right)^{\xi}\right\}, \end{aligned} \tag{14.16}$$

where $\mu\sigma$ and ξ, the location, scale, and shape parameters, respectively, are greater than zero. Figs. 14.12 and 14.13 below give a pictorial view of $g(z; \mu, \sigma, \xi)$ and $G(z)$ for specific values of the parameters.

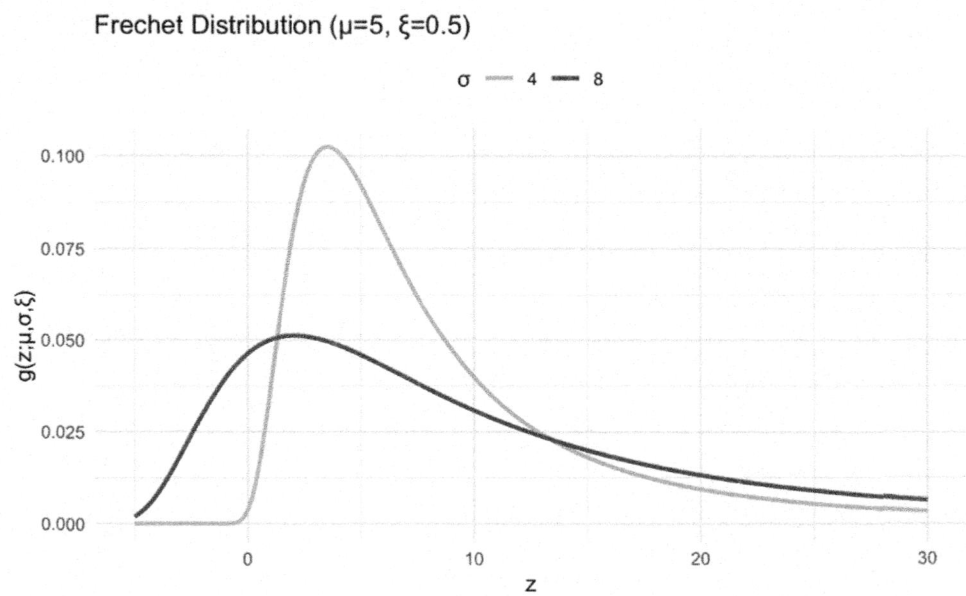

FIGURE 14.12 The Fréchet pdf.

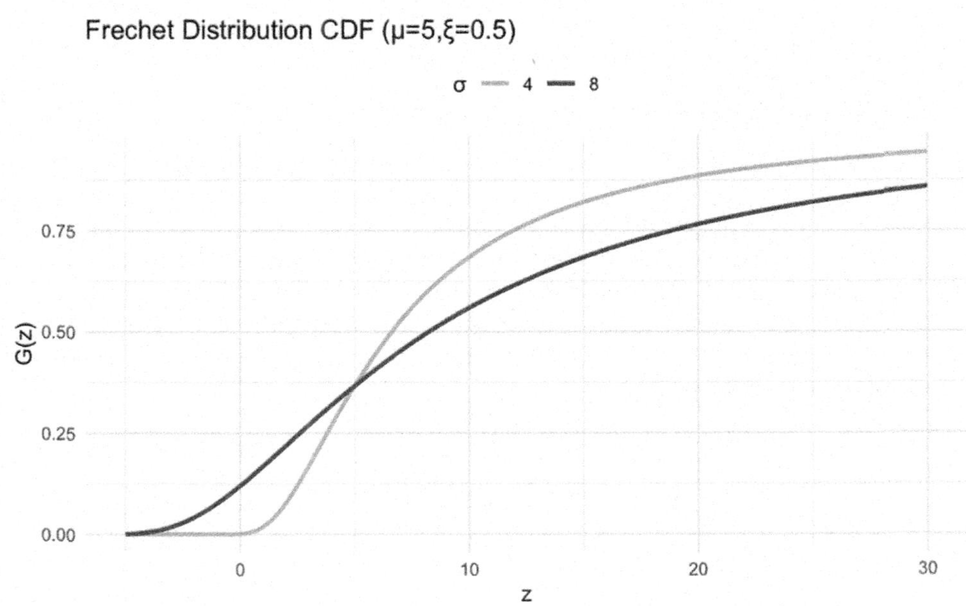

FIGURE 14.13 The Fréchet cumulative distribution.

To apply the Fréchet distribution to real data, we need to estimate the parameters that drive the subject distribution. We obtain the maximum likelihood estimator of μ, σ, ξ in a similar manner as the Gumbel pdf. That is, we form the likelihood estimate of (14.15), the pdf of the Fréchet distribution, $L[g(z;\ \mu,\ \sigma,\ \xi)]$ and $L[g(z;\ \mu,\ \sigma,\ \xi)]$. Then we take the partial derivatives of $\ln L[g(z;\ \mu,\ \sigma,\ \xi)]$, with respect to μ, σ, and ξ, and setting the resulting three equations to zero, we have the following system of likelihood equations, for $\xi > 0$,

$$\frac{m}{\hat{\xi}} + \frac{m \sum_{i=1}^{m} (z_i - \hat{\mu})^{-\hat{\xi}} \ln (z_i - \hat{\mu})}{\sum_{i=1}^{m} (z_i - \hat{\mu})^{-\hat{\xi}}} = \sum_{i=1}^{m} \ln (z_i - \hat{\mu}),$$

$$\frac{m\hat{\xi} \sum_{i=1}^{m} (z_i - \hat{\mu})^{-(\hat{\xi}+1)}}{\sum_{i=1}^{m} (z_i - \hat{\mu})^{-\hat{\xi}}} = (\hat{\xi} + 1) \sum_{i=1}^{m} \frac{1}{z_i - \hat{\mu}},$$

and

$$\hat{\sigma} = \left\{ \frac{1}{m} \sum_{i=1}^{m} (z_i - \hat{\mu})^{-\hat{\xi}} \right\}^{-\frac{1}{\hat{\xi}}}.$$

These three nonlinear equations cannot be solved to obtain exact values of the parameters. Thus, we use numerical iterative methods to obtain the necessary approximate estimates, $\hat{\mu}$, $\hat{\sigma}^2$, $\hat{\xi} > 0$.

14.5.1 Properties of the Fréchet distribution

Some properties of the Fréchet distribution are as follows:

i. It is useful to analyze datasets with heavy tails, where the extreme values are larger than the mean.
ii. The expected value, $E[Z]$, exists only if $\xi > 1$ and is given by

$$E(Z) = \Gamma\left(1 - \frac{1}{\xi}\right), \ \xi > 1,$$

where $\Gamma(z)$ is the gamma function.
iii. The variance, $V(Z)$, exists only if $\xi > 2$ and is given by

$$V(Z) = \Gamma\left(1 - \frac{2}{\xi}\right) - \left\{\Gamma\left(1 - \frac{1}{\xi}\right)\right\}^2, \ \xi > 2.$$

iv. The mode of $Z = \frac{\xi}{\xi+1}$, $\xi > 0$.

14.5.2 The Fréchet Probability Distribution: Application P-tau Protein

Alzheimer's disease is the most common form of dementia and is a very serious disease affecting the brain. It is an invisible disease that destroys memory and important mental functions so that a patient cannot assume daily tasks. One of the proteins in the brain called **phosphorylated tau protein, P-tau protein**, measured in pg/mL contributes to Alzheimer's disease. From the **Alzheimer's Disease Neuroimaging Initiative, ADNI**, database we obtain 210 measurements of P-tau protein in pg/mL. A histogram of the data is given by Fig. 14.14, below.

Our objective is to identify the probability distribution that best fits the P-tau protein data. Both the Kolmogorov–Smirnov and Anderson–Darling tests confirmed that the data fits the Fréchet probability distribution. The approximate MLE of the three parameters are

$$\text{Location: } \hat{\mu} = 20.64 \text{ pg/mL}$$

$$\text{Scale: } \hat{\sigma} = 9.49 \text{ pg/mL}$$

$$\text{Shape: } \hat{\xi} = 2.13.$$

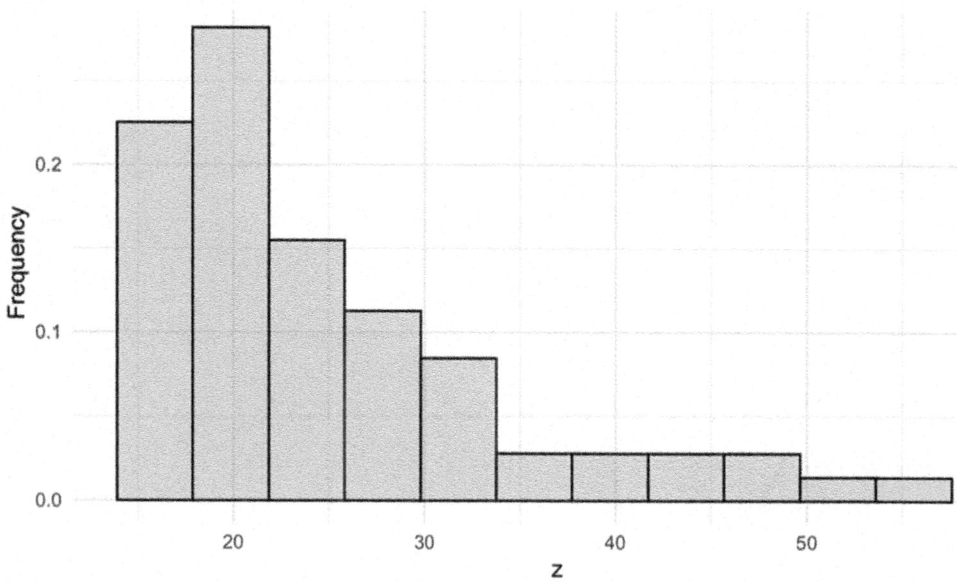

FIGURE 14.14 Histogram of P-tau protein level (pg/mL).

The Fréchet probability density is given by

$$g(z;\ 20.64, 9.49, 2.13) = \frac{2.13}{9.49}\left(\frac{9.49}{z - 20.64}\right)^{3.13} \exp\left\{-\left(\frac{9.49}{z - 20.64}\right)^{2.13}\right\}.$$

The cumulative Fréchet distribution is given by

$$G(z) = \exp\left\{-\left(\frac{9.49}{z - 20.64}\right)^{2.13}\right\}.$$

A graph of $g(z;\ 20.64, 9.49, 2.13)$ is given below by Fig. 14.15.
A graph of $G(z)$ of the P-tau protein is given below by Fig. 14.16.
Thus, for a randomly chosen patient from a given population, the probability of the P-tau protein level will be less than 65 pg/mL is approximately 70%. That is,

$$G(z) = P(z \le 65) \approx 70\%.$$

The expected value of the P-tau protein is

$$E(Z) = \Gamma\left(1 - \frac{1}{2.13}\right) = \Gamma(0.53)$$

$$= 1.77\text{pg/mL}.$$

Thus, a patient selected at random is expected to have 1.77 pg/mL of P-tau protein.
The variance of the subject data exists only if the shape parameter $\hat{\xi} > 2$; thus,

$$V(Z) = \Gamma\left(1 - \frac{2}{\xi}\right) - \left\{\Gamma\left(1 - \frac{1}{\xi}\right)\right\}^2$$

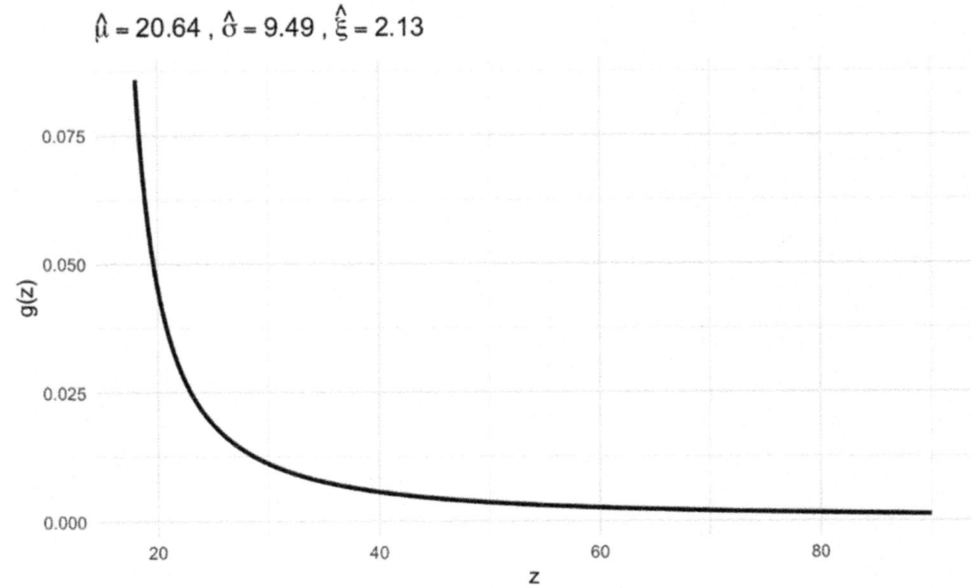

FIGURE 14.15 Fréchet pdf of P-tau protein, pg/mL.

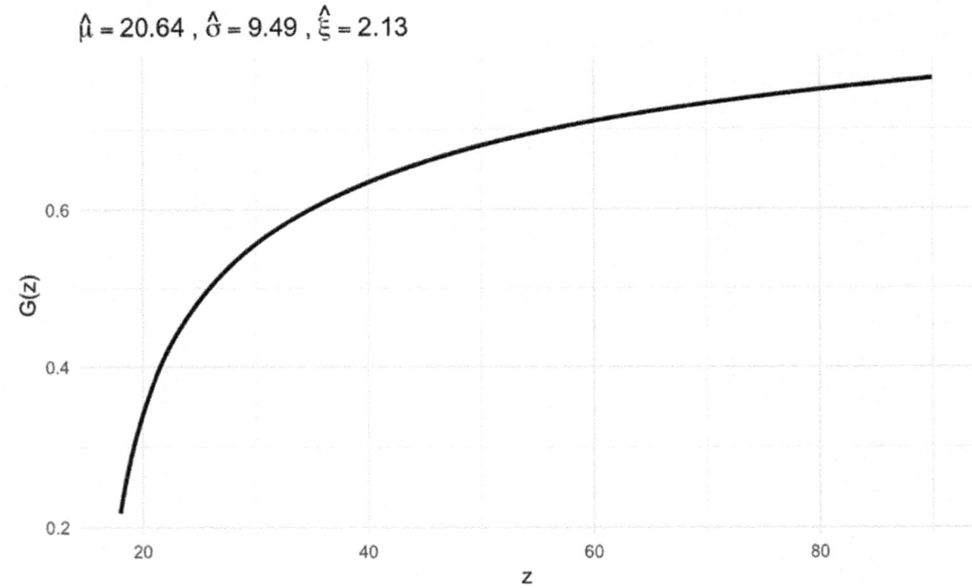

FIGURE 14.16 $G(z)$ for P-tau protein, pg/mL.

$$= \Gamma(0.6) - \{\Gamma(0.53)\}^2$$

$$= 13.65\text{pg/mL}.$$

This high variance in comparison to the mean suggests that individual P-tau measurements differ substantially from the mean, which means that the patients may have much higher or much lower P-tau levels than 1.77.

14.6 THE WEIBULL PROBABILITY DISTRIBUTION

The **Weibull** probability distribution was discovered by **Waloddi Weibull** (1887–1979). Here we consider the Weibull distribution as a special case of the GEV when $\xi < 0$. The Weibull pdf is very important in solving problems in the environment, financial and economic systems, hydrology, insurance, manufacturing, and electrical engineering to decide the overvoltage in electrical systems, but the most important use is identifying the reliability of engineering systems. It is also used in life sciences to determine the survival times of patients with different diseases.

For the shape parameter $\xi < 0$, the GEV distribution $G(z)$ reduces to the Weibull cumulative distribution, that is,

$$G(z) = 1 - \exp\left\{-\left(\frac{z-\mu}{\sigma}\right)^{\xi}\right\},\ z \geq \xi \tag{14.17}$$

and the parameter $\mu > 0$, $\sigma > 0$, and $\xi > 0$.

If we take the derivative of $G(z)$ with respect to z, we obtain the Weibull pdf, given by

$$g(z;\mu,\sigma,\xi) = \frac{\xi}{\sigma}\left(\frac{z-\mu}{\sigma}\right)^{\xi-1} \exp\left\{-\left(\frac{z-\mu}{\sigma}\right)^{\xi}\right\},\ z \geq \mu, \tag{14.18}$$

where $\mu > 0$, $\sigma > 0$, and $\xi > 0$. The Weibull distribution is also known as the **Type III** extreme value distribution.

The graphs of Eqs. (14.17) and (14.18) for different values of the parameters, μ, σ, and ξ, are given by Figs. 14.17 and 14.18 below:

As previously, to obtain approximate MLE of the parameters μ, σ, and ξ, we obtain the maximum likelihood of (14.18), $L[g(z;\mu,\sigma,\xi)]$ and $\ln L[g(z;\mu,\sigma,\xi)]$, then take the partial derivatives with respect to the parameters μ, σ, and ξ, and set the three resulting equations to zero. Thus, we have the three nonlinear maximum likelihood equations for $\xi < 0$, that is,

$$\frac{m}{\hat{\xi}} + \sum_{i=1}^{m} \ln(z_i - \hat{\mu}) = \frac{m\sum_{i=1}^{m}(z_i-\hat{\mu})^{\hat{\xi}}\ln(z_i-\hat{\mu})}{\sum_{i=1}^{m}(z_i-\hat{\mu})^{-\hat{\xi}}},$$

$$\frac{m\hat{\xi}\sum_{i=1}^{m}(z_i-\hat{\mu})^{-(\hat{\xi}-1)}}{\sum_{i=1}^{m}(z_i-\hat{\mu})^{-\hat{\xi}}} = (\hat{\xi}-1)\sum_{i=1}^{m}\frac{1}{z_i-\hat{\mu}},$$

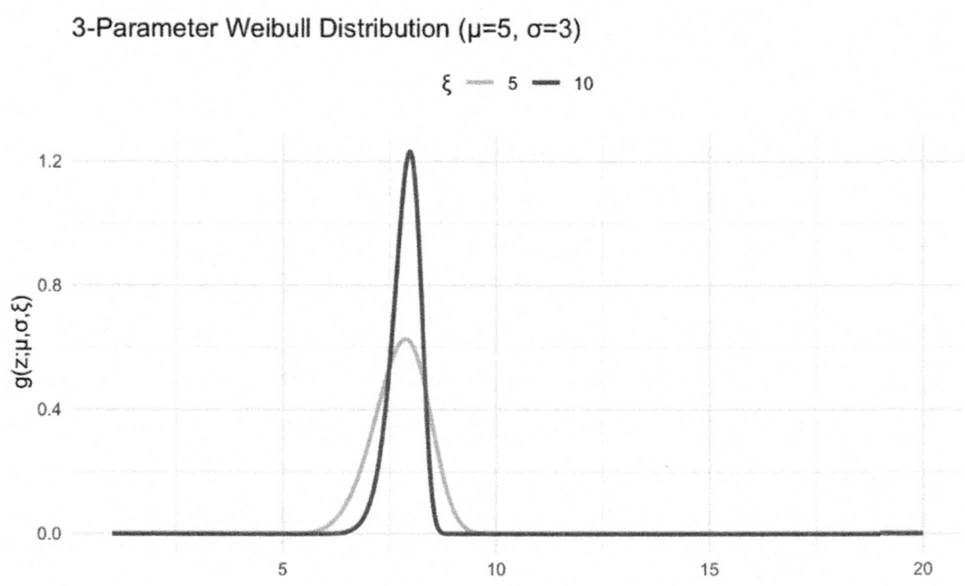

FIGURE 14.17 The Weibull pdf.

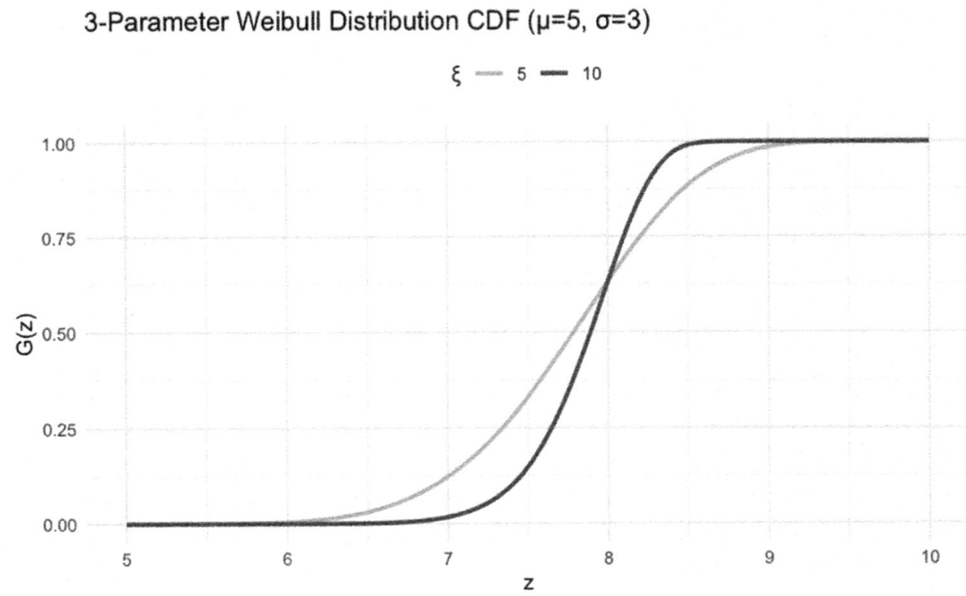

FIGURE 14.18 The Weibull cumulative distribution.

and

$$\hat{\sigma} = \left\{ \frac{1}{m} \sum_{i=1}^{m} (z_i - \hat{\mu})^{-\xi} \right\}^{\frac{1}{\xi}}.$$

We cannot solve these systems of equations to obtain exact estimates of μ, σ, and ξ, but we can use various iterative numerical methods to obtain approximate estimates of the parameters.

The Weibull pdf is commonly used in applications with two parameters because of the difficulties involving estimating three parameters. However, the three parameters will give you a much better fit of a given data set that follows the Weibull pdf than the two parameters and thus a better probabilistic characterization of the phenomenon of interest.

14.6.1 Properties of the Weibull Distribution

Some interesting properties of the Weibull pdf are as follows:

i. The location parameter, μ shifts the behavior of the distribution along the z-axis.
ii. The scale parameter, σ influences the spread of distribution.
iii. The shape parameter, ξ identifies the overall behavior of the distribution.
iv. The expected value of Z is given by

$$E(Z) = \mu + \sigma\Gamma\left(1 + \frac{2}{\xi}\right) - \left\{\Gamma\left(1 + \frac{1}{\xi}\right)\right\}^2,$$

where $\Gamma(.)$ is the gamma function.

v. The variance of Z is given by

$$Var(Z) = \mu^2\left\{\Gamma\left(1 + \frac{2}{\xi}\right) - \left\{\Gamma\left(1 + \frac{1}{\xi}\right)\right\}^2\right\}.$$

vi. The commonly used two-parameter Weibull pdf is obtained by setting the shape parameter $\xi = 0$, and it depends only on μ and σ.

The estimated values of μ, σ, and ξ will be obtained by solving numerically the systems of three maximum likelihood equations.

14.6.2 The Three-Parameter Weibull PD: Application to Carbon Dioxide, CO_2, in the Atmosphere

Carbon Dioxide, CO_2, in the atmosphere, along with the atmospheric temperature, constitutes **Global Warming**. Thus, knowing the probabilistic behavior of CO_2 will give us useful information. CO_2 data was collected by the **Scripps Institution of Oceanography**, **SIO**, on a monthly basis in values and expressed in parts per million, ppm, on the coast of Hawaii, from 1985 to 2004. A histogram of the CO_2 data is given by the histogram below in Fig. 14.19.

Utilizing two goodness-of-fit tests, Kolmogorov–Smirnov and Anderson–Darling, we have identified that the best fit of the CO_2 data is the three-parameter **Weibull** probability distribution.

The approximate maximum likelihood estimate of the three parameters are

$$\text{Location: } \hat{\mu} = 343.7 \text{ ppm}$$

$$\text{Scale: } \hat{\sigma} = 23.0 \text{ ppm}$$

$$\text{Shape: } \hat{\xi} = 2.78.$$

Thus, the probabilistic behavior of CO_2 in the United States is characterized by the **Weibull** PD, given by

$$g(z; 343.7, 23, 2.78) = \frac{2.78}{23}\left(\frac{z - 343.7}{23}\right)^{1.78} \exp\left\{-\left(\frac{z - 343.7}{23}\right)^{2.78}\right\},$$

and the cumulative Weibull PD is given by

$$G(z) = 1 - \exp\left\{-\left(\frac{z - 343.7}{23}\right)^{2.78}\right\}.$$

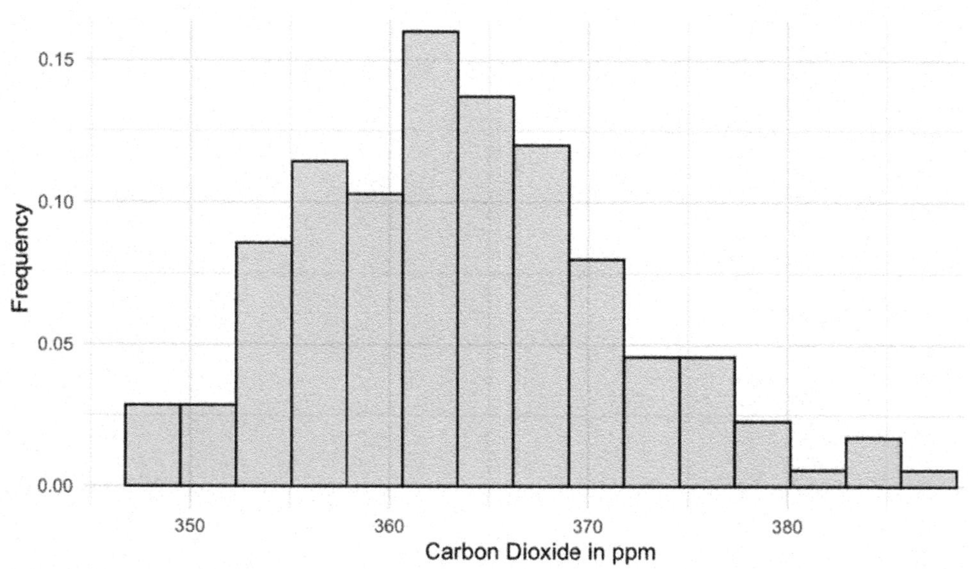

FIGURE 14.19 Histogram of CO_2 in the atmosphere in the United States.

A graph of the $g(z;\, 343.7, 23, 2.78)$ and $G(z)$ is given below by Figs. 14.20 and 14.21, respectively.

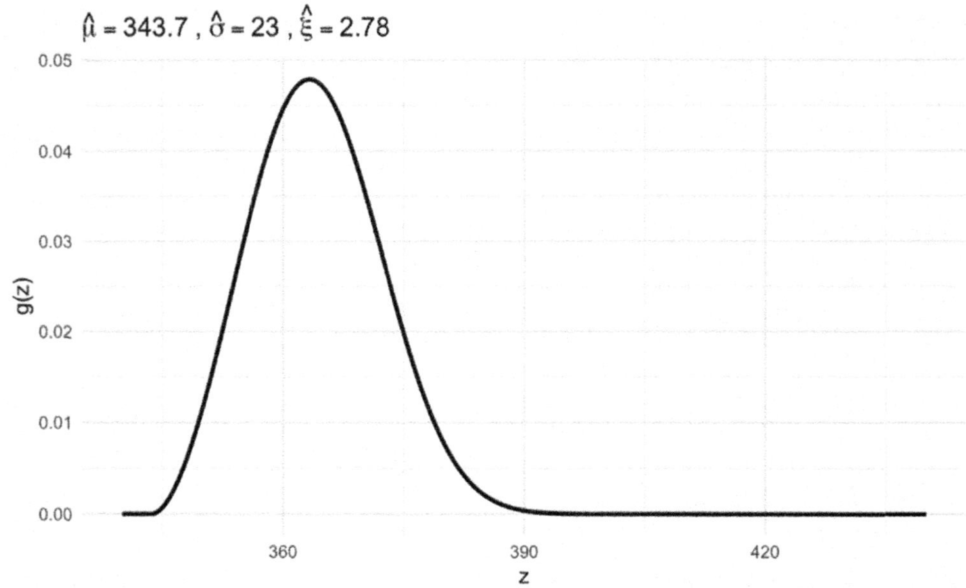

FIGURE 14.20 Probability density of Weibull distribution.

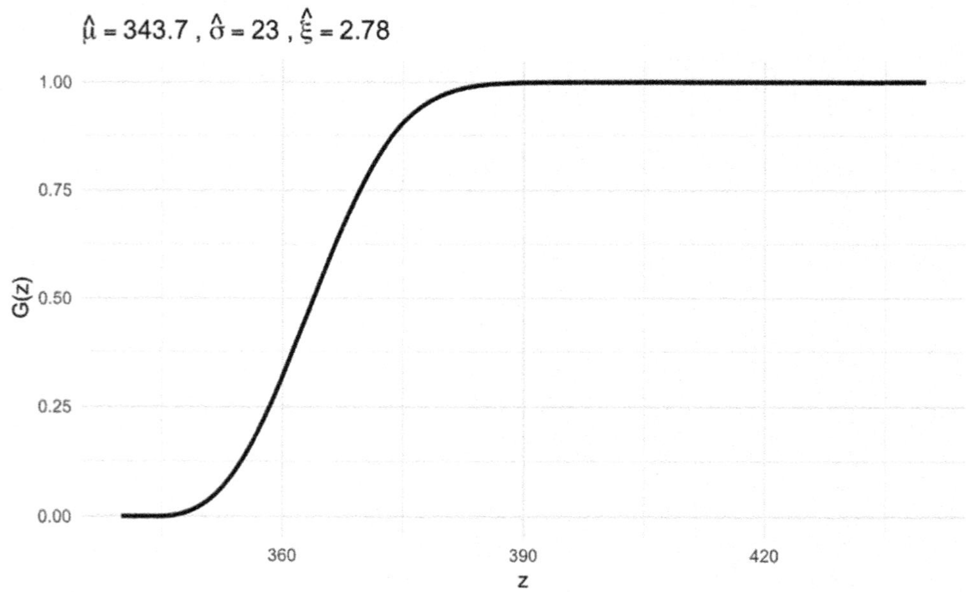

FIGURE 14.21 Weibull cumulative distribution.

We can use the above figures to obtain information, such as what is the probability that the CO_2 in the atmosphere will be less than 364 ppm? That is,

$$P(z \le 364) \approx 59\%,$$

or the probability of CO_2 in the atmosphere will be between 377.67 and 377.88 ppm, that is,

$$P(377.67 \le z \le 377.88) \approx 0.013.$$

The expected value of CO_2 in the atmosphere is

$$E(Z) = \hat{\mu} + \hat{\sigma}\Gamma\left(1 + \frac{2}{\hat{\xi}}\right) - \left\{\Gamma\left(1 + \frac{1}{\hat{\xi}}\right)\right\}^2$$
$$= 343.7 + 23\Gamma(1.72) - \{\Gamma(1.36)\}^2$$
$$= 354.8\text{ppm},$$

and the variance is

$$V(Z) = \mu^2\left\{\Gamma\left(1 + \frac{2}{\xi}\right) - \left\{\Gamma\left(1 + \frac{1}{\xi}\right)\right\}^2\right\}$$
$$= (343.7)^2\{\Gamma(1.72) - \{\Gamma(1.36)\}^2\}$$
$$= 62.68.$$

The results we have found are important in monitoring carbon dioxide in the atmosphere in the United States.

14.7 THE GENERALIZED PARETO PROBABILITY DISTRIBUTION

In probability and statistics, we deal with the **Pareto** probability distribution, **Ppdf**, and the **Generalized Pareto** probability distribution, **GPpdf**. The **Ppdf** is a special case of **GPpdf** with a special parameter, ζ_m, that we call the threshold, which represents the minimum possible value of a phenomenon of interest, whereas the **GPpdf** can be considered a special case of the GEV theory that concentrates on the minimum extreme. The **Ppdf** is used in several areas in social sciences to analyze phenomenon such as income distribution, population growth, actuarial sciences, and many other observable phenomena. It was applied to characterize the wealth of our society, that is, a large portion of the wealth is held by a small (minimum) fraction of the population of our society. The **GPpdf** can be considered as an extension of the **Ppdf** by introducing an additional parameter, ξ, the shape parameter. The **GPpdf** is used to characterize the distributional behavior of data beyond the threshold point, ζ_m, and also used to characterize a broader range behavior of the tails.

The **GPpdf** is used to analyze extreme events in environmental problems, in engineering, in economics and finance, in life and social sciences, among others. We conclude that the **Ppdf** is used to study problems with heavy-tailed data, while the **GPpdf** offers more flexible behavior of the tails, light, heavy, bounded in line with EVT.

14.7.1 The Two-Parameter Pareto Probability Distribution

Let $z_1, z_2, \ldots, z_n$ be the observations of a random variable Z, and $z_m > 0$ be the threshold, the minimum possible value of the observations, we can define the **Ppdf** bywhere z_m is the threshold, $\xi > 0$ is the shape parameter, also known as the **Pareto** index. A small ξ indicates a heavy tail and large ξ indicates a light tail.

$$g(z; z_m, \xi) = \begin{cases} \frac{\xi z^{\xi}}{z^{\xi+1}}, & z_m \le \xi \\ 0, & \xi < z_m \end{cases}$$

The two-parameter **Pareto** cumulative probability distribution is given by

$$G(z) = P(Z \ge z_m) = 1 - P(Z \le z_m)$$

$$= \begin{cases} 1 - \left(\frac{z_m}{z}\right)^{\xi}, & z_m \le \xi \\ 0, & \xi < z_m. \end{cases}$$

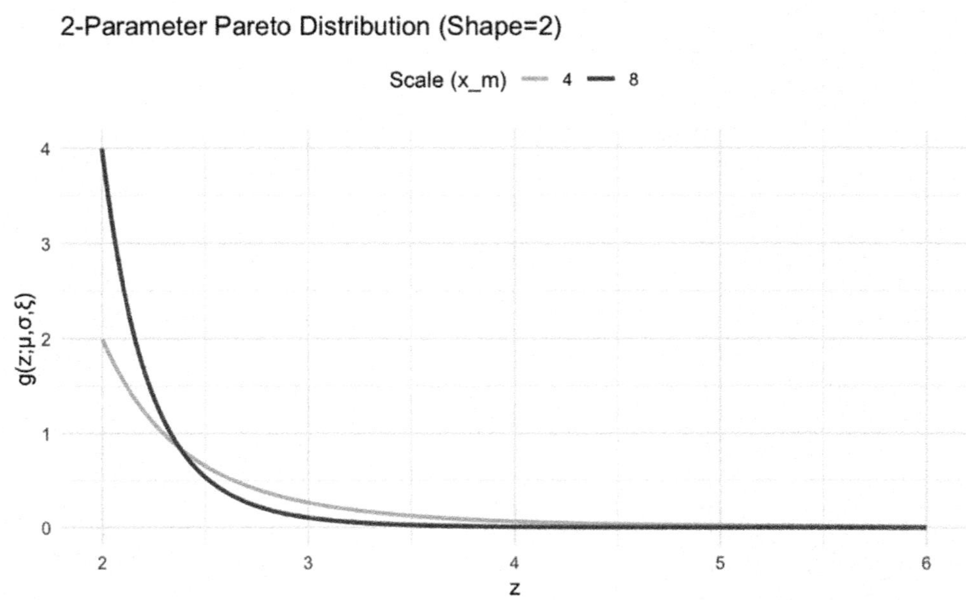

FIGURE 14.22 The Pareto pdf.

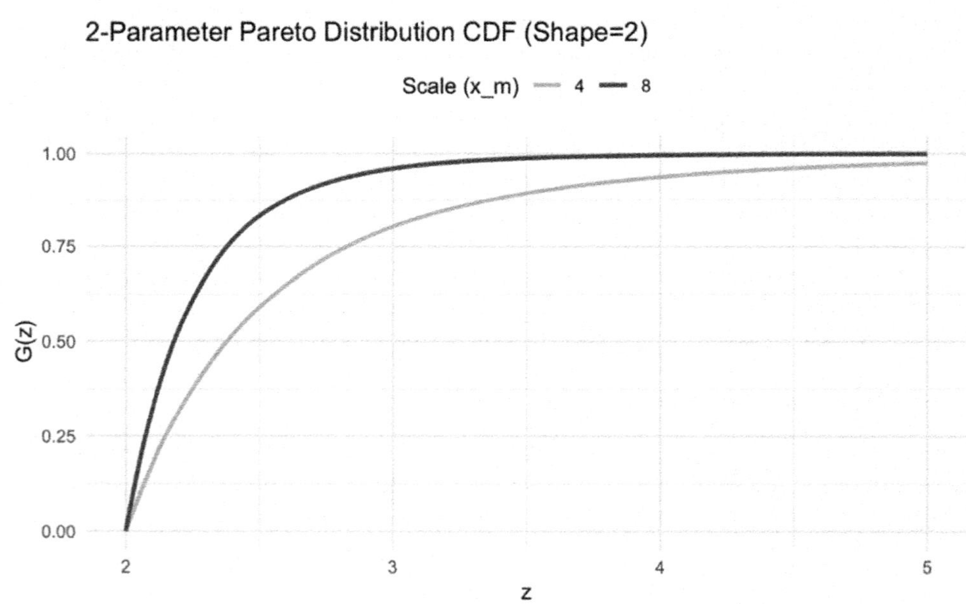

FIGURE 14.23 The Pareto cumulative distribution.

Figs. 14.22 and 14.23 below give a graphical view of **Ppdf** and the cumulative **Ppdf**, for various values of the parameter z_m and ξ.

The expected value and the variance of the **Ppdf** is given by

$$E[Z] = \frac{\xi z_m}{(\xi - 1)}, \; \xi > 1,$$

and

$$Var[Z] = \frac{\xi z_m^{\ 2}}{\{(\xi - 1)^2(\xi - 2)\}}, \ \xi > 2.$$

The maximum likelihood estimate of ξ, the shape parameter, is obtained by taking the ln of the likelihood function of the **Ppdf**, that is, $\ln L\,[g\,(z;z_m,\,\xi)]$ and taking the derivative with respect to z, results in the equation

$$\frac{n}{\xi} + n \ln(z_m) - \sum_{i=1}^{n} \ln(z_i) = 0.$$

Simplifying the above equation, we have

$$\hat{\xi} = \frac{n}{\sum_{i=1}^{n} \ln\left(\frac{z_i}{z_m}\right)},$$

where n is the random sample size, that is, given, $z_1, z_2, \ldots, z_n$ and z_m is the threshold, minimum observation.

14.7.2 The Three-Parameter Generalized Pareto Probability Distribution

The Generalized Pareto probability distribution, **GPpdf**, plays an important role in analyzing and modeling extreme events in various fields. It is used to analyze large insurance claims, precipitation data, maximum rainfall, wind speed, breaking strength of materials, reliability engineering, finance and economics, among many others. The **GPpdf** is an extension of the **Ppdf** by including an additional parameter, μ, the location parameter. Let $z_1, z_2, \ldots, z_n$ be a random sample of the random variable Z, with z_m being the threshold. The **GPpdf** of Z is given by

$$g(z; z_m, \sigma, \xi) = \frac{1}{\sigma}\left(1 + \frac{\xi(z - z_m)}{\sigma}\right)^{\left(-\frac{1}{\xi - 1}\right)}, \ z \geq z_m, \ \xi \geq 0 \text{ and } \sigma > 0.$$

Note that the location parameter μ is now z_m the threshold, where the GPpdf begins.
The cumulative **GP distribution** is given by

$$G(z) = 1 - \left(1 + \frac{\xi(z - z_m)}{\sigma}\right)^{-\frac{1}{\xi}}, \ \xi \neq 0, \ z \geq z_m \text{ and } \sigma > 0.$$

If $\xi = 0$, $G(z)$ reduces to

$$G(z) = 1 - \exp\left(-\frac{z - z_m}{\sigma}\right).$$

The **GPpdf** is very useful in analyzing and modeling distribution with **light**, **heavy**, and **bounded tails** depending on the value of the shape parameter ξ.
Figs. 14.24 and 14.25, below, illustrate graphically **GPpdf** and the cumulative **GPdf** for various values of the parameters $\mu = z_m$, $\sigma > 0$, and ξ.
The expected value of the **GPpdf** is given by

$$E[Z] = z_m + \frac{\sigma}{1 - \xi}, \ \xi < 1.$$

The expected value does not exist for $\xi \geq 1$.
The variance of the **GPpdf** is given by

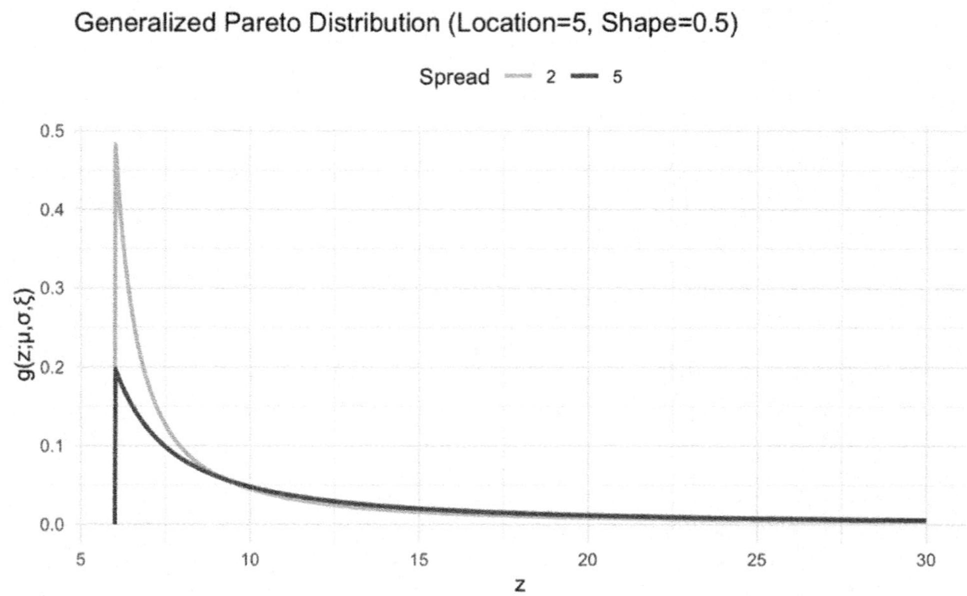

FIGURE 14.24 The Generalized Pareto pdf.

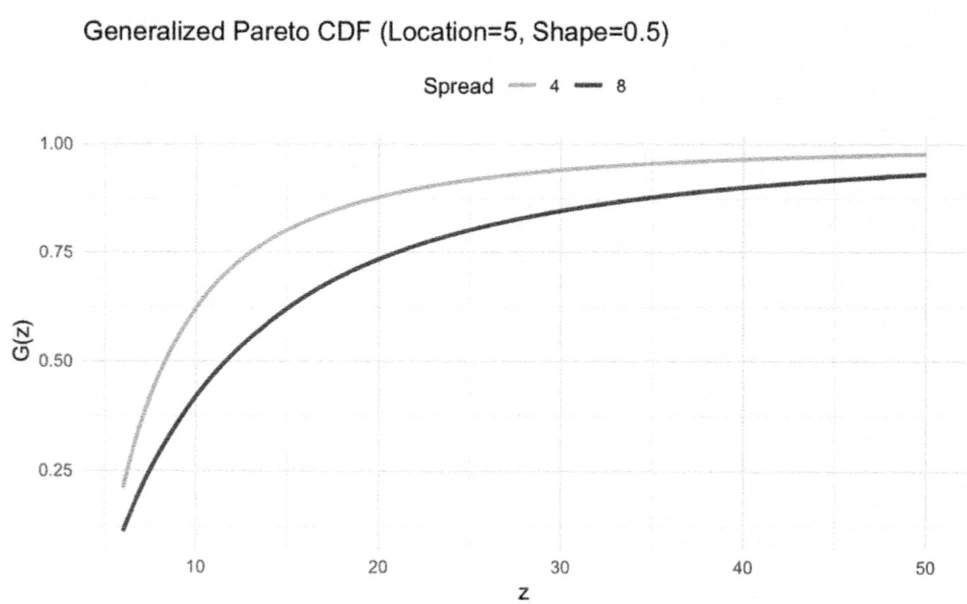

FIGURE 14.25 The Generalized Pareto cumulative distribution.

$$Var\,[Z] = \frac{\sigma^2}{(1-\xi)^2(1-2\xi)},\ \xi < \frac{1}{2}.$$

The variance value does not exist for $\xi \geq \frac{1}{2}$.

To obtain estimates of the parameters of the **GPpdf**, as previously, we use the maximum likelihood method. The likelihood function for a random sample $z_1, z_2, \ldots, z_n$ for $z_i \geq z_m$ is given by

$$L\,[g(z; z_m, \sigma, \xi)] = \prod_{i=1}^{n} \frac{1}{\sigma}(1 + \xi\,(z_i - z_m))^{-\xi-1},$$

for

$$1 + \frac{\xi(z_i - z_m)}{\sigma} > 0,$$

and taking the ln of the likelihood function, we have

$$\ln L[g(z; z_m, \sigma, \xi)] = -n \ln \sigma - \left(1 + \frac{1}{\xi}\right) \sum_{i=1}^{n} \ln\left(1 + \frac{\xi(z_i - z_m)}{\sigma}\right).$$

Taking the partial derivative with respect to z_m, σ, and ξ of the $\ln L[g(z;z_m, \sigma, \xi)]$ and setting them equal to zero, we have the three maximum likelihood equations, that is,

$$\frac{\partial \ln L[g(z; z_m, \sigma, \xi)]}{\partial z_m} = \left(\frac{\xi + 1}{\sigma}\right) \sum_{i=1}^{n} \frac{1}{1 + \frac{\xi(z_i - z_m)}{\sigma}} = 0,$$

$$\frac{\partial \ln L[g(z; z_m, \sigma, \xi)]}{\partial \sigma} = \frac{n}{\sigma} + (\xi + 1) \sum_{i=1}^{n} \frac{(z_i - \xi)}{[\sigma^2 + \sigma\xi(z_i - z_m)]} = 0,$$

and

$$\frac{\partial \ln L[g(z; z_m, \sigma, \xi)]}{\partial \xi} = \frac{1}{\xi^2} + \sum_{i=1}^{n} \ln\left(1 + \frac{\xi(z_i - z_m)}{\sigma}\right) - \left(1 + \frac{1}{\xi}\right) \sum_{i=1}^{n} \frac{(z_i - z_m)}{1 + \frac{\xi(z_i - z_m)}{\sigma}} = 0.$$

To solve these systems of nonlinear equations, we rely on numerical iterative methods. The estimates that we will obtain are approximate estimates of the parameters.

14.7.3 Properties of the GPpdf

Some of the key properties of the **GPpdf** are as follows:

i. The shape parameter ξ identifies the behavior of the tails.
ii. $\xi = 0$, indicates light tail and $z \geq z_m$, where the distribution begins.
iii. $\xi > 0$, indicates heavy tails and $z \geq z_m$.
iv. $\xi < 0$, indicates bounded tails and $z_m \leq z \leq z_m - \frac{\sigma}{\xi}$.
v. Some special cases of the **GPpdf** are that when $\xi > 0$, $\sigma = z_m\xi$, $\mu = z_m$, it results in the **Pareto** probability distribution, and when the shape parameter $\xi = 0$, it results in the **exponential** probability distribution.

14.7.4 The Three-Parameter Generalized Pareto Distribution: Application to Brain Cancer

There are two categories of brain cancer, identified as primary and metastatic. Primary brain cancer is composed of glial cells, and it is called a **gliomas** cancer. Gliomas is the most cancerous tumor arising in the brain. Approximately 78% of malignant brain tumors are gliomas. We obtained the survival times of 6426 patients with gliomas cancer from the National Cancer Institute, NCI. The survival time was given in months, that is, the time that the patient with gliomas cancer survived after the patient was diagnosed with the disease. Our objective is to identify the probability distribution that probabilistically characterizes the survival times of gliomas cancer patients.

The histogram of the survival times data is presented in the following histogram, Fig. 14.26, below:

Both the Kolmogorov–Smirnov and Anderson–Darling goodness-of-fit tests with P-values of 0.4019 and 0.4781, respectively support that the best fit of the data is the three-parameter Generalized **Pareto** distribution, **GPD**. Thus, we proceeded in obtaining approximate estimates of the three parameters of the **GPD** using the maximum likelihood equations, that is,

$$\text{Location: } \hat{\mu} = 0.662 \text{ months}$$

$$\text{Scale: } \hat{\sigma} = 14.395 \text{ months}$$

$$\text{Shape: } \hat{\xi} = -0.292.$$

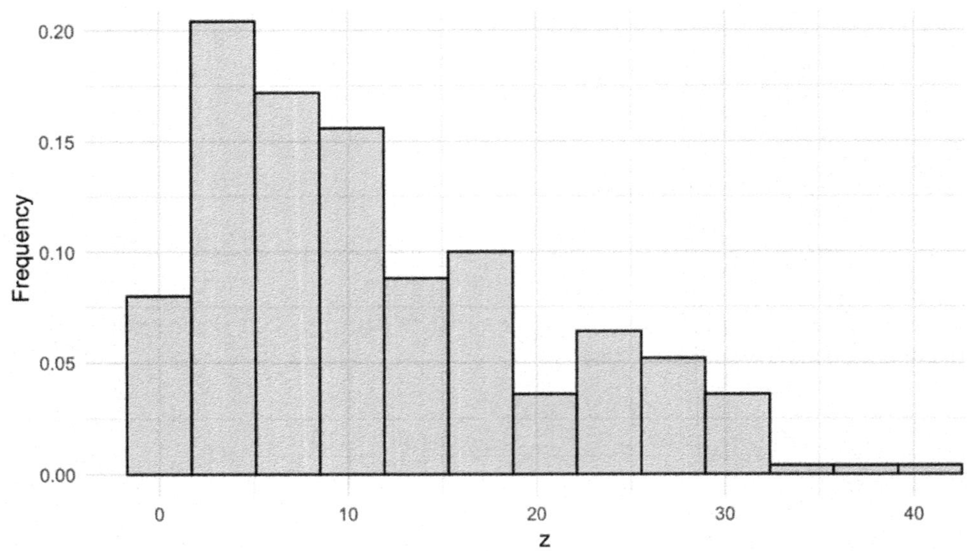

FIGURE 14.26 Histogram of the survival times of gliomas cancer patients.

The probability density that characterizes the probabilistic behavior of the survival times is given by

$$g(t;\hat{\mu},\hat{\sigma},\hat{\xi}) = \begin{cases} \frac{1}{14.395}\left(1+\frac{-0.292(t-0.662)}{14.395}\right)^{\left(-\frac{1}{-0.292}-1\right)}, & \xi \neq 0 \\ \frac{1}{14.395}\exp^{\left(-\frac{-t-0.662}{14.395}\right)}, & \xi = 0. \end{cases}$$

A graph of $g(t;\ 0.662,\ 14.395,\ -0.292)$ is given below in Fig. 14.27.

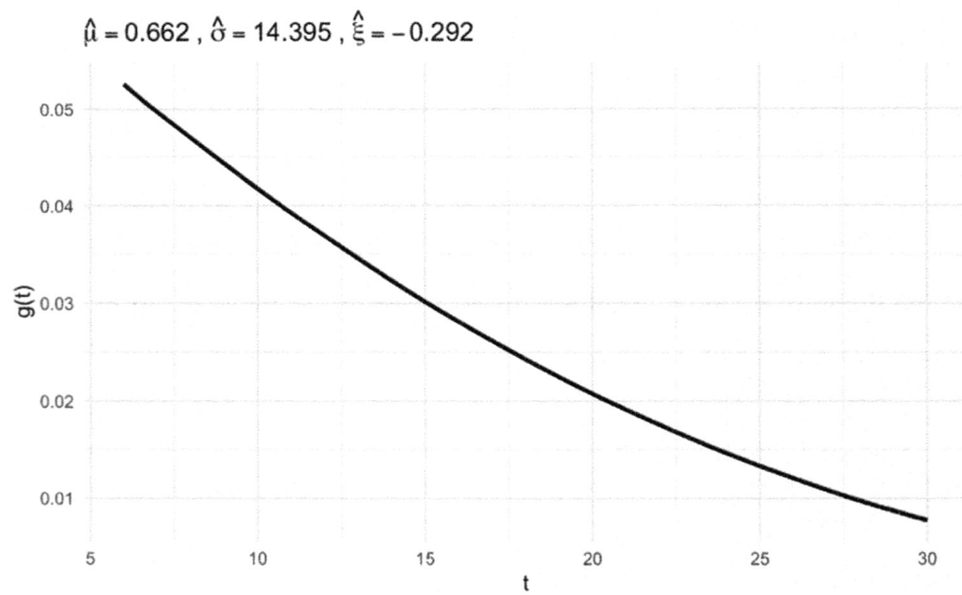

FIGURE 14.27 Probability density of GPD.

The cumulative GPD is given by

$$G(t) = 1 - \left(1 + \frac{-0.292(t-0.662)}{14.395}\right)^{\left(-\frac{1}{-0.292}\right)}.$$

A graph of $G(t)$ for the given parameter estimates is given by Fig. 14.28.

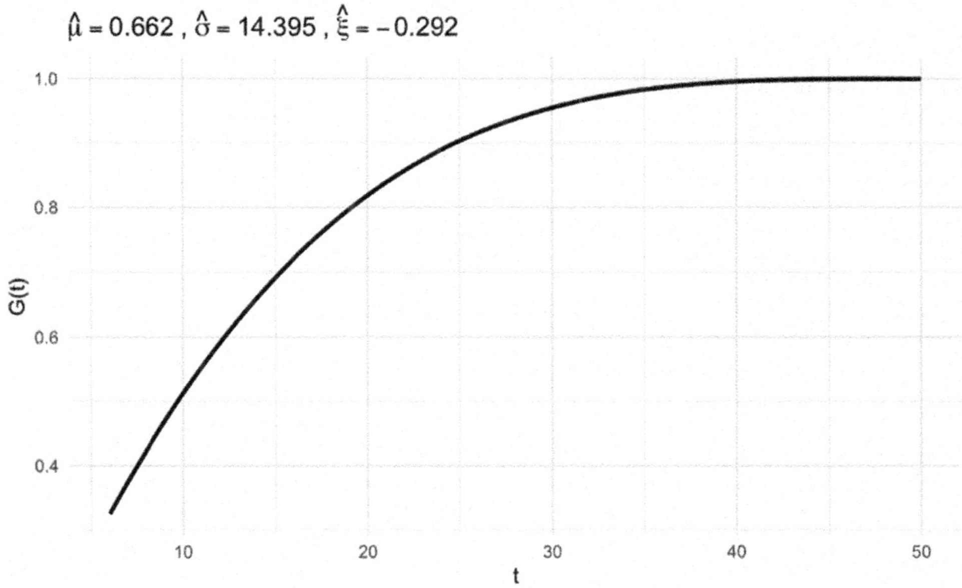

FIGURE 14.28 The cumulative GPD.

We can use the $G(t)$ to obtain various probabilities that a patient once diagnosed with gliomas cancer will survive up to a particular time. For example, what is the probability that a patient with gliomas cancer will survive up to 12 months ($t = 12$) or 25 months ($t = 25$)? Their probabilities are approximately 0.6 (60%) and 0.9 (90%), respectively.

We can also calculate the expected survival time, that is,

$$\begin{aligned} E[T] &= \hat{z}_m + \frac{\hat{\sigma}}{1-\hat{\xi}} \\ &= 0.662 + \frac{14.395}{1-(-0.292)} \\ &= 11.80 \text{ months.} \end{aligned}$$

Thus, we expect an individual diagnosed with gliomas cancer will live 11.80 months.

The variance of the given survival times is

$$\begin{aligned} V[T] &= \frac{\hat{\sigma}^2}{(1-\hat{\xi})^2(1-2\hat{\xi})} \\ &= \frac{(14.395)^2}{(1-(-0.292))^2(1-2(-0.292))} \\ &= 78.37 \text{ months.} \end{aligned}$$

14.8 EXERCISES

14.8.1 For a given phenomenon, Z, we have a random sample of size 100, and it is shown that it follows the **GEV** distribution, with location parameter 15, scale parameter 8, and shape parameter 0.3.

i. State and graph the probability density of the **GEV** distribution.
ii. Find the approximate probabilities that Z assumes a value less than 14.
iii. Obtain the expected value and variance of Z.

iv. State and graph the cumulative distribution of **GEV**.
v. Obtain the approximate probability that Z assumes a variance greater than 16.

14.8.2 It has been verified that a random sample of size n that characterizes the random variable Z follows the **Gumbel** distribution, with location parameter 40, scale parameter 15, and shape parameter 0.

i. State and graph the **Gumbel** probability density of the given data.
ii. Find the approximate probability $P[30 \le Z \le 50]$.
iii. State and graph the cumulative distribution of the **Gumbel**.
iv. Obtain an approximate estimate of $P[Z \ge 55]$.

14.8.3 The monthly rainfall data in the state of Florida for the past 5 years was shown to follow the Fréchet probability, with location parameter 4.5 in, scale parameter 5.1 in, and shape parameter 2.2.

i. State and graph the probability density of the **Fréchet** distribution.
ii. Find and interpret the $P[2 \text{ in} \le R \le 4 \text{ in}]$.
iii. State and graph the cumulative **Fréchet** distribution.
iv. Find and interpret $P[R \le 5 \text{ in}]$.

14.8.4 The monthly data for the past 6 years of carbon dioxide in the atmosphere, CO_2, in EU countries was shown to follow the three-parameter Weibull distribution, with location parameter 300 ppm, scale parameter 18 ppm, and shape parameter 2.9 ppm.

i. State and graph the probability density of the **Weibull** distribution.
ii. Calculate and interpret $P[290 \text{ ppm} \le CO_2 \le 320 \text{ ppm}]$.
iii. Find and interpret $E[CO_2]$ and $Var[CO_2]$.
iv. State and graph the cumulative distribution of **Weibull**.

14.8.5 From the NIH database, we obtain a random sample of 100 patients with breast cancer. From the sample, we obtained the malignant tumor size of each of the patients in mm. It was shown that the data fits the **Generalized Pareto** distribution, with location parameter 45 mm, scale parameter 15 mm, and shape parameter –0.32.

i. State and graph the probability density of the **Generalized Pareto** distribution.
ii. Obtain and interpret the approximate $P[35\text{mm} \le Z \le 50\text{mm}]$.
iii. Find and interpret $E[Z]$ and $Var[Z]$.
iv. State and graph the cumulative distribution of **GPD**.

14.9 CHAPTER SUMMARY

In this chapter, we introduced **EVT**, which is useful in addressing real-world problems where the largest observation, **maximum extreme**, or the smallest observation, **minimum extreme**, is of importance to be studied. It allows us to speak objectively and quantitatively about events which, even though rare, tend to have the largest impacts. We mentioned a number of areas that such problems arise in our daily life. We introduced **Generalized Extreme Value Probability Distribution**, **GEV pdf**, and some of its properties followed by a real-world application. From the GRV pdf, we derived four probability distributions, namely,

- The **Gumbel Probability Distribution**, also known as the **Double Exponential Distribution**.
- The **Fréchet Probability Distribution**.
- The **Weibull Probability Distribution**.
- The **Generalized Pareto Probability Distribution**.

The properties of the above probability distributions are given along with a real application to illustrate their importance.

To estimate the parameters and fit any of the extreme value distribution in R, follow the three-step procedure given below:

1. Load the packages

i. ismev
ii. evd
iii. extRemes

2. Load the data
3. Fit the Particular Probability Distribution

14.10 R-CODES

We will now give general R-codes for the four different types of extreme value distribution. Example 14.3.1 gives how these codes may be modified to particular data set.

EXAMPLE 14.10.1 The Gumbel Probability Distribution

```
loglik_gumbel <- function(params, x) {
  mu <- params[1] # location
  beta <- params[2] # scale (> 0)
  if (beta <= 0) return(-Inf) # invalid parameter
  z <- (x - mu) / beta
  n <- length(x)
  loglik <- -n * log(beta) - sum(z) - sum(exp(-z))
  return(loglik)
  }
  # If we want to generate example data
  > set.seed(123)
  > n <- 100
  > mu_true <- 0
  > beta_true <- 2
  > x <- mu_true - beta_true * log (-log (runif(n))) # simulate Gumbel
  > loglik_gumbel(c(0, 2), x)
  # If we want to estimate parameters using optim
  fit <- optim (par = c(mean (x), sd (x)), # initial guesses
  fn = gumbel_loglik,
  data = x,
  method = "L-BFGS-B,"
  lower = c (-Inf, 1e-6)) # ensure beta > 0
  mu_hat <- fit$par [1]
  beta_hat <- fit$par [2]
  cat ("Estimated mu:," mu_hat, "\n" )
  cat ("Estimated beta:," beta_hat, "\n" )
```

EXAMPLE 14.10.2 The Fréchet Probability Distribution

```
# Fréchet negative log-likelihood
  frechet_loglik <- function (params, data) {
  alpha <- params [1] # shape
  s <- params [2] # scale
  mu <- params [3] # location
  if (alpha <= 0 || s <= 0) return (Inf)
  if (any (data <= mu)) return (Inf)
  z <- (data - mu) / s
  loglik <- sum (log (alpha) - log (s) - (1+ alpha) * log (z) – z ^ ( -alpha))
  return (-loglik)
  }
  # Example data
  set.seed(123)
  n <- 100
```

```
alpha_true <- 2
s_true <- 1
mu_true <- 0
x <- mu_true + s_true * (runif (n) ^ ( -1/alpha_true) - 1) # simulate Fréchet
# Estimate parameters
Fit_frechet <- optim (c (2,1,0), frechet_loglik, data = x,
method = "L-BFGS-B,"
lower = c (le-6, le-6, -Inf ))
fit_frechet$par
```

EXAMPLE 14.10.3 The Weibull Probability Distribution

```
weibull_loglik <- function (params, data) {
  k <- params [1] # shape
  lambda <- params [2] # scale
  if (k <= 0 || lambda <= 0) return (Inf)
  loglik <- sum (log (k) - log(lambda ) + (k - 1) * log (data/lambda) - (data/lambda) ^ k)
  return (-loglik)
  }
  fit_weibull2 <- optim (c(1,1), weibull_loglik, data = x, method= "L-BFGS-B," lower=c(1,1))
  fit_weibull2$par
```

EXAMPLE 14.10.4 The Generalized Pareto Probability Distribution

```
pareto_mle <- function (data) {
  xm <- min (data)
  alpha <- length (data) / sum (log (data/xm))
  return (list(alpha = alpha, xm = xm))
  }
  # Example data
  set.seed(123)
  xm_true <- 1
  alpha_true <- 3
  x <- xm_true * (runif (100) ^ ( -1/alpha_true))
  # Estimate
  fit_ pareto <- pareto_mle (x)
  fit_ pareto
```

Chapter 15

Modeling Issues in Statistical Applications: An Overview

Chapter outline

OBJECTIVE

In this chapter, we discuss some general concepts and useful methods with applications to real-world problems.

Florence Nightingale
(Source: ***http://commons.wikimedia.org/wiki/File:Florence_Nightingale_1920_reproduction.jpg)***

Florence Nightingale (1820–1910) is most remembered as a pioneer of nursing and a reformer of hospital sanitation methods. Her statistical contributions caused Karl Pearson to acknowledge Nightingale as a "prophetess" in the development of applied statistics. Nightingale used data as a tool for improving medical and surgical practices During the Crimean War, she plotted the incidence of preventable deaths in the military and introduced polar-area charts to demonstrate the unnecessary deaths due to unsanitary conditions. With her analysis, Florence Nightingale showed the need for reform and revolutionized the idea that social phenomena could be objectively measured and subjected to mathematical analysis. In addition, she developed a Model Hospital Statistical Form for hospitals to collect and generate data and statistics. She became a Fellow of the Royal Statistical Society in 1858 and an honorary member of the American Statistical Association in 1874.

Mathematical Statistics with Applications in R. DOI: https://doi.org/10.1016/B978-0-443-27548-7.00015-3

15.1 INTRODUCTION

Basically, there can be three major problems in applying the statistical methods that we have studied in the previous chapters to real-world problems. These involve sources of bias, errors in methodology, and the interpretation of the analytical results. Bias occurs in situations or conditions that affect the validity of statistical results. In order for the statistical inferences to be valid, the observed sample must be representative of the target population, and the observed variables must conform to assumptions that underlie the statistical procedures to be used. Of course, the statistical methodology chosen must be also appropriate for the problem under study. We must be careful with the interpretation of the statistical results. For example, in a regression problem, a cause-and-effect relationship may not be warranted, or in a hypothesis testing problem, we may not accept the null hypothesis without exploring the probability of a Type II error. If we present the results graphically, the graphs should be accurate and should reflect the data variations clearly.

In this textbook, we have assumed that a data set is available to us: Either it is a small data set that we can handle without much effort or it is in a computer-readable file. In practical situations, the proper handling of a statistical data set is not an easy task. Going from a stack of disorganized hard copy to online data that are trustworthy, that is, to input, debug, and manipulate the data, is a problem one will face even before one starts the statistical analysis. Here, we will not be dealing with these issues. Interested readers should refer to the references at the end of this book for further study on these aspects.

It is not our aim to discuss comprehensively all the problems that come up in applications. Most of the material presented in this chapter has already been discussed in various parts of the book. One of the problems we face when we study a book of this sort is that the problems of each chapter, say, Chapter 6 on hypothesis testing, we know that we only need to use the techniques of that section, at most of that chapter. For parametric analysis, in Chapter 11, we gave ways to do goodness-of-fit for choosing a particular distribution. In a real-world situation, we will not be able to look at the data analysis in a chapter-by-chapter manner. The purpose of this chapter is to present some methods in a unified way and to discuss generally the various ways in which the techniques developed in previous chapters could be applied to real-world data. Because the material in this chapter is a collection of available techniques, we will not follow the more rigorous pattern of previous chapters, and no proof will be given.

It is very important to mention that every parametric statistical method and also some nonparametric methods are subject to certain assumptions, and when we apply them to real-world problems, we should make every effort to justify these assumptions. If you cannot, it is necessary, when you conclude your analysis and make decisions, that you state that the results are subject to certain assumptions that you could not justify.

15.2 GRAPHICAL METHODS

We first present some useful graphical methods that were not introduced in Chapter 1 on descriptive statistics. Graphical analysis is a very important aspect of any statistical study. Before attempting a complex statistical analysis, summarize the data with a graph. Graphical displays of data analysis help in data exploration, analysis, and presentation and in communication of results. In data analysis, one of the significant steps is to summarize and plot the data. Graphs help in the communication of final results and recommendations inferred from quantitative models. A statistical model is often suggested by an initial graphical analysis. Adequacy of statistical models depends on the model conditions. Because the violations of these model assumptions may sometimes occur as nonlinearities, graphical methods provide an easy and perhaps very effective method of detection. Some examples of graphical displays are the histograms, dotplots, box plots, and scatterplots. Methods of graphing multivariate data are more complex and include scatterplot matrices and icon plots. These are beyond the level of this book.

If we have a data set with one variable (univariate), we first create a dotplot and summary of basic statistics. In a dotplot, we plot the data as dots (one dot for each observation) above the horizontal axis that covers the entire range of observations (see Fig. 15.1). The dotplot will provide us with an idea of the distribution of the data and any unusual behavior of the data that may not be apparent from summary statistics such as mean, median, or standard deviation. The dotplots allow us to visualize the entire distribution of the data set by listing each possible outcome and the frequency of the variable. Other ways of summarizing univariate data, such as histograms, have been discussed in Chapter 1. The histogram differs from the dotplot in that it groups data into categories. We illustrate these problems with several examples.

EXAMPLE 15.2.1

The following data gives the lifetime of 30 light bulbs (rounded to the nearest hour) of a particular type.

1122	922	1146	1120	1079	905	1095	977	1138	966
1150	977	1137	1088	1139	1055	1082	1053	1048	1132
1088	996	1102	1028	1130	1002	990	1052	1116	1135

Construct a dotplot.

Solution

Fig. 15.1 *is the dotplot for these data.*

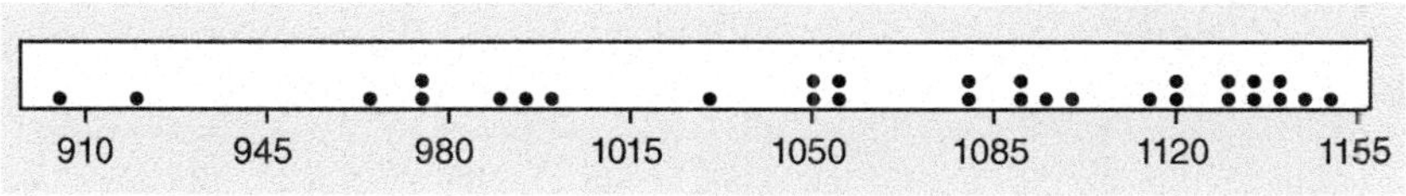

FIGURE 15.1 Dotplot for lifetime of light bulbs.

The dotplot suggests a distribution that is skewed toward the right, because most of the observations are located to the right.

Some of the graphing methods can also be applied to compare two variables, for example, their frequency distributions. For instance, dotplots could also be used to compare bivariate (two variables) or multivariate (many variables) data. When we have independent samples, *side-by-side box plots* could be used for comparing two sample distributions in terms of their centers, dispersions, and skewnesses.

When there are two variables, a *scatterplot* is used as one of the basic graphic tools to examine the relationship between two variables.

The scatterplot in Fig. 15.2 for two variables x and y indicates a possible linear relation between x and y. The strength of the relationship between two variables is often represented through correlation statistics. It should be noted that the correlation coefficient is a single number that is easy to calculate and comprehend, though it only measures the strength of a linear relationship and hence is often used as the primary statistic of interest. However, scatterplots provide information about the strength of association, not necessarily linear, between variables. In addition, scatterplots help us understand other aspects of the data, such as the range. Given n observations on two variables, X and Y, we plot a character or symbol at n points representing (x_i, y_i). If two or more observations in a scatterplot are identical, the plotted symbols will coincide, masking possibly important information.

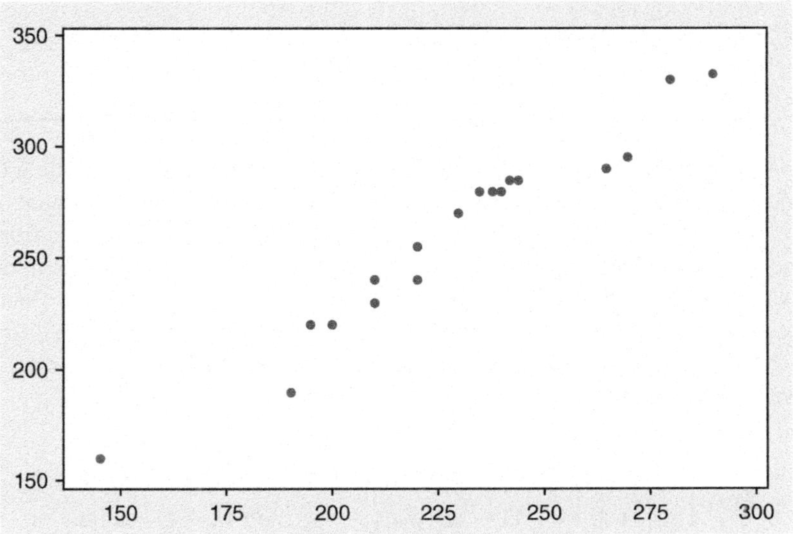

FIGURE 15.2 Scatterplot. A scatter plot showing a positive correlation between two variables.

EXAMPLE 15.2.2

The following data give the cholesterol levels before a certain treatment and after 4 months of treatment.

Before	235	212	277	262	162	212	226	252	185	276
	216	315	289	283	234	223	275	282	311	285
After	233	214	200	266	146	212	238	284	191	247
	244	268	241	289	220	202	221	196	212	247

Draw a scatterplot. Also find the correlation between before- and after-treatment values.

Solution

Fig. 15.3 *is a scatterplot of the data.*

Looking at the scatterplot in Fig. 15.3, *we see a trend in the cholesterol levels before and after the treatment. Correlation of before- and after-treatment data is measured by r, the correlation coefficient, and is given by*

$$r = \frac{\sum_{i=1}^{n}(x_i - \bar{x})(y_i - \bar{y})}{\sqrt{\sum(x_i - \bar{x})^2 \sum(y_i - \bar{y})^2}}.$$

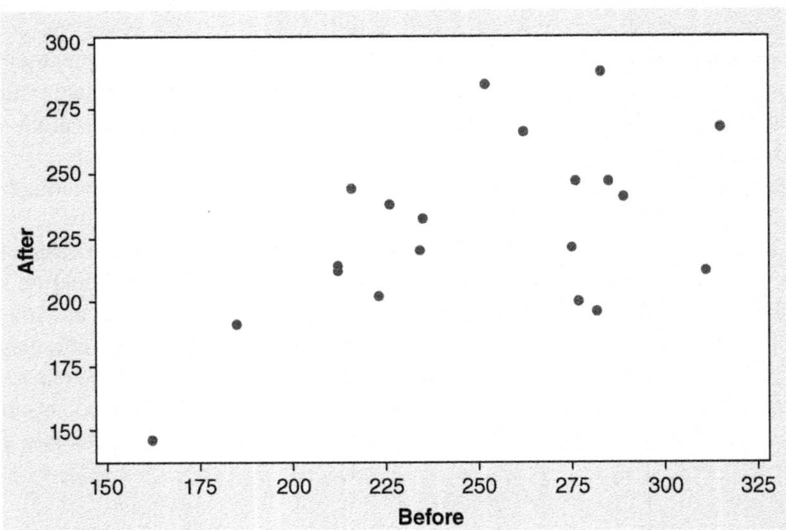

FIGURE 15.3 Scatterplot for cholesterol levels.

The *quantile-quantile* (QQ) *plot* is another useful technique in comparing bivariate data. In a QQ plot, the quantiles of the two samples are plotted against each other. For two distributions that are almost the same, their quantiles would be nearly equal. As a result, the quantiles would plot along the 45-degree line. Deviation of plots from this line can be used to draw inferences about how the two samples differ from one another. If the two sample sizes n_1 and n_2 are equal, then we can draw the QQ plot by graphing the order statistics $x_{(i)}$ and $y_{(i)}$ against each other. If the two samples are not of the same size, then we can use the following procedure to create the QQ plot. If $n_1 > n_2$, then draw the $(1/(n_i + 1))$th quantiles of the two samples against each other. For a large sample, they are the order statistics, $x_{(1)} < \ldots < x_{(n_1)}$. For the smaller sample sizes, the pth quantile value is obtained by using the following formula:

$$\tilde{x}_p = \begin{cases} x_{p(n+1)}, \text{ if } p(n+1) \text{ is an integer} \\ x_{(m)} + [p(n+1) - m](x_{(m+1)} - x_{(m)}) \text{ if } p(n+1) \text{ is a fraction} \end{cases} \quad (15.1)$$

where m denotes the integer part of $p(n\ 1)$. It should be noted that a QQ plot is not useful for paired data because the same quantiles based on the ordered observations do not, in general, come from the same pair.

EXAMPLE 15.2.3

Draw a QQ plot for the data given in Example 15.2.2.

Solution

Here $n_1 = n_2 = 20$. First, sort the data in ascending order.

Before	162	185	212	212	216	223	226	234	235	252
	262	275	276	277	282	283	285	289	311	315
After	146	191	196	200	202	212	212	214	220	221
	233	238	241	244	247	247	266	268	284	289

Because the QQ plot points lie mostly below the 45-degree line, we may conjecture that the cholesterol level before is generally higher than that after Fig. 15.4.

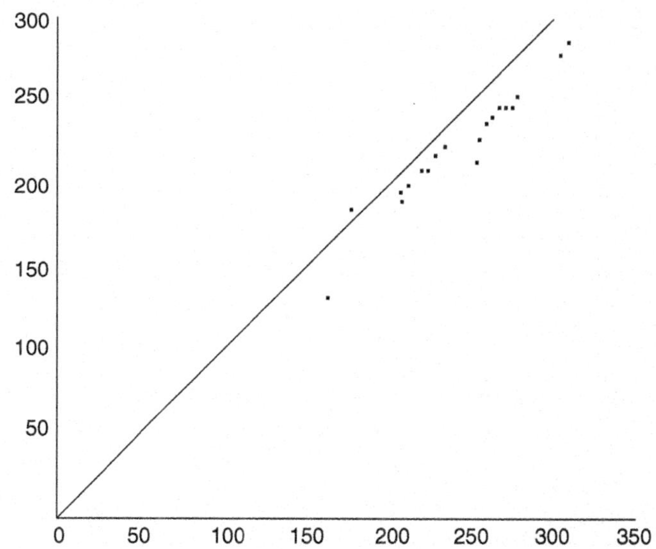

FIGURE 15.4 Q-Q plot for cholesterol levels.

We saw in Chapter 1 that box plots could be used for the identification of outliers. To summarize, we emphasize that graphical procedures, although preliminary, are an integral part of any statistical analysis.

EXERCISES 15.2

15.2.1 In order to study any possible relationship between expense and return, the following data give the percentage of expense ratio and total 1-year return for randomly selected stock mutual funds for the year 2000 (source: *Money*, February 2000).

% expense ratio	1.03	1.80	1.90	1.53	1.03	2.06	3.20	0.49	1.10	1.07
	1.48	1.30	1.23	1.22	1.60	1.50	1.81	1.75	0.97	1.28
% return	7.3	9.5	32.2	11.0	19.5	7.3	25.1	10.2	1.5	7.9
	18.9	26.1	3.4	3.7	23.5	2.9	14.5	14.9	22.7	21.9

Draw a scatterplot. Also, find the sample correlation of percentage expense ratio and percentage return.

15.2.2 In order to study any possible relationship between age and change in systolic blood pressure (BP) (mm Hg) in 24 hours in response to a treatment, the following data were obtained from 11 individuals.

Age	70	51	65	70	48	70	45	48	35	48	30
Systolic BP change	−28	−10	−8	−15	−8	−10	−12	3	1	−5	5

(a) Draw a scatterplot.
(b) Find the sample correlation of age and systolic BP.
(c) Fit a least-squares regression line.
(d) Interpret (a), (b), and (c).

15.2.3 The following data represents 15 randomly selected state finances: revenue and expenditures (in millions of dollars) for the fiscal year 1997 (source: *The World Almanac and Book of Facts 2000*).

Revenue:	9,439 26,538	8,845 5,537	14,520 6,494	24,028 2,818	39,038 49,318	5,215 4,229	20,128 7,724	7,467
Expenditure:	5,722 25,791	7,685 4,808	13,862 5,130	21,975 2,426	35,302 39,296	4,441 4,002	16,200 6,818	7,145

(a) Draw a scatterplot.
(b) Find the sample correlation between revenue and expenditure.
(c) Draw a QQ plot.
(d) Interpret (a), (b), and (c).

15.2.4 The following data give birth rates (per 1000 population) for 20 selected states in 1998 (source: *The World Almanac and Book of Facts 2000*).

14.4	16.3	13.5	14.6	13.7	15.6	10.9	12.8	13.0	14.2
13.4	13.9	15.9	13.3	14.1	15.7	15.2	13.9	15.4	11.3

Construct a dotplot, and interpret.

15.2.5 The following data give the median prices (rounded to nearest $1000) of single-family homes for 18 randomly selected US cities in 1998 (source: *The World Almanac and Book of Facts 2000*).

128	146	109	90	105	152	79	89	109
93	108	128	188	158	93	78	123	137

Construct a dotplot, and interpret.

15.3 OUTLIERS

All statistical procedures make assumptions about a population and the sample values obtained from the population. Before we proceed to analyze the data, we must check to see if there are any outliers, that is, data points that do not belong in the data set or are not in line with the rest of the data.

Outliers are observations that appear to have an abnormal value as compared with the rest of the values in the data set; that is, the value of an outlier is either much higher or significantly lower than any other value in the data set. An outlier could be a discordant observation or a contaminant. A discordant observation is one that appears surprising or discrepant to the investigator and is to some extent subjective. A contaminant is an observation that is from a different distribution than the rest of the data. Outliers may occur because of some limitations on measuring techniques or recording errors. They may also be due to the sample not being entirely from the same population. Extreme values in a data set could also be due to a skewed population. It should be noted that sometimes a data point that is labeled as an outlier may really be indicative of a novel phenomenon. In these cases, an extreme observation may not be classified as an outlier.

The presence of outliers can dramatically affect the estimate of the mean and variance of the sample, especially if the sample size is small. As a result, any test statistic computed from such data would be unreliable and so would be the statistical inferences. For example, the presence of outliers might lead to an incorrect conclusion that the variances of two samples are not equal if the outlier is the result of a recording or measurement error.

In a controlled experiment, such as in a laboratory setting, good record keeping with a clear understanding of the phenomenon under investigation and information about all the data will minimize the occurrence of outliers due to recording errors.

What to do with outliers? As long as these points remain observations, we cannot throw them out on a whim. There are basically two methods that are employed in dealing with outliers. One method is to use statistical testing procedures to detect outliers, possibly removing them from the data set if we know that these are measurement errors, incorrectly entered values, or impossible values in real life, and letting the analysis deal only with the rest of the data. The second method is to use statistical procedures, such as nonparametric tests or data transformations, that are immune or only minimally sensitive to the presence of outliers. Of course, we could run the analysis both with and without the outliers and report both results. We now present some commonly used tests for labeling outliers.

In data analysis, it is necessary to label suspected outliers for further study. For normally distributed data, we give three simple methods to identify an outlier: z-score, modified z-score, and box plot.

In a **z-test**, first find the z-scores of the entire data set, and label any observation with a z-score greater than 3 or less than −3 as an outlier. Recall that for the observed values $x_1,\ldots,x_n$, the z-score is defined by

$$z_i = \frac{x_i - \overline{x}}{s}$$

where s is the sample standard deviation of the sample, that is,

$$s = \sqrt{\frac{1}{n-1}\sum_{i=1}^{n}(x_i - \overline{x})^2}\,.$$

Because both the sample mean and the sample standard deviation are affected by the outliers, this labeling method is not very reliable.

In a *modified z-score test*, the median of absolute deviation about the median (MAD) is used. Let

$$MAD = median\,(|x_i - m|)$$

where m is the median of the observations. Then

$$z_i = \frac{(x_i - \overline{x})}{MAD}.$$

An observation is labeled as an outlier if the corresponding modified z-score is greater than 3.5. A normal plot may be used for testing normality for the data.

If we want a reasonably robust *distribution-free test*, observation x_0 is labeled as an outlier if

$$\frac{|x_0 - m|}{MAD} > 5.$$

Here, the choice of 5 is somewhat arbitrary.

A box plot (also called box-and-whisker plot) gives a method of labeling outliers through a graphical representation. We have seen the method of construction of box plots in Chapter 1. A box plot consists of a box, whiskers, and outliers. We draw a line across the box at the median. For example, in software, the bottom of the box is at the first quartile ($Q1$), and the top is at the third quartile ($Q3$). The whiskers are the lines that extend from the top and bottom of the box to the adjacent values, the lowest and highest observations still inside the region defined by the lower limit $Q1$–$1.5(Q3 - Q1)$ and the upper limit $Q1 + 1.5(Q3 - Q1)$. Outliers are points outside the lower and upper limits, plotted with asterisks (*).

EXAMPLE 15.3.1

The following data give the hours worked by 25 employees of a company in a randomly selected week.

45	40	39	36	42	40	55	58	42	41
48	50	47	54	40	34	18	40	60	56
42	43	46	43	54					

Label all possible outliers using:

(a) *z*-score test, distribution-free test, and modified *z*-score test
(b) Box plot

Solution

(a) *We can create* Table 15.1, *where dfree z stands for the distribution-free scores and modified stands for the modified z-scores.*

TABLE 15.1 The Distribution of Z-scores, Dfree Z-scores, and Modified Z-scores.

Data	*z*-Score	dfree *z*	Modified
45	0.05355	0.12	0.12
40	−0.50427	1.13	−1.13
39	−0.61583	1.38	−1.38
36	−0.95053	2.13	−2.13
42	−0.28114	0.63	−0.63
40	−0.50427	1.13	−1.13
55	1.16919	2.62	2.62
58	1.50389	3.75	3.37
42	−0.28114	0.63	−0.63
41	−0.39271	0.88	−0.88
48	0.38824	0.87	0.87
50	0.61137	1.37	1.37
47	0.27668	0.62	0.62
54	1.05763	2.37	2.37
40	−0.50427	1.13	−1.13
34	−1.17366	2.63	−2.63
18	−2.95868	6.63	−6.63
40	−0.50427	1.13	−1.13
60	1.72701	3.87	3.87
56	1.28076	2.87	2.87
42	−0.28114	0.63	−0.63
43	−0.16958	0.38	−0.38
46	0.16512	0.37	0.37
43	−0.16958	0.38	−0.38
54	1.05763	2.37	2.37

By the z-score test, there are no outliers. Using the distribution-free test, the 18 is the only outlier. By the modified z-score test, 18 and 60 are possible outliers.

(b) *The box plot is given in* Fig. 15.5.

Hence, observation 18 is identified as an outlier using the box plot.

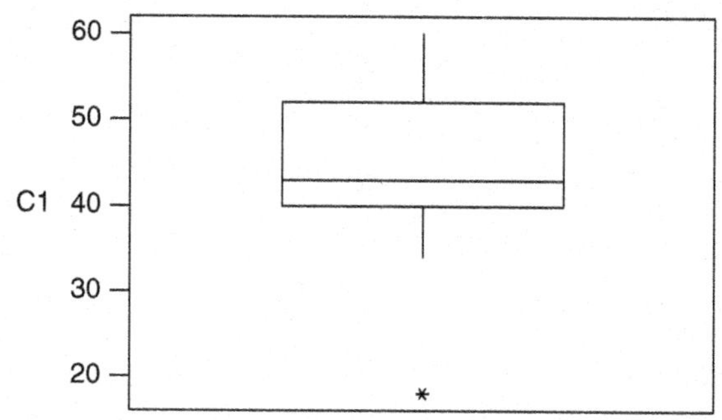

FIGURE 15.5 Box plot for hours of work per week.

Once we identify the outliers, then the question is what to do with them. If we can rule out recording errors as the source of outliers, the situation becomes more difficult. It is often impossible to say whether an outlier is really an extreme value within a skewed population or whether it represents a value drawn from a different population. As we indicated earlier, an outlier can be a legitimate observation representing a special feature of the sample population. In those cases, discarding the outliers may simplify the statistical analysis, although it also reduces the usefulness of such analysis. Understanding the experiment that generated the data might help in determining whether to discard or keep the outliers.

Once we decide to include the outliers, there are two possible ways to deal with them. One is to transform the data, such as by taking the natural logarithm, so as to reduce the undue influence of the outliers. Another possibility is to perform the analysis twice, with and without outliers, and report both results.

If we have bivariate data, a scatterplot may reveal any possible outliers; see Fig. 15.27. There are other methods available to detect outliers in multivariate data.

EXERCISES 15.3

15.3.1 Motor vehicle thefts are a big problem in cities.Table 15.2 displays data on motor vehicle thefts per 100,000 population in the year 1997 for 15 randomly selected large US cities (source: *Statistical Abstracts of the United States*, 1999).

TABLE 15.2 **Motor Vehicle Thefts Per 100,000 Population In The Year 1997 For 15 Randomly Selected Large U.S. Cities.**

Chicago, IL	1215.1	San Antonio, TX	830.0
Columbus, OH	1109.9	Charlotte, NC	780.1
Nashville, TN	1536.5	Tucson, AZ	1403.3
Albuquerque, NM	1797.8	Atlanta, GA	1869.7
Sacramento, CA	1630.5	St. Louis, MO	2152.8
Toledo, OH	939.7	Tampa, FL	1410.0
Birmingham, AL	1219.7	Anchorage, AK	532.8
Norfolk, VA	519.9		

Label all possible outliers using:
(a) (i) z-Score test, (ii) distribution-free test, and (iii) modified z-score test
(b) Box plot

15.3.2 For the data of Example 15.2.1, label all possible outliers using:
(a) (i) z-Score test, (ii) distribution-free test, and (iii) modified z-score test
(b) Box plot

15.3.3 The following data represents test scores of 36 randomly selected students from a large mathematics class.

67	63	39	80	64	95	90	93	21	36	44	66
100	66	72	34	78	66	68	98	74	81	71	100
60	50	81	66	90	89	86	49	77	63	58	43

Label all possible outliers using:
(a) (i) z-Score test, (ii) distribution-free test, and (iii) modified z-score test
(b) Box plot

15.3.4 The following data represent the number of days in 1997 on which selected US metropolitan areas failed to meet acceptable air-quality standards at trend sites (source: *The World Almanac and Book of Facts 2000*).

26	55	30	8	9	15	0	12	3	50	16
47	0	63	3	0	19	23	3	32	15	20
106	2	15	1	14	0	1	44	28		

Label all possible outliers using:
(a) (i) z-Score test, (ii) distribution-free test, and (iii) modified z-score test
(b) Box plot

15.4 CHECKING ASSUMPTIONS

With some exceptions, checking data for agreement with assumptions is not a topic that is strongly emphasized in other textbooks at this level. Even in more advanced books, this step is frequently omitted. In order for the inferences to work correctly, the measured variables must conform to assumptions that underlie the statistical procedures, or methods, to be applied. In hypothesis testing such as the t-tests and ANOVA, we made some fundamental assumptions that the random samples need to satisfy for the tests to yield correct results.

As an example, the basic assumptions underlying a t-test are:

(i) The sample comes from a normal population and usually small, $n < 30$.
(ii) The sample is random. In case of two-sample tests (excluding paired tests), the measurements in one sample are independent of those in the other sample.
(iii) When we are given two random samples, most of the results assume the equality of population variances, that is, $\sigma_1^2 = \sigma_2^2$. This assumption is called the homogeneity of variances. The test for equality of variance may have to be performed first if we doubt the equality of the variance.

Likewise, analysis of variance is based on a model that requires the following three primary assumptions:

(i) The samples come from normal populations.
(ii) Each of the samples is randomly selected from each group, and the samples are independent of each other.
(iii) The population variances for all the samples are equal. That is, if we have k populations with variances $\sigma_i^2, i = 1, 2, \ldots, k$, then $\sigma_1^2 = \sigma_2^2 = \ldots = \sigma_k^2$.

When we say we have a random sample, we implicitly assume that the data are identically distributed. The presence of outliers in an observed sample may affect such an assumption. We now explain a few tests for checking these assumptions such as the assumptions of normality, data transformations, and equality of variances.

15.4.1 Checking the Assumption of Normality

We start with the assumption of normality. Let us consider the example of randomly selected scores of 28 calculus students.

EXAMPLE 15.4.1

Given in the following table are the test scores of 28 randomly selected students from a calculus 1 class.

86	95	82	53	98	85	87	80	49	71	99	40	96	97
94	89	69	23	72	76	78	91	96	77	77	91	35	47

Construct a dotplot and a histogram, and compute the percentage of observations that fall in the intervals $\bar{x} \pm s$, $\bar{x} \pm 2s$, and $\bar{x} \pm 3s$.

Solution

The dotplot is shown in Fig. 15.6.

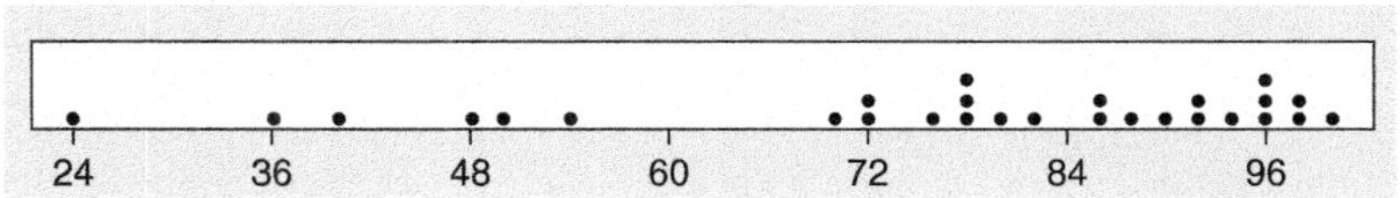

FIGURE 15.6 Dotplot of student scores.

The histogram is shown in Fig. 15.7.

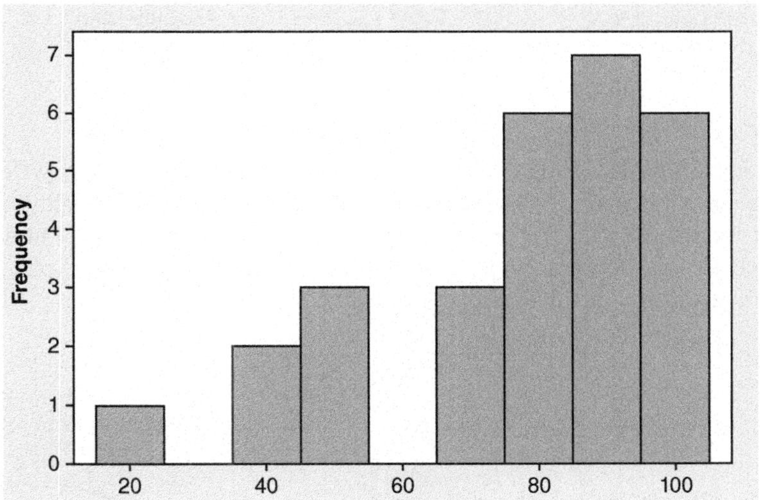

FIGURE 15.7 Histogram for student scores.

We have $\bar{x} = 71.18$ *and* $s = 20.99$. *Also, 57% of the random sample (i.e., 16 observations) fall in the interval* $71.18 \pm 20.99 = (50.19, 92.17)$. *There are 27 observations, or about 96%, that fall in* $71.18 \pm 41.98 = (29.2, 113.16)$, *and all the observations fall in* $71.18 \pm 62.97 = (8.21, 134.94)$. *This suggests that the data set is approximately normally distributed. This procedure is the empirical rule.*

For the previous example, we have seen that the dotplot does not suggest any normality. A histogram also does not suggest any normality (see Fig. 15.7). However, if we used the empirical rule as a test for normality, the data suggests normality. Clearly this leads to a conflicting situation with a simple theoretical check suggesting normality, while visual displays suggest nonnormality. In this case, more sophisticated procedures are warranted.

Sometimes, skewness and kurtosis can be used to test for tilt in and peakedness of a distribution. After getting skewness and kurtosis from the descriptive statistics, divide these by the standard errors. If both skew and kurtosis are within the ±2 range, the data can be considered normal.

We mention some sophisticated testing procedures for two of the most important of the parametric assumptions when running single-factor trials, namely, normality and homogeneity of variance. We have already seen in Project 4C how to construct a normal probability plot and check for normality. In this chapter, we will use software for a normal plot to check for normality. Fig. 15.8 graphs a normal probability plot (using software) for Example 15.4.1.

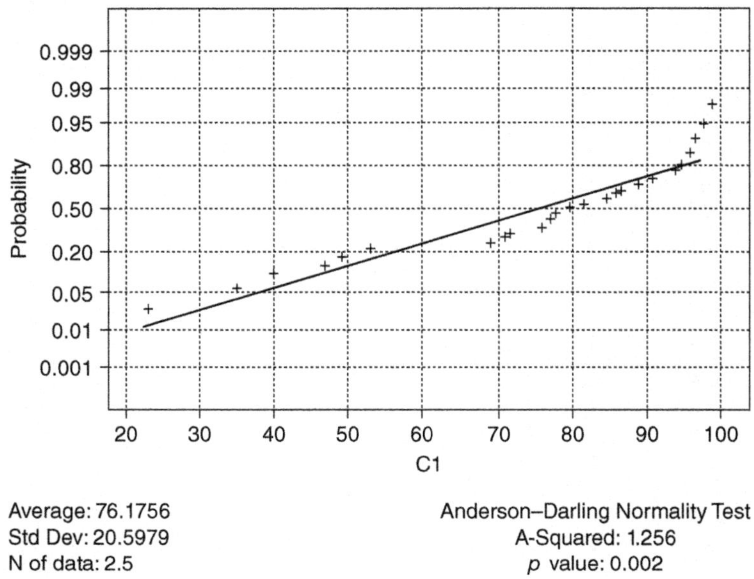

FIGURE 15.8 Normal probability plot of student scores.

We see that the test scores follow the straight line on the normal probability plot pretty well. Serious departures occur for the last four scores, because the values fall well above the line. This suggests normality with possible outliers.

It should be noted that for skewed data, in the normal probability plot, positively skewed data fall below the straight line, whereas the negatively skewed data rise above the straight line. A normal probability plot for the lifetime of 30 light bulbs in Example 15.2.1 is given in Fig. 15.9.

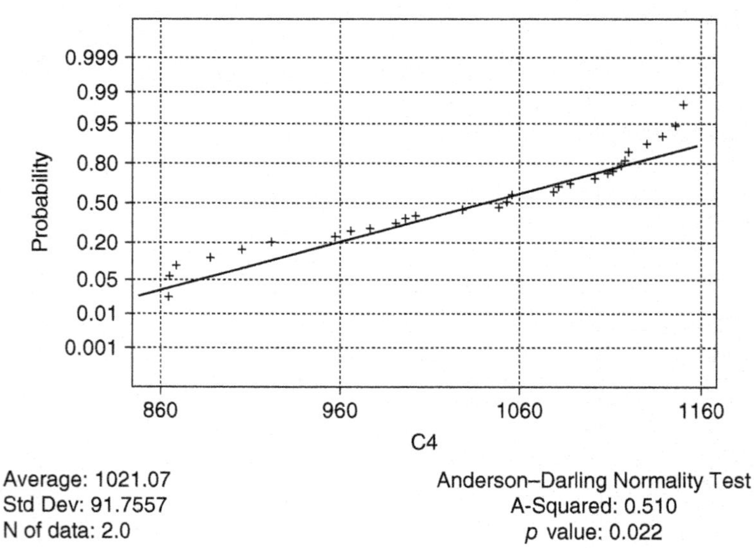

FIGURE 15.9 Normal probability plot for the lifetime of light bulbs.

This graph suggests that the data may not be normal and are more negatively skewed. Fig. 15.10 is a normal probability plot for 30 data points generated from a standard normal distribution.

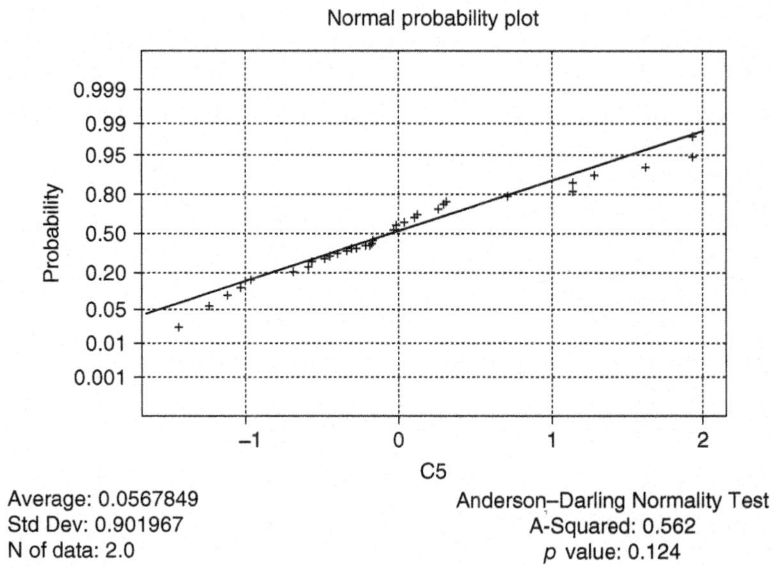

FIGURE 15.10 Normal probability plot of data from a standard normal distribution.

In this chapter, we have presented only simple graphical tests for testing of normality. We should mention that in the literature, a variety of procedures for testing for normality are available, including the Kolmogorov–Smirnov test, the Shapiro–Wilks W test, and the Lilliefors test. In Chapter 11, we have learned how to use Kolmogorov–Smirnov test, Anderson–Darling test, and Chi-square test. Some of these tests are incorporated in statistical software packages such as R and could be performed as easily as the graphical tests. If the sample size is very small, with any of these tests, it may be difficult to detect assumption violations. It is important to keep in mind that these tests are only rough indicators of assumption violations. For small sample sizes, even when the tests show that none of the test assumptions are violated, a normality test may not have sufficient power to detect a significant departure from normality, though it is present.

15.4.2 Data Transformation

Many data in real life do not meet the assumptions of parametric statistical tests: they may not be normally distributed, the variances may not be homogeneous, or both. Using most parametrical tests on those data may give a misleading result. Data transformation uses mathematical operations (filters) on each of the observations, transforming the original scores into a new set of scores. An appropriate transformation may (i) reduce the influence of outliers; (ii) make data, from a nonnormal distribution, more normal; and/or (iii) make the variances of different data sets more homogeneous. Some of the more commonly used transformations are (i) power transformations such as square root, (ii) logarithm, (iii) reciprocal, and (iv) arcsine. Used correctly, data transformation can be a useful tool for the practitioner. Some of these transformations can be put into a popular class of transformations called the Box–Cox power law transformation

$$y = \frac{x^\lambda - 1}{\lambda}$$

where λ can be optimally adjusted from 0 to 1. For example, as $\lambda \to 0$, we obtain the $y = \ln x$ (logarithmic filter) transformation, and when $\lambda = 1/2$, we get the square root transformation.

Even though we have done a statistical test on a transformed variable, it is not a good idea to report the summary statistics such as mean, standard errors, etc. in transformed units. We should back-transform by doing the opposite of the mathematical function we used in the data transformation. For instance, if we had originally used the natural logarithm, we should use exponential transformation as the back transformation. For instance, if we got a symmetric confidence interval for transformed mean as in Chapter 5, which are symmetric for a ln transformed data, we should take exponentials of the lower and upper limits. In the process, we may lose the symmetry of the confidence interval.

As we have seen in Project 9A, it is sometimes possible to use appropriate data transformations to transform nonnormal data into approximately normal data. Then we can use this normality property to perform statistical analysis on these transformed values. For instance, if the distribution of data has a long tail (which could be seen by drawing a histogram of observations) or a few laggards on the right (which could be seen by drawing a dotplot of observations), the $\sqrt{x}$ or $\ln x$ transforms will pull larger values down further than they pull the smaller or center values. Sometimes it is necessary to try several different transformations (trial and error) in order to find one that is more appropriate.

EXAMPLE 15.4.2

Consider the following data from an experiment.

1.15	3.84	0.01	2.06	3.28	2.61	0.59	3.19	1.32	1.07
7.80	1.74	0.25	0.21	3.42	4.52	0.43	0.38	0.07	1.26
4.03	7.28	0.85	3.24	0.62					

(a) Draw a histogram and normal plot.
(b) Take the transform $y = \sqrt{x}$, and draw a histogram and normal plot for the transformed data.

Solution

(a) *The histogram and normal plots for the data are shown in* Figs. 15.11 and 15.12.

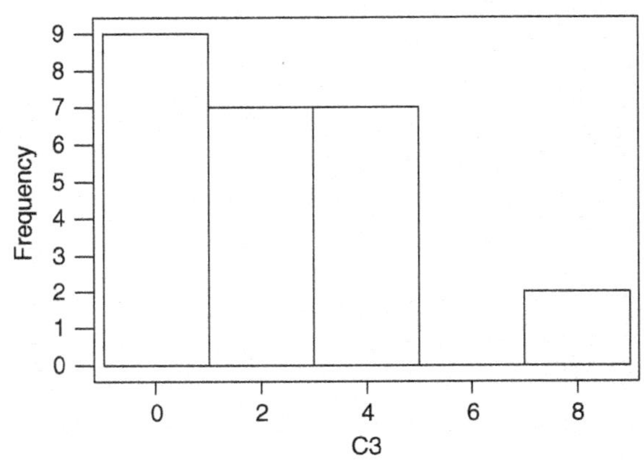

FIGURE 15.11 A histogram of the data.

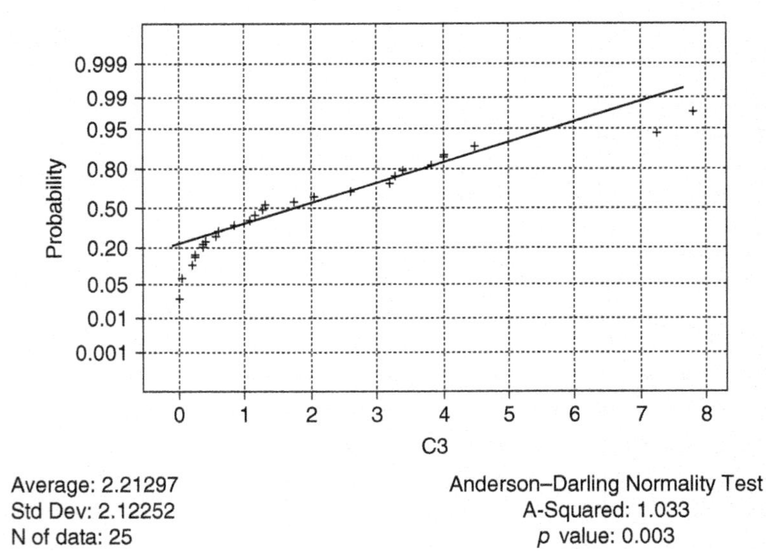

FIGURE 15.12 Normal probability plot of the data.

These graphs clearly show that the data does not follow a normal distribution.

(b) *The histogram and normal plot for the transformed data are shown in* Figs. 15.13 and 15.14. *With this transformation (filter), we can see that the filtered data follows normality.*

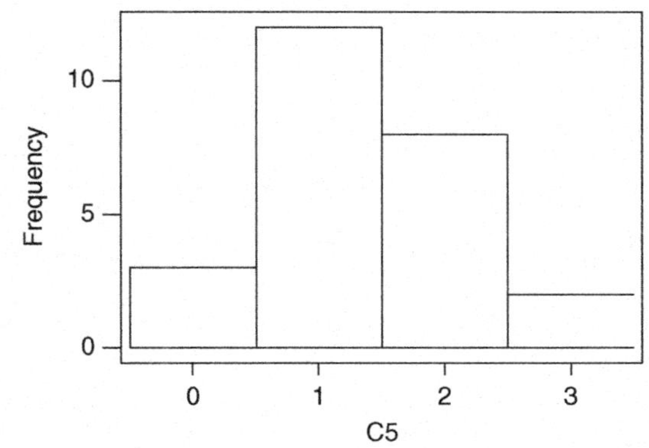

FIGURE 15.13 Histogram of the transformed data.

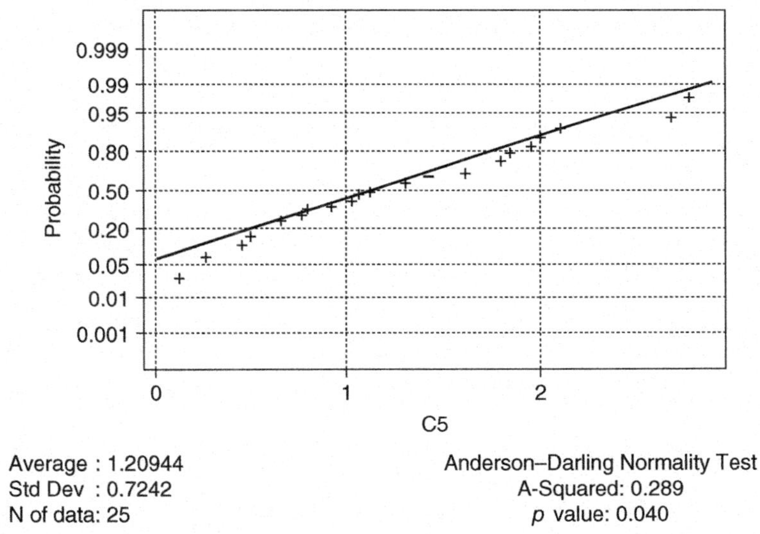

FIGURE 15.14 Normal probability plot of the transformed data.

We have only pointed out transformations in single-variable cases. The transformation methods are also useful in multivariable and multifactor studies; however, these involve more difficult analysis.

15.4.3 Test for Equality of Variances

Now we will discuss the tests for equality of variances, that is, the tests for heteroscedasticity. Our recommendation is that, in a real-world problem, after accounting for outliers, one should conduct tests for normality and heterogeneity of variance routinely before analyzing any data. Here, we give two tests. One, for the two-sample case, is based on the F-test, and for the multisampling case, we give Levene's test based on analysis of variance procedures. Albert Madansky's book Prescriptions for Working Statisticians (Springer-Verlag, 1988) gives various other tests for normality and heteroscedasticity.

Testing Equality of Variances for Two Normal Populations

The following procedure has already been discussed in the hypothesis testing Chapter 6. For the sake of completeness, here we again briefly discuss this procedure. Let $X_{11}, \ldots, X_{1n_1}$ be a random sample from an $N\left(\mu_1, \sigma_1^2\right)$ distribution and $X_{21}, \ldots, X_{2n_2}$ be a random sample from an $N\left(\mu_2, \sigma_2^2\right)$ distribution. Assume that the X_{1i}' s and X_{2j}' s are independent of each other for all i, j. Let

$$\bar{x}_i = \frac{1}{n_i} \sum_{j=1}^{n_i} x_{ij}, \; i = 1,2.$$

Assuming that μ_1 and μ_2 are unknown, we can test the hypothesis that $\sigma_1^2 = \sigma_2^2$ based on the ratio

$$F = \frac{s_1^2}{s_2^2} = \frac{\sum_{j=1}^{n_1} (x_{1j} - \bar{x}_1)^2/(n_1 - 1)}{\sum_{j=1}^{n_2} (x_{2j} - \bar{x}_2)^2/(n_2 - 1)}.$$

We know that $(n_1 - 1)s_1^2/\sigma_1^2$ has a $\chi^2(n_1-1)$ distribution and $(n_2 - 1)s_2^2/\sigma_2^2$ has a $\chi^2(n_2-1)$ distribution. Therefore, under the null hypothesis H_0: $\sigma_1^2 = \sigma_2^2$, the F-statistic has an $F(n_1-1, n_2-1)$ distribution.

Based on the alternate hypothesis, we will reject the equality of variance assumption if the test statistic falls into the appropriate tail of the F-distribution. For example, if H_a: $\sigma_1^2 > \sigma_2^2$ with α 0.05, we would reject H_0 when $F > F_{0.95}(n_1-1, n_2-1)$, and if H_a: $\sigma_1^2 < \sigma_2^2$ with $\alpha = 0.05$, we would reject H_0 when $F \le F_{0.05}(n_i - 1, n_2-1)$. When H_a: $\sigma_1^2 \ne \sigma_2^2$ with α 0.05, we would reject H_0 when $F \ge F_{0975}(n_1-1, n_2-1)$ or $F \le F_{0.025}(n_1-1, n_2-1)$. It should be noted that in the case of a two-sided alternative, this procedure is not the best one in the sense of minimizing the Type II error. However, for simplicity, we will not discuss the optimal two-tailed procedure.

EXAMPLE 15.4.3

An aquaculture farm takes water from a stream and returns it after it has circulated through the fish tanks. Suppose the owner thinks that, because the water circulates rather quickly through the tank, there is little organic matter in the effluent. To find out, some samples of the water are taken at the intake, and other samples are taken at the downstream outlet, and tests are performed for biochemical oxygen demand (BOD). If BOD increases, it can be said that the effluent contains more organic matter than the stream can handle. Table 15.3 gives the data for this problem.

(a) Using normal plots, check for normality of each sample.
(b) Test for the equality of variances of the BOD for the downstream and upstream samples at $\alpha = 0.05$.

Solution

(a) *The normal plots are shown in* Figs. 15.15 and 15.16.

The BOD data for the downstream and upstream samples are approximately normal.

TABLE 15.3 Downstream And Upstream BOD Data.

Upstream	Downstream
7.863	8.132
5.714	9.128
5.871	7.574
6.479	8.678
7.124	9.336
7.539	8.798
6.682	8.457

(Continued)

TABLE 15.3 (Continued)

Upstream	Downstream
5.877	9.756
6.227	8.548
6.771	7.992

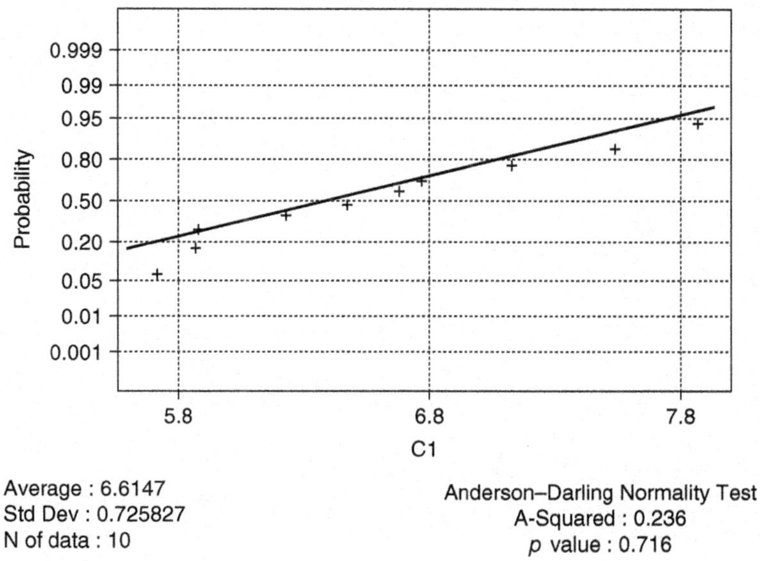

FIGURE 15.15 Normal plot of upstream data.

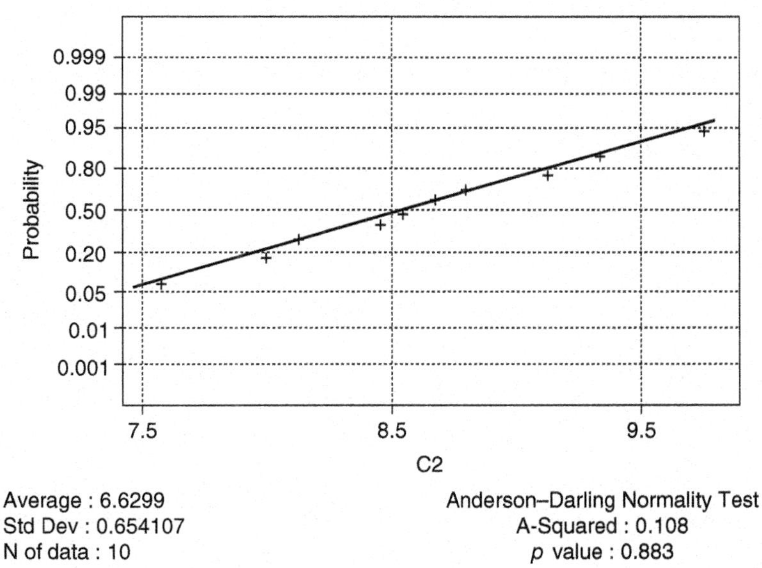

FIGURE 15.16 Normal plot of downstream data.

(b) *We test H_0: $\sigma_1^2 = \sigma_2^2$ versus H_a: $\sigma_1^2 \neq \sigma_2^2$. We have $n_1 = n_2 = 10$, and $\alpha = 0.05$. Because the normal plots of each sample conform with the normality assumption, we can use the F-statistic:*

$$F = \frac{s_1^2}{s_2^2} = \frac{(0.729)^2}{(0.654)^2} = 1.2425.$$

From the F-table, the rejection region is $\{F \le F_{0.025}(9, 9) = 0.248\}$ or $\{F > F_{0.975}(9, 9) = 4.03\}$. Because the observed value of the test statistic does not fall in the rejection region, we conclude based on the sample evidence that the variances of the two populations are equal.

Test for Equality of Variances, k ≥ 2 Populations

Generalizing to k populations, let $X_{i1}, X_{i2},\ldots,X_{i_{i}}, i\ 1, 2,\ldots,k$, be k random samples from $N\left(\mu_i, \sigma_i^2\right)$ distributions, with both $\mu'_i\,s$ and $\sigma'_i\,s$ unknown. Also assume that X_{ij}, X_{kl} are independent for all (i, j), (k, l). We wish to test the hypothesis H_0: $\sigma_1^2 = \sigma_2^2 = \ldots = \sigma_k^2$ against H_a: At least one of the σ_i^2 is different. There are many tests available. One of the basic graphical procedures is to use side-by-side box plots (see Example 9.3.2). We describe Levene's test based on the analysis of variance (source: Levene, 1960).

Let $y_{ij} = |x_{ij} - \bar{x}_i|$. Now perform an analysis of variance test for equality of the means of the y_{ij}. Let

$$n = \sum_{i=1}^{k} n_i,\ \bar{y}_i = \sum_{j=1}^{n_i} y_{ij}/n_i \quad \text{and} \quad \bar{y}_{..} = \sum_{i=1}^{k}\sum_{j=1}^{n_i} y_{ij}/\sum_{i=1}^{k} n_i.$$

The analysis of variance statistic is given by

$$z = \frac{\sum_{i=1}^{k} n_i(\bar{y}_i - \bar{y}_{..})^2/(k-1)}{\sum_{i=1}^{k}\sum_{j=1}^{n_i}(y_{ij} - \bar{y}_{i.})^2/(n-k)} = \frac{MST}{MSE}.$$

Recall that MST (mean square for treatments) and MSE (mean square error) were defined in Section 9.3; the MST is a measure of the variability between the sample means of the groups, and the MSE is a measure of variability within the groups. For a 95% confidence level, the rejection region is $\{z > F_{0.95}\ (k-1, n-k)\}$.

It should be noted that the y_{ij} is not independent, but the analysis of variance method is found to be robust against the deviation from this assumption of independence.

EXAMPLE 15.4.4

The three random samples in Table 15.4 are independently obtained from three different normal populations.

TABLE 15.4 Random Samples From Three Different Normal Populations.

Sample 1	Sample 2	Sample 3
64	56	81
84	74	92
75	69	84
77		
80		

At the $\alpha = 0.05$ level of significance, test for equality of variances.

Solution

We test H_0: $\sigma_1^2 = \sigma_2^2 = \sigma_3^2$ *versus* H_a: *Not all the* σ_i^2 *are equal. For this sample,* $\bar{x}_1 = 76$, $\bar{x}_2 = 66.33$, *and* $\bar{x}_3 = 85.67$. *Also* $n = 11$, *and* $k = 3$. *Letting* $y_{ij} = |x_{ij} - \bar{x}_i|$, *we obtain the following* y_{ij} *values:*

12	10.33	4.67000
8	7.67	6.33000
1	2.67	1.67000
1		
4		

The test statistic is

$$z = \frac{\sum_{i=1}^{k} n_i (\overline{y_i.} - \overline{y..})^2 / (k-1)}{\sum_{i=1}^{k} \sum_{j=1}^{n_i} (y_{ij} - \overline{y_i.})^2 / (n-k)}$$

$$= \frac{MST}{MSE} = \frac{5.5}{16.5} = 0.33.$$

From the F-table, the 95% point is $F_{0.05}(2, 8) = 4.46$. *Hence, the rejection region is* $\{z > 4.46\}$. *Because the observed value of* $z = 0.33$ *does not fall in the rejection region, the null hypothesis is not rejected, and we conclude that the assumption of equality of variances seems to be justified.*

Through our testing, if we find that the homogeneity of variance of the data is violated significantly, then nonparametric tests are more appropriate. Another popular test for equality of variance is Bartlett's test.

15.4.4 Test of Independence

Almost all the results in this book assume that we have independent random samples. In the situation where we suspect that the sample data may not be independent, perform a run test as described in Project 12B to test for independence. There are parametric procedures available to test independence; however, the run test is independent of the distributional assumptions and simpler to perform. In general, whether the two samples are independent of each other is decided by the structure of the experiment from which they arise. In case of correlated samples, such as a set of pre- and posttest observations on the same subject that are not independent, a two-sample paired test may be more appropriate. Another popular method used to check for independence is the chi-squared test of independence; see Section 11.4.2. For time series data, the Durbin–Watson test (http://www.alchemygroup.net/Permutation%20Durbin-Watson%20Final.pdf) is effective.

In practical sampling situations, the underlying populations are unlikely to be exactly normally distributed with homogeneity of variances. Both *t*-tests and ANOVA are robust for reasonable departures in some of these assumptions. However, these tests may not be robust with respect to certain other assumption violations. For example, ANOVA is quite sensitive to the violation of the independence assumption. These factors need to be given special attention in data analysis.

EXERCISES 15.4

15.4.1 The scores of 25 randomly selected students from a large calculus class are given below.

47	73	90	22	68	86	94	32	88	86
80	97	48	70	61	82	67	73	78	55
63	59	42	46	90					

(a) Test the data for normality.
(b) If the data are not normal, try a suitable transformation (filter) to make the transformed data normal.

15.4.2 Refer to Example 15.3.1. Suppose we use the transformation $y_i = \ln x_i$ for each observation.
(a) Test whether the transformed data is normal.
(b) Determine whether the data value 18 is still an outlier in the transformed data set.

15.4.3 The data shown in the following table relates to the concealed weapons permits issued in 13 randomly selected Florida counties in 1996.

31,603	20,873	15,963	10,294	8,956	7,901	6,820
5,695	5,485	4,827	3,969	3,278	1,731	

(a) Test whether the data is normal.
(b) If not, try a suitable transformation to make the transformed data normal.

15.4.4 The following table represents a summary by state for Medicare enrollment (in thousands) for 15 randomly selected states in 1998 (source: *Statistical Abstracts of the United States*, 1999).

665	3,757	623	757	541	448	478	2,728	103	771
224	86	623	1,373	713					

(a) Test to determine whether the data is normal.
(b) If not, try a suitable transformation to make the transformed data approximately normal.
(c) Test for outliers. If an observation is extreme, would you classify it as an outlier?

15.4.5 Given in the following table are 15 randomly selected state expenditures (in millions of dollars) for the fiscal year 1997 (source: *The World Almanac and Book of Facts* 2000).

5,722	7,685	13,862	21,975	35,302	4,441	16,200	25,791
4,80	5,130	2,426	39,296	4,002	6,818	7,145	

(a) Test the data for normality.
(b) If the data is not normal, try a suitable transformation to make the transformed data approximately normal.

15.4.6 For the data of Exercise 15.3.4,
(a) Test whether the data is normal.
(b) If not, try a suitable transformation to make the transformed data approximately normal.

15.4.7 The following data give in-city mileage per gallon for 25 small and midsize cars (source: *Money Magazine*, March 2001).

25	23	20	20	27	26	20	32	25	22
24	21	28	20	22	19	21	29	23	32
23	52	24	24	22					

(a) Test to determine whether the data is normal.
(b) If not, try a suitable transformation to make the transformed data approximately normal.
(c) Test for outliers. If an observation is extreme, would you classify it as an outlier?

15.4.8 The following table gives in-state tuition costs (in dollars) for 15 randomly selected colleges taken from a list of the 100 best values in public colleges (source: *Kiplinger's Magazine*, October 2000).

3788	4065	2196	7360	5212	4137	4060	3956	3975	7395
4058	3683	3999	3156	4354					

(a) Test for outliers.
(b) Test whether the data is normal.

15.4.9 For the data given in Exercise 15.2.1, test for equality of variances.
15.4.10 For the data given in Exercise 15.2.3, test for equality of variances.
15.4.11 The following data represents a random sample of end-of-year bonuses for lower-level managerial personnel employed by a large firm. Bonuses are expressed as a percentage of yearly salary.

Female	6.2	9.2	8.0	7.7	8.4	9.1	7.4	6.7
Male	8.9	10.0	9.4	8.8	12.0	9.9	11.7	9.8

Test for equality of variances. State any assumptions you have made, and interpret your result. Use $\alpha = 0.05$.

15.4.12 In an effort to investigate the premium charged by insurance companies for auto insurance, an agency randomly selects a few drivers who are insured by three different companies. These individuals have similar cars, driving records, and level of coverage. Table 15.5 gives the premiums paid per 6 months by these drivers with these three companies.

TABLE 15.5 Premiums Paid Per 6 Months By Similar Drivers With Three Different Companies.

Company I	Company II	Company III
396	348	378
438	360	330
336	522	294
318		474
432		

Test for equality of variances. State any assumptions you have made, and interpret your result. Use $\alpha = 0.01$.

15.4.13 Three classes in elementary statistics are taught by three different persons, a regular faculty member, a graduate teaching assistant, and an adjunct from outside the university. At the end of the semester, each student is given a standardized test. Five students are randomly picked from each of these classes, and their scores are as shown in Table 15.6.

TABLE 15.6 Scores of Five Randomly Picked Students From Each Class.

Faculty	Teaching Assistant	Adjunct
93	88	86
61	90	56
87	76	73
75	82	90
92	58	47

Test for equality of variances. State any assumptions you have made, and interpret your result. Use $\alpha = 0.05$.

15.5 MODELING ISSUES

A model is a theoretical description in the language of mathematical statistics of a physical phenomenon. Even though interpretations can be developed by analogy, past experience, or intuition, the scientific approach requires a model for the phenomenon of interest. Models are simplifications (or approximations) of real-world situations and are designed to make it easier to identify and to understand relationships among variables. A good model is crucial for accurate estimation, forecasting, or prediction. If the observed data show a good fit to the estimates obtained through the model, we consider the model to be an adequate representation of the real-world phenomenon. If not, the model must be improved to incorporate additional variables or modify the equations defining relationships. In statistical modeling, it is important not to lose perspective on the essential purpose of the modeling effort. The emphasis should be on making these models work on real data sets in lieu of spending a large amount of time on the capabilities of the models. Even though the study of properties and abilities of models is important, equally important is an ability to know when and how to fit models to a particular data set. A regression line is a two-parameter model that depicts a linear dependence of one variable on another. Again, it is not our objective to discuss all the issues related to statistical modeling. We will only discuss briefly some simple issues relevant to modeling.

15.5.1 A Simple Model for Univariate Data

Suppose that we have a data set that characterizes a phenomenon of interest. Suppose our problem is to create a statistical model for the data set in the form of a probability distribution from which the data set came. First, we create a dotplot and summary of the basic statistics. The dotplot will provide us with an idea of the probability distribution of the data and any unusual behavior of the data that will not be apparent from the basic statistics, such as sample mean and sample standard deviation. Having identified the probability distribution of the sample statistic, we can proceed to obtain 95% confidence limits on parameters such as the mean and variance. In addition, we can obtain a 95% prediction interval of the next observation using the expression

$$\bar{y} \pm (t - \text{value}) s \sqrt{1 + \frac{1}{n}}.$$

Note that the prediction interval is always wider than the corresponding confidence interval. The confidence interval provides a measure of reliability for estimating a parameter. The prediction interval provides a measure of reliability for the prediction of an observation. Thus, the prediction interval needs to account for estimation error as well as the natural variability of a single observation. These steps can be considered as the first modeling effort for univariate data. Note that if we have a small sample size, using a t-value in the confidence interval and/or prediction interval supposes a modeling assumption of normality for the corresponding population. The preliminary verification of this is done by the dotplot. For more detailed verification of this modeling assumption, use the normal plots.

EXAMPLE 15.5.1

Consider the following data from an experiment:

0.15	0.14	0.15	0.14	0.26	0.00	0.00	0.47	0.35	0.16
0.15	0.15	0.23	0.13	0.19	0.15	0.22	0.53	0.17	0.23
0.22	0.16	0.12	0.13	0.11	0.14	0.18	0.15	0.14	0.21
0.13	0.12	0.13	0.13	0.21	0.22	0.18	0.20	0.22	0.16
0.17	0.00	0.23	0.21	0.18	0.05	0.16	0.13	0.23	0.18
0.14	0.29	0.21	0.22	0.11	0.16	0.23	0.13	0.07	0.17
0.08	0.14	0.06	0.08	0.07	0.11	0.12	0.14	0.16	0.12
0.10	0.27	0.19	0.13	0.27	0.16	0.07	0.09	0.04	0.53
0.29	0.15	0.12	0.11	0.10	0.14	0.14	0.16	0.16	0.17
0.36	0.46	1.21	0.39	0.01	0.52	0.09	0.18	0.16	0.16
0.14	0.15	0.09	0.09	0.13	0.13	0.08	0.14	0.20	0.09
0.09	0.16	0.08	0.10	0.34	0.24	0.15	0.44	0.08	0.08
0.16	0.14	0.18	0.23	0.19	0.11	0.19	0.10	0.14	0.11
0.14	0.17	0.17	0.17	0.05	0.12	0.14	0.11	0.20	0.14
0.23	0.03	0.10	0.29	0.13	0.26	0.13	0.15	0.27	0.14
0.50	0.16	0.15	0.18	0.16	0.14	0.13	0.08	0.20	0.17

0.17	0.16	0.15	0.11	0.13	0.76	0.18	0.19	0.09	0.12
0.11	0.12	0.08	0.26	0.23	0.20	0.19	0.19	0.16	0.11
0.12	0.13	0.32	0.05	0.18	0.12	0.13	0.50	0.13	0.04
0.00	−0.11	0.18	0.15	0.14	0.15	0.02	0.20		

(a) Obtain a dotplot.
(b) Calculate the basic statistics, sample mean, sample median, and sample standard deviation.
(c) Obtain a 95% confidence interval for the true mean.
(d) Obtain a 95% prediction interval.

Solution

(a) *Each dot in* Fig. 15.17 *represents three points.*

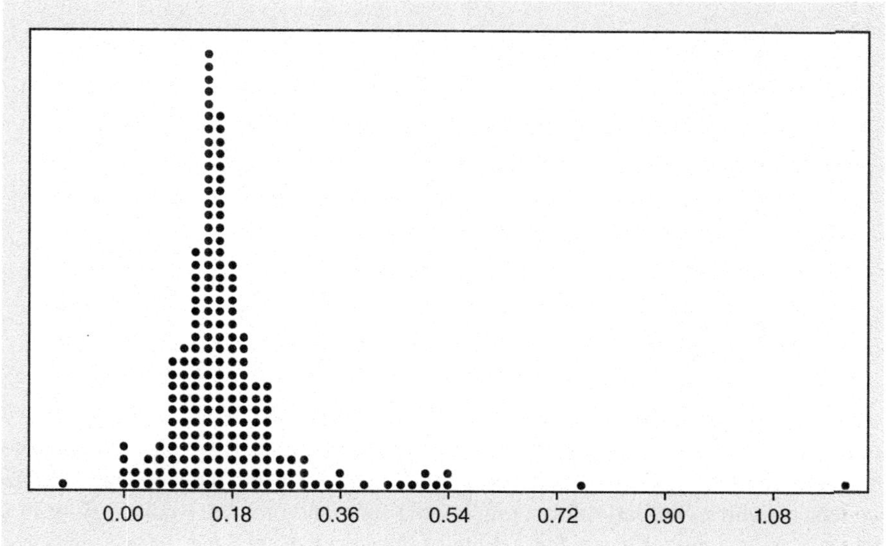

FIGURE 15.17 Dotplot of the data.

(b) *We can use software's* ***describe*** *command to obtain the following.*

	N	Mean	Median	TrMean	StDev	SE Mean
C1	198	0.17038	0.15121	0.15982	0.13610	0.00967
	MIN	MAX	Q1	Q3		
	−0.39575	1.22076	0.12059	0.19284		

(c) *Again using software commands, we can obtain (where data are stored in* ***C1****), MTB > ZInterval 95 0 0.136 c1.*

The Assumed Sigma 0.136

	N	Mean	StDev	SE Mean	95.0 Percent CI
C1	198	0.17038	0.13610	0.00967	(0.15143, 0.18933)

(d) *For the prediction interval, use the large sample formula* $\bar{y} \pm (z_{\alpha/2})s\sqrt{1+\frac{1}{n}}$, *to obtain the 95% prediction interval for the true mean as (0.097, 0.4387).*

15.5.2 Modeling Bivariate Data

When a scatterplot of bivariate data exhibits a linear pattern, the modeling is usually done using linear regression to study their linear relationship as explained in Chapter 8. Clearly a linear relationship is desirable because it is easy to interpret, departure from linearity is easy to detect, and predicting dependent values from independent variables is straightforward. However, when a scatterplot shows a curved nonlinear pattern, then finding a "good" model that fits the observed data may not be very easy. Sometimes, instead of fitting a curve, we may be able to transform the data so as to make the scatterplots of the transformed data look more linear.

A popular statistical method used to straighten a plot is the so-called power transformation. Power transformation is defined by specifying an exponent, k, which could be a positive or negative real number, and then computing each transformed value as the original value to the power k. Note that $k = 1/2$ gives the square root transformation. When $k = 0$, every transformed value is equal to 1. Instead, it is customary to think of $k = 0$ as corresponding to a logarithmic transformation so as to unify the transformation concept. The power $k = 1$ corresponds to no transformation at all. Observe that these are the same transformations we have explained in Subsection 15.4.2 to transform nonnormal data into normal transformed data. The shape of the scatterplots should suggest an appropriate transformation. The four curves in Fig. 15.18 represent possible shapes of scatterplots that are usually encountered in practice.

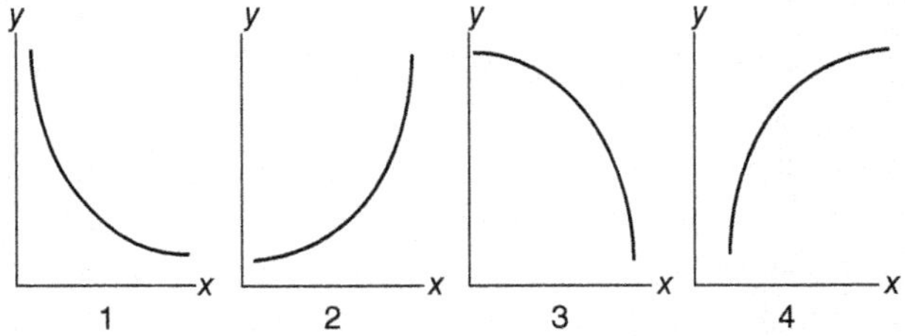

FIGURE 15.18 Possible shapes of a scatterplot.

We can use the following as a general guideline for making transformations. If we have a scatterplot that looks like plot 1 of Fig. 15.18, then to straighten the plot, we should use a power $k < 1$ for x (the independent variable) and/or use a power $k > 1$ for y (the dependent variable). Similarly, for curve 2, $k > 1$ for x and/or $k < 1$ (such as $\sqrt{y}$ or ln y) for y. For curve 3, take $k > 1$ for x (such as x^2 or x^3) and/or $k > 1$ for y. Finally, for curve 4, take $k < 1$ for x and/or $k > 1$ for y. Once we straighten the data through transformations, obtain the least-squares equation of the line as explained in Chapter 7. By reversing the transformation (or solving for y in the transformed equation), we can obtain the original nonlinear relationship between x and y.

EXAMPLE 15.5.2

For the following bivariate data:

x	0	4	8	10	15	18	20	25
y	2.4	2.6	3.1	3.6	4.1	4.2	4.6	4.7

(a) Draw a scatterplot.
(b) Use appropriate transformation (if necessary) to linearize the scatterplot.
(c) Fit the data to an appropriate curve.

Solution

(a) *The scatterplot is shown in* Fig. 15.19.

This looks more like curve 4.

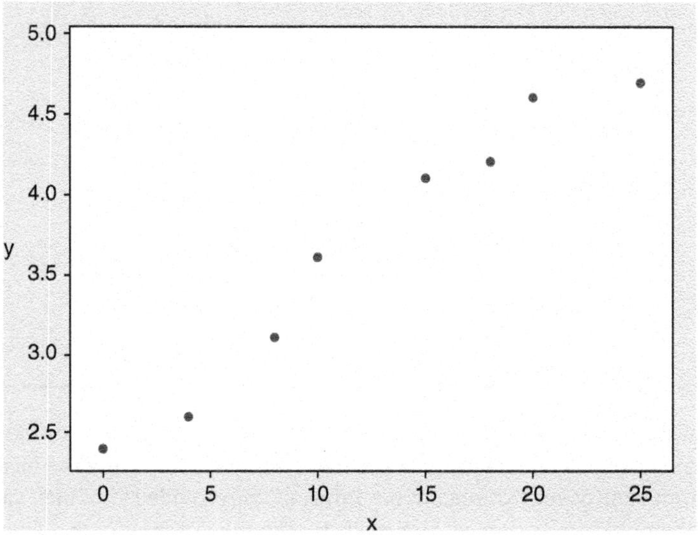

FIGURE 15.19 Scatterplot of the data.

(b) *Let us use the transformation* $x' = \ln x$ *and* $y' = y^2$. *We will get the scatterplot shown in* Fig. 15.20.

This looks more linear.

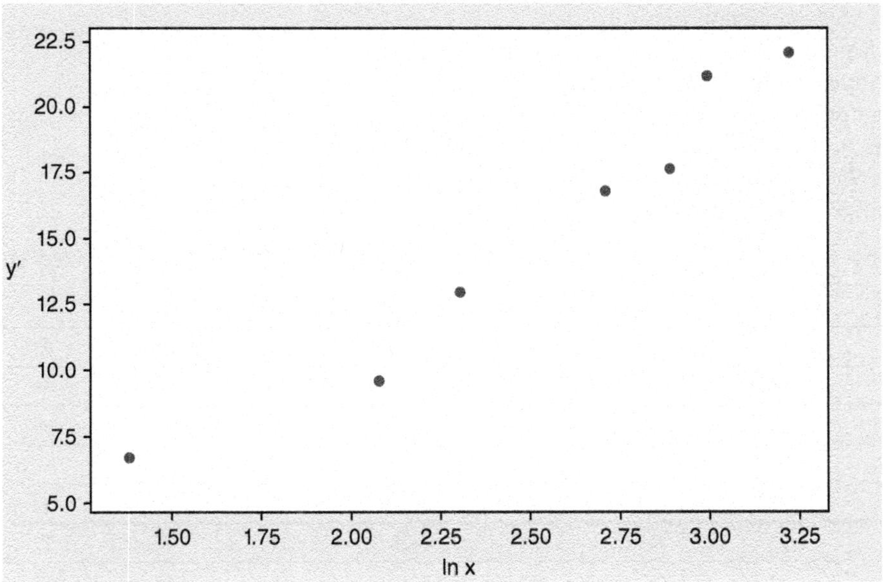

FIGURE 15.20 Scatterplot of the transformed data.

(c) *The regression line for the transformed data is* $y' = 8.86x' - 6.96$. *Therefore, for the original data,* $y^2 = 8.86 \ln x - 6.96$. *The fitted curve is shown in* Fig. 15.21.

Looking at Fig. 15.21, we can see that the data are only slightly nonlinear. In addition, using the equation, for a given value of x, we can predict the value of the response variable y. For instance, if $x = 1.5$, we estimate y^2 to be −3.3676.

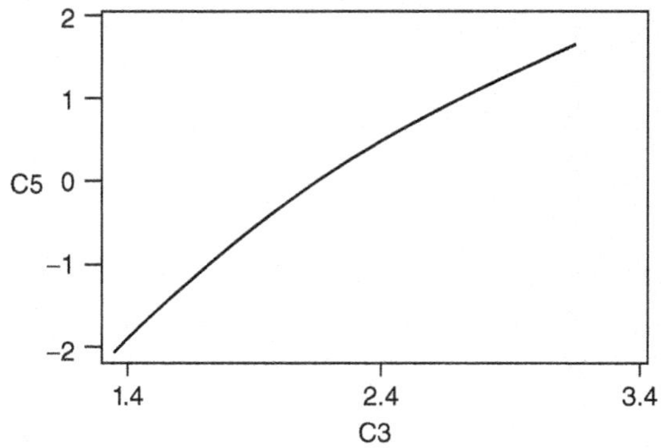

FIGURE 15.21 Fitted curve.

There are various other modeling issues that one may encounter in applications. For example, in multiple regression modeling, an investigator may have data on a number of predictor variables that might be incorporated into a model. Some of these variables may be irrelevant or may duplicate the information provided by other variables. The problem then is how to detect and eliminate the duplicate variables. However, for the sake of brevity and level of presentation, we will not go into the difficulty of these issues of model selection.

EXERCISES 15.5

15.5.1 For the data of Exercise 15.4.5:
- **(a)** Obtain a dotplot.
- **(b)** Describe the data, such as mean, median, and standard deviation.
- **(c)** Obtain a 95% confidence interval for the mean.
- **(d)** Obtain a 95% prediction interval.
- **(e)** Explain your solutions, and state any assumptions.

15.5.2 For the gas mileage data of Exercise 15.4.7:
- **(a)** Obtain a dotplot.
- **(b)** Describe the data, such as mean, median, and standard deviation.
- **(c)** Obtain a 95% confidence interval for the mean.
- **(d)** Obtain a 95% prediction interval.

15.5.3 The following represents the midterm and final exam scores for 35 randomly selected students from a large mathematics class.

Midterm:	67	63	39	80	64	95	90	93	21	36
	44	66	66	72	34	78	66	68	98	43
	74	81	71	100	60	50	81	66	90	89
	86	49	77	63	58					
Final:	29	33	100	33	55	20	10	5	67	64
	71	25	34	66	28	34	16	27	32	20
	14	21	16	62	50	14	61	11	14	41
	52	35	37	51	43					

- **(a)** Draw a scatterplot.
- **(b)** Use appropriate transformation (if necessary) to linearize the scatterplot.
- **(c)** Fit the data to an appropriate curve and explain the usefulness.

15.5.4 For the state finance data of Exercise 15.2.3:
- **(a)** Draw a scatterplot.
- **(b)** Fit a least-squares line.
- **(c)** Explain your solutions, and state any assumptions.

15.5.5 Table 15.7 gives in-state tuition costs (in dollars) and 4-year graduation rate (%) for 15 randomly selected colleges, taken from a list of the 100 best values in public colleges (source: *Kiplinger's Magazine*, October 2000).

TABLE 15.7 Instate-Tuition Costs (In Dollars) and 4-Year Graduation Rate (Percentage) For 15 Randomly Selected Colleges.

In-state tuition	3788	4065	2196	7360	5212	4137	4060	4354
Graduation rate	45	64	40	58	38	20	39	48
In-state tuition	3956	3975	7395	4058	3683	3999	3156	
Graduation rate	40	20	45	39	39	20	9	48

(a) Draw a scatterplot.
(b) Fit a least-squares line, and graph it.
(c) Looking at the scatterplot of part (a), do you think the least-squares line is a good choice? Discuss.

15.6 PARAMETRIC VERSUS NONPARAMETRIC ANALYSIS

Up until Chapter 11, we basically assumed that random variables belong to specific probability distributions, such as a normal distribution or binomial distribution. The members of those distributions are associated by different parameters such as means or variances. Most of our efforts were concentrated on making some inferences about the unknown parameters. In this vein, we looked at point estimators, confidence intervals, and hypothesis testing problems. In practice, the assumption that observations come from a particular family of distributions, such as normal or exponential, may be quite sensible. As we have already mentioned, slight violations of these assumptions in many practical cases may not significantly affect statistical inferences. However, this is not always true. Furthermore, sometimes we may want to make inferences that have nothing to do with parameters. We may not even have precise measurement data, but only the rank order of observations. For example, if we want to study the performance of students at an institution, we may not have the precise scores the students obtained; instead, we may only have their letter grades, such as A, B, C, D, and F. Even if we have precise measurements, we may not be able to assume a distribution, such as normality. Still, we may be able to say that the distribution is symmetric and skewed or has some other characteristics. Basically, if there is doubt about parametric assumptions or the data are not suitable for parametric inference or we are not interested in inference about parameters, a nonparametric test that is valid under weaker assumptions is preferable. It should be noted that weaker assumptions do not mean that nonparametric methods are assumption free. The inference that can be made depends on valid assumptions being made.

When using nonparametric tests, a common question is: "Why substitute a set of nonnormal numbers, such as ranks, for the original data?" Rank tests are often useful in circumstances when we have no idea about the population distribution. We suspect that the data are not normal, and either we cannot transform the data to make them more normal or we do not wish to do so. Few data are truly normal, despite the robustness of common parametric tests; unless we are quite sure that the nonnormality is a minor problem and will not affect the conclusions, we may often be better off using a rank test. However, there is a small penalty for using rank tests. If the original data are really normal, in the long run, the rank tests will be about 95.5% as efficient as a Student t-test would have been. This means that in such situations, the t-test will require about 95 samples compared to 100 for the rank test. But when data is far from normal, the rank tests will require fewer samples than the t-test; in fact, we should not use the t-test in such cases.

Basically, if we know the distribution of the underlying population, we can use parametric tests. Otherwise, for a given data set, we first perform the normality test as explained in Section 15.3. If normality fails, try transformations. If that fails, we can use nonparametric methods for data analysis.

Another situation in which we can use nonparametric tests is when the data contains some outliers. A box plot or a normal plot, as explained in Section 15.3, will reveal the existence of outliers. However, in many applied areas, such as in most bioavailability data, there will appear to be outliers. It is not feasible to determine whether these are skewed or contaminated distributions. They are not errors. In those situations, a conservative approach will be to use nonparametric methods. For example, because the statistic for the rank sum test is resistant to outliers, it will not be seriously affected by the presence of outliers unless the number of outliers becomes large relative to the sample size.

It should be noted that we ought to be careful even when we use nonparametric tests. For example, if the data for one or both of the samples to be analyzed by a rank sum test comes from a population whose distribution violates the assumption that the distributional shapes are the same, then the rank sum test on the original data may provide misleading results or may not be the most powerful test available. Transforming the data (e.g., a logarithmic transformation pulls in long tails) to obtain normality and then performing a two-sample t-test, or using another nonparametric test, may be more appropriate for the analysis. In general, nonparametric methods are appropriate when the sample sizes are small. When the data set is large, say $n > 100$, it often makes little sense to use nonparametric methods.

Finally, we must conclude that we do not perform nonparametric tests on a given set of data unless it is necessary, that is, if we cannot assume a classical probability distribution that characterizes the given data. Also, parametric statistical analysis is, in general, more powerful than nonparametric analysis. We will end this section with a quote from W. J. Conover: "Non-parametric methods use approximate solutions to exact problems, while parametric methods use exact solutions to approximate problems."

EXERCISES 15.6

15.6.1 Consider the following data.

0.01	0.012	0.016	0.018	0.036	0.042	0.036	0.048
0.072	0.042	0.22	0.096	0.76	0.055	0.13	0.016

(a) Test for normality, and comment whether a parametric or nonparametric test is appropriate.

(b) Try a suitable transformation (filter) to make the transformed data normal, if possible, and then use a parametric procedure.

15.6.2 For the Medicare data in Exercise 15.4.4, if parametric procedures are not appropriate, use a nonparametric procedure.

15.7 TYING IT ALL TOGETHER

Now we will give some real-world problems that we will use standard methods to analyze the given data. Software reliability is a major aspect in any kind of software development. One of the ways to do this is to observe time to failure and/or time between failure (TBF). If the defects are fixed, we would expect, on average, the TBF to increase. Based on that data, one studies reliability of the software. There are a variety of methods to analyze software reliability problems. Here we will not dwell on reliability issues. We will only do some simple data analysis on a set of software failure data. The following data represents software failure times in the Apollo 8 software system. They were obtained from http://www.dacs.dtic.mil/databases/sled/swrel.shtml. It is assumed that these failure times are random.

EXAMPLE 15.7.1

The following data set consists of 26 software failure times taken from testing of the Apollo 8 software system.

T:	9	21	32	36	43	45	50	58	63
	70	71	77	78	87	91	92	95	98
	104	105	116	149	156	247	249	250	
TBF:	9	12	11	4	7	2	5	8	5
	7	1	6	1	9	4	1	3	3
	6	1	11	33	7	91	2	1	

(a) Obtain a dotplot, and describe the TBF data.

(b) Identify any outliers and test for normality with and without outliers for TBF data. If the data are not normal, does any simple transformation make the data normal?

(c) Obtain a 95% confidence interval for TBF.

(d) For estimation problems, does a parametric or nonparametric method seem more appropriate for the data?

(e) Obtain a scatterplot between T and TBF, and discuss its usefulness.

Solution

(a) *The dotplot for the TBF data is shown in* Fig. 15.22.

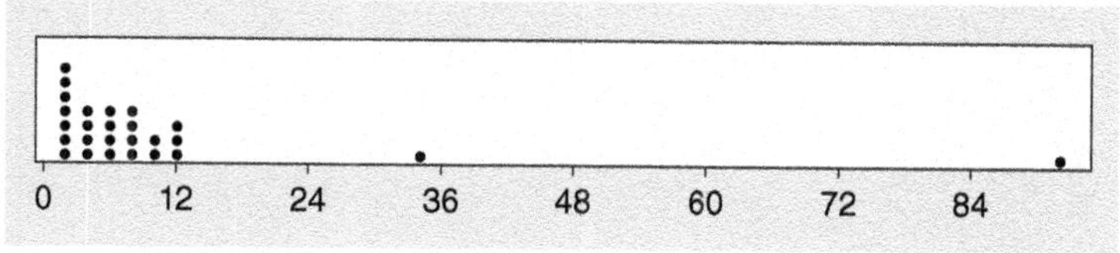

FIGURE 15.22 Dotplot of TBF data.

The following is the result of the describe command from software.

TBF	N	Mean	Median	TrMean	StDev	SE Mean
	26	9.62	5.50	6.58	17.79	3.49
TBF	MIN	MAX	Q1	Q3		
	1.00	91.00	2.00	9.00		

(b) *We will use the box plot shown in* Fig. 15.23 *to identify the outliers.*

From the box plot the observations 33 and 91 are outliers.
Figs. 15.24 and 15.25 *show the normal plots with and without outliers.*

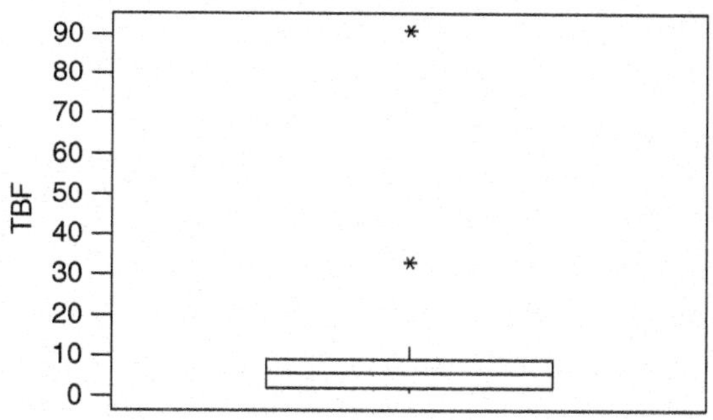

FIGURE 15.23 Box plot of TBF data.

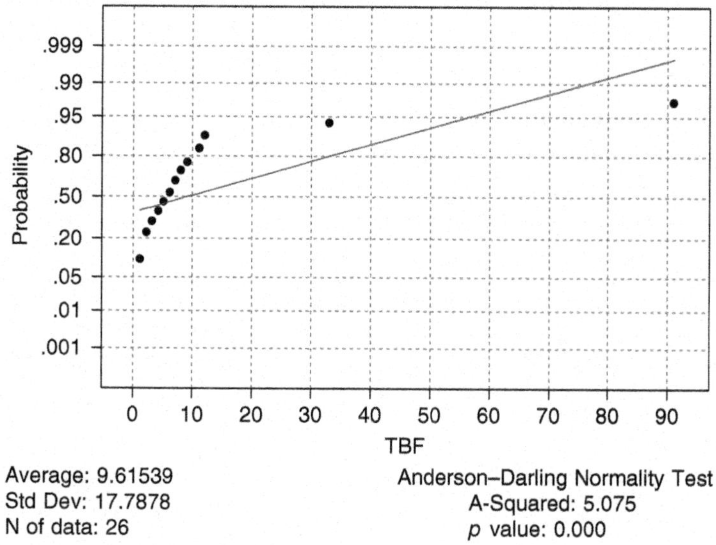

FIGURE 15.24 Normal probability plot of TBF data with outliers.

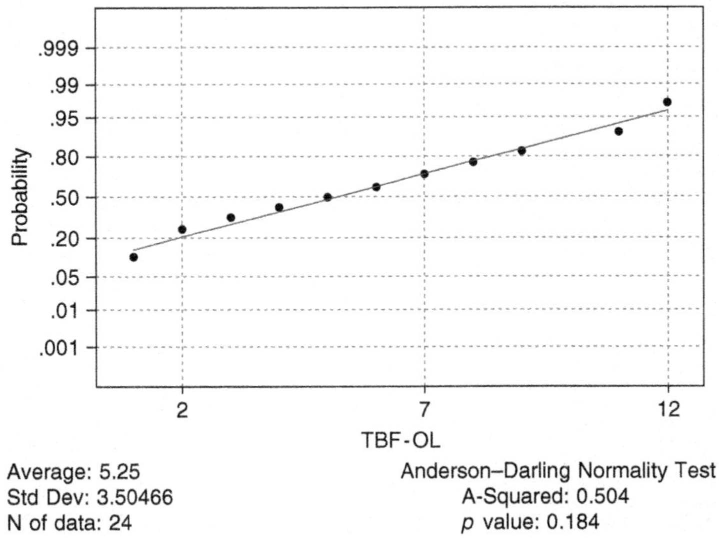

FIGURE 15.25 Normal probability plot of TBF data without outliers.

It is clear that the data with outliers are not normal, whereas if we remove the outliers, the data becomes normal.
Fig. 15.26 *gives the normal plot by taking the natural log of the TBF data with outliers. The figure shows that the data becomes approximately normal.*

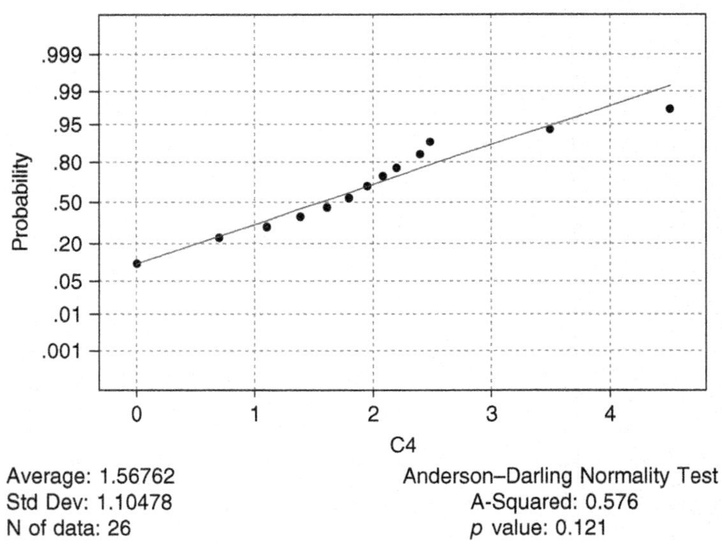

FIGURE 15.26 Normal probability plot of transformed TBF data with outliers.

(c) *It is clear that to obtain a small sample confidence interval, to satisfy the assumption of normality, we need to take the data without the outliers. Hence a 95% confidence interval for TBF with the outliers removed is (3.77, 6.73). Running a nonparametric Wilcoxon test in software for the 95% confidence interval with outliers gave the following.*

		ESTIMATED	ACHIEVED	
TBF	N	MEDIAN	CONFIDENCE	CONFIDENCE INTERVAL
	26	6.00	94.9	(4.00, 8.00)

(d) *If we are analyzing the data without outliers or the log-transformed data, parametric methods are better. With the original data, because the normality assumption may not be appropriate, we need to use nonparametric methods.*

(e) Fig. 15.27 *gives the scatterplot of T and TBF.*

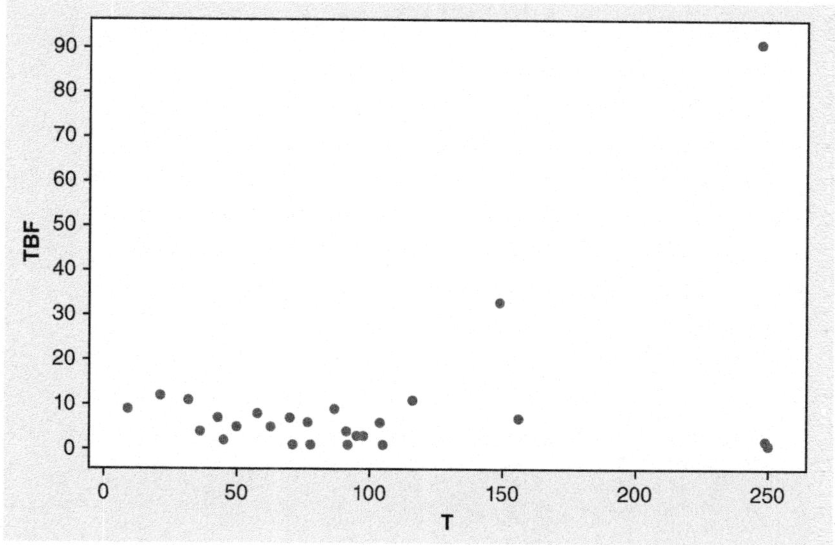

FIGURE 15.27 Scatterplot of T and TBF.

EXAMPLE 15.7.2

Table 15.8 gives dealer cost and sticker price for four-door base models of 25 small and midsize cars (source: *Money Magazine*, March 2001).

(a) Obtain a dotplot, and describe the sticker price data.

(b) Identify any outliers and test for normality with and without outliers for sticker price data. If the data is not normal, does any simple transformation make the data normal?

(c) Obtain a 95% confidence interval for sticker price.

(d) For estimation problems, do parametric or nonparametric methods seem more appropriate for the data?

(e) Obtain a scatterplot between dealer cost and sticker price.

(f) Fit a least-squares regression line, and run a residual model diagnostic using software.

TABLE 15.8 Dealer Cost and Sticker Price (In Dollars) For Select Models of 25 Small and Mid-size Cars.

Model	Dealer Cost (in Dollars)	Sticker Price (in Dollars)
Acura Integra GS	19,479	21,600
Chevy Cavalier	12,398	13,260
Chevy Impala LS	21,251	23,225
Chrysler Concord LX	20,834	22,510
Dodge Neon SE	11,856	12,715
Ford Escort	12,277	12,970

(Continued)

TABLE 15.8 (Continued)

Model	Dealer Cost (in Dollars)	Sticker Price (in Dollars)
Ford Taurus SE	17,606	19,035
Honda Civic DX	11,723	12,960
Honda Accord 2.3 LX	16,727	18,790
Hyundai Sonata	13,805	14,999
Kia Sephia	9,914	10,595
Mazda 626 LX V6	18,181	19,935
Mitsubishi Mirage ES	12,534	13,627
Mercury Sable GS	17,777	19,185
Nissan Maxima GXE	19,430	21,249
Oldsmobile Intrigue GL	22,097	24,150
Pontiac Grand Am GT	18,790	20,535
Saturn SL	9,936	10,570
Subaru Impreza L	14,695	15,995
Toyota Corolla LE	12,042	13,383
Toyota Camry LE	18,169	20,415
Toyota Prius	18,793	19,995
VW Jetta GLS	15,347	16,500
VW Passat GLS	19,519	21,450
Volvo S40	22,090	23,500

Solution

(a) *The dotplot for the sticker price is shown in* Fig. 15.28.

The following summary statistics are obtained by the describe command in software.

	N	Mean	Median	TrMean	StDev	SE Mean
St.price	25	17726	19035	17758	4278	856
	MIN	MAX	Q1	Q3		
St.price	10570	24150	13322	21350		

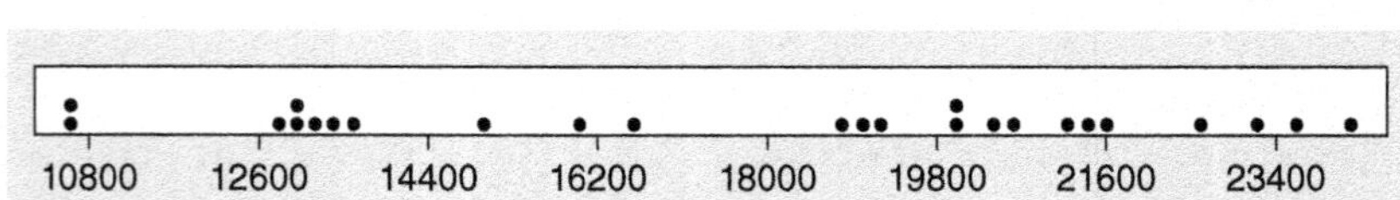

FIGURE 15.28 Dotplot for the sticker price.

(b) *The box plot for the sticker price is shown in* Fig. 15.29.

According to this, there are no outliers. The normal plot is shown in Fig. 15.30. *This is approximately normal.*

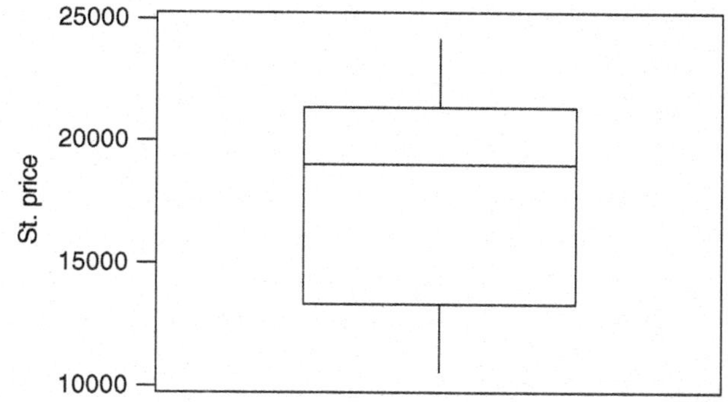

FIGURE 15.29 Box plot for the sticker price.

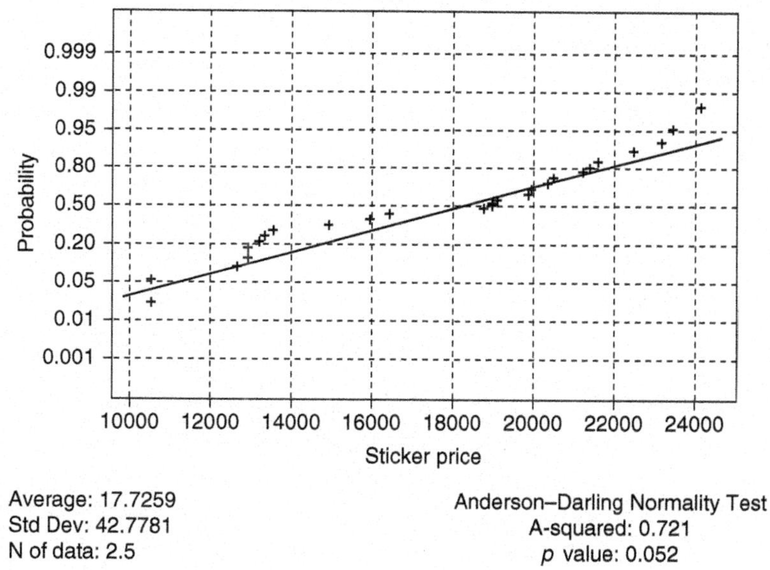

FIGURE 15.30 Normal plot for the sticker price.

(c) *The 95% confidence interval for the sticker price is*

	N	Mean	StDev	SE Mean	95.0 Percent CI
St.price	25	17726	4278	856	(15960, 19492)

(d) *Because there are no outliers and the data look approximately normal, parametric tests seem to be appropriate for this data.*
(e) *The scatterplot for dealer cost versus sticker price is shown in* Fig. 15.31.
(f) Fig. 15.32 *shows the fitted regression line.*

An analysis of residuals by software gives Fig. 15.33.

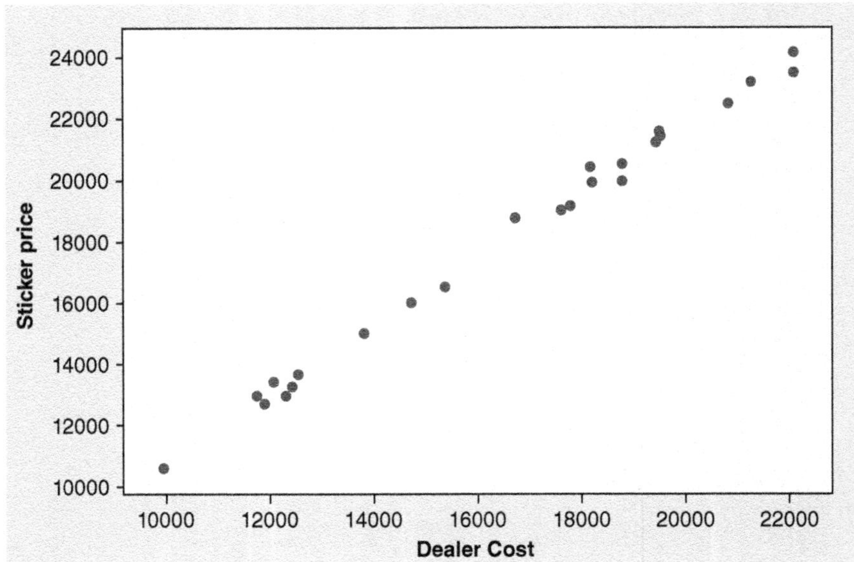

FIGURE 15.31 Scatterplot for dealer cost versus sticker price.

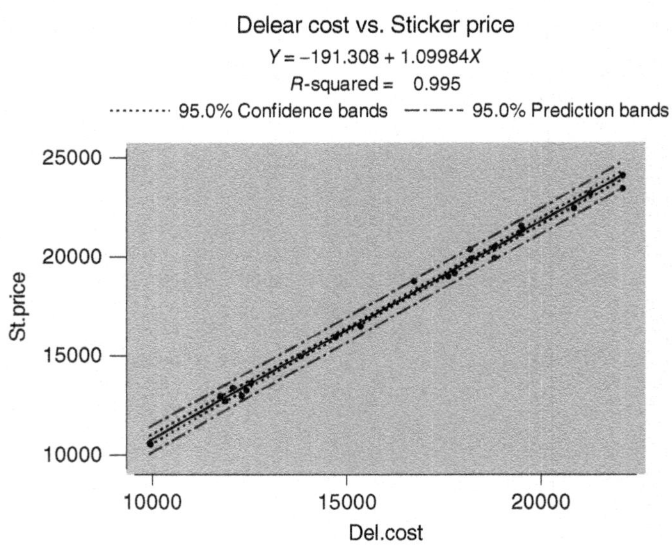

FIGURE 15.32 Regression line for dealer cost versus sticker price.

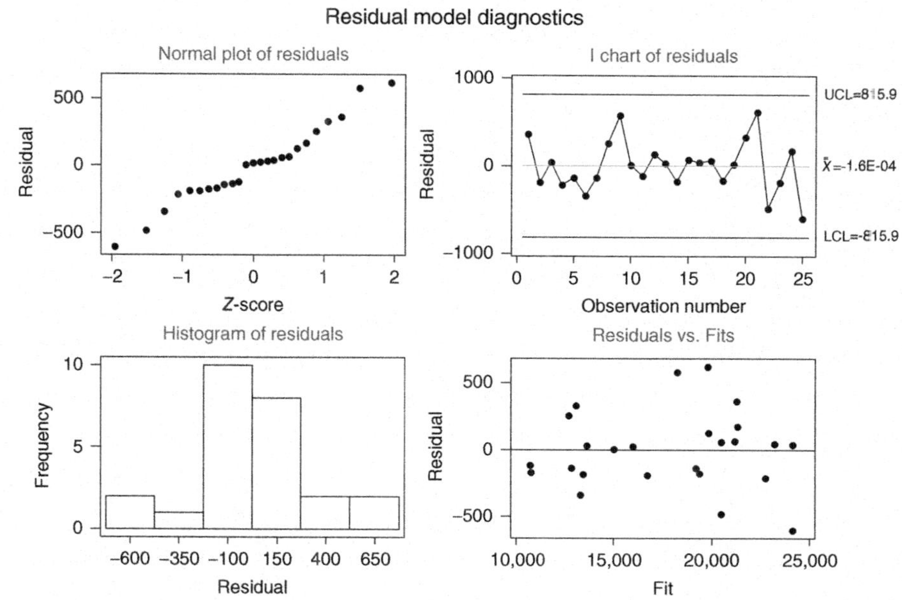

FIGURE 15.33 Residuals versus fit.

By looking at the residuals versus fits, we can see that we have a good fit, and hence the model seems to be appropriate.

EXERCISES 15.7

15.7.1 Table 15.9 gives revenue (in thousands) for public elementary and secondary schools, by state, for 1997–98 and corresponding pupils per teacher for that state for 20 randomly selected states (source: The World Almanac and Book of Facts 2000).

(a) Obtain a dotplot, and describe the pupils per teacher data.

(b) Identify any outliers, and test for normality with and without outliers for the pupils per teacher data. If the data are not normal, does any simple transformation make the data normal?

(c) Obtain a 95% confidence interval for pupils per teacher.

(d) Obtain a scatterplot between total revenue and pupils per teacher.

(e) Fit a regression line between total revenue and pupils per teacher.

TABLE 15.9 Revenue (In Thousands) For Public Elementary and Secondary Schools by State and The Corresponding Pupils Per Teacher.

State	Total Revenue	Pupils per Teacher
Arizona	4,388,915	19.8
Connecticut	5,112,950	14.2
Alabama	4,030,356	16.3
Indiana	7,006,752	17.2
Kansas	3,090,829	14.9
Oregon	3,119,028	20.1
Nebraska	1,688,662	14.5
New York	27,690,556	15.0

(Continued)

TABLE 15.9 (Continued)

State	Total Revenue	Pupils per Teacher
Virginia	6,661,612	14.7
Washington	6,722,916	20.2
Illinois	13,649,628	16.8
North Carolina	7,127,549	15.9
Georgia	8,579,628	16.2
Nevada	1,754,717	18.5
Ohio	12,694,407	16.7
New Hampshire	1,365,391	15.6

15.7.2 Table 15.10 gives the dealer cost and sticker price for luxury cars and sports utility vehicles with popular options (source: *Money Magazine*, March 2001).

(a) Obtain a dotplot, and describe the sticker price data.

(b) Identify any outliers, and test for normality with and without outliers for sticker price data. If the data is not normal, does any simple transformation make the data normal?

(c) Obtain a 95% confidence interval for sticker price.

(d) Do parametric or nonparametric methods seem more appropriate for the data?

(e) Obtain a scatterplot between dealer cost and sticker price.

(f) Fit a least-squares regression line, and run residual model diagnostics using software.

15.7.3 For the college tuition data of Exercise 15.5.5, fit a least-squares regression line, and run residual model diagnostics using R.

TABLE 15.10 Dealer Cost and Sticker Price (In Dollars) For Luxury Cars and Sports Vehicles.

Model	Dealer Cost (in Dollars)	Sticker Price (in Dollars)
Acura TL 3.2	26,218	29,030
Audi A6 4.2	45,385	50,754
BMW 525i	33,800	37,245
Cadillac DeVille DHS4	43,825	47,603
Infiniti I30 Touring	28,604	32,065
Jaguar XJ8	52,535	58,171
Lexus GS430	41,881	48,581
Mercedes-Benz C320	35,067	36,950
SAAB 9–3 Viggen	35,270	38,690
Volvo S80T-6	39,315	41,768
BMW X5 4.4i	45,994	50,774
Chevrolet Blazer LT	26,958	29,725

(Continued)

TABLE 15.10 (Continued)

Model	Dealer Cost (in Dollars)	Sticker Price (in Dollars)
Dodge Durango	26,845	29,370
GMC Jimmy SLE	26,637	29,370
Honda CR-V LX	17,578	19,190
Isuzu Trooper LS	27,901	31,285
Jeep Cherokee SE	21,392	23,130
Lexus LX470	54,785	63,474
Mercedes-Benz ML430	42,243	45,337
Nissan Pathfinder SE	27,203	29,869
Pontiac Aztek GT	22,912	24,995
Subaru Forester S	21,990	24,190
Suzuki Vitara JS	16,063	17,079
Toyota RAV4	18,786	20,630

15.7.4 The following data give the area (in square feet) and the sale prices (approximated to the nearest \$1000) of homes that were sold in a particular city in a 6-week period of 2003.

Area:	1123	1028	1490	2172	2300	1992	3200	3063	3720
	7228	720	943	904	912	1031	1152	1482	1426
	1491	1184	1650	1392	1755	2062	2495	3253	5152
	1270	1723	1161	1220	837	1446	2442	2300	2518
Price:	75	75	102	149	152	154	327	425	625
	775	775	57	66	68	75	86	90	93
	95	95	104	105	135	159	169	253	725
	67	67	110	65	74	95	156	183	207

(a) Obtain a dotplot, and describe the home price data.
(b) Identify any outliers and test for normality with and without outliers for home price data. If the data is not normal, does any simple transformation make the data normal?
(c) Obtain a 95% confidence interval for home price.
(d) Do parametric or nonparametric methods seem more appropriate for the data?
(e) Obtain a scatterplot between the square-foot area of a home and its price.
(f) Fit a least-squares regression line, and run residual model diagnostics using R.

15.8 SOME REAL-WORLD PROBLEMS: APPLICATIONS

In this section, we will use the goodness-of-fit methods discussed in the previous sections to identify the probability distribution that characterizes the behavior of some real-world problems that our society is facing. All the data sets used in this section are available at http://booksite.elsevier.com/9780124171138.

15.8.1 Global Warming

The concept of "Global Warming" consists of two interacting entities, the atmospheric temperature and carbon dioxide (CO_2) in the atmosphere. The United States collects annual data for both of these variables in our observatories in Alaska and Hawaii. The actual data can be found on the website: http://scrippsco2.ucsd.edu/data/atmosperic-co2.html.

Our objective is to identify the probability distribution function, pdf, that follows the CO_2 data that is given in thousands of metric tons annually for 31 years. Once we know the pdf that fits the CO_2 data, we can obtain useful information, such as probabilistic characterization of its behavior, the expected value of CO_2 (theoretical average), and obtain confidence limits on the true amount of CO_2, among other interesting information.

We begin our process of identifying the pdf by structuring a histogram of the 31 randomly selected measurements of CO_2. The histogram of the subject data will give us some idea about the possible pdf that we should be testing. After some preliminary testing of some pdfs, we proceeded to test the following hypothesis:

H_0: The CO_2 data follows the gamma pdf.
Vs
H_a: The CO_2 data does not follow the gamma pdf.

To test this hypothesis, we applied the Kolmogorov–Smirnov, Anderson–Darling, and the Chi-square tests with a level of significance $\alpha = 0.05$. The test statistic results of the three goodness-of-fit tests are given below:

Kolmogorov–Smirnov test:	$D = 0.08771$, P-value = .954
Anderson–Darling test:	$A = 0.3627$, P-value = .883
Chi-square test:	$\chi^2 = 0.95844$, P-value = .811

All three goodness-of-fit tests strongly support the null hypothesis that the CO_2 measurements follow the gamma pdf. We obtained the maximum likelihood estimates of the two parameters α and β of the gamma pdf as $\hat{\alpha} = 635.29$ and $\hat{\beta} = 0.557$. Thus, we can write the estimated gamma pdf for the subject data. That is,

$$f(x) = \frac{x^{635.29-1}}{(0.557)^{635.29}\Gamma(635.29)} \exp\left(\frac{-x}{0.557}\right), \quad x > 0.$$

We can use $f(x)$ to determine various probabilities of interest concerning the behavior of X, the amount of CO_2 in the atmosphere. Also, we can calculate cumulative distribution function $F(x)$, the expected amount that we would find in the atmosphere, confidence limits, among other interesting questions about the behavior of CO_2 using procedures previously explained in this book.

15.8.2 Hurricane Katrina

One of the most devastating hurricanes in the last 100 years to hit the United States was Hurricane Katrina. The Atlantic-based hurricane, category 5 (most devastating), lasted 9 days, August 23–31, 2005. The wind velocity of Katrina is one of the most important variables, and we wish to identify the pdf that characterizes its behavior; that is, perform goodness-of-fit testing to determine the pdf that follows the pressure wind data that was obtained from http://weather.unisys.com/hurricane/atlantic/2005H/KATRINA/track.dat

We have 63 observations of the wind velocity, mph, that reached a maximum wind velocity of 150 miles per hour. After looking at the histogram of the data, we believe that the wind velocity of Katrina follows the two parameters ($\delta = 0$) Weibull pdf. Thus, we proceed to test the following hypothesis:

H_0: The wind velocity data of Hurricane Katrina follows the two-parameter Weibull pdf.
Versus
H_a: The wind velocity data of Hurricane Katrina does not follow the Weibull pdf.

To test this hypothesis, we apply the Kolmogorov–Smirnov, Anderson–Darling, and the Chi-square tests. All these tests strongly support the acceptance of the null hypothesis. The test results are given below:

Kolmogorov–Smirnov test:	$D = 0.0792$, P-value = .795
Anderson–Darling test:	$A = 0.5949$, P-value = .863
Chi-square test	$\chi^2 = 3.4031$, P-value = .638.

Thus, the wind velocity measurements of Hurricane Katrina follow the two-parameter Weibull pdf with the maximum likelihood estimates of the parameter given by $\widehat{\alpha} = 2.1281$ and $\widehat{\beta} = 86.376$. The pdf of the subject data is given by:

$$f(x) = \begin{cases} \frac{86.376}{2.1281}\left(\frac{x}{2.1281}\right)^{85.376} \exp\left(\frac{x}{2.1281}\right)^{86.376}, & x > 0 \\ 0 & \text{elsewhere.} \end{cases}$$

A graphical display of $f(x)$ is given by Fig. 15.34, below:

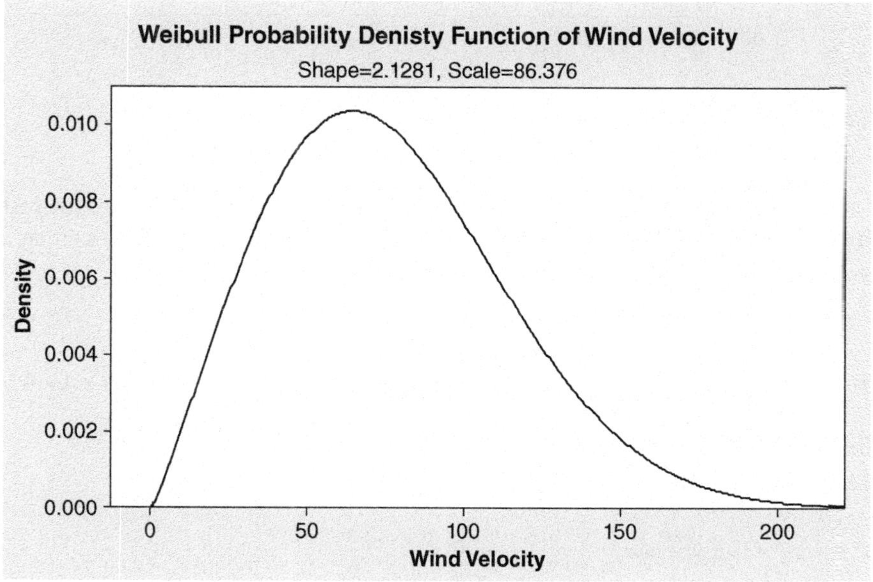

FIGURE 15.34 Weibull probability density function of wind velocity of Hurricane Katrina.

Knowing the pdf that characterizes the two probabilistic behaviors of the wind velocity of Katrina, we can calculate the expected wind velocity and confidence limits. That is,

$$E(X) = 76 \text{ miles/hour}.$$

and the 95% confidence limits of the true mean of the wind velocity are between 23.8 and 150.6 mph.

That is, we are at least 95% certain that the true wind velocity of Hurricane Katrina or similar hurricanes will be between 23.8 and 150.6 miles per hour.

Also, the cumulative probability distribution, $F(x)$, of the wind velocity in its analytical and graphical form (Fig. 15.35) is given below:

$$F(x) = P(X \le x) = \int_0^x \frac{\beta}{\alpha}\left(\frac{t}{\alpha}\right)^{\beta-1} \exp\left(-\left(\frac{t}{\alpha}\right)^{\beta}\right) dt, \; x > 0, \; \alpha, \; \beta > 0$$

and for the given data we will get

$$F(x) = \int_0^x \frac{86.37}{2.128}\left(\frac{t}{2.128}\right)^{85.37} \exp\left(-\left(\frac{t}{2.128}\right)^{86.37}\right) dt.$$

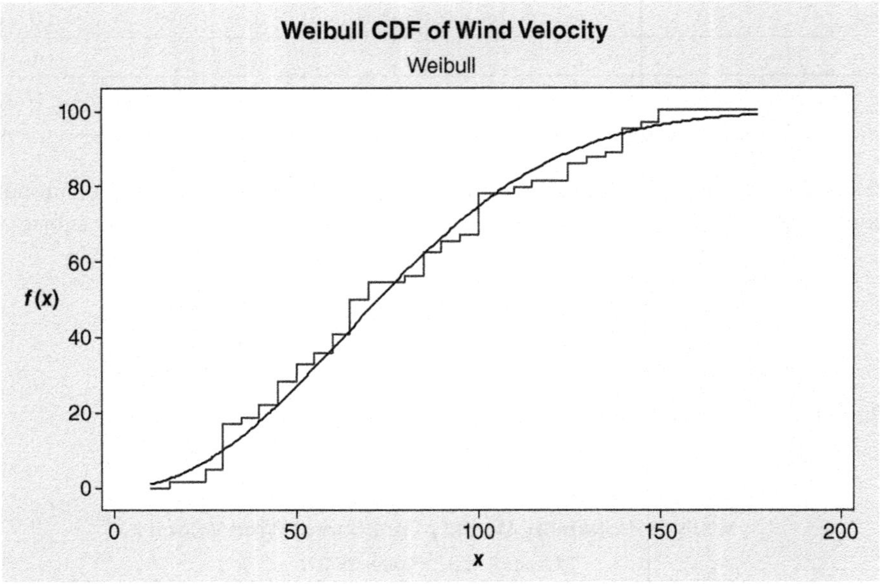

FIGURE 15.35 Weibull cumulative distribution function of wind velocity of Hurricane Katrina.

Thus, we can use the above graph to obtain various probabilities. For example, if we are interested in the probability that the wind velocity of a category 5 hurricane is less than 150 miles per hour, we can obtain an approximate estimate from the graph above, that is,

$$F(150) = P(X \leq 150) \approx 0.93.$$

This means that based on the given data, we are approximately 93% certain that the wind velocity will be less than 150 miles per hour.

Given below is the R code for the goodness-to-fit tests for the Katrina data.

```
HK<-read.delin("~/Documents/Hurricane Katrina.txt")
View(HK)
summary(HK) #### descriptive Stat###
xkHK$WIND
hist(xk) ### Histogram ####
library(lessR)
dens(xk,typec("both,""normal"),xlab"Wind,"ylab"f(x)")
color.density(xk)
mmean(xk);m
stdsqrt(var(xk));std hist(xk),density12,breaks8,probT,col"plum4,"xlab"Wind,"xlimc(0,200),main"Hi
stogram of Wind velocity of Hurricane Katrina")
library(vcd) ## Goodness of fit test
fitdistr(xk,'weibull') ### estimate the parameters using MLE ks.test(xk,"pweibull,"shape1.805,s-
cale52.323) #Kolmogorov-Smirnov test
ad.test(xk)#Anderson-Darling
```

15.8.3 National Unemployment

The aim of the present problem is to identify the probability distribution that characterizes the rates of unemployment in the United States. The subject data was obtained from the US Bureau of Labor Statistics, http://www.bls.gov/, under Database & Tools. We will consider the data on the annual average unemployment rate in the United States from 1957 to 2008. Initially, we looked at the histogram of the data, and it gave us a visual interpretation that it may follow the gamma pdf. Initially we tested for the two-parameter gamma pdf and obtained a fairly good fit, but when we tried the three-parameter gamma pdf, we obtained a better fit. That is,

H_0: The annual average rates of unemployment in the United States follow the three-parameter gamma pdf.
Versus

H_a: The subject data does not fit the three-parameter gamma pdf.

Given below is the value of the goodness-of-fit test statistics for a sample of 51 data points:

Kolmogorov–Smirnov test:	$D = 0.0847$, P-value = 0.8276
Anderson–Darling test:	$A = 0.3424$, P-value = 0.7916
Chi-square test:	$\chi^2 = 2.2353$, P-value = .8172

All three goodness-of-fit tests strongly support that the three-parameter gamma pdf is probabilistically the best to characterize the behavior of the US annual average of unemployment, with the maximum likelihood estimate of the parameter $\hat{\alpha} = 5.5871$, $\hat{\beta} = 0.5954$, and $\hat{\gamma} = 2.5008$. Thus, the subject pdf is given by

$$f(x) = \frac{1}{(0.5954)^{5.5971}\Gamma(0.5954)}(x - 2.5008)^{4.5871}\exp - \left(\frac{x - 2.5008}{0.5954}\right), \quad x > 0.$$

The expected value of the subject pdf is

$$E(X) = \hat{\delta} + \hat{\alpha}(\hat{\beta}) = 2.5008 + (5.5871)(0.5954) = 5.83.$$

Thus, one will expect the unemployment rate to be approximately 5.83% based on the actual data we analyzed. A graphical form of $f(x)$ over the initial histogram that guided us to the three-parameter gamma pdf is given in Fig. 15.36 below:

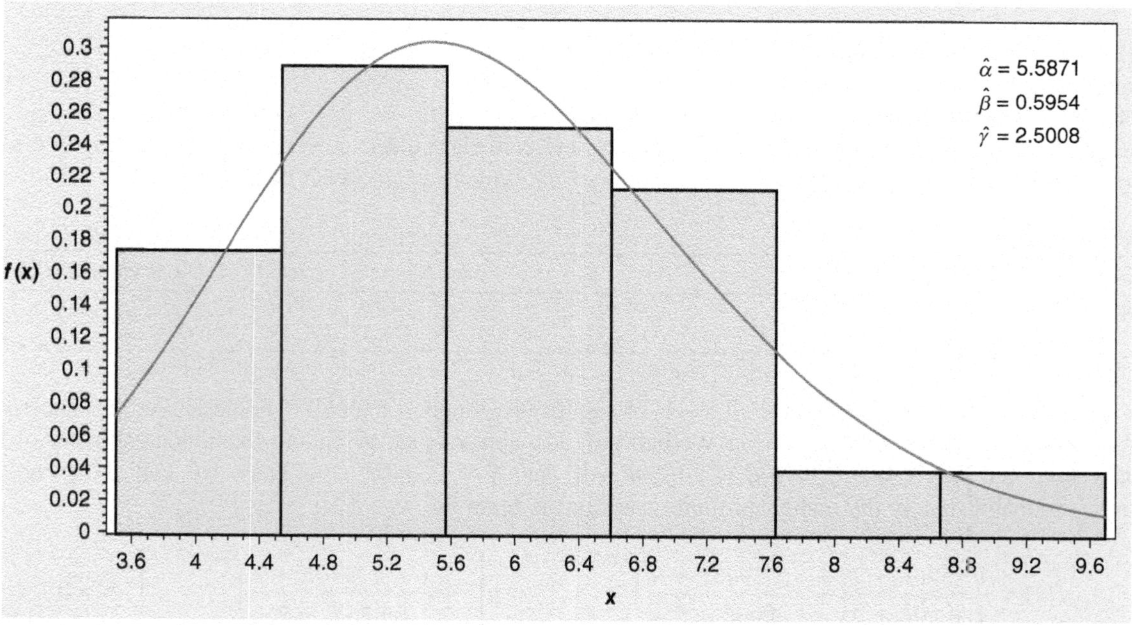

FIGURE 15.36 Three-parameter gamma probability density function for unemployment in the United States.

We can use the pdf to obtain confidence limits on the true rate of unemployment, the cumulative pdf, $F(x)$, to obtain various probabilities of interest on the subject problem, among other useful information.

15.8.4 Brain Cancer

A brain tumor is an abnormal growth of cells within the brain, which can be cancerous (malignant) or benign. It is estimated that there have been more than 43,800 new cases of brain cancerous tumors in the United States during the last few years. In this application, we are interested in studying the behavior of the malignant tumor sizes in the brain. The subject data was obtained from the Surveillance Epidemiology and End Results (SEER) database. We have taken a random sample of 200 brain cancer patients from a large database with their cancerous tumor size measured in millimeters. Our aim is to find the probability distribution that characterizes the behavior of the tumor sizes. Thus, after testing several pdfs and looking at the histogram, we believe that the three-parameter Weibull pdf is a prime candidate. The table below contains the actual data for 50 patients (Table 15.11).

TABLE 15.11 Brain Cancer Data

Frequency	1	3	3	1	1	2	3	2	4	2	1	7	1	1	2
Tumor size(mm)	7	8	10	11	12	14	15	19	20	23	24	25	26	27	28
Frequency	1	7	1	34	1	7	1	1	27	3	1	1	2	11	2
Tumor size (mm)	34	35	37	40	41	45	46	48	50	55	56	57	59	60	63
Frequency	3	2	9	1	1	1	2	6	1	1	1	2	1	1	2
Tumor size (mm)	65	67	70	72	73	74	75	80	83	85	86	90	94	100	120
Frequency	1	1	1	2	21										
Tumor size (mm)	150	160	250	21	30										

Now, we proceed to test our belief.

H_0: The sizes of the malignant tumors in the brain fit the three-parameter Weibull pdf.

Versus

H_a: The subject data does not follow the three-parameter Weibull pdf.

We are applying the most commonly used goodness-of-fit tests to make a decision concerning accepting or rejecting the stated hypothesis for say, α, = .01, 05, 0.10. The results of the three tests are given below:

Kolmogorov–Smirnov test:	$D = 0.0502$, P-value = 0.6746
Anderson–Darling test:	$A = 0.6948$, P-value = 0.7321
Chi-square test:	$\chi^2 = 9.6143$, P-value = .2115.

Thus, all three goodness-of-fit tests, for all levels of significance, support the null hypothesis that the cancerous tumor sizes of the brain follow the three-parameter Weibull pdf. The approximate maximum likelihood estimates of the three parameters used are $\hat{\alpha} = 9.4826E + 7$, $\hat{\beta} = 1.4060E + 9$, and $\hat{\gamma} = 1.3940E + 9$. Thus, we can write the pdf that characterizes probabilistically the malignant tumor sizes in the brain by

$$f(x) = \frac{9.4826E + 7}{1.4060E + 9}\left(\frac{x + 1.3940E + 9}{1.4060E + 9}\right)^{9.4826E+7-1} \exp\left[-\left(\frac{x + 1.3940E + 9}{1.4060E + 9}\right)^{9.4826E+7}\right], \quad x > 0,$$

and the cumulative pdf is given by

$$F(x) = 1 - \exp\left[-\left(\frac{x + 1.3940E + 9}{1.4060E + 9}\right)^{9.482E+7}\right], \quad x \geq 0.$$

A graphical illustration of the three-parameter Weibull pdf, along with a frequency histogram of the data, is given in Fig. 15.37 below:

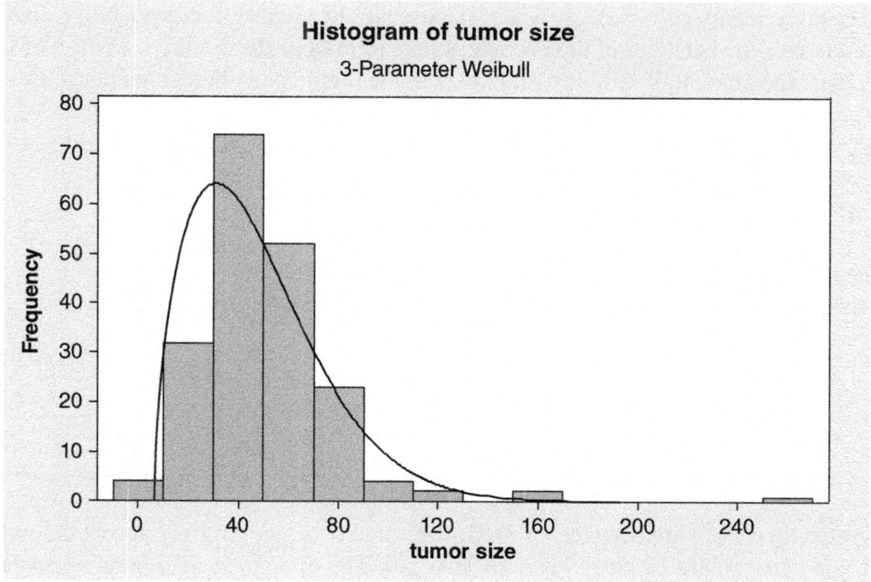

FIGURE 15.37 Three-parameter Weibull pdf for tumor size data.

We can use the above diagram to obtain approximate probabilities of the behavior of the cancerous tumor sizes. For example, the probability that the tumor size is less 60 mm is approximately 0.25, that is,

$$P(X \leq 60\text{mm}) \approx 0.25, \ P(X \leq 48\text{mm}) \approx 0.25,$$

And the probability that the tumor size is larger than 48 mm is approximately 74%, that is,

$$P(X > 48\text{mm}) = 1 - P(X \leq 48\text{mm}) \approx 0.74.$$

We also can proceed to obtain the expected value of the tumor size and approximate confidence limits on the true size of the tumor.

15.8.5 Rain Fall Data Analysis

For the Southern Region of the United States, we have the average annual rainfall data in inches from 1975 to 2007. The actual data is given in Table 15.12 below:

TABLE 15.12 Rainfall Data

Year	Rain	Year	Rain	Year	Rain	Year	Rain	Year	Rain	Year	Rain
1975	3.957	1981	3.68	1987	4.465	1993	4.175	1999	4.103	2005	5.22
1976	4.031	1982	5.224	1988	4.487	1994	4.889	2000	2.737	2006	3.526
1977	3.918	1983	5.639	1989	3.612	1995	5.468	2001	4.104	2007	3.211
1978	4.299	1984	3.563	1990	3.322	1996	3.668	2002	5.037		
1979	4.942	1985	3.592	1991	4.463	1997	5.029	2003	4.633		
1980	3.921	1986	4.307	1992	4.514	1998	4.73	2004	5.219		

Using thirty-three measurements ($n = 33$), we wish, if possible, to identify the probability distribution function that probabilistically characterizes the behavior of the average annual rainfall of the Southern District of United States. Having such pdf, we can calculate the amount of rain we will expect in the region and obtain confidence limits of the true amount of annual rainfall.

From a preliminary view of the histogram of the data, we believe that the rainfall data follows the beta pdf.

Thus, let us proceed to test our belief:

H_0: We believe that the rainfall data follows the beta pdf.

Versus

H_a: The subject data does not follow the beta pdf.

Given below are the goodness-of-fit results applying the three commonly used statistical tests:

Kolmogorov–Smirnov test:	D = 0.0773, *P*-value = 0.9806
Anderson–Darling test:	A = 0.2098, *P*-value = 0.8836
Chi-square test:	$\chi^2 = 0.2888$, *P*-value = 0.9905.

For all commonly used levels of significance, $\alpha = 0.01$, 0.05, and 0.10, we strongly accept the null hypothesis that our belief is true, that is, the given rainfall data follows the beta pdf. The maximum likelihood estimates of the parameter α and β of the beta pdf are $\hat{\alpha} = 2.2823$ and $\hat{\beta} = 1.8754$. Thus, the beta pdf of the rainfall data is given by

$$f(x) = \frac{\Gamma(2 + \widehat{\beta})}{\Gamma(2)\Gamma(\widehat{\beta})} x^{\hat{\alpha}-1}(1 - x)^{\widehat{\beta}-1}, \quad x \geq 0,$$

or

$$f(x) = \frac{2.5237}{5.7593} x^{0.2823}(1 - x)^{0.8754}, \quad x \geq 0,$$

where

$\Gamma(2.2823 + 1.8754) = 2.5237$, and $\Gamma(2.2823)\Gamma(1.87554) = 5.7593$.

The graph of the subject pdf over the histogram of the data is given in Fig. 15.38 below:

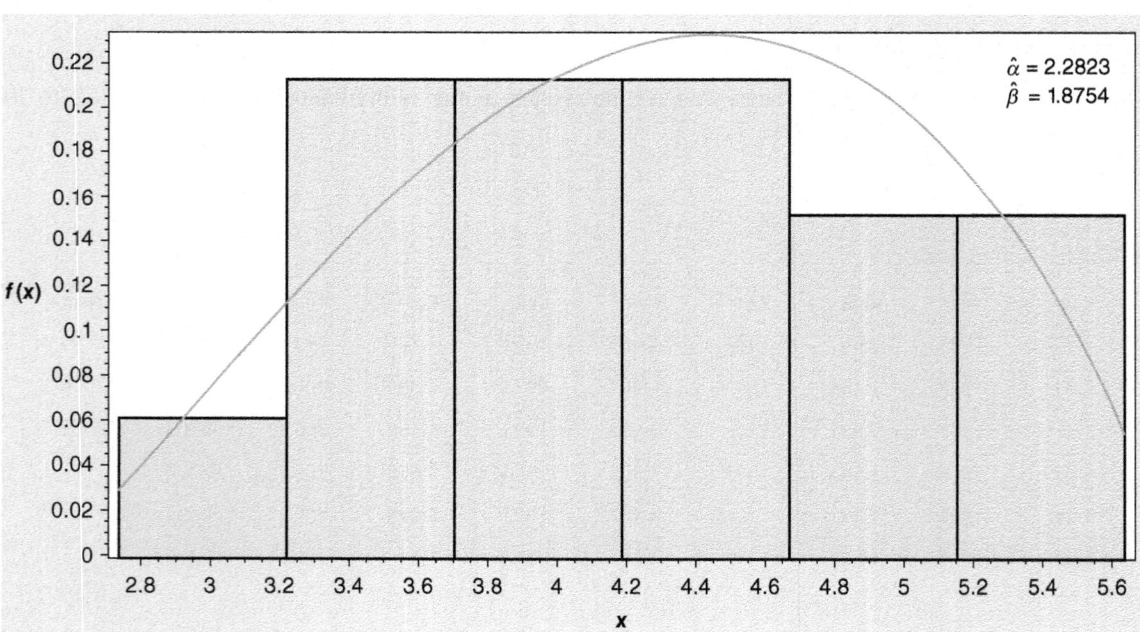

FIGURE 15.38 Beta probability density function for rainfall data.

The expected amount of average rainfall in the Southern Region is 4.2998 inches, that is,

$$E(X) = \int_0^\infty xf(x)dx = 4.2998 \ inches.$$

We can also calculate confidence limits around the true value of the annual average rainfall. For example, we are at least 95% confident that the true annual average rainfall in the Southern District is between 4.0579 and 4.5167 inches.

15.8.6 Prostate Cancer

In this application, we will study the behavior of the cancerous tumor in prostate cancer patients. We shall use real prostate cancer data for white men from 1973 to 2007 from the Surveillance Epidemiology and End Results known as the SEER Program. The tumor size is the random variable of interest for 20,645 prostate cancer patients. Our primary objective is to identify the pdf that characterizes probabilistically the cancerous tumor size in mm. From the initial structure of the histogram, we believe that the two-parameter Weibull pdf may fit the subject data. Thus, we set up our hypothesis to test our belief.

H_0: The prostate cancerous tumor sizes follow the Weibull pdf.

Versus

H_a: The subject data does not follow the Weibull pdf.

Applying the Kolmogorov–Smirnov, Anderson–Darling, and Chi-squared tests, they all support the null hypothesis that the subject data follows the two-parameter Weibull pdf. The maximum likelihood estimates of the parameter α and β that drive the Weibull pdf are $\widehat{\alpha} = 0.8704$ and $\widehat{\beta} = 12.4403$.

Thus, the two-parameter Weibull pdf is given by:

$$f(x) = \frac{0.8704}{12.4403}\left(\frac{x}{12.4403}\right)^{-0.1296} \exp\left[-\left(\frac{x}{12.4403}\right)^{-0.8704}\right], \ x \geq 0,$$

where x represents the size of the cancerous tumor in mm. The cumulative Weibull pdf is useful in obtaining various probabilities of the size of the tumor and is given by

$$F(x) = P(X \leq x) = 1 - \exp\left[-\left(\frac{x}{12.4403}\right)^{-8704}\right], \ x \geq 0.$$

Given in Fig. 15.39 below is the Weibull pdf over the initial histogram along the cumulative pdf.

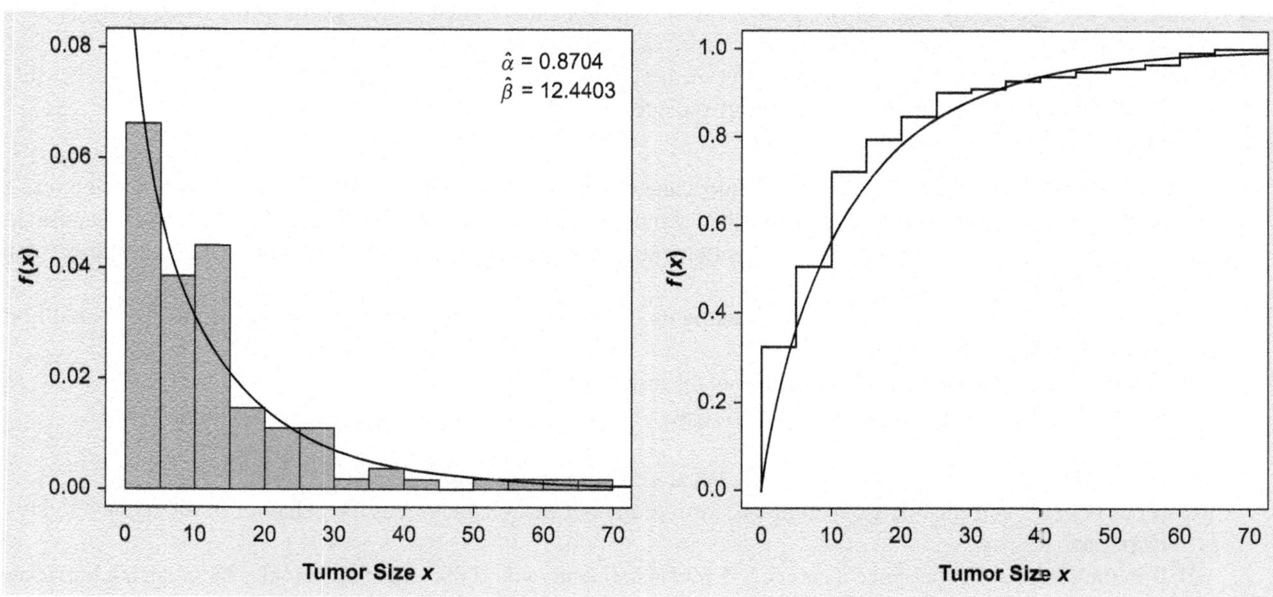

FIGURE 15.39 Weibull pdf and cdf for prostate tumor sizes.

Thus, for an individual patient drawn at random from the subject population, we expect their cancer tumor size to be 13.341 mm, that is,

$$E(X) = \int_0^\infty xf(x)dx = 13.341\text{mm}.$$

Furthermore, we can calculate confidence limits around the true unknown size of the prostate tumor, that is, a 90% confidence interval for the true mean size is (0.410, 43.81). We can conclude that we are at least 90% certain that the true size of the tumor will be between 0.410 and 43.81 mm for an individual that falls in the subject population.

EXERCISES 15.8

15.8.1 Global Warming

Carbon dioxide CO_2 data in the United States is collected in Point Barrow, Alaska, and Mauna Lao, Hawaii. This data is given http://booksite.elsevier.com/9780124171138. Using the CO_2 data collected in Point Barrow from 1974 to 2004, perform the following analysis:

(a) Structure a histogram of the data, and interpret its visual behavior.
(b) Apply the Chi-square goodness-of-fit test to prove or disprove that the CO_2 data follows the exponential power probability distribution, using $\alpha = 0.05$.
(c) If you have proven that the CO_2 data follows the exponential power pdf, proceed to calculate and interpret the expected value of the subject pdf.

15.8.2 Answer the same questions stated in Exercise 15.8.1 using the CO_2 data that was collected at Mauna Lao.

15.8.3 Rainfall Data

In http://booksite.elsevier.com/9780124171138, you will find the average yearly rainfall data in inches for the North, Central, and Southern Region of the United States from 1975 to 2007. Using the Northern Region data, perform the following analysis:

(a) Construct a histogram of the yearly average rainfall for the Northern Region. Does the histogram give you any visual indication of the type of pdf that the data follows?
(b) Using the Kolmogorov–Smirnov goodness-of-fit test, verify if the subject data follows the normal pdf for $\alpha = 0.05$.
(c) If you have proven that the data follows the normal pdf, what is the expected rainfall for a given year? Also, calculate a 95% confidence limit for the true average rainfall, and interpret its meaning.
(d) Calculate P-P plot, and interpret its visual meaning with respect to part (b).

15.8.4 Using the average yearly rainfall from the Central Region of the United States, perform the same analysis as the Northern Region, and in place of the normal pdf, use the gamma pdf.

15.8.5 Using the data given for the Southern Region of the United States, perform the same analysis as for the Northern Region, Exercise 15.8.3, with the normal pdf replaced with the beta pdf.

15.8.6 Hurricane Katrina

Hurricane Katrina was the most devastating hurricane to hit the United States in the last 100 years. Katrina was an Atlantic-based hurricane category 5 that reached wind velocity of more than 160 mph. In http://booksite.elsevier.com/9780124171138, you will find 63 measurements of the wind velocity of Katrina. Using the subject data, perform the following analysis:

(a) We believe that the wind velocity measurements of Hurricane Katrina follow the three-parameter Weibull pdf. Test this belief using:
  i) The Kolmogorov–Smirnov goodness-of-fit test
  ii) The Anderson–Darling goodness-of-fit test
  Using $\alpha = 0.05$:
(b) Discuss the results of (i) and (ii) above. What conclusion have you reached about our belief?
(c) If our belief is correct, write the complete form of the pdf that characterizes the behavior of the wind velocity of Hurricane Katrina.
(d) If in the future we experience a category 5 hurricane, what would the expected velocity be of such a hurricane?

15.8.7 With respect to Exercise 15.8.6, there is a group of scientists that believe that the Rayleigh pdf is a better fit of the wind speed measurements of Hurricane Katrina. Follow the same questions posed in Exercise 15.8.6 using Rayleigh pdf. What do you conclude by comparing the results of Exercise 15.8.6 with those of Exercise 15.8.7?

15.8.8 National Unemployment

In http://booksite.elsevier.com/9780124171138, you will find the annual percent average of unemployment data for the United States from 1957 to 2007. Using this data, perform the following analysis:

(a) Structure a histogram of the data. Does this histogram convey any useful information concerning the behavior of the data?

(b) Using the goodness-of-fit test of your choice, can you identify the pdf that characterizes the behavior of the data, that is, the pdf that the subject data was drawn from using $\alpha = 0.05$?

(c) Once you have found the subject pdf of the unemployment data, calculate the expected value of the annual average percentage of unemployment rate.

15.8.9 Breast Cancer

In http://booksite.elsevier.com/9780124171138, we have the malignant breast tumor size in mm of 250 breast cancer patients. In the database, draw a random sample of the tumor sizes of $n = 50$ breast cancer patients. For the 50 tumor sizes in mm, perform the following analysis:

(a) Structure a histogram of 50 tumor sizes. Discuss any visual information you might obtain concerning the possible pdf that characterizes the data behavior.

(b) Identify, if possible, a pdf that you believe may characterize the given data, using one or more of the goodness-of-fit tests, using $\alpha = 0.05$

(c) If you were not able to identify the pdf, why not? If you were successful, identify completely the pdf, with appropriate parameter estimates.

(d) If you have identified correctly the pdf, calculate and interpret the expected value of the subject data.

15.8.10 In http://booksite.elsevier.com/9780124171138, we have the survival times (in years) of 250 breast cancer patients, that is, the age that they died due to breast cancer. From this database, draw a random sample of $n = 50$ survival times. Use these survival times to perform the following analysis:

(a) Structure a histogram to possibly guide you in identifying the pdf of the subject data.

(b) Use any of the goodness-of-fit test to search in identifying the correct pdf that characterizes the behavior of the given survival times for $\alpha = 0.05$.

(c) State completely the pdf you have identified, and discuss its usefulness in obtaining information about the subject data.

(d) Obtain the cumulative distribution function $F(t)$ of the pdf $f(t)$ you have found. If you take one minus the $F(t)$ you will obtain the survival function, $S(t)$, of the given data. That is, $S(t) = 1 - F(t)$. The survival function, $S(t)$, gives you the probability that a given patient drawn from the database of 250 breast cancer patients will survive a specified year.

(e) Write the survival function of the given data set, and graph it, that is, $S(t)$ versus t. Discuss the useful information that the graph gives concerning breast cancer patients.

15.8.11 Lung Cancer

In http://booksite.elsevier.com/9780124171138, we have the malignant tumor sizes for male and female lung cancer patients. We also include the survival times of both genders, that is, the age in years that they died due to lung cancer. From the male database, draw a random sample of $n = 60$ malignant tumor sizes, and perform the following analysis:

(a) Structure a histogram of the 60 measurements of the tumor sizes in mm.

(b) Let part (a) guide you, if possible, in performing goodness-of-fit testing at $\alpha = 0.05$ to identify the best possible pdf that characterizes the probabilistic behavior of the tumor sizes.

(c) Write the pdf completely with appropriate parameter estimates, and obtain and interpret its expected value.

15.8.12 Proceed to obtain a random sample of $n = 60$ from the female database and perform the same analysis (a) – (c) as in Exercise 15.8.11.

15.8.13 Give a precise comparison of males and females for each of the analysis you performed (a)–(c) in Exercises 15.8.11 and 15.8.12. Discuss your comparison findings.

15.8.14 In the lung cancer database, we have also given information about the survival times of male and female lung cancer patients. Take a random sample of n = 50 of the survival times of male lung patients, and proceed to perform the same analysis for the survival times as in Exercise 15.8.10.

15.8.15 Similarly as in Exercise 15.8.4, proceed to take a random sample of n = 50 of the survival times of female lung patients, and perform the same analysis as you did for the male patients in Exercise 15.8.14.

15.8.16 Give a precise comparison of the analysis of the findings of male and female lung patients that you have found in Exercises 15.8.14 and 15.8.15, respectively. Discuss your comparison findings.

15.8.17 Colon Cancer

In http://booksite.elsevier.com/9780124171138, we have the malignant tumor sizes of male and female colon cancer patients. From this database, draw a random sample of $n = 50$ tumor sizes of the male colon cancer patients. Using this data, proceed to perform the same analysis that you did for the lung cancer data in Exercise 15.8.11.

15.8.18 Proceed to draw a random sample of n = 50 from the female database that gives the malignant colon tumor size. Perform the same analysis for the females that you did for the males in Exercise 15.8.17.

15.8.19 In the colon cancer database, we also give the survival times for both male and female patients. From the male database, draw a random sample of n = 60 survival times, and proceed to perform the same analysis as you did in Exercise 15.8.11.

15.8.20 From the survival times of female colon cancer patients, draw a random sample of n = 60, and proceed with the same analysis that you did for the male patients in Exercise 15.8.19.

15.8.21 Give a precise comparison of the survival times analysis (a)–(c) for males and females.

15.9 CONCLUSION

We have briefly discussed some of the real-world problems that arise in applied data analysis. However, this discussion is not exhaustive. There are various other special problems that can arise in applied data analysis. For example, if one or both of the sample sizes are small, it may be hard to detect violations of some of the assumptions. For small samples, violation of assumptions such as inequalities of variances is hard to discover. Also, for small sample sizes, possible outliers whose detection may be in doubt may have undue influence on the inferences. It is better to avoid such problems in the design stage of an experiment, when suitable sample sizes can be determined before we start collecting data.

Differences in distributional shapes can influence the testing procedures of two or more samples. In those cases, utilizing a transformation may settle that problem and may also promote normality as well as correct the problem of inequality of variances. There are also many issues related to simulation that are discussed in Chapter 13 in the utilization of empirical methods—for instance, in the application of MCMC methods, the issues of burn-in, choice of the correct proposal function, and convergence. These are beyond the scope of this book.

Combining the issues discussed in this chapter with the rest of the material of this textbook should give the student a good footing in the theory of statistics as well as the ability to deal with many real-world problems.

Appendix A

Set Theory

In this appendix, we present some basic ideas and concepts of set theory essential for a modern introduction to probability and statistics. The origin of set theory is attributed to Georg Cantor, when he proved the uncountability of the real line in 1873. A ***set*** is defined as a collection of well-defined, distinct objects. These objects of a set are called ***elements*** or ***members***. The elements of a set can be anything: the alphabet, numbers, people, other sets, and so forth. Sets are conventionally denoted with capital letters, A, B, C, and so on. A ***universal set***, denoted by S, is the collection of all possible elements under consideration. If a is an element of a set A, we write $a \in A$. If a is not an element of A, we write $a \notin A$.

A set is described either by listing its elements or by stating the properties that characterize the elements of the set. For example, to specify the set A of all positive integers less than 12, we may write

$$A = \begin{cases} \{1, 2, 3, 4, 5, 6, 7, 8, 9, 10, 11\} \\ \{\text{all positive integres less than } 12\} \\ \{x\text{: } x < 12, \quad a \text{ a positive integer}\}. \end{cases}$$

Sets are classified as finite or infinite. A set is ***finite*** if it contains exactly n objects, where n is a nonnegative integer. A set is ***infinite*** if it is not finite. For example, if A is a set containing all positive integers less than or equal to 50, then A is a finite set. If B is a set containing all the positive integers, it is an infinite set.

Describing a set by stating its properties is the practical way to represent a set with a large or infinite number of elements.

A set B is a ***subset*** of a set A if every element of B is also an element of A. We denote this by writing $B \subseteq A$, which is read "A contains B " or "B is contained in A." For example, if A is the set of real numbers and

$$B = \{x\text{: } x \leq 5, \quad x \text{ a positive integer}\},$$

it is clear that B is a subset of A. Also, every subset is a subset of itself. Two sets A and B are ***equal***, $A = B$, if and only if $A \subseteq B$ and $B \subseteq A$. Thus, two sets A and B are said to be equal if they have the same members. A set B is a ***proper subset*** of a set A if every element of B is an element of A and A contains at least one element that is not an element of B. We denote this relationship by $B \subset A$. In the previous example, we have $B \subset A$. The set, which contains no elements, is called the ***empty set*** (or ***null set***) and is denoted by ϕ. The null set ϕ is a subset of every set.

A ***Venn diagram*** is used for the visual representation of sets. In the Venn diagram, the universal set, S, is represented by a rectangle. The subsets are represented by circles inside this rectangle.

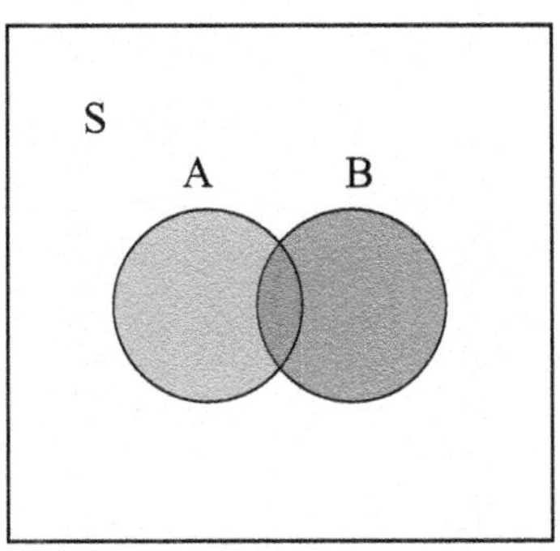

FIGURE A.1 A Venn diagram.

Set Operations

Union, ∪: The union of two sets A and B is the *set* of all elements that belong to A or B (or both; elements that belong to both sets are included only once) and is denoted by $A \cup B$.

$$A \cup B = \{x:\ x \in A \text{ or } x \in B\}$$

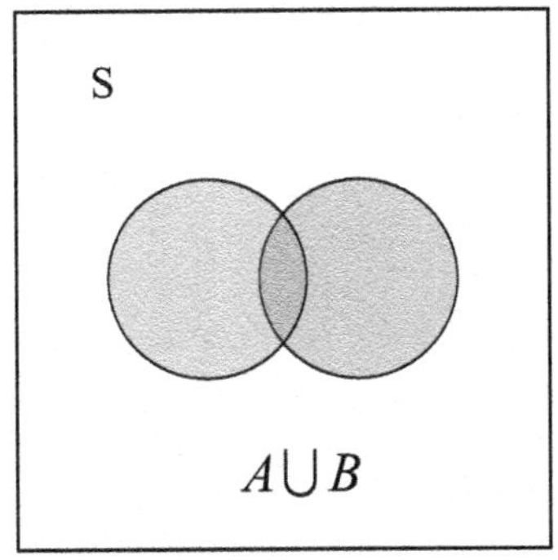

FIGURE A.2 Union of two sets.

Intersection, ∩: The intersection of two sets A and B is the set of all elements that belong to both A and B and is denoted by $A \cap B$. $A \cap B = \{x \in S:\ x \in A \text{ and } x \in B\}$.

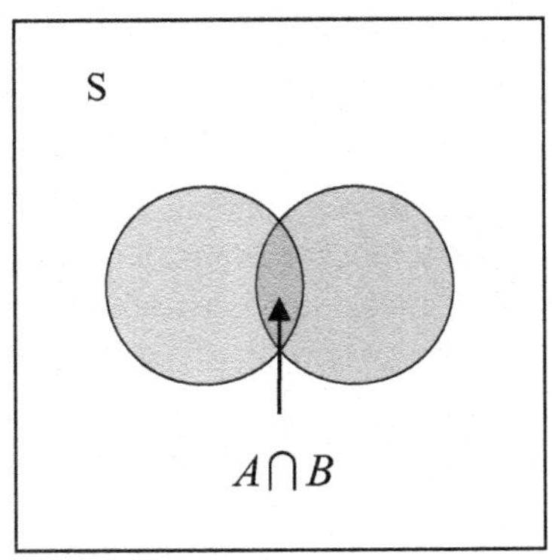

FIGURE A.3 Intersection of two sets.

If $A \cap B = \varnothing$, then the sets A and B are said to be ***disjoint*** or ***mutually exclusive*** sets.

Complement: The complement of a set A is the set of all elements that belong to S but not to A.

$$A^c = \{x: x \in S;\ x \notin A\}.$$

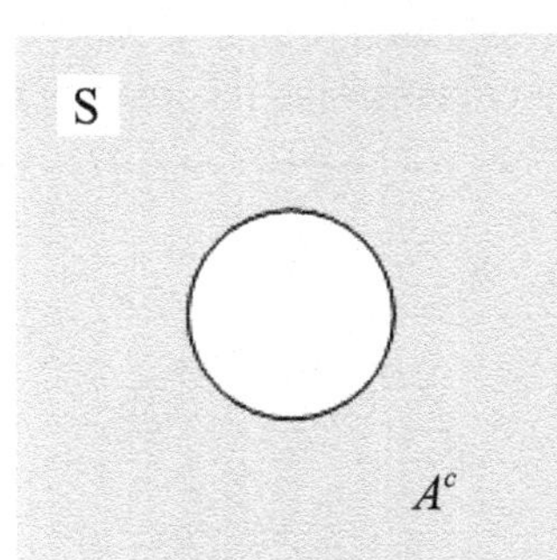

FIGURE A.4 Complement of a set.

The ***difference*** of any two sets, A and B, denoted by $A\backslash B$, is equal to $A \cap B^c$. Thus, $A^c = S\backslash A$. It should be noted that $(A^c)^c = A$. The ***symmetric difference*** between any two sets, A and B, denoted by $A\Delta B$, is the set of elements in A or B, but not both, that is, $(A\backslash B) \cup (B\backslash A)$.

Properties of Sets

If A, B, and C are the subsets of the universal set S, then they satisfy the following properties.

Commutative law

$$A \cup B = B \cup A$$

$$A \cap B = B \cap A$$

Associative law

$$A \cup (B \cup C) = (A \cup B) \cup C = A \cup B \cup C$$

$$A \cap (B \cap C) = (A \cap B) \cap C$$

Distributive law

$$A \cup (B \cap C) = (A \cup B) \cap (A \cup C),$$

$$A \cap (B \cup C) = (A \cap B) \cup (A \cap C)$$

Idempotent law

$$A \cup A = A, \ \ A \cap A = A$$

Identity law

$$A \cup S = S, \ \ A \cap S = A;$$

$$A \cup \varnothing = A, \ \ A \cap \varnothing = \varnothing$$

Complement law

$$A \cup A^c = S, \ \ A \cap A^c = \varnothing$$

De Morgan's laws

$$(A \cup B)^c = A^c \cap B^c$$

$$(A \cap B)^c = A^c \cup B^c$$

The two sets A and B are said to be in ***one-to-one correspondence*** (denoted by 1:1) if each element $a \in A$ is paired with one and only one element $b \in B$ in such a manner that each element of B is paired with exactly one element of A. For example, if $A = \{a_1, a_2, a_3, a_4\}$ and $B = \{1, 2, 3, 4\}$, then A and B are in a 1:1 correspondence.

A set whose elements can be put into a one-to-one correspondence with the set of all positive integers is referred to as being a ***countably infinite*** set. Also, a set is said to be ***countable***, ***denumerable***, or ***enumerable*** if it is finite or countably infinite. The product or Cartesian product of sets A and B is denoted by $A \times B$ and consists of all ordered pairs (a, b), where $a \in A$ and $b \in B$, that is,

$$A \times B = \{(a, b): a \in A, b \in B\}.$$

For example, if $A = \{a_1, a_2, a_3\}$ and $B = \{1, 2\}$, then

$$A \times B = \{(a_1, 1), (a_1, 2), (a_2, 1), (a_2, 2), (a_3, 1), (a_3, 2)\}.$$

The notion of a Cartesian product can be extended to any finite number of sets; that is, $A_1 \times A_2 \times \ldots \times A_n$ is the set of all ordered n-tuples, $(a_1, a_2, \ldots, a_n)$, where

$$a_1 \in A_1, a_2 \in A_2, \ldots, a_n \in A_n$$

Appendix B

Review of Markov Chains

A ***stochastic*** or ***random process*** is defined as a family of random variables, $\{X(t)\}$, describing an empirical process, the development of which in time is governed by probabilistic laws. The ***state space***, S, of the stochastic process is the set of all possible values that the random variable $X(t)$ can take. The parameter t is often interpreted as time and may be either discrete or continuous. When the set of possible values of t forms a countable set, the process $\{X(t),\ t = 0, 1, 2, \ldots\}$ is ***discrete.*** If t forms an interval of the real line, the process $\{X(t),\ t \geq 0\}$ is said to be ***continuous.*** In the discrete case, the state space can be finite or infinite.

Among many different discrete stochastic processes, we are interested in a special class called Markov chains. The basic concepts of Markov chains were introduced in 1907 by the Russian mathematician A. A. Markov.

Let $i_1, i_2, \ldots$ represent the states of the chain. The sequence of random variables $X_1, X_2, \ldots$ is called a ***Markov chain*** if

$$P(X_n = i_{k_n}|X_1 = i_{k_1}, \ldots, X_{n-1} = i_{k_{n-1}}) = P(X_n = i_{k_n}|X_n = i_{k_{n-1}}).$$

An intuitive interpretation is that a stochastic process $\{X(t)\}$ has the Markov property if the conditional probability of any future state, given the present and past states, is independent of the past states and depends only on the present state. Thus a Markov chain can be used to model the position of an object in a discrete set of possible states over time, where the subsequent position is chosen at random from a distribution that depends only on the current location of the chain and not on any previous locations of the chain.

The conditional probabilities that the chain moves to the state j at time n, given that it is in the state i at time $n - 1$, are called ***transition probabilities*** and are denoted by p_{ij},

$$p_{ij} = P(X_n = j|X_{n-1} = i)$$

with the subscript ij of p indicating the direction of transition $i \to j$. Sometimes, p_{ij} may also be represented by $p(i, j)$, and if we need to represent the time points, then we use the notation $p_{n-1,n}(i, j) = P(X_n = j|X_{n-1} = i)$.

Two basic assumptions we make are that (i) $p_{ij} \geq 0$ for all i and j; the transition probabilities are nonnegative. Also, (ii) for every i,

$$\sum_{j=1}^{\infty} p_{ij} = 1 \left(\sum_{j=1}^{n} p_{ij} = 1 \text{ if the state space is finite} \right),$$

that is, the chain makes a transition to some state in the state space.

If the transition probabilities p_{ij} depend only on the states i and j and not on the time n, then the conditional probabilities are called ***stationary***. Markov chains with stationary probabilities are called (time) ***homogeneous Markov chains***. We shall consider only homogeneous Markov chains.

The behavior of homogeneous Markov chains is described by the transition or stochastic matrices of the processes, where the transition probabilities are arranged as elements of a matrix. The ***transition*** or ***stochastic matrix*** of a chain having transition probabilities $i, j = 1, 2, \ldots n$ is

$$P = \begin{pmatrix} p_{11} & \cdots & p_{1n} \\ \vdots & \ddots & \vdots \\ p_{n1} & \cdots & p_{nn} \end{pmatrix}.$$

In the infinite state space case, we represent the transition matrix in the following manner:

$$\begin{pmatrix} p_{11} & \cdots & p_{1n} \cdots \\ \vdots & \ddots & \vdots \\ p_{m1} & \cdots & p_{mn} \cdots \\ \vdots & & \vdots \end{pmatrix}.$$

Each element of the matrix is nonnegative, and each row sums to 1. If we look at any particular row, say the mth row, then we can see the probabilities of going from state m to the various other states including state m.

Example B.1. Four quarterbacks are warming up by throwing a football to one another. Let 1, 2, 3, and 4 denote the four quarterbacks. It has been observed that 1 is as likely to throw the ball to 2 as to 3 and 4. Player 2 never throws to 3 but splits his throws between 1 and 4. Quarterback 3 throws twice as many passes to 1 as to 4 and never to 2, but 4 throws only to 1. This process forms a Markov chain because the player who is about to throw the ball is not influenced by the player who had the ball before him. The one-step transition matrix is

$$\begin{pmatrix} 0 & 1/3 & 1/3 & 1/3 \\ 1/2 & 0 & 0 & 1/2 \\ 2/3 & 0 & 0 & 1/3 \\ 1 & 0 & 0 & 0 \end{pmatrix}.$$

The following is a standard example of a chain with an infinite state space.

Example B.2. Consider a chain with state space $S = (0, 1, 2, 3, \ldots)$ and transition matrix

$$P = \begin{pmatrix} r_0 & p_0 & 0 & 0 & \cdots \\ q_1 & r_1 & p_1 & 0 & \cdots \\ 0 & q_2 & r_2 & p_2 & \cdots \\ 0 & 0 & q_3 & r_3 & p_3 \cdots \\ \vdots & \vdots & \vdots & \vdots & \vdots \end{pmatrix},$$

where $p_i, q_i, r_i \geq 0$ for all $i \geq 0$, $p_0 + r_0 = 1$, and $p_i + q_i + r_i = 1$ for all $i \geq 1$. Thus, for this Markov chain, the transition probabilities are $p_{00} = r_0$, $p_{01} = p_0$, and for $i, j \neq 0$,

$$p_{ij} = \begin{cases} p_i, & j = i + 1 \\ r_i, & j = i \\ q_i, & j = i - 1 \\ 0, & \text{otherwise.} \end{cases}$$

This chain is known as the random walk chain (with barrier at 0).

The following example gives a transition matrix for the random walk chain in a special case. We can think of this as a chain resulting from tossing a fair coin. If we are not at state 0, then if heads comes up, we take a step to the right, and if tails comes up, we take a step to the left. If at state 0, we remain at 0 for a tails outcome and move a step to the right for heads.

Example B.3. Consider a Markov chain with state space $S = (0, 1, 2, 3, \ldots)$ and the transition probabilities given by

$$p_{00} = 1/2, \quad p_{ij} = \begin{cases} 1/2, & j = i - 1 \\ 1/2, & j = i + 1 \\ 0, & \text{otherwise.} \end{cases}$$

This results in the symmetric transition matrix with elements

$$A = \begin{bmatrix} 1/2 & 1/2 & 0 & 0 & 0 & \dots \\ 1/2 & 0 & 1/2 & 0 & 0 & \dots \\ 0 & 1/2 & 0 & 1/2 & 0 & \dots \\ 0 & 0 & 1/2 & 0 & 1/2 & \cdots \\ \vdots & \vdots & \vdots & \vdots & \vdots & \vdots \end{bmatrix}.$$

The ***n*-step transition probability**, $p_{ij}^{(n)}$, is defined as the probability that the chain is in state i and will go to the state j in n steps. If p_{ij} is the one-step transition probability, $p_{ij}^{(n)}$ can be obtained as follows. Let i be the state of the process at time, m, that is, $X_m = i$. Then, the n-step transition probability is

$$\begin{aligned} p_{ij}^{(n+m)} &= P(X_{n+m} = j | X_0 = i) \\ &= \sum_{k=0}^{\infty} P(X_{n+m} = j, X_n = k | X_0 = i) \\ &= \sum_{k=0}^{\infty} P(X_{n+m} = j | X_n = k, X_0 = i) P(X_n = k | X_0 = i) \\ &= \sum_{k=0}^{\infty} p_{kj}^{m} p_{ik}^{n}. \end{aligned}$$

This can be rewritten in the matrix notation as

$$P^{(n+m)} = P^{(m)} P^{(n)} = P^{(n)} P^{(m)}.$$

This is known as the Chapman-Kolmogorov equation.

The following example shows how to compute an n-step transition matrix.

Example B.4. Consider the one-step transition matrix given in Example 1,

$$\begin{pmatrix} 0 & 1/3 & 1/3 & 1/3 \\ 1/2 & 0 & 0 & 1/2 \\ 2/3 & 0 & 0 & 1/3 \\ 1 & 0 & 0 & 0 \end{pmatrix}.$$

The two-step transition matrix, P^2, is

$$P^2 = P.\,P = \begin{pmatrix} 0 & 1/3 & 1/3 & 1/3 \\ 1/2 & 0 & 0 & 1/2 \\ 2/3 & 0 & 0 & 1/3 \\ 1 & 0 & 0 & 0 \end{pmatrix} \begin{pmatrix} 0 & 1/3 & 1/3 & 1/3 \\ 1/2 & 0 & 0 & 1/2 \\ 2/3 & 0 & 0 & 1/3 \\ 1 & 0 & 0 & 0 \end{pmatrix}$$

$$= \begin{pmatrix} 13/18 & 0 & 0 & 5/18 \\ 1/2 & 1/6 & 1/6 & 1/6 \\ 1/3 & 2/9 & 2/9 & 2/9 \\ 0 & 1/3 & 1/3 & 1/3 \end{pmatrix}.$$

The three-step transition matrix, P^3, is

$$P^3 = P^2 P = \begin{pmatrix} 5/18 & 13/54 & 13/54 & 13/54 \\ 13/36 & 1/6 & 1/6 & 11/36 \\ 13/27 & 1/9 & 1/9 & 8/27 \\ 13/18 & 0 & 0 & 5/18 \end{pmatrix}.$$

For instance, the third row of P^3,

$$(13/27 \;\; 1/9 \;\; 1/9 \;\; 8/27),$$

denotes that, after three throws, the ball is in the hands of players 1, 2, 3, and 4, with respective probabilities 13/27, 1/9, 1/9, and 8/27.

A transition matrix, P, all entries of which are positive, is called a ***positive transition matrix***. A state j of a Markov chain is ***accessible*** from a state i if $p_{ij}^{(n)} > 0$ for some $n \geq 0$. If state j is accessible from the state i, and state i is accessible from the state j, the states are said to ***communicate***. If all the states communicate, then the Markov chain is called ***irreducible***. A state i is ***periodic*** (of period d) if the only way to revisit it is through steps of length $k.\,d$ for some value of k

and a fixed value of $d > 1$. Thus the period, d, is the greatest common divisor of the number of steps n needed for the chain, starting at the state i, to revisit the state i:

$$d = GCD\{n \geq 1 | p_{ii}^n > 0\}.$$

If a state is not periodic, then it is called ***aperiodic***. A state i is ***recurrent*** if it will be revisited by the chain with probability 1. That is,

$$P(X_n = i \text{ for infinitely many } n | X_0 = i) = 1.$$

If a state is not recurrent, it is called ***transient***. Recurrent, aperiodic states are called ***ergodic***. It is necessary to impose an extra condition for ergodicity, that the expected recurrence time be finite. This is satisfied for recurrent states in a finite-state Markov chain. A Markov chain is called ***ergodic*** if every state is ergodic. It is clear that a finite-state Markov chain with a positive transition matrix is ergodic.

The following result is of fundamental importance.

Theorem B.1. For an ergodic Markov chain, $\lim_{n\to\infty} p_{ij}^{(n)} = \pi_j$ exists, and this limit is independent of the initial state i. Let the vector π with elements (π_j) be the limiting or the stationary distribution of the chain. Then, this stationary probability vector is the unique solution of the equation

$$\pi = \pi P$$

and satisfies the normalization condition

$$\sum_{j \in S} \pi_j = 1.$$

If, at any transition step n, the distribution of the chain is the same as π obtained in Theorem AII.1, we say that the chain has reached the ***steady state***. Thus the vector π would be the unique steady-state probability vector of the Markov chain.

Analogous to the law of large numbers for a sequence of independent random variables, we can obtain the following ***ergodic theorem*** for Markov chains.

Theorem B.2. For any ergodic Markov chain $\{X_n\}$ with a stationary distribution π: $\frac{1}{n}\sum_{k=1}^{n} f(X_k) \to \sum_{i \in S} f(i)\pi_i = Ef(X)$ $w.p.1$.

The validity of the Markov chain Monte Carlo method lies in this ergodic theorem.

Appendix C

Common Probability Distributions

In this appendix, we present some common probability distributions useful in the statistical methods that we have used in this book. There is a much greater variety of distributions that are very important in a particular area of applications. A good reference is available at http://www.causascientia.org/math_stat/Dists/Compendium.pdf. We give the density function, mean, variance, and moment-generating function (mgf). For some distribution functions, if the mgf is complicated, we just leave it and refer the reader to one of the references in the book.

Name	pdf	Mean	Variance	mgf
Bernoulli distribution	$f(x, p) = \begin{cases} p, & x = 1 \\ 1 - p, & x = 0 \\ 0, & \text{otherwise} \end{cases}$ $0 \le p \le 1$	p	$p(1 - p)$	$q + pe^t$, $q = 1 - p$
Binomial	$f(x, n, p) = \binom{n}{x} p^x q^{n-x}$, $x = 0, 1, \ldots, n$	np	npq	$(q + pe^t)^n$
Geometric	$f(x, p) = q^{x-1}p, \ x = 1, 2, \ldots,$ $0 \le p \le 1$	$\frac{1}{p}$	$\frac{q}{p^2}$	$\frac{pe^t}{1 - qe^t}$
Hypergeometric	$f(x, N, m, n)$ $= \frac{\binom{m}{x}\binom{N-m}{n-x}}{\binom{N}{n}}$, $N = 0, 1, 2, \ldots, \ m = 0, 1, \ldots, N,$ $n = 0, 1, \ldots, N, \ x = 0, 1, \ldots, n$	$\frac{mn}{N}$	$\frac{n\left(\frac{m}{N}\right)\left(1 - \frac{m}{N}\right)\left(1 - \frac{n}{N}\right)}{N-1}$	No closed-form mgf
Negative binomial	$f(x, N, m, n) = \binom{x + r - 1}{x} p^r q^x$, $x = 0, 1, 2, \ldots$	$r\frac{q}{p}$	$r\frac{q}{p^2}$	$\left(\frac{p}{1 - qe^t}\right)^r$
Poisson	$f(x, \lambda) = \frac{\lambda^x e^{-\lambda}}{x!}$, $x = 0, 1, 2, \ldots$	λ	λ	$\exp(\lambda(e^t - 1)$
Beta	$f(x, \alpha, \beta)$ $= \left(\frac{\Gamma(\alpha + \beta)}{\Gamma(\alpha)\Gamma(\beta)}\right) x^{\alpha-1}(1 - x)^{\beta-1}$, $0 < x < 1$	$\frac{\alpha}{\alpha + \beta}$	$\frac{\alpha\beta}{(\alpha + \beta)^2(\alpha + \beta + 1)}$	
Chi-square	$f(x, n) = \frac{x^{(n/2)-1} e^{-x/2}}{\Gamma(n/2) 2^{n/2}}$, $x \ge 0, \ n > 0$ (degrees of freedom)	n	$2n$	$\frac{1}{(1 - 2t)^{n/2}}$, $t < \frac{1}{2}$
Exponential	$f(x, \beta)$ $= \begin{cases} \frac{1}{\beta} e^{-x/\beta}, & \beta > 0, \ 0 \le x < \infty \\ 0, & \text{otherwise} \end{cases}$	β	β^2	$\frac{1}{(1 - \beta t)}$, $t < \frac{1}{\beta}$
Gamma	$f(x, \alpha, \beta)$ $= \begin{cases} \frac{1}{\beta^\alpha \Gamma(\alpha)} x^{\alpha-1} e^{-x/\beta}, & x > 0, \ \alpha, \beta > 0 \\ 0, & \text{otherwise} \end{cases}$	$\alpha\beta$	$\alpha\beta^2$	$\frac{1}{(1 - \beta t)^\alpha}$, $t < \frac{1}{\beta}$
Laplace	$f(x, \mu, \sigma) = \frac{1}{2\sigma} \exp\left(-\frac{\lvert x - \mu \rvert}{\sigma}\right)$, $x, \mu > -\infty$	μ	$2\sigma^2$	

Normal	$f(x, \mu, \sigma^2)$ $=\frac{1}{\sigma\sqrt{2\pi}}\exp\left(-\frac{(x-\mu)^2}{2\sigma^2}\right)$, $-\infty < x, \mu < \infty,\ \sigma > 0$	μ	σ^2	$e^{t\mu+\frac{1}{2}t^2\sigma^2}$
Uniform	$f(x, a, b) = \frac{1}{b-a}$	$\frac{a+b}{2}$	$\frac{(b-a)^2}{12}$	$\frac{e^{tb}-e^{ta}}{t(b-a)}$
Three-parameter gamma	$f(x{:}\ \alpha, \beta, \gamma)$ $=\begin{cases}\frac{1}{\beta^{\alpha}\Gamma(\alpha)}(x-\gamma)^{\alpha-1}\exp-\frac{(x-\gamma)}{\beta}, & \gamma < x < \infty,\ \alpha, \beta > 0 \\ 0, & \text{otherwise}\end{cases}$	$\gamma + \alpha\beta$	$\alpha\beta^2$	$\frac{\exp(\gamma t)}{(1-\beta t)^{\alpha}}.$
Two-parameter Weibull	$f(x, \alpha, \beta)$ $=\frac{\alpha}{\beta}\left(\frac{x}{\beta}\right)^{\alpha-1}e^{-\left(\frac{x}{\beta}\right)^{\alpha}}$, $x \geq 0, \alpha, \beta > 0$	$\beta\Gamma\left(1+\frac{1}{\alpha}\right)$	$\beta^2\left[\Gamma\left(1+\frac{2}{\alpha}\right)-(\Gamma\left(1+\frac{1}{\alpha}\right))^2\right]$	$\Sigma_{n=0}^{\infty}\ \frac{t^n\beta^n}{n!}\Gamma\left(1+\frac{n}{\alpha}\right)$
Three-parameter Weibull	$f(x, \alpha, \beta, \gamma)$ $=\frac{\beta}{\alpha}(x-\gamma)^{\beta-1}e^{-((x-\gamma)\frac{\beta}{\alpha})}$, $x > \gamma,\ \alpha, \beta, \gamma > 0$	$\alpha^{\frac{1}{\beta}}\Gamma\left(1+\frac{1}{\beta}\right)+\gamma$	$\alpha^{\frac{2}{\beta}}\left\{\Gamma\left(1+\frac{2}{\beta}\right)-(\Gamma\left(1+\frac{1}{\beta}\right))^2\right\}$	$e^{t\gamma}\int_0^{\infty} e^{-y+t(\alpha y)^{1/\beta}}dy$

Appendix D

What is R?

R is a language and environment for statistical computing and graphics. It provides a wide variety of statistical methods, including basic statistical tests, regression models, and many other classical and graphical methods/techniques. R can be easily used to produce technical plots of quality along with mathematical symbols and formulas. R is available as free software under the terms of the Free Software Foundation's GNU General Public License in source code form. R can be downloaded from https://www.r-project.org/.

R is similar to the S language and environment. The S language is considered the choice in statistical methods, and R provides an open-source route to work with statistical methods, among others.

R is also designed like S around a true computer language, with the flexibility in that it allows users to add additional functions.

Some users of R think of it as a statistics system; others think of R as an environment within which statistical methods and graphics are implemented.

Finally, **RStudio** is an integrated development environment for the R programming language. It makes it easier to write, run, visualize, and debug R code—everything in one window. We can think of it as a workspace for data science and artificial intelligence in R. After installing R, go to https://posit.co/download/rstudio/. Download the free version and install it; it will automatically detect the R installation. Now, when you open RStudio, you'll see four panels (Source, Console, Environment, and Plots).

Appendix E

Probability Tables

Table E.1 Cumulative Binomial Probabilities, $P(X \le x) = \sum_{i=0}^{x} p(i)$

$n = 2$	$p =$												
	0.01	**0.05**	**0.10**	**0.20**	**0.30**	**0.40**	**0.50**	**0.60**	**0.70**	**0.80**	**0.90**	**0.95**	**0.99**
$x = 0$	0.980	0.903	0.810	0.640	0.490	0.360	0.250	0.160	0.090	0.040	0.010	0.003	0.000
1	1.000	0.998	0.990	0.960	0.910	0.840	0.750	0.640	0.510	0.360	0.190	0.098	0.020
2	1.000	1.000	1.000	1.000	1.000	1.000	1.000	1.000	1.000	1.000	1.000	1.000	1.000
$n = 3$													
$x = 0$	0.970	0.857	0.729	0.512	0.343	0.216	0.125	0.064	0.027	0.008	0.001	0.000	0.000
1	1.000	0.993	0.972	0.896	0.784	0.648	0.500	0.352	0.216	0.104	0.028	0.007	0.000
2	1.000	1.000	0.999	0.992	0.973	0.936	0.875	0.784	0.657	0.488	0.271	0.143	0.030
3	1.000	1.000	1.000	1.000	1.000	1.000	1.000	1.000	1.000	1.000	1.000	1.000	1.000
$n = 4$													
$x = 0$	0.961	0.815	0.656	0.410	0.240	0.130	0.063	0.026	0.008	0.002	0.000	0.000	0.000
1	0.999	0.986	0.948	0.819	0.652	0.475	0.313	0.179	0.084	0.027	0.004	0.000	0.000
2	1.000	1.000	0.996	0.973	0.916	0.821	0.688	0.525	0.348	0.181	0.052	0.014	0.001
3	1.000	1.000	1.000	0.998	0.992	0.974	0.938	0.870	0.760	0.590	0.344	0.185	0.039
4	1.000	1.000	1.000	1.000	1.000	1.000	1.000	1.000	1.000	1.000	1.000	1.000	1.000

(Continued)

Table E.1 (Continued)

***n* = 5**													
***x* = 0**	0.951	0.774	0.590	0.328	0.168	0.078	0.031	0.010	0.002	0.000	0.000	0.000	0.000
1	0.999	0.977	0.919	0.737	0.528	0.337	0.188	0.087	0.031	0.007	0.000	0.000	0.000
2	1.000	0.999	0.991	0.942	0.837	0.683	0.500	0.317	0.163	0.058	0.009	0.001	0.000
3	1.000	1.000	1.000	0.993	0.969	0.913	0.813	0.663	0.472	0.263	0.081	0.023	0.001
4	1.000	1.000	1.000	1.000	0.998	0.990	0.969	0.922	0.832	0.672	0.410	0.226	0.049
***n* = 6**													
***x* = 0**	0.941	0.735	0.531	0.262	0.118	0.047	0.016	0.004	0.001	0.000	0.000	0.000	0.000
1	0.999	0.967	0.886	0.655	0.420	0.233	0.109	0.041	0.011	0.002	0.000	0.000	0.000
2	1.000	0.998	0.984	0.901	0.744	0.544	0.344	0.179	0.070	0.017	0.001	0.000	0.000
3	1.000	1.000	0.999	0.983	0.930	0.821	0.656	0.456	0.256	0.099	0.016	0.002	0.000
4	1.000	1.000	1.000	0.998	0.989	0.959	0.891	0.767	0.580	0.345	0.114	0.033	0.001
5	1.000	1.000	1.000	1.000	0.999	0.996	0.984	0.953	0.882	0.738	0.469	0.265	0.059
***n* = 7**													
***x* = 0**	0.932	0.698	0.478	0.210	0.082	0.028	0.008	0.002	0.000	0.000	0.000	0.000	0.000
1	0.998	0.956	0.850	0.577	0.329	0.159	0.063	0.019	0.004	0.000	0.000	0.000	0.000
2	1.000	0.996	0.974	0.852	0.647	0.420	0.227	0.096	0.029	0.005	0.000	0.000	0.000
3	1.000	1.000	0.997	0.967	0.874	0.710	0.500	0.290	0.126	0.033	0.003	0.000	0.000
4	1.000	1.000	1.000	0.995	0.971	0.904	0.773	0.580	0.353	0.148	0.026	0.004	0.000
5	1.000	1.000	1.000	1.000	0.996	0.981	0.938	0.841	0.671	0.423	0.150	0.044	0.002
6	1.000	1.000	1.000	1.000	1.000	0.998	0.992	0.972	0.918	0.790	0.522	0.302	0.068
***n* = 8**													
***x* = 0**	0.923	0.663	0.430	0.168	0.058	0.017	0.004	0.001	0.000	0.000	0.000	0.000	0.000
1	0.997	0.943	0.813	0.503	0.255	0.106	0.035	0.009	0.001	0.000	0.000	0.000	0.000
2	1.000	0.994	0.962	0.797	0.552	0.315	0.145	0.050	0.011	0.001	0.000	0.000	0.000
3	1.000	1.000	0.995	0.944	0.806	0.594	0.363	0.174	0.058	0.010	0.000	0.000	0.000
4	1.000	1.000	1.000	0.990	0.942	0.826	0.637	0.406	0.194	0.056	0.005	0.000	0.000
5	1.000	1.000	1.000	0.999	0.989	0.950	0.855	0.685	0.448	0.203	0.038	0.006	0.000
6	1.000	1.000	1.000	1.000	0.999	0.991	0.965	0.894	0.745	0.497	0.187	0.057	0.003
7	1.000	1.000	1.000	1.000	1.000	0.999	0.996	0.983	0.942	0.832	0.570	0.337	0.077
***n* = 9**													
***x* = 0**	0.914	0.630	0.387	0.134	0.040	0.010	0.002	0.000	0.000	0.000	0.000	0.000	0.000
1	0.997	0.929	0.775	0.436	0.196	0.071	0.020	0.004	0.000	0.000	0.000	0.000	0.000
2	1.000	0.992	0.947	0.738	0.463	0.232	0.090	0.025	0.004	0.000	0.000	0.000	0.000
3	1.000	0.999	0.992	0.914	0.730	0.483	0.254	0.099	0.025	0.003	0.000	0.000	0.000
4	1.000	1.000	0.999	0.980	0.901	0.733	0.500	0.267	0.099	0.020	0.001	0.000	0.000
5	1.000	1.000	1.000	0.997	0.975	0.901	0.746	0.517	0.270	0.086	0.008	0.001	0.000
6	1.000	1.000	1.000	1.000	0.996	0.975	0.910	0.768	0.537	0.262	0.053	0.008	0.000
7	1.000	1.000	1.000	1.000	1.000	0.996	0.980	0.929	0.804	0.564	0.225	0.071	0.003
8	1.000	1.000	1.000	1.000	1.000	1.000	0.998	0.990	0.960	0.866	0.613	0.370	0.086
***n* = 10**													
***x* = 0**	0.904	0.599	0.349	0.107	0.028	0.006	0.001	0.000	0.000	0.000	0.000	0.000	0.000
1	0.996	0.914	0.736	0.376	0.149	0.046	0.011	0.002	0.000	0.000	0.000	0.000	0.000

(Continued)

Table E.1 (Continued)

2	1.000	0.988	0.930	0.678	0.383	0.167	0.055	0.012	0.002	0.000	0.000	0.000	0.000
3	1.000	0.999	0.987	0.879	0.650	0.382	0.172	0.055	0.011	0.001	0.000	0.000	0.000
4	1.000	1.000	0.998	0.967	0.850	0.633	0.377	0.166	0.047	0.006	0.000	0.000	0.000
5	1.000	1.000	1.000	0.994	0.953	0.834	0.623	0.367	0.150	0.033	0.002	0.000	0.000
6	1.000	1.000	1.000	0.999	0.989	0.945	0.828	0.618	0.350	0.121	0.013	0.001	0.000
7	1.000	1.000	1.000	1.000	0.998	0.988	0.945	0.833	0.617	0.322	0.070	0.012	0.000
8	1.000	1.000	1.000	1.000	1.000	0.998	0.989	0.954	0.851	0.624	0.264	0.086	0.004
9	1.000	1.000	1.000	1.000	1.000	1.000	0.999	0.994	0.972	0.893	0.651	0.401	0.096
$n = 15$													
$x = 0$	0.860	0.463	0.206	0.035	0.005	0.000	0.000	0.000	0.000	0.000	0.000	0.000	0.000
1	0.990	0.829	0.549	0.167	0.035	0.005	0.000	0.000	0.000	0.000	0.000	0.000	0.000
2	1.000	0.964	0.816	0.398	0.127	0.027	0.004	0.000	0.000	0.000	0.000	0.000	0.000
3	1.000	0.995	0.944	0.648	0.297	0.091	0.018	0.002	0.000	0.000	0.000	0.000	0.000
4	1.000	0.999	0.987	0.836	0.515	0.217	0.059	0.009	0.001	0.000	0.000	0.000	0.000
5	1.000	1.000	0.998	0.939	0.722	0.403	0.151	0.034	0.004	0.000	0.000	0.000	0.000
6	1.000	1.000	1.000	0.982	0.869	0.610	0.304	0.095	0.015	0.001	0.000	0.000	0.000
7	1.000	1.000	1.000	0.996	0.950	0.787	0.500	0.213	0.050	0.004	0.000	0.000	0.000
8	1.000	1.000	1.000	0.999	0.985	0.905	0.696	0.390	0.131	0.018	0.000	0.000	0.000
9	1.000	1.000	1.000	1.000	0.996	0.966	0.849	0.597	0.278	0.061	0.002	0.000	0.000
10	1.000	1.000	1.000	1.000	0.999	0.991	0.941	0.783	0.485	0.164	0.013	0.001	0.000
11	1.000	1.000	1.000	1.000	1.000	0.998	0.982	0.909	0.703	0.352	0.056	0.005	0.000
12	1.000	1.000	1.000	1.000	1.000	1.000	0.996	0.973	0.873	0.602	0.184	0.036	0.000
13	1.000	1.000	1.000	1.000	1.000	1.000	1.000	0.995	0.965	0.833	0.451	0.171	0.010
14	1.000	1.000	1.000	1.000	1.000	1.000	1.000	1.000	0.995	0.965	0.794	0.537	0.140
$n = 20$													
$x = 0$	0.818	0.358	0.122	0.012	0.001	0.000	0.000	0.000	0.000	0.000	0.000	0.000	0.000
1	0.983	0.736	0.392	0.069	0.008	0.001	0.000	0.000	0.000	0.000	0.000	0.000	0.000
2	0.999	0.925	0.677	0.206	0.035	0.004	0.000	0.000	0.000	0.000	0.000	0.000	0.000
3	1.000	0.984	0.867	0.411	0.107	0.016	0.001	0.000	0.000	0.000	0.000	0.000	0.000
4	1.000	0.997	0.957	0.630	0.238	0.051	0.006	0.000	0.000	0.000	0.000	0.000	0.000
5	1.000	1.000	0.989	0.804	0.416	0.126	0.021	0.002	0.000	0.000	0.000	0.000	0.000
6	1.000	1.000	0.998	0.913	0.608	0.250	0.058	0.006	0.000	0.000	0.000	0.000	0.000
7	1.000	1.000	1.000	0.968	0.772	0.416	0.132	0.021	0.001	0.000	0.000	0.000	0.000
8	1.000	1.000	1.000	0.990	0.887	0.596	0.252	0.057	0.005	0.000	0.000	0.000	0.000
9	1.000	1.000	1.000	0.997	0.952	0.755	0.412	0.128	0.017	0.001	0.000	0.000	0.000
10	1.000	1.000	1.000	0.999	0.983	0.872	0.588	0.245	0.048	0.003	0.000	0.000	0.000
11	1.000	1.000	1.000	1.000	0.995	0.943	0.748	0.404	0.113	0.010	0.000	0.000	0.000
12	1.000	1.000	1.000	1.000	0.999	0.979	0.868	0.584	0.228	0.032	0.000	0.000	0.000
13	1.000	1.000	1.000	1.000	1.000	0.994	0.942	0.750	0.392	0.087	0.002	0.000	0.000
14	1.000	1.000	1.000	1.000	1.000	0.998	0.979	0.874	0.584	0.196	0.011	0.000	0.000
15	1.000	1.000	1.000	1.000	1.000	1.000	0.994	0.949	0.762	0.370	0.043	0.003	0.000
16	1.000	1.000	1.000	1.000	1.000	1.000	0.999	0.984	0.893	0.589	0.133	0.016	0.000
17	1.000	1.000	1.000	1.000	1.000	1.000	1.000	0.996	0.965	0.794	0.323	0.075	0.001
18	1.000	1.000	1.000	1.000	1.000	1.000	1.000	0.999	0.992	0.931	0.608	0.264	0.017

(*Continued*)

Table E.1 (Continued)

19	1.000	1.000	1.000	1.000	1.000	1.000	1.000	1.000	0.999	0.988	0.878	0.642	0.182
$n = 25$													
$x = 0$	0.778	0.277	0.072	0.004	0.000	0.000	0.000	0.000	0.000	0.000	0.000	0.000	0.000
1	0.974	0.642	0.271	0.027	0.002	0.000	0.000	0.000	0.000	0.000	0.000	0.000	0.000
2	0.998	0.873	0.537	0.098	0.009	0.000	0.000	0.000	0.000	0.000	0.000	0.000	0.000
3	1.000	0.966	0.764	0.234	0.033	0.002	0.000	0.000	0.000	0.000	0.000	0.000	0.000
4	1.000	0.993	0.902	0.421	0.090	0.009	0.000	0.000	0.000	0.000	0.000	0.000	0.000
5	1.000	0.999	0.967	0.617	0.193	0.029	0.002	0.000	0.000	0.000	0.000	0.000	0.000
6	1.000	1.000	0.991	0.780	0.341	0.074	0.007	0.000	0.000	0.000	0.000	0.000	0.000
7	1.000	1.000	0.998	0.891	0.512	0.154	0.022	0.001	0.000	0.000	0.000	0.000	0.000
8	1.000	1.000	1.000	0.953	0.677	0.274	0.054	0.004	0.000	0.000	0.000	0.000	0.000
9	1.000	1.000	1.000	0.983	0.811	0.425	0.115	0.013	0.000	0.000	0.000	0.000	0.000
10	1.000	1.000	1.000	0.994	0.902	0.586	0.212	0.034	0.002	0.000	0.000	0.000	0.000
11	1.000	1.000	1.000	0.998	0.956	0.732	0.345	0.078	0.006	0.000	0.000	0.000	0.000
12	1.000	1.000	1.000	1.000	0.983	0.846	0.500	0.154	0.017	0.000	0.000	0.000	0.000
13	1.000	1.000	1.000	1.000	0.994	0.922	0.655	0.268	0.044	0.002	0.000	0.000	0.000
14	1.000	1.000	1.000	1.000	0.998	0.966	0.788	0.414	0.098	0.006	0.000	0.000	0.000
15	1.000	1.000	1.000	1.000	1.000	0.987	0.885	0.575	0.189	0.017	0.000	0.000	0.000
16	1.000	1.000	1.000	1.000	1.000	0.996	0.946	0.726	0.323	0.047	0.000	0.000	0.000
17	1.000	1.000	1.000	1.000	1.000	0.999	0.978	0.846	0.488	0.109	0.002	0.000	0.000
18	1.000	1.000	1.000	1.000	1.000	1.000	0.993	0.926	0.659	0.220	0.009	0.000	0.000
19	1.000	1.000	1.000	1.000	1.000	1.000	0.998	0.971	0.807	0.383	0.033	0.001	0.000
20	1.000	1.000	1.000	1.000	1.000	1.000	1.000	0.991	0.910	0.579	0.098	0.007	0.000
21	1.000	1.000	1.000	1.000	1.000	1.000	1.000	0.998	0.967	0.766	0.236	0.034	0.000
22	1.000	1.000	1.000	1.000	1.000	1.000	1.000	1.000	0.991	0.902	0.463	0.127	0.002
23	1.000	1.000	1.000	1.000	1.000	1.000	1.000	1.000	0.998	0.973	0.729	0.358	0.026
24	1.000	1.000	1.000	1.000	1.000	1.000	1.000	1.000	1.000	0.996	0.928	0.723	0.222

Table E.2 Standard Norms Table.

Area between 0 and *z*

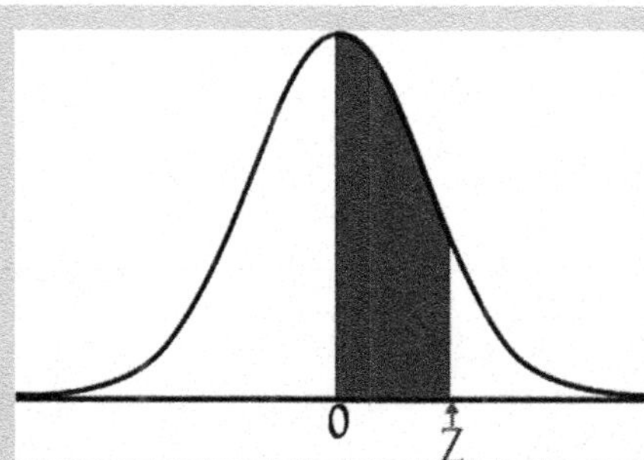

z	0.00	0.01	0.02	0.03	0.04	0.05	0.06	0.07	0.08	0.09
0.0	0.0000	0.0040	0.0080	0.0120	0.0160	0.0199	0.0239	0.0279	0.0319	0.0359
0.1	0.0398	0.0438	0.0478	0.0517	0.0557	0.0596	0.0636	0.0675	0.0714	0.0753
0.2	0.0793	0.0832	0.0871	0.0910	0.0948	0.0987	0.1026	0.1064	0.1103	0.1141
0.3	0.1179	0.1217	0.1255	0.1293	0.1331	0.1368	0.1406	0.1443	0.1480	0.1517
0.4	0.1554	0.1591	0.1628	0.1664	0.1700	0.1736	0.1772	0.1808	0.1844	0.1879
0.5	0.1915	0.1950	0.1985	0.2019	0.2054	0.2088	0.2123	0.2157	0.2190	0.2224
0.6	0.2257	0.2291	0.2324	0.2357	0.2389	0.2422	0.2454	0.2486	0.2517	0.2549
0.7	0.2580	0.2611	0.2642	0.2673	0.2704	0.2734	0.2764	0.2794	0.2823	0.2852
0.8	0.2881	0.2910	0.2939	0.2967	0.2995	0.3023	0.3051	0.3078	0.3106	0.3133
0.9	0.3159	0.3186	0.3212	0.3238	0.3264	0.3289	0.3315	0.3340	0.3365	0.3389
1.0	0.3413	0.3438	0.3461	0.3485	0.3508	0.3531	0.3554	0.3577	0.3599	0.3621
1.1	0.3643	0.3665	0.3686	0.3708	0.3729	0.3749	0.3770	0.3790	0.3810	0.3830
1.2	0.3849	0.3869	0.3888	0.3907	0.3925	0.3944	0.3962	0.3980	0.3997	0.4015
1.3	0.4032	0.4049	0.4066	0.4082	0.4099	0.4115	0.4131	0.4147	0.4162	0.4177
1.4	0.4192	0.4207	0.4222	0.4236	0.4251	0.4265	0.4279	0.4292	0.4306	0.4319
1.5	0.4332	0.4345	0.4357	0.4370	0.4382	0.4394	0.4406	0.4418	0.4429	0.4441
1.6	0.4452	0.4463	0.4474	0.4484	0.4495	0.4505	0.4515	0.4525	0.4535	0.4545
1.7	0.4554	0.4564	0.4573	0.4582	0.4591	0.4599	0.4608	0.4616	0.4625	0.4633
1.8	0.4641	0.4649	0.4656	0.4664	0.4671	0.4678	0.4686	0.4693	0.4699	0.4706
1.9	0.4713	0.4719	0.4726	0.4732	0.4738	0.4744	0.4750	0.4756	0.4761	0.4767
2.0	0.4772	0.4778	0.4783	0.4788	0.4793	0.4798	0.4803	0.4808	0.4812	0.4817
2.1	0.4821	0.4826	0.4830	0.4834	0.4838	0.4842	0.4846	0.4850	0.4854	0.4857
2.2	0.4861	0.4864	0.4868	0.4871	0.4875	0.4878	0.4881	0.4884	0.4887	0.4890
2.3	0.4893	0.4896	0.4898	0.4901	0.4904	0.4906	0.4909	0.4911	0.4913	0.4916
2.4	0.4918	0.4920	0.4922	0.4925	0.4927	0.4929	0.4931	0.4932	0.4934	0.4936
2.5	0.4938	0.4940	0.4941	0.4943	0.4945	0.4946	0.4948	0.4949	0.4951	0.4952
2.6	0.4953	0.4955	0.4956	0.4957	0.4959	0.4960	0.4961	0.4962	0.4963	0.4964
2.7	0.4965	0.4966	0.4967	0.4968	0.4969	0.4970	0.4971	0.4972	0.4973	0.4974
2.8	0.4974	0.4975	0.4976	0.4977	0.4977	0.4978	0.4979	0.4979	0.4980	0.4981
2.9	0.4981	0.4982	0.4982	0.4983	0.4984	0.4984	0.4985	0.4985	0.4986	0.4986
3.0	0.4987	0.4987	0.4987	0.4988	0.4988	0.4989	0.4989	0.4989	0.4990	0.4990
3.1	0.4990	0.4991	0.4991	0.4991	0.4992	0.4992	0.4992	0.4992	0.4993	0.4993
3.2	0.4993	0.4993	0.4994	0.4994	0.4994	0.4994	0.4994	0.4995	0.4995	0.4995
3.3	0.4995	0.4995	0.4995	0.4996	0.4996	0.4996	0.4996	0.4996	0.4996	0.4997
3.4	0.4997	0.4997	0.4997	0.4997	0.4997	0.4997	0.4997	0.4997	0.4997	0.4998

Table E.3 *t*-Table.

Right Tail Probabilities

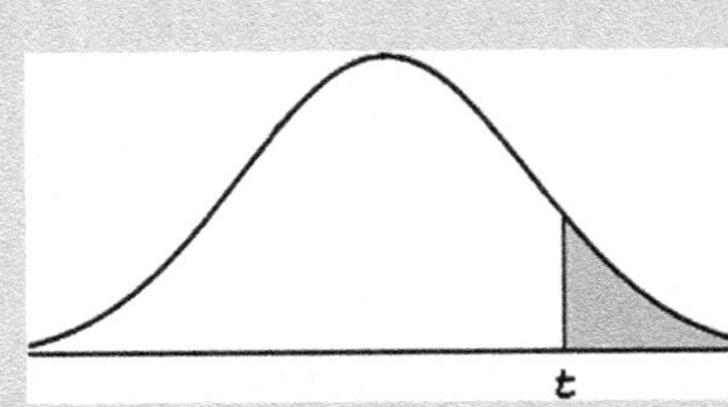

df\p	0.40	0.25	0.10	0.05	0.025	0.01	0.005	0.0005
1	0.324920	1.000000	3.077684	6.313752	12.70620	31.82052	63.65674	636.6192
2	0.288675	0.816497	1.885618	2.919986	4.30265	6.96456	9.92484	31.5991
3	0.276671	0.764892	1.637744	2.353363	3.18245	4.54070	5.84091	12.9240
4	0.270722	0.740697	1.533206	2.131847	2.77645	3.74695	4.60409	8.6103
5	0.267181	0.726687	1.475884	2.015048	2.57058	3.36493	4.03214	6.8688
6	0.264835	0.717558	1.439756	1.943180	2.44691	3.14267	3.70743	5.9588
7	0.263167	0.711142	1.414924	1.894579	2.36462	2.99795	3.49948	5.4079
8	0.261921	0.706387	1.396815	1.859548	2.30600	2.89646	3.35539	5.0413
9	0.260955	0.702722	1.383029	1.833113	2.26216	2.82144	3.24984	4.7809
10	0.260185	0.699812	1.372184	1.812461	2.22814	2.76377	3.16927	4.5869
11	0.259556	0.697445	1.363430	1.795885	2.20099	2.71808	3.10581	4.4370
12	0.259033	0.695483	1.356217	1.782288	2.17881	2.68100	3.05454	4.3178
13	0.258591	0.693829	1.350171	1.770933	2.16037	2.65031	3.01228	4.2208
14	0.258213	0.692417	1.345030	1.761310	2.14479	2.62449	2.97684	4.1405
15	0.257885	0.691197	1.340606	1.753050	2.13145	2.60248	2.94671	4.0728
16	0.257599	0.690132	1.336757	1.745884	2.11991	2.58349	2.92078	4.0150
17	0.257347	0.689195	1.333379	1.739607	2.10982	2.56693	2.89823	3.9651
18	0.257123	0.688364	1.330391	1.734064	2.10092	2.55238	2.87844	3.9216
19	0.256923	0.687621	1.327728	1.729133	2.09302	2.53948	2.86093	3.8834
20	0.256743	0.686954	1.325341	1.724718	2.08596	2.52798	2.84534	3.8495
21	0.256580	0.686352	1.323188	1.720743	2.07961	2.51765	2.83136	3.8193
22	0.256432	0.685805	1.321237	1.717144	2.07387	2.50832	2.81876	3.7921
23	0.256297	0.685306	1.319460	1.713872	2.06866	2.49987	2.80734	3.7676
24	0.256173	0.684850	1.317836	1.710882	2.06390	2.49216	2.79694	3.7454
25	0.256060	0.684430	1.316345	1.708141	2.05954	2.48511	2.78744	3.7251
26	0.255955	0.684043	1.314972	1.705618	2.05553	2.47863	2.77871	3.7066
27	0.255858	0.683685	1.313703	1.703288	2.05183	2.47266	2.77068	3.6896
28	0.255768	0.683353	1.312527	1.701131	2.04841	2.46714	2.76326	3.6739
29	0.255684	0.683044	1.311434	1.699127	2.04523	2.46202	2.75639	3.6594
30	0.255605	0.682756	1.310415	1.697261	2.04227	2.45726	2.75000	3.6460
∞	0.253347	0.674490	1.281552	1.644854	1.95996	2.32635	2.57583	3.2905

Table E.4 Chi-Square Probabilities.

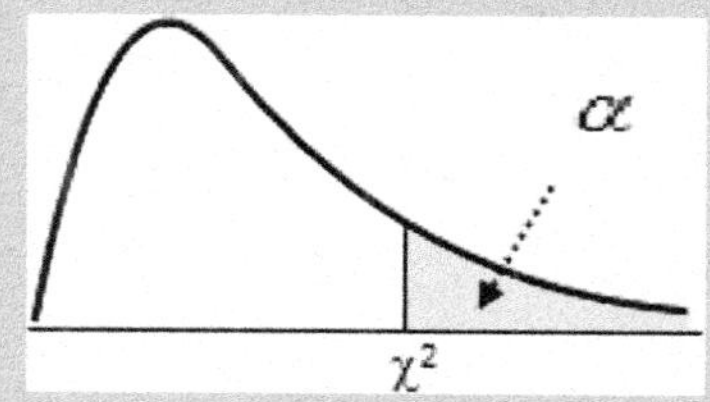

df\p	0.995	0.99	0.975	0.95	0.90	0.10	0.05	0.025	0.01	0.005
1	4×10^{-5}	16×10^{-5}	0.001	0.004	0.016	2.706	3.841	5.024	6.635	7.879
2	0.010	0.020	0.051	0.103	0.211	4.605	5.991	7.378	9.210	10.597
3	0.072	0.115	0.216	0.352	0.584	6.251	7.815	9.348	11.345	12.838
4	0.207	0.297	0.484	0.711	1.064	7.779	9.488	11.143	13.277	14.860
5	0.412	0.554	0.831	1.145	1.610	9.236	11.070	12.833	15.086	16.750
6	0.676	0.872	1.237	1.635	2.204	10.645	12.592	14.449	16.812	18.548
7	0.989	1.239	1.690	2.167	2.833	12.017	14.067	16.013	18.475	20.278
8	1.344	1.646	2.180	2.733	3.490	13.362	15.507	17.535	20.090	21.955
9	1.735	2.088	2.700	3.325	4.168	14.684	16.919	19.023	21.666	23.589
10	2.156	2.558	3.247	3.940	4.865	15.987	18.307	20.483	23.209	25.188
11	2.603	3.053	3.816	4.575	5.578	17.275	19.675	21.920	24.725	26.757
12	3.074	3.571	4.404	5.226	6.304	18.549	21.026	23.337	26.217	28.300
13	3.565	4.107	5.009	5.892	7.042	19.812	22.362	24.736	27.688	29.819
14	4.075	4.660	5.629	6.571	7.790	21.064	23.685	26.119	29.141	31.319
15	4.601	5.229	6.262	7.261	8.547	22.307	24.996	27.488	30.578	32.801
16	5.142	5.812	6.908	7.962	9.312	23.542	26.296	28.845	32.000	34.267
17	5.697	6.408	7.564	8.672	10.085	24.769	27.587	30.191	33.409	35.718
18	6.265	7.015	8.231	9.390	10.865	25.989	28.869	31.526	34.805	37.156
19	6.844	7.633	8.907	10.117	11.651	27.204	30.144	32.852	36.191	38.582
20	7.434	8.260	9.591	10.851	12.443	28.412	31.410	34.170	37.566	39.997
21	8.034	8.897	10.283	11.591	13.240	29.615	32.671	35.479	38.932	41.401
22	8.643	9.542	10.982	12.338	14.041	30.813	33.924	36.781	40.289	42.796
23	9.260	10.196	11.689	13.091	14.848	32.007	35.172	38.076	41.638	44.181
24	9.886	10.856	12.401	13.848	15.659	33.196	36.415	39.364	42.980	45.559
25	10.520	11.524	13.120	14.611	16.473	34.382	37.652	40.646	44.314	46.928
26	11.160	12.198	13.844	15.379	17.292	35.563	38.885	41.923	45.642	48.290
27	11.808	12.879	14.573	16.151	18.114	36.741	40.113	43.195	46.963	49.645
28	12.461	13.565	15.308	16.928	18.939	37.916	41.337	44.461	48.278	50.993
29	13.121	14.256	16.047	17.708	19.768	39.087	42.557	45.722	49.588	52.336
30	13.787	14.953	16.791	18.493	20.599	40.256	43.773	46.979	50.892	53.672
40	20.707	22.164	24.433	26.509	29.051	51.805	55.758	59.342	63.691	66.766
50	27.991	29.707	32.357	34.764	37.689	63.167	67.505	71.420	76.154	79.490
60	35.534	37.485	40.482	43.188	46.459	74.397	79.082	83.298	88.379	91.952
70	43.275	45.442	48.758	51.739	55.329	85.527	90.531	95.023	100.425	104.215
80	51.172	53.540	57.153	60.391	64.278	96.578	101.879	106.629	112.329	116.321
90	59.196	61.754	65.647	69.126	73.291	107.565	113.145	118.136	124.116	128.299
100	67.328	70.065	74.222	77.929	82.358	118.498	124.342	129.561	135.807	140.169

Table E.5 Percentage Point of *F*-Distributions.

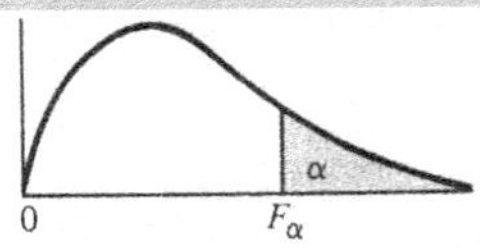

Denominator d.f.	α	Numerator d.f. 1	2	3	4	5	6	7	8	9
1	0.100	39.86	49.50	53.59	55.83	57.24	58.20	58.91	59.44	59.86
	0.050	161.4	199.5	215.7	224.6	230.2	234.0	236.8	238.9	240.5
	0.025	647.8	799.5	864.2	899.6	921.8	937.1	948.2	956.7	963.3
	0.010	4052	4999.5	5403	5625	5764	5859	5928	5982	6022
	0.005	16211	20000	21615	22500	23056	23437	23715	23925	24091
2	0.100	8.53	9.00	9.16	9.24	9.29	9.33	9.35	9.37	9.38
	0.050	18.51	19.00	19.16	19.25	19.30	19.33	19.35	19.37	19.38
	0.025	38.51	39.00	39.17	39.25	39.30	39.33	39.36	39.37	39.39
	0.010	98.50	99.00	99.17	99.25	99.30	99.33	99.36	99.37	99.39
	0.005	198.5	199.0	199.2	199.2	199.3	199.3	199.4	199.4	199.4
3	0.100	5.54	5.46	5.39	5.34	5.31	5.28	5.27	5.25	5.24
	0.050	10.13	9.55	9.28	9.12	9.01	8.94	8.89	8.85	8.81
	0.025	17.44	16.04	15.44	15.10	14.88	14.73	14.62	14.54	14.47
	0.010	34.12	30.82	29.46	28.71	28.24	27.91	27.67	27.49	27.35
	0.005	55.55	49.80	47.47	46.19	45.39	44.84	44.43	44.13	43.88
4	0.100	4.54	4.32	4.19	4.11	4.05	4.01	3.98	3.95	3.94
	0.050	7.71	6.94	6.59	6.39	6.26	6.16	6.09	6.04	6.00
	0.025	12.22	10.65	9.98	9.60	9.36	9.20	9.07	8.98	8.90
	0.010	21.20	18.00	16.69	15.98	15.52	15.21	14.98	14.80	14.66
	0.005	31.33	26.28	24.26	23.15	22.46	21.97	21.62	21.35	21.14
5	0.100	4.06	3.78	3.62	3.52	3.45	3.40	3.37	3.34	3.32
	0.050	6.61	5.79	5.41	5.19	5.05	4.95	4.88	4.82	4.77
	0.025	10.01	8.43	7.76	7.39	7.15	6.98	6.85	6.76	6.68
	0.010	16.26	13.27	12.06	11.39	10.97	10.67	10.46	10.29	10.16
	0.005	22.78	18.31	16.53	15.56	14.94	14.51	14.20	13.96	13.77
6	0.100	3.78	3.46	3.29	3.18	3.11	3.05	3.01	2.98	2.96
	0.050	5.99	5.14	4.76	4.53	4.39	4.28	4.21	4.15	4.10
	0.025	8.81	7.26	6.60	6.23	5.99	5.82	5.70	5.60	5.52
	0.010	13.75	10.92	9.78	9.15	8.75	8.47	8.26	8.10	7.98
	0.005	18.63	14.54	12.92	12.03	11.46	11.07	10.79	10.57	10.39
7	0.100	3.59	3.26	3.07	2.96	2.88	2.83	2.78	2.75	2.72
	0.050	5.59	4.74	4.35	4.12	3.97	3.87	3.79	3.73	3.68
	0.025	8.07	6.54	5.89	5.52	5.29	5.12	4.99	4.90	4.82
	0.010	12.25	9.55	8.45	7.85	7.46	7.19	6.99	6.84	6.72
	0.005	16.24	12.40	10.88	10.05	9.52	9.16	8.89	8.68	8.51

(*Continued*)

Table E.5 (Continued)

F_{α}

Numerator d.f.											
10	*12*	*15*	*20*	*24*	*30*	*40*	*60*	*120*	∞	α	d.f.
60.19	60.71	61.22	61.74	62.00	62.26	62.53	62.79	63.06	63.33	0.100	1
241.9	243.9	245.9	248.0	249.1	250.1	251.1	252.2	253.3	254.3	0.050	
968.6	976.7	984.9	993.1	997.2	1001	1006	1010	1014	1018	0.025	
6056	6106	6157	6209	6235	6261	6287	6313	6339	6366	0.010	
24224	24426	24630	24836	24940	25044	25148	25253	25359	25465	0.005	
9.39	9.41	9.42	9.44	9.45	9.46	9.47	9.47	9.48	9.49	0.100	2
19.40	19.41	19.43	19.45	19.45	19.46	19.47	19.48	19.49	19.50	0.050	
39.40	39.41	39.43	39.45	39.46	39.46	39.47	39.48	39.49	39.50	0.025	
99.40	99.42	99.43	99.45	99.46	99.47	99.47	99.48	99.49	99.50	0.010	
199.4	199.4	199.4	199.4	199.5	199.5	199.5	199.5	199.5	199.5	0.005	
5.23	5.22	5.20	5.18	5.18	5.17	5.16	5.15	5.14	5.13	0.100	3
8.79	8.74	8.70	8.66	8.64	8.62	8.59	8.57	8.55	8.53	0.050	
14.42	14.34	14.25	14.17	14.12	14.08	14.04	13.99	13.95	13.90	0.025	
27.23	27.05	26.87	26.69	26.60	26.50	26.41	26.32	26.22	26.13	0.010	
43.69	43.39	43.08	42.78	42.62	42.47	42.31	42.15	41.99	41.83	0.005	
3.92	3.90	3.87	3.84	3.83	3.82	3.80	3.79	3.78	3.76	0.100	4
5.96	5.91	5.86	5.80	5.77	5.75	5.72	5.69	5.66	5.63	0.050	
8.84	8.75	8.66	8.56	8.51	8.46	8.41	8.36	8.31	8.26	0.025	
14.55	14.37	14.20	14.02	13.93	13.84	13.75	13.65	13.56	13.46	0.010	
20.97	20.70	20.44	20.17	20.03	19.89	19.75	19.61	19.47	19.32	0.005	
3.30	3.27	3.24	3.21	3.19	3.17	3.16	3.14	3.12	3.10	0.100	5
4.74	4.68	4.62	4.56	4.53	4.50	4.46	4.43	4.40	4.36	0.050	
6.62	6.52	6.43	6.33	6.28	6.23	6.18	6.12	6.07	6.02	0.025	
10.05	9.89	9.72	9.55	9.47	9.38	9.29	9.20	9.11	9.02	0.010	
13.62	13.38	13.15	12.90	12.78	12.66	12.53	12.40	12.27	12.14	0.005	
2.94	2.90	2.87	2.84	2.82	2.80	2.78	2.76	2.74	2.72	0.100	6
4.06	4.00	3.94	3.87	3.84	3.81	3.77	3.74	3.70	3.67	0.050	
5.46	5.37	5.27	5.17	5.12	5.07	5.01	4.96	4.90	4.85	0.025	
7.87	7.72	7.56	7.40	7.31	7.23	7.14	7.06	6.97	6.88	0.010	
10.25	10.03	9.81	9.59	9.47	9.36	9.24	9.12	9.00	8.88	0.005	
2.70	2.67	2.63	2.59	2.58	2.56	2.54	2.51	2.49	2.47	0.100	7
3.64	3.57	3.51	3.44	3.41	3.38	3.34	3.30	3.27	3.23	0.050	
4.76	4.67	4.57	4.47	4.42	4.36	4.31	4.25	4.20	4.14	0.025	
6.62	6.47	6.31	6.16	6.07	5.99	5.91	5.82	5.74	5.65	0.010	
8.38	8.18	7.97	7.75	7.65	7.53	7.42	7.31	7.19	7.08	0.005	

(*Continued*)

Table E.5 (Continued)

F_α

Denominator d.f.	α	*Numerator d.f.* 1	2	3	4	5	6	7	8	9
8	0.100	3.46	3.11	2.92	2.81	2.73	2.67	2.62	2.59	2.56
	0.050	5.32	4.46	4.07	3.84	3.69	3.58	3.50	3.44	3.39
	0.025	7.57	6.06	5.42	5.05	4.82	4.65	4.53	4.43	4.36
	0.010	11.26	8.65	7.59	7.01	6.63	6.37	6.18	6.03	5.91
	0.005	14.69	11.04	9.60	8.81	8.30	7.95	7.69	7.50	7.34
9	0.100	3.36	3.01	2.81	2.69	2.61	2.55	2.51	2.47	2.44
	0.050	5.12	4.26	3.86	3.63	3.48	3.37	3.29	3.23	3.18
	0.025	7.21	5.71	5.08	4.72	4.48	4.32	4.20	4.10	4.03
	0.010	10.56	8.02	6.99	6.42	6.06	5.80	5.61	5.47	5.35
	0.005	13.61	10.11	8.72	7.96	7.47	7.13	6.88	6.69	6.54
10	0.100	3.29	2.92	2.73	2.61	2.52	2.46	2.41	2.38	2.35
	0.050	4.96	4.10	3.71	3.48	3.33	3.22	3.14	3.07	3.02
	0.025	6.94	5.46	4.83	4.47	4.24	4.07	3.95	3.85	3.78
	0.010	10.04	7.56	6.55	5.99	5.64	5.39	5.20	5.06	4.94
	0.005	12.83	9.43	8.08	7.34	6.87	6.54	6.30	6.12	5.97
11	0.100	3.23	2.86	2.66	2.54	2.45	2.39	2.34	2.30	2.27
	0.050	4.84	3.98	3.59	3.36	3.20	3.09	3.01	2.95	2.90
	0.025	6.72	5.26	4.63	4.28	4.04	3.88	3.76	3.66	3.59
	0.010	9.65	7.21	6.22	5.67	5.32	5.07	4.89	4.74	4.63
	0.005	12.23	8.91	7.60	6.88	6.42	6.10	5.86	5.68	5.54
12	0.100	3.18	2.81	2.61	2.48	2.39	2.33	2.28	2.24	2.21
	0.050	4.75	3.89	3.49	3.26	3.11	3.00	2.91	2.85	2.80
	0.025	6.55	5.10	4.47	4.12	3.89	3.73	3.61	3.51	3.44
	0.010	9.33	6.93	5.95	5.41	5.06	4.82	4.64	4.50	4.39
	0.005	11.75	8.51	7.23	6.52	6.07	5.76	5.52	5.35	5.20
13	0.100	3.14	2.76	2.56	2.43	2.35	2.28	2.23	2.20	2.16
	0.050	4.67	3.81	3.41	3.18	3.03	2.92	2.83	2.77	2.71
	0.025	6.41	4.97	4.35	4.00	3.77	3.60	3.48	3.39	3.31
	0.010	9.07	6.70	5.74	5.21	4.86	4.62	4.44	4.30	4.19
	0.005	11.37	8.19	6.93	6.23	5.79	5.48	5.25	5.08	4.94
14	0.100	3.10	2.73	2.52	2.39	2.31	2.24	2.19	2.15	2.12
	0.050	4.60	3.74	3.34	3.11	2.96	2.85	2.76	2.70	2.65
	0.025	6.30	4.86	4.24	3.89	3.66	3.50	3.38	3.29	3.21
	0.010	8.86	6.51	5.56	5.04	4.69	4.46	4.28	4.14	4.03
	0.005	11.06	7.92	6.68	6.00	5.56	5.26	5.03	4.86	4.72

(Continued)

Table E.5 (Continued)

F_α

Numerator d.f.											
10	*12*	*15*	*20*	*24*	*30*	*40*	*60*	*120*	∞	α	d.f.
2.54	2.50	2.46	2.42	2.40	2.38	2.36	2.34	2.32	2.29	0.100	8
3.35	3.28	3.22	3.15	3.12	3.08	3.04	3.01	2.97	2.93	0.050	
4.30	4.20	4.10	4.00	3.95	3.89	3.84	3.78	3.73	3.67	0.025	
5.81	5.67	5.52	5.36	5.28	5.20	5.12	5.03	4.95	4.86	0.010	
7.21	7.01	6.81	6.61	6.50	6.40	6.29	6.18	6.06	5.95	0.005	
2.42	2.38	2.34	2.30	2.28	2.25	2.23	2.21	2.18	2.16	0.100	9
3.14	3.07	3.01	2.94	2.90	2.86	2.83	2.79	2.75	2.71	0.050	
3.96	3.87	3.77	3.67	3.61	3.56	3.51	3.45	3.39	3.33	0.025	
5.26	5.11	4.96	4.81	4.73	4.65	4.57	4.48	4.40	4.31	0.010	
6.42	6.23	6.03	5.83	5.73	5.62	5.52	5.41	5.30	5.19	0.005	
2.32	2.28	2.24	2.20	2.18	2.16	2.13	2.11	2.08	2.06	0.100	10
2.98	2.91	2.85	2.74	2.77	2.70	2.66	2.62	2.58	2.54	0.050	
3.72	3.62	3.52	3.42	3.37	3.31	3.26	3.20	3.14	3.08	0.025	
4.85	4.71	4.56	4.41	4.33	4.25	4.17	4.08	4.00	3.91	0.010	
5.85	5.66	5.47	5.27	5.17	5.07	4.97	4.86	4.75	4.64	0.005	
2.25	2.21	2.17	2.12	2.10	2.08	2.05	2.03	2.00	1.97	0.100	11
2.85	2.79	2.72	2.65	2.61	2.57	2.53	2.49	2.45	2.40	0.050	
3.53	3.43	3.33	3.23	3.17	3.12	3.06	3.00	2.94	2.88	0.025	
4.54	4.40	4.25	4.10	4.02	3.94	3.86	3.78	3.69	3.60	0.010	
5.42	5.24	5.05	4.86	4.76	4.65	4.55	4.44	4.34	4.23	0.005	
2.19	2.15	2.10	2.06	2.04	2.01	1.99	1.96	1.93	1.90	0.100	12
2.75	2.69	2.62	2.54	2.51	2.47	2.43	2.38	2.34	2.30	0.050	
3.37	3.28	3.18	3.07	3.02	2.96	2.91	2.85	2.79	2.72	0.025	
4.30	4.16	4.01	3.86	3.78	3.70	3.62	3.54	3.45	3.36	0.010	
5.09	4.91	4.72	4.53	4.43	4.33	4.23	4.12	4.01	3.90	0.005	
2.14	2.10	2.05	2.01	1.98	1.96	1.93	1.90	1.88	1.85	0.100	13
2.67	2.60	2.53	2.46	2.42	2.38	2.34	2.30	2.25	2.21	0.050	
3.25	3.15	3.05	2.95	2.89	2.84	2.78	2.72	2.66	2.60	0.025	
4.10	3.96	3.82	3.66	3.59	3.51	3.43	3.34	3.25	3.17	0.010	
4.82	4.64	4.46	4.27	4.17	4.07	3.97	3.87	3.76	3.65	0.005	
2.10	2.05	2.01	1.96	1.94	1.91	1.89	1.86	1.83	1.80	0.100	14
2.60	2.53	2.46	2.39	2.35	2.31	2.27	2.22	2.18	2.13	0.050	
3.15	3.05	2.95	2.84	2.79	2.73	2.67	2.61	2.55	2.49	0.025	
3.94	3.80	3.66	3.51	3.43	3.35	3.27	3.18	3.09	3.00	0.010	
4.60	4.43	4.25	4.06	3.96	3.86	3.76	3.66	3.55	3.44	0.005	

(Continued)

Table E.5 (Continued)

F_α

Denominator d.f.	α	1	2	3	4	5	6	7	8	9
		Numerator d.f.								
15	0.100	3.07	2.70	2.49	2.36	2.27	2.21	2.16	2.12	2.09
	0.050	4.54	3.68	3.29	3.06	2.90	2.79	2.71	2.64	2.59
	0.025	6.20	4.77	4.15	3.80	3.58	3.41	3.29	3.20	3.12
	0.010	8.68	6.36	5.42	4.89	4.56	4.32	4.14	4.00	3.89
	0.005	10.80	7.70	6.48	5.80	5.37	5.07	4.85	4.67	4.54
16	0.100	3.05	2.67	2.46	2.33	2.24	2.18	2.13	2.09	2.06
	0.050	4.49	3.63	3.24	3.01	2.85	2.74	2.66	2.59	2.54
	0.025	6.12	4.69	4.08	3.73	3.50	3.34	3.22	3.12	3.05
	0.010	8.53	6.23	5.29	4.77	4.44	4.20	4.03	3.89	3.78
	0.005	10.58	7.51	6.30	5.64	5.21	4.91	4.69	4.52	4.38
17	0.100	3.03	2.64	2.44	2.31	2.22	2.15	2.10	2.06	2.03
	0.050	4.45	3.59	3.20	2.96	2.81	2.70	2.61	2.55	2.49
	0.025	6.04	4.62	4.01	3.66	3.44	3.28	3.16	3.06	2.98
	0.010	8.40	6.11	5.18	4.67	4.34	4.10	3.93	3.79	3.68
	0.005	10.38	7.35	6.16	5.50	5.07	4.78	4.56	4.39	4.25
18	0.100	3.01	2.62	2.42	2.29	2.20	2.13	2.08	2.04	2.00
	0.050	4.41	3.55	3.16	2.93	2.77	2.66	2.58	2.51	2.46
	0.025	5.98	4.56	3.95	3.61	3.38	3.22	3.10	3.01	2.93
	0.010	8.29	6.01	5.09	4.58	4.25	4.01	3.84	3.71	3.60
	0.005	10.22	7.21	6.03	5.37	4.96	4.66	4.44	4.28	4.14
19	0.100	2.99	2.61	2.40	2.27	2.18	2.11	2.06	2.02	1.98
	0.050	4.38	3.52	3.13	2.90	2.74	2.63	2.54	2.48	2.42
	0.025	5.92	4.51	3.90	3.56	3.33	3.17	3.05	2.96	2.88
	0.010	8.18	5.93	5.01	4.50	4.17	3.94	3.77	3.63	3.52
	0.005	10.07	7.09	5.92	5.27	4.85	4.56	4.34	4.18	4.04
20	0.100	2.97	2.59	2.38	2.25	2.16	2.09	2.04	2.00	1.96
	0.050	4.35	3.49	3.10	2.87	2.71	2.60	2.51	2.45	2.39
	0.025	5.87	4.46	3.86	3.51	3.29	3.13	3.01	2.91	2.84
	0.010	8.10	5.85	4.94	4.43	4.10	3.87	3.70	3.56	3.46
	0.005	9.94	6.99	5.82	5.17	4.76	4.47	4.26	4.09	3.96
21	0.100	2.96	2.57	2.36	2.23	2.14	2.08	2.02	1.98	1.95
	0.050	4.32	3.47	3.07	2.84	2.68	2.57	2.49	2.42	2.37
	0.025	5.83	4.42	3.82	3.48	3.25	3.09	2.97	2.87	2.80
	0.010	8.02	5.78	4.87	4.37	4.04	3.81	3.64	3.51	3.40
	0.005	9.83	6.89	5.73	5.09	4.68	4.39	4.18	4.01	3.88

(Continued)

Table E.5 (Continued)

F_α											
Numerator d.f.											
10	*12*	*15*	*20*	*24*	*30*	*40*	*60*	*120*	∞	α	d.f.
2.06	2.02	1.97	1.92	1.90	1.87	1.85	1.82	1.79	1.76	0.100	15
2.54	2.48	2.40	2.33	2.29	2.25	2.20	2.16	2.11	2.07	0.050	
3.06	2.96	2.86	2.76	2.70	2.64	2.59	2.52	2.46	2.40	0.025	
3.80	3.67	3.52	3.37	3.29	3.21	3.13	3.05	2.96	2.87	0.010	
4.42	4.25	4.07	3.88	3.79	3.69	3.58	3.48	3.37	3.26	0.005	
2.03	1.99	1.94	1.89	1.87	1.84	1.81	1.78	1.75	1.72	0.100	16
2.49	2.42	2.35	2.28	2.24	2.19	2.15	2.11	2.06	2.01	0.050	
2.99	2.89	2.79	2.68	2.63	2.57	2.51	2.45	2.38	2.32	0.025	
3.69	3.55	3.41	3.26	3.18	3.10	3.02	2.93	2.84	2.75	0.010	
4.27	4.10	3.92	3.73	3.64	3.54	3.44	3.33	3.22	3.11	0.005	
2.00	1.96	1.91	1.86	1.84	1.81	1.78	1.75	1.72	1.69	0.100	17
2.45	2.38	2.31	2.23	2.19	2.15	2.10	2.06	2.01	1.96	0.050	
2.92	2.82	2.72	2.62	2.56	2.50	2.44	2.38	2.32	2.25	0.025	
3.59	3.46	3.31	3.16	3.08	3.00	2.92	2.83	2.75	2.65	0.010	
4.14	3.97	3.79	3.61	3.51	3.41	3.31	3.21	3.10	2.98	0.005	
1.98	1.93	1.89	1.84	1.81	1.78	1.75	1.72	1.69	1.66	0.100	18
2.41	2.34	2.27	2.19	2.15	2.11	2.06	2.02	1.97	1.92	0.050	
2.87	2.77	2.67	2.56	2.50	2.44	2.38	2.32	2.26	2.19	0.025	
3.51	3.37	3.23	3.08	3.00	2.92	2.84	2.75	2.66	2.57	0.010	
4.03	3.86	3.68	3.50	3.40	3.30	3.20	3.10	2.99	2.87	0.005	
1.96	1.91	1.86	1.81	1.79	1.76	1.73	1.70	1.67	1.63	0.100	19
2.38	2.31	2.23	2.16	2.11	2.07	2.03	1.98	1.93	1.88	0.050	
2.82	2.72	2.62	2.51	2.45	2.39	2.33	2.27	2.20	2.13	0.025	
3.43	3.30	3.15	3.00	2.92	2.84	2.76	2.67	2.58	2.49	0.010	
3.93	3.76	3.59	3.40	3.31	3.21	3.11	3.00	2.89	2.78	0.005	
1.94	1.89	1.84	1.79	1.77	1.74	1.71	1.68	1.64	1.61	0.100	20
2.35	2.28	2.20	2.12	2.08	2.04	1.99	1.95	1.90	1.84	0.050	
2.77	2.68	2.57	2.46	2.41	2.35	2.29	2.22	2.16	2.09	0.025	
3.37	3.23	3.09	2.94	2.86	2.78	2.69	2.61	2.52	2.42	0.010	
3.85	3.68	3.50	3.32	3.22	3.12	3.02	2.92	2.81	2.69	0.005	
1.92	1.87	1.83	1.78	1.75	1.72	1.69	1.66	1.62	1.59	0.100	21
2.32	2.25	2.18	2.10	2.05	2.01	1.96	1.92	1.87	1.81	0.050	
2.73	2.64	2.53	2.42	2.37	2.31	2.25	2.18	2.11	2.04	0.025	
3.31	3.17	3.03	2.88	2.80	2.72	2.64	2.55	2.46	2.36	0.010	
3.77	3.60	3.43	3.24	3.15	3.05	2.95	2.84	2.73	2.61	0.005	

(Continued)

Table E.5 (Continued)

F_α

Denominator d.f.	α	Numerator d.f. 1	2	3	4	5	6	7	8	9
22	0.100	2.95	2.56	2.35	2.22	2.13	2.06	2.01	1.97	1.93
	0.050	4.30	3.44	3.05	2.82	2.66	2.55	2.46	2.40	2.34
	0.025	5.79	4.38	3.78	3.44	3.22	3.05	2.93	2.84	2.76
	0.010	7.95	5.72	4.82	4.31	3.99	3.76	3.59	3.45	3.35
	0.005	9.73	6.81	5.65	5.02	4.61	4.32	4.11	3.94	3.81
23	0.100	2.94	2.55	2.34	2.21	2.11	2.05	1.99	1.95	1.92
	0.050	4.28	3.42	3.03	2.80	2.64	2.53	2.44	2.37	2.32
	0.025	5.75	4.35	3.75	3.41	3.18	3.02	2.90	2.81	2.73
	0.010	7.88	5.66	4.76	4.26	3.94	3.71	3.54	3.41	3.30
	0.005	9.63	6.73	5.58	4.95	4.54	4.26	4.05	3.88	3.75
24	0.100	2.93	2.54	2.33	2.19	2.10	2.04	1.98	1.94	1.91
	0.050	4.26	3.40	3.01	2.78	2.62	2.51	2.42	2.36	2.30
	0.025	5.72	4.32	3.72	3.38	3.15	2.99	2.87	2.78	2.70
	0.010	7.82	5.61	4.72	4.22	3.90	3.67	3.50	3.36	3.26
	0.005	9.55	6.66	5.52	4.89	4.49	4.20	3.99	3.83	3.69
25	0.100	2.92	2.53	2.32	2.18	2.09	2.02	1.97	1.93	1.89
	0.050	4.24	3.39	2.99	2.76	2.60	2.49	2.40	2.34	2.28
	0.025	5.69	4.29	3.69	3.35	3.13	2.97	2.85	2.75	2.68
	0.010	7.77	5.57	4.68	4.18	3.85	3.63	3.46	3.32	3.22
	0.005	9.48	6.60	5.46	4.84	4.43	4.15	3.94	3.78	3.64
26	0.100	2.91	2.52	2.31	2.17	2.08	2.01	1.96	1.92	1.88
	0.050	4.23	3.37	2.98	2.74	2.59	2.47	2.39	2.32	2.27
	0.025	5.66	4.27	3.67	3.33	3.10	2.94	2.82	2.73	2.65
	0.010	7.72	5.53	4.64	4.14	3.82	3.59	3.42	3.29	3.18
	0.005	9.41	6.54	5.41	4.79	4.38	4.10	3.89	3.73	3.60
27	0.100	2.90	2.51	2.30	2.17	2.07	2.00	1.95	1.91	1.87
	0.050	4.21	3.35	2.96	2.73	2.57	2.46	2.37	2.31	2.25
	0.025	5.63	4.24	3.65	3.31	3.08	2.92	2.80	2.71	2.63
	0.010	7.68	5.49	4.60	4.11	3.78	3.56	3.39	3.26	3.15
	0.005	9.34	6.49	5.36	4.74	4.34	4.06	3.85	3.69	3.56
28	0.100	2.89	2.50	2.29	2.16	2.06	2.00	1.94	1.90	1.87
	0.050	4.20	3.34	2.95	2.71	2.56	2.45	2.36	2.29	2.24
	0.025	5.61	4.22	3.63	3.29	3.06	2.90	2.78	2.69	2.61
	0.010	7.64	5.45	4.57	4.07	3.75	3.53	3.36	3.23	3.12
	0.005	9.28	6.44	5.32	4.70	4.30	4.02	3.81	3.65	3.52

(Continued)

Table E.5 (Continued)

F_α											
Numerator d.f.											
10	*12*	*15*	*20*	*24*	*30*	*40*	*60*	*120*	∞	α	d.f.
1.90	1.86	1.81	1.76	1.73	1.70	1.67	1.64	1.60	1.57	0.100	22
2.30	2.23	2.15	2.07	2.03	1.98	1.94	1.89	1.84	1.78	0.050	
2.70	2.60	2.50	2.39	2.33	2.27	2.21	2.14	2.08	2.00	0.025	
3.26	3.12	2.98	2.83	2.75	2.67	2.58	2.50	2.40	2.31	0.010	
3.70	3.54	3.36	3.18	3.08	2.98	2.88	2.77	2.66	2.55	0.005	
1.89	1.84	1.80	1.74	1.72	1.69	1.66	1.62	1.59	1.55	0.100	23
2.27	2.20	2.13	2.05	2.01	1.96	1.91	1.86	1.81	1.76	0.050	
2.67	2.57	2.47	2.36	2.30	2.24	2.18	2.11	2.04	1.97	0.025	
3.21	3.07	2.93	2.78	2.70	2.62	2.54	2.45	2.35	2.26	0.010	
3.64	3.47	3.30	3.12	3.02	2.92	2.82	2.71	2.60	2.48	0.005	
1.88	1.83	1.78	1.73	1.70	1.67	1.64	1.61	1.57	1.53	0.100	24
2.25	2.18	2.11	2.03	1.98	1.94	1.89	1.84	1.79	1.73	0.050	
2.64	2.54	2.44	2.33	2.27	2.21	2.15	2.08	2.01	1.94	0.025	
3.17	3.03	2.89	2.74	2.66	2.58	2.49	2.40	2.31	2.21	0.010	
3.59	3.42	3.25	3.06	2.97	2.87	2.77	2.66	2.55	2.43	0.005	
1.87	1.82	1.77	1.72	1.69	1.66	1.63	1.59	1.56	1.52	0.100	25
2.24	2.16	2.09	2.01	1.96	1.92	1.87	1.82	1.77	1.71	0.050	
2.61	2.51	2.41	2.30	2.24	2.18	2.12	2.05	1.98	1.91	0.025	
3.13	2.99	2.85	2.70	2.62	2.54	2.45	2.36	2.27	2.17	0.010	
3.54	3.37	3.20	3.01	2.92	2.82	2.72	2.61	2.50	2.38	0.005	
1.86	1.81	1.76	1.71	1.68	1.65	1.61	1.58	1.54	1.50	0.100	26
2.22	2.15	2.07	1.99	1.95	1.90	1.85	1.80	1.75	1.69	0.050	
2.59	2.49	2.39	2.28	2.22	2.16	2.09	2.03	1.95	1.88	0.025	
3.09	2.96	2.81	2.66	2.58	2.50	2.42	2.33	2.23	2.13	0.010	
3.49	3.33	3.15	2.97	2.87	2.77	2.67	2.56	2.45	2.33	0.005	
1.85	1.80	1.75	1.70	1.67	1.64	1.60	1.57	1.53	1.49	0.100	27
2.20	2.13	2.06	1.97	1.93	1.88	1.84	1.79	1.73	1.67	0.050	
2.57	2.47	2.36	2.25	2.19	2.13	2.07	2.00	1.93	1.85	0.025	
3.06	2.93	2.78	2.63	2.55	2.47	2.38	2.29	2.20	2.10	0.010	
3.45	3.28	3.11	2.93	2.83	2.73	2.63	2.52	2.41	2.29	0.005	
1.84	1.79	1.74	1.69	1.66	1.63	1.59	1.56	1.52	1.48	0.100	28
2.19	2.12	2.04	1.96	1.91	1.87	1.82	1.77	1.71	1.65	0.050	
2.55	2.45	2.34	2.23	2.17	2.11	2.05	1.98	1.91	1.83	0.025	
3.03	2.90	2.75	2.60	2.52	2.44	2.35	2.26	2.17	2.06	0.010	
3.41	3.25	3.07	2.89	2.79	2.69	2.59	2.48	2.37	2.25	0.005	

(Continued)

Table E.5 (Continued)

F_α

Denominator d.f.	α	1	2	3	4	5	6	7	8	9
		Numerator d.f.								
29	0.100	2.89	2.50	2.28	2.15	2.06	1.99	1.93	1.89	1.86
	0.050	4.18	3.33	2.93	2.70	2.55	2.43	2.35	2.28	2.22
	0.025	5.59	4.20	3.61	3.27	3.04	2.88	2.76	2.67	2.59
	0.010	7.60	5.42	4.54	4.04	3.73	3.50	3.33	3.20	3.09
	0.005	9.23	6.40	5.28	4.66	4.26	3.98	3.77	3.61	3.48
30	0.100	2.88	2.49	2.28	2.14	2.05	1.98	1.93	1.88	1.85
	0.050	4.17	3.32	2.92	2.69	2.53	2.42	2.33	2.27	2.21
	0.025	5.57	4.18	3.59	3.25	3.03	2.87	2.75	2.65	2.57
	0.010	7.56	5.39	4.51	4.02	3.70	3.47	3.30	3.17	3.07
	0.005	9.18	6.35	5.24	4.62	4.23	3.95	3.74	3.58	3.45
40	0.100	2.84	2.44	2.23	2.09	2.00	1.93	1.87	1.83	1.79
	0.050	4.08	3.23	2.84	2.61	2.45	2.34	2.25	2.18	2.12
	0.025	5.42	4.05	3.46	3.13	2.90	2.74	2.62	2.53	2.45
	0.010	7.31	5.18	4.31	3.83	3.51	3.29	3.12	2.99	2.89
	0.005	8.83	6.07	4.98	4.37	3.99	3.71	3.51	3.35	3.22
60	0.100	2.79	2.39	2.18	2.04	1.95	1.87	1.82	1.77	1.74
	0.050	4.00	3.15	2.76	2.53	2.37	2.25	2.17	2.10	2.04
	0.025	5.29	3.93	3.34	3.01	2.79	2.63	2.51	2.41	2.33
	0.010	7.08	4.98	4.13	3.65	3.34	3.12	2.95	2.82	2.72
	0.005	8.49	5.79	4.73	4.14	3.76	3.49	3.29	3.13	3.01
120	0.100	2.75	2.35	2.13	1.99	1.90	1.82	1.77	1.72	1.68
	0.050	3.92	3.07	2.68	2.45	2.29	2.17	2.09	2.02	1.96
	0.025	5.15	3.80	3.23	2.89	2.67	2.52	2.39	2.30	2.22
	0.010	6.85	4.79	3.95	3.48	3.17	2.96	2.79	2.66	2.56
	0.005	8.18	5.54	4.50	3.92	3.55	3.28	3.09	2.93	2.81
∞	0.100	2.71	2.30	2.08	1.94	1.85	1.77	1.72	1.67	1.63
	0.050	3.84	3.00	2.60	2.37	2.21	2.10	2.01	1.94	1.88
	0.025	5.02	3.69	3.12	2.79	2.57	2.41	2.29	2.19	2.11
	0.010	6.63	4.61	3.78	3.32	3.02	2.80	2.64	2.51	2.41
	0.005	7.88	5.30	4.28	3.72	3.35	3.09	2.90	2.74	2.62

(Continued)

Table E.5 (Continued)

					F_{α}						
					Numerator d.f.						
10	*12*	*15*	*20*	*24*	*30*	*40*	*60*	*120*	∞	α	d.f.
1.83	1.78	1.73	1.68	1.65	1.62	1.58	1.55	1.51	1.47	0.100	29
2.18	2.10	2.03	1.94	1.90	1.85	1.81	1.75	1.70	1.64	0.050	
2.53	2.43	2.32	2.21	2.15	2.09	2.03	1.96	1.89	1.81	0.025	
3.00	2.87	2.73	2.57	2.49	2.41	2.33	2.23	2.14	2.03	0.010	
3.38	3.21	3.04	2.86	2.76	2.66	2.56	2.45	2.33	2.21	0.005	
1.82	1.77	1.72	1.67	1.64	1.61	1.57	1.54	1.50	1.46	0.100	30
2.16	2.09	2.01	1.93	1.89	1.84	1.79	1.74	1.68	1.62	0.050	
2.51	2.41	2.31	2.20	2.14	2.07	2.01	1.94	1.87	1.79	0.025	
2.98	2.84	2.70	2.55	2.47	2.39	2.30	2.21	2.11	2.01	0.010	
3.34	3.18	3.01	2.82	2.73	2.63	2.52	2.42	2.30	2.18	0.005	
1.76	1.71	1.66	1.61	1.57	1.54	1.51	1.47	1.42	1.38	0.100	40
2.08	2.00	1.92	1.84	1.79	1.74	1.69	1.64	1.58	1.51	0.050	
2.39	2.29	2.18	2.07	2.01	1.94	1.88	1.80	1.72	1.64	0.025	
2.80	2.66	2.52	2.37	2.29	2.20	2.11	2.02	1.92	1.80	0.010	
3.12	2.95	2.78	2.60	2.50	2.40	2.30	2.18	2.06	1.93	0.005	
1.71	1.66	1.60	1.54	1.51	1.48	1.44	1.40	1.35	1.29	0.100	60
1.99	1.92	1.84	1.75	1.70	1.65	1.59	1.53	1.47	1.39	0.050	
2.27	2.17	2.06	1.94	1.88	1.82	1.74	1.67	1.58	1.48	0.025	
2.63	2.50	2.35	2.20	2.12	2.03	1.94	1.84	1.73	1.60	0.010	
2.90	2.74	2.57	2.39	2.29	2.19	2.08	1.96	1.83	1.69	0.005	
1.65	1.60	1.55	1.48	1.45	1.41	1.37	1.32	1.26	1.19	0.100	120
1.91	1.83	1.75	1.66	1.61	1.55	1.50	1.43	1.35	1.25	0.050	
2.16	2.05	1.94	1.82	1.76	1.69	1.61	1.53	1.43	1.31	0.025	
2.47	2.34	2.19	2.03	1.95	1.86	1.76	1.66	1.53	1.38	0.010	
2.71	2.54	2.37	2.19	2.09	1.98	1.87	1.75	1.61	1.43	0.005	
1.60	1.55	1.49	1.42	1.38	1.34	1.30	1.24	1.17	1.00	0.100	∞
1.83	1.75	1.67	1.57	1.52	1.46	1.39	1.32	1.22	1.00	0.050	
2.05	1.94	1.83	1.71	1.64	1.57	1.48	1.39	1.27	1.00	0.025	
2.32	2.18	2.04	1.88	1.79	1.70	1.59	1.47	1.32	1.00	0.010	
2.52	2.36	2.19	2.00	1.90	1.79	1.67	1.53	1.36	1.00	0.005	

Table E.6 Wilcoxon Signed Rank Test: $P(W+ \le c)$.

	n								
c	3	4	5	6	7	8	9	10	11
0	0.125	0.062	0.031	0.016	0.008	0.004	0.002	0.001	0.000
1	0.250	0.125	0.062	0.031	0.016	0.008	0.004	0.002	0.001
2	0.375	0.188	0.094	0.047	0.023	0.012	0.006	0.003	0.001
3	0.625	0.312	0.156	0.078	0.039	0.020	0.01	0.005	0.002
4	0.750	0.438	0.219	0.109	0.055	0.027	0.014	0.007	0.003
5	0.875	0.562	0.312	0.156	0.078	0.039	0.020	0.01	0.005
6	1.000	0.688	0.406	0.219	0.109	0.055	0.027	0.014	0.007
7		0.812	0.500	0.281	0.148	0.074	0.037	0.019	0.009
8		0.875	0.594	0.344	0.188	0.098	0.049	0.024	0.012
9		0.938	0.688	0.422	0.234	0.125	0.064	0.032	0.016
10		1.000	0.781	0.500	0.289	0.156	0.082	0.042	0.021
11			0.844	0.578	0.344	0.191	0.102	0.053	0.027
12			0.906	0.656	0.406	0.230	0.125	0.065	0.034
13			0.938	0.719	0.469	0.273	0.150	0.080	0.042
14			0.969	0.781	0.531	0.320	0.180	0.097	0.051
15			1.000	0.844	0.594	0.371	0.213	0.116	0.062
16				0.891	0.656	0.422	0.248	0.138	0.074
17				0.922	0.711	0.473	0.285	0.161	0.087
18				0.953	0.766	0.527	0.326	0.188	0.103
19				0.969	0.812	0.578	0.367	0.216	0.120
20				0.984	0.852	0.629	0.410	0.246	0.139
21				1.000	0.891	0.680	0.455	0.278	0.160
22					0.922	0.727	0.500	0.312	0.183
23					0.945	0.770	0.545	0.348	0.207
24					0.961	0.809	0.590	0.385	0.232
25					0.977	0.844	0.633	0.423	0.260
26					0.984	0.875	0.674	0.461	0.289
27					0.992	0.902	0.715	0.500	0.319
28					1.000	0.926	0.752	0.539	0.350
29						0.945	0.787	0.577	0.382
30						0.961	0.820	0.615	0.416
31						0.973	0.850	0.652	0.449
32						0.980	0.875	0.688	0.483
33						0.988	0.898	0.722	0.517
34						0.992	0.918	0.754	0.551
35						0.996	0.936	0.784	0.584
36						1.000	0.951	0.812	0.618
37							0.963	0.839	0.650
38							0.973	0.862	0.681
39							0.980	0.884	0.711
40							0.986	0.903	0.740
41							0.990	0.920	0.768
42							0.994	0.935	0.793
43							0.996	0.947	0.817
44							0.998	0.958	0.840
45							1.000	0.968	0.861
46								0.976	0.880
47								0.981	0.897
48								0.986	0.913

(Continued)

Table E.6 (Continued)

49								0.990	0.926
50								0.993	0.938
51								0.995	0.949
52								0.997	0.958
53								0.998	0.966
54								0.999	0.973
55								1.000	0.979
56									0.984
57									0.988
58									0.991
59									0.993
60									
61									0.997
62									0.998
63									0.999
64									0.999
65									1.000
									1.000

(continued)

	n								
c	**12**	**13**	**14**	**15**	**16**	**17**	**18**	**19**	**20**
0	0.000	0.000	0.000	0.000	0.000	0.000	0.000	0.000	0.000
1	0.000	0.000	0.000	0.000	0.000	0.000	0.000	0.000	0.000
2	0.001	0.000	0.000	0.000	0.000	0.000	0.000	0.000	0.000
3	0.001	0.001	0.000	0.000	0.000	0.000	0.000	0.000	0.000
4	0.002	0.001	0.000	0.000	0.000	0.000	0.000	0.000	0.000
5	0.002	0.001	0.001	0.000	0.000	0.000	0.000	0.000	0.000
6	0.003	0.002	0.001	0.000	0.000	0.000	0.000	0.000	0.000
7	0.005	0.002	0.001	0.001	0.000	0.000	0.000	0.000	0.000
8	0.006	0.003	0.002	0.001	0.000	0.000	0.000	0.000	0.000
9	0.008	0.004	0.002	0.001	0.001	0.000	0.000	0.000	0.000
10	0.010	0.005	0.003	0.001	0.001	0.000	0.000	0.000	0.000
11	0.013	0.007	0.003	0.002	0.001	0.000	0.000	0.000	0.000
12	0.017	0.009	0.004	0.002	0.001	0.001	0.000	0.000	0.000
13	0.021	0.011	0.005	0.003	0.001	0.001	0.000	0.000	0.000
14	0.026	0.013	0.007	0.003	0.002	0.001	0.000	0.000	0.000
15	0.032	0.016	0.008	0.004	0.002	0.001	0.001	0.000	0.000
16	0.039	0.020	0.010	0.005	0.003	0.001	0.001	0.000	0.000
17	0.046	0.024	0.012	0.006	0.003	0.002	0.001	0.000	0.000
18	0.055	0.029	0.015	0.008	0.004	0.002	0.001	0.000	0.000
19	0.065	0.034	0.018	0.009	0.005	0.002	0.001	0.001	0.000
20	0.076	0.040	0.021	0.011	0.005	0.003	0.001	0.001	0.000
21	0.088	0.047	0.025	0.013	0.007	0.003	0.002	0.001	0.000
22	0.102	0.055	0.029	0.015	0.008	0.004	0.002	0.001	0.001
23	0.117	0.064	0.034	0.018	0.009	0.005	0.002	0.001	0.001
24	0.133	0.073	0.039	0.021	0.011	0.005	0.003	0.001	0.001
25	0.151	0.084	0.045	0.024	0.012	0.006	0.003	0.002	0.001
26	0.170	0.095	0.052	0.028	0.014	0.007	0.004	0.002	0.001

(Continued)

Table E.6 (Continued)

27	0.190	0.108	0.059	0.032	0.017	0.009	0.004	0.002	0.001
28	0.212	0.122	0.068	0.036	0.019	0.010	0.005	0.003	0.001
29	0.235	0.137	0.077	0.042	0.022	0.012	0.006	0.003	0.002
30	0.259	0.153	0.086	0.047	0.025	0.013	0.007	0.004	0.002
31	0.285	0.170	0.097	0.053	0.029	0.015	0.008	0.004	0.002
32	0.311	0.188	0.108	0.060	0.033	0.017	0.009	0.005	0.002
33	0.339	0.207	0.121	0.068	0.037	0.020	0.010	0.005	0.003
34	0.367	0.227	0.134	0.076	0.042	0.022	0.012	0.006	0.003
35	0.396	0.249	0.148	0.084	0.047	0.025	0.013	0.007	0.004
36	0.425	0.271	0.163	0.094	0.052	0.028	0.015	0.008	0.004
37	0.455	0.294	0.179	0.104	0.058	0.032	0.017	0.009	0.005
38	0.485	0.318	0.195	0.115	0.065	0.036	0.019	0.010	0.005
39	0.515	0.342	0.213	0.126	0.072	0.040	0.022	0.011	0.006
40	0.545	0.368	0.232	0.138	0.080	0.044	0.024	0.013	0.007
41	0.575	0.393	0.251	0.151	0.088	0.049	0.027	0.014	0.008
42	0.604	0.420	0.271	0.165	0.096	0.054	0.030	0.016	0.009
43	0.633	0.446	0.292	0.180	0.106	0.060	0.033	0.018	0.01
44	0.661	0.473	0.313	0.195	0.116	0.066	0.037	0.020	0.011
45	0.689	0.500	0.335	0.211	0.126	0.073	0.041	0.022	0.012
46	0.715	0.527	0.357	0.227	0.137	0.080	0.045	0.025	0.013
47	0.741	0.554	0.380	0.244	0.149	0.087	0.049	0.027	0.015
48	0.765	0.580	0.404	0.262	0.161	0.095	0.054	0.030	0.016
49	0.788	0.607	0.428	0.281	0.174	0.103	0.059	0.033	0.018
50	0.810	0.632	0.452	0.300	0.188	0.112	0.065	0.036	0.020
51	0.830	0.658	0.476	0.319	0.202	0.122	0.071	0.040	0.022
52	0.849	0.682	0.500	0.339	0.217	0.132	0.077	0.044	0.024
53	0.867	0.706	0.524	0.360	0.232	0.142	0.084	0.048	0.027
54	0.883	0.729	0.548	0.381	0.248	0.153	0.091	0.052	0.029
55	0.898	0.751	0.572	0.402	0.264	0.164	0.098	0.057	0.032
56	0.912	0.773	0.596	0.423	0.281	0.176	0.106	0.062	0.035
57	0.924	0.793	0.620	0.445	0.298	0.189	0.114	0.067	0.038
58	0.935	0.812	0.643	0.467	0.316	0.202	0.123	0.072	0.041
59	0.945	0.830	0.665	0.489	0.334	0.215	0.132	0.078	0.045
60	0.954	0.847	0.687	0.511	0.353	0.229	0.142	0.084	0.049
61	0.961	0.863	0.708	0.533	0.372	0.244	0.152	0.091	0.053
62	0.968	0.878	0.729	0.555	0.391	0.259	0.162	0.098	0.057
63	0.974	0.892	0.749	0.577	0.410	0.274	0.173	0.105	0.062
64	0.979	0.905	0.768	0.598	0.430	0.290	0.185	0.113	0.066
65	0.983	0.916	0.787	0.619	0.450	0.306	0.196	0.121	0.071
66	0.987	0.927	0.805	0.640	0.470	0.322	0.209	0.129	0.077
67	0.990	0.936	0.821	0.661	0.490	0.339	0.221	0.138	0.082
68	0.992	0.945	0.837	0.681	0.510	0.356	0.234	0.147	0.088
69	0.994	0.953	0.852	0.700	0.530	0.373	0.248	0.156	0.095
70	0.995	0.960	0.866	0.719	0.550	0.391	0.261	0.166	0.101
71	0.997	0.966	0.879	0.738	0.570	0.409	0.275	0.176	0.108
72	0.998	0.971	0.892	0.756	0.590	0.427	0.290	0.187	0.115
73	0.998	0.976	0.903	0.773	0.609	0.445	0.305	0.198	0.123
74	0.999	0.980	0.914	0.789	0.628	0.463	0.320	0.209	0.131
75	0.999	0.984	0.923	0.805	0.647	0.482	0.335	0.221	0.139
76	1.000	0.987	0.932	0.820	0.666	0.500	0.351	0.233	0.147
77	1.000	0.989	0.941	0.835	0.684	0.518	0.367	0.245	0.156
78	1.000	0.991	0.948	0.849	0.702	0.537	0.383	0.258	0.165

(Continued)

Table E.6 (Continued)

79		0.993	0.955	0.862	0.719	0.555	0.399	0.271	0.174
80		0.995	0.961	0.874	0.736	0.573	0.416	0.284	0.184
81		0.996	0.966	0.885	0.752	0.591	0.433	0.297	0.194
82		0.997	0.971	0.896	0.768	0.609	0.449	0.311	0.205
83		0.998	0.975	0.906	0.783	0.627	0.466	0.325	0.215
84		0.998	0.979	0.916	0.798	0.644	0.483	0.340	0.226
85		0.999	0.982	0.924	0.812	0.661	0.500	0.354	0.237
86		0.999	0.985	0.932	0.826	0.678	0.517	0.369	0.249
87		0.999	0.988	0.940	0.839	0.694	0.534	0.384	0.261
88		1.000	0.990	0.947	0.851	0.710	0.551	0.399	0.273
89		1.000	0.992	0.953	0.863	0.726	0.567	0.414	0.285
90		1.000	0.993	0.958	0.874	0.741	0.584	0.430	0.298
91		1.000	0.995	0.964	0.884	0.756	0.601	0.445	0.311
92			0.996	0.968	0.894	0.771	0.617	0.461	0.324
93			0.997	0.972	0.904	0.785	0.633	0.476	0.337
94			0.997	0.976	0.912	0.798	0.649	0.492	0.351
95			0.998	0.979	0.920	0.811	0.665	0.508	0.364
96			0.998	0.982	0.928	0.824	0.680	0.524	0.378
97			0.999	0.985	0.935	0.836	0.695	0.539	0.392
98			0.999	0.987	0.942	0.847	0.710	0.555	0.406
99			0.999	0.989	0.948	0.858	0.725	0.570	0.420
100			1.000	0.991	0.953	0.868	0.739	0.586	0.435
101			1.000	0.992	0.958	0.878	0.752	0.601	0.449
102			1.000	0.994	0.963	0.888	0.766	0.616	0.464
103			1.000	0.995	0.967	0.897	0.779	0.631	0.478
104			1.000	0.996	0.971	0.905	0.791	0.646	0.493
105			1.000	0.997	0.975	0.913	0.804	0.660	0.507
106				0.997	0.978	0.920	0.815	0.675	0.522
107				0.998	0.981	0.927	0.827	0.689	0.536
108				0.998	0.983	0.934	0.838	0.703	0.551
109				0.999	0.986	0.940	0.848	0.716	0.565
110				0.999	0.988	0.946	0.858	0.729	0.580
111				0.999	0.989	0.951	0.868	0.742	0.594
112				0.999	0.991	0.956	0.877	0.755	0.608
113				1.000	0.992	0.960	0.886	0.767	0.622
114				1.000	0.993	0.964	0.894	0.779	0.636
115				1.000	0.995	0.968	0.902	0.791	0.649
116				1.000	0.995	0.972	0.909	0.802	0.663
117				1.000	0.996	0.975	0.916	0.813	0.676
118				1.000	0.997	0.978	0.923	0.824	0.689
119				1.000	0.997	0.980	0.929	0.834	0.702
120				1.000	0.998	0.983	0.935	0.844	0.715
121					0.998	0.985	0.941	0.853	0.727
122					0.999	0.987	0.946	0.862	0.739
123					0.999	0.988	0.951	0.871	0.751
124					0.999	0.990	0.955	0.879	0.763
125					0.999	0.991	0.959	0.887	0.774
126					0.999	0.993	0.963	0.895	0.785
127					1.000	0.994	0.967	0.902	0.795
128					1.000	0.995	0.970	0.909	0.806
129					1.000	0.995	0.973	0.916	0.816
130					1.000	0.996	0.976	0.922	0.826

(Continued)

Table E.6 (Continued)

131					1.000	0.997	0.978	0.928	0.835
132					1.000	0.997	0.981	0.933	0.844
133					1.000	0.998	0.983	0.938	0.853
134					1.000	0.998	0.985	0.943	0.861
135					1.000	0.998	0.987	0.948	0.869
136					1.000	0.999	0.988	0.952	0.877
137						0.999	0.990	0.956	0.885
138						0.999	0.991	0.960	0.892
139						0.999	0.992	0.964	0.899
140						0.999	0.993	0.967	0.905
141						1.000	0.994	0.970	0.912
142						1.000	0.995	0.973	0.918
143						1.000	0.996	0.975	0.923
144						1.000	0.996	0.978	0.929
145						1.000	0.997	0.980	0.934
146						1.000	0.997	0.982	0.938
147						1.000	0.998	0.984	0.943
148						1.000	0.998	0.986	0.947
149						1.000	0.998	0.987	0.951
150						1.000	0.999	0.989	0.955
151						1.000	0.999	0.990	0.959
152						1.000	0.999	0.991	0.962
153						1.000	0.999	0.992	0.965
154							0.999	0.993	0.968
155							0.999	0.994	0.971
156							1.000	0.995	0.973
157							1.000	0.995	0.976
158							1.000	0.996	0.978
159							1.000	0.996	0.980
160							1.000	0.997	0.982
161							1.000	0.997	0.984
162							1.000	0.998	0.985
163							1.000	0.998	0.987
164							1.000	0.998	0.988
165							1.000	0.999	0.989
166							1.000	0.999	0.990
167							1.000	0.999	0.991
168							1.000	0.999	0.992
169							1.000	0.999	0.993
170							1.000	0.999	0.994
171							1.000	1.000	0.995
172								1.000	0.995
173								1.000	0.996
174								1.000	0.996
175								1.000	0.997
176								1.000	0.997
177								1.000	0.998
178								1.000	0.998
179								1.000	0.998
180								1.000	0.998
181								1.000	0.999
182								1.000	0.999

(Continued)

Table E.6 (Continued)

183								1.000	0.999
184								1.000	0.999
185								1.000	0.999
186								1.000	0.999
187								1.000	0.999
188								1.000	1.000

Table E.7 Wilcoxon Rank Sum Test.

$$P(W \leq a)$$

$$H_0 : \tilde{\mu}_1 = \tilde{\mu}_2$$

$$n_1 = n, \quad n_2 = m, \quad \text{where } n \leq m$$

$$W = \sum_{i=1}^{m} R_{i2} - \frac{1}{2}m(m+1)$$

$n = 3$:

	m							
a	3	4	5	6	7	8	9	10
0	0.0500	0.0286	0.0179	0.0119	0.0083	0.0061	0.0045	0.0035
1	0.1000	0.0571	0.0357	0.0238	0.0167	0.0121	0.0091	0.0070
2	0.2000	0.1143	0.0714	0.0476	0.0333	0.0242	0.0182	0.0140
3	0.3500	0.2000	0.1250	0.0833	0.0583	0.0424	0.0318	0.0245
4	0.5000	0.3143	0.1964	0.1310	0.0917	0.0667	0.0500	0.0385
5	0.6500	0.4286	0.2857	0.1905	0.1333	0.0970	0.0727	0.0559
6	0.8000	0.5714	0.3929	0.2738	0.1917	0.1394	0.1045	0.0804
7	0.9000	0.6857	0.5000	0.3571	0.2583	0.1879	0.1409	0.1084
8	0.9500	0.8000	0.6071	0.4524	0.3333	0.2485	0.1864	0.1434
9	1.0000	0.8857	0.7143	0.5476	0.4167	0.3152	0.2409	0.1853
10		0.9429	0.8036	0.6429	0.5000	0.3879	0.3000	0.2343
11		0.9714	0.8750	0.7262	0.5833	0.4606	0.3636	0.2867
12		1.0000	0.9286	0.8095	0.6667	0.5394	0.4318	0.3462
13			0.9643	0.8690	0.7417	0.6121	0.5000	0.4056
14			0.9821	0.9167	0.8083	0.6848	0.5682	0.4685
15			1.0000	0.9524	0.8667	0.7515	0.6364	0.5315
16				0.9762	0.9083	0.8121	0.7000	0.5944
17				0.9881	0.9417	0.8606	0.7591	0.6538
18				1.0000	0.9667	0.9030	0.8136	0.7133
19					0.9833	0.9333	0.8591	0.7657
20					0.9917	0.9576	0.8955	0.8147
21					1.0000	0.9758	0.9273	0.8566
22						0.9879	0.9500	0.8916
23						0.9939	0.9682	0.9196
24						1.0000	0.9818	0.9441
25							0.9909	0.9615
26							0.9955	0.9755
27							1.0000	0.9860
28								0.9930
29								0.9965
30								1.0000

(Continued)

Table E.7 (Continued)

$P(W \le a)$

$H_0 : \tilde{\mu}_1 = \tilde{\mu}_2$

$n_1 = n, \quad n_2 = m, \quad \text{where } n \le m$

$$W = \sum_{i=1}^{m} R_{i2} - \frac{1}{2}m(m+1)$$

	n = 4:						
	m						
a	4	5	6	7	8	9	10
0	0.0143	0.0079	0.0048	0.0030	0.0020	0.0014	0.0010
1	0.0286	0.0159	0.0095	0.0061	0.0040	0.0028	0.0020
2	0.0571	0.0317	0.0190	0.0121	0.0081	0.0056	0.0040
3	0.1000	0.0556	0.0333	0.0212	0.0141	0.0098	0.0070
4	0.1714	0.0952	0.0571	0.0364	0.0242	0.0168	0.0120
5	0.2429	0.1429	0.0857	0.0545	0.0364	0.0252	0.0180
6	0.3429	0.2063	0.1286	0.0818	0.0545	0.0378	0.0270
7	0.4429	0.2778	0.1762	0.1152	0.0768	0.0531	0.0380
8	0.5571	0.3651	0.2381	0.1576	0.1071	0.0741	0.0529
9	0.6571	0.4524	0.3048	0.2061	0.1414	0.0993	0.0709
10	0.7571	0.5476	0.3810	0.2636	0.1838	0.1301	0.0939
11	0.8286	0.6349	0.4571	0.3242	0.2303	0.1650	0.1199
12	0.9000	0.7222	0.5429	0.3939	0.2848	0.2070	0.1518
13	0.9429	0.7937	0.6190	0.4636	0.3414	0.2517	0.1868
14	0.9714	0.8571	0.6952	0.5364	0.4040	0.3021	0.2268
15	0.9857	0.9048	0.7619	0.6061	0.4667	0.3552	0.2697
16	1.0000	0.9444	0.8238	0.6758	0.5333	0.4126	0.3177
17		0.9683	0.8714	0.7364	0.5960	0.4699	0.3666
18		0.9841	0.9143	0.7939	0.6586	0.5301	0.4196
19		0.9921	0.9429	0.8424	0.7152	0.5874	0.4725
20		1.0000	0.9667	0.8848	0.7697	0.6448	0.5275
21			0.9810	0.9182	0.8162	0.6979	0.5804
22			0.9905	0.9455	0.8586	0.7483	0.6334
23			0.9952	0.9636	0.8929	0.7930	0.6823
24			1.0000	0.9788	0.9232	0.8350	0.7303
25				0.9879	0.9455	0.8699	0.7732
26				0.9939	0.9636	0.9007	0.8132
27				0.9970	0.9758	0.9259	0.8482
28				1.0000	0.9859	0.9469	0.8801
29					0.9919	0.9622	0.9061
30					0.9960	0.9748	0.9291
31					0.9980	0.9832	0.9471
32					1.0000	0.9902	0.9620
33						0.9944	0.9730
34						0.9972	0.9820

(Continued)

Table E.7 (Continued)

$$P(W \leq a)$$
$$H_0 : \tilde{\mu}_1 = \tilde{\mu}_2$$
$$n_1 = n, \quad n_2 = m, \quad \text{where } n \leq m$$
$$W = \sum_{i=1}^{m} R_{i2} - \frac{1}{2}m(m+1)$$

	$n = 4$:						
	m						
a	4	5	6	7	8	9	10
35						0.9986	0.9880
36						1.0000	0.9930
37							0.9960
38							0.9980
39							0.9990
40							1.0000

	$n = 5$:					
	m					
a	5	6	7	8	9	10
0	0.0040	0.0022	0.0013	0.0008	0.0005	0.0003
1	0.0079	0.0043	0.0025	0.0016	0.0010	0.0007
2	0.0159	0.0087	0.0051	0.0031	0.0020	0.0013
3	0.0278	0.0152	0.0088	0.0054	0.0035	0.0023
4	0.0476	0.0260	0.0152	0.0093	0.0060	0.0040
5	0.0754	0.0411	0.0240	0.0148	0.0095	0.0063
6	0.1111	0.0628	0.0366	0.0225	0.0145	0.0097
7	0.1548	0.0887	0.0530	0.0326	0.0210	0.0140
8	0.2103	0.1234	0.0745	0.0466	0.0300	0.0200
9	0.2738	0.1645	0.1010	0.0637	0.0415	0.0276
10	0.3452	0.2143	0.1338	0.0855	0.0559	0.0376
11	0.4206	0.2684	0.1717	0.1111	0.0734	0.0496
12	0.5000	0.3312	0.2159	0.1422	0.0949	0.0646
13	0.5794	0.3961	0.2652	0.1772	0.1199	0.0823
14	0.6548	0.4654	0.3194	0.2176	0.1489	0.1032
15	0.7262	0.5346	0.3775	0.2618	0.1818	0.1272
16	0.7897	0.6039	0.4381	0.3108	0.2188	0.1548
17	0.8452	0.6688	0.5000	0.3621	0.2592	0.1855
18	0.8889	0.7316	0.5619	0.4165	0.3032	0.2198
19	0.9246	0.7857	0.6225	0.4716	0.3497	0.2567
20	0.9524	0.8355	0.6806	0.5284	0.3986	0.2970
21	0.9722	0.8766	0.7348	0.5835	0.4491	0.3393
22	0.9841	0.9113	0.7841	0.6379	0.5000	0.3839
23	0.9921	0.9372	0.8283	0.6892	0.5509	0.4296
24	0.9960	0.9589	0.8662	0.7382	0.6014	0.4765
25	1.0000	0.9740	0.8990	0.7824	0.6503	0.5235
26		0.9848	0.9255	0.8228	0.6968	0.5704
27		0.9913	0.9470	0.8578	0.7408	0.6161
28		0.9957	0.9634	0.8889	0.7812	0.6607
29		0.9978	0.9760	0.9145	0.8182	0.7030
30		1.0000	0.9848	0.9363	0.8511	0.7433
31			0.9912	0.9534	0.8801	0.7802
32			0.9949	0.9674	0.9051	0.8145
33			0.9975	0.9775	0.9266	0.8452
34			0.9987	0.9852	0.9441	0.8728
35			1.0000	0.9907	0.9585	0.8968
36				0.9946	0.9700	0.9177
37				0.9969	0.9790	0.9354
38				0.9984	0.9855	0.9504
39				0.9992	0.9905	0.9624
40				1.0000	0.9940	0.9724
41					0.9965	0.9800
42					0.9980	0.9860
43					0.9990	0.9903
44					0.9995	0.9937
45					1.0000	0.9960
46						0.9977
47						0.9987
48						0.9993
49						0.9997
50						1.0000

(Continued)

Table E.7 (Continued)

$$P(W \leq a)$$
$$H_0 : \tilde{\mu}_1 = \tilde{\mu}_2$$
$$n_1 = n, \quad n_2 = m, \quad \text{where } n \leq m$$
$$W = \sum_{i=1}^{m} R_{i2} - \frac{1}{2}m(m+1)$$

	$n = 6$:				
	m				
a	6	7	8	9	10
0	0.0011	0.0006	0.0003	0.0002	0.0001
1	0.0022	0.0012	0.0007	0.0004	0.0002
2	0.0043	0.0023	0.0013	0.0008	0.0005
3	0.0076	0.0041	0.0023	0.0014	0.0009
4	0.0130	0.0070	0.0040	0.0024	0.0015
5	0.0206	0.0111	0.0063	0.0038	0.0024
6	0.0325	0.0175	0.0100	0.0060	0.0037
7	0.0465	0.0256	0.0147	0.0088	0.0055
8	0.0660	0.0367	0.0213	0.0128	0.0080
9	0.0898	0.0507	0.0296	0.0180	0.0112
10	0.1201	0.0688	0.0406	0.0248	0.0156
11	0.1548	0.0903	0.0539	0.0332	0.0210
12	0.1970	0.1171	0.0709	0.0440	0.0280
13	0.2424	0.1474	0.0906	0.0567	0.0363
14	0.2944	0.1830	0.1142	0.0723	0.0467
15	0.3496	0.2226	0.1412	0.0905	0.0589
16	0.4091	0.2669	0.1725	0.1119	0.0736
17	0.4686	0.3141	0.2068	0.1361	0.0903
18	0.5314	0.3654	0.2454	0.1638	0.1099
19	0.5909	0.4178	0.2864	0.1942	0.1317
20	0.6504	0.4726	0.3310	0.2280	0.1566
21	0.7056	0.5274	0.3773	0.2643	0.1838
22	0.7576	0.5822	0.4259	0.3035	0.2139
23	0.8030	0.6346	0.4749	0.3445	0.2461
24	0.8452	0.6859	0.5251	0.3878	0.2811
25	0.8799	0.7331	0.5741	0.4320	0.3177
26	0.9102	0.7774	0.6227	0.4773	0.3564
27	0.9340	0.8170	0.6690	0.5227	0.3962
28	0.9535	0.8526	0.7136	0.5680	0.4374
29	0.9675	0.8829	0.7546	0.6122	0.4789
30	0.9794	0.9097	0.7932	0.6555	0.5211

	$n = 6$:				
	m				
a	6	7	8	9	10
31	0.9870	0.9312	0.8275	0.6965	0.5626
32	0.9924	0.9493	0.8588	0.7357	0.6038
33	0.9957	0.9633	0.8858	0.7720	0.6436
34	0.9978	0.9744	0.9094	0.8058	0.6823
35	0.9989	0.9825	0.9291	0.8362	0.7189
36	1.0000	0.9889	0.9461	0.8639	0.7539
37		0.9930	0.9594	0.8881	0.7861
38		0.9959	0.9704	0.9095	0.8162
39		0.9977	0.9787	0.9277	0.8434
40		0.9988	0.9853	0.9433	0.8683
41		0.9994	0.9900	0.9560	0.8901
42		1.0000	0.9937	0.9668	0.9097
43			0.9960	0.9752	0.9264
44			0.9977	0.9820	0.9411
45			0.9987	0.9872	0.9533
46			0.9993	0.9912	0.9637
47			0.9997	0.9940	0.9720
48			1.0000	0.9962	0.9790
49				0.9976	0.9844
50				0.9986	0.9888
51				0.9992	0.9920
52				0.9996	0.9945
53				0.9998	0.9963
54				1.0000	0.9976
55					0.9985
56					0.9991
57					0.9995
58					0.9998
59					0.9999
60					1.0000

(Continued)

Table E.7 (Continued)

$$P(W \le a)$$
$$H_0 : \tilde{\mu}_1 = \tilde{\mu}_2$$
$$n_1 = n, \quad n_2 = m, \quad \text{where } n \le m$$
$$W = \sum_{i=1}^{m} R_{i2} - \frac{1}{2}m(m+1)$$

	$n = 7$:			
	m			
a	7	8	9	10
0	0.0003	0.0002	0.0001	0.0001
1	0.0006	0.0003	0.0002	0.0001
2	0.0012	0.0006	0.0003	0.0002
3	0.0020	0.0011	0.0006	0.0004
4	0.0035	0.0019	0.0010	0.0006
5	0.0055	0.0030	0.0017	0.0010
6	0.0087	0.0047	0.0026	0.0015
7	0.0131	0.0070	0.0039	0.0023
8	0.0189	0.0103	0.0058	0.0034
9	0.0265	0.0145	0.0082	0.0048
10	0.0364	0.0200	0.0115	0.0068
11	0.0487	0.0270	0.0156	0.0093
12	0.0641	0.0361	0.0209	0.0125
13	0.0825	0.0469	0.0274	0.0165
14	0.1043	0.0603	0.0356	0.0215
15	0.1297	0.0760	0.0454	0.0277
16	0.1588	0.0946	0.0571	0.0351
17	0.1914	0.1159	0.0708	0.0439
18	0.2279	0.1405	0.0869	0.0544
19	0.2675	0.1678	0.1052	0.0665
20	0.3100	0.1984	0.1261	0.0806
21	0.3552	0.2317	0.1496	0.0966
22	0.4024	0.2679	0.1755	0.1148
23	0.4508	0.3063	0.2039	0.1349
24	0.5000	0.3472	0.2349	0.1574
25	0.5492	0.3894	0.2680	0.1819
26	0.5976	0.4333	0.3032	0.2087
27	0.6448	0.4775	0.3403	0.2374
28	0.6900	0.5225	0.3788	0.2681
29	0.7325	0.5667	0.4185	0.3004
30	0.7721	0.6106	0.4591	0.3345
31	0.8086	0.6528	0.5000	0.3698
32	0.8412	0.6937	0.5409	0.4063
33	0.8703	0.7321	0.5815	0.4434
34	0.8957	0.7683	0.6212	0.4811
35	0.9175	0.8016	0.6597	0.5189
36	0.9359	0.8322	0.6968	0.5566
37	0.9513	0.8595	0.7320	0.5937
38	0.9636	0.8841	0.7651	0.6302
39	0.9735	0.9054	0.7961	0.6655
40	0.9811	0.9240	0.8245	0.6996
41	0.9869	0.9397	0.8504	0.7319
42	0.9913	0.9531	0.8739	0.7626
43	0.9945	0.9639	0.8948	0.7913
44	0.9965	0.9730	0.9131	0.8181
45	0.9980	0.9800	0.9292	0.8426
46	0.9988	0.9855	0.9429	0.8651
47	0.9994	0.9897	0.9546	0.8852
48	0.9997	0.9930	0.9644	0.9034
49	1.0000	0.9953	0.9726	0.9194
50		0.9970	0.9791	0.9335
51		0.9981	0.9844	0.9456
52		0.9989	0.9885	0.9561
53		0.9994	0.9913	0.9649
54		0.9997	0.9942	0.9723
55		0.9998	0.9961	0.9785
56		1.0000	0.9974	0.9835
57			0.9983	0.9875
58			0.9990	0.9907
59			0.9994	0.9932
60			0.9997	0.9952
61			0.9998	0.9966
62			0.9999	0.9977
63			1.0000	0.9985
64				0.9990
65				0.9994
66				0.9996
67				0.9998
68				0.9999
69				0.9999
70				1.0000

(Continued)

Table E.7 (Continued)

$$P(W \le a)$$

$$H_0 : \tilde{\mu}_1 = \tilde{\mu}_2$$

$$n_1 = n, \quad n_2 = m, \quad \text{where } n \le m$$

$$W = \sum_{i=1}^{m} R_{i2} - \frac{1}{2}m(m+1)$$

	$n = 8$:		
	m		
a	8	9	10
0	0.0001	0.0000	0.0000
1	0.0002	0.0001	0.0000
2	0.0003	0.0002	0.0001
3	0.0005	0.0003	0.0002
4	0.0009	0.0005	0.0003
5	0.0015	0.0008	0.0004
6	0.0023	0.0012	0.0007
7	0.0035	0.0019	0.0010
8	0.0052	0.0028	0.0015
9	0.0074	0.0039	0.0022
10	0.0103	0.0056	0.0031
11	0.0141	0.0076	0.0043
12	0.0190	0.0103	0.0058
13	0.0249	0.0137	0.0078
14	0.0325	0.0180	0.0103
15	0.0415	0.0232	0.0133
16	0.0524	0.0296	0.0171
17	0.0652	0.0372	0.0217
18	0.0803	0.0464	0.0273
19	0.0974	0.0570	0.0338
20	0.1172	0.0694	0.0416
21	0.1393	0.0836	0.0506
22	0.1641	0.0998	0.0610
23	0.1911	0.1179	0.0729
24	0.2209	0.1383	0.0864
25	0.2527	0.1606	0.1015
26	0.2869	0.1852	0.1185
27	0.3227	0.2117	0.1371
28	0.3605	0.2404	0.1577
29	0.3992	0.2707	0.1800
30	0.4392	0.3029	0.2041
31	0.4796	0.3365	0.2299
32	0.5204	0.3715	0.2574
33	0.5608	0.4074	0.2863
34	0.6008	0.4442	0.3167
35	0.6395	0.4813	0.3482
36	0.6773	0.5187	0.3809
37	0.7131	0.5558	0.4143
38	0.7473	0.5926	0.4484
39	0.7791	0.6285	0.4827
40	0.8089	0.6635	0.5173
41	0.8359	0.6971	0.5516
42	0.8607	0.7293	0.5857
43	0.8828	0.7596	0.6191
44	0.9026	0.7883	0.6518
45	0.9197	0.8148	0.6833
46	0.9348	0.8394	0.7137
47	0.9476	0.8617	0.7426
48	0.9585	0.8821	0.7701
49	0.9675	0.9002	0.7959
50	0.9751	0.9164	0.8200
51	0.9810	0.9306	0.8423
52	0.9859	0.9430	0.8629
53	0.9897	0.9536	0.8815
54	0.9926	0.9628	0.8985
55	0.9948	0.9704	0.9136
56	0.9965	0.9768	0.9271
57	0.9977	0.9820	0.9390
58	0.9985	0.9863	0.9494
59	0.9991	0.9897	0.9584
60	0.9995	0.9924	0.9662
61	0.9997	0.9944	0.9727
62	0.9998	0.9961	0.9783
63	0.9999	0.9972	0.9829
64	1.0000	0.9981	0.9867
65		0.9988	0.9897
66		0.9992	0.9922
67		0.9995	0.9942
68		0.9997	0.9957
69		0.9998	0.9969
70		0.9999	0.9978
71		1.0000	0.9985
72		1.0000	0.9990
73			0.9993
74			0.9996
75			0.9997
76			0.9998
77			0.9999
78			1.0000
79			1.0000
80			1.0000

(Continued)

Table E.7 (Continued)

$$P(W \leq a)$$

$$H_0 : \tilde{\mu}_1 = \tilde{\mu}_2$$

$$n_1 = n, \quad n_2 = m, \quad \text{where } n \leq m$$

$$W = \sum_{i=1}^{m} R_{i2} - \frac{1}{2}m(m + 1)$$

	$n = 9$:	
	m	
a	9	10
0	0.0000	0.0000
1	0.0000	0.0000
2	0.0001	0.0000
3	0.0001	0.0001
4	0.0002	0.0001
5	0.0004	0.0002
6	0.0006	0.0003
7	0.0009	0.0005
8	0.0014	0.0007
9	0.0020	0.0011
10	0.0028	0.0015
11	0.0039	0.0021
12	0.0053	0.0028
13	0.0071	0.0038
14	0.0094	0.0051
15	0.0122	0.0066
16	0.0157	0.0086
17	0.0200	0.0110
18	0.0252	0.0140
19	0.0313	0.0175
20	0.0385	0.0217
21	0.0470	0.0267
22	0.0567	0.0326
23	0.0680	0.0394
24	0.0807	0.0474
25	0.0951	0.0564
26	0.1112	0.0667
27	0.1290	0.0782
28	0.1487	0.0912
29	0.1701	0.1055
30	0.1933	0.1214
31	0.2181	0.1388
32	0.2447	0.1577
33	0.2729	0.1781
34	0.3024	0.2001
35	0.3332	0.2235
36	0.3652	0.2483
37	0.3981	0.2745
38	0.4317	0.3019
39	0.4657	0.3304
40	0.5000	0.3598
41	0.5343	0.3901
42	0.5683	0.4211
43	0.6019	0.4524
44	0.6348	0.4841
45	0.6668	0.5159
46	0.6976	0.5476
47	0.7271	0.5789
48	0.7553	0.6099
49	0.7819	0.6402
50	0.8067	0.6696
51	0.8299	0.6981
52	0.8513	0.7255
53	0.8710	0.7517
54	0.8888	0.7765
55	0.9049	0.7999
56	0.9193	0.8219
57	0.9320	0.8423
58	0.9433	0.8612
59	0.9530	0.8786
60	0.9615	0.8945
61	0.9687	0.9088
62	0.9748	0.9218
63	0.9800	0.9333
64	0.9843	0.9436
65	0.9878	0.9526
66	0.9906	0.9606
67	0.9929	0.9674
68	0.9947	0.9733
69	0.9961	0.9783
70	0.9972	0.9825
71	0.9980	0.9860
72	0.9986	0.9890
73	0.9991	0.9914
74	0.9994	0.9934
75	0.9996	0.9949
76	0.9998	0.9962
77	0.9999	0.9972
78	0.9999	0.9979
79	1.0000	0.9985
80	1.0000	0.9989
81	1.0000	0.9993
82		0.9995
83		0.9997
84		0.9998
85		0.9999
86		0.9999
87		1.0000
88		1.0000
89		1.0000
90		1.0000

(Continued)

Table E.7 (Continued)

$$P(W \le a)$$

$$H_0 : \tilde{\mu}_1 = \tilde{\mu}_2$$

$$n_1 = n, \quad n_2 = m, \quad \text{where } n \le m$$

$$W = \sum_{i=1}^{m} R_{i2} - \frac{1}{2}m(m+1)$$

	$n = 10$:		$n = 10$:		$n = 10$:
	m		m		m
a	10	a	10	a	10
0	0.0000	35	0.1399	70	0.9385
1	0.0000	36	0.1575	71	0.9474
2	0.0000	37	0.1763	72	0.9554
3	0.0000	38	0.1965	73	0.9624
4	0.0001	39	0.2179	74	0.9685
5	0.0001	40	0.2406	75	0.9738
6	0.0002	41	0.2644	76	0.9784
7	0.0002	42	0.2894	77	0.9823
8	0.0004	43	0.3153	78	0.9856
9	0.0005	44	0.3421	79	0.9884
10	0.0008	45	0.3697	80	0.9907
11	0.0010	46	0.3980	81	0.9927
12	0.0014	47	0.4267	82	0.9943
13	0.0019	48	0.4559	83	0.9955
14	0.0026	49	0.4853	84	0.9966
15	0.0034	50	0.5147	85	0.9974
16	0.0045	51	0.5441	86	0.9981
17	0.0057	52	0.5733	87	0.9986
18	0.0073	53	0.6020	88	0.9990
19	0.0093	54	0.6303	89	0.9992
20	0.0116	55	0.6579	90	0.9995
21	0.0144	56	0.6847	91	0.9996
22	0.0177	57	0.7106	92	0.9998
23	0.0216	58	0.7356	93	0.9998
24	0.0262	59	0.7594	94	0.9999
25	0.0315	60	0.7821	95	0.9999
26	0.0376	61	0.8035	96	1.0000
27	0.0446	62	0.8237	97	1.0000
28	0.0526	63	0.8425	98	1.0000
29	0.0615	64	0.8601	99	1.0000
30	0.0716	65	0.8763	100	1.0000
31	0.0827	66	0.8912		
32	0.0952	67	0.9048		
33	0.1088	68	0.9173		
34	0.1237	69	0.9284		

Table E.8 Friedman Test.

$$H_0 : \tilde{\mu}_1 = \tilde{\mu}_2 = \cdots = \tilde{\mu}_k$$

$$P = P(S \leq s)$$

$$S = \frac{12km}{k(k+1)} \sum_{i=1}^{k} \left[R_{i\cdot} - \frac{1}{2}(k+1) \right]^2$$

Note: k = **number of treatment levels,** m = **number of blocks.**

s	P
$k = 3,\ m = 2$	
0.000	0.167
1.000	0.500
3.000	0.833
4.000	1.000

s	P
$k = 3,\ m = 3$	
0.000	0.056
0.667	0.472
2.000	0.639
2.667	0.806
4.667	0.972
6.000	1.000

s	P
$k = 3,\ m = 4$	
0.000	0.069
0.500	0.347
1.500	0.569
2.000	0.727
3.500	0.875
4.500	0.931
6.000	0.958
6.500	0.995
8.000	1.000

s	P
$k = 3,\ m = 5$	
0.000	0.046
0.400	0.309
1.200	0.478
1.600	0.633
2.800	0.818
3.600	0.876
4.800	0.907
5.200	0.961
6.400	0.976
7.600	0.992
8.400	0.999
10.000	1.000

s	P
$k = 3,\ m = 6$	
0.000	0.044
0.333	0.260
1.000	0.430
1.333	0.570
2.333	0.748
3.000	0.816
4.000	0.858
4.333	0.928
5.333	0.948
6.333	0.971
7.000	0.988
8.333	0.992
9.000	0.994
9.333	0.998
10.333	1.000
12.000	1.000

s	P
$k = 3,\ m = 7$	
0.000	0.036
0.286	0.232
0.857	0.380
1.143	0.514
2.000	0.695
2.571	0.763
3.429	0.808
3.714	0.888
4.571	0.915
5.429	0.949
6.000	0.973
7.143	0.979
7.714	0.984
8.000	0.992
8.857	0.996
10.286	0.997
10.571	0.999
11.143	1.000
12.286	1.000
14.000	1.000

s	P
$k = 3,\ m = 8$	
0.000	0.033
0.250	0.206
0.750	0.346
1.000	0.469
1.750	0.645
2.250	0.715
3.000	0.764
3.250	0.851
4.000	0.880
4.750	0.921
5.250	0.953
6.250	0.962
6.750	0.970
7.000	0.982
7.750	0.990
9.000	0.992
9.250	0.995
9.750	0.998
10.750	0.999
12.000	0.999
12.250	1.000
13.000	1.000
14.250	1.000
16.000	1.000

s	P
$k = 3,\ m = 9$	
0.000	0.029
0.222	0.186
0.667	0.315
0.889	0.431
1.556	0.602
2.000	0.672
2.667	0.722
2.889	0.813
3.556	0.846
4.222	0.893
4.667	0.931
5.556	0.943
6.000	0.952
6.222	0.969
6.889	0.981
8.000	0.984
8.222	0.990
8.667	0.994
9.556	0.996
10.667	0.997
10.889	0.999
11.556	0.999
12.667	1.000
13.556	1.000
14.000	1.000
14.222	1.000
14.889	1.000
16.222	1.000
18.000	1.000

s	P
$k = 3,\ m = 10$	
0.000	0.026
0.200	0.170
0.600	0.290
0.800	0.399
1.400	0.564
1.800	0.632
2.400	0.684
2.600	0.778
3.200	0.813
3.800	0.865
4.200	0.908
5.000	0.922
5.400	0.934
5.600	0.954
6.200	0.970
7.200	0.974
7.400	0.982
7.800	0.988
8.600	0.993
9.600	0.994
9.800	0.997
10.400	0.998
11.400	0.999
12.200	0.999
12.600	0.999
12.800	1.000
13.400	1.000
14.600	1.000
15.000	1.000
15.200	1.000
15.800	1.000
16.200	1.000
16.800	1.000
18.200	1.000
20.000	1.000

(Continued)

Table E.8 (Continued)

$$H_0 : \tilde{\mu}_1 = \tilde{\mu}_2 = \cdots = \tilde{\mu}_k$$

$$P = P(S \leq s)$$

$$S = \frac{12km}{k(k+1)} \sum_{i=1}^{k} \left[R_{i\cdot} - \frac{1}{2}(k+1) \right]^2$$

Note: k = number of treatment levels, m = number of blocks.

s	P
k = 3, m = 11	
0.000	0.024
0.182	0.156
0.545	0.268
0.727	0.371
1.273	0.530
1.636	0.597
2.182	0.649
2.364	0.744
2.909	0.781
3.455	0.837
3.818	0.884
4.545	0.900
4.909	0.913
5.091	0.938
5.636	0.957
6.545	0.962
6.727	0.973
7.091	0.981
7.818	0.987
8.727	0.989
8.909	0.994
9.455	0.996
10.364	0.997
11.091	0.998
11.455	0.999
11.636	0.999
12.182	0.999
13.273	1.000
13.636	1.000
13.818	1.000
14.364	1.000
14.727	1.000
15.273	1.000
16.545	1.000
16.909	1.000
17.636	1.000
18.182	1.000
18.727	1.000
20.182	1.000
22.000	1.000

s	P
k = 3, m = 12	
0.000	0.022
0.167	0.144
0.500	0.249
0.667	0.346
1.167	0.500
1.500	0.566
2.000	0.617
2.167	0.713
2.667	0.751
3.167	0.809
3.500	0.859
4.167	0.877
4.500	0.892
4.667	0.920
5.167	0.942
6.000	0.949
6.167	0.962
6.500	0.973
7.167	0.980
8.000	0.983
8.167	0.989
8.667	0.993
9.500	0.995
10.167	0.996
10.500	0.997
10.667	0.998
11.167	0.998
12.167	0.999
12.500	0.999
12.667	0.999
13.167	1.000
13.500	1.000
14.000	1.000
15.167	1.000
15.500	1.000
16.167	1.000
16.667	1.000
17.167	1.000
18.000	1.000
18.167	1.000
18.500	1.000
18.667	1.000
19.500	1.000
20.167	1.000
20.667	1.000
22.167	1.000
24.000	1.000

s	P
k = 3, m = 13	
0.000	0.020
0.154	0.134
0.462	0.233
0.615	0.325
1.077	0.473
1.385	0.537
1.846	0.588
2.000	0.684
2.462	0.722
2.923	0.783
3.231	0.835
3.846	0.855
4.154	0.871
4.308	0.902
4.769	0.927
5.538	0.935
5.692	0.950
6.000	0.963
6.615	0.972
7.385	0.975
7.538	0.984
8.000	0.988
8.769	0.991
9.385	0.993
9.692	0.995
9.846	0.996
10.308	0.997
11.231	0.998
11.538	0.998
11.692	0.999
12.154	0.999
12.462	0.999
12.923	0.999
14.000	1.000
14.308	1.000
14.923	1.000
15.385	1.000
15.846	1.000
16.615	1.000
16.769	1.000
17.077	1.000
17.231	1.000
18.000	1.000
18.615	1.000
19.077	1.000
19.538	1.000
19.846	1.000
20.462	1.000
21.385	1.000
22.154	1.000
22.615	1.000
24.154	1.000
26.000	1.000

s	P
k = 3, m = 14	
0.000	0.019
0.143	0.126
0.429	0.219

s	P
k = 3, m = 10	
0.571	0.306
1.000	0.449
1.286	0.511
1.714	0.562
1.857	0.656
2.286	0.695
2.714	0.758
3.000	0.812
3.571	0.833
3.857	0.850
4.000	0.884
4.429	0.910
5.143	0.920
5.286	0.937
5.571	0.952
6.143	0.963
6.857	0.967
7.000	0.978
7.429	0.983
8.143	0.987
8.714	0.990
9.000	0.992
9.143	0.993
9.571	0.995
10.429	0.996
10.714	0.997
10.857	0.998
11.286	0.998
11.571	0.998
12.000	0.999
13.000	0.999
13.286	1.000
13.857	1.000
14.286	1.000
14.714	1.000
15.429	1.000
15.571	1.000
15.857	1.000

(Continued)

Table E.8 (Continued)

$$H_0 : \tilde{\mu}_1 = \tilde{\mu}_2 = \cdots = \tilde{\mu}_k$$

$$P = P(S \leq s)$$

$$S = \frac{12km}{k(k+1)} \sum_{i=1}^{k} \left[R_{i\cdot} - \frac{1}{2}(k+1) \right]^2$$

Note: k = number of treatment levels, m = number of blocks.

s	P
16.000	1.000
16.714	1.000
17.286	1.000
17.714	1.000
18.143	1.000
18.429	1.000
19.000	1.000
19.857	1.000
20.571	1.000
21.000	1.000
21.143	1.000
21.571	1.000
22.286	1.000
22.429	1.000
23.286	1.000
24.143	1.000
24.571	1.000
26.143	1.000
28.000	1.000

s	P
$k = 3$, $m = 15$	
0.000	0.018
0.133	0.118
0.400	0.206
0.533	0.289
0.933	0.427
1.200	0.487
1.600	0.537
1.733	0.631
2.133	0.670
2.533	0.733
2.800	0.789
3.333	0.811
3.600	0.830
3.733	0.865
4.133	0.894
4.800	0.904
4.933	0.923
5.200	0.940
5.733	0.953
6.400	0.958
6.533	0.970
6.933	0.977
7.600	0.982
8.133	0.986
8.400	0.989
8.533	0.990
8.933	0.993
9.733	0.994
10.000	0.995
10.133	0.996
10.533	0.997
10.800	0.997
11.200	0.998
12.133	0.999
12.400	0.999
12.933	0.999
13.333	0.999
13.733	1.000
14.400	1.000
14.533	1.000
14.800	1.000
14.933	1.000
15.600	1.000
16.133	1.000
16.533	1.000
16.933	1.000
17.200	1.000
17.733	1.000
18.533	1.000
19.200	1.000
19.600	1.000
19.733	1.000
20.133	1.000
20.800	1.000
20.933	1.000
21.733	1.000
22.533	1.000
22.800	1.000
22.933	1.000
23.333	1.000
24.133	1.000
24.400	1.000
25.200	1.000
26.133	1.000
26.533	1.000
28.133	1.000
30.000	1.000

s	P
$k = 4$, $m = 2$	
0.000	0.042
0.600	0.167
1.200	0.208
1.800	0.375
2.400	0.458
3.000	0.542
3.600	0.625
4.200	0.792
4.800	0.833
5.400	0.958
6.000	1.000

s	P
$k = 4$, $m = 3$	
0.200	0.042
0.600	0.090
1.000	0.273
1.800	0.392
2.200	0.476
2.600	0.554
3.400	0.658
3.800	0.700
4.200	0.793
5.000	0.825
5.400	0.852
5.800	0.925
6.600	0.946
7.000	0.967
7.400	0.983
8.200	0.998
9.000	1.000

s	P
$k = 4$, $m = 4$	
0.000	0.008
0.300	0.072
0.600	0.100
0.900	0.200
1.200	0.246
1.500	0.323
1.800	0.351
2.100	0.476
2.400	0.492
2.700	0.568
3.000	0.611
3.300	0.645
3.600	0.676
3.900	0.758
4.500	0.800
4.800	0.810
5.100	0.842
5.400	0.859
5.700	0.895
6.000	0.906
6.300	0.923
6.600	0.932
6.900	0.946
7.200	0.948
7.500	0.964
7.800	0.967

s	P
$k = 3$, $m = 10$	
8.100	0.981
8.400	0.986
8.700	0.988
9.300	0.993
9.600	0.994
9.900	0.997
10.200	0.998
10.800	0.999
11.100	1.000
12.000	1.000

s	P
$k = 4$, $m = 5$	
0.120	0.025
0.360	0.056
0.600	0.143
1.080	0.229
1.320	0.291
1.560	0.348
2.040	0.439
2.280	0.479
2.520	0.555
3.000	0.592
3.240	0.628
3.480	0.702
3.960	0.740
4.200	0.774
4.440	0.790
4.920	0.838
5.160	0.849
5.400	0.877
5.880	0.893
6.120	0.907
6.360	0.925
6.840	0.933
7.080	0.945

(Continued)

Table E.8 (Continued)

$$H_0 : \tilde{\mu}_1 = \tilde{\mu}_2 = \cdots = \tilde{\mu}_k$$

$$P = P(S \leq s)$$

$$S = \frac{12km}{k(k+1)} \sum_{i=1}^{k} \left[R_{i\cdot} - \frac{1}{2}(k+1) \right]^2$$

Note: k = number of treatment levels, *m* = number of blocks.

s	*P*
7.320	0.956
7.800	0.966
8.040	0.969
8.280	0.977
8.760	0.980
9.000	0.983
9.240	0.988
9.720	0.991
9.960	0.993
10.200	0.995
10.680	0.997
10.920	0.998
11.160	0.998
11.640	0.998
11.880	0.999
12.120	0.999
12.600	1.000
12.840	1.000
13.080	1.000
13.560	1.000
14.040	1.000
15.000	1.000

s	*P*
***k* = 4, *m* = 6**	
0.000	0.004
0.200	0.043
0.400	0.060
0.600	0.126
0.800	0.156
1.000	0.211
1.200	0.228
1.400	0.321
1.600	0.332
1.800	0.391
2.000	0.426
2.200	0.459
2.400	0.488
2.600	0.569
3.000	0.614
3.200	0.625
3.400	0.662
3.600	0.683
3.800	0.730
4.000	0.744
4.200	0.770
4.400	0.782
4.600	0.803
4.800	0.806
5.000	0.837
5.200	0.845
5.400	0.873
5.600	0.886
5.800	0.892
6.200	0.911
6.400	0.912
6.600	0.927
6.800	0.934
7.000	0.940
7.200	0.944
7.400	0.957
7.600	0.959
7.800	0.963
8.000	0.965
8.200	0.968
8.400	0.971
8.600	0.977
8.800	0.978
9.000	0.983
9.400	0.986
9.600	0.987
9.800	0.990
10.000	0.990
10.200	0.991
10.400	0.993
10.600	0.994
10.800	0.994
11.000	0.996
11.400	0.997
11.600	0.997
11.800	0.998
12.000	0.998
12.200	0.999
12.600	0.999
12.800	0.999
13.000	0.999
13.200	0.999
13.400	1.000
13.600	1.000
13.800	1.000
14.000	1.000
14.400	1.000
14.600	1.000
14.800	1.000
15.000	1.000
15.200	1.000
15.400	1.000
15.800	1.000
16.000	1.000
16.200	1.000
16.400	1.000
17.000	1.000
18.000	1.000

s	*P*
***k* = 4, *m* = 7**	
0.086	0.016
0.257	0.037
0.429	0.094
0.771	0.155
0.943	0.200
1.114	0.243
1.457	0.315
1.629	0.348
1.800	0.410
2.143	0.443
2.314	0.476
2.486	0.544
2.829	0.582
3.000	0.618
3.171	0.634
3.514	0.690
3.686	0.703
3.857	0.738
4.200	0.761
4.371	0.780
4.543	0.805
4.886	0.820
5.057	0.839
5.229	0.857
5.571	0.878
5.743	0.882
5.914	0.900
6.257	0.907
6.429	0.915
6.600	0.927
6.943	0.937
7.114	0.944
7.286	0.948
7.629	0.959
7.800	0.962
7.971	0.965
8.314	0.967
8.486	0.970
8.657	0.977
9.000	0.980
9.171	0.983
9.343	0.985
9.686	0.987
9.857	0.988
10.029	0.990
10.371	0.991
10.543	0.992
10.714	0.993
11.057	0.995
11.229	0.996
11.400	0.996
11.743	0.997
11.914	0.997
12.086	0.998
12.429	0.998
12.600	0.998
12.771	0.999
13.114	0.999
13.286	0.999
13.457	0.999
13.800	0.999
13.971	0.999
14.143	1.000
14.486	1.000
14.657	1.000
14.829	1.000
15.171	1.000
15.343	1.000
15.514	1.000
15.857	1.000
16.029	1.000
16.200	1.000
16.543	1.000
16.714	1.000
16.886	1.000
17.229	1.000
17.400	1.000
17.571	1.000
17.914	1.000
18.257	1.000
18.771	1.000
18.943	1.000
19.286	1.000
19.971	1.000
21.000	1.000

(Continued)

Table E.8 (Continued)

$$H_0 : \tilde{\mu}_1 = \tilde{\mu}_2 = \cdots = \tilde{\mu}_k$$

$$P = P(S \leq s)$$

$$S = \frac{12km}{k(k+1)} \sum_{i=1}^{k} \left[R_{i\cdot} - \frac{1}{2}(k+1) \right]^2$$

Note: k = number of treatment levels, m = number of blocks.

s	P
$k = 4$, $m = 8$	
0.000	0.002
0.150	0.029
0.300	0.041
0.450	0.088
0.600	0.110
0.750	0.151
0.900	0.163
1.050	0.235
1.200	0.243
1.350	0.290
1.500	0.319
1.650	0.346
1.800	0.371
1.950	0.442
2.250	0.483
2.400	0.493
2.550	0.529
2.700	0.550
2.850	0.596
3.000	0.611
3.150	0.638
3.300	0.650
3.450	0.674
3.600	0.677
3.750	0.713
3.900	0.722
4.050	0.758
4.200	0.774
4.350	0.781
4.650	0.807
4.800	0.809
4.950	0.832
5.100	0.842
5.250	0.852
5.400	0.859
5.550	0.879
5.700	0.883
5.850	0.890
6.000	0.894
6.150	0.900
6.300	0.906
6.450	0.919
6.600	0.921
6.750	0.932
7.050	0.940
7.200	0.942
7.350	0.949
7.500	0.951
7.650	0.954
7.800	0.958
7.950	0.962
8.100	0.963
8.250	0.969
8.550	0.972
8.700	0.975
8.850	0.977
9.000	0.978
9.150	0.981
9.450	0.984
9.600	0.985
9.750	0.986
9.900	0.986
10.050	0.989
10.200	0.989
10.350	0.991
10.500	0.991
10.650	0.992
10.800	0.992
10.950	0.994
11.100	0.994
11.250	0.995
11.400	0.995
11.550	0.996
11.850	0.996
12.000	0.996
12.150	0.997
12.300	0.997
12.450	0.998
12.600	0.998
12.750	0.998
12.900	0.998
13.050	0.998
13.200	0.999
13.350	0.999
13.500	0.999
13.650	0.999
13.800	0.999
13.950	0.999
14.250	0.999
14.400	0.999
14.550	0.999
14.700	1.000
14.850	1.000
15.000	1.000
15.150	1.000
15.300	1.000
15.450	1.000
15.600	1.000
15.750	1.000
15.900	1.000
16.050	1.000
16.200	1.000
16.350	1.000
16.650	1.000
16.800	1.000
16.950	1.000
17.100	1.000
17.250	1.000
17.400	1.000
17.550	1.000
17.700	1.000
17.850	1.000
18.150	1.000
18.300	1.000
18.450	1.000
18.600	1.000
18.750	1.000
19.050	1.000
19.200	1.000
19.350	1.000
19.500	1.000
19.650	1.000
19.800	1.000
19.950	1.000
20.250	1.000
20.400	1.000
20.550	1.000
20.700	1.000
20.850	1.000
21.150	1.000
21.600	1.000
21.750	1.000
21.900	1.000
22.200	1.000
22.950	1.000
24.000	1.000

s	P
$k = 5$, $m = 2$	
0.000	0.008
0.400	0.042
0.800	0.067
1.200	0.117
1.600	0.175
2.000	0.225
2.400	0.258
2.800	0.342
3.200	0.392
3.600	0.475
4.000	0.525
4.400	0.608
4.800	0.658
5.200	0.742
5.600	0.775
6.000	0.825
6.400	0.883
6.800	0.933
7.200	0.958
7.600	0.992
8.000	1.000

s	P
$k = 5$, $m = 3$	
0.000	0.000
0.267	0.012
0.533	0.028
0.800	0.059
1.067	0.086
1.333	0.155
1.600	0.169
1.867	0.232
2.133	0.280
2.400	0.318
2.667	0.351
2.933	0.405
3.200	0.441
3.467	0.507
3.733	0.525
4.000	0.568
4.267	0.594
4.533	0.653
4.800	0.674
5.067	0.709
5.333	0.747
5.600	0.764
5.867	0.787
6.133	0.828
6.400	0.837
6.667	0.873

(Continued)

Table E.8 (Continued)

$$H_0 : \tilde{\mu}_1 = \tilde{\mu}_2 = \cdots = \tilde{\mu}_k$$

$$P = P(S \leq s)$$

$$S = \frac{12km}{k(k+1)} \sum_{i=1}^{k} \left[R_{i\cdot} - \frac{1}{2}(k+1) \right]^2$$

Note: k **= number of treatment levels,** m **= number of blocks.**

s	P
6.933	0.883
7.200	0.904
7.467	0.920
7.733	0.937
8.000	0.944
8.267	0.955
8.533	0.962
8.800	0.972
9.067	0.974
9.333	0.983
9.600	0.985
9.867	0.992
10.133	0.995
10.400	0.996
10.667	0.997
10.933	0.999
11.467	1.000
12.000	1.000

s	P
$k = 5$, $m = 4$	
0.000	0.001
0.200	0.009
0.400	0.020
0.600	0.041
0.800	0.060
1.000	0.094
1.200	0.105
1.400	0.150
1.600	0.185
1.800	0.215
2.000	0.241
2.200	0.285
2.400	0.315
2.600	0.370
2.800	0.388
3.000	0.421
3.200	0.448
3.400	0.500
3.600	0.521
3.800	0.558
4.000	0.587
4.200	0.605
4.400	0.630
4.600	0.671
4.800	0.683
5.000	0.714
5.200	0.725
5.400	0.751
5.600	0.773
5.800	0.795
6.000	0.803
6.200	0.822
6.400	0.839
6.600	0.857
6.800	0.864
7.000	0.879
7.200	0.887
7.400	0.905
7.600	0.914
7.800	0.920
8.000	0.928
8.200	0.937
8.400	0.940
8.600	0.951
8.800	0.957
9.000	0.962
9.200	0.965
9.400	0.972
9.600	0.975
9.800	0.979
10.000	0.981
10.200	0.983
10.400	0.986
10.600	0.989
10.800	0.990
11.000	0.992
11.200	0.993
11.400	0.994
11.600	0.995
11.800	0.996
12.000	0.996
12.200	0.997
12.400	0.998
12.600	0.998
12.800	0.999
13.000	0.999
13.200	0.999
13.400	0.999
13.600	1.000
13.800	1.000
14.000	1.000
14.200	1.000
14.400	1.000
14.600	1.000
14.800	1.000
15.200	1.000
15.400	1.000
16.000	1.000

Table E.9 Studentized Range q Table.

The following tables provide the critical value (upper quantiles) for q(k, df, α) for α = .10, .05 and .01.

Level of Significance $\underline{\alpha = 0.10}$

	k-->																		
df	2	3	4	5	6	7	8	9	10	11	12	13	14	15	16	17	18	19	20
1	8.929	13.437	16.358	18.488	20.150	21.504	22.642	23.621	24.477	25.237	25.918	26.536	27.100	27.618	28.097	28.542	28.958	29.347	29.713
2	4.129	5.733	6.772	7.538	8.139	8.633	9.049	9.409	9.725	10.006	10.259	10.488	10.698	10.891	11.070	11.237	11.392	11.538	11.676
3	3.328	4.467	5.199	5.738	6.162	6.511	6.806	7.062	7.287	7.487	7.667	7.831	7.982	8.120	8.248	8.368	8.479	8.584	8.683
4	3.015	3.976	4.586	5.035	5.388	5.679	5.926	6.139	6.327	6.494	6.645	6.783	6.909	7.025	7.132	7.233	7.326	7.414	7.497
5	2.850	3.717	4.264	4.664	4.979	5.238	5.458	5.648	5.816	5.965	6.100	6.223	6.336	6.439	6.536	6.626	6.710	6.788	6.863
6	2.748	3.558	4.065	4.435	4.726	4.966	5.168	5.344	5.499	5.637	5.762	5.875	5.979	6.075	6.164	6.247	6.325	6.398	6.466
7	2.679	3.451	3.931	4.280	4.555	4.780	4.971	5.137	5.283	5.413	5.530	5.637	5.735	5.826	5.910	5.988	6.061	6.130	6.195
8	2.630	3.374	3.834	4.169	4.431	4.646	4.829	4.987	5.126	5.250	5.362	5.464	5.558	5.644	5.724	5.799	5.869	5.935	5.997
9	2.592	3.316	3.761	4.084	4.337	4.545	4.721	4.873	5.007	5.126	5.234	5.333	5.423	5.506	5.583	5.655	5.722	5.786	5.845
10	2.563	3.270	3.704	4.018	4.264	4.465	4.636	4.783	4.913	5.029	5.134	5.229	5.316	5.397	5.472	5.542	5.607	5.668	5.726
11	2.540	3.234	3.658	3.965	4.205	4.401	4.567	4.711	4.838	4.951	5.053	5.145	5.231	5.309	5.382	5.450	5.514	5.573	5.630
12	2.521	3.204	3.621	3.921	4.156	4.349	4.511	4.652	4.776	4.886	4.986	5.076	5.160	5.236	5.308	5.374	5.436	5.495	5.550
13	2.504	3.179	3.589	3.885	4.116	4.304	4.464	4.602	4.724	4.832	4.930	5.019	5.100	5.175	5.245	5.310	5.371	5.429	5.483
14	2.491	3.158	3.563	3.854	4.081	4.267	4.424	4.560	4.679	4.786	4.882	4.969	5.050	5.124	5.192	5.256	5.316	5.372	5.426
15	2.479	3.140	3.540	3.828	4.052	4.235	4.390	4.524	4.641	4.746	4.841	4.927	5.006	5.079	5.146	5.209	5.268	5.324	5.376
16	2.469	3.124	3.520	3.804	4.026	4.207	4.360	4.492	4.608	4.712	4.805	4.890	4.968	5.040	5.106	5.169	5.227	5.282	5.333
17	2.460	3.110	3.503	3.784	4.003	4.182	4.334	4.464	4.579	4.681	4.774	4.857	4.934	5.005	5.071	5.133	5.190	5.244	5.295
18	2.452	3.098	3.487	3.766	3.984	4.161	4.310	4.440	4.553	4.654	4.746	4.829	4.905	4.975	5.040	5.101	5.158	5.211	5.262
19	2.445	3.087	3.474	3.751	3.966	4.142	4.290	4.418	4.530	4.630	4.721	4.803	4.878	4.948	5.012	5.072	5.129	5.182	5.232
20	2.439	3.077	3.462	3.736	3.950	4.124	4.271	4.398	4.510	4.609	4.699	4.780	4.855	4.923	4.987	5.047	5.103	5.155	5.205
21	2.433	3.069	3.451	3.724	3.936	4.109	4.255	4.380	4.491	4.590	4.678	4.759	4.833	4.901	4.965	5.024	5.079	5.131	5.180
22	2.428	3.061	3.441	3.712	3.923	4.095	4.239	4.364	4.474	4.572	4.660	4.740	4.814	4.882	4.944	5.003	5.058	5.109	5.158
23	2.424	3.054	3.432	3.701	3.911	4.082	4.226	4.350	4.459	4.556	4.644	4.723	4.796	4.863	4.926	4.984	5.038	5.089	5.138
24	2.420	3.047	3.423	3.692	3.900	4.070	4.213	4.336	4.445	4.541	4.628	4.707	4.780	4.847	4.909	4.966	5.020	5.071	5.119
25	2.416	3.041	3.416	3.683	3.890	4.059	4.201	4.324	4.432	4.528	4.614	4.693	4.765	4.831	4.893	4.950	5.004	5.055	5.102
26	2.412	3.036	3.409	3.675	3.881	4.049	4.191	4.313	4.420	4.515	4.601	4.680	4.751	4.817	4.878	4.936	4.989	5.039	5.086
27	2.409	3.030	3.402	3.667	3.873	4.040	4.181	4.302	4.409	4.504	4.590	4.667	4.739	4.804	4.865	4.922	4.975	5.025	5.072
28	2.406	3.026	3.396	3.660	3.865	4.032	4.172	4.293	4.399	4.493	4.579	4.656	4.727	4.792	4.853	4.909	4.962	5.012	5.058
29	2.403	3.021	3.391	3.654	3.858	4.024	4.163	4.284	4.389	4.484	4.568	4.645	4.716	4.781	4.841	4.897	4.950	4.999	5.046
30	2.400	3.017	3.386	3.648	3.851	4.016	4.155	4.275	4.381	4.474	4.559	4.635	4.706	4.770	4.830	4.886	4.939	4.988	5.034
31	2.398	3.013	3.381	3.642	3.845	4.009	4.148	4.268	4.372	4.466	4.550	4.626	4.696	4.760	4.820	4.876	4.928	4.977	5.023
32	2.396	3.010	3.376	3.637	3.839	4.003	4.141	4.260	4.365	4.458	4.541	4.617	4.687	4.751	4.811	4.866	4.918	4.967	5.013
33	2.393	3.006	3.372	3.632	3.833	3.997	4.135	4.253	4.357	4.450	4.533	4.609	4.679	4.743	4.802	4.857	4.909	4.957	5.003
34	2.391	3.003	3.368	3.627	3.828	3.991	4.129	4.247	4.351	4.443	4.526	4.602	4.671	4.734	4.794	4.848	4.900	4.949	4.994
35	2.389	3.000	3.364	3.623	3.823	3.986	4.123	4.241	4.344	4.436	4.519	4.594	4.663	4.727	4.786	4.841	4.892	4.940	4.986
36	2.388	2.998	3.361	3.619	3.819	3.981	4.117	4.235	4.338	4.430	4.512	4.588	4.656	4.720	4.778	4.833	4.884	4.932	4.978
37	2.386	2.995	3.357	3.615	3.814	3.976	4.112	4.230	4.332	4.424	4.506	4.581	4.650	4.713	4.771	4.826	4.877	4.925	4.970
38	2.384	2.992	3.354	3.611	3.810	3.972	4.107	4.224	4.327	4.418	4.500	4.575	4.643	4.706	4.765	4.819	4.870	4.918	4.963
39	2.383	2.990	3.351	3.608	3.806	3.967	4.103	4.220	4.322	4.413	4.495	4.569	4.637	4.700	4.758	4.812	4.863	4.911	4.956
40	2.381	2.988	3.348	3.605	3.802	3.963	4.099	4.215	4.317	4.408	4.490	4.564	4.632	4.694	4.752	4.806	4.857	4.904	4.949
48	2.372	2.973	3.330	3.583	3.778	3.937	4.070	4.185	4.285	4.375	4.455	4.528	4.593	4.656	4.713	4.766	4.816	4.863	4.907
60	2.363	2.959	3.312	3.562	3.755	3.911	4.042	4.155	4.254	4.342	4.421	4.493	4.558	4.619	4.675	4.727	4.775	4.821	4.864
80	2.353	2.945	3.294	3.541	3.731	3.885	4.014	4.125	4.223	4.309	4.387	4.457	4.521	4.581	4.636	4.687	4.735	4.780	4.822
120	2.344	2.930	3.276	3.520	3.707	3.859	3.986	4.096	4.191	4.276	4.353	4.422	4.485	4.543	4.597	4.647	4.694	4.738	4.779
240	2.335	2.916	3.258	3.499	3.684	3.834	3.959	4.066	4.160	4.244	4.319	4.386	4.448	4.505	4.558	4.607	4.653	4.696	4.737
inf	2.326	2.902	3.240	3.478	3.661	3.808	3.931	4.037	4.129	4.211	4.285	4.351	4.412	4.468	4.519	4.568	4.612	4.654	4.694

(Continued)

Table E.9 (Continued)

Level of Significance $\alpha = 0.10$

	k-->																		
df	2	3	4	5	6	7	8	9	10	11	12	13	14	15	16	17	18	19	20
1	17.969	26.976	32.819	37.082	40.408	43.119	45.397	47.357	49.071	50.592	51.957	53.194	54.323	55.361	56.320	57.212	58.044	58.824	59.558
2	6.085	8.331	9.798	10.881	11.734	12.435	13.027	13.539	13.988	14.389	14.749	15.076	15.375	15.650	15.905	16.143	16.365	16.573	16.769
3	4.501	5.910	6.825	7.502	8.037	8.478	8.852	9.177	9.462	9.717	9.946	10.155	10.346	10.522	10.686	10.838	10.980	11.114	11.240
4	3.926	5.040	5.757	6.287	6.706	7.053	7.347	7.602	7.826	8.027	8.208	8.373	8.524	8.664	8.793	8.914	9.027	9.133	9.233
5	3.635	4.602	5.218	5.673	6.033	6.330	6.582	6.801	6.995	7.167	7.323	7.466	7.596	7.716	7.828	7.932	8.030	8.122	8.208
6	3.460	4.339	4.896	5.305	5.628	5.895	6.122	6.319	6.493	6.649	6.789	6.917	7.034	7.143	7.244	7.338	7.426	7.508	7.586
7	3.344	4.165	4.681	5.060	5.359	5.606	5.815	5.997	6.158	6.302	6.431	6.550	6.658	6.759	6.852	6.939	7.020	7.097	7.169
8	3.261	4.041	4.529	4.886	5.167	5.399	5.596	5.767	5.918	6.053	6.175	6.287	6.389	6.483	6.571	6.653	6.729	6.801	6.869
9	3.199	3.948	4.415	4.755	5.024	5.244	5.432	5.595	5.738	5.867	5.983	6.089	6.186	6.276	6.359	6.437	6.510	6.579	6.643
10	3.151	3.877	4.327	4.654	4.912	5.124	5.304	5.460	5.598	5.722	5.833	5.935	6.028	6.114	6.194	6.269	6.339	6.405	6.467
11	3.113	3.820	4.256	4.574	4.823	5.028	5.202	5.353	5.486	5.605	5.713	5.811	5.901	5.984	6.062	6.134	6.202	6.265	6.325
12	3.081	3.773	4.199	4.508	4.750	4.950	5.119	5.265	5.395	5.510	5.615	5.710	5.797	5.878	5.953	6.023	6.089	6.151	6.209
13	3.055	3.734	4.151	4.453	4.690	4.884	5.049	5.192	5.318	5.431	5.533	5.625	5.711	5.789	5.862	5.931	5.995	6.055	6.112
14	3.033	3.701	4.111	4.407	4.639	4.829	4.990	5.130	5.253	5.364	5.463	5.554	5.637	5.714	5.785	5.852	5.915	5.973	6.029
15	3.014	3.673	4.076	4.367	4.595	4.782	4.940	5.077	5.198	5.306	5.403	5.492	5.574	5.649	5.719	5.785	5.846	5.904	5.958
16	2.998	3.649	4.046	4.333	4.557	4.741	4.896	5.031	5.150	5.256	5.352	5.439	5.519	5.593	5.662	5.726	5.786	5.843	5.896
17	2.984	3.628	4.020	4.303	4.524	4.705	4.858	4.991	5.108	5.212	5.306	5.392	5.471	5.544	5.612	5.675	5.734	5.790	5.842
18	2.971	3.609	3.997	4.276	4.494	4.673	4.824	4.955	5.071	5.173	5.266	5.351	5.429	5.501	5.567	5.629	5.688	5.743	5.794
19	2.960	3.593	3.977	4.253	4.468	4.645	4.794	4.924	5.037	5.139	5.231	5.314	5.391	5.462	5.528	5.589	5.647	5.701	5.752
20	2.950	3.578	3.958	4.232	4.445	4.620	4.768	4.895	5.008	5.108	5.199	5.282	5.357	5.427	5.492	5.553	5.610	5.663	5.714
21	2.941	3.565	3.942	4.213	4.424	4.597	4.743	4.870	4.981	5.081	5.170	5.252	5.327	5.396	5.460	5.520	5.576	5.629	5.679
22	2.933	3.553	3.927	4.196	4.405	4.577	4.722	4.847	4.957	5.056	5.144	5.225	5.299	5.368	5.431	5.491	5.546	5.599	5.648
23	2.926	3.542	3.914	4.180	4.388	4.558	4.702	4.826	4.935	5.033	5.121	5.201	5.274	5.342	5.405	5.464	5.519	5.571	5.620
24	2.919	3.532	3.901	4.166	4.373	4.541	4.684	4.807	4.915	5.012	5.099	5.179	5.251	5.319	5.381	5.439	5.494	5.545	5.594
25	2.913	3.523	3.890	4.153	4.358	4.526	4.667	4.789	4.897	4.993	5.079	5.158	5.230	5.297	5.359	5.417	5.471	5.522	5.570
26	2.907	3.514	3.880	4.141	4.345	4.511	4.652	4.773	4.880	4.975	5.061	5.139	5.211	5.277	5.339	5.396	5.450	5.500	5.548
27	2.902	3.506	3.870	4.130	4.333	4.498	4.638	4.758	4.864	4.959	5.044	5.122	5.193	5.259	5.320	5.377	5.430	5.480	5.528
28	2.897	3.499	3.861	4.120	4.322	4.486	4.625	4.745	4.850	4.944	5.029	5.106	5.177	5.242	5.302	5.359	5.412	5.462	5.509
29	2.892	3.493	3.853	4.111	4.311	4.475	4.613	4.732	4.837	4.930	5.014	5.091	5.161	5.226	5.286	5.342	5.395	5.445	5.491
30	2.888	3.486	3.845	4.102	4.301	4.464	4.601	4.720	4.824	4.917	5.001	5.077	5.147	5.211	5.271	5.327	5.379	5.429	5.475
31	2.884	3.481	3.838	4.094	4.292	4.454	4.591	4.709	4.812	4.905	4.988	5.064	5.134	5.198	5.257	5.313	5.365	5.414	5.460
32	2.881	3.475	3.832	4.086	4.284	4.445	4.581	4.698	4.802	4.894	4.976	5.052	5.121	5.185	5.244	5.299	5.351	5.400	5.445
33	2.877	3.470	3.825	4.079	4.276	4.436	4.572	4.689	4.791	4.883	4.965	5.040	5.109	5.173	5.232	5.287	5.338	5.386	5.432
34	2.874	3.465	3.820	4.072	4.268	4.428	4.563	4.680	4.782	4.873	4.955	5.030	5.098	5.161	5.220	5.275	5.326	5.374	5.420
35	2.871	3.461	3.814	4.066	4.261	4.421	4.555	4.671	4.773	4.863	4.945	5.020	5.088	5.151	5.209	5.264	5.315	5.362	5.408
36	2.868	3.457	3.809	4.060	4.255	4.414	4.547	4.663	4.764	4.855	4.936	5.010	5.078	5.141	5.199	5.253	5.304	5.352	5.397
37	2.865	3.453	3.804	4.054	4.249	4.407	4.540	4.655	4.756	4.846	4.927	5.001	5.069	5.131	5.189	5.243	5.294	5.341	5.386
38	2.863	3.449	3.799	4.049	4.243	4.400	4.533	4.648	4.749	4.838	4.919	4.993	5.060	5.122	5.180	5.234	5.284	5.331	5.376
39	2.861	3.445	3.795	4.044	4.237	4.394	4.527	4.641	4.741	4.831	4.911	4.985	5.052	5.114	5.171	5.225	5.275	5.322	5.367
40	2.858	3.442	3.791	4.039	4.232	4.388	4.521	4.634	4.735	4.824	4.904	4.977	5.044	5.106	5.163	5.216	5.266	5.313	5.358
48	2.843	3.420	3.764	4.008	4.197	4.351	4.481	4.592	4.690	4.777	4.856	4.927	4.993	5.053	5.109	5.161	5.210	5.256	5.299
60	2.829	3.399	3.737	3.977	4.163	4.314	4.441	4.550	4.646	4.732	4.808	4.878	4.942	5.001	5.056	5.107	5.154	5.199	5.241
80	2.814	3.377	3.711	3.947	4.129	4.277	4.402	4.509	4.603	4.686	4.761	4.829	4.892	4.949	5.003	5.052	5.099	5.142	5.183
120	2.800	3.356	3.685	3.917	4.096	4.241	4.363	4.468	4.560	4.641	4.714	4.781	4.842	4.898	4.950	4.998	5.043	5.086	5.126
240	2.786	3.335	3.659	3.887	4.063	4.205	4.324	4.427	4.517	4.596	4.668	4.733	4.792	4.847	4.897	4.944	4.988	5.030	5.069
Inf	2.772	3.314	3.633	3.858	4.030	4.170	4.286	4.387	4.474	4.552	4.622	4.685	4.743	4.796	4.845	4.891	4.934	4.974	5.012

(*Continued*)

Table E.9 (Continued)

Level of Significance $\alpha = 0.01$

df	2	3	4	5	6	7	8	9	10	11	12	13	14	15	16	17	18	19	20
1	90.024	135.041	164.258	185.575	202.210	215.769	227.166	236.966	245.542	253.151	259.979	266.165	271.812	277.003	281.803	286.263	290.426	294.328	297.997
2	14.036	19.019	22.294	24.717	26.629	28.201	29.530	30.679	31.689	32.589	33.398	34.134	34.806	35.426	36.000	36.534	37.034	37.502	37.943
3	8.260	10.619	12.170	13.324	14.241	14.998	15.641	16.199	16.691	17.130	17.526	17.887	18.217	18.522	18.805	19.068	19.315	19.546	19.765
4	6.511	8.120	9.173	9.958	10.583	11.101	11.542	11.925	12.264	12.567	12.840	13.090	13.318	13.530	13.726	13.909	14.081	14.242	14.394
5	5.702	6.976	7.804	8.421	8.913	9.321	9.669	9.971	10.239	10.479	10.696	10.894	11.076	11.244	11.400	11.545	11.682	11.811	11.932
6	5.243	6.331	7.033	7.556	7.972	8.318	8.612	8.869	9.097	9.300	9.485	9.653	9.808	9.951	10.084	10.208	10.325	10.434	10.538
7	4.949	5.919	6.542	7.005	7.373	7.678	7.939	8.166	8.367	8.548	8.711	8.860	8.997	9.124	9.242	9.353	9.456	9.553	9.645
8	4.745	5.635	6.204	6.625	6.959	7.237	7.474	7.680	7.863	8.027	8.176	8.311	8.436	8.552	8.659	8.760	8.854	8.943	9.027
9	4.596	5.428	5.957	6.347	6.657	6.915	7.134	7.325	7.494	7.646	7.784	7.910	8.025	8.132	8.232	8.325	8.412	8.495	8.573
10	4.482	5.270	5.769	6.136	6.428	6.669	6.875	7.054	7.213	7.356	7.485	7.603	7.712	7.812	7.906	7.993	8.075	8.153	8.226
11	4.392	5.146	5.621	5.970	6.247	6.476	6.671	6.841	6.992	7.127	7.250	7.362	7.464	7.560	7.648	7.731	7.809	7.883	7.952
12	4.320	5.046	5.502	5.836	6.101	6.320	6.507	6.670	6.814	6.943	7.060	7.166	7.265	7.356	7.441	7.520	7.594	7.664	7.730
13	4.260	4.964	5.404	5.726	5.981	6.192	6.372	6.528	6.666	6.791	6.903	7.006	7.100	7.188	7.269	7.345	7.417	7.484	7.548
14	4.210	4.895	5.322	5.634	5.881	6.085	6.258	6.409	6.543	6.663	6.772	6.871	6.962	7.047	7.125	7.199	7.268	7.333	7.394
15	4.167	4.836	5.252	5.556	5.796	5.994	6.162	6.309	6.438	6.555	6.660	6.756	6.845	6.927	7.003	7.074	7.141	7.204	7.264
16	4.131	4.786	5.192	5.489	5.722	5.915	6.079	6.222	6.348	6.461	6.564	6.658	6.744	6.823	6.897	6.967	7.032	7.093	7.151
17	4.099	4.742	5.140	5.430	5.659	5.847	6.007	6.147	6.270	6.380	6.480	6.572	6.656	6.733	6.806	6.873	6.937	6.997	7.053
18	4.071	4.703	5.094	5.379	5.603	5.787	5.944	6.081	6.201	6.309	6.407	6.496	6.579	6.655	6.725	6.791	6.854	6.912	6.967
19	4.046	4.669	5.054	5.334	5.553	5.735	5.889	6.022	6.141	6.246	6.342	6.430	6.510	6.585	6.654	6.719	6.780	6.837	6.891
20	4.024	4.639	5.018	5.293	5.510	5.688	5.839	5.970	6.086	6.190	6.285	6.370	6.449	6.523	6.591	6.654	6.714	6.770	6.823
21	4.004	4.612	4.986	5.257	5.470	5.646	5.794	5.924	6.038	6.140	6.233	6.317	6.395	6.467	6.534	6.596	6.655	6.710	6.762
22	3.986	4.588	4.957	5.225	5.435	5.608	5.754	5.882	5.994	6.095	6.186	6.269	6.346	6.417	6.482	6.544	6.602	6.656	6.707
23	3.970	4.566	4.931	5.195	5.403	5.573	5.718	5.844	5.955	6.054	6.144	6.226	6.301	6.371	6.436	6.497	6.553	6.607	6.658
24	3.955	4.546	4.907	5.168	5.373	5.542	5.685	5.809	5.919	6.017	6.105	6.186	6.261	6.330	6.394	6.453	6.510	6.562	6.612
25	3.942	4.527	4.885	5.144	5.347	5.513	5.655	5.778	5.886	5.983	6.070	6.150	6.224	6.292	6.355	6.414	6.469	6.522	6.571
26	3.930	4.510	4.865	5.121	5.322	5.487	5.627	5.749	5.856	5.951	6.038	6.117	6.190	6.257	6.319	6.378	6.432	6.484	6.533
27	3.918	4.495	4.847	5.101	5.300	5.463	5.602	5.722	5.828	5.923	6.008	6.087	6.158	6.225	6.287	6.344	6.399	6.450	6.498
28	3.908	4.481	4.830	5.082	5.279	5.441	5.578	5.697	5.802	5.896	5.981	6.058	6.129	6.195	6.256	6.314	6.367	6.418	6.465
29	3.898	4.467	4.814	5.064	5.260	5.420	5.556	5.674	5.778	5.871	5.955	6.032	6.103	6.168	6.228	6.285	6.338	6.388	6.435
30	3.889	4.455	4.799	5.048	5.242	5.401	5.536	5.653	5.756	5.848	5.932	6.008	6.078	6.142	6.202	6.258	6.311	6.361	6.407
31	3.881	4.443	4.786	5.032	5.225	5.383	5.517	5.633	5.736	5.827	5.910	5.985	6.055	6.119	6.178	6.234	6.286	6.335	6.381
32	3.873	4.433	4.773	5.018	5.210	5.367	5.500	5.615	5.716	5.807	5.889	5.964	6.033	6.096	6.155	6.211	6.262	6.311	6.357
33	3.865	4.423	4.761	5.005	5.195	5.351	5.483	5.598	5.698	5.789	5.870	5.944	6.013	6.076	6.134	6.189	6.240	6.289	6.334
34	3.859	4.413	4.750	4.992	5.181	5.336	5.468	5.581	5.682	5.771	5.852	5.926	5.994	6.056	6.114	6.169	6.220	6.268	6.313
35	3.852	4.404	4.739	4.980	5.169	5.323	5.453	5.566	5.666	5.755	5.835	5.908	5.976	6.038	6.096	6.150	6.200	6.248	6.293
36	3.846	4.396	4.729	4.969	5.156	5.310	5.439	5.552	5.651	5.739	5.819	5.892	5.959	6.021	6.078	6.132	6.182	6.229	6.274
37	3.840	4.388	4.720	4.959	5.145	5.298	5.427	5.538	5.637	5.725	5.804	5.876	5.943	6.004	6.061	6.115	6.165	6.212	6.256
38	3.835	4.381	4.711	4.949	5.134	5.286	5.414	5.526	5.623	5.711	5.790	5.862	5.928	5.989	6.046	6.099	6.148	6.195	6.239
39	3.830	4.374	4.703	4.940	5.124	5.275	5.403	5.513	5.611	5.698	5.776	5.848	5.914	5.974	6.031	6.084	6.133	6.179	6.223
40	3.825	4.367	4.695	4.931	5.114	5.265	5.392	5.502	5.599	5.685	5.764	5.835	5.900	5.961	6.017	6.069	6.118	6.165	6.208
48	3.793	4.324	4.644	4.874	5.052	5.198	5.322	5.428	5.522	5.606	5.681	5.750	5.814	5.872	5.926	5.977	6.024	6.069	6.111
60	3.762	4.282	4.594	4.818	4.991	5.133	5.253	5.356	5.447	5.528	5.601	5.667	5.728	5.784	5.837	5.886	5.931	5.974	6.015
80	3.732	4.241	4.545	4.763	4.931	5.069	5.185	5.284	5.372	5.451	5.521	5.585	5.644	5.698	5.749	5.796	5.840	5.881	5.920
120	3.702	4.200	4.497	4.709	4.872	5.005	5.118	5.214	5.299	5.375	5.443	5.505	5.561	5.614	5.662	5.708	5.750	5.790	5.827
240	3.672	4.160	4.450	4.655	4.814	4.943	5.052	5.145	5.227	5.300	5.366	5.426	5.480	5.530	5.577	5.621	5.661	5.699	5.735
Inf	3.643	4.120	4.403	4.603	4.757	4.882	4.987	5.078	5.157	5.227	5.290	5.348	5.400	5.448	5.493	5.535	5.574	5.611	5.645

Table E.10 Critical values of the Kolmogorov–Smirnov one-sample test statistics.

One-sided test α *two-sided test* α *n*	.10 .20	.05 .10	.025 .05	.01 .02	.005 .01	*n*	.10 .20	.05 .10	.025 .05	.01 .02	.005 .10
1	.900	.950	.975	.990	.995	21	.226	.259	.287	.321	.344
2	.684	.776	.842	.900	.929	22	.221	.253	.281	.314	.337
3	.565	.636	.708	.785	.829	23	.216	.247	.275	.307	.330
4	.493	.565	.624	.689	.734	24	.212	.242	.269	.301	.323
5	.447	.509	.563	.627	.669	25	.208	.238	.264	.295	.317
6	.410	.468	.519	.577	.617	26	.204	.233	.259	.290	.311
7	.381	.436	.483	.538	.576	27	.200	.229	.254	.284	.305
8	.358	.410	.454	.507	.542	28	.197	.225	.250	.279	.300
9	.339	.387	.430	.480	.513	29	.193	.221	.246	.275	.295
10	.323	.369	.409	.457	.489	30	.190	.218	.242	.270	.290
11	.308	.352	.391	.437	.468	31	.187	.214	.238	.266	.285
12	.296	.338	.375	.419	.449	32	.184	.211	.234	.262	.281
13	.285	.325	.361	.404	.432	33	.182	.208	.231	.258	.277
14	.275	.314	.349	.390	.418	34	.179	.205	.227	.254	.273
15	.266	.304	.338	.377	.404	35	.177	.202	.224	.251	.269
16	.258	.295	.327	.366	.392	36	.174	.199	.221	.247	.265
17	.250	.286	.318	.355	.381	37	.172	.196	.218	.244	.262
18	.244	.279	.309	.346	.371	38	.170	.194	.215	.241	.258
19	.237	.271	.301	.337	.361	39	.168	.191	.213	.238	.255
20	.232	.265	.294	.329	.352	40	.165	.189	.210	.235	.252

This table gives the values of $D^{+}_{n,\alpha}$ and $D_{n,\alpha}$ for which $\alpha \geq P\{D^{+}_{n} > D^{+}_{n,\alpha}\}$ and $\alpha \geq P\{D_{n} > D_{n,\alpha}\}$ for some selected values of n and α.

Bibliography

References

Aggarwal, C.C., 2013. Outlier Analysis. Springer.
Agresti, A., 2013. Categorical Data Analysis, 3rd ed. Wiley.
Albert, J., 2009. Bayesian Computation with R, 2nd ed. Springer.
Antony, J., 2014. Design of Experiments for Engineers and Scientists, 2nd ed. Elsevier.
Atkinson, A.C., 1988. Recent developments in the methods of optimum and related experimental designs. Int. Statist. Rev. 56, 99–115.
Bain, L.J., Engelhardt, M., 2000. Introduction to Probability and Mathematical Statistics, 2nd ed. Duxbury Classic.
Balakrishnan, N., Nevzorov, V.B., 2003. A Primer on Statistical Distributions. Wiley, New York.
Barker, T.B., 1986. Quality engineering by design: Taguchi's philosophy. Qual. Prog. 19, 32–42.
Barnett, V., Lewis, T., 1995. Outliers in Statistical Data, 3rd ed. Wiley, New York.
Berinstein, P., 1998. Finding statistics online: how to locate the elusive numbers you need. In: Bjorner, S. (Ed.), Information Today. Independent Pub Group, Medford, NJ.
Berk, K.N., Carey, P., 2009. Data analysis with Microsoft Excel: updated for Office 2007, 3nd ed. Brooks/Cole Cengage Learning.
Box, G.E.P., Hunter, J.S., Hunter, W.G., 2005. Statistics for experimenters: design, innovation, and discovery, 2nd ed. Wiley, New York.
Bratcher, T.L., Moran, M.A., Zimmer, W.J., 1970. Tables of sample sizes in the analysis of variance. J. Qual. Technol. 2 (3), 156–164.
Brereton, R.G., 1990. Chemometrics: Application of Mathematics and Statistics to Laboratory Systems. Ellis Horwood, Chichester, UK.
Byrne, D.M., Taguchi, S., 1986. The Taguchi approach to parameter design. ASQC Quality Cong. Trans. Anaheim, CA.
Bush, L.B., Unal, R., 1992. Preliminary structural design of a lunar transfer vehicle aerobrake. Paper presented at the AIAA 1992 Aerospace Design Conference, Irvine, CA, 3–6 February, AIAA-92-1108.
Casella, G., 1985. An introduction to Empirical Bayes data analysis. Am. Statistician 39 (2).
Chiang, C.L., 2003. Statistical Methods of Analysis. World Scientific, Singapore.
Cobb, B.D., Clarkson, J.M., 1994. A simple procedure for optimizing the polymerase chain reaction (PCR) using modified Taguchi methods. Nucleic Acids Res. 22 (18), 3801–3805.
Cochran, W.G., 1977. Sampling Techniques, 3rd ed. Wiley, New York.
Cody, R.P., 2005. Applied Statistics and the SAS Programming Language, 5th ed. Prentice Hall, Upper Saddle River, NJ.
Coles, S., 2001. An Introduction to Statistical Modeling of Extreme Values. Springer.
Condra, L.W., 2001. Reliability Improvement with Design of Experiments, 2nd ed. Marcel Dekker, New York.
Conover, W.J., 1998. Practical Nonparametric Statistics. Wiley, New York.
Converse, P.E., Traugott, M.W., 1986. Assessing the accuracy of polls and surveys. Science 234, 1094–1098.
Crossfield, R.T., Dale, B.G., 1991. Applying Taguchi methods to the design improvement process of turbochargers. Qual. Eng. 3 (4), 501–516.
Dalgard, P., 2008. Introductory Statistics with R. Springer, New York.
Daniel, W.W., 1978. Applied Nonparametric Statistics. Houghton Mifflin, New York.
Davidson, F., 1996. Principles of Statistical Data Handling. Sage, Thousand Oaks, CA.
Davies, O.L., 1960. The Design and Analysis of Industrial Experiments. Oliver and Boyd, London.
Davison, A.C., Hinkley, D.V., 2003. Bootstrap Methods and Their Application. Cambridge University Press, Cambridge, UK.
Dempster, A.P., Laird, N.M., Rubin, D.B., 1977. Maximum likelihood from incomplete data via the EM algorithm. J. R. Statist. Soc. Ser. B, Methodol. 39, 1–38.
Dertouzos, M.L., Lester, R.K., Solow, R.M., 1989. Made in America: Regaining the Productive Edge. HarperPerennial, New York.
Edgington, E.S., 1987. Randomization Tests, 2nd ed. Marcel Dekker, New York.
Efron, B., 1979. Bootstrap methods: another look at the Jackknife. Ann. Statist. (No. 1), 1–26.
Efron, B., Tibshirani, R.J., 1993. An Introduction to the Bootstrap. Chapman and Hall, New York.
Fisher, R.A., 1926. The arrangement of field experiments. J. Ministry Agriculture Gt. Br. 33, 503–513.
Fisher, R.A., 1971. The Design of Experiments, 9th ed. Hafner Press (Macmillan), New York.
Fisher, R.A., Balmukand, B., 1928. The estimation of linkage from the offspring of selfed heterozygotes. J. Genet. 20, 70–92.
Fisher, R.A., Yates, F., 1963. Statistical Tables for Biological, Agricultural and Medical Research, 6th ed. Hafner Press (Macmillan), New York.
Frees, E.W., 1996. Data Analysis Using Regression Models: The Business Perspective. Prentice Hall, Upper Saddle River, NJ.

Freund, J.E., 1992. Mathematical Statistics, 5th ed. Prentice Hall, Upper Saddle River, NJ.

Geman, S., Geman, D., 1984. Stochastic relaxation, Gibbs distributions, and Bayesian restoration of images. IEEE Trans. Pattern Anal. Mach. Intell. 6, 721–741.

Casella, G., George, E.I., 1992. Explaining the Gibbs Sampler. Am. Statistician Vol. 46 (Issue 3), 167–174.

Gilbert, R.O., 1987. Statistical Methods for Environmental Pollution Monitoring. Van Nostrand Reinhold, New York.

Gilks, W.R., Richardson, S., Spiegelhalter, D.J. (Eds.), 1996. Markov Chain Monte Carlo in Practice. Chapman & Hall, London.

de Haan, L., Ferreira, A., 2006. Extreme Value Theory: An Introduction. Springer.

Hald, A., 1990. A History of Probability and Statistics and Their Applications before 1750. Wiley, New York.

Hand, D.J., Daly, F., McConway, K., Lunn, D., Ostrowski, E., 1993. A Handbook of Small Data Sets. Chapman & Hall, London.

Harwell, M.R., 1988. Choosing between parametric and nonparametric tests. J. Counseling Dev. 67, 35–38.

Hastings, W.K., 1970. Monte Carlo sampling methods using Markov chains and their applications. Biometrika 57, 97–109.

Hegsted, D.M., Nicolosi, R.J., 1987. Individual variation in serum cholesterol levels. Proc. Nattl. Acad. Sci. USA.

Hinkley, D.V., 1983. Jackknife methods. Encycl. Statist. Sci. 4, 280–287.

Hoenig, J.M., Heisey, D.M., 2001. The abuse of power: the pervasive fallacy of power calculations for data analysis. Am. Statistician 55, 19–24.

Hogg, R.V., Tanis, E.A., 1993. Probability and Statistical Inference. Macmillan, New York.

Hogg, R.V., Craig, A.T., 1978. Introduction to Mathematical Statistics, 4th ed. Macmillan, New York.

Hollander, M., Wolfe, D.A., 1999. Nonparametric Statistical Methods. Wiley, New York.

Iglewicz, B., Hoaglin, D.C., 1993. How to Detect and Handle Outliers. American Society for Quality Control, Milwaukee, WI.

Iman, R.L., 1994. A Data-Based Approach to Statistics. Duxbury, Pacific Grove, CA.

Johnson, R.A., Miller, I., Freund, J.E., 2010. Miller and Freund's Probability and Statistics for Engineers, 8th ed. Prentice Hall, Upper Saddle River, NJ.

Lee, J.-S.R., McNickle, D.C., Pawlikowski, K., 1998. A Survey of Confidence Interval Formulae for Coverage Analysis. Technical Report TR-COSC 04/98. Christchurch: University of Canterbury.

Kiefer, J., 1959. Optimum experimental designs (with discussions). J. R. Statist. Soc. Ser. B 21, 272–319.

Kuehl, R.O., 2000. Design of Experiments: Statistical Principles of Research Design and Analysis, 2nd ed. Brooks/Cole, Pacific Grove, CA.

Larsen, R.J., Marx, M.L., 2011. An Introduction to Mathematical Statistics and Its Applications, 5th ed. Prentice Hall, Upper Saddle River, NJ.

Lehmann, E.L., 1975. Nonparametrics: Statistical Methods Based on Ranks. Holden-Day, San Francisco.

Lenth, R.V., 2001. Some practical guidelines for effective sample size determination. Am. Statistician 55, 187–193.

Levene, H., 1960. Robust tests for the equality of variances. In: Olkin, I. (Ed.), Contributions to Probability and Statistics: Essays in Honor of Harold Hotelling. Stanford University Press, Palo Alto, CA, pp. 278–292.

Liang, F., Liu, C., Carroll, R.J., 2010. Advanced Markov Chain Monte Carlo Methods: Learning from Past Samples. Wiley.

Madansky, A., 2012. Prescriptions for Working Statisticians. Springer-Verlag, New York.

Maghsoodloo, S., 1990. The exact relation of Taguchi's signal-to-noise ratio to his quality loss function. J. Qual. Technol. 22 (1), 57–67.

Manly, B.F.J., 1991. Randomization and Monte Carlo Methods in Biology. Chapman Hall, London.

Maritz, J.S., 1970. Empirical Bayes Methods. Methuen, London.

McLachlan, G.J., Krishnan, T., 1997. The EM Algorithm and Extensions. Wiley, New York.

Mendenhall, W., Sincich, T., 1996. A Second Course in Statistics: Regression Analysis, 5th ed. Prentice Hall, Upper Saddle River, NJ.

Mendenhall, W., Wackerly, D.D., Scheaffer, R.L., 2007. Mathematical Statistics with Applications, 7th ed. PWS-Kent, Boston.

Metropolis, N., Rosenbluth, A.W., Rosenbluth, M.N., Teller, A.H., Teller, E., 1953. Equation of state calculations by fast computing machines. J. Chem. Phys. 21, 1087–1092.

Miller, B.O., 2001. Beyond Statistics: A Practical Guide to Data Analysis. Allyn and Bacon, Boston.

Morris, C.N., 1983. Parametric empirical bayes inference: theory and applications (with discussion). J. Am. Stat. Assoc. 78, 47–65.

Murphy, K.R., Myors, B., 1998. Statistical Power Analysis: A Simple and General Model for Traditional and Modern Hypothesis Tests. L. Erlbaum Associates, London.

Nair, V.N. (Ed.), Abraham, B., MacKay, J., Nelder, J.A., Box, G., Phadke, M.S., Kacker, R.N., Lorenzen, T.J., Lucas, J.M., Myers, R.H., Vining, G.G., Sacks, J., Welch, W.J., Shoemaker, A.C., Tsui, K.L., Taguchi, S., Wu, C.F.J. 1992. Taguchi's parameter design: a panel discussion (Ed.). Technometrics 34 (2), 127–161.

Nguyen, H.T., Rogers, G.S., 1989. Fundamentals of Mathematical Statistics: Statistical Inference Vol. 2 Springer-Verlag, New York.

Peace, G.S., 1993. Taguchi Methods: A Hands-On Approach. Addison-Wesley, Reading, MA.

Pearson, E.S., 1939. "Student" as statistician. Biometrika 30, 210–250.

Pearson, K., 1900. On the criterion that a given system of deviations from the probable in the case of a correlated system of variables is such that it can be reasonably supposed to have arisen from random sampling. Lond. Edinb. Dubl. Phil. Mag. & J. Sci. 50 (302), 157–175.

Peterson, G.E., St. Clair, D.C., Aylward, S.R., Bond, W.E., 1995. Using Taguchi's method of experimental design to control errors in layered perceptrons. IEEE Trans. Neural Netw. 6 (4), 949–961.

Phadke, M.S., 1989. Quality Engineering Using Robust Design. Prentice Hall, Englewood Cliffs, NJ.

Porter, T., 1986. The Rise of Statistical Thinking, 1820–1900. Princeton University Press, Princeton, NJ.

Raudenbush, S.W., 1997. Statistical analysis and optimal design for cluster randomized trials. Psychol. Methods 2 (2), 173–185.

Rinaman, W.C., 1993. Foundations of Probability and Statistics. Saunders College Publishing, Fort Worth.

Robbins, H., 1956. An empirical Bayes approach to statistics. In: Neyman, J. (Ed.), Proceedings of the Third Berkeley Symposium on Mathematical Statistics and Probability, 1954–1955, Vol. I: Contributions to the Theory of Statistics. University of California Press, Berkeley, CA, pp. 157–163.

Robert, C.P., Casella, G., 2004. Monte Carlo Statistical Methods, 2nd ed. Springer.

Robert, C.P., Casella, G., 2009. Introducing Monte Carlo Methods with R. Springer.

Robinson, G.K., 2000. Practical Strategies for Experimenting. Wiley, New York.

Ross, P.J., 1988. Taguchi Techniques for Quality Engineering: Loss Function, Orthogonal Experiments, Parameter and Tolerance Design. McGraw-Hill, New York.

Rousseeuw, P.J., Leroy, A.M., 1987. Robust Regression and Outlier Detection. Wiley, New York.

Roy, R.K., 1990. A Primer on the Taguchi Method. Van Nostrand Reinhold, New York.

Ryan, B.F., Joiner, B.L., Cryer, J.D., 2013. Minitab Handbook: Updated for Release 16, 6th ed. Thomson Brooks/Cole.

Sahai, H., Ageel, M.I., 2000. The Analysis of Variance: Fixed, Random and Mixed Models. Birkhauser, Boston.

Salsburg, D., 2001. The Lady Tasting Tea: How Statistics Revolutionized Science in the Twentieth Century. W.H. Freeman, New York.

Savage, L.J., 1954. The Foundations of Statistics. Wiley, New York.

Shao, J., Tu, D., 1995. The Jackknife and Bootstrap. Springer-Verlag, New York.

Shina, S.G., 1991. The successful use of the Taguchi method to increase manufacturing process capability. Qual. Eng. 3 (3), 333–349.

Snedecor, G.W., Cochran, W.G., 1980. Statistical Methods, 7th ed. Iowa State University Press, Ames.

Sprent, P., 1998. Data Driven Statistical Methods. Chapman & Hall, New York.

Sullivan, L.P., 1987. The power of Taguchi methods. Qual. Prog. 76–79.

Taguchi, G. (1986). *Introduction to Quality Engineering: Designing Quality into Products and Processes.* Asian Productivity Organization. Available in the United States from American Supplier Institute, Dearborn, MI.

Taguchi, G., 1987. System of Experimental Design: Engineering Methods to Optimize Quality and Minimize Costs, Vols. 1 and 2. UNIPUB/Kraus International, White Plains, NY.

Taguchi, G., 1993. Taguchi on Robust Technology Development: Bringing Quality Engineering Upstream. ASME, New York.

Taguchi, G., Clausing, D., 1990. Robust quality. Harv. Bus. Rev.

Taguchi, G., Konishi, S., 1987. Taguchi Methods: Orthogonal Arrays and Linear Graphs. American Supplier Institute, Dearborn, MI.

Taguchi, G., Yokoyama, Y., 1994. Taguchi Methods: Design of Experiments. American Supplier Institute, Dearborn, MI in conjunction with the Japanese Standards Association, Tokyo.

Tamhane, A.C., Dunlop, D.D., 2000. Statistics and Data Analysis: From Elementary to Intermediate. Prentice Hall, Upper Saddle River, NJ.

Tankard, J., 1984. The Statistical Pioneers. Schenkman Books, New York.

Tanur, J.M., Mosteller, F., Kruskal, W., Link, R., Pieters, R., Rising, G., 1972. Statistics: A Guide to the Unknown. Holden-Day, New York.

Thompson, S.K., 1992. Sampling. Wiley, New York.

Thompson, S.K., Seber, G.A.F., 1996. Adaptive Sampling. Wiley, New York.

Tsao, H., Wright, T., 1983. On the maximum ratio: a tool for assisting inaccuracy assessment. Am. Statistician 37 (4), 339–342.

Walters, F.H., Parker Jr., L.R., Morgan, S.L., Deming, S.N., 1993. Sequential Simplex Optimization: A Technique for Improving Quality and Productivity in Research, Development, & Manufacturing. CRC Press, Boca Raton, FL.

Wei, G.C.G., Tanner, M.A., 1990. A Monte Carlo implementation of the EM algorithm and the poor man's data augmentation algorithm. J. Am. Statist. Assoc. 85, 699–704.

Wooding, W.M., 1994. Planning Pharmaceutical Clinical Trials: Basic Statistical Principles. Wiley, New York.

Ya-lun, C., 1989. Statistical Analysis for Business and Economics. Elsevier, New York.

Yang, M.C.K., Robinson, D.H., 1986. Understanding and Learning Statistics by Computer. World Scientific, Singapore.

Zar, J.H., 1996. Biostatistical Analysis, 3rd ed. Prentice Hall, Upper Saddle River, NJ.

Index

Note: Page numbers followed by "*f*" indicate figures, "*t*" indicate tables and "*b*" indicate boxes.

O

P

Q

R

S

T

U

V

W

Z

Printed and bound by CPI Group (UK) Ltd, Croydon, CR0 4YY
15/09/2026
02217083-0001